# SCANNING ELECTRON MICROSCOPY
## STUDIES OF
# EMBRYOGENESIS

Edited by
**Gary C. Schoenwolf**

Based on symposia organized principally by
**S. Robert Hilfer, Gary C. Schoenwolf, Robert E. Waterman**

Published By
**Scanning Electron Microscopy, Inc.**
P.O. Box 66507
AMF O'Hare, IL 60666 U.S.A.

Copyright © 1986, Scanning Electron Microscopy, Inc. except for contributions in the public domain.

All rights reserved.

Individual readers of this volume and non-profit libraries acting for them are freely permitted to make fair use of the material herein, such as to copy an article for use in teaching or research. Permission is granted to quote from this volume in scientific works with the customary acknowledgment of the source. To print a table, figure, micrograph or other excerpt requires, in addition, the consent of one of the original authors and notification to SEM, Inc. Republication or systematic or multiple reproduction of any material in this volume (including the abstracts) is permitted only after obtaining written approval from SEM, Inc.; and, in addition, SEM, Inc. will require that permission also be obtained from one of the original authors.

Every effort has been made to trace the ownership of all copyrighted material in this volume and to obtain permission for its use.

The articles in this book are reprinted from the Journal:
  Scanning Electron Microscopy, Vol. 1978, Part II
  Scanning Electron Microscopy, Vol. 1980, Part II
  Scanning Electron Microscopy, Vol. 1981, Part II
  Scanning Electron Microscopy, Vol. 1982, Part I
  Scanning Electron Microscopy, Vol. 1983, Part III
  Scanning Electron Microscopy, Vol. 1984, Part I, II, III & IV
  Scanning Electron Microscopy, Vol. 1985, Part I, II, & IV

The paginations from these publications are given at the top left of the first page of each article.

In quoting the papers in this book, it is strongly recommended that the original pagination be used in the following format:
  Scanning Electron Microsc. year; part: p. no.

SEM, Inc. is a **not-for-profit organization** with the following goals:
a. Promotion of advancement of science of SEM and related characterization techniques;
b. Promotion of application of these techniques in existing and new areas of applications;
c. Promotion of these techniques so that their users obtain the best information of the highest quality from their instruments.

SEM, Inc. sponsors annual meetings on Scanning Electron Microscopy and Pfefferkorn Conferences on basic subjects related to SEM. SEM, Inc. publishes the quarterly international journal, "**Scanning Electron Microscopy.**"

For more information or other inquiries contact:
  Dr. Om Johari — phone 312/529-6677
  P.O. Box 66507
  AMF O'Hare, IL 60666, U.S.A.

ISBN: 0-931288-36-3

Printed in the United States of America

International Journal **Scanning Electron Microscopy**

| | |
|---|---|
| **Editor** | **Om Johari** |
| Guest Editors | Godfried M. Roomans |
| | Robert P. Becker |
| Associate Editors | Sudha A. Bhatt |
| | Joseph Staschke |

**Scanning Electron Microscopy, Inc.**

| | |
|---|---|
| President | John D. Fairing |
| Vice President | Robert P. Becker |
| Secretary-Treasurer | John D. Fairing |

## FOREWORD

This monograph consists of invited and contributed papers that were presented principally at the developmental biology sessions of the 1982, 1983, 1984, and 1985 Scanning Electron Microscopy meetings. (A few additional papers were selected from earlier meetings.) The theme of these sessions was the use of scanning electron microscopy (SEM) in developmental biology, with emphasis on the importance of SEM as a tool in the teaching of developmental biology and embryology. Authors were asked to provide a concise text written in a teaching style geared to the undergraduate student. The overall goal was to provide a collection of high quality scanning electron micrographs on a wide variety of developing systems to help students visualize dynamic three-dimensional changes in morphology over time. It was hoped that such micrographs would speak largely for themselves, and that the text would serve to guide the student from one topic to the next. A second goal was to provide a bibliography for further in-depth, individual study. These goals have been realized in large part by the present monograph. Appreciation is due to the many active researchers who believe that it is important to communicate their fields to beginning students and who were willing to take the time to do exactly that. Hence, this monograph is for the student -- enjoy learning about the way of the embryo!

G. C. Schoenwolf
The University of Utah
School of Medicine
Salt Lake City, UT

REVIEWING PROCEDURE
AND
DISCUSSION WITH REVIEWERS

Each paper in this volume contains a Discussion with Reviewers. This discussion follows the text and should be read with the paper. Each paper submitted to SEM, Inc. for publication is reviewed by at least three, up to an average of five, reviewers. The reviewers are asked to separate their comments from their questions. The comments are useful in determining the acceptability of the papers as submitted. Although the comments require no written response, in several cases, the authors have included responses to comments, or to questions phrased from, or based on, comments (either as a result of editorial suggestions or on the author's own initiative). Based on these comments approximately 15% of the submitted papers were not accepted for publication, while almost all of the others were asked to make changes involving from minor to major revisions.

The questions, for the most part, originate as a result of statements included in our cover letter accompanying each paper sent to the reviewers. The reviewers are asked to suppose they are attendees at a conference where this paper, as written, is being presented, and then ask relevant questions which would occur to them resulting from the presentation. From the questions so asked, some are not included with the published paper because the authors attended to them by text revisions. In some cases, editorial and/or space considerations may exclude inclusion of all questions asked by reviewers. The authors are asked to prepare their Discussion with Reviewers section in a camera-ready format. In some instances the authors edit the questions and/or combine several similar questions from different reviewers to provide one answer. While all efforts are made to check that the questions in the printed version faithfully follow the views of the specific reviewer, the editors apologize, if in some instances, the actual meaning and/or emphasis may have been changed by the author.

The cover letter to the reviewers states:
"1. Your name will be conveyed to the author with your review UNLESS YOU ASK US NOT TO.
2. The questions published in the Journal will be identified as originating from you UNLESS YOU ADVISE OTHERWISE..."

In all cases sincere efforts are made to respect the reviewer's wishes to remain anonymous; however, in nearly 95% of the cases, the reviewers have given permission to be identified; so their names are conveyed to the authors and are included with the questions printed with each paper. An overall list of reviewers is provided in the opening pages of each SEM part. We apologize for any error/omissions which may occur.

Finally, readers are urged to be cautious regarding the weight they attach to the authors' replies, since the answers to the questions represent the authors' unchallenged views--except for minor editorial changes--the authors generally have the last word. Also, please consider that the questions were, in most cases, relevant to the originally submitted paper, and they may not have the same significance for the revised paper published in this volume.

If you disagree with the results, conclusions or approaches in a paper, please send your comments, as a Letter to Editor, typed in a column format (each column is 4-1/8 inches wide and 11-1/2 inches long; i.e., 10.5 by 29.3 cm.). Your comments along with author's response will be published in a subsequent issue.

The editor gratefully thanks the authors and reviewers (see p. $xiii$) for their contributions, invites your comments on ways to improve this procedure and seeks qualified volunteers to assist with reviewing papers in the future.

ERRATA: Despite the best efforts of authors, reviewers and editors, errors may remain. Please help by pointing out errors that you notice. Please provide enough information to locate each error (volume, page, column, line, etc.) and indicate suitable correction.

The Editors

FERTILIZATION AND EARLY DEVELOPMENT OF SEA URCHINS

G. Schatten* and H. Schatten

Center for the Study of Cells, Reproduction and Development, Department of Biological Science, Florida State University, Tallahassee, Florida

(Paper received January 27 1983, Complete manuscript received June 27 1983)

## Abstract

Scanning electron microscopy (SEM) has been successfully employed for the study of several surface-mediated events during fertilization and early development in sea urchins. In addition to basic morphological descriptions of the sperm, the extrusion of the acrosomal process has been documented with SEM. During sperm incorporation, short microvilli are found to elongate around the successful sperm. In eggs denuded of vitelline layers, in which the elevation and hardening of this fertilization coat is prevented, numerous long microvilli have been shown to cluster around and elongate over the entering sperm during sperm incorporation. Following sperm incorporation and the elevation of the fertilization coat, scanning electron microscopy has been utilized to study the bursts of elongation of the previously short egg microvilli. These microvilli appear to undergo two bursts in length due primarily to the new assembly of microfilaments in the egg cortex. Cytokinesis occurs shortly after the second burst of microvillar elongation. The morula stage is characterized by loosely attached cells which become more closely apposed in subsequent cell divisions to result in the hollow blastula. The ciliated blastula hatches from the fertilization coat, whereupon gastrulation occurs, resulting in a free-swimming, feeding larval stage. This paper reviews the surface alterations and the contribution of scanning electron microscopy to the study of these surface alterations, from fertilization through early development.

Key words: fertilization, development, sea urchin, scanning electron microscopy, sperm, egg, plasma membrane, microvilli, microfilaments, microtubules.

*Department of Biological Science, Florida State University, Tallahassee, Florida 32306

Phone No. (904) 644-5825.

## Introduction

Fertilization and early development of sea urchins is characterized by numerous surface changes, which can be admirably investigated with scanning electron microscopy(SEM). In the sperm, the most dramatic change is the extension of the acrosomal process, which must be considered the first event of fertilization. In the egg at fertilization, sperm attachment, fusion and incorporation have been documented by SEM, as have the elevation and hardening reactions of the fertilization coat. Shortly after fertilization, the egg undergoes bursts in microvillar elongations, which were discovered using scanning electron microscopy. These microvillar bursts may be central to the division cycles required to transform the singled cell egg into a multicellular organism.

Finally, the cells undergo cytoskeletal transformations and locomotion to transform the irregular shaped morula into first a blastula and later a gastrula. The scope of this article will be to review the contributions of scanning electron microscopy to the understanding of the behavior of the gametes during fertilization and of the zygotes and developing embryos.

### Sperm and the Acrosome Reaction

The basic morphological features of the sperm can be well resolved in scanning electron microscopy (Figs. 1, 2). At the anterior of the sperm is the acrosome (arrow), which can extend a filamentous process and is responsible for the first contact between the sperm and the egg. The acrosome reaction will be considered in the next paragraph. The conical sperm head contains the paternal genome; the activation of the incorporated sperm nucleus has been recently reviewed by Poccia (1982). The acrosome and nucleus together comprise the sperm head. Directly posterior to the sperm head and closely apposed to it is the cylindrical mid-piece (arrowhead), which contains a ring-shaped mitochondrion and a pair of centrioles. The significance of this pair of sperm centrioles should not be underestimated. The unfertilized sea urchin egg lacks these microtubule organizing centers: it is the contribution of the sperm-derived centrioles that establish the two poles for the first cell division.

Directly attached to one of these centrioles, which serves as a basal body, is the sperm tail, which is responsible for sperm motility. The molecular aspect of the manner in which the nine outer doublet and two inner singlet microtubules of the sperm tail are capable of propelling the sperm through its suspending medium is now a classic model in the understanding of ciliary motion and has been recently reviewed by Gibbons (1981). Surrounding the entire sperm is a highly diversified plasma membrane (reviewed by Friend, 1982). The acrosome reaction of the sperm, triggered by diffusible factors from the egg surface, involves both the secretion of the acrosomal vesicle and the polymerization of actin in the periacrosomal region of the sperm head (Tilney, 1978). In Fig. 2, the extension of the acrosomal process (arrow) is well displayed in scanning electron microscopy. At the apex of this process, a species-specific egg-binding protein, called bindin, has been characterized by Vacquier and Moy (1977) and appears to be the crucial adhesive between the gametes.

## Sperm Incorporation

Perhaps the greatest contribution of scanning electron microscopy to the study of fertilization has been in its utility for investigating aspects of sperm adhesion, sperm egg-membrane fusion and sperm incorporation. The surface features of sperm incorporation in sea urchins have been studied by Schatten and Mazia (1976), Schatten and Schatten (1980), Tegner and Epel (1976) and Usui et al. (1980). The mammalian egg surface during sperm incorporation has been studied by Shalgi and Phillips (1980a, b), and Yanagimachi (1978). Figure 1 is a relatively low magnification image of insemination of a sea urchin egg as observed by scanning electron microscopy. Numerous sperm are attached to the unfertilized egg. In Figure 2, the initial attachment of the acrosome-reacted sperm to the egg surface is demonstrated. The unfertilized egg surface is characterized by numerous short microvilli, which appear here as papillae (arrow) in the vitelline layer; the vitelline layer is an extracellular layer draped over the unfertilized egg surface. The successful sperm invariably attaches in a perpendicular fashion (Fig. 3) whereupon membrane fusion follows (Fig. 4). Microvilli adjacent to the successful sperm elongate and cluster about the sperm head to form into the fertilization cone (Fig. 5; arrow), and shortly thereafter the sperm during incorporation is obscured by the elevation of the fertilization coat (Fig. 6).

In addition to signaling the egg to begin the process of sperm incorporation, the successful fusion of the sperm with the egg initiates the cortical reaction (see below), which results in the elevation and hardening of the fertilization coat. The elevation around the successful sperm of the vitelline layer, which then hardens into the electron-opaque fertilization coat, precludes direct observations with SEM at the plasma membrane surface of the events during sperm incorporation (Fig. 6). To overcome this steric interference to electron observations, eggs can be denuded of their vitelline layers with disulfide-reducing agents such as dithiothreitol (Epel et al. 1970) and then studied with scanning electron microscopy (Schatten and Schatten, 1980).

The basic ultrastructural features of sperm incorporation are analogous in these denuded eggs to those in untreated controls. The initial contact again is via the extended acrosomal process; sperm incorporation starts with the localized elongation of microvilli around the sperm head (Fig. 7). These microvilli elongate around the

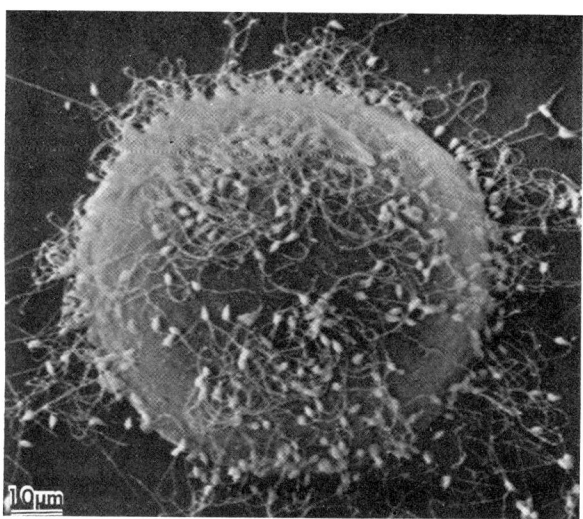

Fig. 1. Insemination observed by scanning electron microscopy. An early stage of insemination of an egg glued to a polylysine-coated slide. Only the tops and sides of the egg are available for sperm binding. <u>Strongylocentrotus purpuratus</u>. Bar: 10 μm. Reprinted, with permission, from Schatten and Mazia, 1976.

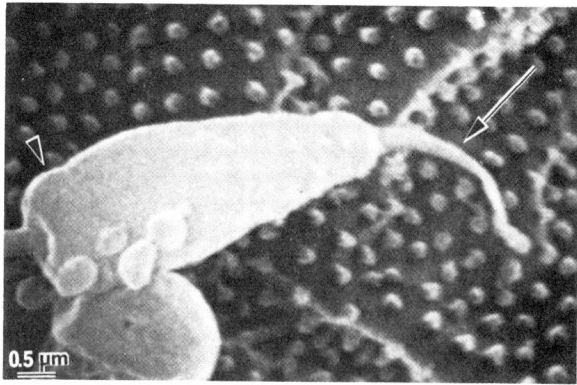

Fig. 2. The Acrosome Reaction of the Sperm. The acrosomal process observed with scanning electron microscopy appears as an elongated fiber (arrow) and establishes the initial contact with the surface of the unfertilized egg. <u>S. purpuratus</u>. Bar: 500 nm. Reprinted, with permission, from Schatten and Mazia, 1976.

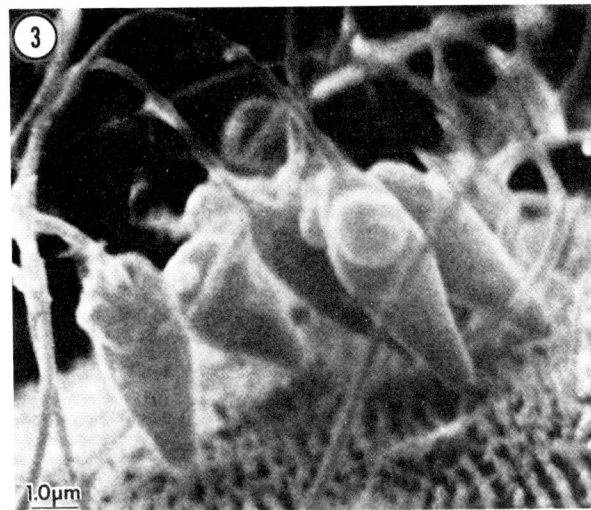

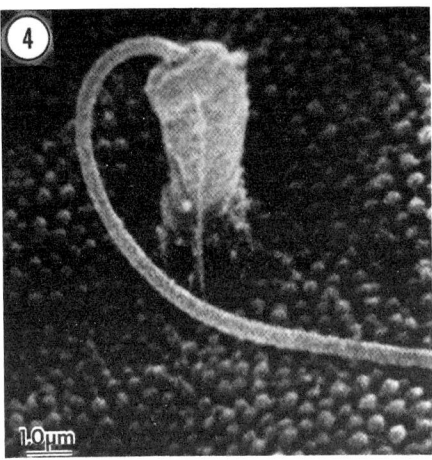

Figs. 3 to 6 are reprinted, with permission, from Schatten and Mazia, 1976.

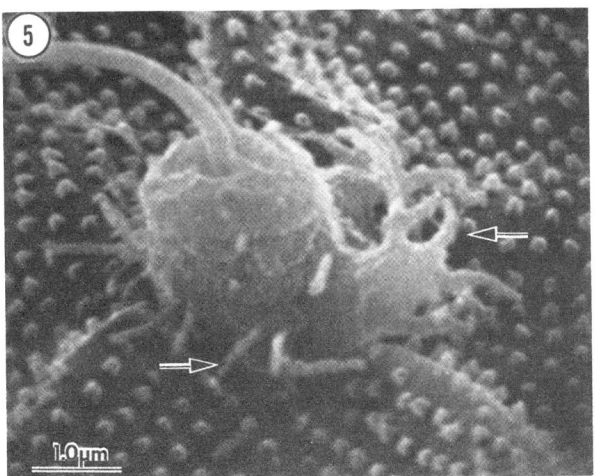

Fig. 5. Sperm Incorporation: The egg membrane continues to rise around the sperm head. Microvilli (arrow) elongate around the spermatozoon as the membrane derived from the sperm appears slack and convoluted. S. purpuratus.

Fig. 3. Sperm-Egg Attachment: The sperm attach perpendicularly to the vitelline sheet of the egg surface. The short, arrayed microvilli are characteristic of an unfertilized egg. The sperm always adhere to the egg surface by the apical tip of the sperm head. S. purpuratus. Bar: 1 µm.

Fig. 4. Sperm-Egg Membrane Fusion. A later stage of membrane fusion. The membrane derived from the egg now surrounds the anterior portion of the sperm head. S. purpuratus. Bar: 1 µm.

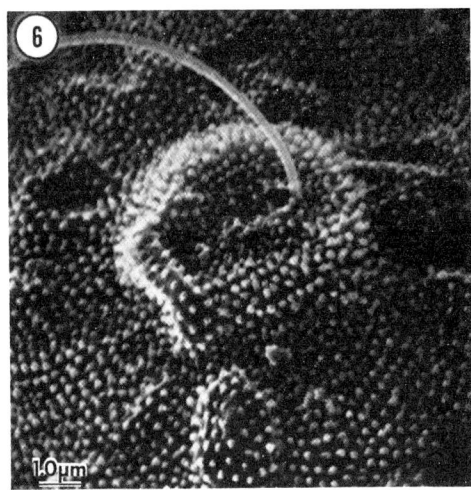

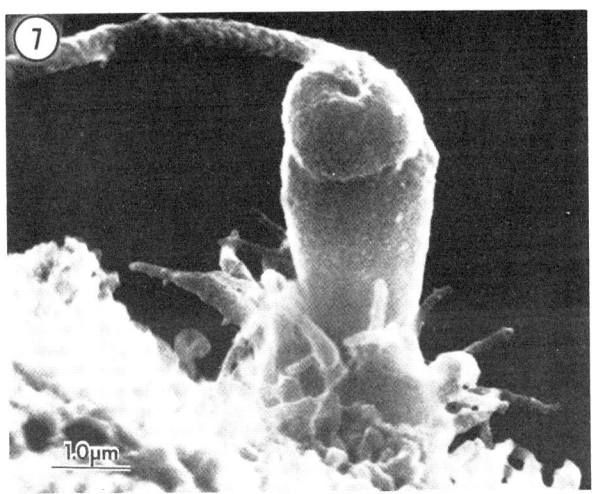

Fig. 6. Sperm Incorporation. The sperm head and midpiece are pulled from view, leaving only the sperm tail protruding as the fertilization coat starts to elevate. S. purpuratus. Bar: 1 µm.

Fig. 7. Sperm Incorporation Viewed at the Egg Plasma Membrane. In eggs devoid of their vitelline layers the activity of the egg surface in engulfing the sperm is clearly apparent. Microvilli have elongated, to 1.2 µm, to completely surround the successful sperm. These microvilli will continue to elongate to form the fertilization cone. L. variegatus. Bar: 1 µm. Reprinted, with permission, from Schatten and Schatten, 1980.

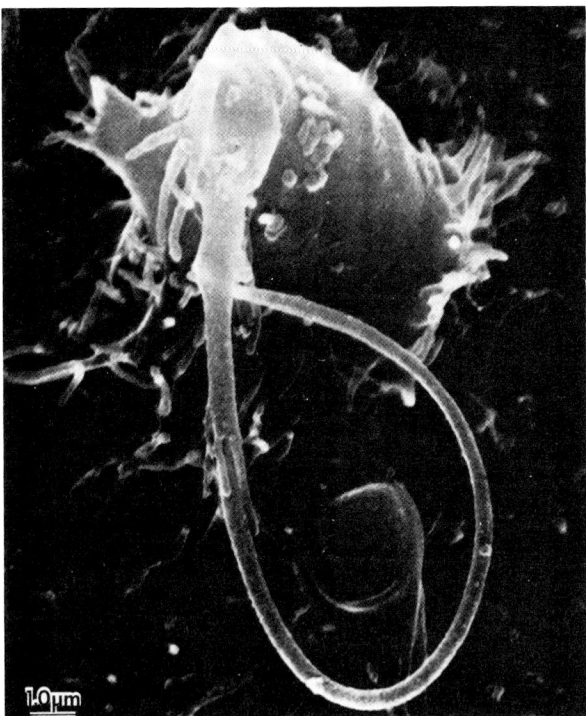

Fig. 8. Sperm Incorporation Viewed at the Egg Plasma Membrane Surface. The fertilization cone forms from these elongating microvilli, which surround the base of the fertilization cone and which continue to engulf the sperm. Note the microvilli surrounding the sperm tail. L. variegatus. Bar: 1 μm. Reprinted, with permission, from Schatten and Schatten, 1980.

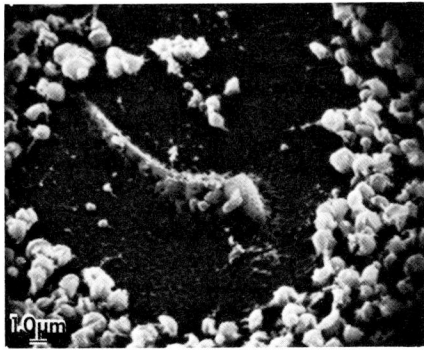

Fig. 10. Scanning Electron Microscopy of the Cortical Surface During Insemination. The intimate association between these cortical elements and the membraneless spermatozoon can be observed as the spermatozoon continues to rotate through the egg surface. S. purpuratus. Bar: 1 μm. Reprinted, with permission, from Schatten and Mazia, 1976.

sperm head, the midpiece, and, surprisingly, the sperm tail; engulf the entire spermatozoan; and form into the fertilization cone (Fig. 8), which at this stage is apparent in light microscopy. A short time later, the sperm head and midpiece rotate on the egg surface to lie parallel with the egg cortex during the subsequent events of incorporation (Schatten, 1981).

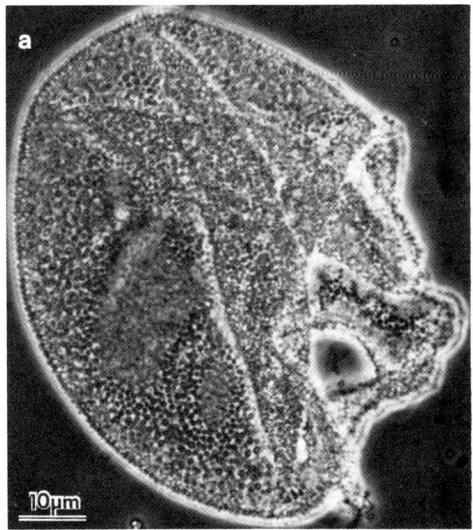

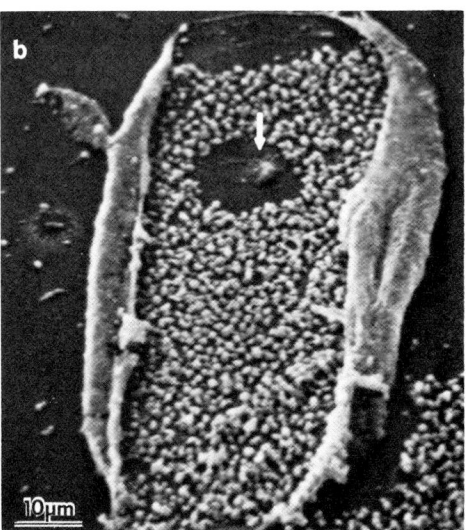

Fig. 9. The Cortical View of Insemination. a. A phase-contrast micrograph of an isolated surface cortex. Although empty, these surfaces tend to maintain the spherical shape of the egg. The cortical granules can be observed as the small granules underlying the egg surface. S. purpuratus. Bar: 10 μm. Reprinted, with permission, from Schatten and Mazia, 1976. b. Scanning electron microscopy of a surface isolated shortly after fertilization. The cortical granules are discharged around the site of sperm entry(↓). The outer surface can be observed at the right and left where the surface is folded over. S. purpuratus. Bar: 10 μm. Reprinted, with permission, from Schatten and Mazia, 1976.

Since scanning electron microscopy is limited to the study of cell surfaces, it is not possible to follow the sperm through the plasma membrane during incorporation. The only means by which investigators could exploit the resolving power and depth of field of the scanning electron microsocope was to study isolated cellular structures, such as the isolated egg surface.

Advances in understanding the surface events at the inner face of the egg surface during sperm incorporation are reviewed by the study of the cortical (inner) face of isolated egg surfaces immediately following insemination. Vacquier (1975) developed an ingenious method for isolating the unfertilized egg surface by affixing eggs to cationic surfaces and shearing the tops of the eggs off in a calcium-free environment. These cortical "lawns" could be induced to undergo secretion of the cortical granules in vitro by the addition of calcium ions. By a modification of this method and the development of an isolation medium that mimicked the intracellular environment of the unfertilized surface, Schatten and Mazia (1976) were able to isolate the entire egg surface in suspension. When the egg surface was isolated moments after insemination, the ultrastructural features of the sperm moving through the egg surface could be captured with scanning electron microscopy. In Figure 9a, the phase contrast image of an unfertilized egg surface isolated in suspension is depicted. Figure 9b is a low-magnification scanning electron micrograph of an egg surface isolated one minute after insemination in which a sperm is captured within the egg surface (arrow). In this scanning electron micrograph, the outer faces of the egg surface are apparent at the left and right. The egg cortex with its adherent cortical granules is apparent in the middle of the image, and towards the top of the image a patch of cortical granules through which the sperm is entering is documented. The isolation medium employed a calcium-free environment; had calcium been present, the exocytosis of the cortical granules would have continued to radiate from the sperm-egg fusion site.

Figure 10, the higher-magnification image of a sperm during incorporation viewed from the interface of the egg cortex, demonstrates an intricate array of fibrous netting. This netting is extractable with 0.6 M KI, and the extract contains an electrophoretic band that comigrates with rabbit muscle actin, inviting speculations about the role of microfilaments during sperm incorporation.

To evaluate the role of microfilaments during sperm incorporation, microfilament inhibitors have been used on both gametes and the results assayed with the scanning electron microscope. Sanger and Sanger (1975) demonstrated that the actin polymerization occurring during the acrosome reaction of the sperm was not sensitive to cytochalasin B. Gould-Somero et al. (1977), Longo (1980), Byrd and Perry (1980) and Schatten and Schatten (1979, 1980, 1981) present evidence that eggs treated with cytochalasin B are unable to incorporate the sperm even though sperm-induced cortical activation has occurred. Banzhaf et al. (1980) have demonstrated the cortical sensitivity of fertilized eggs to cytochalasin and have determined the speed of cytochalasin permeability. The finding that the fertilization cone and sperm incorporation are blocked by the cytochalasins when they are added before insemination (Fig. 11), and that the fertilization cone is rapidly resorbed when the cytochalasins are added subsequent to sperm-egg fusion (Schatten and

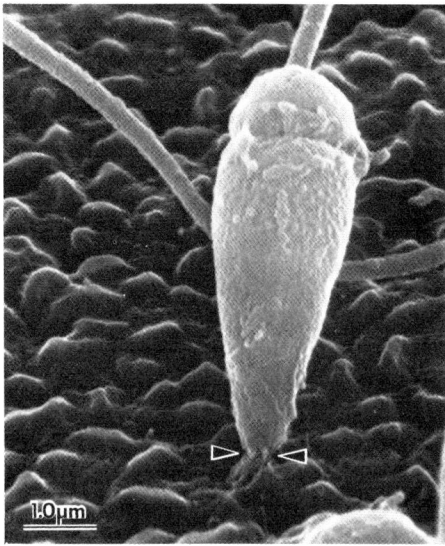

Fig. 11. The Surface Events During Cytochalasin Fertilization. Sperm binding occurs at the plasma membrane (arrowheads), though the egg-mediated elongation of microvilli and the formation of the fertilization cone are prevented by cytochalasin B. L. variegatus. Bar: 1 μm. Reprinted, with permission, from Schatten and Schatten, 1980.

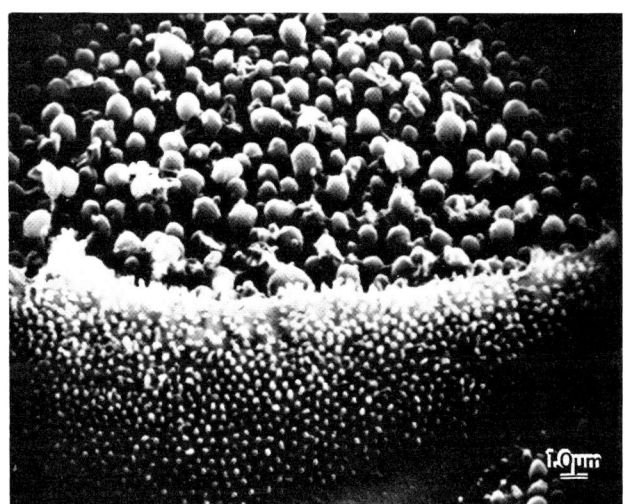

Fig. 12. The Unfertilized Egg Surface. In this isolated egg surface, observed by scanning electron microscopy, the various layers are depicted. At the bottom, the casts of microvilli forming papillae in the vitelline layer are apparent. The vitelline layer drapes over the egg plasma membrane and conforms to its topography. At the top of the image, viewing the inner cortical face of the egg surface, are numerous cortical granules, which appear attached to the plasma membrane by fibrous elements. Immediately following sperm-egg fusion, these cortical granules fuse their membranes with the plasma membrane and the resultant secretory event elevates the fertilization coat. S. purpuratus. Bar: 1 μm. Reprinted, with permission, from Schatten, 1975.

Schatten, 1981), argue strongly for an important role of microfilaments within the fertilization cone during sperm incorporation.

## Cortical Restructuring, Microvilli Elongation, and Fertilization Coat Elevation

The egg surface at fertilization undergoes a complex series of modifications and alterations, which are crucial in converting the unfertilized egg to a fertilized zygote and which affect the manner in which the egg interacts with its environment. The various components have been isolated and characterized: the vitelline layer (Glabe and Vacquier, 1977), the plasma membrane with attached cortical granules (Detering et al., 1977) and the entire unfertilized cortex (Vacquier, 1975). The biochemical features of this surface restructuring has recently been reviewed by Shapiro and Eddy (1980).

To understand the surface events at fertilization it is first essential to review the layers covering the unfertilized egg. Beneath the jelly coat, which likely plays an important role in triggering the acrosome reaction of the sperm, is the vitelline layer. The vitelline layer is draped over the plasma membrane of the unfertilized egg and conforms to its topography. Beneath the plasma membrane of the unfertilized egg and attached to it are a monolayer of 1-µm secretory granules, the cortical granules (Fig. 12; reviewed by Anderson, 1968, Anderson, 1974, and Schuel, 1978). At the moment of sperm-egg fusion, the cortical granules fuse their membranes with the plasma membrane, in a wave-like motion starting at the site of sperm-egg fusion and radiating to encompass the entire egg surface. The fusion of the cortical granules with the plasma membrane, the cortical reaction, externalizes the contents of the cortical granules into the space between the vitelline layer and the plasma membrane, the perivitelline space. The contents of the cortical granules attach subjacent to the vitelline layer, elevating it and hardening it to form the fertilization coat (Chandler and Heuser, 1979, 1980, 1981; Foerder and Shapiro, 1977). The membrane added during the cortical reaction may well be resorbed by clathrin-coated vesicles (Fischer and Rebhun, 1981). The fertilization coat serves as a barrier between the developing embryo and the external environment, protecting it from supernumerary sperm and bacterial infestation (M. Daniels, unpublished results).

In addition to the secretion of the cortical granules the surface undergoes a series of motile events following that of sperm incorporation (discussed in the previous section). The egg cortex has recently been carefully reviewed by Vacquier (1981). The microvilli on the unfertilized egg are short, stubby and very well aligned. Following fertilization, the microvilli adjacent to the sperm elongate to form the fertilization cone and are essential for proper sperm incorporation. Additionally, the remaining microvilli on the egg surface undergo two bursts in microvillar elongation (Schroeder, 1979). The first burst occurs within five minutes of the sperm-egg fusion and is associated with the addition of the cortical granule membrane to the plasma membrane. The second burst (Figure 13) occurs prior to mitosis, perhaps resulting in the increase in surface area necessary for cell division. Cytokinesis, of course, is another cyclical event mediated by cortical motility.

Loeb (1913) was one of the earliest workers recognizing the crucial importance of the cortical rearrangement to the onset of development. He regarded the cortical reaction as a cytolytic process, an analogy that is not unfair considering the dramatic changes in physiognomy between the surfaces of unfertilized and fertilized eggs. The requirements for first rapid (Jaffe, 1976) and then permanent blocks to polyspermy (Vacquier et al., 1972a, b, reviewed by Dale and Monroy, 1981) are perhaps obvious reasons for this dramatic alteration of surface features as the unfertilized egg progresses to the fertilized state. Still unexplained surface changes, shown to be correlated with fertilization or artificial activation (Mazia et al., 1975; Spiegel and Spiegel, 1977) and required for the proper completion of the first cell cycle (Schatten and Schatten, 1981), are the changes in microvillar length. The changing configurations of the surface microvilli have been studied by scanning electron microscopy (Eddy and Shapiro, 1976; Mazia et al., 1975; Spiegel and Spiegel, 1977), and the presence of microfilaments as the underlying substructure has been demonstrated by transmission electron microscopy (Kidd et al., 1976; Burgess and Schroeder, 1977; Longo and Anderson, 1968; Tilney and Jaffe, 1980). Interestingly, the sperm-induced cortical reaction appears insensitive to cytochalasin B, an inhibitor of microfilament assembly, though cytochalasin B alone has been reported to cause limited exocytosis of individual cortical granules (Longo, 1978; Schatten and Schatten, 1980). However, the elongation of the egg microvilli is sensitive to this inhibitor (Eddy and Shapiro, 1976; Longo, 1980; Schatten and Schatten, 1980). Additionally, the progression of the fertilized egg through the first cell cycle requires the proper restructuring of the egg cortex, which itself is sensitive to cytochalasin B (Schatten and Schatten, 1981).

In summary then, the surface modifications occurring at fertilization include, of course, sperm incorporation and then the elevation of the fertilization coat resulting from the cortical reaction. Two bursts of microvillar elongation are noted during the first cell cycle, and changes in total surface area have been reviewed by Schroeder (1981). It appears likely that one function of the alteration in microvillar length is to provide a means first to secure and later to resorb the membrane added by the cortical granules following insemination.

## Cell Division and Later Development

Once fertilization has been successfully completed, signalled by the proper fusion of the male pronucleus with the female pronucleus (reviewed by Schatten, 1982), the task of the

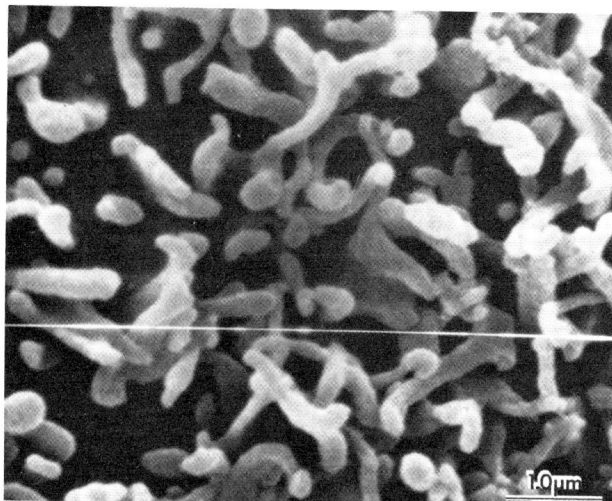

Fig. 13. Elongated Microvilli. Following fertilization, the egg microvilli undergo microfilament-mediated bursts in microvillar elongation. In this scanning electron micrograph of an egg denuded of all surface layers at forty minutes following insemination, the elongated pattern of microvilli is striking when compared to the previous image of the unfertilized egg surface. L. variegatus. Bar: 1 µm. From Schatten et al., 1981.

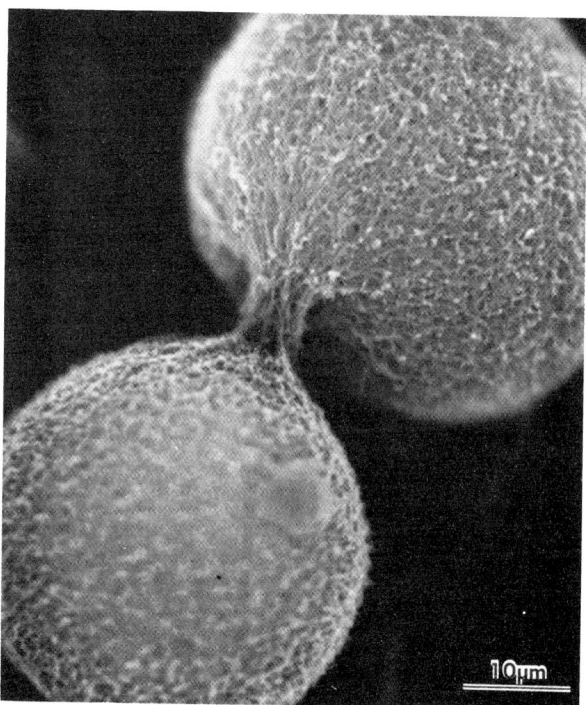

Fig. 14. First Division. The division of the spherical egg into two cells, a process referred to as cytokinesis, involves the contraction of the equatorial region of the dividing egg. Note the presence of numerous microvilli on the egg surface. The fertilization coat elevation was prevented with dithiothreitol and the hyaline layer was removed with a calcium-free wash before fixation. Arbacia punctulata. Bar: 10 µm.

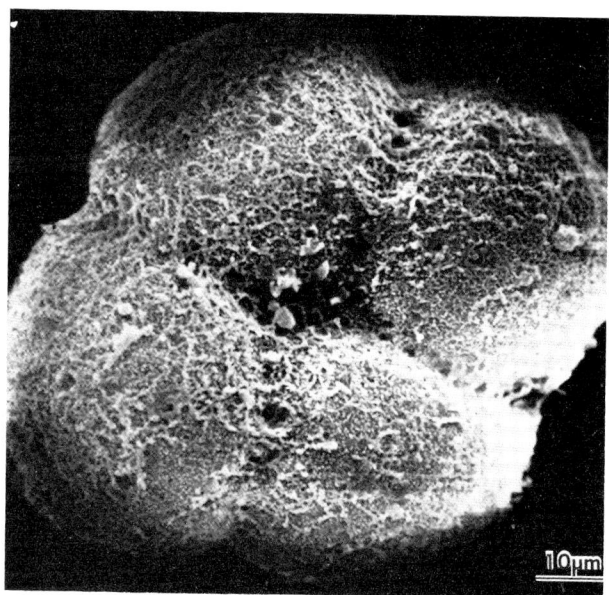

Fig. 15. Second Division. The second and subsequent division planes occur perpendicular to the previous axis for division. Here remnants of the hyaline layer are apparently draped over the egg surface. A. punctulata. Bar: 10 µm.

zygote is to initiate the cell divisions that will permit the morphogenetic motions leading to embryogenesis and later development. A full discussion of the theories concerning the processes of mitosis and cytokinesis are beyond the scope of this section. Interested readers are directed to the classic reviews (Mazia, 1961; Wilson, 1925) and to the recent reviews (Inoue, 1981; Forer and Zimmerman, 1982).

To appreciate the events leading to first division it is important to analyze the contribution of each gamete to the resultant zygote. Of course, each has a haploid genome that will be required for later development. The egg, containing a large pool of microfilament and microtubule precursors, does not appear to have any existing centrioles; a pair are contributed by the sperm at insemination. This point is underscored by the problems confronting a polyspermic egg, which, instead of having a pair of centrioles contributed by one sperm and leading to a bipolar mitotic apparatus, has one contributed by each supernumerary sperm, leading to a multipolar mitotic apparatus and an abortive cleavage attempt. Each contributes a pair of centrioles, all of which attempt to organize asters at the first division.

The pair of sperm centrioles is directed toward the egg center when the sperm rotates during sperm incorporation. At first, a monopolar structure, the sperm aster, forms and plays crucial roles in the pronuclear migrations (Bestor and Schatten, 1981; Balczon and Schatten, 1983). During the final stages of the growth of the sperm aster this monastral structure develops two focal points because of the separation, and perhaps even replication, of the pair of sperm-contributed centrioles. This separation of the pair of sperm centrioles occurs perpendicular to the direction of the movement of the pronuclei to the egg center and will be crucial in the establishment of the first embryonic axis. The sperm aster typically disassembles prior to syngamy, and as the cell progresses through the first cell cycle a planar apparatus forms immediately prior to prophase. This structure is referred to as the "streak," the "interim apparatus" and the "interphase asters" and is rather transitory in nature. The axis of the streak is typically perpendicular to the final motion of the sperm aster and is usually parallel with the mitotic axis. The streak distorts the swollen spherical zygote nucleus and typically disassembles to permit the zygote nucleus to reform its spherical shape immediately prior to nuclear envelope breakdown. Mitosis involves the well known scheme of astral formation at the poles and chromosome condensation moving the paired chromosomes to the metaphase plate when first the anaphase movements of the chromosomes and later the separation of the poles occur.

In addition to these cytoplasmic events during the first cell cycle, there are also cortical changes. Tracing the surface alterations from fertilization, microvilli form into the fertilization cone surrounding the successful sperm when the cortical reaction propagates over the egg surface. The remaining egg microvilli undergo their first burst of elongation within five minutes of insemination. Following this first burst of elongation, the net surface area is reduced, presumably because of the resorption of the added cortical granule membrane. Prior to prophase and around the streak stage the microvilli undergo a second burst of elongation. Following mitosis, the cell surface undergoes perhaps the most impressive cortical change, namely first cleavage, when the contractile ring divides the fertilized egg into two (Fig. 14) and, at second division (Fig. 15) four, cells.

Scanning electron microscopy has provided a number of concrete answers to old problems of cell division. With this optimistic statement, let us dispel the notion that we understand the mechanics of cell division; we only understand why certain hypotheses are wrong. SEM, which has characterized the bursts of microvillar elongation, has been utilized to demonstrate that gross changes in microvillar elongation and distribution are not central to the question of cytokinesis (reviewed by Schroeder, 1981).

In Figures 14 and 15 numerous microvilli are seen at all regions of egg surface. The impression of increased microvillar density in the cleavage furrow has been refuted by morphometric measurements (cited in Schroeder, 1981). The attractive hypothesis that a contractile ring, specifically a band of microfilaments at the equatorial region during cell division, is responsible for cytokinesis (Schroeder, 1981) appears as an unlikely mechanism to account fully for the surface changes during cell division. Recent fluorescence labelling of the cortical microfilaments argues for an involvement of the entire surface during cytokinesis (Cline et al., 1983). A more likely theory is that the entire egg cortex plays a coordinate role with polar expansion and equatorial contraction resulting in the observed motions during division.

The later stages of development are characterized by repeated rounds of cell division to form a somewhat irregular aggregate of loosely apposed cells called the morula because of their mulberry-like appearance figure (Fig. 16). Following the rapid cell division cycle phase, the cells become more closely apposed and begin to interact as an organism. At this stage, the blastula (Fig. 17) is characterized by a smooth spherical appearance; the cells have long cilia protruding on their outer surfaces. The coordinated beating of these cilia is responsible in part for the hatching of this organism from the fertilization coat. Within the interior of this hollow ball, a fluid-filled cavity, the blastocoel, has formed into which the primary mesenchymal cells will invade as the first step leading towards gastrulation (Fig. 18). The gastrula, which now has the beginnings of a digestive tract, is a free-swimming, feeding larva capable of leading an independent existence.

## Future Directions

The contribution of scanning electron microscopy to our understanding of fertilization and early development, all performed within the last decade, has resulted in leaps in our understanding of the surface events crucial for development, reproduction and cell biology generally. Though higher resolution scanning electron microscopy will provide additional answers especially in the fields of cell division and intracellular interactions, a truly bright future exists for the more sophisticated electron imaging and analytical instrumentation. Scanning transmission electron microscopy and high voltage electron microscopy are liable to provide the resolution necessary to answer many of the remaining questions. X-ray microanalysis and electron loss spectroscopy are likely to solve questions concerning the presence of crucial elements suspected to regulate the events of fertilization, division and early development.

The contributions of the past several years, many of these involving scanning electron microscopy, will have no doubt provided a stable foundation from which future investigations will lead to solutions of these basic problems in cell and developmental biology.

## Acknowledgements

The support of the authors' research by the National Institutes of Health (Research Grant HD 12913; Research Career Development Award HD363; Analytical Scanning Transmission Electron Microscope Instrumentation Award RR 1466) is gratefully acknowledged.

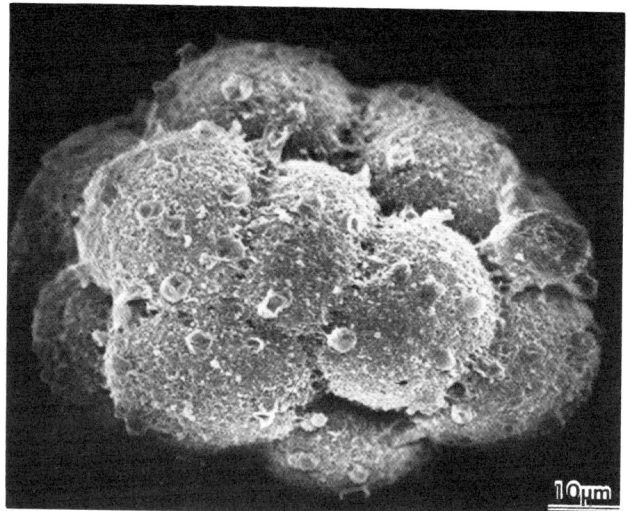

Fig. 16. Morula. The morula stage is characterized by the loose association of numerous cells. A. punctulata. Bar: 10 μm.

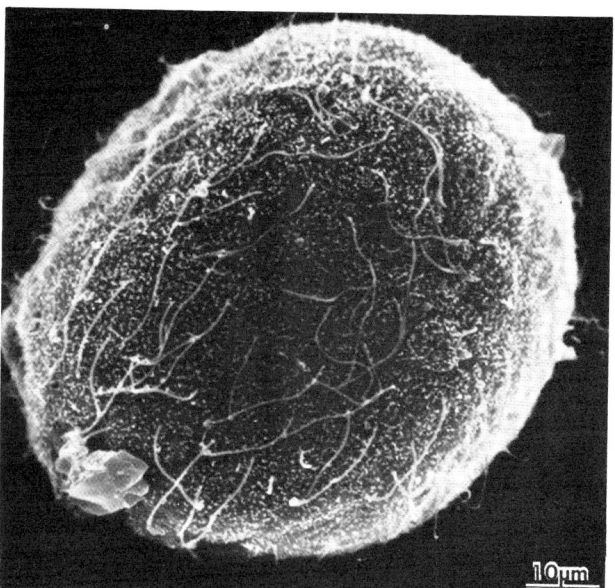

Fig. 17. Blastula. The blastula is a hollow ball of closely apposed ciliated cells. The cilia permit the blastula to hatch from the fertilization coat and to swim freely. A. punctulata. Bar: 10 μm.

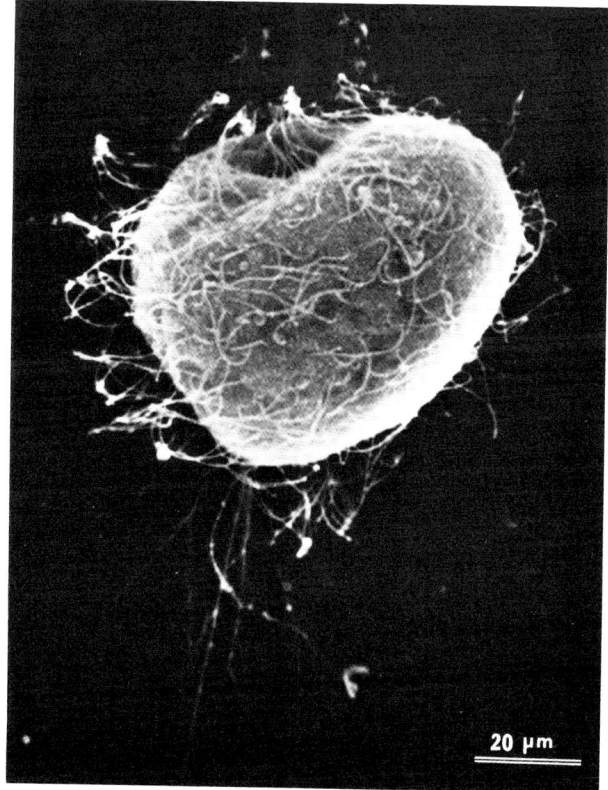

Fig. 18. Gastrula. With the invasion of the primary mesenchymal cells into the blastocoel of the blastula, the gastrula stage forms. This feeding stage is capable of dispersing the species. A. punctulata. Bar: 20 μm.

## References

Anderson E. (1968). Oocyte differentiation in the sea urchin, Arbacia punctulata, with particular reference to the origin of cortical granules and their participation in the cortical reaction. J. Cell Biol. 37, 514-539.

Anderson E. (1974). Comparative aspects of the ultrastructure of the female gamete. Int. Rev. Cytol. Suppl. 4, 1-70.

Balczon R, Schatten G. (1983). Microtubule-containing detergent-extracted cytoskeletons during sea urchin fertilization. Cell Motility 3, in press.

Banzhaf WC, Warren RH, McClay DR. (1980). Cortical reorganization following fertilization of sea urchin eggs: sensitivity to cytochalasin B. Dev. Biol. 80, 506-515.

Bestor TH, Schatten G. (1981). Anti-tubulin immunofluorescence microscopy of microtubules present during the pronuclears of sea urchin fertilization. Develop. Biol. 88, 80-91.

Burgess DR, Schroeder TE. (1977). Polarized bundles of actin filaments within microvilli of fertilized sea urchin eggs. J. Cell Biol. 74, 1032-1037.

Byrd W, Perry G (1980). Cytochalasin B blocks sperm incorporation but allows activation of the sea urchin egg. Exp. Cell Res. 126, 333-342.

Chandler DE, Heuser J. (1979). Membrane fusion during secretion. J. Cell Biol. 83, 91-108.

Chandler DE, Heuser J. (1980). The vitelline layer of the sea urchin egg and its modification during fertilization. J. Cell Biol. 84, 618-632.

Chandler DE, Heuser J. (1981). Postfertilization growth of microvilli in the sea urchin egg: New views from eggs that have been quick-frozen, freeze-fractured, and deeply etched. Dev. Biol. 82, 393-400.

Cline C, Schatten H, Balczon R, Schatten G. (1983). Actin-mediated surface motility during sea urchin fertilization. Cell Motility, in press.

Dale B, Monroy A. (1981). How is polyspermy prevented? Gamete Res. 4, 151-169.

Detering NK, Decker GL, Schmell ED, Lennarz WJ. (1977). Isolation and characterization of plasma membrane associated cortical granules from sea urchin eggs. J. Cell Biol. 75, 899-914.

Eddy EM, Shapiro BM. (1976). Changes in the topography of the sea urchin egg after fertilization. J. Cell Biol. 71, 35-48.

Epel D, Weaver AM, Mazia D. (1970). Methods for removal of the vitelline membrane of sea urchin eggs. Exp. Cell Res. 61, 64-68.

Fischer GW, Rebhun LI. (1981). Turn on of endocytotic processes in response to sea urchin egg activation accompanies restructuring of the egg surface. J. Cell Biol. 91, 185a.

Foerder CA, Shapiro BM. (1977). Release of ovoperoxidase from sea urchin eggs hardens the fertilization membrane with tyrosine crosslinks. Proc. Natl. Acad. Sci. USA 74, 4214-4218.

Forer A, Zimmerman A. (1982). Mitosis and Cytokinesis, Academic Press, New York, 1-479.

Friend DS. (1982). Plasma-membrane diversity in a highly polarized cell. J. Cell Biol. 93, 243-250.

Gibbons IR. (1981). Cilia and flagella of eukaryotes. J. Cell Biol. 91, 107S-124S.

Glabe CG, Vacquier VD. (1977). Isolation and characterization of the vitelline layer of sea urchin eggs. J. Cell Biol. 75, 410-421.

Gould-Somero M, Holland L, Paul M. (1977). Cytochalasin B inhibits sperm penetration into eggs of Urechis caupo. Develop. Biol. 58, 11-22.

Inoue S. (1981). Cell division and the mitotic spindle. J. Cell Biol. 91, 131S-147S.

Jaffe LA. (1976). Fast block to polyspermy in sea urchin eggs is electrically mediated. Nature (London) 261, 68-71.

Kidd P, Schatten G, Grainger J, Mazia D. (1976). Microfilaments in the sea urchin egg at fertilization. Biophys. J. 16, 117a.

Loeb J. (1913). Artificial Parthenogenesis and Fertilization, University of Chicago Press, 1-113.

Longo F. (1978). Effects of cytochalasin B on sperm-egg interactions. Develop. Biol. 62, 271-291.

Longo F. (1980). Organization of microfilaments in sea urchin (Arbacia punctulata) eggs at fertilization: Effects of cytochalasin B. Develop. Biol. 74, 422-431.

Longo F, Anderson E. (1968). The fine structure of pronuclear development in the sea urchin, Arbacia punctulata. J. Cell Biol. 39, 339-368.

Mazia D. (1961). Mitosis and physiology of cell division, in: The Cell, Vol. 3, J. Brachet and A. E. Mirsky (eds), Academic Press, New York, 77-424.

Mazia D, Schatten G, Steinhardt R. (1975). Turning on of activities in unfertilized sea urchin eggs: Correlation with changes of the surface. Proc. Natl. Acad. Sci. USA 72, 4469-4473.

Poccia DL. (1982). Biochemical aspects of sperm nucleus activation by egg cytoplasm. J. Wash. Acad. Sci. 72, 24-33.

Sanger JW, Sanger JM. (1975). Polymerization of sperm actin in the presence of cytochalasin-B(1). J. Exp. Zool. 193, 441-447.

Schatten G. (1975). The Cell Surface Complex of the Ovum. University of California Berkeley.

Schatten G. (1981). Sperm incorporation, the pronuclear migrations, and their relation to the establishment of the first embryonic axis: Time-lapse video microscopy of the movements during fertilization of the sea urchin Lytechinus variegatus. Develop. Biol. 86, 426-437.

Schatten G. (1982). Motility during fertilization. Int. Rev. Cytl. 79, 59-164.

Schatten G, Mazia D. (1976). The surface events at fertilization: The movements of the spermatozoon through the sea urchin egg surface and the roles of the surface layers. J. Supramolec. Struct. 5, 343-369.

Schatten G, Schatten H. (1979). Sperm-egg membrane fusion and interactions in denudated sea urchin eggs. Scanning Electron Microsc. 1979; III: 299-305.

Schatten G, Schatten H. (1981). Effects of motility inhibitors during sea urchin fertilization. Exp. Cell Res. 135, 311-330.

Schatten G, Bestor T, Cline C, Schatten H. (1981). Regulation of motility during fertilization. J. Cell Biol. 91, 184a.

Schatten H, Schatten G. (1980). Surface activity at the egg plasma membrane during sperm incorporation and its cytochalasin B sensitivity. Develop. Biol. 78, 435-449.

Schroeder T. (1979). Surface area change at fertilization: resorption of the mosaic membrane. Develop. Biol. 70, 306-326.

Schroeder T. (1981). Interrelations between the cell surface and the cytoskeleton in clearing sea urchin eggs, in: Cytoskeletal Elements and Plasma Membrane Organization, G. Poste and G. L. Nicolson (eds), Elsevier-North Holland, 170-216.

Schuel H. (1978). Secretory functions of egg cortical granules in fertilization and development: A critical review. Gamete Res. 1, 299-382.

Shalgi R, Phillips DM. (1980a). Mechanics of in vitro fertilization in the hamster. Biol. Reprod. 23, 433-444.

Shalgi R, Phillips DM. (1980b). Mechanics of sperm entry in cyclic hamster. J. Ultrastruct. Res. 71, 154-161.

Shapiro BM, Eddy EM. (1980). When sperm meets egg: Biochemical mechanisms of gamete interactions. Intl. Rev. Cytol. 66, 257-302.

Spiegel E, Spiegel M. (1977). Microvilli in sea urchin eggs (differences in their formation and type). Exp. Cell Res. 109, 452-465.

Tegner M, Epel D. (1976). Scanning electron microscope studies of sea urchin fertilization. I. Eggs with vitelline layers. J. Exp. Zool. 197, 31-58.

Tilney LG. (1978). Polymerization of actin (V. A new organelle, the actomere, that initiates the assembly of actin filaments in Thyone sperm). J. Cell Biol. 77, 551-564.

Tilney LG, Jaffe LA. (1980). Actin, microvilli, and the fertilization cone of sea urchin eggs. J. Cell Biol. 87, 771-782.

Usui N, Sano K, Mohri H. (1980). The surface events at fertilization of the sea urchin egg. Dev., Growth Diff. 22, 461-473.

Vacquier VD. (1975). The isolation of the intact cortical granule from sea urchin eggs: Calcium ions trigger granule discharge. Develop. Biol. 43, 62-74.

Vacquier VD. (1981). Dynamic changes of the egg cortex. Develop. Biol. 84, 1-26.

Vacquier VD, Moy GW. (1977). Isolation of bindin: The protein responsible for adhesion of sperm to sea urchin eggs. Proc. Natl. Acad. Sci. USA 74, 2456-2460.

Vacquier VD, Epel D, Douglas L. (1972a). Sea urchin eggs release protease activity at fertilization. Nature (London) 237, 34-36.

Vacquier VD, Tegner MJ, Epel D. (1972b). Protease activity establishes the block against polyspermy in sea urchin eggs. Nature (London) 240, 352-353.

Wilson EB. (1925). The Cell in Development and Heredity, The MacMillan Co., New York, 1-1232.

Yanagimachi R. (1978). Sperm-egg association in mammals. Current Top. Dev. Biol. 12, 83-105.

## Discussion with Reviewers

P. Kenemans: Please indicate the extent to which the descriptions you have given for fertilization and early development in sea urchins apply to other non-mammalian, mammalian, primate, and human beings, respectively.
Authors: The images and descriptions in this article apply specifically to those of sea urchins, but generally to fertilization per se. There are wide variations in the basic themes; for example, in other non-mammalian organisms such as horseshoe crabs, the acrosome reaction can elicit a process that is longer than the sperm tail. However, our best evidence for the surface events during mammalian and human fertilization includes the salient features described here: an acrosome reaction occurs in the mammalian sperm albeit with a considerably different morphological change. Microvilli on the mammalian oocyte surface appear to engulf the equatorial region of the mammalian sperm, and interestingly the mammalian sperm has actin at that site. Additionally, there appears to be a cortical reaction at mammalian fertilization that is central to the block to polyspermy.

P. Kenemans: What criteria should be satisfied before fertilization can be said to be unequivocally demonstrated?
Authors: Strictly speaking, fertilization is the process that culminates when the parental genomes merge. Therefore, for fertilization to be successful, the DNA of the sperm and egg must meet, either during fusion of the pronuclei or when the chromosomes align at the metaphase plate.

P. Kenemans: After successful sperm incorporation, what is the fate of the sperm flagellum components and sperm mitochondrial DNA? Is the latter incorporated into the egg mitochondria, thus giving a paternal contribution to the cytoplasmic inheritance of the new embryo?
Authors: It appears that the sperm flagellum is retained during the first cell divisions and ultimately breaks down. It may well contribute its tubulin to the cytoplasmic tubulin pool for use in later embryogenesis. The dogma regarding the sperm mitochondrion is that it is broken down and lost and therefore does not provide a paternal contribution for the developing embryo. However, this point ought to be reinvestigated with modern molecular techniques.

E. M. Eddy: What are the relative time courses for the processes considered (acrosome reaction, cortical reaction, fertilization envelope formation, sperm incorporation, microvillar elongation, fusion of the pronuclei, cell division) in relation to the time of fertilization and the initiation of development?
Authors: The acrosome reaction is a very rapid event and is concluded within seconds. The cortical reaction starts roughly twenty seconds after the sperm triggers the bioelectric events of fertilization and the cortical reaction leads directly to the fertilization envelope formation. The fertilization envelope is fully formed by three minutes. Sperm incorporation starts roughly twelve seconds after the onset of the bioelectrical events and is concluded within two minutes. The microvilli typically elongate within ten minutes after fertilization. The fusion of the pronuclei occurs at roughly fifteen minutes, and cell division occurs by an hour. We should point out that these values are obtained on animals living in the relatively warm, tropical Gulf of Mexico.

Additional discussion with reviewers of the paper "A Scanning Electron Microscopic Surface and Cryofracture Study of Development in the Planulae of the Hydrozoan, Pennaria Tiarella" by A.E. Hotchkiss, V.J. Martin and R.P. Apkarian continued from page 22.

V.C. Barber: Could the authors elaborate on the methods used to obtain measurements in the SEM? Without the use of stereological methods, measurements of the accuracy presented in the text must be highly suspect.
Authors: Although stereological methods to obtain accurate structural measurements were not employed in this study, approximate dimensions are provided in order that relative developmental changes in shape and size could be more easily assessed.

J.A. Westfall: I fail to see the distinction between interstitial and gastrodermal cell nuclei. Also, I cannot see close contact between interstitial cells in these preparations. Are interstitial cell boundaries visible by cryofracture?
Authors: There is no real distinction between interstitial and gastrodermal cell nuclei on the level of our study. We attempted to distinguish between the two cell types by relative positions of the cells in the endoderm at various stages of development. Interstitial cells were observed in close contact with one another (Fig. 5). At present, however, we have not observed junctional complexes between cell membranes in cryofracture SEM preparations.

J.A. Westfall: How do you know that interstitial cells and nematoblasts migrate from endoderm to ectoderm? Have you observed them in the process of crossing the mesoglea?
Authors: We have not observed interstitial cells or nematoblasts in the process of crossing the mesoglea in this study. It has been suggested by other investigators (11, 12, 13, 15) that this migration occurs. We did observe interstitial cells and nematoblasts in the ectoderm by 17 hours and prior to this stage these cell types were found only in the endoderm. Thus, our observations seem to support this previously suggested migratory phenomenon.

Reviewer III: How commonly was the ganglionic nerve cell plexus observed? Was it seen only once? Was it seen only in 24 hr. planulae? Was there any anterior or posterior concentration?
J.A. Westfall: How can you tell from these preparations that the nerve plexus is running both longitudinally and transversely?
Authors: The nerve plexus composed of ganglionic cells was first detected by SEM-cryofracture at 24 hours post-fertilization and also observed with TEM in older planulae (e.g., 12). An analysis of anterior or posterior concentrations of ganglionic cells was not performed in the present study. In these preparations, you can only observe the transverse orientation of the nerve plexus. TEM micrographs demonstrate both the longitudinal and transverse orientations of the neurites composing the nerve plexus (compare Figs 12 and 15).

J.A. Westfall: Why are the sensory cell microvilli so different in Figs. 10 and 16? Could the microvillar protrusions in Fig. 16 belong to mucous cells? Also, could the cone-shaped ciliary structure near the magnification bar in Fig. 16 belong to a nematocyte?
Authors: Figure 10 demonstrates the early appearance of the sensory cell surface specialization at 24 hours, while Fig. 16 demonstrates development of these specializations by 48 hours. The change in the appearance of the sensory cell microvilli from a simple collar configuration to a "button" formation may possibly coincide with differentiation of the sensory cell. It is not likely that the microvillar protrusions in Fig. 16 belong to mucous cells since this type of association containing a collar of tall microvilli was not characteristic of mucous cells in younger planulae. It is very possible that the cone-shaped ciliary structure seen in Fig. 16 belongs to a cnidocyte. The 48 hour planula contained differentiating cnidocytes in its ectodermal regions. Surface specializations of cnidocytes which resemble the cone-shaped ciliary structure on the 48 hour planulae in Figure 16 have been seen on the surfaces of the 60 hour planulae.

J.A. Westfall: What is the function of the mesoglea-associated granules?
Authors: The function of these granules is unknown.

A SCANNING ELECTRON MICROSCOPIC SURFACE AND CRYOFRACTURE STUDY OF DEVELOPMENT IN THE PLANULAE OF THE HYDROZOAN, PENNARIA TIARELLA

Anne E. Hotchkiss,[1]* Vicki J. Martin,[2] and Robert P. Apkarian[3]

[1]Department of Biology, University of Louisville, Louisville, KY 40292
[2]Department of Biology, University of Notre Dame, Notre Dame, IN 46556
[3]Graduate School SEM Facility, University of Louisville, Louisville, KY 40292

(Paper received January 16, 1984, Completed manuscript received June 22, 1984)

## Abstract

Combined techniques of scanning electron microscopy (SEM) and cryofracture were used to study cellular morphology and cellular interactions during development of the planula larva of the marine hydrozoan, Pennaria tiarella. 10, 17, 24, and 48 hour planulae were prepared for either surface or cryofracture analysis. Specimens to be cryofractured were dehydrated to 100% ethanol, wrapped in parafilm, frozen in liquid Freon-22 and delicately fractured under liquid nitrogen using a modified tissue-chopper followed by critical point drying.

SEM-cryofracture produced finely preserved three-dimensional views of ectodermal and endodermal cell layers and the acellular mesogleal layer. Examination of the surfaces of planulae at these developmental stages revealed the presence of at least four surface specializations: 1) microvilli of ectodermal surface cells; 2) cilia of epitheliomuscular and mucous cells; 3) a single cilium of a sensory cell surrounded by a collar of long microvilli; and 4) depressions in the ectoderm containing a cilium surrounded by microvilli. This work complements and extends earlier light and transmission electron microscopic studies of the development of planula larvae.

KEY WORDS: Cnidaria, Cryofracture, Hydrozoa, Invertebrate, Pennaria tiarella, planula, ultrastructure, Scanning Electron Microscopy

*Address for correspondence:
Anne E. Hotchkiss c/o Robert P. Apkarian
Scanning Electron Microscope Facility
Yerkes Regional Primate Research Center
Emory University
954 Gatewood Road
Atlanta, GA 30322
Phone no.: (404) 727-7766

## Introduction

Most Cnidarians have a planula larval stage at some time during their life cycle (Fig. 1). Planulae are characteristically elliptical in shape and constructed of an outer ectoderm and inner endoderm separated by a thin mesoglea. The histological composition of planulae is simple. Pre-metamorphic anthozoan planulae typically possessed 7-10 types of ectodermal cells and 2-6 types of endodermal cells (2, 3). Mature hydrozoan planulae possessed 7 ectodermal cell types and 2 endodermal cell types (10, 11, 12, 13). Scyphozoan planulae possessed 2-3 ectodermal cell types and 2 endodermal cell types (9). Due to their simplicity and ready availability for embryological studies, planulae have been employed in numerous studies focusing on phylogenetic classification, mechanisms and patterns of cellular differentiation, metamorphosis, polarity, and morphogenesis (4, 9, 10, 11, 12 13, 14, 15).

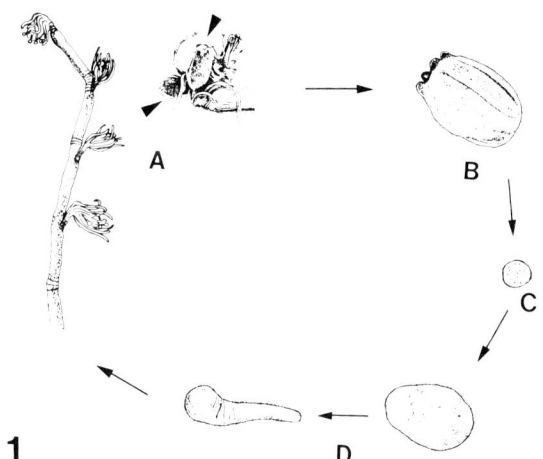

Figure 1. Life-cycle of Pennaria tiarella. A, Polyp stage is colonial and possesses gastrozooids. Free-swimming medusae (B) are formed by asexual buds (arrows) on gastrozooids. The medusa releases egg or sperm into the water. Fertilized egg (C) develops into a free-crawling planula (D) which attaches to a substrate and undergoes metamorphosis to form the polyp stage.

Previous studies of Cnidarian development primarily utilized the techniques of light microscopy (LM) and transmission electron microscopy (TEM) (9, 10, 11, 12, 13). These studies concentrated on the internal morphology of developing planulae. The present study was undertaken to specifically examine the surface morphology of developing planulae using scanning electron microscopy (SEM). A SEM-cryofracture technique which would correlate with TEM was developed in order to examine the internal morphology of planular cells. Due to the 3-dimensional nature of SEM surface analysis and the internal views of cells afforded by SEM-cryofracture, these techniques provided information on the external morphological changes of planulae during development, on cellular morphology, and on cell to cell and cell to mesoglea interactions.

Planulae of the marine hydrozoan Pennaria tiarella were examined in this study. LM and TEM studies have provided detailed characteristics of the internal anatomy of this planula (11, 12, 13, 14). The ectoderm of a planula was composed of epitheliomuscular, mucous, sensory, cnidocyte, ganglionic, and interstitial cell types. The endoderm was composed of digestive, glandular, nematoblast (differentiating cnidocyte), and interstitial cell types. Most interstitial cells differentiate into ganglionic cells and cnidocytes after migration from the endoderm to the ectoderm, however, some interstitial cells begin to differentiate into cnidocytes while still in the endoderm and complete differentiation in the ectoderm. Cnidocytes, the stinging cells of Cnidarians, function in capture and entanglement of prey and defense (7, 8). Epitheliomuscular cells possess the functional capabilities of both epithelial and muscular cells. Apically, they possess single cilia, prominent nuclei, and numerous granules. Basal cytoplasmic extensions (foot processes), known to contain myofibrils (5, 11, 17) organized into a "myoneme," form an intricate network of contractile processes that construct a continuous, "muscular" sheet onto the mesoglea thus aiding in movement of the animal (5, 16). The functions of sensory and ganglionic cells in Pennaria tiarella are still being resolved; however, a strong similarity exists between these cell types and cells which are involved in a proposed simple reflex pathway for longitudinal muscles in the tentacles of hydra (17). The precise function of mucous cells is unresolved. Mucous cells possess granules that contain mucopolysaccharide. It is suggested that the products of these cells are released to aid attachment of the planula to a substrate (11). The gastrodermal cells within the endoderm of the planulae of Pennaria tiarella line the later gastrovascular cavity of the polyp and participate in intracellular digestion. The relatively few cell types in this hydrozoan made it a suitable subject for the study of embryological patterns by SEM.

## Materials and Methods

During late afternoon in July and August, colonies of Pennaria tiarella were collected from floating docks in Wrightsville Beach, North Carolina. Fronds from mature male and female colonies were placed together in large glass bowls filled with seawater. The bowls were placed in the dark at 5:30 p.m. This time was recorded as time 0 for our notations. At 8:30 p.m. the bowls were re-exposed to the light. Early cleavage stages were observed in the bottoms of the bowls. These stages were transferred by mouth pipet into small finger-bowls of seawater and allowed to develop. Embryos were collected at 10, 17, 24, and 48 hours of development and fixed for electron microscopy. Specimens were fixed for 1 hour at 25°C in 2% glutaraldehyde in 0.1 M sodium-cacodylate buffered seawater, pH 7, containing 0.44 M sucrose and 0.01 M $CaCl_2$. Specimens were rinsed for three 10 minute changes in the above buffer solution without fixative. Specimens were post-fixed in 1% osmium tetroxide in 0.1 M sodium-cacodylate, pH 7, for 1 hour and then rinsed for three ten minute changes in 0.1 M sodium-cacodylate buffer, pH 7.

In preparation for SEM, fixed specimens from different developmental stages (10, 17, 24, and 48 hours) were rinsed in distilled water for 5 minutes and then dehydrated through a graded series of ethanols at 5 minute intervals until the specimens were in 100% ethanol. Specimens to be cryofractured were wrapped in absolute ethanol filled parafilm squares. Specimens were initially quick-frozen in Freon-22 (-160°C) and then transferred into liquid nitrogen (-196°C). The specimens within the parafilm packets were cracked under liquid nitrogen using a modified tissue-chopper (1). A Sorvall Smith-Farquhar tissue-chopper was adapted for cryofracture, as was previously described, by the installation of a $LN_2$ trough which was constructed either of brass or styrofoam walls and a copper base. The trough surrounded a center stage for placement of the specimen. The chopper arm was adjusted to stop the blade 3 mm from the trough base at the bottom of its stroke. This prevented mechanical damage to the fracture faces of the tissue by limiting penetration of the blade into the tissue. Both the blade and stage were pre-cooled with $LN_2$ prior to cryofracture. Fractured specimens were thawed to 0-4°C in petri-dishes containing chilled, absolute ethanol. The specimens were pipetted into fine-mesh tissue baskets, critical point dried from $CO_2$ using an exhaust flow monitor, mounted on stubs, and coated with 6 nm of gold in a Polaron E-3000 DC sputter-coater. Observations were made with an ISI-40 SEM operated at 10-15 kV. Individual cell types observed in cryofracture sections were identified according to developmental age of the planula, germ layer location, and morphological characteristics of the nucleus, granules, cellular boundaries, and surface specializations previously described in TEM studies (11, 12, 13, 14). Fixed planulae that were not cryofractured were dehydrated to 100% ethanol and critical point dried from $CO_2$. These samples were mounted on stubs, coated with 6 nm of gold and examined by SEM.

## Results

The cryofracture method for the preparation of planulae for SEM differed slightly from previously used techniques (1, 6). Due to size, delicacy, and physical properties of the planulae tissue, a delicate handling procedure was developed. The initial distilled water rinse was limited to 5 minutes. Exposures longer than this led to swelling and bursting of the planulae. During the dehydration step, specimens were also retained in each concentration of ethanol for 5 minute intervals. Longer periods of exposure led to distorted cracked specimens with separated germ layers. Initial attempts to examine internal surfaces and cells of planulae by razor cutting produced rough, debris-covered surfaces and displaced cells making interpretation difficult and warranted the implementation of a cryofracture technique. Smooth, shiny faces of fractured planulae were produced under $LN_2$ on a modified tissue-chopper. Planulae were also fractured under $LN_2$ using a hand-held blade and hammer. This process, due to the uncontrolled strike force, produced less planar fracture faces that contained several fracture steps making interpretation difficult.

Depending upon the orientation of planulae during the cryofracture procedure, cross-sections or longitudinal sections were produced (Figs. 2, 3). Distinct ectoderm, mesoglea, and endoderm layers were easily discernible in the cross-sectional fractures while longitudinal fractures afforded less distinct views.

## 10 Hour Post-fertilization Planula

At 10 hours post-fertilization, embryos were 230 μm in length (Fig. 4). Polarity was established as evidenced by a broad anterior end (150 μm wide) and narrower posterior end (90 μm wide). They have gastrulated and possessed a distinct ectoderm, endoderm, and mesoglea (Fig. 2). Tall columnar epitheliomuscular cells comprise the ectoderm. Surface analysis showed a single cilium (average length and width of 2.9 μm and 0.2 μm respectively) and numerous short microvilli (average length and width of 0.3 μm and 0.1 μm respectively) which projected from the surface of each epitheliomuscular cell. These cells extended from the free surface of the planula to the mesoglea. Each cell was characterized by the presence of numerous granules (1 μm in diameter) in the cytoplasm, a medially-located nucleus, and basal extensions of the cell cytoplasm (basal foot processes) which inserted on the mesoglea. Also in the basal region of the cell were clusters of granules (2.8 μm in diameter). The epitheliomuscular cells were tightly packed together and few spaces were present between cells. The endoderm consisted of an outer layer of gastrodermal cells and an inner core of interstitial cells (Figs. 2, 3, and 5). Gastrodermal cells rested on the mesoglea and in close association with interstitial cells. The gastrodermal cell was characterized by a vacuolated cytoplasm, as indicated by the irregularly shaped holes in the cells, a centrally located nucleus, and basally located cytoplasmic granules (Fig. 2). The interstitial cells were in close contact with each other and contained a centrally located nucleus.

## 17 Hour Post-fertilization Planula

The ectoderm of the 17 hour post-fertilization planula consisted of epitheliomuscular cells, mucous cells, interstitial cells, and developing nematoblasts (Figs. 6 and 7a). The epitheliomuscular cells were similar to those described at 10 hours with two exceptions. The foot processes were well-developed, tortuous, and appeared to overlap each other as they inserted onto the mesoglea (Fig. 7a, 7b). Also the average length of the epitheliomuscular cell cilium had increased to 5.9 μm (Fig. 6). Mucous cells were interspersed among the epitheliomuscular cells (Fig. 6). The mucous cell had a short cilium and microvilli which projected from its surface (personal communication, Martin and Thomas). Located in the apical region of the mucous cell were clusters of small granules (0.6 μm in diameter). The granules of the mucous cells were smaller than those of the epitheliomuscular cells, thus enabling easy identification of the two cell types (Fig. 6). By this stage interstitial cells and developing nematoblasts had migrated to the ectoderm and nematoblasts with their developing capsules were observed just apical to the epitheliomuscular cell foot processes (Fig. 7a).

The endoderm of the 17 hour planula had differentiated sufficiently to discern cellular organization (Fig. 3). Gastrodermal cells and associated granules bordered the mesoglea and encircled an endodermal core of interstitial cells. In the regions where the gastrodermal cells were in contact with the mesoglea, granules (1.5-2 μm in diameter) were often present (Fig. 8). These granules either rested on the mesoglea or were situated in close proximity to other granules in contact with the mesoglea.

## 24 Hour Post-fertilization Planula

At 24 hours post-fertilization, the planula was 500-600 μm in length and possessed a distinct anterior end (150 μm wide) and a narrower posterior end (75 μm wide) (Fig. 9). The planula exhibited three surface specializations in both its anterior and posterior regions: 1) randomly scattered microvilli (average length and width of 0.4 μm and 0.1 μm respectively) found on all ectodermal cells; 2) a solitary cilium (average length and width of 7.4 μm and 0.3 μm respectively) of either epitheliomuscular or mucous cells; and 3) a single cilium of a sensory cell (average length and width of 5.0 μm and 0.3 μm respectively) surrounded by a collar of 15-32 tall microvilli (average length and width of 1.7 μm and 0.1 μm respectively) (Fig. 10). SEM observations of 24 hour planulae ectoderm demonstrated the presence of epitheliomuscular cells, mucous cells, sensory cells, nematoblasts, and ganglionic cells. Epitheliomuscular and mucous cells comprised the majority of the ectodermal cells. The apical regions of both cells were filled with clusters of granules. Occasional cryofracture preparations, which only grazed the planula, removed the superficial membranous layer from the apical regions of

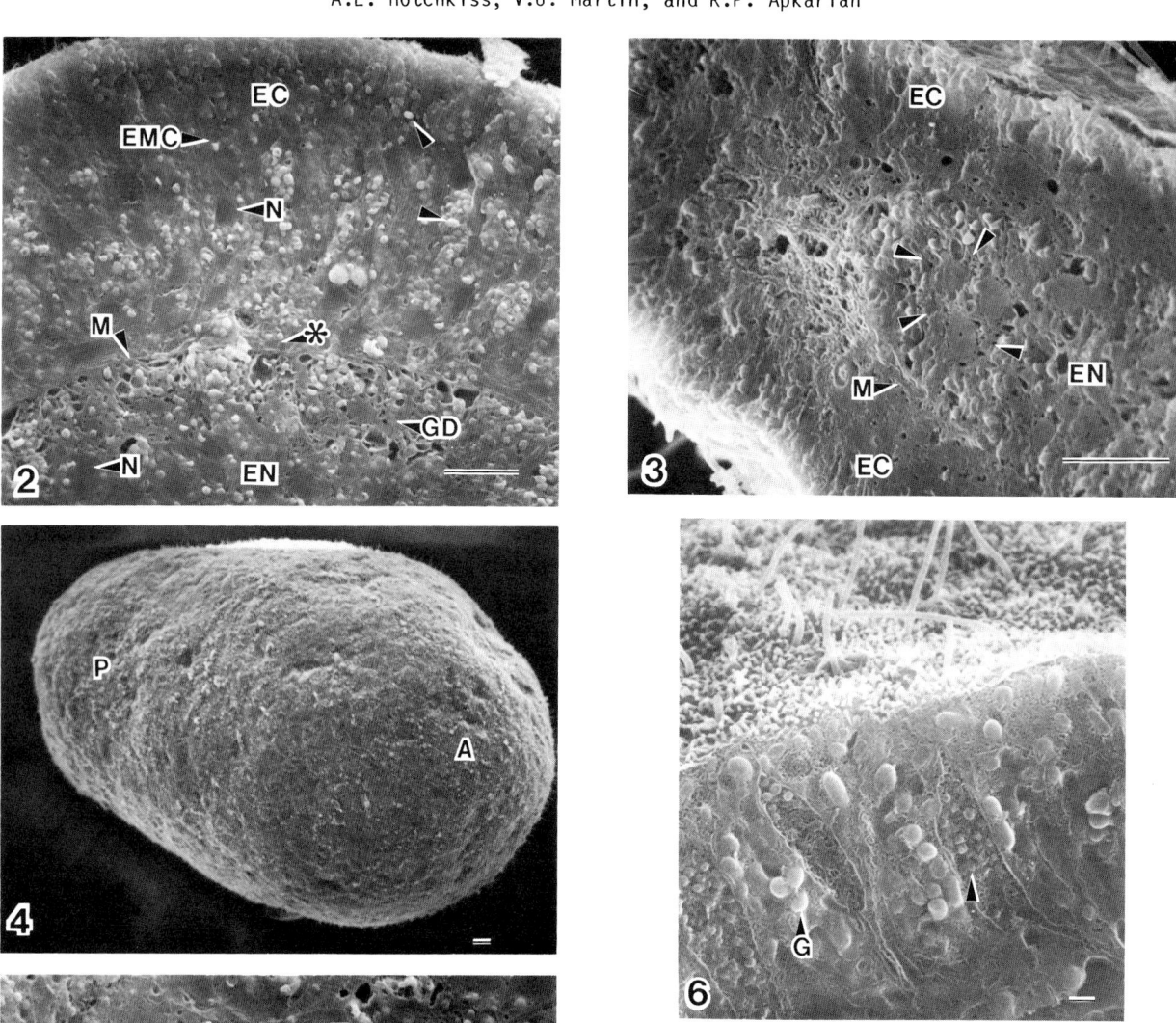

Figure 2: A cross-sectional cryofracture through a 10 hour post-fertilization planula. Note the well-defined ectoderm (EC), mesoglea (M), and endoderm (EN). Columnar epitheliomuscular cells (EMC) were located in the ectoderm and extended from the planula surface onto the mesoglea via newly formed basal foot processes (*). Oval nuclei (N) and granules 1.0 μm in diameter (arrows) were characteristic of these cells. The endoderm contained gastrodermal cells (GD) and interstitial cells. Interstitial cells have prominent nuclei (N). Bar = 10 μm.

Figure 3: A longitudinal cryofracture section through a 10 hour post-fertilization planula. Note the ectoderm (EC), mesoglea (M), and endoderm (EN). The endoderm contained a central core of interstitial cells (arrows). Bar = 10 μm.

Figure 4: Micrograph of surface view of 10 hour post-fertilization planula. Polarity was established as evidenced by a broad anterior end (A) and narrower posterior end (P). Bar = 10 μm.

Figure 5: A cross-sectional cryofracture through the endoderm of a 10 hour post-fertilization planula. Interstitial cells located in the endoderm were characterized by centrally located nuclei (N). Note boundaries between interstitial cells (arrows). Bar = 10 μm.

Figure 6: A cross-sectional cryofracture through the apical region of the ectoderm in a 17 hour post-fertilization planula. Apical regions of epitheliomuscular cells were characterized by large granules (G), while adjacent regions of mucous cells contained smaller granules (arrows). Bar = 1 μm.

# SEM-Cryofracture Study of Planulae of Pennaria tiarella

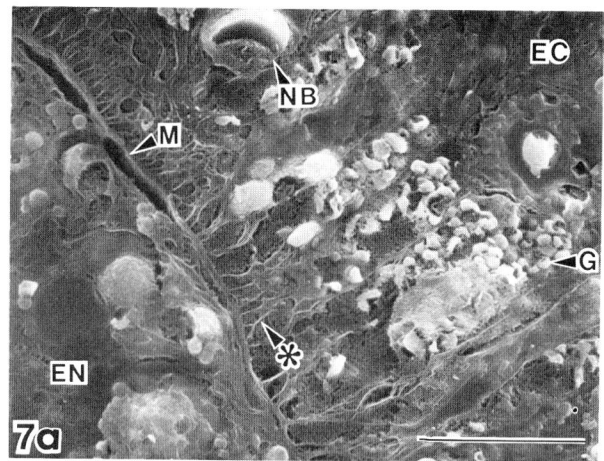

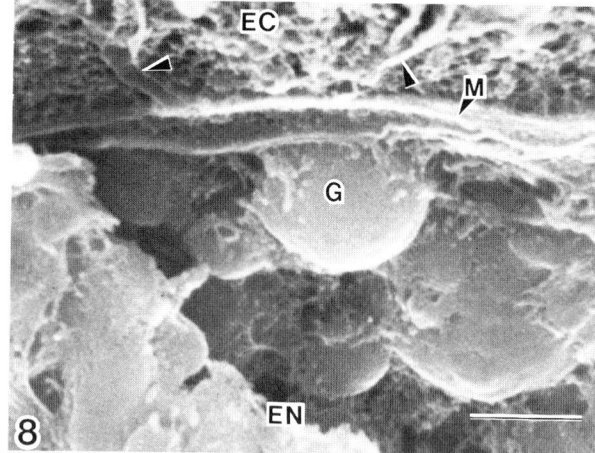

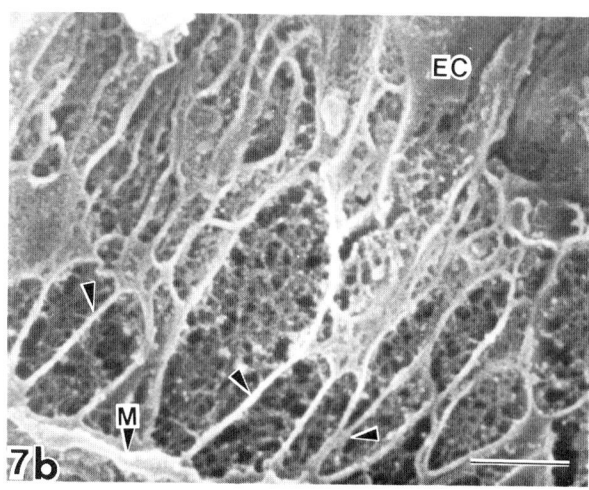

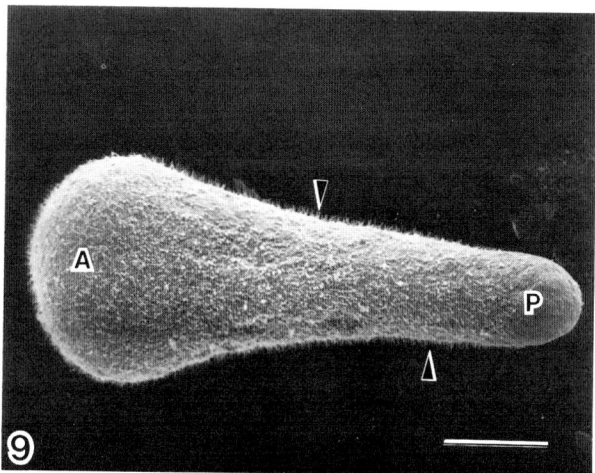

Figure 7a: A cross-sectional cryofracture through the basal region of the ectoderm (EC) in a 17 hour post-fertilization planula. Note mesoglea (M) and endoderm (EN). Columnar epitheliomuscular cells contained granules (G) and complex basal foot processes which inserted onto the mesoglea (*). A developing nematoblast (NB) with capsule has migrated from the endoderm and can be seen above the foot processes. Bar = 10 μm.

Figure 7b: A higher magnification micrograph of the basal foot processes (arrows) of epitheliomuscular cells. Note mesoglea (M) and ectoderm (EC). Bar = 10 μm.

Figure 8: A cross-sectional cryofracture of a 17 hour post-fertilization planula showing the acellular mesoglea (M) separating ectoderm (EC) from endoderm (EN). Note distinct basal foot processes of epitheliomuscular cells after they had inserted onto the mesoglea (arrows). Granules (G) in a digestive cell are seen in the endoderm. Bar = 1 μm.

Figure 9: An overview of a 24 hour post fertilization planula which possessed a characteristically broad anterior end (A) tapered to a narrow posterior end (P). Note the numerous cilia projecting from the surface of the planula (arrows). Bar = 100 μm.

Figure 10: Surface view of 24 hour post-fertilization planula with the three basic cell surface specializations. Microvilli (M) were uniformly scattered over the surfaces of all ectodermal cell types. Cilia (C) of epitheliomuscular cells and mucous cells are indistinguishable at this stage. Apical regions of sensory cells were characterized by single cilia (arrow) projecting from a basal collar of tall microvilli (B). Bar = 1 μm.

A.E. Hotchkiss, V.J. Martin, and R.P. Apkarian

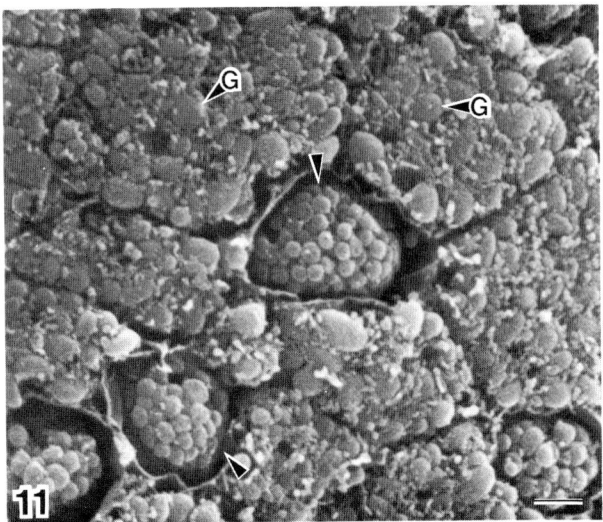

Figure 11: Cytoplasmic surfaces of epitheliomuscular and mucous cells in which the membrane was removed during the cryofracture procedure. Epitheliomuscular cells were characterized by numerous large granules (G) while mucous cells contained clusters of smaller granules (arrows). Bar = 1 μm.

the ectodermal cells (Fig. 11). These exposed apical surfaces contained cells which possessed granules characteristic of either epitheliomuscular or mucous cells. The basal regions of sensory cells were extremely tortuous, therefore positive identification of these cells by cryofracture was inconclusive. A nerve plexus composed of ganglionic cells and their neurites was prominent at the base of the ectoderm just above the foot processes of the epitheliomuscular cells (Fig. 12). This plexus ran transversely around basal foot processes. The endoderm of 24 hour post-fertilization planulae was similar to that of the 17 hour post-fertilization planula. A large number of interstitial cells were seen closely apposed to the mesoglea (Fig. 13). Deep cavities often containing cytoplasmic projections were also present in the endoderm.

48 Hour Post-fertilization Planula

By 48 hours post-fertilization, planulae had an average length of 700 μm, a broad anterior end (200 μm wide), and a narrow posterior end (75 μm wide) (Fig. 14). They possessed the same ectodermal and endodermal cell types as were previously described for 24 hour post-fertilization planulae. A thin section TEM preparation demonstrates the ganglionic cell nerve plexus and the longitudinal and transverse orientation of the neurites among the epitheliomuscular cell basal foot processes of a 48 hour planula (Fig. 15). Four types of surface specializations were found on 48 hour planulae (Fig. 16). Microvilli (average length and width of 0.4 μm and 0.1 μm respectively) were randomly scattered over the surfaces of all epitheliomuscular, mucous, and sensory cells. Solitary cilia of epitheliomuscular and mucous cells averaged 8.1 μm in length and 0.4 μm in width. The tall collar microvilli (average length and width

of 1.0 μm and 0.13 μm respectively) associated with the single cilia (average length and width of 5.42 μm and 0.25 μm respectively) of sensory cell surfaces were clustered in button-like, bulbous projections above the planula surface (compare Figs. 10 and 16). The button-like clusters of microvilli were approximately 3-5 μm in diameter. Scattered over the surface of the planula were small depressions, 3-6.5 μm in diameter (Figs. 16 and 17). Projecting from each depression was a single cilium (average length and width of 4.5 μm and 0.3 μm respectively) surrounded by 30-35 microvilli (average length and width of 1.0 μm and 0.1 μm respectively). Sensory cell surface specializations with button-like clusters of microvilli and the ectodermal surface depressions were present on both the anterior and posterior regions of the planula. Surface specializations of cnidocytes were not detected at 48 hours post-fertilization.

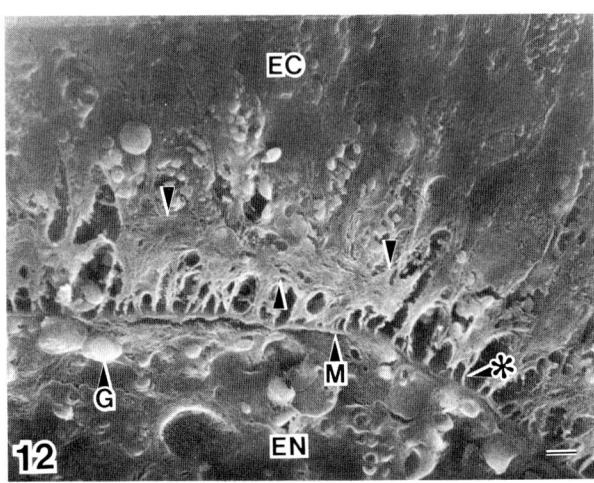

Figure 12: A cross-sectional cryofracture through a 24 hour post-fertilization planula. Basal foot processes (*) of the epitheliomuscular cells are tortuous and closely abut the mesoglea (M). Note the nerve plexus (arrows) seen at the base of the ectoderm (EC). Granules (G) of digestive cells in the endoderm (EN) contact the mesoglea. Bar = 2 μm.

Discussion

Surface specializations of epitheliomuscular, mucous, and sensory cells have been partially characterized by TEM and LM in hydrozoan planulae (11, 12). This study represented the first detailed description, using SEM, of the nature of surface specializations during development of planulae larvae. Martin and Thomas (11, 12) used TEM to describe the hydrozoan planula of Pennaria tiarella. They reported that epitheliomuscular cells were present in planulae by 10 hours post-fertilization; mucous cells appeared shortly after 10 hours; and sensory cells were present by 48 hours post-fertilization. SEM surface analysis demonstrated the presence of at least three surface specializations that corresponded to the

18

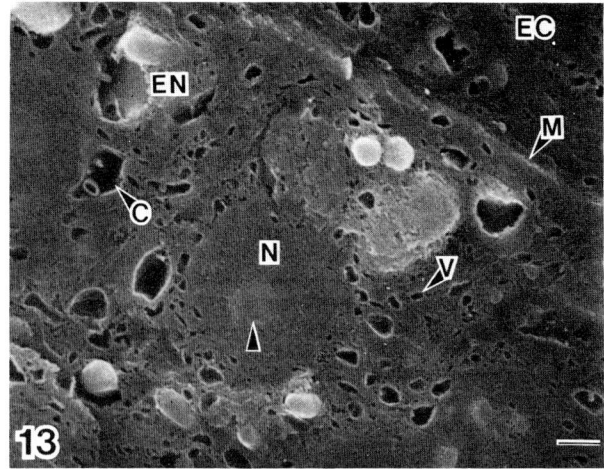
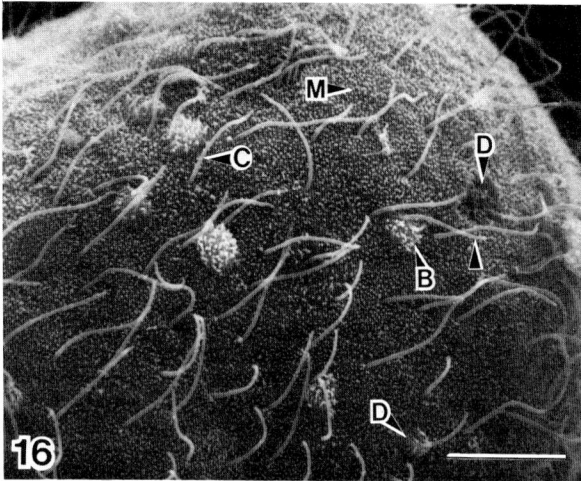
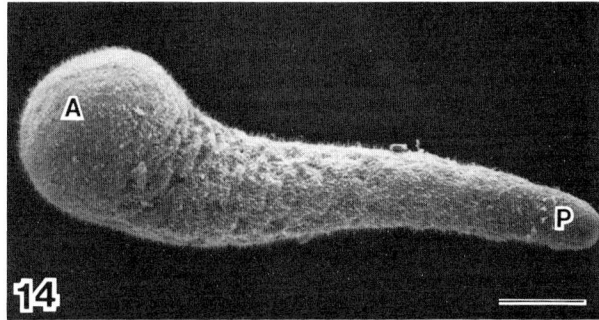
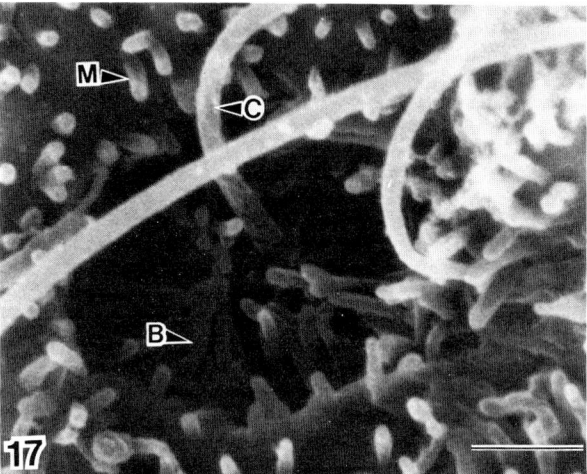
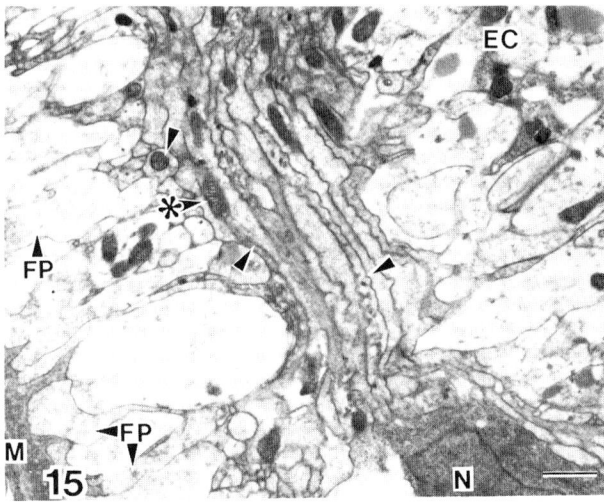

Figure 13: An interstitial cell within the endoderm (EN) of a 24 hour post-fertilization planula prior to its migration across the mesoglea (M) into the ectoderm (EC). The cell possesses a centrally positioned nucleus (N), a prominent nucleolus (arrow) and abundant scattered vacuoles (V). Deep cavities (C) containing cytoplasmic projections were present in the endoderm. Bar = 1 μm.

Figure 14: An overview of 48 hour post-fertilization planula. A characteristically broad anterior end (A) tapering to a narrow posterior end (P) demonstrates the increase in polarity. Bar = 100 μm.

Figure 15: Transmission electron micrograph of a nerve plexus of a 48 hour post-fertilization planula. Neurites (arrows) of ganglionic cells are oriented parallel to the mesoglea (M) in the basal region of the ectoderm (EC). Note mitochondrion (*) in neurite, nucleus (N) of a ganglionic cell, and basal foot processes (FP) of epitheliomuscular cells. Bar = 1 μm.

Figure 16: Surface view of 48 hour post-fertilization planula showing 4 types of cell surface specializations. Microvilli (M) were uniformly scattered over all ectodermal cell surfaces. Apical regions of epitheliomuscular and mucous cells were characterized by solitary cilia (C). Apical regions of sensory cells were characterized by single cilia (arrow) surrounded by a bulbous cluster of tall microvilli (B). Depressions in ectoderm layer contained a single cilium surrounded by a basal collar of tall microvilli (D). Bar = 10 μm.

Figure 17: A higher magnification micrograph of an ectodermal surface depression on a 48 hour post-fertilization planula. Note the single cilium (C) with an associated collar of tall microvilli (B) and scattered microvilli (M) on the ectodermal surface. Bar = 1 μm.

appearance of the epitheliomuscular, mucous, and sensory cells during development (Figs. 10 and 16).

Cilia of epitheliomuscular cells appear on the surface of the embryo during gastrulation. 17 hour post-fertilization planulae possess two classes of cilia (long and short) on their surfaces. Short cilia are thought to be attached to developing mucous cells; one cilium per cell (personal communication, Martin and Thomas). During planula development, the long cilia of the epitheliomuscular cells and the cilia of the mucous cells grow in length until they are indistinguishable from one another. Specializations of the surfaces of sensory cells from 24 hour post-fertilization planulae were observed in cryofracture SEM preparations. Observations of the microvillar collar, which is associated with the sensory cell cilium, demonstrated that it was much denser than previous TEM studies had suggested. By 48 hours post-fertilization, an unknown surface structure is present. The specialization, which we described in this paper, consists of a depression in the surface of the planula containing a single cilium projecting from the center of the depression and surrounded by a collar of tall microvilli. The dimensions of the cilia and microvilli projecting from the depression corresponded to those of the sensory cell surface specializations. Further correlative microscopy is needed to confirm the identity of this cell surface specialization. The sunken disposition of this structure may be characteristic of the sensory cell type in the final stages of development. Such a surface specialization has not been previously observed by other microscopic techniques.

Previous TEM studies have shown that ganglionic cells are derived from interstitial cells and are found in the ectoderm adjacent to the forming foot processes of the epitheliomuscular cells (12, 13, 14). Characteristically, they possessed a nucleus with a prominent nucleolus and a complex of longitudinally and transversely oriented neurites. Observations of 24 hour post-fertilization planulae revealed a broad fibrillar layer of transversely oriented cytoplasmic processes in close association with the epitheliomuscular cell basal foot processes. This structure closely resembled a nerve plexus of ganglionic cells and their neurites (Fig. 12).

SEM-cryofracture of planulae results in excellent preservation of germ layers, acellular membranes, and cell boundaries. Distinct cell types are easily recognized in cryofractured material and correspond well with TEM studies of planular cells. The SEM-cryofracture procedure for planulae allowed for 3-dimensional imaging of the internal morphology of planular cells. Such a view is not possible with LM or TEM. The cryofracture of planulae allows for a number of new ideas concerning morphology of the planulae. When cross-sectional cryofractured fragments from different stages of planulae were viewed in the SEM, the epitheliomuscular cell basal foot processes exhibited an increase in complexity as they intermingled with the mesoglea and neighboring cells.

The foot processes appeared well-developed and more tortuous than previously suggested in TEM studies. Micrographs of cryofractured planulae also show the 3-dimensional nature of granules in epitheliomuscular, mucous, and gastrodermal cells.

Cryofractured planulae have a well-preserved endoderm, attesting to the quality of fixation and delicate handling procedures employed in this SEM study (Fig. 5). Previous TEM studies on planula larvae have concentrated on the organization of the ectoderm (3, 11, 12). The close association of the interstitial cells is highlighted in cryofractured material. In cryofractured planulae, SEM micrographs revealed endodermal cavities containing cytoplasmic processes. These cavities which have not been previously reported, warrant future correlative microscopic studies so as to allude to the nature of cellular interactions in developing planulae.

### Summary

A modified cryofracture technique was applied to the planula larval system of _Pennaria tiarella_ and results correlated with previous LM and TEM studies. Solitary cilia and characteristic granules of epitheliomuscular and mucous cells, and single cilia with a basal collar of microvilli characteristic of sensory cells, are described during planula development. The development of the basal foot processes of epitheliomuscular cells and their attachment on to the mesoglea was also assessed. This suggests the usefulness of SEM-cryofracture procedures for embryological studies in invertebrate larval systems.

### Acknowledgments

This work was supported by a grant from the Academic Excellence Committee of the University of Louisville Graduate School.

### References

1. Apkarian R, Curtis JC. (1981). SEM cryofracture study of ovarian follicles of immature rats. Scanning Electron Microsc., 1981; IV:165-172.

2. Chia FS, Crawford B. (1977). Comparative fine structural studies of planulae and primary polyps of identical age of the sea pen, _Ptilosarcus gurneyi_. J. Morphol., 151:131-158.

3. Chia F, Koss R. (1979). Apical organ in the planula larva of the sea anemone _Anthopleura elegantissima_. J. Morphol., 160:275-298.

4. Freeman G. (1981). The role of polarity in the development of the hydrozoan planula larva. Wilhelm Roux Arch., 190:168-184.

5. Haynes JF, Burnett AL, Davis, LE. (1968). Histological and ultrastructural study of the muscular and nervous systems in hydra. I. The muscular system and the mesoglea. J. Exp. Zool., 167:283-294.

6. Humphreys WJ, Spurlock BO, Johnson JS. (1974). Critical point drying of ethanol-infiltrated cryofractured biological specimens for scanning electron microscopy. Scanning Electron Microsc., 1974: 275-282.

7. Hyman LH. (1940). The Invertebrates: Protozoa through Ctenophora. McGraw-Hill Book Company, Inc., New York:382-392.

8. Mariscal RN. (1977). Nematocysts. In: Coelenterate Biology, Reviews and New Perspectives. Academic Press, Inc., New York:129-167.

9. Martin VJ, Chia FS. (1982). Fine structure of a scyphozoan planula, Cassiopeia xamachana. Biol. Bull., 163:320-328.

10. Martin VJ, Chia FS, Koss R. (1983). A finestructural study of metamorphosis of the marine hydrozoan Mitrocomella polydidemata. J. Morphol., 176:261-287.

11. Martin VJ, Thomas MB. (1977). A fine structural study of embryonic and larval development in the marine hydrozoan Pennaria tiarella. Biol. Bull., 153:198-218.

12. Martin VJ, Thomas MB. (1980). Nerve elements in the planula of the hydrozoan Pennaria tiarella. J. Morph., 166:27-36.

13. Martin VJ, Thomas MB. (1981). The origin of the nervous system in Pennaria tiarella, as revealed by treatment with colchicine. Biol. Bull., 160:303-310.

14. Martin VJ, Thomas MB. (1983). Establishment and maintenance of morphological polarity in epithelial planulae. Trans. Am. Microsc. Soc., 102:18-24.

15. Summers RG, Haynes JF. (1969). The ontogeny of interstitial cells in Pennaria tiarella. J. Morphol., 129:81-88.

16. West DL. (1978). The epitheliomuscular cell of hydra: its fine structure, three-dimensional architecture and relation to morphogenesis. Tissue and Cell, 10(4):629-646.

17. Westfall, JA. (1973). Ultrastructural evidence for neuromuscular systems in coelenterates. Amer. Zool., 13:237-246.

## Discussion with Reviewers

V.C. Barber: Would the authors like to comment on the preparation artifacts that will be inherent in their preparative techniques? For example, ice crystal damage will occur on freezing, and shrinkage will occur on critical point drying. Did the authors re-embed their SEM material to try to reevaluate the state of preservation by TEM? Did the authors try to assess the amount of shrinkage by CPD (as per the papers by Boyde et al in previous SEM meetings)? Why did the authors use Freon 22 instead of other freons such as Freon-12? Would the authors like to comment on the results obtained by "their" cryofracture method as compared with other similar techniques used in the past? Please provide some general references to these other methods.

It is clear that an SEM examination of the surface of the planulae gave new information that would not be possible to obtain by any other method. However, it appears to the reviewer that this was not the case with the cryofracture technique, particularly in view of the artifacts that might have been produced. Could the authors comment on any particular advantage that the cryofracture method may have over TEM and LM preparations?

J.A. Westfall: Do you believe accurate information can be obtained by cryofracture methods without prior TEM?

Authors: Delicate handling procedures were employed during cryofracture and critical point drying in order to minimize the possibility of artifacts. To eliminate the formation of icecrystals during freezing, we removed water by absolute ethanol-infiltration of specimens prior to cryofracture (6). The large volume of absolute ethanol used during the thawing process would effectively remove any water that may be condensed on the specimen surfaces during the fracturing process.

Boyde (1978) determined 26% linear and 60% volume reduction in embryonic tissue after critical point drying. Planulae specimens were fixed in glutaraldehyde, post-fixed in osmium-tetroxide, dehydrated in ethanol, and critical point dried from $CO_2$. Reductions in tissue size were, hopefully, less than that estimated by Boyde because artifacts inherent to CPD routines were lessened by our delicate handling procedures. It is known that all critical point drying specimens, except those with monovalent or divalent cations added to either the fixative, washing solution, or one of the ethanol series, will swell during treatment with dehydration solvents below 70% (18). Mechanical deformation caused by excessive swelling (followed by shrinkage during CPD) was avoided during planulae preparation by addition of 0.01 M $CaCl_2$ to the 0.1 M Na-cacodylate buffer used in the glutaraldehyde fixative. Our Polaron E-3000 critical point dryer with exhaust flow monitor maintained a continuous flow of 1.2 L/min during purging and decompression procedures which resulted in excellent preservation of fragile cell structures such as microvilli and cilia. A long period of purging ensured complete exchange of $CO_2$ for ethanol within the drying chamber since it is known that residual amounts of intermediate fluid lead to excessive shrinkage when "wet" specimens are removed from the drying chamber and exposed to the air (19, 22).

We did not re-embed out SEM material; however, Humphreys et al. (6) did re-embed ethanolinfiltrated, cyrofractured, critical point dried tissues and re-examined these with TEM. Results showed no damage that could be attributed to cryofracturing, thawing, or critical point drying procedures, thereby demonstrating adequate tissue preservation. The same specimen fixation procedures used in previous TEM studies of Pennaria tiarella (11) were employed in this SEM study. The quality of specimen preservation and image contrast achieved by this cryofracture preparation provided a direct correlation with previously reported TEM (compare text Fig. 12 with Fig. 1 from Martin and Thomas (12). Delicate structures such as epitheliomuscular cell basal foot processes and ganglionic cell nerve plexuses exhibited comparable preservation in SEM-cryofracture and TEM.

The delicate cryofracture procedures employed in this study avoided excessive mechanical damage that normally occurs in specimens cryofractured after critical point drying (21). Freon-22, chosen because of its ready availability, was used in the cryofracture procedure. Due to its high thermal capacity, Freon-22 (chlorodifluoromethane) was used to rapidly freeze tissues as is routinely done in freeze fracture TEM preparations. Initial quick freezing in Freon-22 slush followed by immersion in $LN_2$ avoided the formation of Leidenfrost bubbles in the tissue that are caused by localized boiling of $LN_2$. The blade of the modified tissue-chopper used in this study was prepositioned to strike only the top of the frozen specimen packet, thus avoiding shear force damage to the fracture faces common to the hand-held hammer and chisel method (1).

Modifications of the cryofracture technique were developed by Haggis et al. (20) and Tanaka (23) to improve observations of intracellular structures. Haggis et al. employed a method where unfixed tissues were initially impregnated with glycerol or 20% DMSO (dimethylsulphoxide) as cryoprotectants; the tissues were then rapidly frozen, fractured under $LN_2$, thawed into fixative, dehydrated to 100% ethanol, and critical point dried. In this procedure soluble proteins were washed out of the fracture face before they had a chance to be cross-linked and made insoluble by the fixative thus creating spaces around cellular organelles which were once filled with soluble proteins. Tanaka used an osmium-DMSO-osmium (O-D-O) method where specimens were first osmicated, cryoprotected with DMSO, frozen, fractured, thawed in 50% DMSO, rinsed in buffer, and then macerated for 24-72 hours in a 0.1% osmium tetroxide solution during which the excess cytoplasmic matrices were removed from the fracture face. Although the cryofracture methods utilized by Haggis et al. and Tanaka were relatively successful, the use of DMSO in both studies created the potential for shrinkage artifacts. DMSO has a low vapor pressure and therefore exhibits a low volatility as a cryoprotectant and is retained by the specimens (18). Specimens treated with DMSO cannot be processed directly by CPD but must undergo further substitution with a more volatile intermediate fluid for the DMSO, and it is known that DMSO causes quite substantial shrinkage due to its low dielectric constant. There may be overall shrinkage of the tissue during alcohol dehydration or during passage into liquid $CO_2$, but structural relationships are well preserved. Samples treated with DMSO undergo greater deformations due to differential shrinkage in mixed consistency samples than those in tissues of a uniform consistency (18). Thus, in our study, in which tissue and cellular orientation are integral to understanding embryonic relationships, the possible creation of spaces once filled with soluble proteins was avoided.

Ethanol-infiltrated cryofractured specimens provided an entire sectional view through the planulae in which ectodermal, mesogleal, and endodermal layers could be qualitatively assessed for embryonic relationships with greater resolution than that afforded by light microscopy and greater sample size than that afforded by TEM. We felt that the use of the ethanol-infiltrated cryofracture procedure was preferable to that of techniques utilizing DMSO because specimens infiltrated with ethanol could be directly critical point dried, whereas those processed with DMSO must first be substituted with a more volatile intermediate fluid such as ethanol to replace the DMSO prior to critical point drying.

J.A. Westfall: Why did longitudinal fractures provide less distinct views than cross-sectional ones? Is this a problem of the method or the specimen?
Authors: Although the ectodermal, mesogleal, and endodermal layers were observed in longitudinal fractures, the topographic contrast created at the periphery of individual cells was lower than the contrast produced at cell boundaries in cross-sections. We feel that the heightened contrast observed in cross-sections, which lead to more distinct views, is a function of topographic relief produced in the specimen when cryofractured in this manner.

Discussion References

18. Boyde A. (1978). Pros and cons of critical point drying and freeze drying for SEM. Scanning Electron Microsc., 1978; II:303-314.

19. Cohen AL. (1977). A critical look at critical point drying theory, practice, and artifacts. Scanning Electron Microsc., 1977; I:525-536.

20. Haggis GH, Bond EF, and Phipps B. (1976). Visualization of mitochondrial cristae and nuclear chromatin by SEM. Scanning Electron Microsc., 1976; I:281-285

21. Nemanic MK. (1972). Critical point drying, cryofracture, and serial sectioning. Scanning Electron Microsc., 1972: 297-304.

22. Peters K-R. (1980). Improved handling of structural fragile cell-biological specimens during electron microscopic preparations by the exchange method. J. Microsc., 118:429-441.

23. Tanaka, K. (1981). Demonstration of intracellular structures by high resolution scanning electron microscopy. Scanning Electron Microsc., 1981; II:1-8.

For additional discussion see page 12.

# DEVELOPMENT OF SURFACE POLARITY IN MOUSE EGGS

F.J. Longo* and D.Y. Chen

The University of Iowa, Department of Anatomy,
Iowa City, Iowa 52240

(Paper received February 18, 1984, Completed manuscript received June 15, 1984)

## Abstract

Investigations were carried out to determine what effects components of the cytoskeletal system and meiotic spindle have on the development and maintenance of surface polarity in mouse ova. The surface of the mature egg possessed numerous microvilli except for a region (microvillus-free area) adjacent to the meiotic spindle. In contrast, the surface of the immature oocyte was covered uniformly with a dense population of microvilli. When cultured in vitro immature oocytes spontaneously underwent maturation; a meiotic spindle formed in the center of the ovum which then moved to the cortex. Coincident with the cortical localization of the meiotic spindle was the formation of a microvillus-free area and subjacent layer of microfilaments. A microvillus-free area did not form when meiotic maturation was inhibited with dibutyryl cyclic AMP or chloroquine. If immature oocytes were incubated in cytochalasin B a meiotic spindle developed, but it did not become localized to the egg cortex and a microvillus-free area failed to form. Oocytes incubated in colchicine underwent germinal vesicle breakdown and chromosome condensation, a meiotic spindle did not form but the chromosomes became localized to the ovum cortex where a microvillus-free area developed. These results and observations of mature ova treated with cytochalasin B or colchicine indicate that mechanisms involving the movement of the meiotic spindle to the oocyte cortex and development and maintenance of surface polarity are cytochalasin B sensitive. Cortical localization of meiotic chromosomes brings about the formation of a microvillus-free area.

KEY WORDS: Cell polarity, microvilli, cytoskeleton, microfilaments, meiotic maturation.

*Address for correspondence:
Frank J. Longo
Department of Anatomy
The University of Iowa
Iowa City, IA 52240    Phone No.: (319) 353-3811

## Introduction

Studies by Johnson et al. (1975) and Eager et al. (1976) showed that the plasmalemma overlying the meiotic spindle of mature mouse eggs has a reduced capacity to bind concanavalin A, owing, in part, to a reduction in the density of microvilli in this region of the ovum (Wolf and Ziomek, 1983). These observations and studies by Nicosia et al. (1977) and Wolf and Ziomek (1983) have demonstrated morphological and macromolecular polarities in the surface of the mouse ovum, which may be involved with processes of meiotic maturation and fertilization.

The question of when and how morphological and macromolecular polarities are established in the egg surface is central to the elucidation of their possible role in processes of oocyte development and fertilization. The area of the cortex overlying the meiotic spindle in the mouse ovum contains a filamentous layer which seems to be involved in polar body formation (Stefanini et al., 1969; Zamboni, 1970, 1971; Nicosia et al., 1977; Van Blerkom and Motta, 1979). Stefanini et al. (1969), Thompson et al. (1974) and Nicosia et al. (1977) have suggested that this filamentous layer may be involved in the abstriction of the polar bodies much in the manner described for the cleavage furrow of dividing cells (Schroeder, 1975). It may also contribute to conditions that maintain this region of egg surface free of microvilli. In addition, it has been noted that sperm incorporation rarely occurs in that region of the mouse egg free of microvilli (Nicosia et al., 1977). Hence, the filamentous layer associated with the meiotic spindle in mouse eggs may be a part of conditions that preclude from certain areas of the ovum surface the necessary juxtaposition of the gametes required for their fusion (Nicosia et al., 1977). Because similar morphological features are also present in eggs of other mammals (Odor and Renninger, 1960; Gulyas, 1976, 1980), it is possible that the origin of surface polarity in mammalian oocytes is temporally and spatially related to meiotic maturation and has a bearing on the site of gamete fusion and the establishment of embryonic gradients and axes (Johnson et al., 1981). Such polarities are common in eggs where sperm incorporation is limited to a specific region of the plasma membrane (Austin, 1965).

We have carried out microscopic investigations in mouse ova in an effort to: (1) characterize surface changes occurring in spontaneously maturing oocytes, and (2) determine what effects components of the cytoskeletal system and meiotic spindle have on the development and maintenance of surface polarity (Chen and Longo, 1984). The present account describes scanning electron microscopic observations of these experiments and reviews previous studies of surface polarity in eggs from other organisms.

## Materials and Methods

Germinal vesicle oocytes were obtained from the ovaries of 10 to 14 wks old CD-1, virgin mice (Charles River). Ovaries were removed and placed in watch-glasses containing 0.5 ml oocyte culture medium (Donahue, 1968) and covered with paraffin oil. Oocytes, liberated from ovaries by puncturing follicles with hypodermic needles, were collected, washed, placed into droplets of culture medium (10-20 oocytes/0.1 ml medium) and covered with paraffin oil. Incubation was carried out in Falcon petri dishes at 37°C under a mixture of 5% $CO_2$ and 95% air for up to 24 hrs. Oocytes were also cultured for varying periods in medium containing colchicine or cytochalasin B and examined or washed and recultured in fresh medium and then prepared for observation. For these investigations, 25 µM colchicine and 20 µM cytochalasin B were employed (Wassarman et al., 1979). Cytochalasin B was dissolved in dimethylsulfoxide (DMSO) as a stock solution and stored frozen. For controls, oocytes were incubated in culture medium containing ten times the concentration of DMSO added to ova treated with cytochalasin B. To inhibit germinal vesicle breakdown, immature oocytes were incubated for up to 17 hrs in 20 µM dibutyryl cyclic AMP or 10 µM chloroquine (Wassarman et al., 1976, 1979).

Mice were superovulated with intraperitoneal injections of pregnant mare serum (5 IU; Gestyl, Organon) followed 40 hrs later with human chorionic gonadotropin (5 IU; Sigma). Ovulated eggs at the second metaphase of meiosis were recovered 12 to 14 hrs after the injection of human chorionic gonadotropin. Eggs, surrounded by cumulus cells, were incubated in a modified Krebs-Ringer bicarbonate medium (Hoppe and Pitts, 1973) containing 0.1% hyaluronidase (Sigma) for approximately 5 min. Following the removal of cumulus cells, eggs were washed in fresh medium and processed for microscopic observation or placed in 0.1 ml medium containing cytochalasin B or colchicine, covered with paraffin oil and incubated for variable periods up to 9 hrs. Some specimens treated with colchicine or cytochalasin B were washed and recultured in fresh medium and then processed for observation.

Specimens for light and electron microscopy were fixed in 3% glutaraldehyde in 0.05 M sodium phosphate buffer (pH 7.4) for 30 min at 4°C, washed overnight in 0.1 M phosphate buffer, incubated in 0.5% $OsO_4$ in phosphate buffer for 30 min, dehydrated in ascending concentrations of ethanol and embedded in Spurr's embedding medium. Thick sections for light microscopy were stained with 1% toluidine blue. Thin sections were stained with uranyl acetate and lead citrate and examined in a Philips 300 electron microscope.

For scanning electron microscopy, zonae pellucidae of eggs and oocytes were removed by incubating specimens in Krebs-Ringer bicarbonate medium containing 0.001% chymotrypsin for 30 to 60 sec. Specimens were washed in culture medium and fixed as described above. Eggs and oocytes were critically point dried in carbon dioxide, coated with 60% gold/40% palladium, and examined in a JEOL 35C scanning electron microscope at 13kV. To determine the size and density of microvilli, micrographs of 3 to 6 specimens were taken at a magnification of $10^4$ times. Counts of microvilli were made of 5-7 areas comprising 100 µm² of the surfaces of 6 eggs and oocytes observed with scanning electron microscopy. Microvillar dimensions were determined from specimens examined with both transmission and scanning electron microscopy; means ± S.D. were determined for the sum of 10 determinations per egg type examined. The diameter of eggs (76µm) and oocytes (78µm) was determined from measurements of 20 zonae-free specimens with a Leitz SM LUX microscope at a magnification of 400X. The total surface area and surface area of specific regions of mature and immature eggs were determined by standard geometric formulas (Diem, 1962).

## Results

### Mature Egg.

The oviductal mouse ovum had a definite polarity that was readily apparent when examined with scanning electron microscopy (Fig. 1). At one pole of the egg was a region approximately 30 µm in diameter (about 740 µm²) that was virtually free of microvilli. The plasma membrane at this region undulated to some extent, but with respect to other areas of the ovum surface, it was relatively smooth (Fig. 1). This region was present on all of the mature eggs examined and is referred to as the microvillus-free area. The remainder of the egg surface was projected into numerous microvilli (3.7/µm²; Table 1) and is referred to as the microvillous area. This region of the egg surface, consisting of the microvilli as well as the surface proper, comprised an area of approximately 35,600 µm² (Table 2).

The microvillus-free area overlaid a cortical layer approximately 800 nm in width consisting of microfilaments (Figs. 2, 3). Mitochondria and endoplasmic reticulum, as well as the meiotic spindle were located subjacent to this layer (Figs. 2, 4). The presence of this layer could be discerned in light microscopic preparations; in such cases the plasma membrane overlying the meiotic spindle appeared much denser than the plasmalemma at other regions of the egg surface (Fig. 4). Microfilaments were also present in the cortex of the microvillous area; however, they were not organized into as dense a layer as in the microvillus-free area (Figs. 4, 5).

As to the composition of the cortical microfilaments, immunocytochemical methods for the demonstration of actin revealed a bright

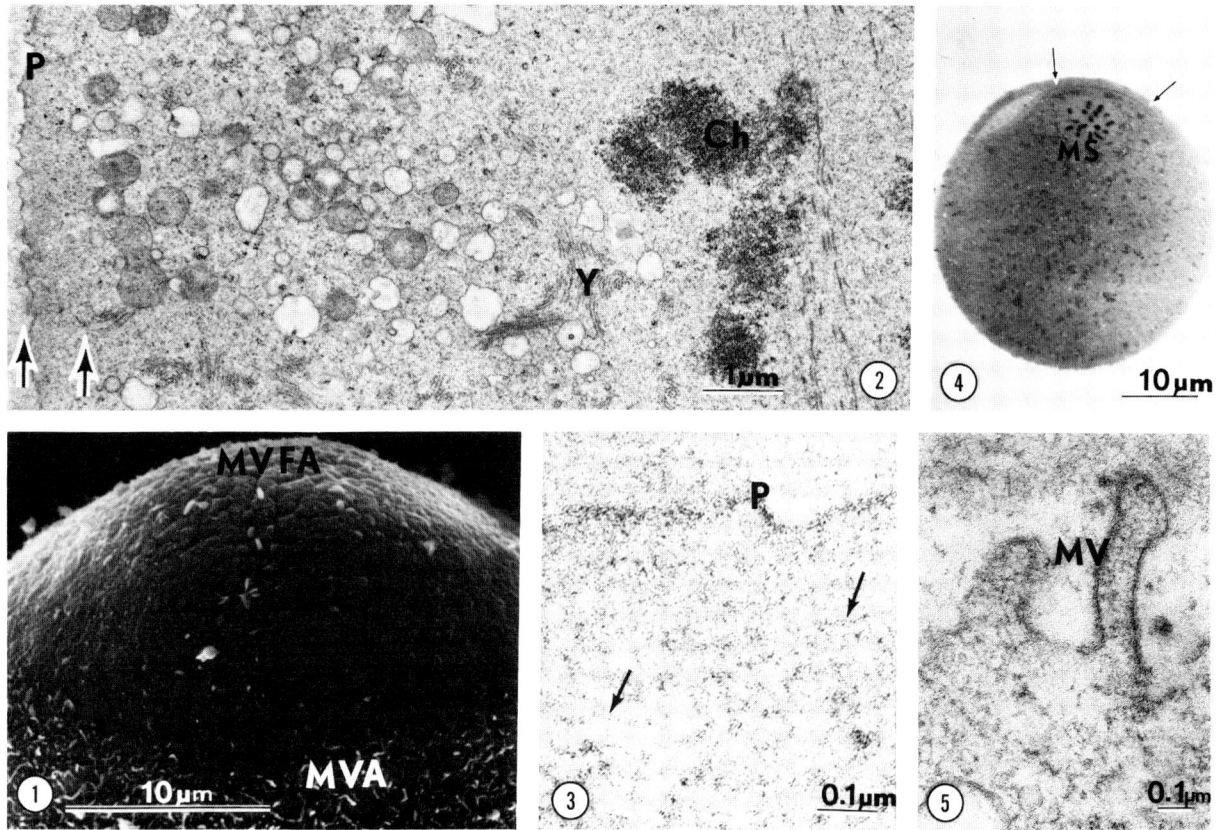

Figure 1. Scanning micrograph of a mature, oviductal mouse egg showing the microvillus-free area (MVFA) and a portion of the microvillous region (MVA).
Figure 2. Section through a portion of the meiotic spindle of a mature mouse egg demonstrating its structural relation to the ovum's surface. Although the plasma membrane (P) in the region of the meiotic spindle undulates, it is not projected into microvilli. The cortex immediately subjacent to the plasmalemma (arrows) is filled with an accumulation of microfilaments. Notice that vesicles and mitochondria are restricted from this cortical region. Ch, chromosomes of the meiotic spindle; Y, yolk plates (Szollosi, 1972).
Figure 3. Cortical region of the egg overlying the meiotic spindle showing an aggregation of microfilaments (arrows). P, plasmalemma.
Figure 4. Mature mouse egg sectioned to show a polar view of the metaphase plate of the meiotic spindle (MS). The plasma membrane that overlies the spindle, and which is located between the two arrows, is denser than the plasmalemma along other portions of the ovum.
Figure 5. Surface of a mature ovum along the microvillous area. MV, microvillus.

fluorescent layer in the egg cortex that was enhanced in the area of the microvillus-free area. Specimens prepared with antibodies to tubulin showed fluorescence in the meiotic spindle (Wassarman et al., 1979; Schatten et al., 1983; Albertini, 1984).

Immature Oocyte.

The surface of all germinal vesicle oocytes examined had a homogeneous distribution of microvilli (Nicosia et al., 1978); a microvillus-free area, characteristic of mature eggs, was not present (Fig. 6). Dense segments of the plasma membrane, characteristically associated with the microvillus-free area of mature ova, were never observed in sections of immature oocytes prepared for light microscopy. The microvilli were comparable in size and shape to those of mature ova ($0.9 \pm 0.2$ μm in length and $0.1 \pm 0.03$ μm in diameter) and had a density of $7.2/\mu m^2$ of oocyte surface (Table 1). Internally, they possessed a small bundle or core of microfilaments. Along the bases of the microvilli was a relatively dense meshwork of microfilaments that formed a cortical layer approximately 200 nm in width (Fig. 7). The surface area of the immature oocyte, including microvilli as well as the surface proper, was about 58,000 $\mu m^2$, approximately 1.6 times that of the mature ovum (Table 2).

Table 1. Density of egg and oocyte microvilli treated with and without colchicine and cytochalasin B.

| Specimen and Treatment | Microvillar Density[1] |
|---|---|
| Mature, oviductal egg | 3.7 ± 0.3 |
|     25 µM colchicine, 5 hrs | 4.2 ± 0.6 |
|     20 µM cytochalasin B, 1-2 hrs | 0.28 ± 0.1 |
|     20 µM cytochalasin B, 5 hrs | 0.33 ± 0.1 |
|     20 µM cytochalasin B, 1-2 hrs; washed and recultured in fresh medium, 4 hrs | 2.6 ± 0.3 |
| Immature oocyte | 7.2 ± 0.4 |
|     cultured, 9 hrs | 5.8 ± 1.3 |
|     cultured (matured _in vitro_), 17 hrs | 4.2 ± 0.6 |
|     25 µM colchicine, 17 hrs _in vitro_ | 3.1 ± 0.7 |
|     20 µM cytochalasin B, 17 hrs _in vitro_ | 1.5 ± 0.8 |

[1]Mean ± S.D. of microvilli present on 1 µm$^2$/egg or oocyte surface. Counts of mature and maturing eggs were taken from scanning electron micrographs of areas other than those overlying the meiotic spindle.

Table 2. Estimated surface areas of immature and mature mouse ova.

| Surface Areas of Mature Egg (µm$^2$) as: | Surface Area of Immature Oocyte (µm$^2$) as: |
|---|---|
| Microvillus-free Area - 740 | Microvillus-free Area - 0 |
| Microvillous Area | Microvillus Area |
|     Surface proper - 17,400 |     Surface Proper - 19,100 |
|     Microvilli - 18,200 |     Microvilli - 38,900 |
| Estimated Total Surface Area - 36,340 | Estimated Total Surface Area - 58,000 |

Procedures employed to calculate surface areas are provided in Materials and Methods.

Figure 6. Scanning electron micrograph of the surface of an immature oocyte which is characterized by numerous microvilli; a microvillus-free area is not present.

Figure 7. Cortex of an immature oocyte demonstrating microvilli containing microfilaments. A layer of microfilaments is located at the base of the microvilli (arrows).

Figure 8. Scanning electron micrograph of an oocyte that was matured _in vitro_. As in the case of mature, oviductal ova, oocytes having undergone maturation _in vitro_ have a polar region lacking microvilli that overlies the second meiotic spindle (MVFA). The remainder of the egg surface possesses microvilli (MVA).

Figure 9. Scanning electron micrograph of a mature oviductal egg treated with 25 µM colchicine for 5 hrs. The microvillus-free area (MVFA) is located at one pole of the egg; the remainder of the ovum's surface (MVA) is reflected into numerous microvilli.

Figures 10 and 11. Sections of mature oviductal egg treated with 25 µM colchicine for 5 hrs. The meiotic spindle of both eggs have disappeared and only the chromosomes (Ch), surrounded by some light staining material, remain. In Fig. 10 the chromosomes are present as a single aggregation closely associated with the plasma membrane. At this site (arrows) the plasma membrane is relatively denser than at other regions of the egg. In Fig. 11 two chromosomal masses have formed and both are closely associated with the plasma membrane. At each site the plasma membrane (arrows) is denser than at other regions of the egg.

Figure 12. Section of a mature egg treated with cytochalasin B showing the second meiotic spindle (MS) that is located in the ovum's cortex. The plasma membrane (arrows) that overlies the meiotic spindle does not demonstrate the increased density found in untreated preparations. C, crystalloids formed as a result of cytochalasin B treatment (cf. Moskalewski et al., 1982).

Figure 13. Scanning electron micrograph of a mature egg treated with 20 µM cytochalasin B for 2 hrs. The microvillus-free area normally associated with second meiotic spindle is difficult to identify in such specimens because of the overall disappearance of microvilli. Arrows indicate blebs that form on cytochalasin B-treated oocytes. Notice that the blebs are restricted from a region (*) of the egg surface that corresponds in size to the microvillus-free area of untreated specimens.

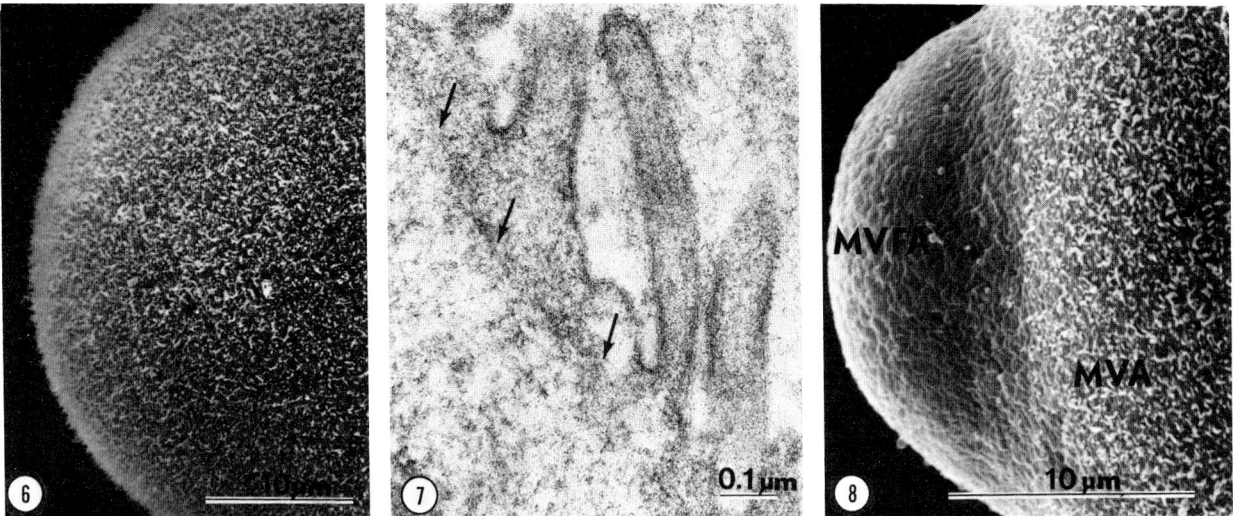

Figure 14. Scanning electron micrograph of a mature egg that was incubated in cytochalasin B for 2 hrs then washed and recultured in fresh medium for 4 hrs. A polar region, free of microvilli (MVFA), is present; the remainder of the egg possesses microvilli.

Figure 15. Section of a mature egg that was incubated in cytochalasin B for 2 hrs and then washed and recultured in medium for 4 hrs, demonstrating a portion of the meiotic spindle (MS) subjacent to the plasma membrane. The plasma membrane (arrows) overlying the meiotic spindle is denser than that along other regions of the egg.

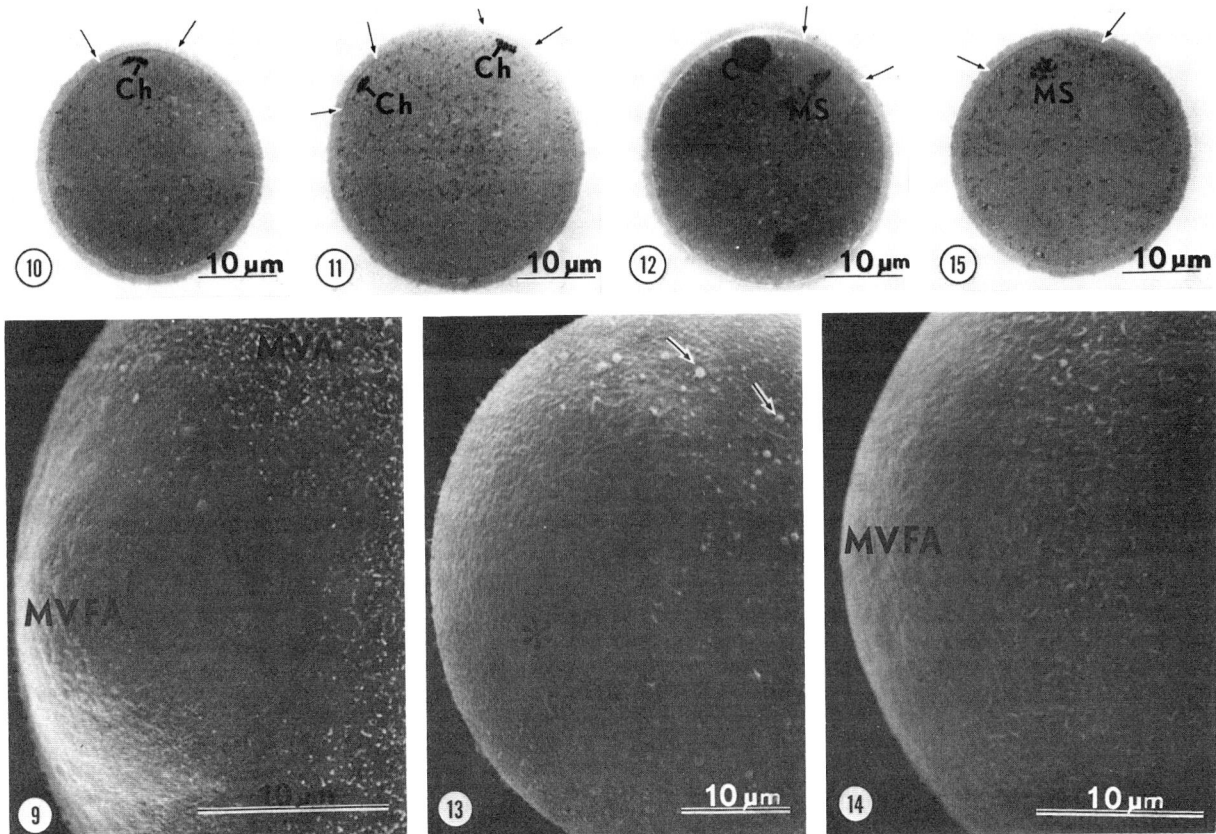

The captions for Figures 6 to 13 are on the facing page.

The sequence of germinal vesicle breakdown, first meiotic division and formation of the second meiotic spindle observed in this study was identical to that previously reported (Donahue, 1968; Wassarman et al., 1979). By 3 hrs in culture germinal vesicle breakdown had occurred, the meiotic spindle developed within the center of the oocyte and then moved peripherally to become localized within the egg cortex by 9 hrs in culture. After 17 hrs in culture greater than 85% of the oocytes examined had matured and formed the first polar body; the second meiotic spindle was localized to the ovum's cortex. By 17 hrs in culture the oocyte cortex was polarized, showing a well-defined microvillus-free area, which was equivalent in size and structure to those of mature, oviductal ova (Fig. 8). This area was associated with the meiotic spindle and a dense aggregation of microfilaments (Chen and Longo, 1984). As in the case of mature, oviductal eggs, the microvillus-free area was sharply demarcated by the appearance of numerous microvilli that projected from the remainder of the egg surface (Table 1; Fig. 8). Structurally, the microvilli were similar to those observed in mature, oviductal eggs; furthermore, the layer of microfilaments found along the bases of microvilli in immature oocytes was no longer present (Chen and Longo, 1984). The cortex of the microvillous area of the spontaneously matured oocyte was structurally identical to that of the mature, oviductal ovum.

When germinal vesicle breakdown was inhibited with chloroquine or dibutyryl cyclic AMP the surfaces of oocytes incubated up to 17 hrs in vitro did not become polarized and demonstrate a microvillus-free area. The surfaces of such oocytes were covered uniformly with microvilli comparable to untreated specimens.

Mature Eggs Treated with Colchicine or Cytochalasin B.

The surface of mature, oviductal eggs treated with 25 μM colchicine was structurally similar to untreated ova (Fig. 9). In the presence of colchicine, microtubules of the meiotic spindle disappeared and the chromosomes, which usually aggregated into a single mass surrounded by some amorphous material, remained confined to the egg's cortex (Table 3; Figs. 10, 11). The cortex and surface of the egg where the chromosomes were located was structurally similar to the microvillus-free area of untreated specimens. Occasionally, colchicine-treated specimens were found in which several groups of chromosomes were localized along the cortex. In these cases all of the chromosomal aggregates were associated with microvillus-free areas. The plasma membrane in such regions possessed a dense accumulation of microfilaments and in light microscopic preparations appeared much denser than membrane delimiting other areas of the ovum's surface (Fig. 11). The number ($4.2/\mu m^2$) and morphology of microvilli along the microvillus areas of colchicine-treated ova were comparable to microvilli of untreated eggs (Table 1; Fig. 9).

In mature, oviductal eggs treated with cytochalasin B for up to 9 hrs, the meiotic spindle remained intact and confined to the cortex, although the spindle in a few specimens did become "detached" from the oolemma (Table 3; Fig. 12). There was a reduction in the number of microvilli on the egg's surface to about $0.3/\mu m^2$ after 1-5 hrs exposure to cytochalasin B (Table 1; Fig. 13). In addition, blebs, saccules of the plasma membrane, projected into the perivitelline space. These structures were distributed over the entire surface of the egg, except for a region approximately 30 μm in diameter that corresponded to the area overlying the meiotic spindle (Fig. 13). The surface of specimens incubated in DMSO was structurally identical to that of untreated ova.

The surface of cytochalasin B-treated eggs overlying the meiotic spindle had a relatively smooth profile and when examined with light microscopy lacked the density seen in untreated specimens (Figs. 12, 13). Microfilaments were present in this area of the egg cortex, however, they were not present in a dense, wide layer as seen in untreated preparations (Chen and Longo, 1984).

When eggs were treated with cytochalasin B for 2 hrs, washed and recultured in medium, there was a reappearance of microvilli along the ovum surface, except for that portion associated with the meiotic spindle (Table 1; Fig. 14). In addition, blebbing of the plasma membrane was no longer apparent. The plasma membrane overlying the meiotic spindle of cytochalasin B-treated specimens that were washed and recultured in medium exhibited a wide aggregation of microfilaments comparable to those observed in untreated ova which rendered it denser than the plasmalemma delimiting other portions of the egg surface (Fig. 15; Chen and Longo, 1984).

Immature Oocyte Cultured in Colchicine or Cytochalasin B.

Examination of maturing oocytes cultured in medium containing 25 μM colchicine for 9 or 17 hrs indicated that a meiotic spindle did not develop following germinal vesicle breakdown

---

Figure 16. Scanning electron micrograph of an oocyte incubated in medium containing 25 μM colchicine for 17 hrs. Polar body formation has been inhibited, however, a microvillus-free area (MVFA) has developed.

Figure 17. Section of oocyte that was incubated in medium containing 25 μM colchicine for 17 hrs. The oocyte underwent germinal vesicle breakdown, however, a meiotic spindle was not formed. Nevertheless, the chromosomes (Ch) moved to the ovum's cortex. The plasma membrane (arrows) associated with the chromosomes is much denser than that associated with other regions of the ovum.

Figure 18. Section of an oocyte incubated in medium containing cytochalasin B for 17 hrs. In such a specimen the germinal vesicle breaks down and a meiotic spindle forms (MS) but it fails to move to the ovum's cortex. Dense segments of the plasma membrane are not found in these preparations.

Table 3. Localization of chromosomes in eggs and oocytes treated with cytochalasin B or cholchicine.

| Specimen and Treatment | Time of Examination (Hours after Culture) | Chromosome Location Central | Peripheral[1] |
|---|---|---|---|
| Oocyte, Untreated | 9 | 2 | 18 (90) |
|  | 17 | 0 | 20 (100) |
| Oocyte, Colchicine (25 μM) | 9 | 8 | 12 (60) |
|  | 17 | 2 | 18 (90) |
| Oocyte, Cytochalasin B (20 μM) | 9 | 18 | 2 (10) |
| Oocyte, Cytochalasin B (20 μM), 9 hrs; washed and recultured, 8 hrs | 17 | 8 | 20 (71) |
| Mature Egg, Untreated | 9 | 0 | 20 (100) |
| Mature Egg, Colchicine (25 μM) | 9 | 3 | 18 (86) |
| Mature Egg, Cytochalasin B (20 μM) | 9 | 4 | 24 (86)[2] |
| Mature Egg, Cytochalasin B (20 μM), 9 hrs, washed and recultured, 4 hrs | 13 | 6 | 20 (76)[2] |

[1]Numbers in parenthesis are percentages of eggs/oocytes with peripherally located chromosomes. Meiotic chromosomes within 12 μm (diameter of the metaphase plate) of the oolemma were considered to be peripherally located.

[2]The chromosomes from 6 mature eggs treated with cytochalasin B and 2 that were then washed and recultured in fresh medium moved from the oolemma but were within 12 μm of the egg surface.

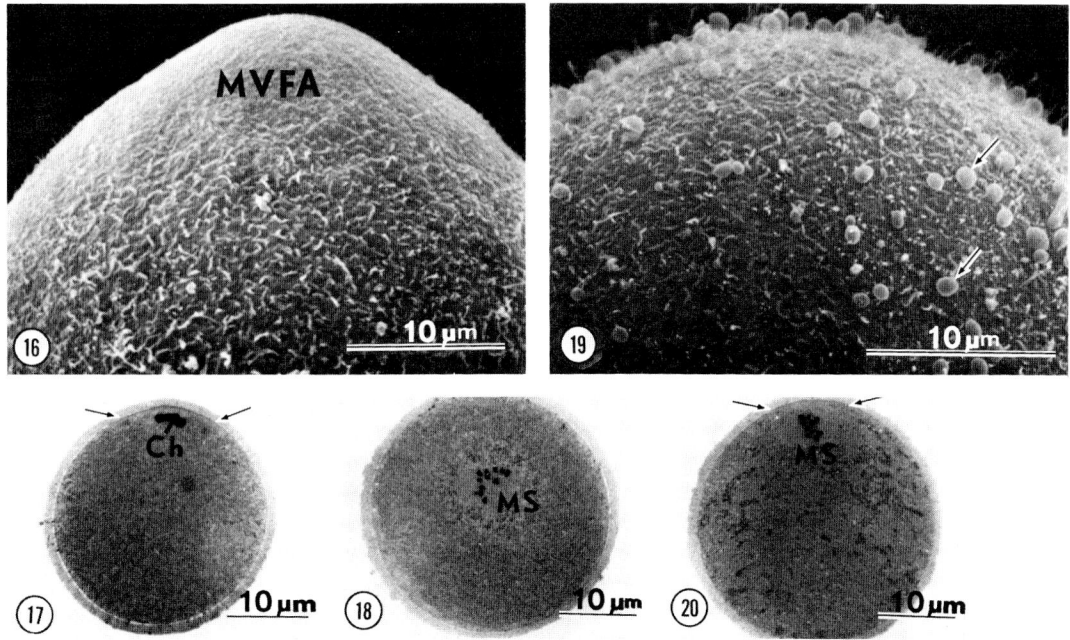

The captions for Figures 16 to 18 are on the facing page.

Figure 19. Scanning electron micrograph of an oocyte incubated in 20 μM cytochalasin B for 17 hrs. Although there is an overall reduction in microvilli and the formation of blebs (arrows), a microvillus-free area does not develop in such specimens.

Figure 20. Section of an oocyte incubated in medium containing cytochalasin B for 9 hrs, washed and recultured in medium for 8 hrs. The plasma membrane (arrows) overlying the meiotic spindle (MS), which has moved to the egg cortex, is denser than the plasmalemma associated with other regions of the ovum.

and chromosome condensation (Donahue, 1968; Wassarman et al., 1976, 1979). Nevertheless, the chromosomes moved peripherally and became positioned within the ovum's cortex (Fig. 16). The cortex and surface of the egg overlying the aggregated chromosomes were organized structurally identical to that of untreated specimens (Figs. 16, 17). The surface of the egg along the microvillous area possessed 5.2 and 4.4 microvilli/$\mu m^2$ following 9 and 17 hrs of culture in colchicine-containing media, respectively (Table 1). Structurally, the microvilli were identical to those of untreated ova.

When immature oocytes were incubated with 20 $\mu M$ cytochalasin B, germinal vesicle breakdown and development of a meiotic spindle occurred; however, the latter failed to move to the egg cortex and polar body formation was inhibited (Table 3; Wassarman et al., 1976). The chromosomes remained in the center of the ovum for up to 17 hrs in culture; during this period a microvillus-free area failed to form (Figs. 18, 19). The surface of the oocyte possessed microvilli (1.5/$\mu m^2$) and numerous blebs, structurally similar to those seen in mature eggs treated with cytochalasin B (Table 1; Fig. 19).

Immature oocytes incubated with 20 $\mu M$ cytochalasin B for 9 hrs and then washed and recultured in medium for 8 hrs failed to form polar bodies. However, in almost all of the specimens examined, the meiotic spindle became positioned within the ovum's cortex by 17 hrs in culture (Table 3; Fig. 20). The surface and cortex overlying the meiotic spindle of oocytes treated with cytochalasin B and then washed were distinguished by the presence of a microvillus-free area and the accumulation of microfilaments which in light microscopic preparations resulted in an increased density of the plasma membrane.

## Discussion

The presence of a gradient, polarity or mosaicism in the organization of the egg cortex and plasma membrane has been observed in ova from a wide variety of animals (Austin, 1965). Such polarities are common in ova where entrance and penetration of the spermatozoon is limited to a specific region of the plasma membrane (Brummett and Dumont, 1979; cf. Austin, 1965) and does not appear to be related to the stage of meiosis at which the egg is inseminated. Eggs of the sea anemone, Actinia fragacea (reportedly fertilized at the completion of meiosis) are covered uniformly with tufts of long microvilli known as cytospines (Larkman and Carter, 1984). Ova of the molluscs Mytilus and Spisula, which are inseminated at the first metaphase of meiosis and at the germinal vesicle stage, respectively, do not appear to be polarized (Longo and Anderson, 1969, 1970). Following germinal vesicle breakdown in Spisula the meiotic spindle moves to the egg cortex. Although cortical granules are moved from the site where the meiotic spindle becomes localized, the egg surface remains structurally identical to that of the unfertilized ovum. Echinoderm eggs have a uniform, microvillous surface with no apparent polarity (cf. Guraya, 1982). However, observations of Schroeder (1980) have demonstrated the reproducible alignment of polar bodies, jelly canal, micromeres, and the vegetal clear zone indicating an antecedent axis of egg organization in ova of the sea urchins, Paracentratus lividus and Arbacia punctulata (both fertilized at the completion of meiosis). Examination of the pattern of cortical organelles did not establish the nature of the underlying axis which may define the animal-vegetal poles in these eggs.

The surface and cortex of unfertilized amphibian eggs has been reviewed (Elinson, 1980; Schmell et al., 1983; cf. also Picheral and Charbonneau, 1982; Campanella and Andreuccetti, 1977; Gardiner and Grey, 1983). Unfertilized amphibian eggs are covered with numerous microvilli; in Rana pipiens those along the animal pole are long and sometimes ridge-like, while those near the vegetal pole are short and stubby. The transition between the two forms of microvilli is gradual. Within the cortex of Xenopus eggs are cisternae of endoplasmic reticulum which form junctions with the plasmalemma. Junctions in the cortex of animal hemisphere are two to three times more abundant as compared with the vegetal hemisphere (Gardiner and Grey, 1983). It has been speculated that these junctions are sites that transduce extracellular events into intracellular calcium release during fertilization and activation of development. In addition to the cortical granule reaction following fertilization, the egg cortex undergoes considerable rearrangements, including cortical contractions and formation of the grey crescent which have been shown to be involved with establishing the symmetry of the developing embryo (cf. Elinson, 1980).

Eggs of the rabbit, hamster and rat possess a cortical polarity similar to that described for the mouse (Odor and Renninger, 1960; Longo, 1974, 1975; Nicosia et al., 1977; Stefanini et al., 1969; Thompson et al., 1974; Yanagimachi and Chang, 1961; Szollosi 1962, 1967, 1976; Gulyas, 1976; Phillips and Shalgi, 1980; Chen and Longo, 1984). Polarization may not be a general feature of mammalian eggs; observations of human ova have not verified the presence of a cortical polarity comparable to that described for mouse eggs (Zamboni, 1972).

Structural comparisons of the cortices of mature and immature mouse ova indicate that there is a concomitant reorganization of the egg surface and cortical cytoskeleton during meiotic maturation (Chen and Longo, 1984). Unlike the mature ovum, the surface of the immature oocyte is apolar and has a uniform, microvillous surface (Nicosia et al., 1978). Within the oocyte cortex, along the bases of the microvilli, is a thin layer of microfilaments. This layer is replaced during meiotic maturation by one that is relatively more diffuse, except in that region associated with the meiotic spindle where it is greatly augmented (Chen and Longo, 1984). Microfilaments are also found within the microvilli of both mature and immature mouse ova, however they are not as well-developed in number and organization as the cores of actin filaments within the microvilli of intestinal epithelial cells and activated echinoderm eggs (Mooseker and Tilney, 1975; Schroeder, 1981; Chen and Longo, 1984). Based on their structural dimensions and immunochemical staining properties, Chen and Longo (1984) have demonstrated that the microfilaments within the microvilli and cortex of mouse ova consist of actin. Contractile proteins have also been demonstrated in the ova of other organisms and are believed to function in many of the dynamic changes of the egg cortex during oogenesis and fertilization (Vacquier, 1981; Schroeder, 1981).

The polar modifications within the cortex and surface of mouse eggs described here, involving cytoskeletal elements and microvilli, may be relevant to mechanisms of gamete interactions, polar body formation and generation of embryonic polarity (Szollosi, 1967; Zamboni, 1970, 1971; Stefanini et al., 1969; Thompson et al., 1974; Campanella, 1975; Elinson, 1975; Burgess, 1977; Shimizu, 1981a,b; Johnson et al., 1981). Nicosia et al. (1978) have indicated that changes in surface morphology seen upon the resumption of meiotic maturation may be related to satisfying nutritional requirements (Zamboni, 1970; Anderson and Albertini, 1976; Epstein et al., 1976), processes of cytoplasmic maturation (Thibault, 1973) or to changes in membrane fluidity or electrical activity (Powers and Biggers, 1976). In addition, the estimated differences in surface areas of immature oocytes vs. mature eggs (Table 2) may be relevant to metabolic studies analyzing processes such as the uptake and incorporation of precursors (Wassarman et al., 1981).

The simultaneous appearance/disappearance of the microvillus-free area and associated layer of microfilaments in mouse ova, under different experimental conditions, are suggestive that the cortical cytoskeleton maintains the conformation of the egg surface, particularly microvillar domains. Chen and Longo (1984) suggested that the cortical aggregation of microfilaments observed in both mouse oocytes and mature ova may function as a special component of the cytoskeleton, comparable to that postulated for red blood cells (Branton et al., 1981). That is, the shape of the cell, as well as its elasticity may be dependent upon cell-surface proteins whose interactions reinforce the egg plasma membrane with a deformable meshwork that constitutes the cytoskeleton. In the case of the maturing ovum this component of the cytoskeleton may be involved in polar body formation (Stefanini et al., 1969; Thompson, et al., 1974) and responsible for changes in cortical rigidity that accompany this process (Vacquier, 1981). It may also function to secure the meiotic spindle to the egg cortex (Chambers, 1917; Conklin, 1917; Shimizu, 1981a; Hamaguchi et al., 1983).

Reduction of the cortical microfilamentous aggregation in mouse eggs treated with cytochalasin B is consistent with previous observations of the effects of this drug in other cells where inhibition of polymerization and/or "breaking" of actin filaments have been described (Yahara et al., 1982; Schliwa, 1982). The disappearance of microvilli and the formation of blebs induced by cytochalasin B have also been reported for mouse eggs and embryos and for other cells (Perry and Snow, 1975; Loor, 1981, Shimizu, 1981a,b; Chen and Longo, 1984). The reduction in microvilli of cytochalasin B-treated eggs reported here stands in contrast to observations of Eager et al. (1976) who reported the formation of microvilli within the area overlying the meiotic spindle. The basis for this discrepancy is unclear, but it may be related to differences in strains of mice, sampling time and/or specimen preparation.

With meiotic maturation and the localization of the meiotic spindle to the egg cortex, there was an overall reduction in the number of microvilli and the development of a microvillus-free area. How modifications in microvillar number were accomplished has not been established. Preferential, asymmetric expression of microvilli has been described for cells such as lymphocytes and embryonic blastomeres (Loor, 1981; Ducibella et al., 1977). Loor (1981) described the development of microvillar asymmetry in lymphocytes as a flow of microvilli towards one pole of the cell. Such a mechanism, if it were to occur, would only partially account for microvillar changes during meiotic maturation in mouse oocytes.

Structural and chemical properties of the surface of mammalian eggs, involving the distribution of cortical granules and the identification of specific enzymatic activities and ligands have been described (Yanagimachi et al., 1973; Nicolson et al., 1975; Solter, 1977; Gulyas, 1980; Longo, 1981a). Detailed accounts have also been given for invertebrate eggs (Longo, 1976, 1981b; Vacquier, 1981; Schroeder, 1981; Carron and Longo, 1983). Although the egg plasmalemma does not demonstrate a wide range of membrane diversity and complex structural mosaicism when compared to a highly polarized cell such as the spermatozoon (Friend, 1982), it does possess areas having specific properties that appear to be spatially related to underlying cortical structures. Johnson et al. (1975) and Eager et al. (1976) described differences in concanavalin A binding to mouse ova that were believed to be related to the distribution of microvilli. Recent investigations by Wolf and Ziomek (1983) showed a true concentration gradient on the surface of the mouse egg beyond those resulting from surface amplification. Variations in membrane protein diffusibility within the microvillous-free and microvillous areas were observed which they speculated may reflect differences in the distribution of cytoskeletal elements. The observations presented here are consistent with their speculation and demonstrate distinct differences in cortical cytoskeletal components that are distributed in a polar fashion (Stefanini et al., 1969; Nicosia et al., 1977; Chen and Longo, 1984).

That immature oocytes were capable of forming microvillus-free areas and associated aggregations of microfilaments only when meiosis was reinitiated and the chromosomes became localized to the egg periphery indicates that cortical changes normally associated with oocyte maturation are not autonomous cytoplasmic events, independent of nuclear structures and events. Investigations with the eggs and zygotes of other organisms demonstrated the presence of cortical changes, often contractile, during maturation and/or fertilization that are autonomous with respect to nuclear activity (Saki and Kubotta, 1981; Yoneda et al., 1978; Shimizu, 1981c; Coffe et al., 1982; Yamamoto and Yoneda, 1983). In these cases the cytoplasm by its own cyclic behavior is believed to regulate the timing of nuclear events during the cell cycle.

Although a meiotic spindle was not formed in oocytes treated with colchicine, when the chromosomes moved to the cortex a microvillus-free area formed, and only in areas associated with the chromosomes. Based on these observations it is compelling to suggest that interaction of the meiotic chromosomes with components in the oocyte cortex brings about the formation of a microvillus-free area. However, the chromosomes of both colchicine-treated eggs and oocytes were also associated with some undefined amorphous material (Chen and Longo, 1984). Therefore, we cannot eliminate the possibility that a substance(s) associated with the chromosomes, and not the chromosomes themselves, mediate changes in the ovum cortex and surface. Hence, reference to the possible role of chromosomes as inducers of cortical and surface changes in mouse ova should be interpreted in light of this possibility.

The appearance of more than one cortical group of chromosomes in colchicine-treated eggs, each with its own microvillus-free area, is consistent with the suggestion that the chromosomes may have an inductive role in the development of cortical modifications of maturing mouse ova. The absence of a microvillus-free area and associated aggregation of microfilaments in colchicine-treated eggs, where the chromosomes were surrounded by cisternae, is also supportive of this proposal (Chen and Longo, 1984). This observation also suggests that the presumed inductive effect is eliminated when the chromosomes are surrounded by membrane. In the fresh water polyp, Pelmatohydra egg, microvilli cover the egg surface except for an area at the animal pole occupied by the female pronucleus (Noda and Kania, 1981). Although this appears to be in contrast to the situation seen in colchicine-treated mouse ova, the presumed inductive effect of nuclear components on the egg cortex may not be a general one; it may be mediated according to local conditions involving the status of the nuclear material. The presence of more than one microvillus-free area in some colchicine-treated eggs also suggests that the capacity of the surface and cortex to differentiate into specialized areas is not restricted to a single site.

Subsequent to its migration to the egg cortex, the meiotic spindle becomes anchored to the plasma membrane (Chambers, 1917; Conklin, 1917; Shimizu, 1981c). In Chaetopterus, attachment of the meiotic spindle to the egg surface was colchicine sensitive and unaffected by cytochalasin B (Hamaguchi et al., 1983). Johnson et al. (1975) indicated that in mature mouse eggs treated with cytochalasin B, the meiotic spindle was slightly displaced from the plasma membrane, although as demonstrated here, it did remain within the cortex. In contrast to what was observed in the eggs of some invertebrates (Longo, 1972; Peaucellier et. al., 1974), cytochalasin B prevented the localization of meiotic chromosomes to the cortex of maturing mouse oocytes, indicating that a cytochalasin B sensitive component of the egg is involved in this movement (Wassarman et al., 1976). Hence, it is possible that movement to and maintenance of the meiotic spindle within the ovum cortex is not modulated in the same manner for eggs of different organisms. Actin has been demonstrated in nuclei and mitotic spindles and has been implicated in forced production of chromosome movements during mitosis (cf. Zimmerman and Forer, 1981). In light of these investigations, the observations reviewed here suggest that an actin based system may also be responsible for the cortical localization of the meiotic spindle in mouse oocytes.

Acknowledgements

The technical assistance of Frederick So is gratefully acknowledged. Investigations supported by funds from the NIH and the Rockefeller Foundation.

References

Albertini DF. (1984). Novel morphological approaches for the study of oocyte maturation. Biol. Reprod. 30, 13-28.

Anderson E, Albertini DF. (1976). Gap junctions between the oocytes and companion follicle cells in the mammalian ovary. J. Cell Biol. 71, 680-686.

Austin CR. (1961). The Mammalian Egg. Blackwell Scientific Publications, Oxford, 183.

Austin CR. (1965). Fertilization. Prentice-Hall, Englewood Cliffs, New Jersey, 145.

Branton D, Cohen CM, Tyler J. (1981). Interaction of cytoskeletal proteins on the human erythrocyte membrane. Cell 24, 24-32.

Brummett AR, Dumont JH. (1979). Initial stages of sperm penetration into the egg of Fundulus heteroclitus. J. Exp. Zool. 210, 417-434.

Burgess DR. (1977). Ultrastucture of meiotic and polar body formation in the egg of the mud snail, Ilyanassa obsoleta, in: Cell Shape and Surface Architecture, JR Revel, U Henning, F Fox, (eds), Alan R. Liss, New York, 569-579.

Campanella C. (1975). The site of spermatozoa entrance in the unfertilized egg of Discoglossus pictus (Anura): An electron microscopic study. Biol. Reprod. 12, 439-447.

Campanella C, Andreuccetti P. (1977). Ultrastructural observations on cortical endoplasmic reticulum and on residual cortical granules in the egg of Xenopus laevis. Dev. Biol. 56, 1-10.

Carron CP, Longo FJ. (1983). Filipin-sterol complexes in fertilized and unfertilized sea urchin egg membranes. Dev. Biol. 99, 482-488.

Chambers R. (1917). Microdissection studies. II. The cell aster: A reversal gelation phenomenon. J. Exp. Zool. 23, 483-505.

Chen DY, Longo FJ. (1984). Involvement of microfilaments and the meiotic apparatus in the development of surface polarity in mouse eggs. Anat. Rec. 208:29A.

Coffe G, Foucault G, Soyer MO, DeBilly F, Pudles J. (1982). State of actin during the cycle of cohesiveness of the cytoplasm in parthenogenetically activated sea urchin egg. Exp. Cell Res. 142, 365-372.

Conklin EG. (1917). Effects of centrifugal force on the structure and development of the eggs of Cripedula. J. Exp. Zool. 22, 311-419.

Diem K. (1962). Scientific Tables. Geigy Pharmaceuticals, New York, 142-143.

Donahue RP. (1968). Maturation of the mouse oocyte in vitro. I. Sequence and timing of nuclear progression. J. Exp. Zool. 169, 237-250.

Ducibella T, Ukena T, Karnovsky M, Anderson E. (1977). Changes in cell surface and cytoplasmic organization during early embryogenesis in the preimplantation mouse embryo. J. Cell. Biol. 74, 153-167.

Eager DD, Johnson MH, Thurley KW. (1976). Ultrastructural studies on the surface membrane of the mouse egg. J. Cell Sci. 22, 345-353.

Elinson RP. (1975). Site of sperm entry and cortical contraction associated with egg activation in the frog, Rana pipiens. Dev. Biol. 47, 257-268.

Elinson RP. (1980). The amphibian egg cortex in fertilization and early development, in: The Cell Surface: Mediator of Developmental Processes, S Subtelny, NK Wessells, (eds), Academic Press, New York, 217-234.

Epstein ML, Beers WH, Gilula NB. (1976). Cell communication between the rat cumulus oophorus and the oocyte. J. Cell Biol. 70, 904a.

Friend DS. (1982). Plasma-membrane diversity in a highly polarized cell. J. Cell Biol. 93, 243-249.

Gardiner DM, Grey RD. (1983). Membrane junction in Xenopus eggs: Their distribution suggests a role in calcium regulation. J. Cell Biol. 96, 1159-1163.

Gulyas B. (1976). Ultrastructural observations on rabbit, hamster and mouse eggs following electrical stimulation in vitro. Am. J. Anat. 147, 203-218.

Gulyas B. (1980). Cortical granules of mammalian eggs. Intl. Rev. Cytol. 63, 357-392.

Guraya SS. (1982). Recent progress in the structure, origin, composition, and function of cortical granules in animal eggs. Intl. Rev. Cytol. 78, 257-360.

Hamaguchi Y, Lutz DA, Inoué S. (1983). Cortical differentiation asymmetric positioning and attachment of the meiotic spindle in Chaetopterus pergamentaceous oocytes. J. Cell Biol. 97, 254a.

Hoppe P, Pitts S. (1973). Fertilization in vitro and development of mouse ova. Biol. Reprod. 8, 420-426.

Johnson MH, Eager D, Muggleton-Harris A. (1975). Mosaicism in organization of concanavalin A receptors on surface membrane of mouse egg. Nature 257, 321-322.

Johnson MH, Pratt HPM, Handyside AH. (1981). The generation and recognition of positional information in the preimplanation mouse embryo, in: Cellular and Molecular Aspects of Implantation, SR Glasser, DW Bullock, (eds), Plenum Press, New York, 55-74.

Larkman AU, Carter MA. (1984). The apparent absence of a cortical reaction after fertilization in a sea anemone. Tissue & Cell 16, 125-130.

Longo FJ. (1972). The effects of cytochalasin B on the events of fertilization in the surf clam, Spisula solidissima. I. Polar body formation. J. Exp. Zool. 182, 321-344.

Longo FJ. (1974). Ultrastructural changes in rabbit eggs aged in vivo. Biol. Reprod. 11, 22-39.

Longo FJ. (1975). Spontaneous activation of the hamster egg in vivo. in: Electron microscopic concepts of secretion. Ultrastructure of Endocrinal and Reproductive Organs, M Hess, (ed), John Wiley & Sons, Inc., New York, 35-51.

Longo FJ. (1976). Ultrastructural aspects of fertilization in spiralian eggs. Amer. Zool. 16, 375-394.

Longo FJ. (1981a). Changes in the zonae pellucidae and plasmalemmae of aging mouse eggs. Biol. Reprod. 25, 399-411.

Longo FJ. (1981b). Morphological features of the surface of the sea urchin (Arbacia punctulata) egg: Oolemma-cortical granule association. Dev. Biol. 84, 173-182.

Longo FJ, Anderson E. (1969). Cytological aspects of fertilization in the lamellibranch, Mytilus edulis. I: Polar body formation and development of the female pronucleus. J. Exp. Zool. 172, 69-96.

Longo FJ, Anderson E. (1970). An ultrastructural analysis of fertilization in the surface clam, Spisula solidissima. I. Polar body formation and development of the female pronucleus. J. Ultrastruct. Res. 33, 495-574.

Loor F. (1981). Cell surface-cell cortex transmembranous interactions with special reference to lymphocyte functions, in: Cytoskeletal Elements and Plasma Membrane Organization, G. Poste, GL Nicolson, (eds), Elsevier/North Holland Biomedical Press, Amsterdam, 255-335.

Mooseker MS, Tilney LG. (1975). Organization of actin filament-membrane complex. Filament polarity and membrane attachment in the microvilli of intestinal epithelial cells. J. Cell Biol. 67, 725-743.

Moskalewski S, Sawicki W, Gabara B, Koprowski H. (1972). Crystalloid formation in unfertilized mouse ova under influence of cytochalasin B. J. Exp. Zool. 180, 1-12.

Nicolson GL, Yanagimachi R, Yanagimachi H. (1975). Ultrastructural localization of lectin-binding sites on the zonae pellucidae and plasma membranes of mammalian eggs. J. Cell Biol 66, 263-274.

Nicosia SV, Wolf DP, Inoue M. (1977). Cortical granule distribution and cell surface characteristics in mouse eggs. Dev. Biol. 57, 56-74.

Nicosia SV, Wolf DP, Mastroianni L. (1978). Surface topography of mouse eggs before and after insemination. Gam. Res. 1, 145-155.

Noda K, Kania C. (1981). Light and electron microscopic studies on fertilization of Pelmatohydra robusta I. Sperm entry to a specialized region of the egg. Develop. Growth Differ. 23, 401-413.

Odor DL, Renninger DF. (1960). Polar body formation in the rat oocyte as observed with the electron microscope. Anat. Rec. 137, 13-23.

Peaucellier G, Guerrier P, Bergerard J. (1974). Effects of cytochalasin B on meiosis and development of fertilized and activated eggs of Sabellaria alveolata (Polychaete Annelid). J. Embryol. Exp. Morph. 31, 61-74.

Perry MM, Snow MHL. (1975). The blebbing response of 2-4 cell stage mouse embryos to cytochalasin B. Dev. Biol. 45, 372-377.

Phillips DM, Shalgi R. (1980). Surface architecture of the mouse and hamster zona pellucida and oocytes. J. Ultrastruct. Res. 72, 1-12.

Picheral B, Charbonneau M. (1982). Anuran fertilization: A morphological reinvestigation of some early events. J. Ultrastruct. Res. 81, 306-321.

Powers DR, Biggers JD. (1976). Inhibition of mouse oocyte maturation by cell membrane potential hyperpolarization. J. Cell Biol. 70, 1054a.

Saki M, Kubotta HY. (1981). Cyclic surface changes in the non-nucleate egg fragment of Xenopus laevis. Develop. Growth Differ. 23, 41-49.

Schatten G, Simerly C, Cline C, Schatten H. (1983). Microtubules and microfilaments in mouse oocytes and sperm during fertilization: Immunofluorescence and fluorescence localization. J. Cell Biol. 97, 27a.

Schliwa M. (1982). Action of cytochalasin D in cytoskeletal networks. J. Cell Biol. 92, 79-91.

Schmell ED, Gulyas BJ, Hedrick JL. (1983). Egg surface changes during fertilization and the molecular mechanism of the block to polyspermy, in: Mechanism and Control of Animal Fertilization. JF Hartmann, (ed), Academic Press, New York, 365-413.

Schroeder TE. (1975). Dynamics of the contractile ring, in: Molecules and Cell Movement. S Inoué and RE Stephens, (eds), Raven Press, New York, 305-334.

Schroeder TE. (1980). Expressions of the prefertilization polar axis in sea urchin eggs. Dev. Biol. 79, 428-443.

Schroeder TE. (1981). Interrelations between the cell surface and the cytoskeleton in cleaving sea urchin eggs, in: Cytoskeletal Elements and Plasma Membrane Organization, G Poste, GL Nicolson, (eds), Elsevier/North Holland Biomedical Press, Amsterdam, 170-221.

Shimizu T. (1981a). Cortical differentiation of the animal pole during maturation division in fertilized eggs of Tubifex (Annelida, Oligochaeta). I. Meiotic apparatus formation. Dev. Biol. 85, 65-76.

Shimizu T. (1981b). Cortical differentiation of the animal pole during maturation division in fertilized eggs of Tubifex (Annelida, Oligochaeta). II. Polar body formation. Dev. Biol. 85, 77-88.

Shimizu T. (1981c). Cyclic changes in shape of a non-nucleate egg fragment of Tubifex (Annelida, Oligochaeta). Develop. Growth Differ. 23, 101-109.

Solter D. (1977). Organization and the antigenic properties of the egg membrane, in: Immunobiology of Gametes, M Edidin, MH Johnson, (eds), Cambridge Univ. Press, Cambridge, 207-234.

Stefanini M, Oura C, Zamboni L. (1969). Meiotic cleavage of the mammalian ovum. J. Cell Biol. 43, 138a.

Szollosi D. (1962). Cortical granules: a general feature of mammalian eggs. J. Reprod. Fertil. 4, 223.

Szollosi D. (1967). Development of cortical granules and the cortical reaction in rat and hamster eggs. Anat. Rec. 159, 431-446.

Szollosi D. (1972). Changes of some organelles during oogenesis in mammals, in: Oogenesis, JD Biggers, AW Schuetz, (eds), University Park Press, Baltimore, 47-64.

Szollosi D. (1976). Oocyte maturation and paternal contribution to the embryo in mammals, in: Current Topics in Pathology, Vol. 62, E Grundman, WH Kirsten, (eds), Springer-Verlag, New York, 9-27.

Thibault C. (1973). In vitro maturation and fertilization of rabbit and cattle oocytes, in: The Regulation of Mammalian Reproduction, SJ Segal, RC Rozier, PA Corfman, PG Condliffe, (eds), CC Thomas Pub., Springfield, 231-240.

Thompson RS, Moore-Smith D, Zamboni L. (1974). Fertilization of mouse ova in vitro: An electron microscopy study. Fert. Steril. 25, 222-249.

Vacquier VD. (1981). Dynamic changes of the egg cortex. Dev. Biol. 84, 1-26.

Van Blerkom J, Motta P. (1979). The Cellular Basis of Mammalian Reproduction, Urban and Schwarzenberg, Inc., Baltimore, 252.

Wassarman PM, Josefowicz WJ, Letourneau GZ. (1976). Meiotic maturation of mouse oocytes in vitro: Inhibition of maturation at specific stages of nuclear progression. J. Cell Sci. 22, 531-545.

Wassarman PM, Schultz RM, Letourneau GE, LeMarca MI, Josefowicz WJ, Bleil JD. (1979). Meiotic maturation of mouse oocytes in vitro, in: Ovarian Follicular and Corpus Luteum Function, CP Channing, JM Marsh, WA Sadler, (eds), Plenum Press, New York, 251-268.

Wassarman PM, Bleil JD, Cascio SM, Lamarca MJ, Letourneau GE, Mrojak SC, Schultz RM. (1981). Programming of gene expression during mammalian oogenesis, in: Bioregulators of Reproduction, G Jagiello, HJ Vogel, (eds), Academic Press, New York, 119-150.

Wolf DE, Ziomek CA. (1983). Regionalization and lateral diffusion of membrane proteins in unfertilized and fertilized mouse eggs. J. Cell Biol. 96, 1786-1790.

Yahara I, Harada F, Sekita S, Yoshihira K, Natori S. (1982). Correlation between effects of 24 different cytochalasins on cellular structures and cellular events and those on actin in vitro. J. Cell. Biol. 92, 69-78.

Yamamoto K, Yoneda M. (1983). Cytoplasmic cycle in meiotic division of starfish oocytes. Dev. Biol. 96, 166-172.

Yanagimachi R, Chang MC. (1961). Fertilizable life of golden hamster ova and their morphological changes at the time of losing fertilizability. J. Exp. Zool. 148, 185-203.

Yanagimachi R, Nicolson GL, Noda YD, Fujimoto M. (1973). Electron microscopic observations of the distribution of acidic anionic residues on hamster spermatozoa and eggs before and during fertilization. J. Ultrastruct. Res. 43, 344-353.

Yoneda M, Ikeda M, Washitani S. (1978). Periodic change in the tension at the surface of activated non-nucleate fragments of sea urchin eggs. Develop. Growth Differ. 20, 329-336.

Zamboni L. (1970). Ultrastructure of mammalian oocytes and ova. Biol. Reprod., Suppl. 2, 44-63.

Zamboni L. (1971). Fine Morphology of Mammalian Fertilization. Harper and Row, New York, 223.

Zamboni L. (1972). Comparative studies on the ultrastructure of mammalian oocytes, in: Oogenesis, JD Biggers, AW Schuetz, (eds), University Park Press, Baltimore, 5-45.

Zimmerman AM, Forer A. (1981). Mitosis/Cytokinesis. Academic Press, New York, 479

## Discussion with Reviewers

G.C. Schoenwolf: Are the undulations of the MVFA real or are they probably shrinkage artifacts?
Authors: We are unsure whether the undulations seen on the MVFA are real or shrinkage artifacts. With the techniques we have employed to study the egg surface they are a constant feature of the microvillus-free area.

G.C. Schoenwolf: Do any changes occur in microvillar length or diameters following treatments with colchicine or cytochalasin B?
Authors: Changes in microvillar morphology of eggs treated with colchicine were not obvious. As indicated in Table 1 the microvilli of eggs treated with cytochalasin B disappeared.

E.M. Eddy: In the colchicine-treated eggs with several groups of chromosomes localized along the cortex, was the total surface area of the microvillus-free areas similar to that of the untreated egg?
Authors: The total surface area of microvillus-free areas in colchicine-treated eggs with several cortically located groups of chromosomes was similar to that of the untreated egg. This is a rather intriguing result which we plan to investigate in greater detail.

S.V. Nicosia: Would you comment on the significance and role of the microfilamentous band which underlies the microvilli-free plasmalemma and overlies the meiotic spindle? Also would you comment on the significance of surface polarity in mouse eggs?
Authors: The microfilamentous layer appears to be involved in several major processes. (1) It may function during polar body formation both in the development of cytoplasmic processes that become the first and second polar bodies and in their cleavage from the egg. In the latter case the microfilaments may be involved in the formation of the cleavage furrow. (2) The microfilamentous layer may also contribute to conditions that maintain this region free of microvilli and nonfusigenic with sperm. Repeated examinations of fertilizing mammalian eggs have demonstrated that sperm incorporation occurs along that region of the egg possessing microvilli.

S.V. Nicosia: As you (see Fig. 8) and others have shown, the microvilli-free portion of mouse eggs has a conical shape and stands out from the remaining eggs circumference. Is this geometrical diversity due to the underlying cytoskeleton or an artifactual result of the removal of the zona pellucida?
Authors: We are not the first to describe this asymmetry in mammalian eggs. It was observed by early light microscopists and is reviewed by Austin (1961). We feel that this geometrical diversity is due to both the presence of the underlying cytoskeleton and is, in part, an artifactual result of the removal of the zona pellucida. Zonae-intact eggs possess a "conical" microvillus-free area that stands out from the remaining eggs circumference, although it is not as prominent as when the zona pellucida is removed. If eggs are treated with cytochalasin B the conical shape of the microvillus-free area and the cortical actin layer associated with the microvillus-free area disappear; the ovum assumes a spherical form. When cytochalasin B-treated eggs are washed in fresh medium a microvillus-free area and cortical actin layer reappear. The microvillus-free areas in such specimens have the same geometrical diversity as untreated ova.

S.V. Nicosia: Does the appearance of surface polarity in mouse oocytes precede or follow the loss of junctional contacts between oocytes and cumulus granulosa cells? Does the interruption of these contacts represent a signal for the resumption of meiosis?
Authors: The appearance of surface polarity in mouse oocytes coincides with the localization of the meiotic spindle to the egg cortex. Germinal vesicle-containing, immature oocytes, lacking cumulus cells, possess a homogeneous, microvillous surface.

S.V. Nicosia: Is the formation of the microfilamentous band due to the inward migration and condensation of microvillar filaments?
Authors: We are unsure how the microfilamentous cytoskeleton is rearranged during oocyte maturation. The microfilamentous layer associated with the microvillus-free area may be formed, in whole or part, as you suggested. We have had difficulty in determining this precisely because of technical aspects involving the size of the eggs and their symmetry.

W. H. Massover: Several of the SEM micrographs show a prominently aspherical shape of the mouse egg. Do the authors consider this to be physiological, or is it an artifact of the procedures utilized to prepare these ova for SEM?
Authors: Eggs with intact zonae have an aspherical shape, however, it is not as prominent in unfixed specimens which have had their zonae removed. Hence, both the zona and the cytoskeleton influence the shape of the mouse ovum.

W. H. Massover: The authors interpret the cytochalasin-induced small protuberance of the surface shown in Fig. 10 as "blebs, saccules of the plasma membrane." In earlier experiments,

Eager et al. (1976, text reference) interpreted such protuberances as "balooning of microvilli"; Merchant and Chang (1971 Anat. Rec. 171:21-28) observed rather similar protuberances in maturing mouse oocytes (not treated with cytochalasin) and interpreted them as "lamella-like processes". Would the authors clarify these varying interpretations of the structural nature of such protuberances?
Authors: The protuberances formed in cytochalasin B-treated eggs consist of two closely apposed membranes and are not simply a "ballooning of microvilli". We have not determined how they are formed and their relation to microvilli.

W. H. Massover: What is the condition of the surface of the polar bodies, and is the microvillar area of the egg surface incorporated into the first polar body?
Authors: Polar body formation occurs at the site of the microvillus-free area such that this particular region of the egg is incorporated into the structure of the polar bodies. The first polar body has a smooth surface reflecting its incorporation of the microvillus-free area.

EMBRYOLOGY OF THE MOUSE FROM OVULATION THROUGH PERI-IMPLANTATION STAGES *IN VITRO*

D. J. Chávez[*]

Department of Anatomy, School of Medicine, Southern Illinois
University, Carbondale, Illinois 62901

(Paper received January 23, 1984, Completed manuscript received June 9, 1984)

## Abstract

Because mammalian embryos derive their nourishment from a placenta their early development is different from that manifested by embryos that develop outside the mother's body. Although development in mammals may be studied by using representative examples obtained by recovery methods, it is very fascinating to watch living embryos develop in the laboratory. Advances in culture techniques have made it possible to observe development of pre-implantation mammalian embryos entirely in vitro. The sequence of in vitro development may be interrupted at any time and selected specimens processed for scanning electron microscopy. Observation with the scanning electron microscope on developing mouse embryos revealed the mosaic nature of the oocyte and the dynamics of the egg membrane as egg activation proceeded from incorporation of the spermatozoon to extrusion of the polar body. During early cleavage the dividing cells remained attached to each other only by ionic attraction but prior to formation of the blastocyst the cells established extensive foci of junctional contacts. Interaction between blastocyst-stage embryos and uterine epithelial cells were approximated by co-culturing embryos with uterine cells and the SEM revealed the cell to cell interactions during peri-implantation development.

KEY WORDS: embryology, oocyte, fertilization, cleavage, morula, blastocyst, mouse embryo, in vitro culture

*Address for correspondence:
D. J. Chávez, Department of Anatomy,
School of Medicine, Southern Illinois University
Carbondale, IL 62901
        Phone No. (618) 536-5513

## Introduction

Typically, students of introductory biology and embryology are first exposed to developing embryos of egg-laying (ooviparous) species. Embryos of frogs or sea urchins are most frequently chosen because they are easily obtainable and develop readily in a teaching laboratory. Another advantage is that they are not covered by an opaque shell. Eggs of ooviparous animals must possess a large amount of yolk in order to provide nourishment for the developing embryo and, therefore, the early cleavage as well as subsequent development is usually quite different from that of species that derive nourishment via a placenta (i.e., eutherian mammals) and subsequently give live birth (viviparous). These differences are normally hidden from casual observation and comparison because, of course, development in viviparous species takes place within the female genital tract. By using techniques that recover the egg from the female genital tract it is possible to obtain representative developmental stages and thus build a picture of the overall embryological process. However, advances in in vitro culture techniques should now make it possible for all students to observe at first hand all developmental processes from egg maturation through peri-implantation development. Pre-implantation embryos of rabbits, mice and guinea pigs are especially easy to maintain in culture. It remains rather difficult to achieve post implantation development in vitro, however, and post-implantation embryogenesis remains hidden within the uterus.

This paper presents a brief sketch of the early embryology of mammals using the laboratory mouse as an example. The manner in which blastocyst stage embryos become implanted within the uterus is presently the object of much active investigation and some examples of the use of the in vitro model of implantation are presented to give the reader a view of peri-implantation embryogenesis.

## Materials and Methods

Examples included in this paper were obtained from female mice that were mated normally with fertile males. Fertilized eggs were flushed from the oviduct a few hours after mating. Pre-blastocyst stage embryos were covered by an acellular mucoprotein coat, the zona pellucida. In order to observe the surface of these embryos by scanning electron microscopy the zona pellucida was removed by chemical dissolution with an acidic Tyrode's solution. In vitro culture was done in plastic culture dishes using tissue culture medium containing an energy source (lactate and pyruvate for pre-blastocyst stages (Brinster, 1967), glucose for blastocyst stages (Eagle's Minimum Essential Medium; Eagle, 1955)), amino acids, salts and trace elements, vitamins, and supplemented with fetal bovine serum. Incubation was at 37°C in an atmosphere of 90% $N_2$, 5% $CO_2$, 5% $O_2$, and saturated humidity. All samples were fixed in an aldehyde/paraformaldehyde fixative. Fixed embryos were allowed to adhere to poly-L-lysine coated cover glasses and were dehydrated in a graded series of ethanol, critical point dried using $CO_2$ as a transition fluid and finally sputter coated with gold. The specimens were observed and photographed using an ISI mini-SEM operated at a constant accelerating voltage of 15kV. Waterman (1980) presents details and gives an excellent discussion of techniques for preparation of embryos for SEM.

## Observations

Ovulatory stage

Ovulation of mammalian oocytes occurs with the oocyte in metaphase of meiosis II. The first polar body has been extruded and occupies a space between the oocyte and the surrounding zona pellucida (peri-vitelline space). The zona pellucida is surrounded by a mass of cells from the ovarian follicle (corona radiata; L., radiant crown; fig. 1). The corona radiata is typically several cells thick and is, in turn surrounded by cumulus fluid from the antrum of the follicle. In polyovular species, such as the mouse, this fluid serves to bind the ovulated oocytes together in a cumulus clot during their initial sojourn through the proximal portion of the oviduct. Dissolution of the cumulus clot and dispersal of the follicle cells normally occurs within a few hours in the oviduct and the oocytes remain surrounded only by the zona pellucida.

Fertilization

Fertilization of a mammalian oocyte generally occurs while the oocyte is still surrounded by the follicle cells and cumulus fluid. The spermatozoa must, therefore, penetrate these investing layers, in addition to the zona pellucida, before coming in contact with the membrane of the oocyte. Fertilization involves fusion of the plasma membranes of oocyte and spermatozoon, thus producing a diploid zygote with a small patch of paternally derived plasma membrane. Normally only a single spermatozoon fuses with and penetrates an oocyte and the events associated with preventing penetration of supernumerary spermatozoa (polyspermy) involve several molecular changes in the cell membrane and the underlying cortex of the oocyte. Numerous supernumerary spermatozoa do, however, reach the perivitelline space and may persist in association with the early embryo through the early cleavage stages. One such supernumerary spermatozoon is shown in figure 2.

Resumption of meiosis

Shortly after fusion of the gametes the oocyte resumes meiosis. The resumption of meiosis is a process called activation and involves changes in the electric potential of the oocyte and changes in permeability of the plasma membrane to ions. Morphological changes in the fertilized oocyte involve extrusion of the second polar body. The location of the male pronucleus in the mouse is identified superficially by a region of the oocyte membrane that is devoid of microvilli (fig. 2). As activation proceeds, this region begins to swell and form a protrusion of the surface of the plasma membrane called the incorporation cone. The final step in meiosis is completion of the second meiotic division and extrusion of the second polar body (fig. 3).

---

Fig. 1. Mature mouse oocyte. This sample was obtained from a mature Graafian follicle. The oocyte (oo) is shown within the follicle. The surface of the oocyte itself is not visible because it is covered by the zona pellucida (ZP). The zona pellucida is, in turn, surrounded by coronal cells (C) derived from the follicle. Some of the follicle cells have been dissected away to show the oocyte and its relation to them. Bar = 15 μm.

Fig. 2. Mouse oocyte in the process of activation. This oocyte has already been penetrated by a spermatozoon and is in the process of resumption of the second meiotic division. The spermatozoon (arrow) that is visible adhering to the surface of the oocyte is supernumerary and did not participate in the fertilization process. The male pronucleus, derived from the fertilizing spermatozoon is located beneath the smooth area of the oocyte that is obviously devoid of microvilli (*). Incorporation of the fertilizing spermatozoon is presumed to be complete in this specimen because no portion of the tail is visible in the micrograph. The zona pellucida has been removed. The specimen shows some shrinkage artifact. Bar = 6.2 μm.

Fig. 3. Abstriction of the second polar body. The plasma membrane incorporating the second polar body (*) is absolutely devoid of microvilli. The rugosities seen on the surface of the oocyte were caused by shrinkage during processing. Nonetheless, the microvillous nature of the oocyte is preserved. The zona pellucida has been removed. Bar = 5 μm.

Fig. 4. Two-cell mouse embryo. The first cleavage division has resulted in the formation of two individual blastomeres of approximately equal size. Shown also is one of the polar bodies (arrow). The zona pellucida has been removed. Bar = 6.2 μm.

---

Cleavage

The first cell division and formation of individual blastomeres from a single-celled zygote marks the culmination of fertilization and the beginning of embryogenesis. The first few cleavage divisions normally result in formation of blastomeres of approximately equal size (fig.

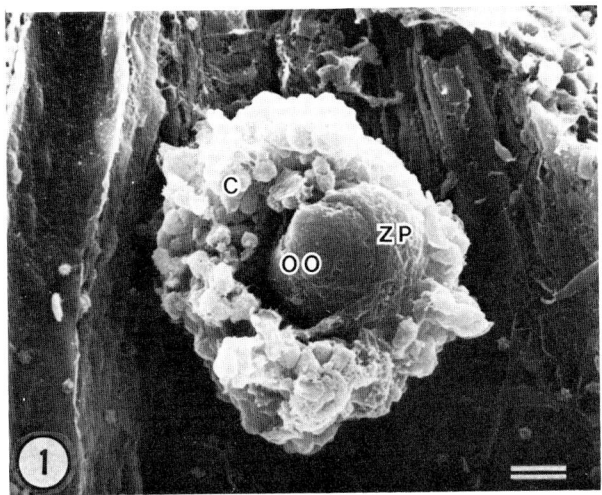
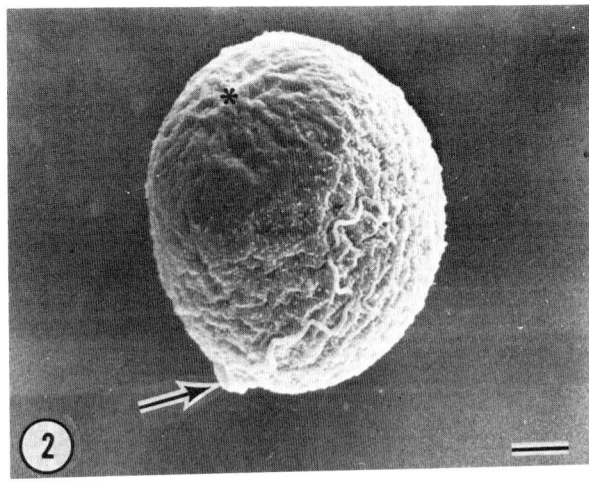
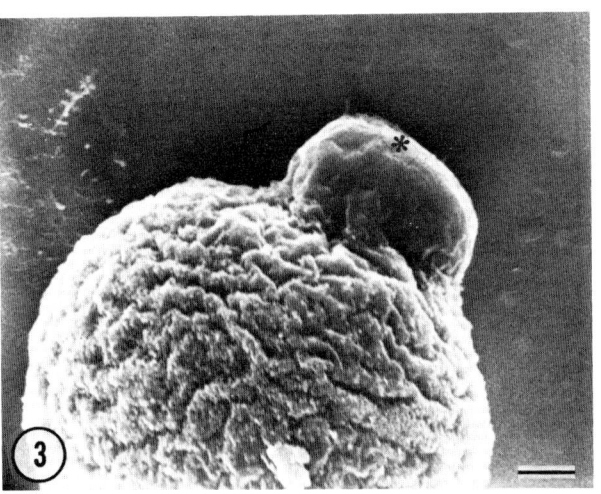
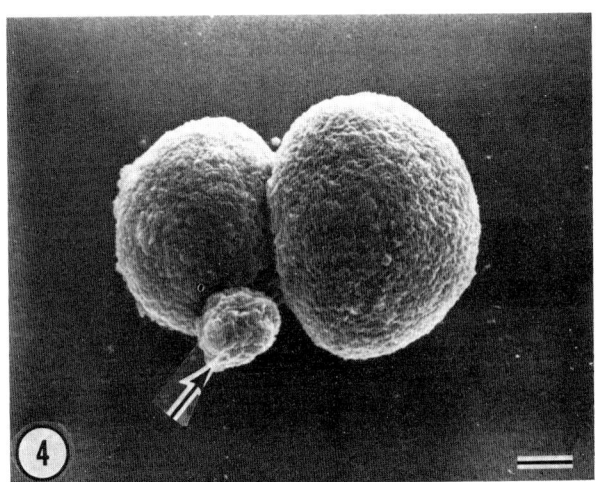

Fig. 5. Stereo view of an eight-cell mouse morula. Cleavage divisions have resulted in eight individual blastomeres of approximately equal size. The cell borders of each blastomere are clearly defined and distinctly separated from neighboring cells. The formation of junctions between blastomeres that precedes formation of the blastocyst cavity has not yet occurred. The zona pellucida has been removed. Bar = 6.2 μm.

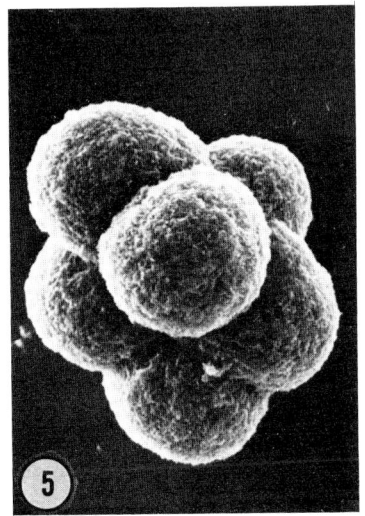
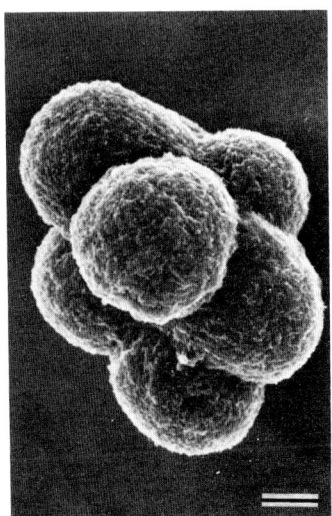

4). Although cytoplasmic bridges can frequently be seen connecting blastomeres, such intercellular connections are transitory and there are usually no junctions established between the individual cells. At this stage, blastomeres are held together by ionic attractions of the plasma membranes and can be easily dissociated by mechanical means. Each blastomere retains the capacity to form a complete embryo (totipotency) and it is at this stage that monozygotic twinning can occur. It must be kept in mind, however, that these embryos are normally bounded by the zona pellucida and that such dissociation is relatively infrequent, in nature.

Subsequent cleavage divisions result in increasingly smaller blastomeres. The embryo resembles a mulberry and at this stage is called a morula from the Latin word for that berry (fig. 5).

Cleavage of eggs will occur following a variety of stimuli and is frequently observed to occur spontaneously. Cleavage of unfertilized eggs is more properly called fragmentation and generally results in a mass of cell fragments of varying sizes. These portions of fragmented oocytes are not analogous to blastomeres for they contain only random portions of cytoplasm and no nuclear material. Fragmentation occurs spontaneously and can proceed in a matter of hours instead of over a period of days as occurs in normal cleavage. It is important for the student of mammalian embryology to recognize these aberrations and not confuse them with normal development for they occur rather frequently. An example of a fragmented oocyte that contains cell fragments of uniform size is presented in figure 6.

### Formation of the blastocyst

The blastocyst stage is characterized by the accumulation of fluid within a blastocyst cavity or blastocoel. For fluid to accumulate within the cavity of the blastocyst there must be established 1) a mechanism of fluid transport and 2) a mechanism of fluid retention. Experiments with rats and mice indicate that fluid is pumped into the blastocyst cavity by active transport and it is retained within the blastocyst by the formation of an epithelial barrier. Blastulation begins to occur in the mouse during the eight-cell stage when the individual blastomeres establish extensive focal contacts with one another. Junctional complexes in the form of tight junctions, gap junctions and desmosomes form at the lateral borders of the outer blastomeres. The outer layer of blastomeres are destined to become the trophoblast of the blastocyst and the inner blastomeres give rise to the inner cell mass. The connections formed between adjacent trophoblast cells can be seen in figure 7.

Shedding of the zona pellucida is normally accomplished during the early blastocyst stage and blastocysts of some mammals begin to expand dramatically in volume at this time. Expansion of blastocysts in some species is several thousand fold and serves to increase the area of contact between the blastocyst and uterus at the time preceding implantation. In comparison, the mouse blastocyst does not expand significantly.

### The peri-implantation stage

Embryos of most mammals initiate the implantation process at the blastocyst stage. Implantation involves the attachment to and penetration of the epithelial lining of the uterus. Implantation stage embryos cannot be easily removed from the uterus for study and there would exist a gap in our knowledge of this stage of embryogenesis were it not for the fact that some embryos will undergo many implantation-related events during in vitro culture. However, many developmental events that an embryo undergoes in vitro are accommodations to tissue culture conditions and are not strictly analogous to embryogenesis in vivo. Nonetheless, many aspects of peri-implantation embryogenesis can be more readily observed in vitro by students of embryology than if a few samples were taken from live mothers.

Attachment of the blastocyst to the surface of the tissue culture dish can be observed at the same time period and developmental stage that it occurs in utero. The actual molecular mechanisms responsible for attachment in vitro are probably not the same as those that adhere the blastocyst to the epithelium of the uterus. Two deviations from normal development that can be observed in figure 7 are the extensive ruffling of the surface of the trophoblast cells and the beginning outgrowth of the trophoblast cells onto the surface of the tissue culture dish.

As peri-implantation embryogenesis proceeds in vitro the trophoblast not only forms an extensive monolayer outgrowth but also migrates away from the inner cell mass (now called an egg cylinder) leaving the underlying embryo exposed to the milieu of the tissue culture medium (fig. 8). The egg cylinder in the mouse is surrounded by a layer of endoderm cells and these cells can clearly be seen in the embryo in figure 8. Of course, a living, transporting epithelial lining of the uterus cannot be duplicated by the surface of a plastic tissue culture dish and several attempts have been made to overcome this serious flaw in the in vitro model. Figure 8 shows a mouse blastocyst that has been placed in co-culture with a monolayer of cells derived from a primary explant of uterine luminal epithelium. Using this technique, it is possible to observe directly the interactions between trophoblast and uterine luminal epithelial cells. These in vitro models cannot duplicate precisely the conditions that are encountered within the uterus (figure 9). The in vitro model does, however, bring this crucial time in the development of all mammals into view of the embryology student or researcher and thus we can gain a better understanding of this fundamental process.

### Summary

Because embryos of Eutherian mammals gain their nourishment via a placenta, and not from massive amounts of yolk stored in the oocyte, their early embryology is different from that of ooviparous species that are typically presented to beginning students of embryology. It has now become routine to grow mammalian embryos in

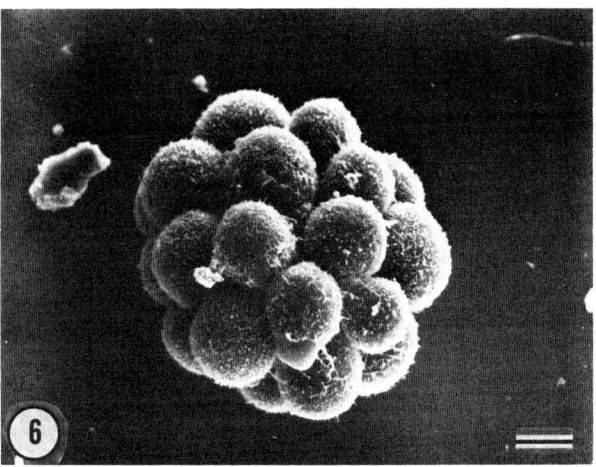

Fig. 6. Fragmenting mouse oocyte. This unviable oocyte has fragmented into several cell fragments that resemble normal blastomeres. This sample is included to remind the reader that confusing abnormalities occur. It is important to recognize these abnormalities, especially with the increased popularity of in vitro insemination and embryo transfer. The zona pellucida has been removed. Bar = 6.2 μm.

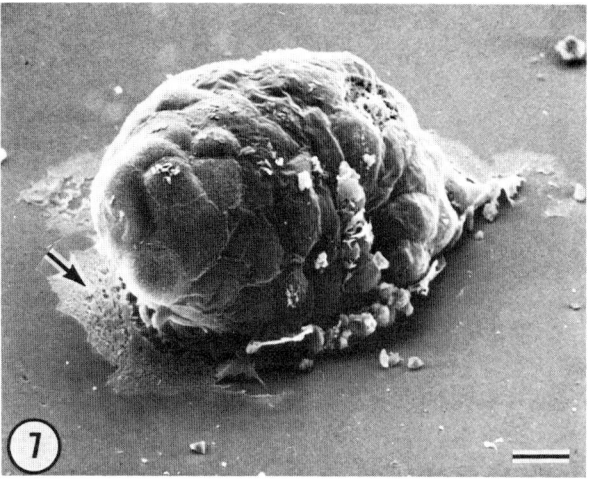

Fig. 7. Mouse blastocyst. This blastocyst stage embryo has escaped from the zona pellucida and has attached to the surface of the tissue culture vessel. It is evident that some of the trophectoderm cells have begun to outgrow onto the culture dish (arrow). The junctions between adjacent trophoblast cells are evident in this micrograph. These junctions are necessary to maintain the fluid filled blastocyst cavity. Bar = 15 μm.

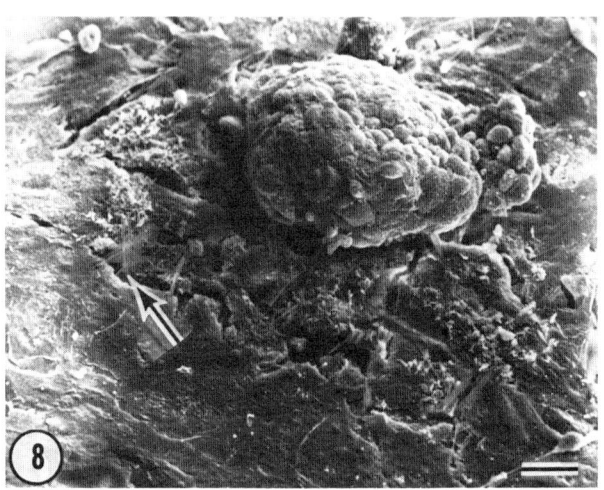

Fig. 8. Peri-implantation mouse embryo in vitro. This embryo was placed in co-culture with a monolayer of cells derived from a primary explant of uterine luminal epithelial cells. The embryonic trophectoderm cells have formed an extensive outgrowth onto some of the uterine cells and have begun to displace the uterine cells (arrow). Trophectoderm cells can easily be distinguished from uterine cells because whereas the former possess extensive ruffles and lamellipodia the latter possess a membrane that is relatively devoid of surface projections. The mass of cells that forms an elongated cylinder that appears to be resting on the monolayer of cells is the developing egg cylinder. The egg cylinder is covered by visceral endoderm. Bar = 25 μm.

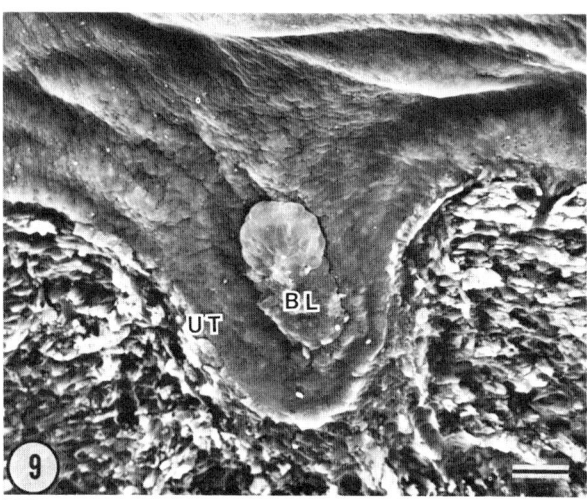

Fig. 9. Peri-implantation mouse embryo in situ. The uterus has been slit longitudinally and split to expose the embryo within the uterus. The uterine lumen extends horizontally at the top of the micrograph and extends downward to form an implantation chamber enclosing the blastocyst. The blastocyst (BL) can be seen in its specific orientation within the uterus (UT). The inner cell mass of the blastocyst is located at the upper pole of the blastocyst and the polar trophoblast cells overlying the inner cell mass appear more refractive to the electron beam. Bar = 50 μm.

tissue culture and it should become common, without significant investment of resources, to grow mammalian embryos in a teaching laboratory, thus making direct observation of mammalian embryogenesis available to all students of embryology.

## Acknowledgements

I thank Ms. Elaine Shriver for assistance in preparation of the electron micrographs and Ms. Candida Trueblood for typing the manuscript. Much of this work was supported by National Institutes of Health Grant HD-16703 and appropriations from the Southern Illinois University School of Medicine.

## Literature Cited

Brinster RL. (1967) Carbon Dioxide produced from glucose by the peri-implantation mouse embryo. Exp. Cell Res. 47, 271-277.

Eagle H. (1955) Amino acid metabolism in mammalian cell cultures. Science 130, 432-437.

Waterman RE (1980) Preparation of embryonic tissues for SEM. Scanning Electron Microsc. 1980; II: 21-44.

## Suggestions for Further Reading

Austin CR. (1961) The Mammalian Egg. Charles C. Thomas, Springfield, IL.

Austin CR. (1968) Ultrastructure of Fertilization. Holt Rinehart and Winston, New York.

Austin CR. (1982) Germ Cells and Fertilization. Cambridge University Press, London.

Austin CR. (1983) Embryonic and Fetal Development, 2nd ed. Cambridge University Press, London.

Balls M, Wild AE., (eds) (1975) The Early Development of Mammals. Cambridge University Press, London.

Biggers JD, Schuetz AW. (1972) Oogenesis. University Park Press, Baltimore.

Blandau RJ. (ed) (1971) The Biology of the Blastocyst. University of Chicago Press, Chicago and London.

Gwatkin RBL. (1977) Fertilization Mechanims in Man and Mammals. Plenum Press, New York and London.

Hadek RF (1969) Mammalian fertilization. An Atlas of Ultrastructure. Academic Press, New York.

Hafez ESE. (ed) (1970) Reproduction and Breeding Techniques for Laboratory Animals. Lea and Febiger Philadelphia, PA.

Hafez ESE. (ed) (1975) Scanning Electron Microscopic Atlas of Mammalian Reproduction. Igaku-Shoin, Ltd., Tokyo.

Hafez ESE, Kenemans P. (eds) (1982) Atlas of Human Reproduction by Scanning Electron Microscopy. MTP Press, Lancaster, U.K.

Markert CL, Papaconstantinou I. (ed) (1975) The Developmental Biology of Reproduction. Academic Press, New York.

Moghissi KS, Hafez ESE. (eds) (1972) Biology of Mammalian Fertilization and Implantation. Charles C. Thomas, Springfield, IL.

Mossman HW, Duke KL. (1973) Comparative Morphology of the Mammalian Ovary. University of Wisconsin Press. Madison, Wisconsin.

Motta PM, Hafez ESE. (eds) (1980) Biology of the ovary. Martinis Nijoff. The Hague, Oxford, Boston.

Poste G, Nicolson GL. (eds) (1977) The Cell Surface in Animal Embryogenesis. North Holland Publishing Co., Amsterdam and New York.

Sherman MI. (ed) (1977) Concepts in Mammalian Embryogenesis. M.I.T. Press, Cambridge, Mass. and London.

Van Blerkom J, Motta PM. (1979) The Cellular Basis of Mammalian Reproduction. Urban and Schwarzenberg, Baltimore and Munich.

Van Blerkom J, Motta PM (eds) (1983) Ultrastructure of Reproduction. Martinus Nijoff, Baltimore, MD.

Waterman E. (1979) Embryonic and foetal tissues of vertebrates. Biomedical Research Applications of Scanning Electron Microscopy. Vol. 1. (Hodges, GM and Hallowes, RC, Eds.) Academic Press, New York, San Francisco, London. pp 1-125.

Wolstenhome GE, O'Conner M. (1965) Preimplantation Stages of Pregnancy. Little, Brown, and Co., Boston, MA.

Zamboni L. (1971) Fine Morphology of Mammalian Fertilization. Harper & Row, New York.

## Discussion with Reviewers

P. Motta: Perusing the pictures of the contribution it seems that a great deal of shrinkage occurred to some preparations (mainly figures 1 to 5) as they were somehow air dried. How long after critical point drying was the material observed? Was it maintained properly under vacuum before observed by SEM? Can you more exactly describe the method used?

Author: Indeed, some shrinkage has occurred and I believe that some air drying occurred during transfer of the specimens from the final ethanol bath for dehydration to the critical point drying apparatus. In other trials, shrinkage of these difficult specimens was minimized by performing all dehydration steps within the holder for the critical point dryer, thus eliminating the necessary transfer step which involves exposure of the samples to air-drying for a few seconds. In most cases, the samples were sputter coated and examined by SEM immediately after critical point drying. In only one case were the specimens kept in a vacuum desiccator overnight before being observed by SEM.

P. Motta: How can you distinguish only by using SEM unfertilized and fertilized eggs? Your figures 5 and 6 only seem two different developmental stages. Magnification is of course a good criterion in not confusing cells with blastomeres but a parallel insert by light and transmission electron microscopy might be more useful.

Author: Regarding fertilized and unfertilized eggs, the smooth area overlying the male pronucleus seems to have a good corelation with fertilization and this has been examined with correlative phase contrast light microscopy (for observing the male pronucleus) and SEM. Regarding figure 6, during collection of 2-cell eggs for SEM I observed spontaneous fragmentation of that specimen through the dissecting microscope. Since students and laoratory personnel have frequently confused fragmented eggs with cleavage-stage embryos it seemed appropriate to include a specimen in this paper. Of course, another criterion for distinguishing fragmenting eggs in the mouse is the number of blastomeres. Mouse embryos typically begin compaction at the 8-cell stage and any egg that consists of more than 12 uncompacted blastomeres is either fragmented or developing abnormally. Unfortunately I have not bothered to prepare these abnormal specimen for TEM; however, figure A shows a phase contrast light micrograph.

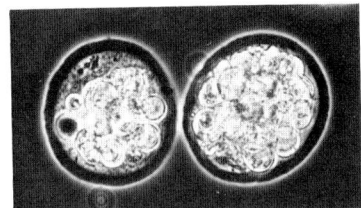

Fig. A: Phase contrast light micrograph of two zona-encased fragmented oocytes. Many of the fragments are of unequal size and no nuclei can be seen.

P. Motta: According to your description of fig. 1 you state that the oocyte is shown in situ. If such is the case I wonder why the granulosa layer is not also shown in the picture and the granulosa cells forming the cumulus and surrounding the oocyte are not attached to the mural granulosa?

Author: The oocyte has somehow been dislocated from its natural position within the follicle.

H. Schatten: The lack of abundant microvilli surrounding the site of sperm incorporation in the mouse provokes questions regarding the mechanism for sperm incorporation in this species. In the invertebrate model, the sea urchin, we have shown (Schatten and Schatten, 1980. Devel. Biol. 78:435-449) that sperm incorporation will be prevented by the microfilament inhibitor cytochalasin B and Shalgi et al. (Shalgi, R., Phillips, D. M., and Kraicer, P. F., 1978. Gamete Res. 27-35) note the cytochalasin sensitivity of the incorporation cone in the rat. Since the mouse seems to contrast with other animals, does the author have any indication of any specific surface alteration responsible for the incorporation of the sperm during fertilization?

Author: The unfertilized mouse egg is structurally mosaic with about one fifth of its surface non-microvillous. The non-microvillous portion overlies the meiotic spindle and the egg cortex at this site is also devoid of cortical granules. Sperm-egg fusion almost always occurs at microvillous regions away from the meiotic apparatus. From evidence currently available, sperm penetration results in a local release of cortical granules and a local swelling in the egg membrane coincident with sperm-egg membrane fusion and formation of the male pronucleus. This local swelling is what I refer to as the incorporation cone and it is transitory. Since I have only monitored sperm penetration by light microscopy followed by SEM I have no data regarding cytochalasin B or other microfilament inhibitors.

G. Schatten: The dramatic surface reorganization during compaction completely reshapes the embryo. Is the transformation from spherical individual cellular aggregates to closely adherent cellular sheets a sudden or slow transformation and can the author speculate on the underlying mechanism?

Author: In a recent publication, Sutherland and Calarco-Gillam (Devel. Biol. 100:328-338; 1983) suggested that compaction is a three step process that began with 1) cell surface recognition (probably mediated by glycoproteins) and adhesion of microvilli of adjacent blastomeres. Initial attachment was followed by 2) microfilament shortening of attached microvilli. 3) Compaction was maintained by microtubules.

E. M. Eddy: What is the source of the zona pellucida and what are its roles in early development?

Author: The origin of the zona pellucida is rather unclear. The majority of the zona pellucida appears to be synthesized and secreted by the granulosa cells. However, some authors have concluded that Golgi vacuoles within the oocyte cortex contain material that may also contribute to the deposition of the zona.

E. M. Eddy: What is the time scale in the mouse of the main steps described for the development from fertilization to implantation?

Author: Although variations are encountered among strains of mice and time of day with regard to the precise occurrence of the events, the various stages occur as follows: Day 1 (the day the vaginal plug is found); oocyte maturation, ovulation, and fertilization. Day 2; 2-cell embryos and early cleavage stages. Day 3; compaction and cavitation. Day 4; early blastocyst and shedding of the zona pellucida. Day 5; blastocyst and adhesion to uterine luminal epithelium. During days 1 through 3 embryos are found in the oviducts. Embryos arrive in the uterus between day 3 and day 4. The above schedule is maintained under in vitro conditions except that the developmental time between stages is somewhat longer. If embryos are kept in culture from fertilization through blastulation, the day 5 embryo is more developmentally related to a day 4 embryo in vivo. Eggs of many strains of mice will not undergo first cleavage divisions in vitro.

P. Calarco: What causes the regional differences in surface morphology (i.e., microvillous vs. smooth) seen during fertilization and cleavage stages?

Author: During fertilization the smooth area overlies the meiotic spindle and is separated from the oocyte with extrusion of the polar body. Also, during incorporation of the spermatozoon there is a transient smooth region as the incorporation cone forms over the decondensing male pronucleus. During the cleavage stages polarity has been noted to exist as the result of intercellular contact.

## GASTRULATION IN THE TELEOST, BRACHYDANIO RERIO

R. G. Thomas and R. E. Waterman*

Department of Anatomy,
School of Medicine
The University of New Mexico
Albuquerque, NM 87131

### Abstract

The morphology of embryos of the fresh water teleost, Brachydania rerio (zebrafish), was examined by SEM techniques from late blastula through gastrula stages. Specimens were fixed in an aldehyde mixture, dehyrated in ethanol and dried from liquid $CO_2$ by the critical point method. Cells within the blastoderm were exposed by several techniques including fracture of dried specimens and ethanolic cryofracture.

The blastoderm consists of several layers of blastomeres (deep cells) covered by a single layer of flattened cells (enveloping layer). The blastoderm is located at the animal pole and is separated from the yolk mass by a layer of uncleaved cytoplasm. The only strong adhesion between the blastoderm and this cytoplasmic layer is at the periphery of the blastoderm. During gastrulation, the blastoderm flattens and its margin (germ ring) progressively moves toward the vegetal pole; finally surrounding the yolk completely. The body of the embryo appears during this period as a small cellular aggregation (embryonic shield) along the germ ring. Cells of the blastoderm converge toward the embryonic shield, enlarging it and reorganizing it into the components of the embryonic body.

Differences in morphology exhibited by the blastomeres during gastrulation as viewed with the SEM appear similar to those previously described in vivo. The deep surface of the blastoderm is readily seen in fractured specimens and differences in the morphology of blastomeres forming the embryonic shield, central region and germ ring were documented. The SEM observations also indicated that the enveloping layer of cells may enlarge in part by continued formation of cells from the syncytial periblast at its margin.

KEY WORDS: Teleost, Zebrafish, Embryogenesis, Gastrulation, Morphogenesis, Locomotion

*Address for correspondence:
R.G. Thomas c/o R.E. Waterman
Address as above. Phone no.: (505) 277-5555.

### Introduction

The morphogenetic movements of gastrulation have been examined with an ever increasing degree of sophistication in several species; particularly echinoderm, amphibian and avian embryos.[1,2] Observation of intact specimens with sufficient magnification is technically difficult or impossible in some cases, however, and attention has therefore been given to the study of gastrulation in the bony fishes, many of which have relatively transparent eggs that allow observation of cellular behavior in a relatively undisturbed condition for considerable periods.

Shortly after fertilization, cytoplasm of the teleost egg accumulates at the animal pole where it surrounds the nucleus of the zygote. Only this portion of the egg cytoplasm, the blastodisc, undergoes cleavage. The first cleavage divisions are incomplete, but eventually result in a blastoderm consisting of a single outer "enveloping layer" of cells covering a number of deeper blastomeres ("deep cells"). The blastoderm next flattens and begins to spread over the yolk by epiboly, eventually surrounding it completely. The margin of the blastoderm thickens during epiboly to form a "germ ring", some cells of which converge toward a point along the ring to form a cellular aggregation, the embryonic shield, which in turn elongates and forms the axis of the early embryo.

Early light microscopic studies of teleost development[3,4] were hampered by difficulties inherent to paraffin histology. These problems, apparently combined with a conceptual bias derived from comparison with studies of amphibian gastrulation led to differing interpretations and a plethora of terminology. Much of this early work has been extensively reviewed[5,6]. The present study deals primarily with more recent studies of gastrulation which have focused on mechanisms of epiboly and the reorganization of blastomeres to form the embryonic body.

Much of our current understanding of epiboly derives from observations made by Trinkaus and co-workers during the past 25 years.[2] Historically, when it was shown that mitotic activity is not necessary for epiboly[7], attention was directed toward the cortex of the uncleaved yolk mass and the cytoplasmic layer (periblast) beneath the blastoderm. The periblast is a

45

syncytium containing nuclei derived from incomplete cytokinesis of the peripheral blastomeres. Nuclei are more prominent around the circumference of the blastoderm (marginal periblast) than immediately beneath the blastoderm (central periblast). In the first extensive studies of epiboly, Lewis proposed that the cortex of the yolk mass, which he termed the "yolk gel layer", exerted a contractile force which both pulled the margin of the blastoderm and periblast down around the yolk and simultaneously forced the yolk upward beneath the animal pole.[8,9] This concept was challenged by Trinkaus who showed experimentally in Fundulus heteroclitus that there is no coordinated contraction of the yolk gel layer during epiboly, and that the marginal periblast begins to spread prior to the blastoderm and will move ventrally even if the connection of the blastoderm to it is experimentally severed.[10]

It was subsequently shown by TEM analysis that the "yolk gel layer" is actually a thin layer of cytoplasm surrounding the yolk mass, and that this yolk cytoplasm layer is continuous with the central and marginal regions of the periblast. The apparent movement of the marginal periblast results, therefore, from a flow of nuclei and organelles from the periblast into the yolk cytoplasmic layer.[11,12] Trinkaus proposed that the marginal periblast may act as a substrate for the advancing edge of the blastoderm. This movement was specifically attributed to the adhesion of ruffled membranes at the free borders of the marginal cells to the underlying marginal periblast.[1,12] The leading edge of the marginal cells was thought to be the prime mover in epiboly by analogy with the spreading of other epithelial sheets.[13]

Demonstration that the blastoderm is not passively pulled down over the yolk suggested that the mechanism of epiboly must be intrinsic to the cells of the blastoderm, and Trinkaus began a series of studies of the behavior of the blastomeres during epiboly in Fundulus.[11,12] Time-lapse cinematography of deep cells in vivo revealed a progressive change in the surface activity of these cells during early development. Undulations of the cell surface are first seen during late cleavage stages and increase until the predominant feature is the production of blunt protuberances of hyaline cytoplasm termed blebs. Both the rapidity of their formation and subsequent retraction, and the number of deep blastomeres exhibiting blebs, increase during early blastula stages. These initial blebs do not appear to form strong adhesive contacts. Some cells begin to move during mid- to late blastula stages, and the number of translocating cells steadily increases as gastrulation progresses. Longer cell processes, lobopodia, become more frequent and many appear to form by cytoplasmic flow into a pre-existing bleb. Some lobopodia flatten to form lamellipodia, and both types of protrusions appear to act as organs of locomotion. Blebs, too, are involved in locomotion; with the cell moving in the direction of the bleb as cytoplasm flows into it.[14]

Observations, primarily of cells in the germ ring, indicate that blastomeres translocate during epiboly using the deep surface of the enveloping layer and adjacent deep cells as substrata. Cells may also use the periblast as a substrate, but they are difficult to view under normal situations. Deep cells left behind after experimental removal of a blastoderm will spread out and move over the periblast surface.[2] These observations of living deep cells have been confirmed in the Medaka (Oryzias latipes)[15], the zebrafish (Brachydanio rerio)[16], and in some Cyprinodonts in which the deep cells of the blastula disperse over the periblast during epiboly, then reaggregate some time later to form a cellular mass analogous to the embryonic shield of other teleosts from which the normal embryonic structures develop.[17,19]

The observed increase in surface activity and adhesiveness between cells and their substrate during late blastula stages is paralleled by the behavior of dissociated blastomeres examined in vitro. Deep cells from Fundulus blastulae grown on glass remain spherical while those from gastrulae flatten rapidly, suggesting an increased adhesiveness.[20] Cells from early gastrulae are also more easily deformed in vitro than cells of early blastulae.[21] A greater frequency of fusion between blastomeres isolated from early blastulae of the Medaka than between cells isolated from early gastrulae may represent membrane changes possibly correlated with observed changes in surface activity in this species.[22]

The ultrastructure of the deep cells in Fundulus is likewise consistent with the proposed increased adhesiveness during late blastula and early gastrula stages.[2,11,12,23] Cells of the enveloping layer are bound together apically by close junctions during blastula stages, and punctate type junctions are added in the apical zone of membrane apposition during epiboly. Primitive desmosomes are also seen.[11,12] Development of similar junctions have been described between enveloping layer cells of trout embryos.[24]

The pathways followed by deep cells as they migrate within the blastoderm to reach their definitive positions in the embryo, and what role the germ ring may play in this process, have been studied in some detail by numerous investigators. Examination of sectioned material reveals a distinction in the germ ring between an external group of blastomeres, the epiblast, and a deeper layer of cells, the hypoblast, which is precursor to the endoderm, notochord and mesoderm of the embryonic shield and body. Several theories were historically suggested to explain the formation of the hypoblast. The most generally accepted concept was that the hypoblast of the germ ring and embryonic shield formed by involution of cells at the surface of the blastoderm at the margin of the germ ring (which was termed the "blastopore").

This long accepted theory has been challenged by Ballard in a series of experimental studies largely involving careful vital marking experiments of Salmo eggs.[25-30] These experiments show that cells of the enveloping layer do not involute to form the hypoblast and do not participate in forming the body of the embryo. By examining the undersurface of blastoderm removed from the yolk mass following fixation, Ballard suggested that the hypoblast is formed by cells which detach from the deep surface of the central region of the blastoderm and migrate toward the margin of the blastoderm where they collect as the hypoblast. The embryonic shield was also shown to form by the continued migration of hypoblast and epiblast cells toward a small cellular thickening termed a "nubbin", which appears prior to the externally observable embryonic shield. The results of these observations have been summarized into a proposed "three-dimensional" fate map for Salmo, which differs in concept as well as detail from those published previously by others.[31]

Because of uncertainties remaining in our understanding of the mechanisms involved in morphogenetic movements during gastrulation, continued study of early teleost development seems warranted. While observation and filming of living cells in situ is an important advantage of the teleost embryo for the study of cell movement in vertebrates, light microscopic observations are often limited by optical problems imposed by the yolk, curvature of the egg, thickness, etc. This is particularly true for the undersurface of the early blastoderm and the embryonic shield region. The present study was undertaken to explore the possibility of using the SEM for examining these regions of the blastoderm during gastrulation in the zebrafish, Brachydanio rerio, a cyprinid which has been extensively used in studies of teleost development.[32]

## Materials and Methods

Adult zebrafish were maintained in a laboratory aquarium with the light-dark cycle controlled to facilitate daily spawning. Eggs were collected from the bottom of the aquarium by means of a dip-tube and incubated in finger bowls at 24-26°C. Embryos were staged according to the classification established by Hisaoka and Battle.[33,34] Embryos from late blastulae (stages 11,12) to closure of the "blastopore" (stage 17) were examined.

Embryos were fixed by immersion in a 0.2M phosphate-buffered 2% paraformaldehyde, 2% glutaraldehyde (pH 7.3-7.4) solution containing 0.01% trinitrophenol[35] for a minimum of one hour. The chorion was removed with fine forceps prior to fixation. Embryos were then washed in buffer, postfixed in 1% $OsO_4$ (30 minutes) and dehydrated in ethanol. Specimens selected for sectioning were placed in two changes of propylene oxide and embedded in Epon. One micrometer sections cut with glass knives were stained with methylene blue:Azure II. For examination with the SEM, specimens were dried by the critical point method using liquid $CO_2$, mounted on stubs with double-stick Scotch brand tape and silver paint, coated with palladium:gold in either a vacuum evaporator equipped with a rotating stage or a Hummer II sputter coater, and viewed with an ETEC "Autoscan"

SEM operated at 10-20 kV.

Several methods were employed to allow visualization of the deep surface of the blastoderm. Blastoderms from late blastulae and early gastrulae were removed essentially intact either by carefully fracturing dried embryos or by lowering the blastoderm onto a piece of double-stick Scotch tape, then gently rolling the egg until the yolk separated from the blastoderm. Intermediate layers of deep cells within the blastoderm were exposed by lowering a piece of double-stick Scotch tape onto the surface and gently lifting the enveloping layer away. It was difficult or impossible to obtain intact blastoderms from older gastrulae (stages 15-17) since they enveloped more than one-half of the yolk sphere of the relatively small zebrafish egg. Best results were obtained at these stages by dissecting the blastoderm from the yolk during initial fixation and prior to osmication. After mechanical removal of the chorion, the embryos were stained with a dilute 1% Nile blue solution made up in 0.4% methocell (methylcellulose) in Holtfreter's solution to increase visibility of the gastrulae. After staining for up to ten minutes in this solution known to allow eggs to develop normally for long periods of time,[14] embryos were placed in fixative solution for a few minutes and the egg gently squeezed until the yolk fractured. Pieces of yolk were then picked away from the undersurface of the blastoderm. Some specimens infiltrated with 100% ethanol were frozen in liquid nitrogen and fractured.[36] The fractured pieces were subsequently critical point dried and processed as described above.

## Observations

### Stages prior to gastrulation

At late blastula stages (11-12) just prior to gastrulation, the blastoderm consists of a group of deep cells covered by a single layer of squamous enveloping cells (Figures 1A,B). The deep cells are generally spheroidal or ovoid. They exhibit blebs but very few filopodia or other longer processes. The interface between the blastoderm and the yolk is flat, and indentations created by the deep blastomeres may be seen on the surface of the central periblast region overlying the yolk mass (Figure 1B).

### Gastrulation

As gastrulation begins (stage 13), the blastoderm thins and the interface between the periblast and blastoderm becomes increasingly curved. Incompletely formed cells at the margin of the enveloping layer are directly continuous with the periblast and yolk cytoplasmic layer (Figures 1C, 2-5). The external surface of the cell membrane over the marginal periblast exhibits numerous microvilli which distinguishes this region from the non-microvillous surface of the yolk cytoplasmic layer (Figure 4).

The margin of the blastoderm formed by deep cells becomes rounded and slightly thickened to form the germ ring (Figure 1C), which is visible in vivo when the blastoderm covers about one-third of the yolk mass (stage 14). This margin is always separated by a small space from the

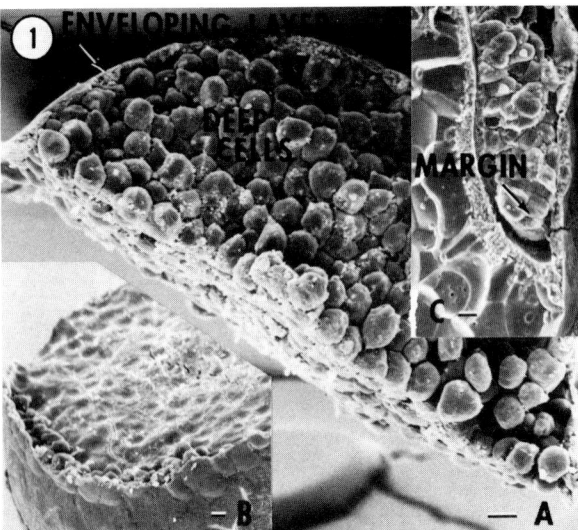

Figure 1. A) Sagittal fracture of Stage 11 blastoderm (Bar=10μm); B) Yolk separated from 1A (Bar=10μm); C) Margin of Stage 14 blastoderm (Bar=10μm).

attachment of the enveloping layer to the periblast. Cells of the early germ ring are rather closely packed, and exhibit blebs, lobopodia and short filopodia which contact adjacent cells. The central region of the blastoderm is fairly uniform in thickness, consisting of approximately three to four layers of deep cells (Figures 6,7). These spheroidal cells exhibit lobopodia and fine filopodial processes, although their morphology varies with their location within the blastoderm. Blastomeres immediately subjacent to the enveloping layer are more rounded and have fewer processes than the deeper layers of blastomeres (Figure 8). Deep cells at the undersurface of blastoderms removed from stage 14 to stage 15 embryos are also rounded and separated by obvious intercellular spaces, but extend numerous long filopodia and lamellipodia which interdigitates with adjacent cells (Figure 9). Individual cells may exhibit more than one bleb or may simultaneously exhibit both blebs and lobopodia or filopodia.

Cells of the germ ring margin become increasingly flattened against the periblast during epiboly. The more centrally located blastomeres do not appear to be pressed as tightly against the yolk, and rarely adhere to the periblast when the blastoderm is removed at early stages. The ease with which the blastoderm can be removed from fixed preparations, and the appearance of the yolk and dome-shaped central periblast from which the blastoderm has been removed (Figure 3), indicate that the deep cells do not form extensive or tight adhesions to the periblast cytoplasm.

Between stages 14 and 15, the location of the body of the embryo appears as a small thickening at one point along the germ ring known as the embryonic shield. This marks the antero-posterior axis of the embryonic body. When first observed, the embryonic shield is a definite, but slight thickening (Figure 10). The profiles of cells at the undersurface of the early embryonic shield are somewhat smaller than surrounding cells. When seen in sagittal fractures of gastrulae in which the blastoderm surrounds one-half of the yolk mass (stage 15), the embryonic shield is composed of a lower group of cells (hypoblast) which exhibit numerous blebs and appear to be incompletely separated from the overlying cells of the germ ring (epiblast) (Figure 6). The epiblast and hypoblast merge together at the germ ring margin, and the hypoblast of the embryonic shield is always more closely opposed to the undersurface of the epiblast than to the underlying periblast. The embryonic shield forms a depression in the periblast which is obvious in fixed and dissected specimens (Figure 3). No other obvious regional differences in the morphology of the surface of the central periblast are seen at this stage.

The embryonic shield lengthens and becomes more distinct as the blastoderm proceeds to encircle between one-half (stage 15) and two-thirds (stage 16) of the yolk mass (Figures 11-16). The central area of the blastoderm thins, and a band of more loosely organized and less flattened cells, the hypoblast of the germ ring, appears immediately centripetal to the compact margin of the germ ring (Figure 11). A slight groove on either side of the embryonic shield demarcates it from the hypoblast cells laterally. Cells forming the embryonic shield exhibit numerous lobopodia, blebs, and occasional lamellipodia (Figures 11,14,16,17). Cells at the anterior end of the embryonic shield initially extend lobopodia and short filopodia to contact cells of the central region (Figure 11), but the anterior end of the shield becomes increasingly distinct and blunt as gastrulation proceeds (Figure 14). By stage 16, the embryonic shield has elongated anteriorly beyond the hypoblast of the germ ring (Figure 16). Cells near the germ ring exhibit numerous blebs (Figure 15).

Cells at the undersurface of the central region of the blastoderm display numerous lobopodia, blebs and filopodial processes which interdigitate extensively with adjacent cells. Some cells appear to project further from the undersurface of the blastoderm during early gastrulation stages (Figure 11), and may be in the process of disengaging from the remainder of the deep blastomeres. The undersurface of the central region, however, always appears more flattened than that of the embryonic shield and germ ring hypoblast as gastrulation proceeds. The thickness of the central area of the blastoderm gradually decreases. There are approximately two layers of deep cells beneath the enveloping layer at stage 16, and only one layer at stage 17. An increasing number of cells remain attached to the periblast surface upon removal of the blastoderm as development proceeds. These are generally elongated or fusiform cells with one or more major processes of the lobopodial or lamellipodial type (Figure 18).

The margins of the germ ring come together beneath the enveloping layer at the vegetal pole at stage 17 at what is generally referred to as

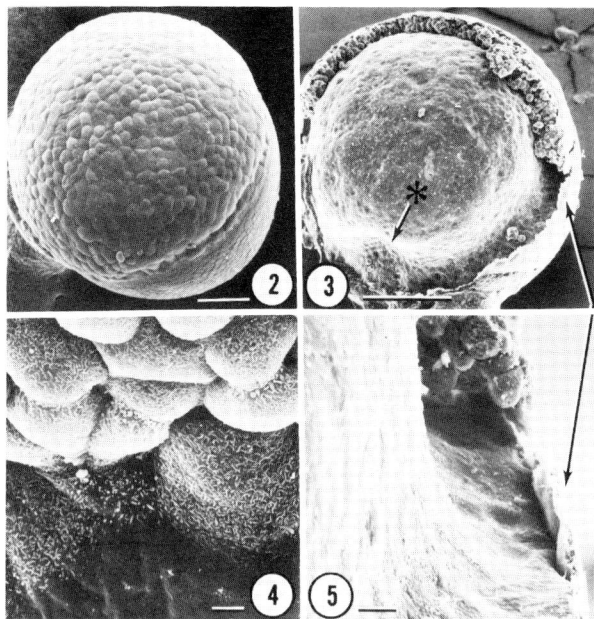

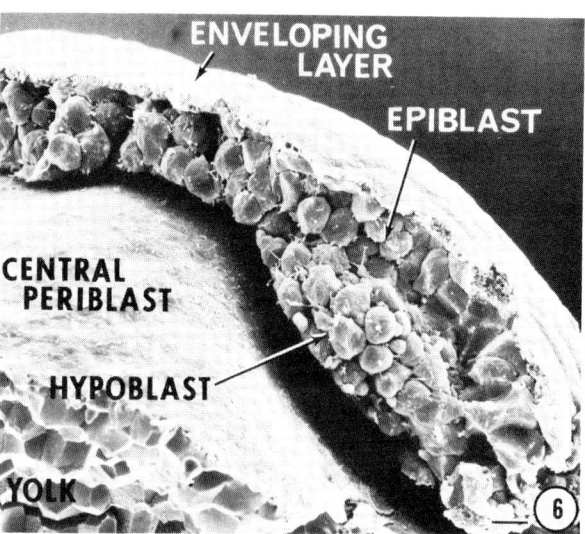

Figure 2. Animal pole of intact specimen. Stage 15. (Bar=100µm).

Figure 3. Blastoderm partially removed. Stage 15. Indentation caused by embryonic shield (*). (Bar=100µm).

Figure 4. Surface of junction between margin of blastoderm and periblast. Stage 14. (Bar=10µm).

Figure 5. Portion of fig. 3 showing marginal enveloping cells apparently forming from marginal periblast. (Bar=10µm).

Figure 6. Sagittal fracture of embryonic shield. Stage 15. (Bar=10µm).

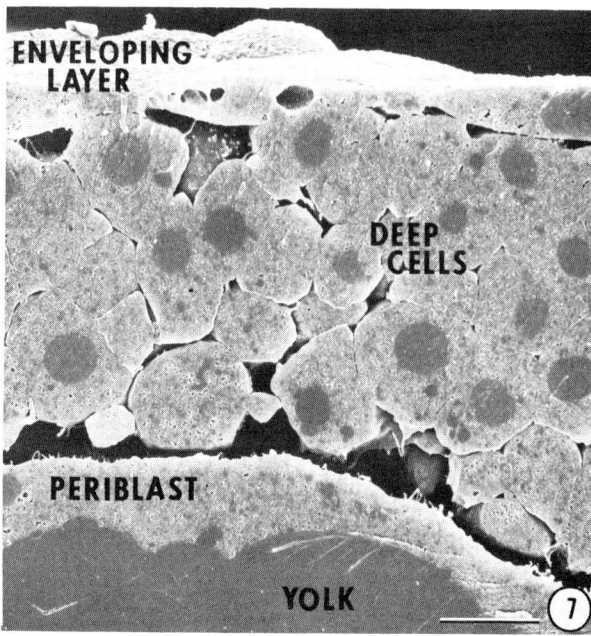

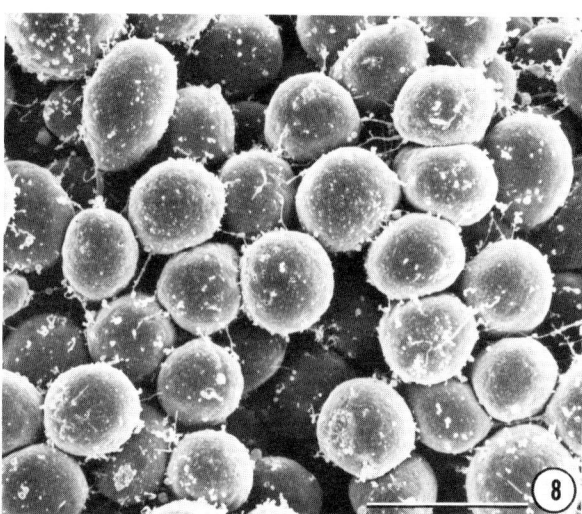

Figure 8. Deep cells immediately beneath enveloping layer. Stage 15. (Bar=10µm).

Figure 7. Cryofracture through central region of blastoderm. Stage 15. (Bar=10µm).

Figure 9. Deep surface of central region of blastoderm. Stage 15. (Bar=10µm).

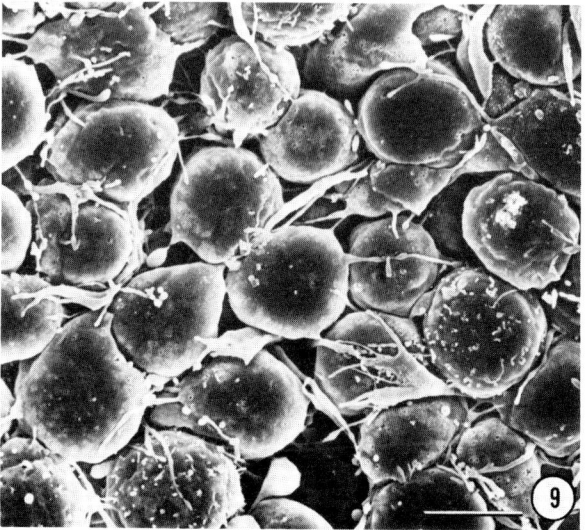

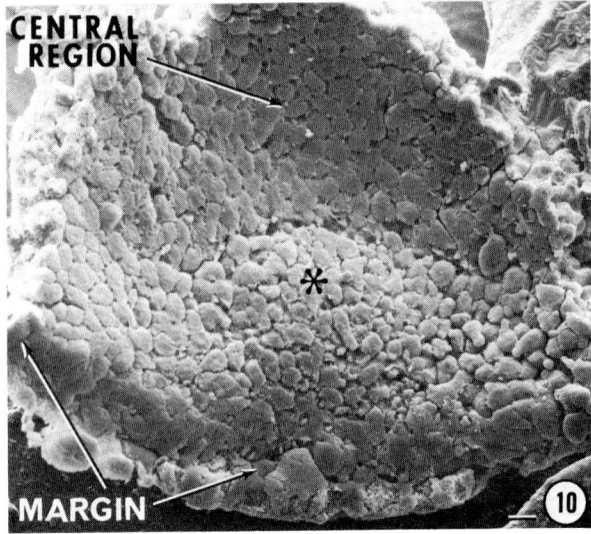

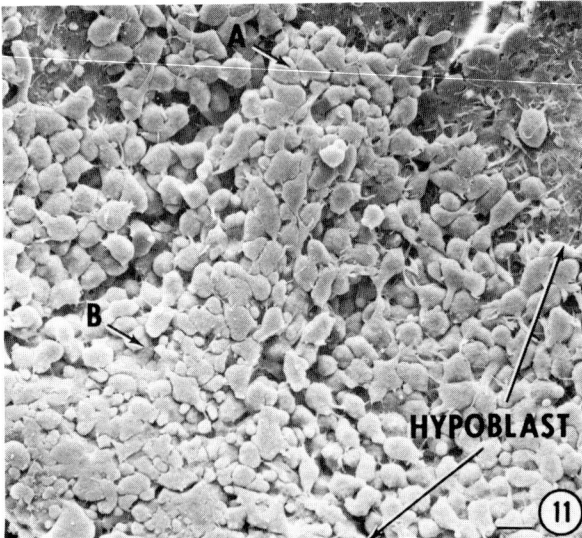

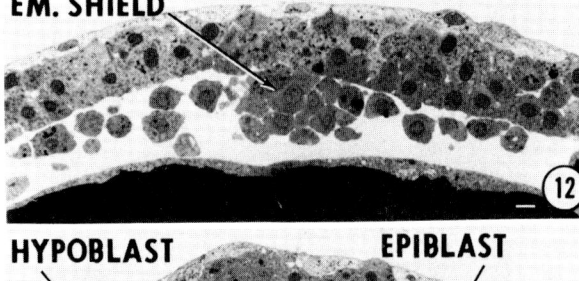

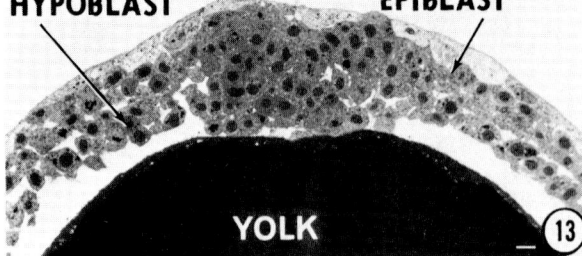

Figure 10. Early embryonic shield (*). Stage 14+. (Bar=10μm).

Figure 11. Embryonic shield and germ ring. Stage 15. Note interdigitation of processes of flattened central cells, hypoblast and germ ring margin. (Bar=10μm).

Figure 12. Cross section through anterior end of an embryonic shield indicated by A in fig. 11. Stage 15. (Bar=10μm).

Figure 13. Cross section through the posterior end of an embryonic shield indicated by B in fig. 11. Stage 15. (Bar=10μm).

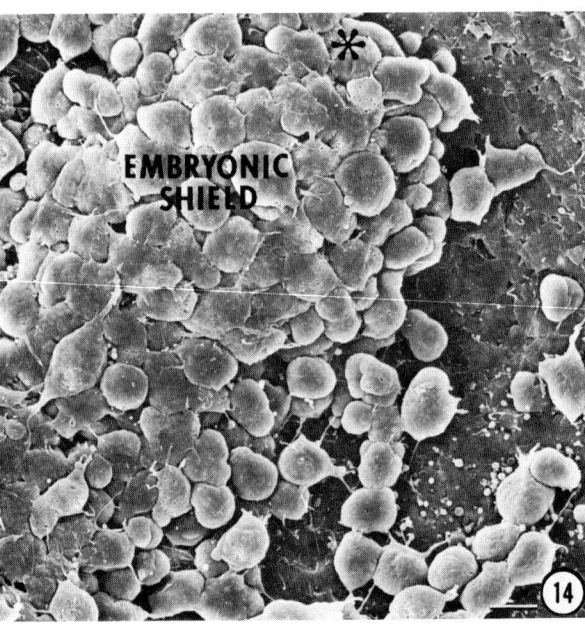

Figure 14. Anterior end of an embryonic shield. Stage 15+. Note flattened cells of deep surface of epiblast, and lack of filopodia at end of embryonic shield (*). (Bar=10μm).

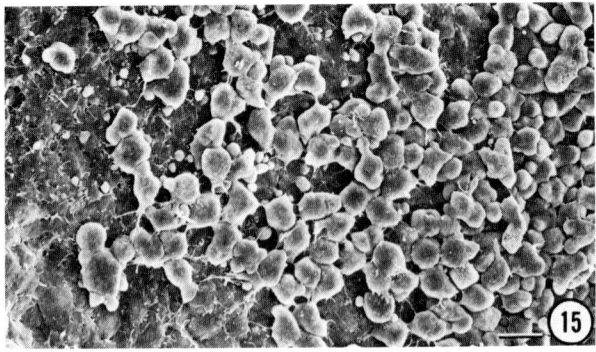

Figure 15. Differences in cell shape from germ ring (right) toward central flattened region (left). Stage 16. (Bar=10μm).

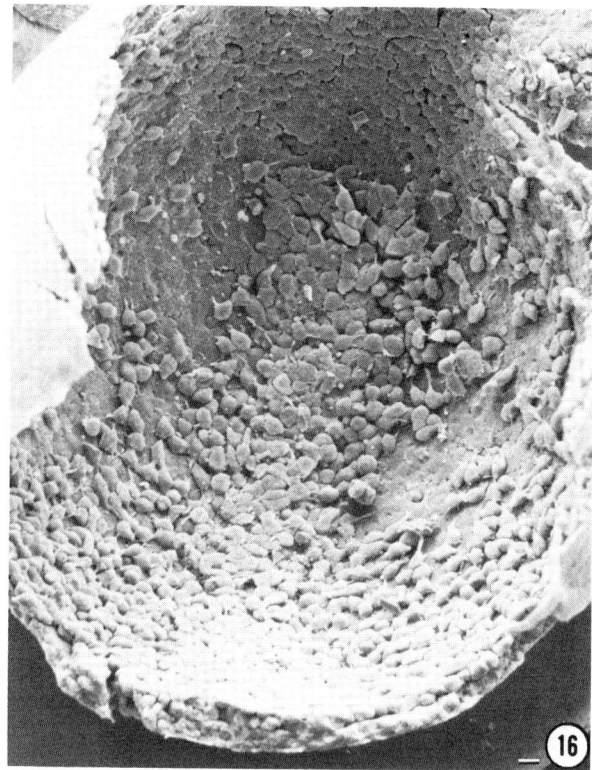

Figure 16. Embryonic shield. Stage 16. Note hypoblast and epiblast at fractured surface. (Bar=10μm).

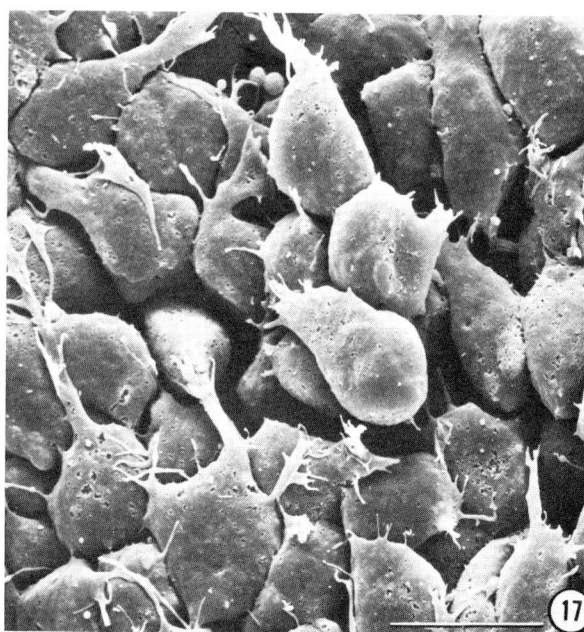

Figure 17. Cells at anterior end of embryonic shield shown in fig. 16. Note lamellipodia. Holes in cell membranes are probably preparation artifact. (Bar=10μm).

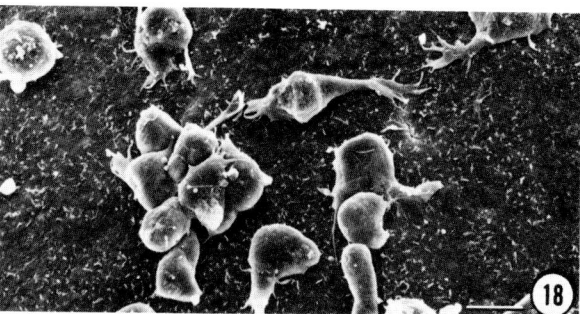

Figure 18. Cells remaining on central periblast after removal of blastoderm. Stage 16. (Bar=10μm).

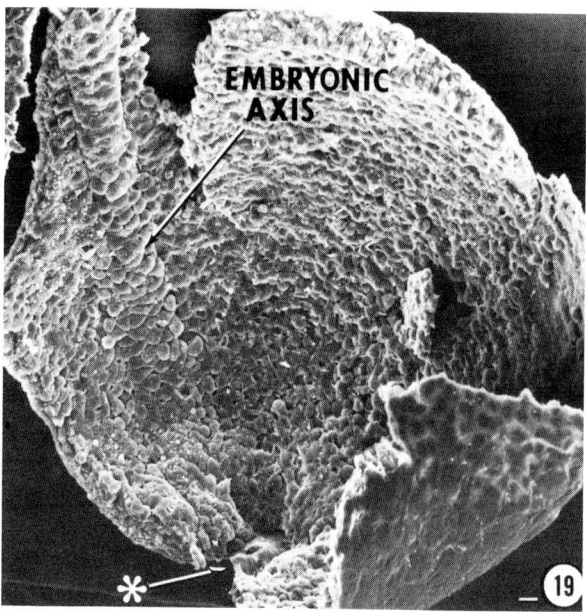

Figure 19. Vegetal half of stage 17 embryo. Embryonic axis extends forward from the closing germ ring (*). (Bar=10μm).

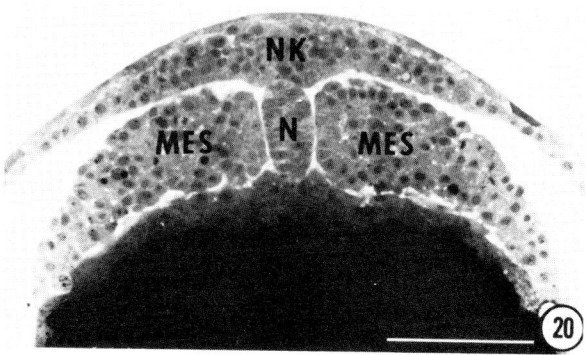

Figure 20. Cross section of embryonic axis. Stage 17+. NK=neural keel; N=notochord; MES=mesoderm. (Bar=100μm).

the "blastopore" (Figure 19). The embryonic shield (or axis) extends anteriorly from the point of closure of the germ ring and encircles approximately one-half of the yolk surface when seen in profile. Cells lateral to the future notochord and neural keel constitute wedges of mesoderm (and perhaps presumptive endoderm) cells. Cells of the shield are more closely associated than the presumptive mesodermal cells which extend numerous filopodia to contact the flattened undersurface of the epiblast as well as adjacent mesodermal cells. The notochord, neural keel and epidermis become increasingly distinct from the mesoderm (Figure 20).

## Discussion

The results of this study clearly indicate that the SEM can be used to examine the morphology of the deeper layers of blastomeres which are not normally visible in the intact specimen. Cellular morphology and differences in cell shape and the number and type of cell processes observed from blastula to late gastrula stages parallel those observed and filmed in vivo. Since analysis of time-lapse motion films indicate that blastomeres move in the direction of their major processes,[14,16] it may be possible to document the direction of cell migration with the SEM. Of special interest is the mapping of patterns of relative cell movements within, and at the undersurface of, the blastoderm during epiboly and formation of the embryonic shield. Such analyses should be closely correlated with in vivo observations of the same sample, however, and the small size, rapid development and transparency of the zebrafish egg offers potential advantages in this regard.

There is a tendency for the zebrafish blastoderm to reorganize so that only a single layer of deep cells remains beneath the enveloping layer. This condition is reached first in the central region near the animal pole and continues until final separation of the epidermis from the neural keel.[37] This appears to be a general feature of teleost development. In the trout embryo, the layer of deep cells subjacent to the enveloping layer gives rise to the basal epidermal cells and the enveloping layer remains as a periderm which is shed at hatching.[24] The mechanics of this thinning process are not clear, however, but might be clarified by SEM observation.

SEM observations may also clarify the historically debated connection between cells of the enveloping layer and the marginal periblast. Trinkaus initially viewed this as an attachment between ruffled membranes at the free margins of the leading edge of the enveloping layer cells and the underlying marginal periblast.[2] Ultrastructural observation of an extensive intercellular junction between the periblast and an enveloping layer cell[12], however, prompted several alternative suggestions.[2]

SEM observations of the zebrafish embryo indicate that there is no free margin of the enveloping layer, and that new cells may be continuously added to the margin of the enveloping layer by cytokinesis of portions of the syncytial periblast cytoplasm. Enlargement of the enveloping layer margin may thus be controlled by the rate of formation of new cells from the marginal periblast.

The continuity between enveloping cells and marginal periblast can be seen in Epon sections of zebrafish embryos,[38] and may have been described at the ultrastructural level in Oryzias by Yamamoto.[39] The external surface of the marginal periblast exhibits numerous microvilli and other surface projections and is distinctly different from the smooth surface of the membrane covering the yolk cytoplasmic layer (see also [40]). Formation of these structures during epiboly may appear as increased surface activity which might be interpreted as ruffling activity when viewed in time-lapse films.[2]

## Acknowledgement

The authors gratefully acknowledge the technical assistance of Mr. G. Minion and the secretarial skills of Ms. A. Kimbrell and Mrs. J. Ivey.

R. Waterman is recipient of a P.H.S. Research Career Development Award DE 00013.

## References

1. J.P. Trinkaus, "Morphogenetic cell movements," In: Major Problems in Developmental Biology. Academic Press, Inc., NY. 1967,125-176.
2. J.P. Trinkaus, "Cells into organs. The forces that shape the embryo," Found. Devel. Biol. Ser. Prentice-Hall, N.J. 1969, 1-237.
3. E.F. Mahon and W.S. Hoar, "The early development of the chum salmon, Oncorhynchus keta (Walbaum)," J. Morph. 98, 1956, 1-47.
4. C. Devillers, "Structural and dynamic aspects of the development of the teleostean egg," Adv. Morph. 1, 1961, 379-428.
5. J. Pasteels, "Études sur la gastrulation des vertébrés méroblastiques, I. Téléostéens," Arch. Biol., 47, 1936, 205-308.
6. J.M. Oppenheimer, "Organization of the teleost blastoderm," Quart. Rev. Biol., 22, 1947, 105-118.
7. R.G. Kessel, "The role of cell division in gastrulation of Fundulus heteroclitus," Exp. Cell Res., 20, 1960, 277-282.
8. W.H. Lewis, "The role of the superficial gel layer in gastrulation of the zebrafish egg." Anat. Rec., 85, 1943, 326.
9. W.H. Lewis, "Mechanics of invagination," Anat. Rec., 97, 1947, 139-156.
10. J.P. Trinkaus, "A study of the mechanism of epiboly in the egg of Fundulus heteroclitus," J. Exp. Zool., 118, 1951, 269-319.
11. T.L. Lentz and J.P. Trinkaus, "A fine structural study of cytodifferentiation during cleavage, blastula and gastrula stages of Fundulus heteroclitus," J. Cell Biol., 32, 1967, 121-138.
12. J.P. Trinkaus and T.L. Lentz, "Surface specializations of Fundulus cells and their relation to cell movements during gastrulation," J. Cell Biol., 32, 1967, 139-154.

13. J.P. Trinkaus, "Mechanisms of morphogenetic movements," In: Organogenesis, R.L. DeHaan and H. Ursprung (eds). Holt, Rinehart and Winston, NY. 1965, Chap. 3, 55-104.
14. J.P. Trinkaus, "Surface activity and locomotion of Fundulus deep cells during blastula and gastrula stages," Devel. Biol., 30, 1973, 68-103.
15. T. Kageyama, "Motility and locomotion of embryonic cells of the Madaka, Oryzias latipes, during early development," Devel., Growth Diff., 19, 1977, 103-110.
16. S. Hamano, "A time-lapse cinematographic study on gastrulation in the zebrafish," Acta Embryol. Morph. Exp., 7, 1964, 42-48.
17. J.P. Wourms, "Developmental biology of annual fishes. I. Stages in the normal development of Austrofundulus myersi Dahl," J. Exp. Zool., 182, 1972, 143-168.
18. J.P. Wourms, "Developmental biology of annual fishes. II. Naturally occurring dispersion and reaggregation of blastomeres during the development of annual fish eggs," J. Exp. Zool., 182, 1972, 169-200.
19. R.J. Lesseps, A.H.M.G. van Kessel and J.M. Denuce, "Cell patterns and cell movements during early development of an annual fish, Nothobranchius neumanni," J. Exp. Zool., 193, 1975, 137-146.
20. J.P. Trinkaus, "The cellular basis of Fundulus epiboly. Adhesivity of blastula and gastrula cells in culture," Devel. Biol. 7, 1963, 513-532.
21. C.A. Tickle and J.P. Trinkaus, "Change in surface extensibility of Fundulus deep cells during early development," J. Cell Sci., 13, 1973, 721-726.
22. S. Mizukami and N. Satoh, "Fusion of dissociated fish embryonic cells," J. Embryol. exp. Morph., 40, 1977, 265-270.
23. J.C. Hogan, Jr. and J.P. Trinkaus, "Intercellular junctions, intramembranous particles and cytoskeletal elements of deep cells of the Fundulus gastrula," J. Embryol. exp. Morph., 40, 1977, 125-141.
24. J. Bouvet, "Enveloping layer and periderm of the trout embryo (Salmo trutta fario L.)," Cell Tiss. Res., 170, 1976, 367-382.
25. W.W. Ballard, "The role of the cellular envelope in the morphogenetic movements of teleost embryos," J. Exp. Zool., 161, 1966, 193-200.
26. W.W. Ballard, "Origin of the hypoblast in Salmo. II. Outward movement of deep central cells," J. Exp. Zool., 161, 1966, 211-220.
27. W.W. Ballard, "Origin of the hypoblast in Salmo. I. Does the blastodisc edge turn inward?" J. Exp. Zool., 161, 1966, 201-210.
28. W.W. Ballard and L.M. Dodes, "The morphogenetic movements at the lower surface of the blastodisc in salmonid embryos," J. Exp. Zool., 168, 1968, 67-84.
29. W.W. Ballard, "History of the hypoblast in Salmo," J. Exp. Zool., 168, 1968, 257-272.
30. W.W. Ballard, "Morphogenetic movements in Salmo gairdneri Richardson," J. Exp. Zool., 184, 1973, 27-48.
31. W.W. Ballard, "A new fate map for Salmo gairdneri," J. Exp. Zool., 184, 1973, 49-74.
32. H.W. Laale, "The biology and use of zebrafish, Brachydanio rerio in fisheries research: A literature review," J. Fish Biol., 10, 1977, 121-173.
33. K.K. Hisaoka and H.I. Battle, "The normal developmental stages of the zebrafish, Brachydanio rerio," J. Morph., 102, 1958, 311-328.
34. K.K. Hisaoka and C.F. Firlit, "Further studies on the embryonic development of the zebrafish, Brachydanio rerio (Hamilton-Buchanan)," J. Morph. 107, 1960, 205-225.
35. S. Ito and M.J. Karnovsky, "Formaldehyde-glutaraldehyde fixatives containing trinitro compounds," J. Cell Biol. 39, 1968, 168a-169a.
36. W.J. Humphreys, B.O. Spurlock and J.S. Johnson, "Critical point drying of ethanol-infiltrated, cryofractured biological specimens for scanning electron microscopy," SEM/1974, O. Johari and I. Corvin, eds. IITRI, Chicago, 1974, 275-282.
37. R.E. Waterman and T.A. McCarty, "SEM observations of the developing teleost central nervous system," SEM/1977/Vol. II, O. Johari and R.P. Becker, eds. IITRI, Chicago, 1977, 387-394.
38. R.J. Thomas, "Yolk distribution and utilization during early development of a teleost embryo (Brachydanio rerio)," J. Embryol. exp. Morph., 19, 1968, 203-215.
39. M. Yamamoto, "Electron microscopy of fish development. V. The fine structure of the periblast in Oryzias latipes eggs," J. Fac. Sci., Univ. Tokyo Sec. IV, 10, 1965, 483-490.
40. H.W. Beams and R.G. Kessel, "Cytokinesis: A comparative study of cytoplasmic division in animal cells," Am. Sci., 63, 1976, 279-290.

## DISCUSSION WITH REVIEWERS

Reviewer II: Would you define more precisely the terms blebs, lobopodia, filopodia and lamellipodia? Does the term "lamellipodia" differ from the term as used in vitro where ruffled membranes are formed?

Authors: A "bleb" is a hemispherical, usually transient, blunt hyaline protrusion whose diameter is between one-fourth and two-fifths of the diameter of the cell. A "lobopodium" is a longer, blunt, finger-like projection. A "filopodium" is a thin, thread-like extension and a "lamellipodium" is a rather large, flattened cell extension. These terms have been more clearly defined and illustrated by several workers, particularly Trinkaus (text reference 14). Lamellipodia in situ within teleost embryos do not appear to be associated with extensive ruffling activity.

Reviewer I: Is there experimental evidence to support the interpretation, based on ultrastructural observations alone, that enlargement of the enveloping layer margin is due to formation of new cells by division from the marginal periblast?

Authors: To our knowledge, there is no direct experimental evidence to support this hypothesis. A series of cinematographic, vital marker and ultrastructural studies designed to examine the relationship between marginal enveloping layer cells and the periblast syncytium is currently under way.

Reviewer II: Why is trinitrophenol added to the fixative? Does it add to the quality of fixation at the SEM level? Does the high osmolarity of the fixative used by the authors account for any of the appreciable cell surface blebbing noted throughout the paper?
Authors: Trinitro compounds used according to the recipe of Ito and Karnovsky (text reference 35) appears to improve the fixation of ultrastructure in certain tissues. It does not, unfortunately, appear to improve the quality of fixation of zebrafish embryos for SEM examination over that obtained with other commonly used fixative systems. The blebs referred to are characteristic of living cells in situ and are not created by the fixative.

Reviewer I: Were the cryofractured embryos fixed prior to ethanol infiltration? If so, how do you account for the poor preservation of the intracellular cytoplasmic structure?
Authors: Embryos were fixed as stated in the Materials and Methods section prior to cryofracture. While nuclei are discernible within the fractured cells, preservation of other cytoplasmic components seems similar to that reported by others using this technique.

Reviewer II: The cell processes shown in Figure 17 are all polarized in one direction, although this remarkable observation drew no specific comment in the text. Do the authors have any thought as to the contact-guidance mechanism directing it, and how these cells are moving? Is there any evidence of extracellular matrix formation in any of the stages observed? If so, was there ever an association of motility appendages with the extracellular macromolecules?
Authors: The problem of what guides the cells during translocation to form the embryonic shield and germ ring is as yet unanswered. Although some flocculent material associated with the blastomeres is occasionally observed, its origin and biochemical composition are unknown, and there is no obvious structural organization to suggest that it functions in directing cell migration. One possible mechanism involves preferential adhesion to the periblast surface. Deep cells have a greater tendency to adhere to the periblast beneath the embryonic shield than that under the germ ring when the blastoderm is experimentally removed (text reference 14). Another possibility for direction of cell movement is suggested by the observation that blastomeres isolated from Fundulus blastulae form new blebs at points other than the site of the recently resorbed bleb. When such cells are experimentally touched or "nudged" in vitro, they form blebs at a point nearly opposite to the point of contact with a micropipette. Similar behavior in vivo might result in directional movement of adjacent cells (C. Tickle and J.P. Trinkaus, "Observations on cells in culture," Nature, 261, 1976, 413). Electrical coupling of cells may also coordinate movements (M.V.L. Bennett and J.B. Trinkaus, "Electrical coupling between embryonic cells by way of extracellular space and specialized junctions," J. Cell Biol., 44, 1970, 592-610).

Reviewer III: You did not mention much about the inner deep cells of the epiblast, probably because they are not easy to visualize. What is your opinion about these cells and their behavior as gastrulation proceeds?
Authors: Some rounded or elongated cells project from the generally flattened undersurface of the zebrafish blastoderm in both histological sections and in SEM preparations during the stages examined, and might be interpreted as disengaging from the blastoderm. Their number does not seem as great, however, as that suggested by light micrographs of Salmo embryos (text references 28,29). The initially recognizable embryonic shield in the zebrafish also does not appear to be formed by a cluster of loosely organized migratory cells as in Salmo. These differences may reflect species variation, or the more rapid development of the zebrafish may have resulted in important stages being missed during this preliminary study.

REVIEWERS:  I  P. B. Bell, Jr.
            II R. R. Markwald

AMPHIBIAN GASTRULATION AS SEEN BY SCANNING ELECTRON MICROSCOPY

Cathy Lundmark, John Shih, Paul Tibbetts, Ray Keller*

Department of Zoology, University of California, Berkeley, California

(Paper received February 4 1983, Completed manuscript received June 28 1984)

## Abstract

The salient events of amphibian gastrulation are readily seen in scanning electron micrographs of dissected Xenopus laevis embryos. Bottle cell formation, involving apical constriction and radial elongation of epithelial cells, initiates blastoporal groove formation first on the dorsal side of the embryo, then laterally and finally ventrally. As bottle cells form in the superficial epithelium, deep cells begin to involute over an internal blastoporal lip, carrying the superficial layer inside to form the archenteron. While material involutes over the blastoporal lip, the blastopore moves toward the vegetal pole as a result of thinning and spreading of the outer layers (epiboly) as well as extension in the dorsal marginal zone. Extension and convergence in the dorsal marginal zone can be accounted for by rearrangement of several layers of deep cells into one layer. These same deep cells are those that then turn over the internal blastoporal lip, migrate toward the animal pole to form the archenteron and ultimately form mesodermal structures. The superficial endoderm is carried inside by the deep cells to line the archenteron. Mesoderm is never found in the superficial layer of Xenopus, but in Ambystoma mexicanum mesoderm makes up the roof of the archenteron for a time. This implies that gastrulation may vary significantly in different amphibians.

KEY WORDS: Ambystoma mexicanum, formation of blastopore, epiboly, extension and convergence, gastrulation in amphibians, invagination, involution, marginal zone in amphibian embryos, morphogenesis, Xenopus laevis
*Address for correspondence:
Department of Zoology
University of California
Berkeley, California  94720
Phone No. (415) 642-8665

## Introduction

Using scanning electron micrographs, we will describe the dramatic changes in cell shape and arrangement occurring during the principal morphogenetic movements of amphibian gastrulation. Various species of amphibians differ greatly in their early morphogenetic movements. We will deal primarily with gastrulation in the African clawed frog, Xenopus laevis, and briefly with gastrulation in the Mexican Axolotl, Ambystoma mexicanum. Other anurans (tailless amphibians) with eggs similar in size to those of Xenopus may gastrulate similarly, but the descriptions that follow should not be extrapolated to other anuran species without caution. A distinctly different pattern of gastrulation is seen in the urodeles (tailed amphibians) studied thus far, though some of the basic features of cell shape and arrangement are similar to those seen in Xenopus.
  The following discussion will also include descriptions of the location of germ layers, the direction and extent of movements, and the mechanical relationships between movements. These are based on extensive experimental studies that will not be discussed in detail but are listed in the references at the end of the paper.
as suggestions for further reading at the end of the paper.

### Overview

The major features of Xenopus gastrulation are best summarized with schematic diagrams (fig. 1) of embryos at the early, early mid-, late mid- and late gastrula stages. The major movements that we will describe are: (1) bottle cell formation; (2) involution of deep and superficial cells of the marginal zone; (3) extension and convergence of the dorsal marginal zone; and (4) epiboly of the animal region. The first sign of gastrulation is the formation of bottle cells about one-third of the way down from the equator to the vegetal pole on the dorsal side of the embryo (fig. 1a). Cells of the superficial epithelium achieve a 'bottle' shape by contracting their apical (surface) ends and elongating toward the interior of the embryo, thereby compacting a large area of the surface into a small darkened area, the blastoporal pigment line (fig. 1a). Such change in shape results in formation of the blastoporal groove by invagination. It is important to note that only the initial blastoporal groove formation is a true invagination--the bending of a sheet of cells inward. Formation of bottle cells, and thus of the blastoporal groove, proceeds laterally on both sides and meet ventrally by the mid-gastrula stage to form a complete circle defining the yolk plug (fig. 1b).

The further increase in depth of the blastoporal groove is due to involution--a rolling of material over the lip of the blastopore. Deep mesodermal cells in the marginal zone--material lying above the blastoporal groove--turn over an internal blastoporal lip as a coherent group of individual cells, and move toward the animal pole by involution (see arrows, fig. 1a). Mesodermal cells which have not yet involuted (preinvolution or prospective mesoderm) are clearly distinct from those that have (postinvolution mesoderm). Involution of deep mesoderm begins in the dorsal marginal zone (fig. 1a) and progresses laterally and ventrally more or less in concert with bottle cell formation (fig. 1b). Postinvolution cells migrate toward the animal pole along the inner surfaces of preinvolution mesoderm and the blastocoel wall to form the mesodermal mantle. The movement of deep mesoderm appears to carry the overlying superficial layer of prospective endoderm, including the bottle cells, with it, so that as the prospective mesoderm turns the lip to become the mesodermal mantle, the superficial prospective endodermal layer involutes with it to form the lining of the roof of the archenteron (figs. 1a-d). The floor of the archenteron is formed from large endodermal cells of the vegetal pole that are pulled dorsally and toward the animal pole by virtue of their continuity with the archenteron roof.

In Xenopus there is no mesoderm in the surface layer of cells, thus the roof of the archenteron is always formed by endoderm. This is not the case in urodele embryos, where prospective chordamesoderm forms the roof of the early archenteron, and is later covered by endoderm moving up from both sides.

As superficial and deep cells involute, more are supplied to the marginal zone by the spreading of uninvoluted material toward the vegetal pole. This spreading, known as epiboly, occurs equally in all directions at the animal pole, while in the dorsal marginal zone it is directed toward the blastopore by extension. Concurrently, the dorsal marginal zone narrows by convergence. Progressively less extension and convergence is shown by the lateral and ventral marginal zones. Thus the dorsal lips move farther across the vegetal region (yolk plug) than the lateral and ventral lips, closing the blastopore on the ventral side of the vegetal pole (figs. 1a-d). It is important to note that movement of external material vegetally (epiboly) is as important a process as movement of involuted material toward the animal pole.

## Bottle Cell Formation

At the onset of gastrulation the Xenopus embryo consists of a superficial epithelial layer one cell thick, and a deep mesenchymal layer ranging from about two cells thick at the animal pole to five or six cells thick at the marginal zone, and more at the vegetal pole (fig. 2). Cells of the vegetal region are much larger than those of the animal region (fig. 2). Bottle cell formation occurs at the boundary of small and large cells and involves both small suprablastoporal and large subblastoporal endodermal cells (fig. 3). Prospective bottle cells in the superficial epithelium contract apically and lengthen radially (figs. 3, 4), forming long, thin necks and bulbous basal ends (fig. 4). Their apical surfaces are thrown into microfolds and microvilli (figs. 5,6). The density of microfolds seen in the scanning electron microscope is roughly proportional to the amount of constriction and the degree of darkening (due to the concentration of melanin granules) seen in the light microscope. As the cells change shape their apices occupy less area than their dilated basal ends, and thus the sheet of cells invaginates to from a deep groove known as the blastoporal groove. The apices of the bottle cells constrict less extensively along the length of the groove than perpendicular to it (fig. 5). This is consistent with the formation of a groove rather than a pit, which is what would be expected if the apices contracted symmetrically. Bottle cells lose their flask shape beginning at the mid-gastrula stage in the dorsal sector, and later laterally and ventrally (fig. 7). When they respread to form the flattened epithelial cells of the anterior portion of the archenteron (figs. 7, 8), their apices flatten and spread, losing their microfolds and increasing their surface area.

## Involution of the Marginal Zone

A sagittal fracture through the dorsal marginal zone (fig. 9) shows the major features of this region during involution. The outer preinvolution mesoderm (PIM) is separated from the involuted mesoderm (IM) by an obvious interface (pointers, fig. 9). Bottle cells (BC) lie anterior to the mesoderm at the end of the archenteron (A). Many of the involuted mesodermal cells have a more rotund morphology than those that have not yet involuted, and bear large protrusions that may be involved in migration toward the animal pole (fig. 10). In contrast, the overlying preinvolution deep cells usually appear more columnar.

The differences between cells of the deep marginal zone before and after involution, seen in sagittal fractures (figs. 9, 10), become more apparent when the two are separated along the interface between them (pointers in fig. 9) and their apposed surfaces are viewed with the SEM. Preinvolution mesodermal cells have flattened, closely packed inner surfaces connected by long, thin filiform protrusions (fig. 11). The postinvolution mesoderm, on the other hand, is less densely and less evenly packed, and consists of

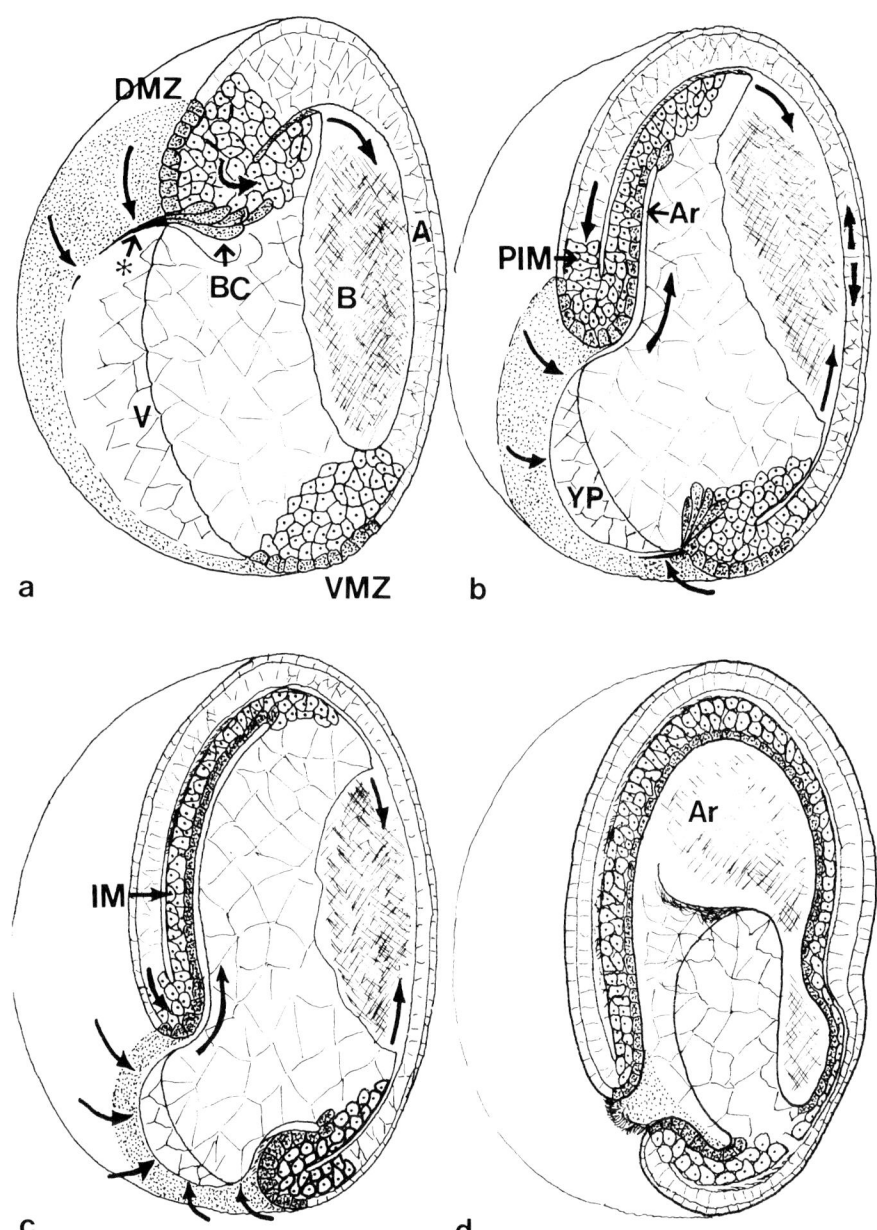

Fig. 1 Schematic diagrams of sagittally fractured halves of X. laevis at the early (a), early middle (b), late middle (c), and late (d) gastrula stages, show the major morphogenetic movements of gastrulation. Shown are the dorsal marginal zone (DMZ), ventral marginal zone (VMZ), the preinvolution mesoderm (PIM), involuted mesoderm (IM), the yolk plug (YP), bottle cells (BC), the animal pole (A), the vegetal pole (V), the archenteron (Ar), the blastocoel (B), and the blastoporal pigment line (*). The prospective mesodermal cells are outlined heavily and the suprablastoporal endodermal cells are stippled. The arrows show direction of movement.

cells separated by larger intercellular spaces bridged by fewer shorter protrusions (fig. 12). Postinvolution cells also extend protrusions (fig. 12) toward the preinvolution cells' inner surfaces (fig. 11) on which they migrate (fig.10). The appearance of the inner surfaces of the preinvolution deep mesodermal cells in the marginal zone is characteristic of the entire blastocoel roof and changes little during gastrulation.

The cells at the leading edge of migrating involuted material exhibit well developed attachments to the blastocoel roof in Xenopus, but more so in urodele embryos (figs. 13, 14). These broad, flat lamelliform protrusions probably function in active migration of these cells on the blastocoel roof.

As involution proceeds, more and more mesoderm moves inside while the preinvolution marginal zone thins (fig. 15). The archenteron deepens as the blastopore advances vegetally and as the involuted mesoderm migrates toward the animal pole, taking the attached epithelium with it (fig. 15). The superficial epithelium which involutes with the deep layer forms the thin, endodermal roof of the archenteron (figs. 15, 16), while the floor is formed by large vegetal endodermal cells (fig. 15). Viewed from the surface of the embryo, superficial cells elongate and narrow as the marginal zone extends toward and turns over the blastoporal lip (fig. 17). The cells become even taller as they pass over the blastoporal lip (open arrow, fig. 15) then become more rounded once inside, except middorsally (fig. 17) where they remain long and narrow. These transient changes in shape are probably due to squeezing of cells laterally as they pass over the lip.

In Xenopus the superficial layer involutes to form a continuous endodermal roof of the archenteron. The three primary germ layers are clearly delineated in the late gastrula (fig. 16). This is not true of all amphibians. The urodeles commonly studied, such as Ambystoma mexicanum, show involution of a superficial layer which consists of mesoderm as well as endoderm. The mesodermal components, somites and notochord, are covered by endoderm prior to and during notochord formation.

### Extension of the Dorsal Marginal Zone

The preinvolution dorsal marginal zone, at the onset of gastrulation, is five or six cell layers of deep cells (fig. 18a). The number of layers of deep cells decreases to one or two layers of elongated cells by the mid-gastrula stage (figs. 18b, c), and finally to one cell layer at the end of gastrulation (fig. 18d). Cells of the deep marginal zone elongate perpendicularly to the surface of the embryo, parallel to its radius. They move between one another in this direction to convert several layers of cells to one layer of tall columnar cells that occupies much greater surface area--a process called radial interdigitation. These columnar cells then shorten, and flatten, thereby forming a thinner layer of even greater surface area. These changes generate the extension seen in the dorsal marginal zone.

The deep cells of the marginal zone are closely packed, in sagittal view, and are connected to one another by many small filiform and lamelliform protrusions (fig. 18b). These may facilitate their rearrangement during extension of the marginal zone.

The narrowing or convergence of the dorsal marginal zone occurs simultaneously with its increase in area and dramatic extension toward the blastopore, yet there is no obvious reflection of this behavior in the morphology or arrangement of the deep cells.

A similar reduction in the number of deep cell layers occurs in the lateral and ventral marginal zones (figs. 19a, b), but in these regions the cells participating in the rearrangement do not show the radial elongation characteristic of their dorsal counterparts.

The superficial cells flatten progressively during the spreading of the marginal zone (figs. 18a-d). Cinemicrography shows that the apices of these cells increase in area, and that they divide during such spreading. They also rearrange in the plane of the epithelium as the dorsal marginal zone narrows and lengthens. They merge toward the dorsal midline much as several lanes of cars might merge on a freeway. These rearrangements cannot be seen in SEM, of course.

### Blastocoel Roof

As in the marginal zone, rearrangement of the deep cells decreases the number of layers seen in sagittal fractures of the blastocoel roof (figs. 20a, b). Cells at the animal pole flatten and the number of cell layers decreases as the blastocoel roof spreads and thins out during epiboly. The deep cell layer decreases from two to one cell layer thick, while the superficial epithelium, though thinner, remains a single cell layer thick (figs. 20a, b). Cinemicrography shows that the superficial cells spread in a manner reminiscent of the dorsal marginal zone, though at half the rate. The inner surfaces of the deep cells of the animal pole are tightly packed and connected by long filiform protrusions (fig. 20c), much like those seen in the marginal zone (fig. 11).

### Tissue Organization

It must be kept in mind that there are two fundamentally different tissue organizations in the amphibian gastrula. The superficial layer of cells (prospective ectoderm and endoderm) consists of epithelial cells which have contiguous borders marked by small raised areas of ridges (fig. 21a). In contrast, the deep cells, in surface view, have small spaces between them and are connected only at localized areas where filiform protrusions lay across adjacent cells' surfaces (fig. 21b). The superficial epithelium forms a physiological barrier isolating the inside of the embryo from the outside. Such demands are not placed on the deep mesenchymal cell population.

---

Fig. 2 A midsagittal fracture of a very early gastrula shows the superficial epithelial layer (S), the deep mesenchymal layer (D), the bottle cells (BC), and the blastocoel (B). The animal pole is at the upper right in the micrograph. Bar = 100 μm

Fig. 3 Midsagittal fracture of the early gastrula shows the cell shape changes and early movements of gastrulation. Bottle cells (BC) can be seen to vary in size above and below the blastopore (BP). Arrows show the direction of movement of deep mesoderm, and the interface (*) between the preinvolution and involuted mesoderm is evident. From Keller (1981), with permission. Bar = 100 μm

Fig. 4 Close up of bottle cells in fig. 3 shows their morphology. The small arrow indicates the long thin necks and the large arrow the rounded basal ends that are the result of apical constriction and radial elongation. Bar = 20 μm

Fig. 5 Surface view of the blastoporal groove shows the apical ends of bottle cells which constrict primarily meridionally (vertically in the micrograph), causing the cells to appear more elongate circumferentially (horizontally in the micrograph). Animal pole is above. From Keller (1981), with permission. Bar = 10 μm

Fig. 6 Highly magnified view of bottle cell apices shows the microfolds and microvilli that form when the apices constrict. Bar = 10 μm

Fig. 7 Lateral view of anterior tip of the archenteron at the late gastrula stage shows (arrows) disappearance of bottle cells which respread to form the epithelial cells of the archenteron. Bar = 100 μm

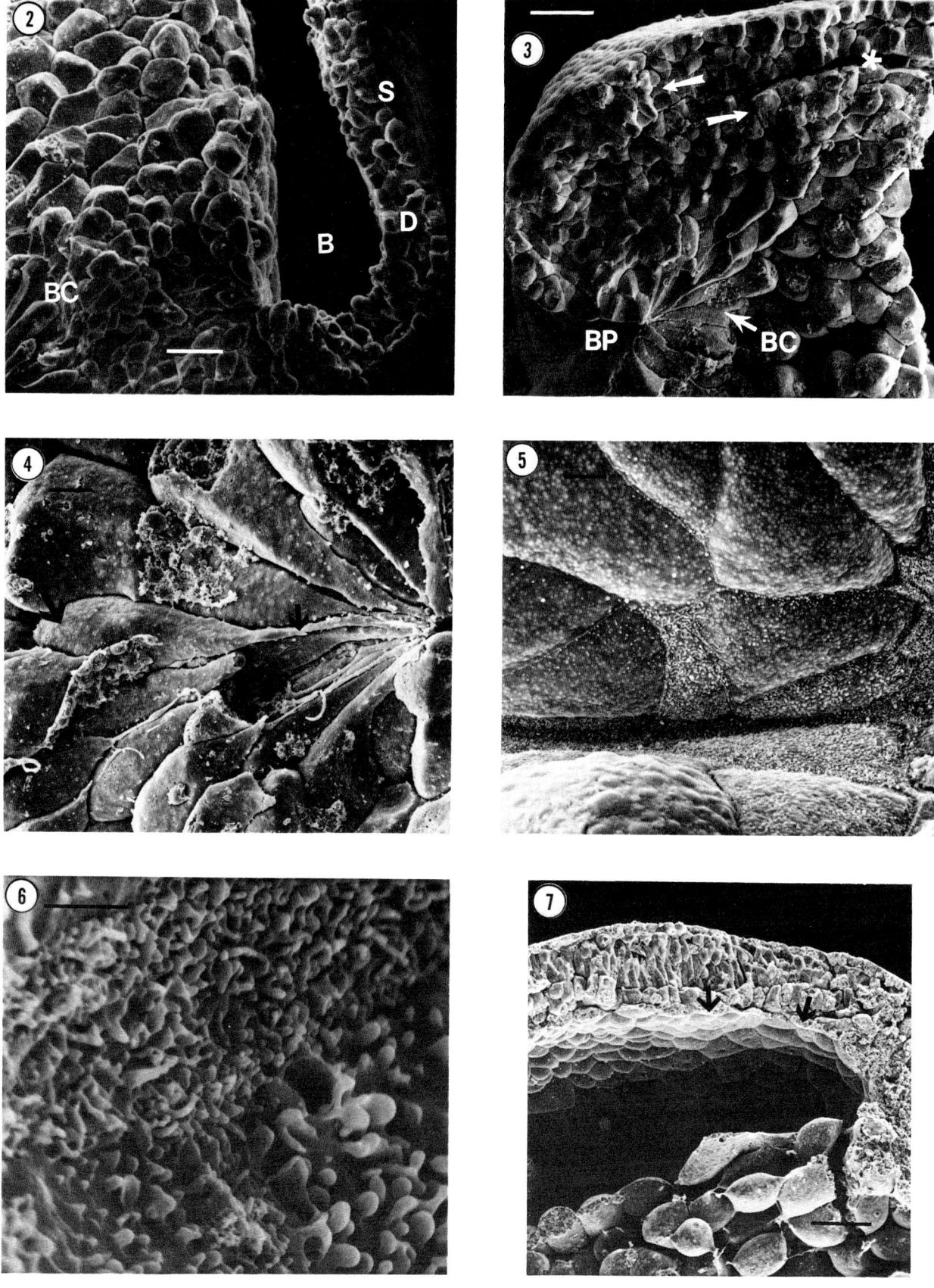

## Discussion

Xenopus gastrulation results from the coordinate operation of several systems of movement occurring in the various cell layers of the embryo. The mesodermal mantle and the archenteron roof are formed by coordinate movement of the deep and superficial layers of the marginal zone toward the blastoporal lip, their involution, and finally, their migration toward the animal pole. Movement toward the lip (epiboly) is accompanied by flattening and spreading of superficial cells and radial interdigitation of several layers of deep cells to form fewer layers of greater area. Involution is thought to involve change in cell behavior to allow migration toward the animal pole. What these changes in behavior are and how they are controlled is not known.

## References

Keller R. (1980). The cellular basis of epiboly: An SEM study of deep-cell rearrangement during gastrulation in Xenopus laevis. Exp.Morph. 60, 201-234.

Keller R. (1981). An experimental analysis of the role of bottle cells and the marginal zone in gastrulation of Xenopus laevis. J. Exp. Zool. 216, 81-101.

Keller R., Schoenwolf G.C. (1977) An SEM study of cellular morphology, contact, and arrangement as related to gastrulation in Xenopus laevis. Roux' Arch. Dev Biol. 182, 165-186.

## Suggestions For Further Reading

Baker P. (1965). Fine structure and morphogenetic movements in the gastrula of the treefrog, Hyla regilla. J. Cell Biol. 24, 95-116.

Holtfreter J. (1939). Gewebeaffinität, ein Mittel der embryonalen Formbildung. Arch. Exp. Zellforsch. Besonders Gewebezücht 23, 169-209.

Holtfreter J. (1943a). Properties and functions of the surface coat in amphibian embryos. J. Exp. Zool. 93, 251-323.

Holtfreter J. (1943b). A study of the mechanics of gastrulation: Part I. J. Exp. Zool. 94: 261-318.

Holtfreter J. (1944). A study of the mechanics of gastrulation: Part II. J. Exp. Zool. 95, 171-212.

Johnson K. (1970). The role of changes in cell contact behavior in amphibian gastrulation. J. Exp. Zool. 175, 391-427.

Keller R. (1975). Vital dye mapping of the gastrula and neurula of Xenopus laevis. I. Prospective areas and morphogenetic movements of the superficial layer. Develop. Biol. 42, 222-241.

Keller R. (1976). Vital dye mapping of the gastrula and neurula of Xenopus laevis. II. Prospective areas and morphogenetic movements of the deep layer. Develop. Biol. 51, 118-137.

Keller R. (1978). Time-lapse cinemicrographic analysis of superficial cell behavior during and prior to gastrulation in Xenopus laevis. J. Morphol. 157, 223-248.

Nakatsutji N. (1974). Studies on the gastrulation of amphibian embryos: pseudopodia in the gastrula of Bufo bufo japonicus and their significance to gastrulation. J. Embryol. Exp. Morphol. 32, 795-804.

Nakatsutji N. (1976). Studies on the gastrulation of amphibian embryos: Ultrastructure of the migrating cells of anurans. Roux' Arch. Dev. Biol. 180, 229-240.

Schechtman A. (1942). The mechanism of amphibian gastrulation. I. Gastrulation-promoting interactions between various regions of an anuran egg (Hyla regilla). Univ. Calif. Publ. Zool. 51, 1-39.

Vogt W. (1929). Gestaltungsanalyse am Amphibienkeim mit örtlicher Vitalfärbung. II Teil. Gastrulation und Mesodermbildung bei Urodelen und Anuren. Wilhelm Roux' Archiv. 120, 384-706.

---

Fig. 8 Surface view of anterior tip of the archenteron shows respread apices of former bottle cells. Bar = 50 μm

Fig. 9 Midsagittal fracture through the dorsal lip of a mid-gastrula stage embryo shows pre- and post-involution mesoderm (arrows show direction of movement) clearly distinguished by an interface (pointer). (A) Archenteron; (BC) bottle cells; (B) blastocoel. Bar = 100 μm

Fig. 10 A postinvolution mesodermal cell attached to the inner surface of the gastrula wall. Protrusions (arrows) attach the cell at both ends to the deep prospective ectoderm, which appears more columnar (upper left). From Keller and Schoenwolf (1977), with permission. Bar = 10 μm

Fig. 11 Inner surfaces of cells of preinvolution mesoderm appear flat, closely packed, and connected to one another by long filiform protrusions. Bar = 10 μm

Fig. 12 A view of the outer surfaces of involuted mesodermal cells shows loose arrangement of cells with larger intercellular spaces (compare to fig. 11). Some cells have protrusions (arrows) where they had been attached to the apposed deep cells of the gastrula wall. Bar = 10 μm

Fig. 13 Cells of the leading edge of the involuted material in Ambystoma mexicanum display broad lamelliform protrusions as they migrate along the inner surfaces of deep cells toward the animal pole (above). Bar = 100 μm

Amphibian Gastrulation as Seen by SEM

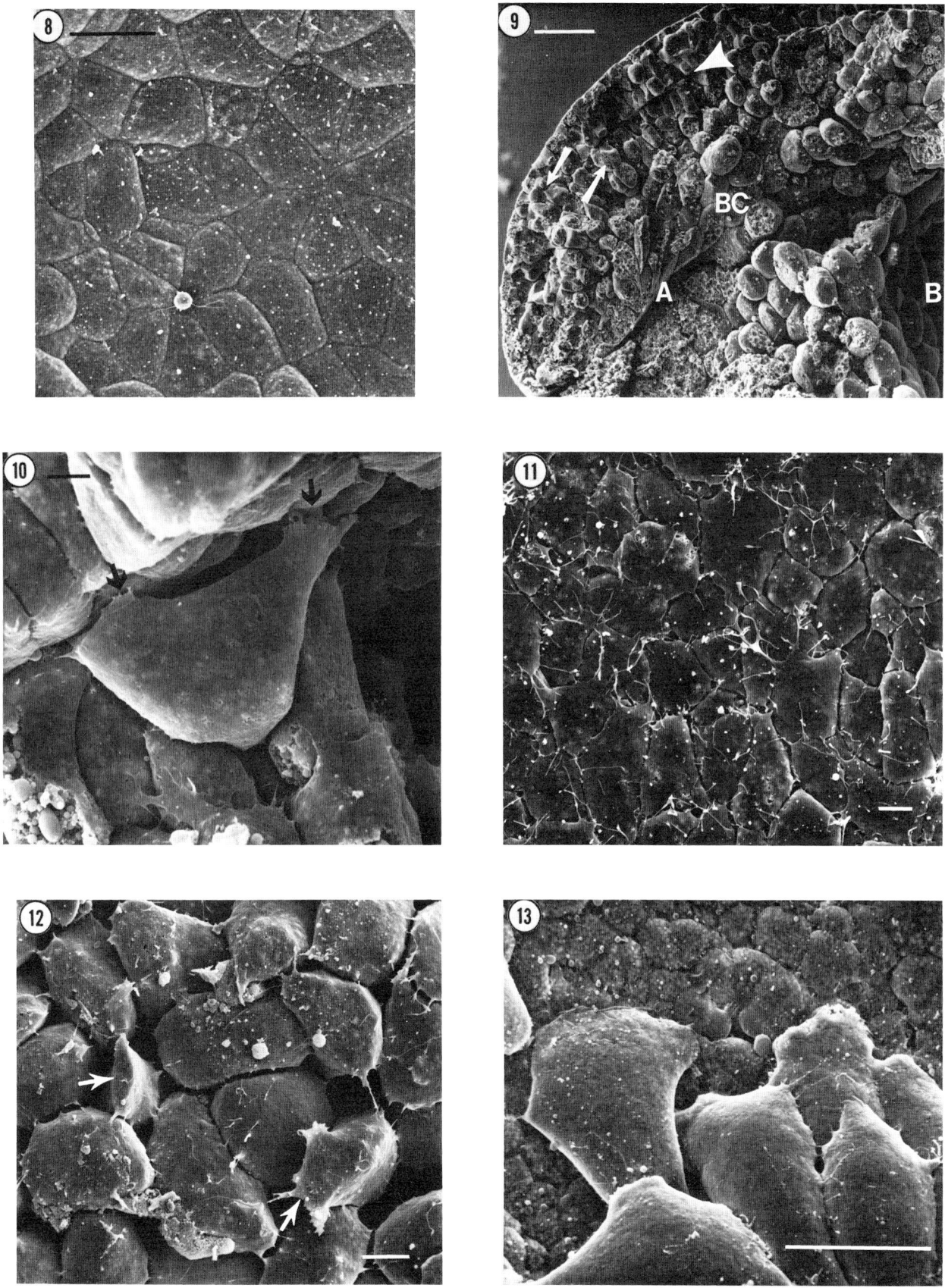

61

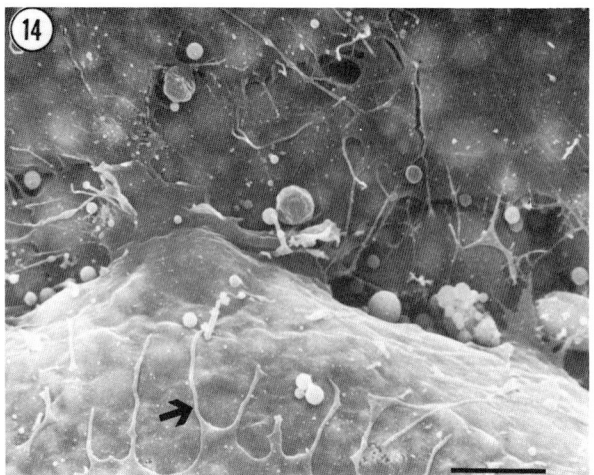

Fig. 14 cells of the leading edge in *Pleurodeles waltl* also bear well-developed protrusions attached to the blastocoel wall as well as to other cells of the leading edge (arrow). Bar = 10 μm

Fig. 18 (Next page) A series of sagittal fractures through the dorsal marginal zone show the changes in morphology and arrangement of the deep marginal zone cells at (a) the beginning of gastrulation, (b) early gastrula, (c) mid-gastrula, and (d) late gastrula stages. Note the decrease in number of deep cell layers and thinning of the superficial layer with concomitant elongation of the cells of the deep layer. The vegetal or blastoporal end of the marginal zone is below in all cases. From Keller (1980), with permission. Bar = 100 μm

Fig. 15 Montage of a midsagittal fracture of a late gastrula shows the involuted mesoderm or mesodermal mantle (thick arrow), the thin preinvolution marginal zone (thin arrow). The superficial layer (pointers) is involuted over the blastoporal lip to form the roof of the archenteron (A). The floor of the archenteron is formed by large endodermal cells in the lower part of the micrograph. The movements of involution are depicted by white arrows while black arrow indicates the wedge-shaped or columnar appearance of superficial cells as they turn over the blastoporal lip. Bar = 100 μm.

Fig. 16. A frontal fracture of a late gastrula-early neurula shows the three germ layers; (E) ectoderm, (M) mesoderm and (En) endoderm. The endoderm forming the room of the archenteron is shown (small pointer indicates area from which fig. 17 was taken). Bar=100 μm.

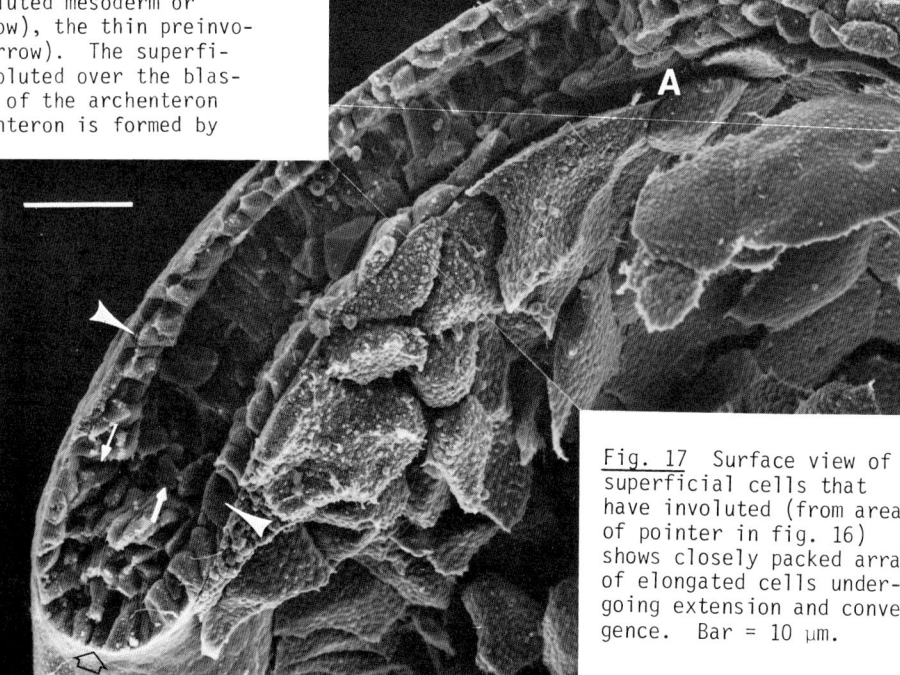

Fig. 17 Surface view of superficial cells that have involuted (from area of pointer in fig. 16) shows closely packed array of elongated cells undergoing extension and convergence. Bar = 10 μm.

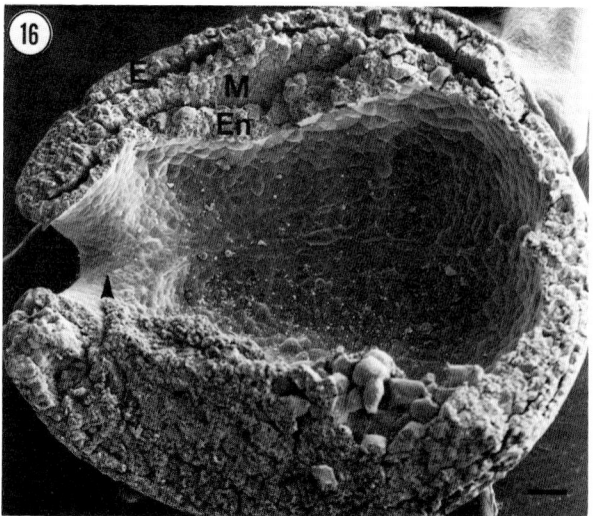

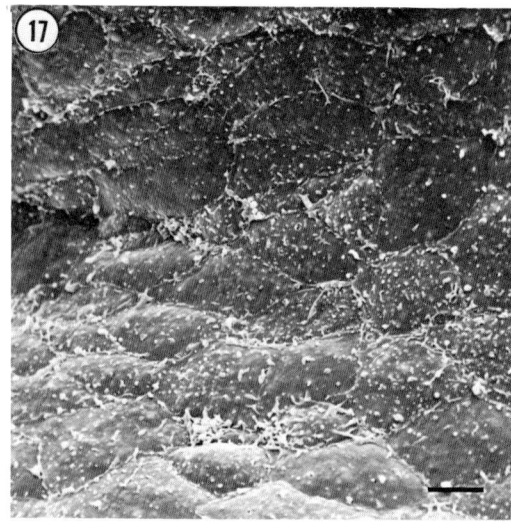

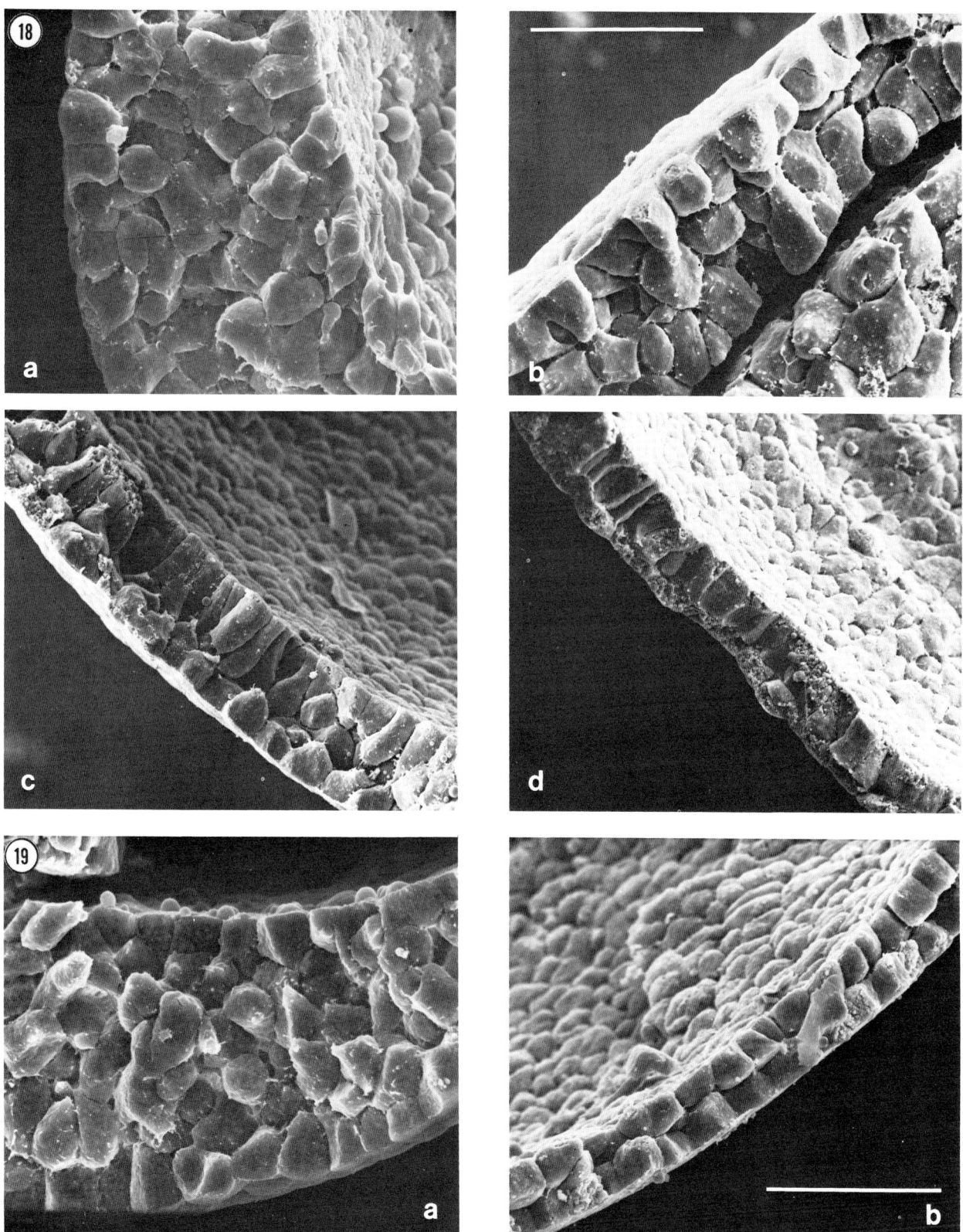

Fig. 19 Sagittal views of the ventral marginal zone at (a) early mid- and (b) late mid-gastrula stages. Note the flattening of cells as well as the reduction of the number of cell layers in the deep region. The vegetal end of the marginal zone is below in both figures. From Keller (1980), with permission. Bar = 100 μm.

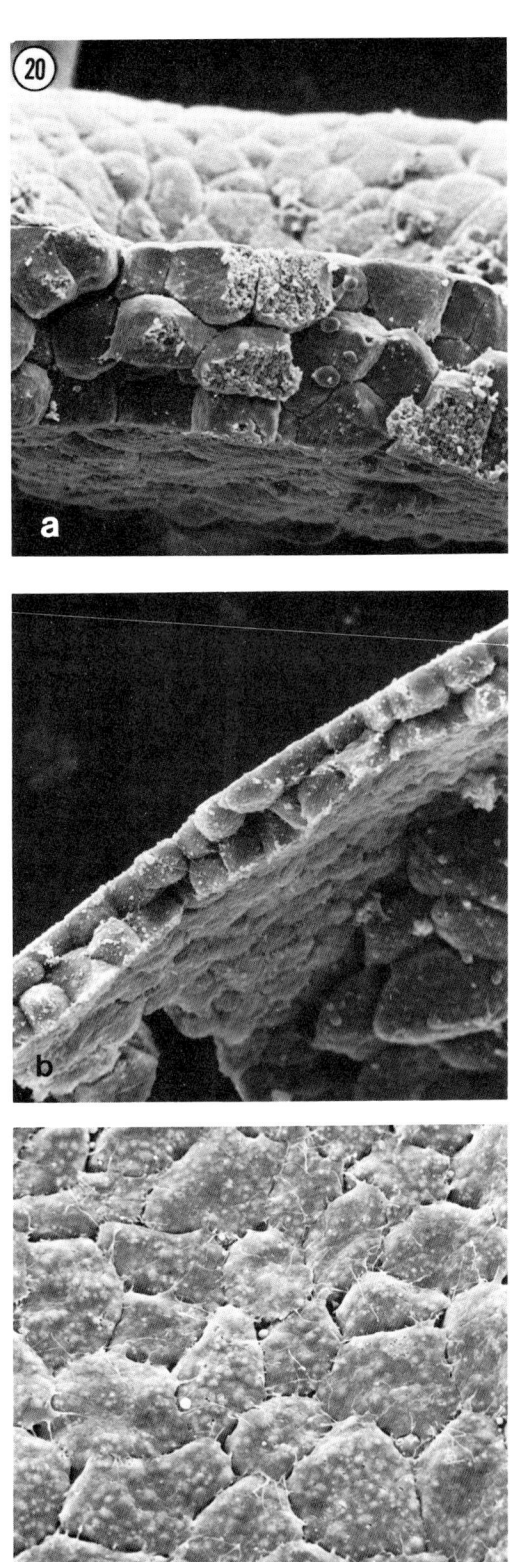

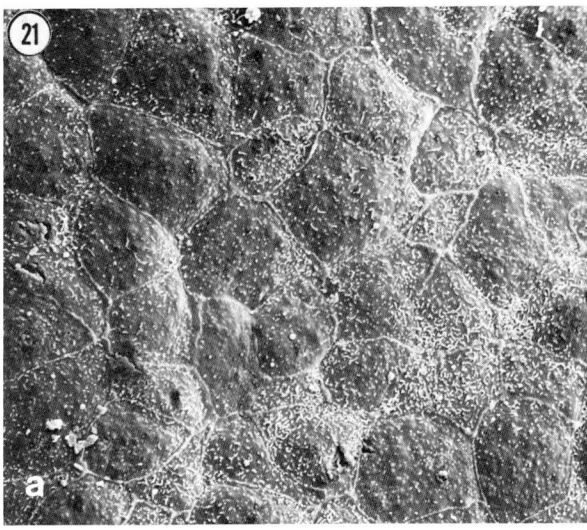

Fig. 21 A surface view of the gastrula shows the contiguous boundaries and raised ridges between superficial epithelial cells (a) in contrast to the intercellular spaces and lack of contiguity of the margins of deep cells (b). Bar = 10 μm

Fig. 20 (Left) Sagittal fractures of animal pole region (a) prior to gastrulation and (b) at late gastrula stage show thinning due to epiboly. Cell size decreases due to cell division. (c) Surface view of the deep layer showing closely apposed cells and the long filiform protrusions which connect them. From Keller (1980), with permission. Bar = 100 μm

Amphibian Gastrulation as Seen by SEM

## Discussion with Reviewers

R. Waterman: For the benefit of beginning students, could you provide a brief historical perspective of how SEM observations have advanced our understanding of gastrulation in amphibians and summarize some of the currently unanswered questions regarding this topic?

Authors: SEM made it possible to see cell shape and relationships with each other directly in three dimensions, and with high resolution for the first time. The high resolution made it possible to see small cellular protrusions possibly used in movement. Stereomicrographs allow a three dimensional view of surface topography not available with the light microscope. The major limitation of SEM is that it deals with fixed tissues. One must still reconstruct the dynamics of gastrulation from a sequence of static pictures and correlate them with the results of other methods, such as cinemicrography.

The major unanswered questions include the following. To what extent do the cells of the mesodermal mantle migrate up the inner surface of the blastocoel roof and to what extent are they displaced by virtue of their active convergence and extension? How do deep mesodermal cells bring about convergence and extension of the dorsal mesodermal mantle?

I. Brick: What is the evidence and/or criteria whereby it is determined that the superficial layer in Ambystoma mexicanum consists of mesoderm as well as endoderm? How are these presumptive tissues deployed in the superficial layer—intermingled or separate—and if separate in what relative positions?

Authors: Current research in our lab and other labs corroborates Vogt's original fate maps showing chordamesoderm on the surface of the axolotl embryo. The chordamesoderm is a coherent dorsal group of cells, separated from endodermal cells by some somitic mesoderm. The latter disappears from the surface by ingression as involution occurs, bringing endoderm next to the chordamesoderm lining the roof of the archenteron.

I. Brick: It is not clear what the relationship is between the "outer preinvolution mesoderm" which "is separated from the involuted mesoderm by an obvious interface". What is the location of the mesoderm cells in the deep marginal zone with respect to the "inner blastopore lip"? What is the basis for the designation, preinvolution and postinvolution, in as much as the postinvolution mesoderm involutes first? When does this happen with respect to the onset of gastrular morphogenetic movements?

Authors: PIM and IM are separated in time and space by involution over the internal blastopore lip. PIM becomes IM by turning over the lip and migrating on the inner surface of the PIM. The difference in cell packing (figs. 11, 12) makes the distinction more obvious. Involution follows on the heels of bottle cell formation.

I. Brick: In what way does the subsuperficial layer constitute a "mesenchymal layer" rather than an epithelium? Mesenchymal may imply mesodermal in origin. Does the mesenchymal layer vary with respect to presumptive germ layer composition in different anterior-posterior and dorso-ventral regions?

Authors: We use the term "mesenchymal" to refer to non-epithelial cells, without any implication of their germ layer origin.

I. Brick: The description of blastoporal groove formation implies that this occurs directly as a consequence of apical constriction resulting in reduced surface area, with concomitant basal expansion. Are other events also involved, for example an apico-basal constriction of microfilaments?

Authors: Experiments in this laboratory show that bottle cell formation from cuboidal epithelial cells is autonomous and active. The role of the cytoskeleton in this process has never been rigorously examined.

G.C. Schoenwolf: Could you provide a more detailed explanation of the movements of convergence and extension? How, exactly, do you define these terms, and what is known about how these movements are generated?

Authors: Convergence means narrowing in the same sense that merging of six lanes of traffic on a freeway to two constitutes narrowing. Likewise, extension means lengthening in the perpendicular direction in the same sense that the merging traffic will occupy about three-fold greater length after merging. Nothing is known about how these movements are generated.

G.C. Schoenwolf: Is there any experimental evidence that cells apically constrict (actively by microfilaments) during formation of bottle cells; deep involuting cells "carry in" superficial cells; the large endodermal cells forming the floor of the archenteron are pulled dorsally and toward the animal pole by virtue of their continuity with the archenteron roof; or postinvolution cells migrate toward the animal pole to form the mesodermal mantle?

Authors: Change in shape of bottle cells is active (see answer above) but the role of microfilaments has never been rigorously examined. The native deep cells are the essential element in involution, whereas the superficial cells of any region can be carried across the lip by native deep cells. The presumptive archenteron roof and floor always maintain continuity at the region of bottle cell formation (Keller, 1975). The former is pulled vegetally over the latter during gastrulation. The contribution of active migration of involuted mesodermal cells to the total force production during gastrulation has never been determined.

G.C. Schoenwolf: Could you provide a picture of bottle cells associated with the lateral lips of the blastopore? Do these cells have the same morphology as dorsal or ventral bottle cells?

Authors: Lateral bottle cells look exactly like those in dorsal and ventral sectors.

G.C. Schoenwolf: Are the highly attenuated necks of bottle cells formed actively by cell elongation (perhaps mediated by microtubules), or passively, by the stretching of bottle cells as the lips of the blastopore converge toward one another (or in the case of the dorsal lip, as this lip moves vegetally)? How are the connections of the necks of bottle cells to surface ultimately severed?

Authors: It is not known how bottle cells change shape. Their connections with the surface layer may not be broken in most cases.

GASTRULATION IN AVIAN EMBRYOS

Marjorie A. England *

Department of Anatomy
University of Leicester,
Leicester LE1 7RH U.K.

(Paper received January 25 1984, Completed manuscript received August 8 1984)

## Abstract

Chick embryo epiblast cells invaginate through the primitive streak and Hensen's node to form the mesoderm and definitive endoblast layers. These invaginating cells lose their epithelial morphology and there is an increase in surface blebbing. There is also a change in the synthesis of extracellular materials.

The mesoderm cells migrate laterally and anteriorly from the primitive streak using fibrils of extracellular material as a contact guidance system. These fibrils form specific patterns at stages 3, 4, and 5 on the epiblast basement membrane. These patterns change with the embryo's further normal development.

KEY WORDS: gastrulation, embryo, mesoderm, primitive streak, primitive node, fibronectin, extracellular materials, basal lamina, contact guidance, cell migration

*Address for correspondence:
  Department of Anatomy,
  University of Leicester,
  Leicester LE1 7RH, U.K.
      Phone No. (0533) 551234

## Introduction

The formation of the third germ layer (mesoderm) is an important morphological event in an embryo's normal development. Many features of this process, however, have only become clear with the advent of the scanning electron microscope. The avian embryo is particularly suitable for study as it is easily obtained and cultured in the laboratory. The following study examines gastrulation in the normal chick embryo by scanning electron microscopy.

## Early Embryo

At the time the egg is laid, the bird or avian embryo consists of a flat disc of cells lying on a large amount of yolk. Both the disc and the yolk are enclosed in a vitelline membrane. Two distinct regions are evident on the disc: the central area pellucida and a surrounding peripheral yolk ring, the area opaca (Fig. 1). The embryo forms in the area pellucida whereas the area opaca gives rise to the yolk sac and other extraembryonic membranes.

The embryo consists of two layers of cells at an early stage (stage 2, Hamburger and Hamilton, 1951) : an upper (dorsal) layer or epiblast, and a lower (ventral) layer or hypoblast. The epiblast is a continuous layer at this stage whereas the hypoblast is continuous only at the future posterior (tail) end. Anteriorly (future head end), the hypoblast is composed of islands of cells. With further development, the hypoblast forms a continuous sheet (stage 3).

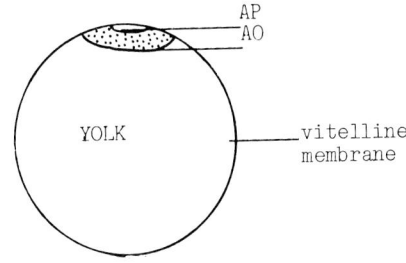

Fig. 1  The relationship of the avian blastoderm to the yolk. AO, area opaca; AP, area pellucida.

The epiblast layer will form the ectoblast or ectoderm layer of the embryo. The hypoblast is largely replaced by the definitive endoblast or endoderm layer during gastrulation.

## Gastrulation

The embryonic process by which the third germ layer (mesoderm) and the definitive endoblast layer form at stages 3 and 4 is called gastrulation.

The epiblast begins to thicken in the midline at the posterior end of the area pellucida at approximately 6-7 hours of incubation (stage 2). This is indicated by the appearance of a dark triangular area. This triangular thickened area is the early primitive streak, which forms as a result of cell migrations within the epiblast layer as well as a high mitotic rate in this region. Migration of epiblast cells posteriorly is induced by the underlying hypoblast layer.

The movements of the epiblast layer have been studied by several workers. Their results are summarized as "fate maps" and whereas they differ in detail, they are generally in agreement on major patterns of movement. The most recent of these studies incorporated tritiated thymidine as a marker to map the development of each region. The general areas based on a study by Rosenquist in 1966 are summarized in Fig. 2. The mesodermal areas marked on the diagram will migrate toward, and invaginate ( or move through the primitive streak) to form internal tissues.

It is possible from an early age of incubation to distinguish by scanning electron microscopy (SEM) those cells in the intact embryo destined to invaginate through the primitive streak from the cells destined to form the epithelium (ectoderm) covering the embryo. The future primitive streak cells possess few microvilli but have many globular and vesicular processes, whereas those cells destined to form the covering of the embryo (Fig. 3) possess many microvilli and few globular and vesicular structures. As development proceeds these distinctions become more apparent. However, if these two cell types are isolated and grown <u>in vitro</u>, there is no discernible difference in <u>their</u> behaviour. This would suggest that the cells are very labile and their definitive characteristics are reversible at this stage.

The hypoblast (Fig.4) is a single-layered sheet of squamous epithelium forming the ventral layer of the embryo. On the ventral surface the perimeters of adjacent cells frequently overlap, and rosettes of cells are present marking areas where cells have been added to the layer (Fig.5).

At stage 3, the primitive streak appears from the dorsal surface as a depression (or primitive groove) in the epiblast marked by small cellular projections (Fig. 6). The anterior end of the streak is a shallow circular depression with some small cellular projections protruding from the surface (Fig. 7). More posteriorly the streak narrows and is studded with numerous small cellular projections (Fig. 8).

As the primitive streak lengthens to two-thirds of the length of the pear-shaped area pellucida by stage 4 (Fig. 9), the amount of cellular projections in the primitive groove increases dramatically (Fig. 10). By stage 4 (Fig. 11), a hillock called Hansen's node or the primitive node (or knot) has formed with a central, usually somewhat deeper, primitive pit at the most anterior end of the groove. The node characteristically contains large quantities of cellular projections that fill and obscure the pit. From the hypoblast side of the embryo, the node is a bulge projecting from the ventral surface (Fig. 12).

The ventral surface of the epiblast layer at stage 3 is covered with a thin basal lamina or membrane through which the cells' perimeters are apparent. By stage 4, the basal lamina has thickened and the cell perimeters are less clearly identifiable. In addition to the basal lamina, fibers radiate from the primitive streak and node. The basal lamina is present on the ventral epiblast except in the region of the primitive streak. Along the borders of the streak in stages 4-5 and around Hensen's node several fibers have collected on the epiblast ventral surface. As the epiblast cells migrate toward the primitive streak, some of their basal lamina and its associated fibers is carried with them and, as the cells invaginate, they leave their fibers behind at the node or primitive streak margins (Figs. 13 and 14).

The cells of the epiblast layer and the area pellucida border generally form a cuboidal epithelium whereas those closer to the streak are more columnar in orientation (Fig. 15). The epiblast cells immediately adjacent to the streak form a pseudo-stratified epithelium. In the primitive streak, many of the cells assume a flask or bottle shape, with a bulbous base ventrally and a long thin neck dorsally. Each cell invaginates (or moves through) the streak by migrating ventrally (Figs. 16 and 17) and then moving laterally between the hypoblast and epiblast layers. The bulbous ventral end of the cell enters the space between the layers while the long, trailing neck separates from the dorsal surface. The leading edges of the cells are characterized by filopodia (Fig. 18).

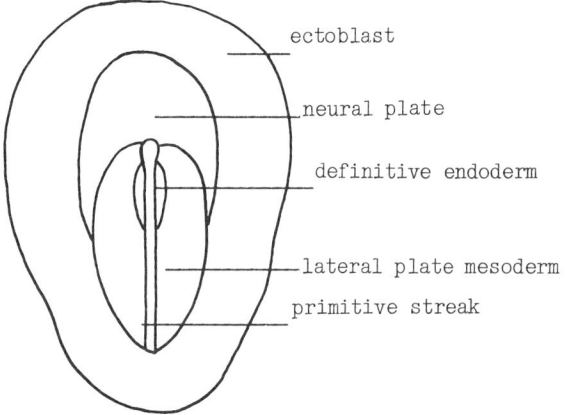

Fig. 2 Presumptive areas of the stage 4 epiblast. After Rosenquist, 1966.

Avian Gastrulation

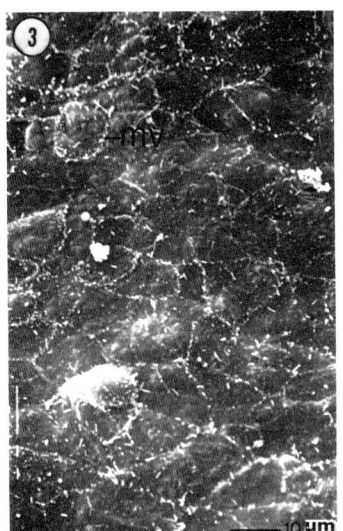

Fig. 3 Stage 4 epiblast cells (dorsal surface) destined to form the embryo's covering. mv, microvilli.

Fig. 4 The ventral surface of the stage 3 hypoblast. Note the flat cells. C, cells.

Fig. 5 Stage 4 hypoblast cells viewed from the ventral surface. R, rosettes.

Fig. 6 Blastoderm at stage 3 fractured transversely anterior to the primitive streak (PS). CP, cellular projections; E, epiblast; H, hypoblast; M, mesoderm.

Fig. 7 The anterior tip of the stage 3 primitive streak looking posteriorly; view of the dorsal epiblast. CP, cellular projections; E, epiblast; T, tip of the primitive streak.

Fig. 8 The middle of the stage 4 primitive streak viewed from the dorsal surface of the epiblast. CP, cellular projections; E, epiblast; PS, primitive streak.

Fig. 9 The stage 4 primitive streak viewed from the dorsal surface of the epiblast. AO, area opaca; AP, area pellucida; HN, Hensen's node; PS, primitive streak.

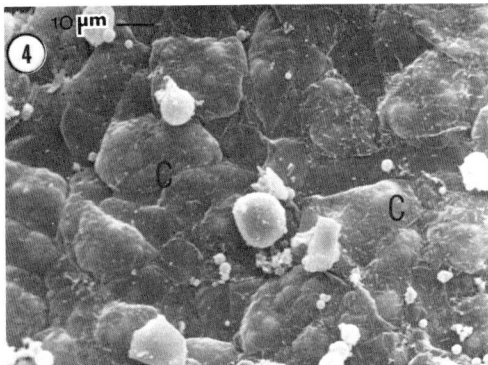

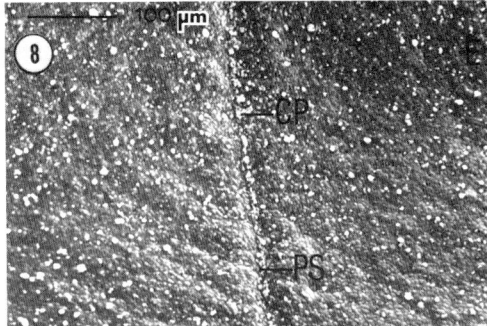

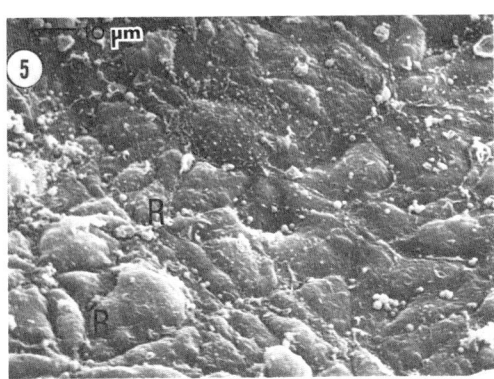

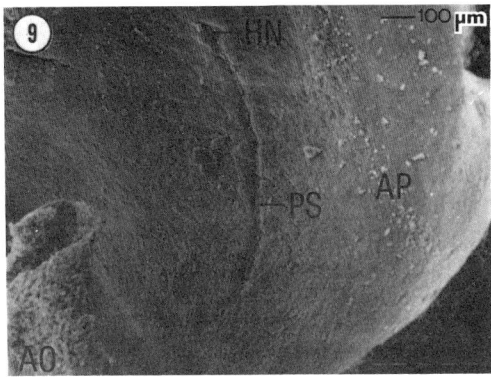

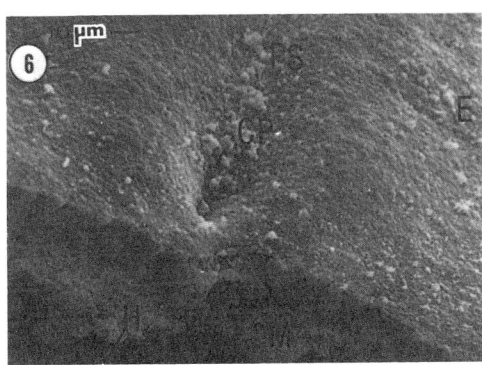

Fig. 10 The stage 4 primitive streak. Note the cellular projections. CP, cellular projections; E, epiblast; Y, yolk.

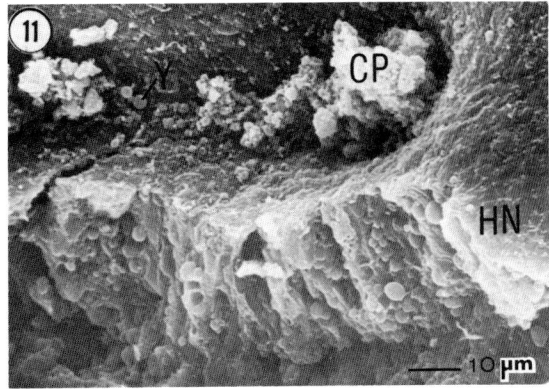

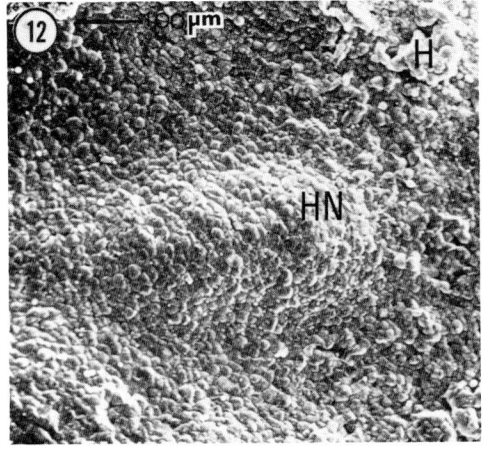

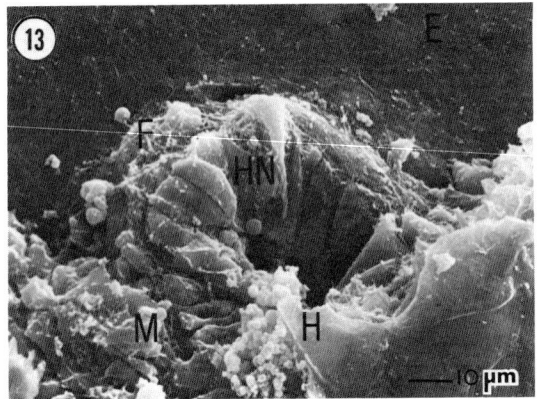

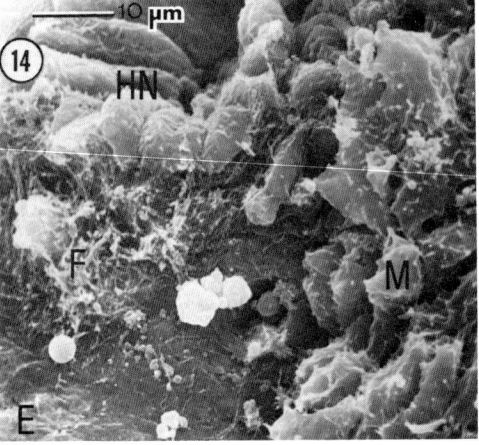

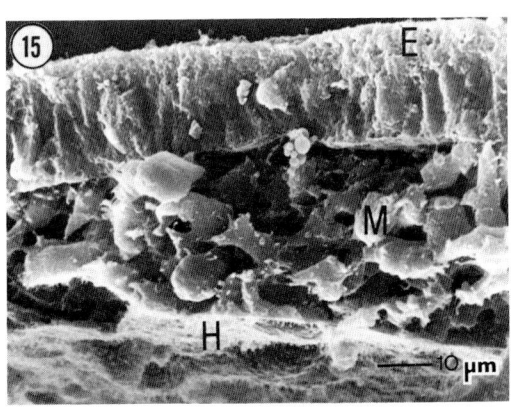

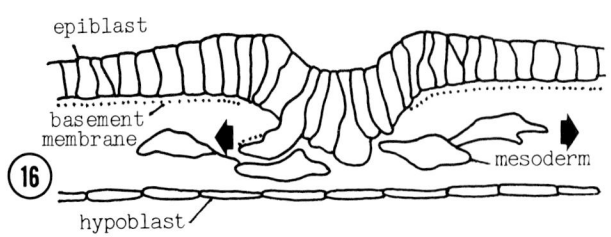

Fig. 11  Hensen's node with cellular projections. CP, cellular projections; HN, Hensen's node; Y, yolk.

Fig. 12  Hensen's node viewed from the ventral surface of the hypoblast. HN, Hensen's node; H, hypoblast.

Fig. 13  Hensen's node viewed from the ventral surface of the epiblast. The hypoblast and mesoderm have been largely dissected away. E, ventral surface of the epiblast; F, fibers; HN, Hensen's node; H, hypoblast; M, mesoderm.

Fig. 14  Hensen's node viewed from the ventral surface of the epiblast. The hypoblast and mesoderm have been dissected away. E, ventral surface of the epiblast; F, fibers; HN, Hensen's node; M, mesoderm.

Fig. 16  Cells invaginating through the primitive streak. Arrows indicate direction of mesoderm cell migration.

Fig. 15  The stage 4 embryo sectioned transversely. Note the columnar epithelium of the epiblast. E, epiblast; H, hypoblast; M, mesoderm.

As is indicated on the fate map (Fig. 2), some of the first cells to invaginate (stages 2-4) penetrate between existing hypoblast cells to contribute to the definitive endoblast while later cells (stage 4) remain in the space between the epiblast and hypoblast to form mesoderm. The separation of these two cell types (i.e. prospective mesoderm and endoblast) is not understood, nor are they identifiable by SEM or other morphological criteria. As cells leave the streak they assume a stellate shape when viewed by SEM and may contact up to eight other cells in the space between the epiblast and hypoblast (Fig. 19). Three regions are identifiable by SEM from the ventral surface of the mesoderm. In the primitive streak the processes are vertically oriented narrow flaps, while immediately adjacent to the streak there are few flap-like processes but numerous filopodia are present. The streak cells are oriented at right angles to the hypoblast, while those adjacent to the streak are oriented parallel to the hypoblast. Especially long filopodia are present in an intermediate zone between the streak cells and cells adjacent to the streak.

Several important changes occur at the surfaces of the cells as they move through the primitive streak. There is an increase in surface blebbing in presumptive streak cells, and they lose their epithelial morphology as they invaginate. There is a decrease in their ability to bind Ca++ and Mg++. In addition, there is a change in the synthesis of extracellular materials in particular hyaluronate and the glycoproteins. Hyaluronate is present in the tissue spaces between the epiblast and hypoblast. It is known to stimulate cell migration and because of its characteristic ability to take up water and swell, it may also open up the tissue spaces to allow mesoderm migration. All of these changes have been suggested as important factors in invagination and subsequent mesoderm migration. Also, mesoderm cells treated with neuraminidase which is reported to remove sialic acid lose their characteristic stellate appearance and reassume the morphological appearance of the streak cells. This would suggest the mesoderm cell shape is partially determined by the surface coat carbohydrates present on the mesoderm cells.

## Mesoderm Migration

Mesoderm cells apparently use the epiblast basal lamina as a contact guidance system in their migration away from the primitive streak. Lying on the basement membrane are long fibrils which form specific patterns in the area pellucida and area opaca at stages 4 and 5 during gastrulation (Figs. 20 and 21). These fibrils have a high concentration of fibronectin, sulphated glycosaminoglycans and collagen type I associated with them (Wakely and England, 1979). The mesoderm cells are in contact with these fibrils (Fig. 22). It has been suggested that a mesoderm cell projects a filopodium which contacts a fibril and pulls the cell towards this contact by contacting the filopodium. The fibrils in the area pellucida form a grid pattern which guides the mesoderm toward the periphery of the area pellucida. A band of parallel fibers (Fig. 23) is present at the anterior border of the area pellucida which guides the primordial germ cells toward the posterior of the embryo. This band also acts to guide the mesoderm cells in their migration. The high concentration of fibronectin (which is known to promote adhesiveness) and the concentration of glycosaminoglycans associated with these fibrils could influence mesoderm migration. The epiblast cells moving to invaginate through the primitive streak between stages 4 and 5 carry their basement membrane and associated fibrils with them. Some of the fibrils of the invaginating cells which will form hypoblast or mesoderm are left at the site of the node. As the normal epiblast morphogenetic movements occur, the grid pattern of fibrils anterior to Hensen's node forms a fan-shaped pattern in early stage 5 which eventually comes together as a band anterior to the node. This band of fibrils could guide the prenotochordal cells into position as a midline structure.

With the appearance of the notochord at the beginning of stage 5 the primitive streak regresses and with further development becomes shorter and shorter. Eventually only a remnant is preserved at the posterior region of the embryo as part of the tailbud (Schoenwolf, 1983).

## Summary

Gastrulation is the process whereby the third germ layer (mesoderm) and the definitive endoblast layer form at stages 2 and 3 in the avian embryo. As the primitive streak cells from the epiblast invaginate and contribute to the mesoderm or definitive endoblast layers.

The invaginating cells change morphologically

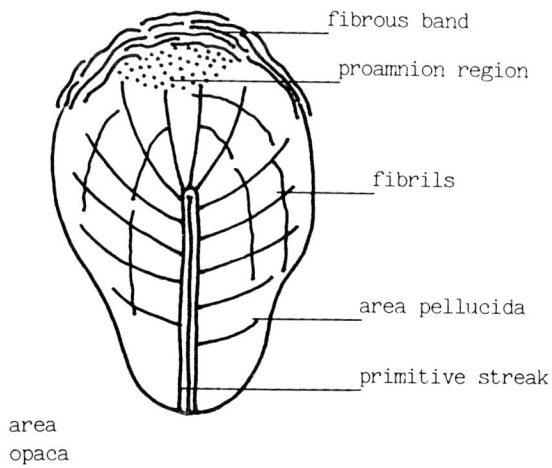

Fig. 20 Distribution of fibrils on the stage 4 ventral epiblast basement membrane.

See next page for figures 17,18,19,21,22 and 23.

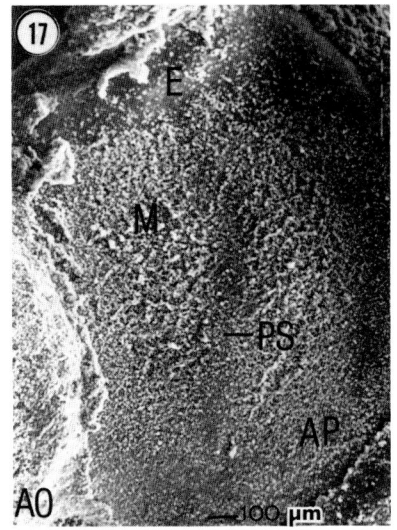

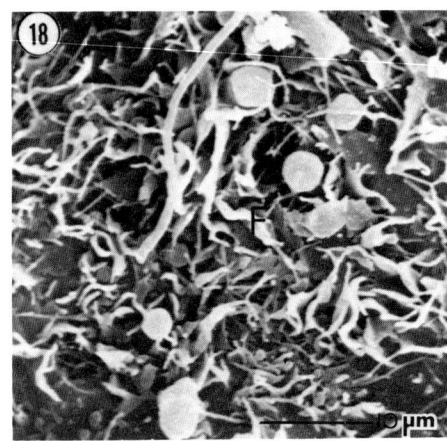

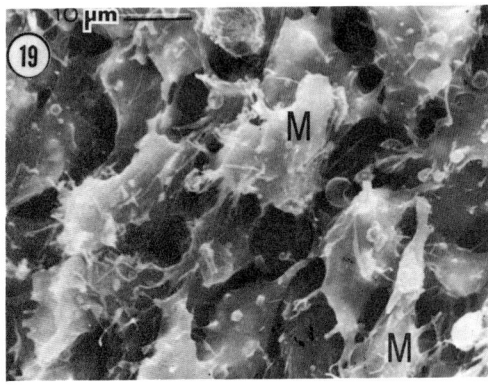

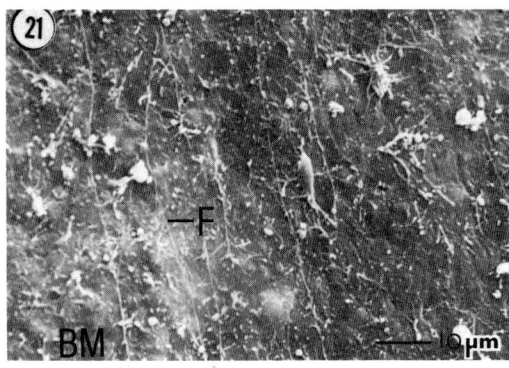

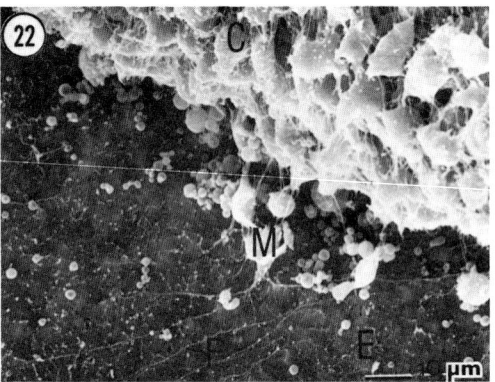

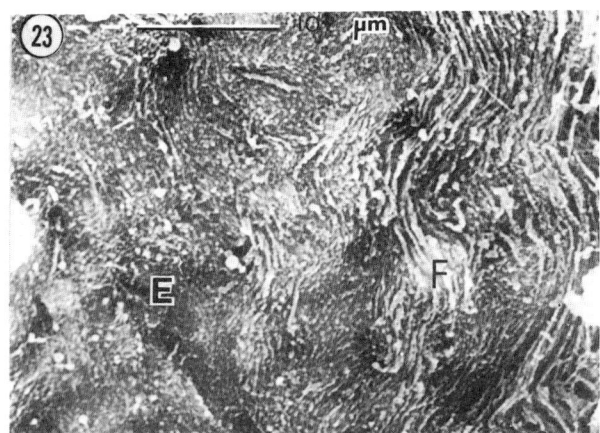

Fig. 17  Mesoderm distribution of the stage 4 embryo. The hypoblast has been dissected away. AO, area opaca; AP, area pellucida; E, epiblast; M, mesoderm; PS, primitive streak.

Fig. 18  Filopodia on the leading, ventral edges of the invaginating primitive streak cells. The hypoblast and mesoderm cells have been dissected away. F, filopodium.

Fig. 19  Mesoderm cells (M) near the primitive streak. The hypoblast has been dissected away.

Fig. 21  Ventral epiblast basement membrane. BM, basement membrane; F, fibrils.

Fig. 22  Mesoderm cells contacting fibrils on the ventral epiblast basement membrane. C, cells at Hensen's node; E, ventral epiblast; F, fibrils; M, mesoderm.

Fig. 23  The fibrous band at the border of the anterior area pellucida. E, ventral epiblast; F, fibrils.

in the streak but little is known as to the factors effecting these changes. Mesoderm cells migrating from the streak region utilize a fibrous network on the epiblast basal lamina as a contact guidance system.

## Acknowledgements

The author is grateful to two colleagues at the University of Leicester who kindly allowed their photographs to be used; Mr. D. Summerton (Fig. 4) and Dr. Jennifer Wakely (Fig. 23). Mr. G.L.C. McTurk and the Leicester University Scanning Electron Microscopy Unit provided excellent technical assistance. The author is also very grateful for the excellent technical assistance of Miss Shemene Uppal.

This work has been supported by the Medical Research Council.

## Literature Cited

Hamburger, V. and Hamilton, H.L. (1951) A series of normal stages in the development of the chick embryo. Journal of Morphology 88, 49-92.

Rosenquist, G. (1966) A radioautographic study of labelled grafts in the chick blastoderm. Development from primitive streak stages to stage 12. Carnegie Institution of Washington Publication 625, Contributions to Embryology 38, 111-121.

Schoenwolf, G. (1983) The Chick Epiblast: A Model for Examining Epithelial Morphogenesis. Scanning Electron Microsc. 1983;III:1371-1385.

Wakely, J. and England, M.A. (1979) Scanning electron microscopical and histochemical study of the structure and function of basement membranes in the early chick embryo. Proceedings of the Royal Society B206, 329-352.

## Suggestions for Further Reading

Bellairs, R. (1982). Gastrulation processes in the chick embryo, in: Cell Behavior (eds. R. Bellairs, A. Curtis, and G. Dunn), Cambridge University Press, Cambridge, pp. 395-427.

England, M.A. (1982). Interactions at basement membranes, in: Functional Integration of Cells in Animal Tissues: British Society for Cell Biology Symposium 5 (eds. J.D. Pitts and M.E. Finbow), Cambridge University Press, Cambridge, pp.209-228.

Hamburger, V. Hamilton, H.L. (1951). A series of normal stages in the development of the chicken embryo. Journal of Morphology 88, 49-92.

Rosenquist, G. (1966). A radioautographic study of labelled grafts in the chick blastoderm. Development from primitive streak stages to stage 12. Carnegie Institution of Washington Publication 625, Contributions to Embryology 38, 111-121.

Solursh, M. (1976). Glycosaminoglycan synthesis in the chick gastrula. Developmental Biology 50, 525-530.

Vanroelen, Ch., Vakaet, L., Andries, L. (1980). Alcian blue staining during the formation of mesoblast in the primitive streak stage chick blastoderm. Anatomy and Embryology 160, 361-367.

Wakely, J., England, M.A. (1979). Scanning electron microscopical and histochemical study of the structure and function of basement membranes in the early chick embryo. Proceedings of the Royal Society B 206, 329-352.

## Discussion with Reviewers

R. Waterman: What evidence is there that the fibrillar extracellular matrix, including the epiblast basement membrane, guides the migration of mesoderm cells?
Author: Several in vivo and in vitro studies have clearly demonstrated the migration of mesoderm cells on the fibrillar extracellular matrix and epiblast basement membrane. SEM studies have been correlated with time lapse video recordings of normal migration patterns. Also, if embryos are treated with cis-hydroxyproline which specifically inhibits the extracellular deposition of procollagen the basement membrane and its associated fibers are not evident and the mesoderm cells invaginating at the primitive streak remain clumped in this region.

R. Waterman: Does gastrulation occur identically in all birds, or are there significant differences?
Author: Gastrulation is very similar in all the birds I have studied including the chick embryo, quail, pheasant and duck. There are minor differences in the thickening of the extracellular fibers and their distribution but generally speaking the systems are similar.

Reviewer II: The following references are relevant to the scope of this study:

Mayer BW, Jr., Packard DS, Jr. (1978). A Study of the expansion of the chick area vasculosa. Develop. Biol. 63, 335-351.

Revel, JP. (1974). Scanning electron microscope studies of cell surface morphology and labeling, in situ and in vitro. Scanning Electron Microsc. 1974:541-548.
Author: Thank you.

Additional discussion of the paper "Ultrastructure of Primary Mesenchyme in Chick and Rat Embryos" by J.P. Revel and M. Solursh continued from page 80.

Reviewer 4: What differences do you observe in cellular movements in vivo and in vitro?
Author: Cells that appear to be moving observed in situ have fewer lamellipodia and more filopodia or microvilli. This does not mean that lamellipodia are not found in various places in the embryo. We have confirmed Pexieder observations on the endocardium (Bull. Assoc. Anat. 61, 63, 1976), but so far as I know there is no evidence that the cells in the endocardium are moving with respect to each other. Lamellipodia-like structures are involved in pinocytosis and other processes, as well as cell movement.

Reviewer 5: Did fixative temperature make any noticeable difference to cell morphology?
Author: We did not make a systematic study of the effect of temperature on the morphology in the case of the chick embryo, since the in vivo morphology of the cell is difficult to establish. However, our experiments with culture cells, including observations on primary chick fibroblast cultures, suggest changes upon cooling.

Reviewer 5: When one sees a short stalk with microvilli extending from it, can one interpret that as a regressing lamellipodium or could this also represent the formation of a lamellipodium by extension of the web of cytoplasm between microvilli, as described for spreading cells in vitro by Rajamaran et al. (Exp. Cell Res. 88, 1974, 327)?
Author: Time lapse cinematography does not support the idea that a lamellipodium forms as a web between a neighboring microvilli. While spreading cells were examined by Rajamaran et al., the state of each individual cell was not examined in vivo. Some of the cells might well have been retracting in some areas while spreading in others.

Reviewer 5: How do the authors distinguish a cytoplasmic flap from a lamellipodium?
Author: Lamellipodia were originally described at the light microscope level and are classically thought to be very large structures extending from the periphery of cells in culture. The structures seen on the neck of the flask itself are relatively small and may not be the same functionally as a true lamellipodium. We therefore decided to use a descriptive word without functional connotations until these structures were better understood.

Reviewer 6: Is there any evidence that the presumptive "retraction fibers" exhibited by mesenchymal cells may be artifactually formed during fixation or specimen preparation?
Author: There is no direct evidence bearing on this subject. It is known from examination of cultures that retraction fibers can form both under normal circumstances and also when the cells are allowed to cool or are treated with trypsin.

Reviewer 6: Although both chick and rat embryos were studied the species of cells illustrated in the figures is not identified.
Author: All of the micrographs are of stage 8 chick cells.

Reviewer 6: The morphology of flask cells observed in the primitive streak of chick embryos is strikingly similar to that of flask cells in a dorsal blastoporal groove of stage 10 + Xenopus embryos (Keller and Schoenwolf, Wilhelm Roux's Arch. 182, 165, 1977) except for the fact that cellular protrusions are usually lacking at the internal ends of these cells in Xenopus. Do you propose a role in exploring the environment for Xenopus flask cells as you do for the chick, or are flask cells likely to have different functions in Xenopus?
Author: The illustrations in Keller and Schoenwolf were taken at relatively low magnification and it is not obvious that they demonstrate the absence of such protrusions; wherever flask cells are found I would expect that their general function would be similar, but there could be flask cells with, as well as flask cells without, cell protrusions.

ULTRASTRUCTURE OF PRIMARY MESENCHYME IN CHICK AND RAT EMBRYOS

J. P. Revel and M. Solursh*

Division of Biology
California Institute of Technology
Pasadena, CA 91125

*Department of Zoology
University of Iowa
Iowa City, IA 52242

## Abstract

Comparisons between the morphology of cells in situ in embryos and cells in culture reveals that there are many similarities between the two. Assuming that one can indeed extrapolate from the in vitro situation some conclusions can be drawn from the shape of the cells observed in situ. But it becomes quite clear that morphology alone will not be sufficient to give any answers. Only by combining scanning electron microscopy and other observational techniques with immunocytochemical, biochemical and histochemical analysis, as well as other experimental approaches, will reasonable answers as to the mechanism of cell movements in embryos become available. We discuss the morphology of cells in migrating mesenchymes and discuss how these could be understood in terms of cellular movement or other behavior.

KEY WORDS: Embryology, Cell Shape, Cell Movement, Flask Cells, Mesenchyme

## Introduction

The scanning electron microscope can provide dramatic three-dimensional views of biological specimens at high resolution and promises to be particularly useful in the study of cellular patterns in embryos (see refs. 1-7). However, it has a major disadvantage over light microscopic approaches[8] in that, at the present, only fixed samples can be used. The morphology of individual cells and their processes is readily studied, but one can only surmise the dynamic nature of the events which were taking place at the moment of fixation. Research protocols in which cells are examined both in the light microscope, by time lapse cinematography, and by scanning electron microscopy, have been useful[4,9,10] in circumventing this problem. Such correlations are relatively easily made in cell cultures, but, except for a few fortunate cases, are impossible to carry out in whole embryos because the thickness or opacity of the material precludes direct observation. This paper addresses itself to the question of what information can be obtained about cell movement in embryogenesis using the scanning electron microscope.

## Morphology of Moving Cells

Cell Movements in Culture. Perhaps the most clearly recognized cell behavior which is associated with cell movement is the extension of lamellipodia,[8,11] thin protoplasmic shelves extending from the cell's periphery and which are believed to serve a role in providing the moving cell with new anchoring points.[10,12] It is often noted that if a lamellipodium is not successful in making an attachment to its substrate, it is retracted upwards and rearwards.[13] There is thus a cycle of extension followed by uplifting which gives rise to the striking "ruffling" activity seen in time lapse movies. Ruffling can only be said to take place if extension of a lamellipodium is followed by uplifting to give the characteristic appearance of "bending back and forth like ruffles on a dress in a slight breeze."[14] "Ruffling" is the rule in vitro but is often not observed in moving cells in situ.[6,15,16] A possible conclusion is that natural substrates are particularly "good," allowing cells to form a strong attachment very readily.[6] It must be noted that the cycle of extension and uplifting of lamellipodia is easily defined, even after fixation, only if cells are moving on a planar surface, as is the rule in cell culture. In the embryo, however, one usually deals with a three-dimensional environment so that "uplifting" is much more difficult to define, especially if live specimens are not transparent enough for direct examination. Other structures which are commonly associated with cellular translocation in culture are lobopodia, or "blebs," and filopodia, very long and thin processes of the diameter of a microvillus which project from the periphery of the cell and come to rest on the substrate.[17,18] While filopodia can be recognized as "active" in cell locomotion in living cultures, they are difficult to distinguish from retraction fibers when fixed specimens are examined. Retraction fibers are long thin cytoplasmic threads which link the main body of a cell (that has moved away or retracted) to small regions of its membrane which stay attached to the substrate at their original position.[10] Presumably, the retraction fibers passively elongate until they snap under tension as they become overly stretched. The interpretation of static images in dynamic terms is obviously fraught with problems and must be approached with a critical eye. Differences between the in vitro and in vivo situations have been noted (see refs. 3, 4, 8, 16), and caution would seem necessary as one tries to analyze cell behavior in embryogenesis by study of fixed samples. Yet, basing oneself on those instances where it has been possible to do correlative studies in vivo,[8,9,15,21,22] one can conclude that, generally speaking, there are more similarities than differences when the cellular behavior associated with cell movement in the in situ and the in vitro situations are compared.

Cell Movements in the Early Embryo. An early step in embryogenesis is the establishment of the three germ layers.[19] First the epiblast forms and then the hypoblast, partially by delamination from the epiblast. The mesoblast, or primary mesenchyme, then becomes wedged between the two. Epithelial cells of the epiblast invaginate at the midline along the primitive streak (Fig. 1), then migrate laterally between the first two germ layers, apparently using either as substrate (Fig. 2). One is dealing not only with the translocation and movement of cells but also with a change in cellular organization. Cells of the epiblast arranged as an epithelium, with close packed columnar cells linked to each other by tight and other junctions,[20] become mesenchymal with rounded or flattened stellate cells separated by large amounts of intercellular space, and quite possibly endowed with a different junctional complement.[20] Our studies have involved observation of the morphology of cells at the level of the primitive streak both in rat and in chick embryos. Embryos obtained from Sprague-Dawley rats on the 9th day of pregnancy, were dissected from the maternal tissues and fixed for 1 hour in 2.5% glutaraldehyde in 0.1 M cacodylate buffer. After rinsing they were dehydrated in a graded series of alcohol. They were sliced through the primitive streak while in 70% alcohol, and dried from liquid $CO_2$ at the critical point from absolute alcohol. The chick embryos used were of Hamburger-Hamilton stages 4-8. They were dissected from the yolk and rinsed in warm (38°C) Earle's Balanced Salt Solution before fixing in warm 2.5% glutaraldehyde in Earle's BSS. The embryos were post-fixed in 1% osmium tetroxide, dehydrated in a graded series of alcohol and dried at the critical point from liquid $CO_2$. The chick embryos were fractured either in alcohol or after critical point drying. The embryos were mounted on stubs with double-sided tape or silver paste and coated with gold or gold-palladium in a sputter coater before examination in an ETEC Autoscan or a Cambridge Stereoscan operated at 20 kV.

## Morphology of Primary Mesenchyme Cells

The shape of cells in the primary mesenchyme varies with the different species as well as with their location in the embryo. The mesenchyme cells are somewhat more rounded in the streak region of rat embryos than they are more laterally. Lateral and anterior to the streak, the cells are flattened somewhat, and in other regions the cells elongate, with many having their long axis along the dorsal ventral plane. Particularly striking both in the chick and the rat is that cells in the regions lateral to the primitive streak do not all seem to migrate in the same direction. Observations made on moving cells in culture, and in those instances where cells can be visualized in the live embryos, lead one to expect that the leading, forward-moving edge of a cell can be identified by the presence of a lamellipodium, or of microvilli or filopodia.[15,16,21,22] Using the same criteria to define the leading edge of cells in the mesenchyme brings one to conclude that primary mesenchyme cells do not, in fact, all have their axis oriented in the expected direction of migration (Fig. 2, Fig. 3). Cells appear to point both towards the periphery and the midline as well as towards the epiblast or hypoblast. This lack of consistent cell orientation may well be a reflection of the fact that one is dealing not with the movement of individual cells all migrating, arrow-like, towards the periphery, but rather with an expanding sheet of cells moving as a unit. As was clearly shown by Trinkaus and his associates (see refs. 21, 23) during epiboly in Fundulus and also in culture systems, when a sheet of cells advances it may be that only the peripheral individual cells exhibit locomotor activity while the cells farther back in the sheet may follow "passively." Thus, while the net result of cell movement may be a displacement of cells from the midline towards the periphery of the embryo, this could be achieved not by a precisely orchestrated, soldier-on-parade-like movement of individuals but rather by a series of displacements and cell divisions, the net effect of which would be the translocation of cells laterally simply because the margin of the mesenchyme keeps on expanding. In fact, observation made by time lapse cinematography of "wounds" in confluent cultures (personal unpublished observation) shows individual cells invading the bare area in what appears a random way, settling down to divide after moving only a few cell diameters before the daughter cells separate to repeat the process.

Besides using their lamellipodia, the cells also appear to make contact with each other by fine processes, which are the diameter of microvilli and of variable length and therefore reminiscent of filopodia. As we have indicated earlier in this paper, such structures could be difficult to differentiate from retraction fibrils, passively anchoring cells rather than active participating in cell movements. Thus, the long thin processes which radiate from the clusters of rounded cells often seen in the primary mesenchyme are, we believe, more easily interpreted as retraction fibrils than as filopodia (Fig. 4), because they often radiate in all directions from rounded cells. The clusters of microvilli seen on some of the cells suggest late stages in the life cycle of a lamellipodium (Fig. 5).[24] Rather than being organs of locomotion per se, they could indicate that in vivo as well as in vitro, a short lived lamellipodium had been extended and has now been withdrawn, leaving a long microvillus behind: in a sense these could also be considered as a type of retraction fibril. The contacts mesenchyme cells make with the base of the

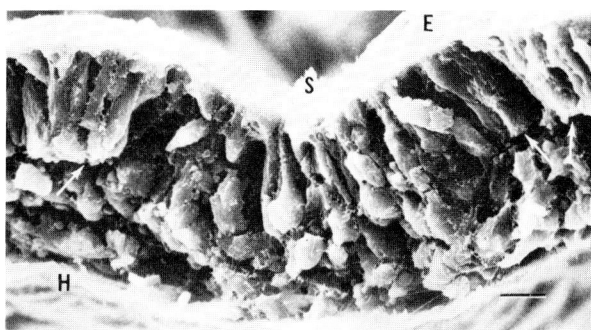

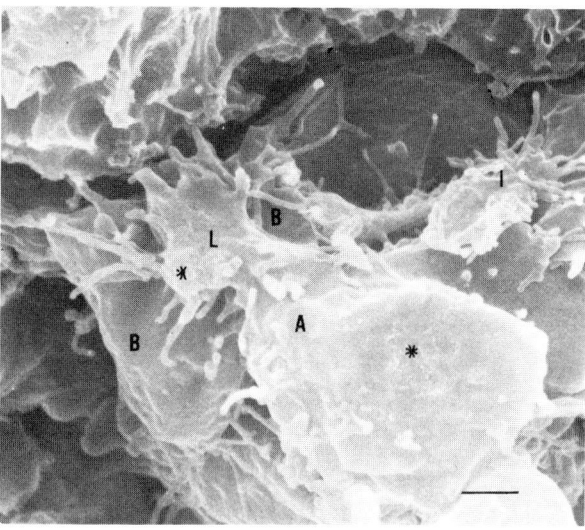

Fig. 1. A stage 8 chick embryo fractured across the primitive streak shows invaginating cells at the midline (S). These cells then migrate laterally between the columnar epiblast cells (E) and the very thin hypoblast (H). The boundary between epiblast and mesenchyme is necessarily ill defined near the streak but becomes sharp more laterally, presumably where a basement membrane exists (→). The diagram below indicates the approximate level at which this particular embryo was fractured. bar: 10 μm

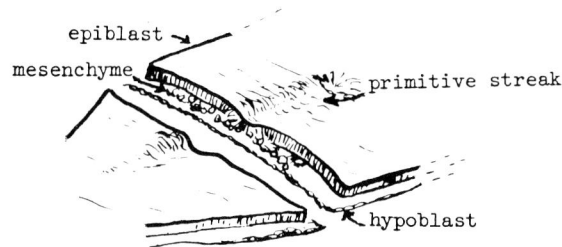

Fig. 3. A view of mesenchyme cells obtained by removing the epiblast. Presumably the portion of the cell exposed here had been in contact with the underside of epiblast or its associated basement lamina. Cell A has a well developed lamellipodium (L). The scars on its surface may show where it had been attached to the epiblast, and damaged during the dissection (*). Cell B is partially obscured, but its lamellipodia (1) seem to extend in a direction some 90° away from that of cell A, suggesting movement in a totally different direction. bar: 1.0 μm

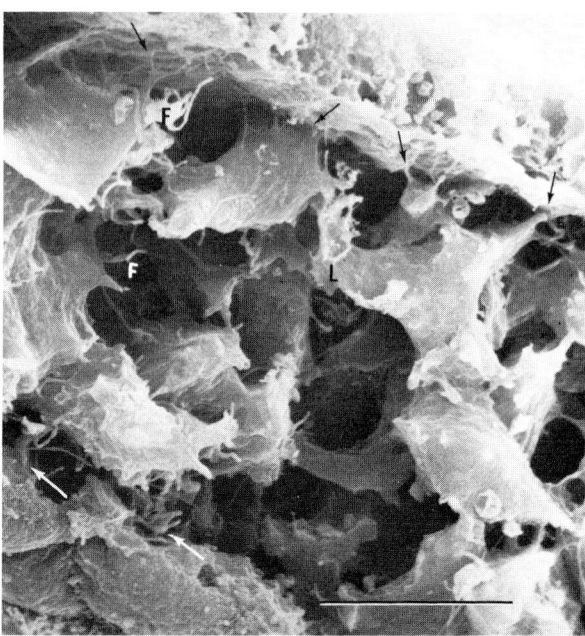

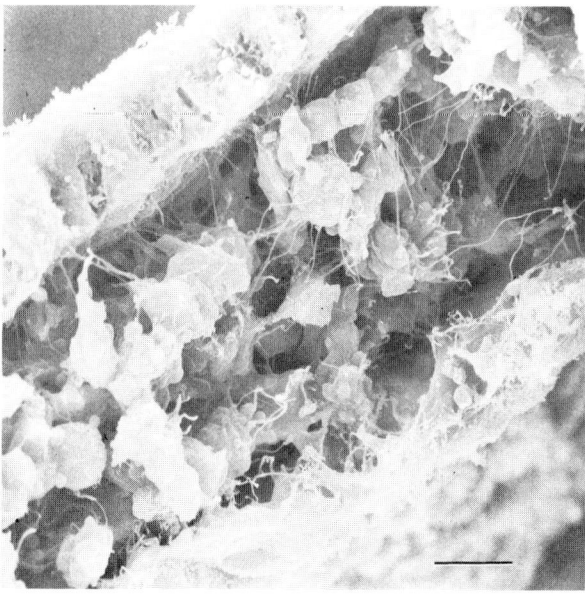

Fig. 2. The migrating mesenchyme cells make contact both with epiblast and hypoblast (upper and lower set of arrows, respectively). The roughly polygonal mesenchyme cells carry structures reminiscent of lamellipodia (L) or filopodia (F). These processes appear oriented at random over the cell surface. bar: 10 μm

Fig. 4. In this region of the mesenchyme the cells are linked to each other and to the undersides of the epi- and hypoblast by long thin processes. While some of these might be filopodia actively extended by the cell and part of the locomotory machinery, most are probably retraction fibers formed when a cell pulls away from its neighbors or substrata without losing its anchorage. bar: 10 μm

epiblast or with the hypoblast are very similar to the contacts that they make with each other and which we have just described. There are no recognizable structures which could act as obvious guides for cell movements (Fig. 2, Fig. 5).[3,6] In summary, one is struck by the fact that throughout the mesenchyme the cells have quite variable orientation with respect to the embryonic axes. If all the cells were straining to move laterally by the most direct route, one would expect that this would be reflected in the orientation and shape of the processes as indeed is the case in cell cultures. This observation is again compatible with the idea that the movements of individual cells in the primary mesenchyme take place with respect to each other and that the expansion of the mesoblast reflects movements of the sheet as a whole rather than of individual components.

### Cells at the Primitive Streak

An interesting cell found at the level of the primitive streak (and in other regions where cells invaginate) is the "flask cell" which characterizes gastrulation movements in a variety of organisms. Flask cells are found near the midline of the embryo both in chick and rat material. Their appearance contrasts dramatically with that of the epithelial cells from which they derive. The epithelial cells are somewhat prismatic, columnar in shape, contacting each other by microvilli or filopodia-like structures. The epithelial cells proper are sometimes a little wider near the apex than they are near their base so that wide intercellular spaces are formed near the bottom of the epithelium. The flask cells, on the contrary, are narrow at their apex and bulge very widely towards their base (Fig. 6a,b). Examination of stereopairs suggests that the basal swelling of the flask cells does not extend only in a dorsoventral direction, but also extends in the anteroposterior direction. In the upper region of the epithelium, the flask cells have a thin neck, at the level of which there are many contacts with the neighboring cells. These contacts are characterized by microvilli-like structures of different lengths, as well as some broader structures, all very similar to those found in the neighboring epithelial cells, although perhaps somewhat slightly more numerous than would be found on the columnar cells. It is as if there were the same number of cell contacts, but, because of the small cell diameter at the neck, the density of the contacts has increased. The bulging portion of the flask cell is relatively free of microvilli, although one does often find at least one cluster of longish microvilli (filopodia) or a lamellipodium. The spatial organization of these structures and their shape rarely convey the impression that they are part of a mechanism by which the cell actively pulls itself out of the epithelium to penetrate the underlying mesenchyme. Rather, the smooth bulb shape of the flask and the fact that many of the filopodia actually terminate without contacting other structures and are oriented in all different directions suggests that they may play a role in exploring the environment much like the lamellipodia of culture cells. In a way, the flask cells are reminiscent of cells in culture (Fig. 6, diagram). Instead of a leading edge flattened by being restricted to a plane, they have a bulbous leading edge crowned by a lamellipodium or filopodia. The "bulge" would represent the flow of cytoplasm towards the mesenchyme and the neck could be the equivalent of the "tail" or uropod of cells in culture; their morphology closely parallels in three dimensions the appearance of moving cells in a planar culture; it becomes tempting to think of the bottle cell as an individually migrating cell in a three-dimensional "space." The unanswered question at the present time is obviously what causes cells to behave as they do and what guides their movement. The scanning electron microscope will have to be combined with other experimental approaches for the embryologist to be able to obtain an answer to these important questions. Clearly, however, it can be an important tool in a detailed analysis of movement and, while not an end in itself, will become increasingly useful in interpreting the events which lead from a fertilized ovum to a self-sufficient organism.

### Acknowledgements

The original work presented here was supported by grants GM-06965 and RR-07003 from the National Institutes of Health.

### References

1. Waterman, R. E. Use of the scanning electron microscope for observation of vertebrate embryos. Devel. Biol. 27, 1972, 276-281.
2. Armstrong, P. B. and Parenti, D. Scanning electron microscopy of the chick embryo. Devel. Biol. 33, 1973, 457-568.
3. Revel, J. P. Some aspects of cellular interactions in development. In The Cell Surface in Development, A. A. Moscona (ed.), John Wiley and Sons, New York, 1974, 51-65.
4. Bard, J. B. L., Hay, E. D. and Meller, S. M. Formation of the endothelium of the avian cornea: a study of cell movement in vivo. Devel. Biol. 42, 1975, 334-361.
5. Bancroft, M. and Bellairs, R. Differentiation of the neural plate and neural tube in the young chick embryo. A study by scanning and transmission electron microscopy. Anat. Embryol. 147, 1975, 309-335.
6. Ebendal, T. Migratory mesoblast cells in the young chick embryo examined by scanning electron microscopy. Zoon 4, 1976, 101-108.
7. Wakely, J. and England, M. A. Scanning electron microscopy (SEM) of the chick embryo primitive streak. Differentiation 7, 1977, 181-186.
8. Trinkaus, J. P. On the mechanism of metazoan cell movements. In The Cell Surface in Animal Embryogenesis and Development. North Holland Publishing Co., New York, 1976, 225-329.
9. Brunk, V., Bell, P., Colling, P., Forsby, N. and Fredriksson, B. A. SEM of in vitro cultivated cells, osmotic effects during fixation. SEM/1975, Ill. Inst. of Tech., Chicago, Ill., 379-386.
10. Revel, J. P., Hoch, P. and Ho, D. Adhesion of culture cells to their substratum. Exp. Cell Res. 84, 1974, 207-218.
11. Ingram, V. A side view of moving fibroblasts. Nature 222, 1969, 641-644.
12. Lochner, L. and Izzard, C. S. Dynamic aspects of cell-substrate contact in fibroblast motility. J. Cell Biol. 59, 1973, 199a.
13. Heaysman, J. E. M. Comment. Locomotion of tissue cells. Ciba Foundation Symposium 14 (new series), Elsevier, North-Holland, Amsterdam, 1973, p. 248.
14. Lewis, W. H. The relation of viscosity changes of protoplasm to amoeboid locomotion and cell division. In The Structure of Protoplasm. W. Seifriz (ed.), Iowa State College Press, Ames, Iowa, 1942, p. 172.

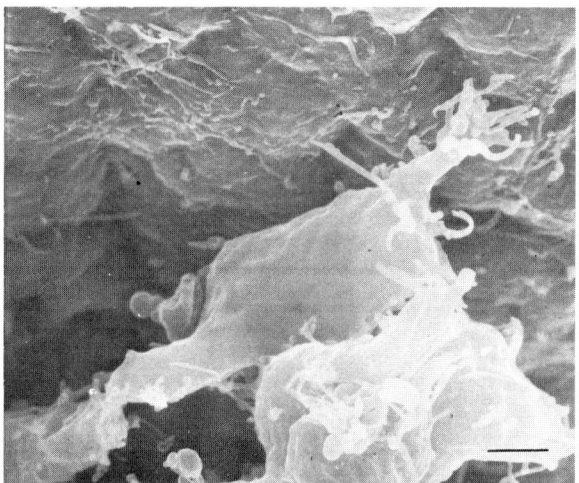

Fig. 5. The underside of the epiblast forms the roof of the space through which the mesenchyme cells are migrating. Two such cells are seen here. One of them extends a small stalk on which numerous microvilli are found. This presumably represents a lamellipodium which had been extended and is now in the process of regressing. The many small cytoplasmic remnants seen plastering the underside of the epiblast presumably represent other cellular processes which have been pulled off during the preparation, or left behind as passing cells moved away. bar: 1.0 μm

Fig. 6(a). A flask cell near the midline in the primitive streak. This cell is characterized by a very long narrow neck and a bulging lower portion, the body of the flask. Numerous microvilli (m) as well as broader cytoplasmic flaps (b) are seen on the neck contacting the neighboring epithelial cells. The body of the cell is relatively free of such specializations, but a number of microvilli are found at the broad end of the cell. Views of the same cell taken at a backward tilt of 55° shows that they are making contacts with neighbors (Fig. 6b). Minute pits presumably the opening of pinocytosis vesicles are found on the surface of the cell. Seen in three-dimensional views, this cell extends forward towards the viewer, i.e., it is elongated in a direction parallel to the long axis of the embryo as well as downward. We argue that the flask cell may represent a three-dimensional counterpart of a single cell moving in culture, as indicated by the diagram.

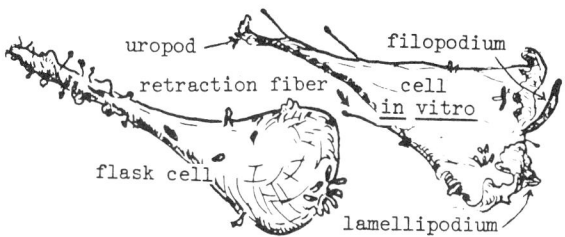

Somehow, even though this cell is in a crowded environment and in contact with its neighbors, it does not seem to obey the general rules of contact inhibition of movement. bar: 1.0 μm

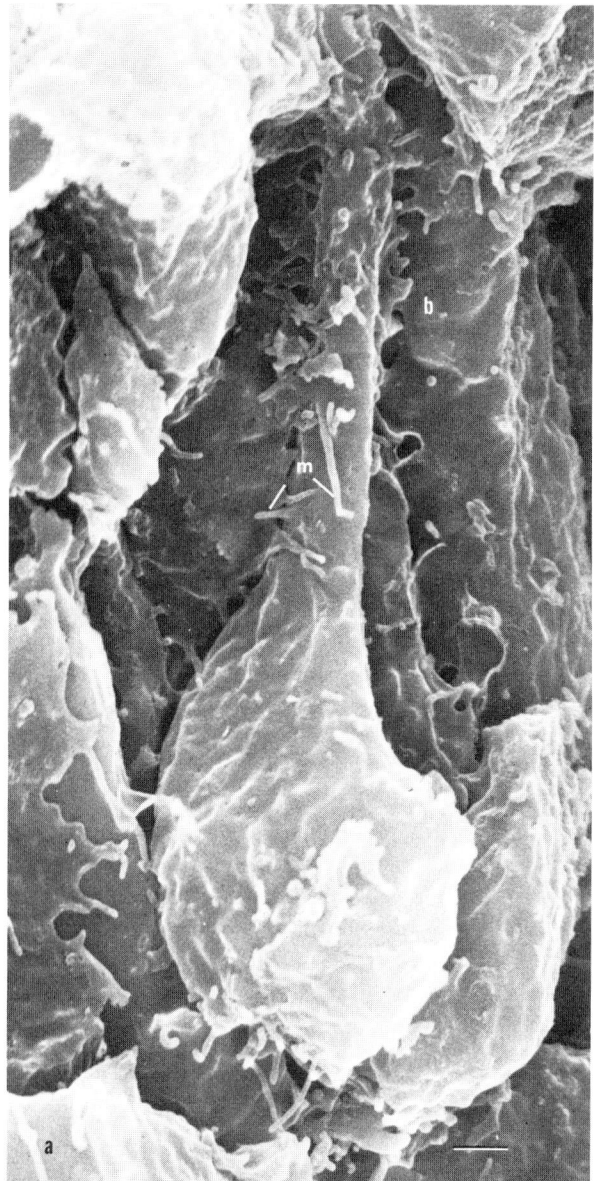

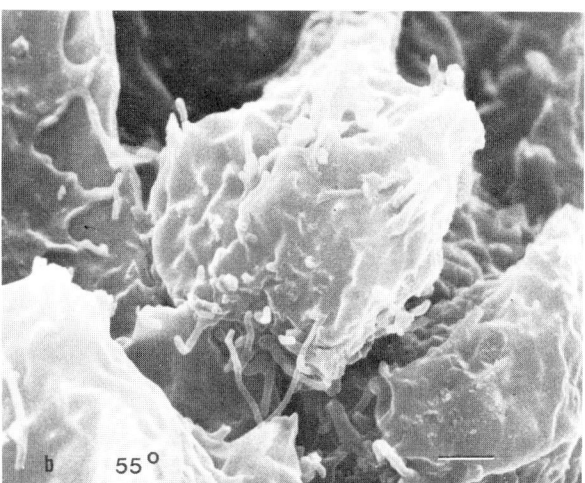

15. Bard, J. B. L. and Hay, E. D. The behavior of fibroblasts from the developing avian cornea morphology and movement in situ and in vitro. J. Cell Biol. 67, 1975, 400-418.

16. Nelson, G. A. and Revel, J. P. Scanning electron microscopic study of cell movements in the corneal endothelium of the avian embryo. Devel. Biol. 42, 1975, 315-333.

17. Taylor, A. C. and Robbins, E. Observation of microextensions from the surface of isolated vertebrate cells. Devel. Biol. 7, 1963, 660-673.

18. Albrecht-Buhler, G. Filopodia of spreading 3T3 cells: do they have a substrate exploring function? J. Cell Biol. 69, 1976, 275-286.

19. Balinsky, B. I. and Walther, H. H. The emigration of presumptive mesoblast from the primitive streak in the chick as studied with the electron microscope. Acta Embryol. Morphol. Exp. 4, 1961, 261-283.

20. Revel, J. P., Yip, P. and Chang, L. L. Cell junctions in the early chick embryo--a freeze etch study. Devel. Biol. 35, 1975, 302-317.

21. Trinkaus, J. P. Surface activity and locomotion of Fundulus deep cells during blastula and gastrula stages. Devel. Biol. 30, 1973, 68-103.

22. Gustafson, T. The role and activities of pseudopodia during morphogenesis of the sea urchin larva. In Primitive Motile Systems in Cell Biology. R. D. Allen and N. Kamiya (eds.), Academic Press, New York, 1964, 333-349.

## DISCUSSION WITH REVIEWERS

**Reviewers 1 and 4:** Did you employ any fixation method which would enhance the preservation of extracellular matrix?
**Author:** No, we were concerned with observing cellular morphology and did not want to take the chance that we might obscure cellular details in attempts at preserving the extracellular material.

**Reviewer 1:** Have you examined gastrulation stages in the sea urchin in which the direction of migration and formation of filopodia are known and have been recorded on film?
**Author:** Yes, we have; filopodia are seen extending from cells in the absence of contact either with the substrate or other cells, and SEM observations closely parallel in vivo events.

**Reviewer 2:** What is the structural nature of the interaction between processes of migrating mesenchymal cells and the adjacent basement lamina?
**Author:** On the basis of our scanning electron micrographs it would appear that microvillus-like structures and filopodia are more commonly found than lamellipodia. Whether this is due to changes induced during fixation or truly represents the in situ situation is not known at the present.

**Reviewer 3:** In view of the proven distortion of embryonic tissues caused by fixation and critical point drying would not freeze drying provide a better correspondence between cells seen by SEM and those seen in vitro?
**Author:** This might be true, but much of the work presented was done before Boyde and his collaborators (SEM/1977, 1, 507) had documented the artifacts caused by dehydration. In any case, I am not certain that it is necessarily wise to trade possible ice crystal artifacts in the center of as large a sample as a whole chick embryo with the better understood pitfalls of shrinkage caused by the standard dehydration procedures.

**Reviewer 3:** Could not the change in cells shape exemplified by the formation of "bottle cells" provide the forces necessary for cellular redistribution?
**Author:** Elongation of cells and constriction in the apical region have been shown to be important factors in the formation of the neural tube, the lens, etc. However, to my knowledge, bottle cells are not found in these situations. Bottle cells are found only in instances where cells are shown to migrate. Without additional evidence it is not possible at the present to decide whether the changes in cell shape seen are causal of morphogenetic movements or merely a consequence of these movements.

**Reviewer 4:** What are the effects of leaving out post-osmication on the morphology of rat mesenchymal cells?
**Author:** The post-osmication steps commonly used stabilize cell membrane structures. At the level of resolution at which our study was carried out lack of post-osmication did not cause any observable changes in morphology.

**Reviewer 4:** How can you distinguish between a filopodium, a retraction fiber and a collagen fiber? These structures can be difficult to distinguish at times.
**Author:** In our paper we tried to indicate that a filopodium is a long and thin process roughly of the diameter of a microvillus, which projects from the periphery of cells and may come to rest on the substrate. Retraction fibers are long thin cytoplasmic threads which are always attached to the substrate at one end. A single retraction fiber can easily be confused for a filopodium attached to the substrate. When cells round up however they remain attached to each other or their substrates by many retraction fibers radiating from their periphery, as can easily be seen in live specimens. When encountering cells linked to each other by many thin taut processes it is therefore tempting to interpret these as retraction fibers rather than filopodia. In very young chick embryos such as we are studying, collagen fibrils are relatively rare, and of a smaller diameter than filopodia or their retraction fibers. Absolute identification of one particular structure may be difficult, and corroborative evidence from other types of observation (thin sections, etc.) has to be used for definitive interpretation.

REVIEWERS:  I   T. P. Fitzharris
II  R. O. Kelley
III A. Tamarin
IV  T. Pexieder
VI  R. E. Waterman
V   J. Wakely

For additional discussion see page 74.

THE CHICK EPIBLAST: A MODEL FOR EXAMINING EPITHELIAL MORPHOGENESIS

G.C. Schoenwolf[*]

Department of Anatomy, University of Utah, School of Medicine,
Salt Lake City, Utah 84132

(Paper received February 12 1983, Complete manuscript received June 10 1983)

## Abstract

The epiblast of early chick embryos is an important model system for examining morphogenesis. Five major morphogenetic processes can be readily examined by scanning electron microscopy of the epiblast: thickening of epithelial sheets by cell elongation, folding of epithelial sheets, fusion of epithelial sheets, cavitation of epithelial cords and dispersal of cell sheets during cell migration. The purpose of this paper is to describe these morphogenetic processes, showing examples of each type. Thickening of epithelial sheets occurs by cell elongation during formation of the neural plate. Folding then ensues to form the neural groove, which is flanked laterally by the neural folds. Fusion of the neural folds closes the neural groove and separates the incipient neural tube from the overlying surface ectoderm. The caudal part of the neural tube develops much differently. Cells derived from the tail bud cluster together as an epithelial medullary cord, the peripheral cells of this cord elongate, and, simultaneously, several small cavities appear at the inner ends of the peripheral cells. All of these cavities eventually coalesce, forming a single lumen. Thus, the neural tube can be formed by markedly different morphogenetic processes depending upon its particular craniocaudal level of origin. Cell migration is exhibited by neural crest cells. In chick embryos, these cells originate from the roof of the closed neural tube. They then migrate laterad to take up residence in a variety of new locations. Scanning electron microscopy has served as an important tool, aiding us greatly in visualizing complex spatial changes that occur during the morphogenesis of epithelia.

KEY WORDS: Body Folds, Cavitation, Cell Elongation, Cell Migration, Epiblast, Epithelium, Fusion, Medullary Cord, Morphogenesis, Neural Crest, Neural Tube

[*]Address for correspondence:
Department of Anatomy
University of Utah, School of Medicine
Salt Lake City, Utah 84132
Phone No. (801) 581-6453

## Introduction

The fertilized egg progresses from a single cell to a multicellular organism during embryonic development. In avian and mammalian embryos, cells formed during cleavage become arranged into two layers, an upper (dorsal or outer) epiblast and a lower (ventral or inner) hypoblast (because the hypoblast is largely replaced by endodermal cells, see below, its development and fate will not be discussed in detail). The epiblast gives rise to the three primary germ (Latin: germen, sprout or bud) layers during gastrulation. These layers, the ectoderm (Greek: ektos, outside; derma, skin), mesoderm (Greek: mesos, middle) and endoderm (Greek: endon, within), are the primitive structural "building blocks" used for the construction of the new organism. The primary germ layers are sculptured during subsequent development, by events called morphogenetic (Greek: morphè, form; genesis, birth or origin) processes, to demarcate the boundaries of the embryonic body, and to form the primitive organ rudiments. Organ rudiments slowly differentiate forming various tissues, organs and organ systems.

The purpose of this paper is to describe some of the early morphogenetic processes that shape the embryo and its rudiments, focussing on the chick epiblast as a model system. Scanning electron micrographs will be used copiously to illustrate morphogenetic processes, aiding visualization of the complex spatial changes that occur over time. Emphasis will be on what happens during morphogenesis, rather than how morphogenesis occurs, because we know considerably more about the former than the latter. I will start with a brief discussion of gastrulation, emphasizing the origin and formation of the germ layers. I will then describe neurulation (i.e., the formation of the neural tube, the rudiment of the entire adult central nervous system). As part of neurulation, I will discuss the following morphogenetic processes: thickening of epithelial sheets by cell elongation, folding or bending of epithelial sheets, fusion of epithelial sheets and cavitation (canalization) of epithelial cords. Finally, I will discuss dispersal of epithelial sheets during cell migration. Spreading of epithelial sheets (epiboly), cell death and differential growth (hypertrophy and hyperplasia) will not be discussed in detail. Spreading of epithelial sheets is discussed elsewhere, as a part of the description of gastrulation in amphibian embryos (Lundmark, Shih, Tibbetts and Keller, 1983), cell death has already been reviewed extensively (Glücksmann, 1951; Saunders, 1966; Saunders and Fallon, 1967) and differential growth is difficult to document by scanning electron microscopy alone. (Studies utilizing morphometric or cell kinetic techniques are needed to document spatial and temporal differences in cell volume or the rate of mitosis.)

A list of suggestions for further reading has been included at the end of the paper. This list consists principally of review articles, articles of historical interest, articles thoroughly illustrated with good-quality scanning electron micrographs or landmark studies. The list is not exhaustive.

## Gastrulation

Gastrulation in chick embryos involves two basic types of cell movement: epiboly and emboly. Epiboly consists of the spreading of an epithelial sheet of cells across the surface of the blastoderm, whereas emboly consists of a surface-to-interior movement of a stream of cells. During gastrulation, epiblast cells spread toward the midline (i.e., undergo epiboly) where they accumulate as a longitudinal thickening called the primitive streak. Epiblast cells then migrate through the streak (i.e., they undergo emboly).

The primitive streak attains its maximal length at stage 4 (all stages are numbered according to the criteria of Hamburger and Hamilton, 1951), occupying the caudal two-thirds of the midline portion of the area pellucida, the central portion of the blastoderm that floats on the sub-blastodermic fluid between the blastoderm and the yolk. (See Fig. B in Watterson, Sweeney and Schoenwolf, 1979, for a drawing of the early chick blastoderm and its subdivisions.) The definitive primitive streak consists of the primitive knot (Hensen's node) at its cranial end, followed caudally by the primitive pit and groove (Fig. 1). The primitive ridges flank the primitive groove laterally. Migration of cells through the streak causes the primitive pit, groove and ridges to form.

Movement of cells through the streak involves an epithelial-mesenchymal transformation. Cells of the epiblast are organized as an epithelium (Fig. 2). An epithelium (especially a columnar epithelium like the epiblast) is a tissue consisting of cells closely apposed over large areas of their surfaces. Typically, epithelia are organized as coherent, sheet-like layers, and little intercellular space separates adjacent epithelial cells. In contrast, a mesenchyme consists of a mass of loosely packed cells. Generally, mesenchymal cells are widely separated by intercellular spaces and materials (see especially right side of Fig. 2). As cells move through the streak they change their positions, morphologies and contacts with neighboring cells. Two general shapes of migrating cells are common: spherical and flask-like (Fig. 2).

The first epiblast cells to move through the streak are prospective endodermal cells. These cells enter the hypoblast to form a new lower layer called the endoderm (as prospective endodermal cells enter the hypoblast they displace hypoblast cells cranially--the ultimate fate of the displaced hypoblast is irrelevant to the present discussion). By stage 4 (the definitive-primitive-streak stage), the formation of the endoderm has been completed, and prospective mesodermal cells are migrating through the primitive streak.

## Neurulation

### Thickening of Epithelial Sheets

The area of the epiblast cranial to the primitive streak is the ectoderm. This area does not exhibit overt regionalism in surface view at stage 4 (Fig. 1). However, by stages 5-6, two structurally different areas are recognizable: the peripheral surface ectoderm and the more central neural ectoderm or neural plate (Figs. 3, 4). Mesoderm has migrated through the primitive streak and completely underlies the neural plate by stage 5. A tongue of mesodermal cells, the so-called head process (i.e., the cranial part of the notochord), lies in the midline beneath the neural plate (Figs. 3, 4). Both the head process and surrounding mesoderm seem to affect the shape of the neural plate as viewed from its dorsal surface: firm attachments between the head process and overlying, midline portion of the neural plate are indicated by a shallow longitudinal groove; and accumulations of loosely packed (i.e., mesenchymal) mesodermal cells (and associated extracellular materials) beneath the lateral and cranial parts of the neural plate are indicated by the dorsal bulging of the neural plate in these areas. A crescentic depression lies just cranial to the neural plate. This area, the so-called proamnion, lacks mesoderm and, hence, consists of only ectoderm and endoderm.

Formation of the neural plate occurs by thickening of the ectoderm. This thickening is due to cell elongation, rather than to cell accumulation with formation of a multilayered area, and is caused or induced by cells that migrate through the cranial end of the primitive streak. (Actually, induction is believed to begin even before cells located in the cranial end of the primitive streak begin their migration.) The neural plate is a pseudostratified (Greek: pseudo, false; Latin: stratum, layer) epithelium. Such an epithelium consists of a single layer of cells that stretch across the entire width of the epithelium (Fig. 5). Due to the varied position of cell nuclei within the epithelium (the position of each nucleus is indicated in scanning electron micrographs by the position of the widest area of each cell), histological sections of a pseudostratified epithelium give the impression that the epithelium is stratified.

Cells of the neural plate have three characteristic shapes (Fig. 5): spherical, spindle-like and flask-like. Spindle-shaped cells have slender necks that extend both apically and basally. The necks of flask-shaped cells extend

## Morphogenetic Processes in the Epiblast

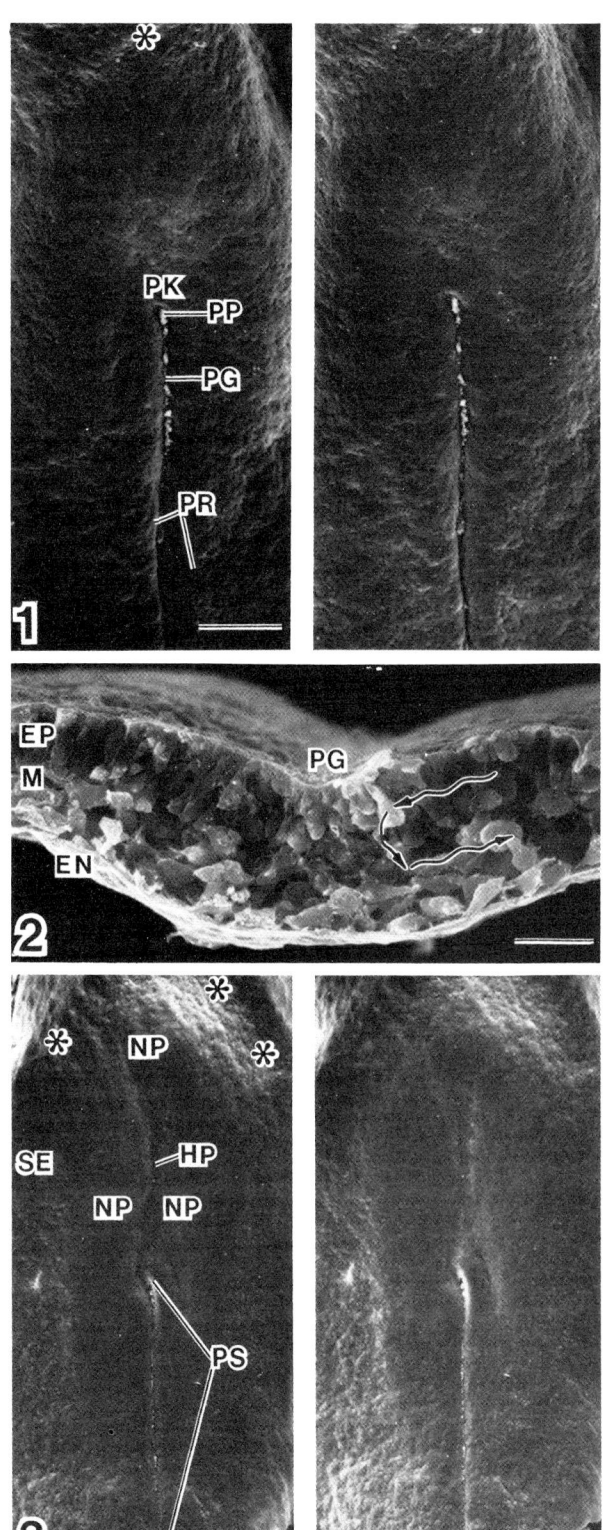

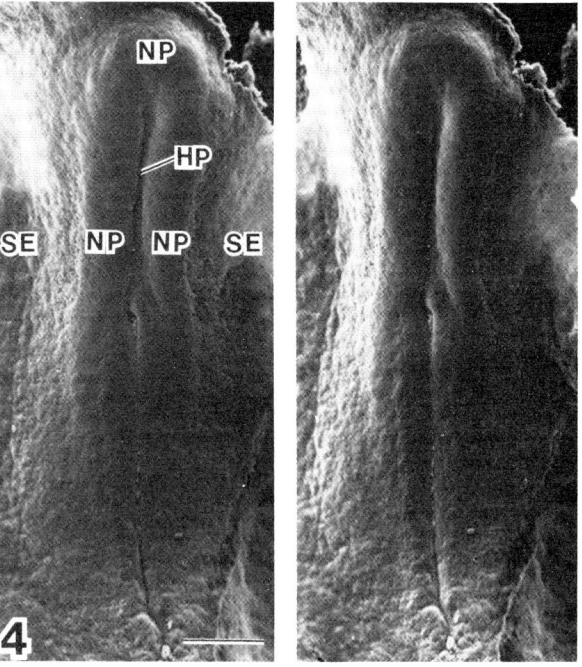

Fig. 1. Stereopair of the dorsal surface of the chick blastoderm at stage 4 (definitive-primitive-streak stage). Prospective mesodermal cells are migrating through the primitive streak. Debris typically occupies the primitive pit (PP) and cranial half of the primitive groove (PG). PK, primitive knot; PR, primitive ridges; asterisk, cranial end of area pellucida. Bar = 200 μm.

Fig. 2. Transverse slice through the primitive streak of a chick embryo at stage 5 (head-process stage). Cells are migrating from the epiblast (EP) through the primitive streak to the mesodermal (M) layer. EN, endoderm; PG, primitive groove; arrows, directions of migration of cells through the primitive streak. Bar = 25 μm.

Fig. 3. Stereopair of the dorsal surface of the chick blastoderm at stage 5. Mesodermal cells have migrated through the cranial end of the primitive streak (PS) forming the head process (HP). The neural plate (NP) has just formed and its boundaries are becoming demarcated. The neural plate overlies the head process in the midline. The close association between these two rudiments results in the formation of a shallow midline groove, bounded by the ectoderm of the neural plate. Mesoderm is lacking cranial to the neural plate in the area of the so-called proamnion (asterisks). SE, surface ectoderm. Bar = 250 μm.

Fig. 4. Stereopair of the dorsal surface of the chick blastoderm at stage 6 (early-head-fold stage). The boundaries of the neural plate (NP) are now well defined. The position of the head process (HP) is still indicated by a shallow midline groove. SE, Surface ectoderm. Bar = 250 μm.

only apically or basally, but not in both directions. The particular shape of a neuroepithelial cell seems to be determined largely by its phase in the mitotic cycle. This is due to the relatively large size of nuclei and a process called interkinetic nuclear migration. During interkinetic nuclear migration, nuclei undergo a to-and-fro movement within the wall of the neuroepithelium: nuclei migrate toward the lumen (i.e., apically) during early prophase; lie next to the lumen during late prophase, metaphase, anaphase and early telophase; and migrate away from the lumen (basally) during late telophase. At present, it is known that spherical cells are in M phase of the mitotic cycle (i.e., metaphase and anaphase), but the exact phase of spindle- or flask-shaped cells is unknown.

Apical (outer or dorsal) ends of neuroepithelial cells are tightly apposed to neighboring cells along their entire circumferences (Fig. 6). Intercellular junctions connect the apices of neuroepithelial cells, probably functioning in cell-to-cell adhesion and possibly also in cell-to-cell communication (Fig. 7). The apices of many neuroepithelial cells each display a single, centrally located cilium. Also present on the apical side of the neuroepithelium are many long, slender cellular processes, each of which bears a single localized dilatation (Fig. 6). These structures, sometimes called beaded threads, are midbodies or remnants of cellular bridges interconnecting daughter cells. Midbodies can be quite long, spanning the apices of several cells (Fig. 6). How cells come to intervene between pairs of daughter cells is unknown. Likewise, the ultimate fate or function of midbodies is unknown. In transmission electron micrographs midbodies appear as tubular processes packed with microtubules (Fig. 7). An electron-dense material fills the area of the dilatation.

Folding of Epithelial Sheets

A common morphogenetic process that occurs in epithelial sheets is folding. After thickening of an epithelial area, such as occurs in the formation of the neural plate, folding ensues. In addition, as the neural plate is forming, the area cranial to the neural plate begins to fold under the plate as the head fold of the body.

The head fold can be readily seen in scanning electron micrographs of the dorsal surface of the blastoderm (Figs. 4, 8, 9) or in midsagittal slices (Figs. 10-13). It consists of two layers (a dorsal or upper layer of ectoderm and a lower or ventral layer of endoderm) and creates two bay-like areas. One of these areas, the subcephalic pocket, lies subjacent to the head and is lined with ectoderm. The other area, the foregut, is lined with endoderm and is contained within the future head region. The foregut opens caudally, via the anterior intestinal portal, into the subblastodermic space--a fluid-filled space between the area pellucida and the yolk.

The lateral body folds appear shortly after formation of the head fold. In surface view, the head fold seems concentric in shape (Figs. 4, 8, 9). This is because the head fold is quickly joined by two lateral folds (Fig. 14). Both the cranial and lateral aspects of the head are formed through the combined action of the head fold and

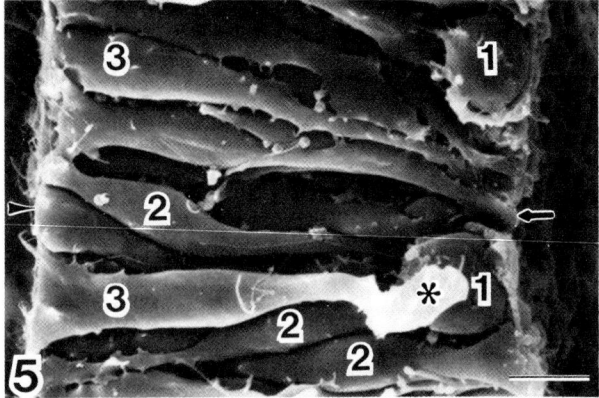

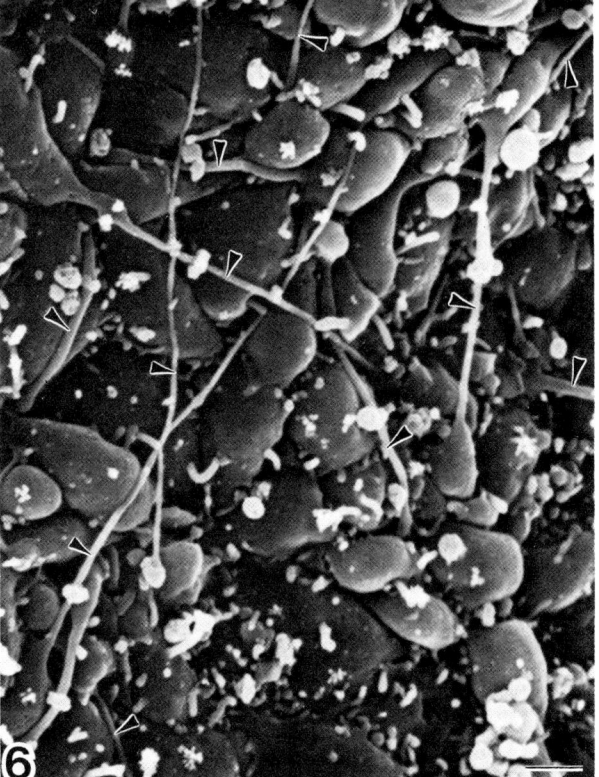

Fig. 5. Enlargement of a transverse slice through the lateral wall of the neural tube (spinal cord) of a chick embryo at stage 10 (10-somite stage). The neural ectoderm is a pseudostratified epithelium. Typically, cells are either spherical (1), or shaped like spindles (2) or flasks (3). Arrow, apical (luminal) side of epithelium; arrowhead, basal side of epithelium; asterisk, cell apex detached during slicing. Bar = 5 μm.

Fig. 6. Apical surface of the wall of the neural tube (spinal cord) from a chick embryo at stage 15 (24- to 27-somite stage). Beaded threads (arrowheads) form long, slender bridges spanning several cells. Bar = 1 μm.

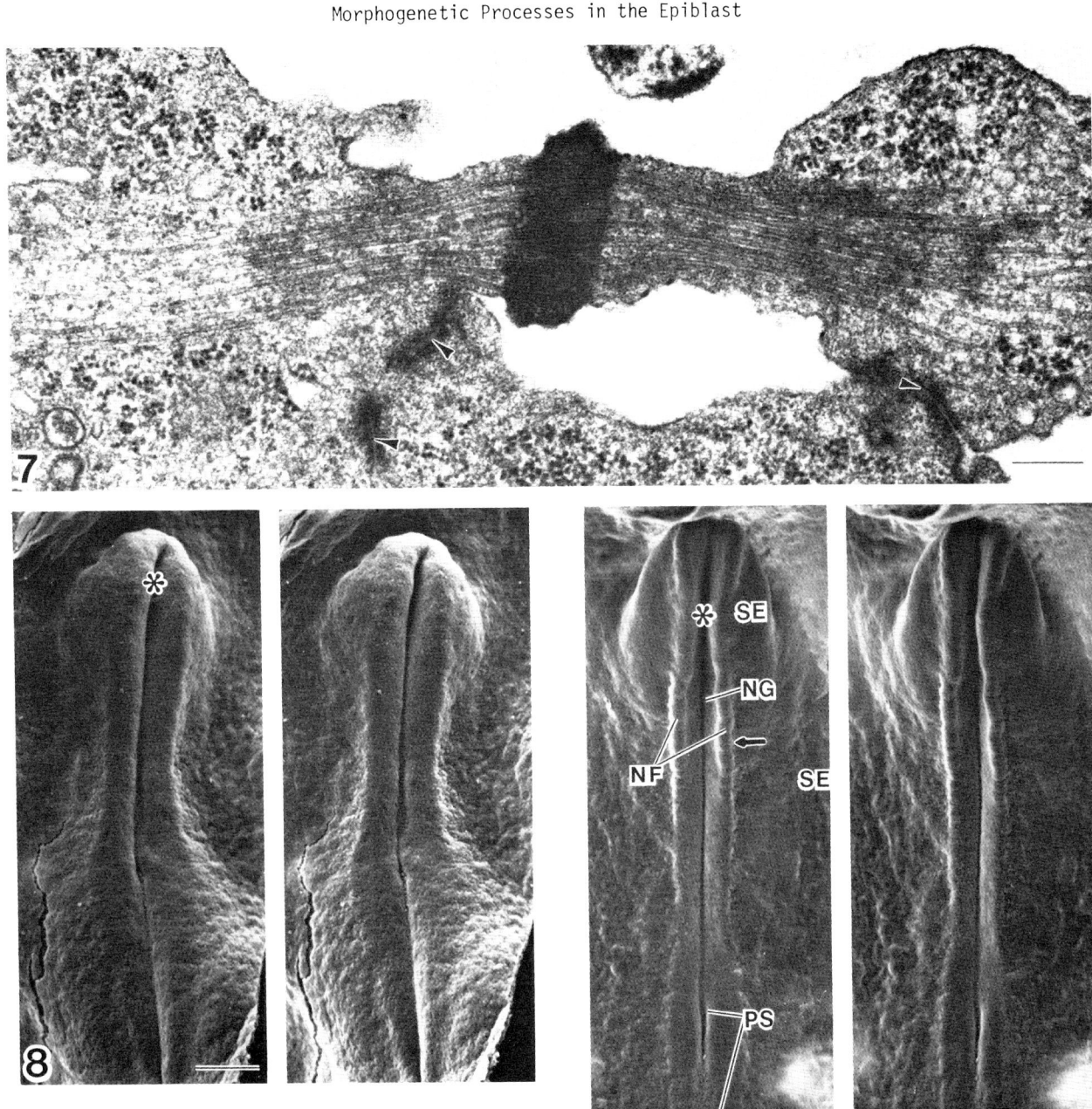

Fig. 7. Transmission electron micrograph of a section through a beaded thread from the spinal cord of a mouse embryo at 10 days of gestation. Numerous microtubules are visible. Arrowheads, apical intercellular junctions. Bar = 0.25 μm.

Fig. 8. Stereopair of the dorsal surface of the chick blastoderm at stage 7 (one-somite stage). The head fold has undercut the cranial end of the neural plate. Folding of the neural plate has just begun cranially, establishing the incipient neural groove (asterisk). Bar = 250 μm.

Fig. 9. Stereopair of the dorsal surface of the chick blastoderm at stage 8- (three-somite stage). The head fold and lateral body folds have demarcated the cranial extent of the head. The neural folds (NF) are about to become approximated in the region of the future midbrain (asterisk). More caudally, a shallow surface ectodermal trough (arrow) lies just lateral to each neural fold. NG, neural groove; PS, primitive streak; SE, surface ectoderm. Bar = 250 μm.

lateral body folds (Fig. 15). Thus, the head is separated from the underlying blastoderm by the ectoderm-lined subcephalic pocket. Simultaneously, the foregut is delimited both cranially and laterally. Caudally, the foregut remains open as the anterior intestinal portal because the body folds have not yet fused across the midline in this area. Note that the lateral body folds contain mesoderm and that this mesoderm is split into two layers (Figs. 14, 16). The layer of mesoderm just adjacent to the surface ectoderm is called somatic mesoderm. The somatic mesoderm and ectoderm collectively constitute the somatopleure. The layer of mesoderm adjacent to the endoderm is called the splanchnic mesoderm. Collectively, the splanchnic mesoderm and endoderm constitute the splanchnopleure. The space between the somatic and splanchnic mesoderm is the coelom.

Fusion of the lateral body folds across the midline occurs in the following sequence. First, the splanchnopleure of each fold fuses to establish the foregut. Next, the somatopleure fuses to separate the head (and, subsequently, the trunk) from the underlying blastoderm. Fusion of somatopleuric and splanchnopleuric components of the lateral body folds establish what is often referred to as the typical tube-within-a-tube vertebrate body plan. Essentially, the embryo consists of two tubes: an outer tube of surface ectoderm and an inner tube of gut endoderm. (As the neural groove is closed, see below, a third tube, the neural tube, is added to this basic body plan.)

A second type of folding, that begins shortly after the head fold, is involved in formation of the brain and spinal cord. The lateral margins of the neural plate begin to elevate, soon after the boundaries of the neural plate are first demarcated, creating a gutter-like space extending down the midline (Figs. 8, 9). This space, the neural groove, is lined with neural plate and is eventually closed as the lateral margins of the neural plate come into contact and fuse across the midline. The lateral margins of the neural plate are loosely designated as the neural folds. I will define these areas more precisely below.

Folding of the neural plate occurs in two distinct steps in chick embryos (Figs. 14-18): elevation and convergence. Elevation consists of the lifting of the lateral aspects of the neural plate upward and the concomitant bending of the neural plate in the midline. The neural groove is distinctly "V" shaped during elevation, both in future brain and spinal cord regions (Figs. 14, 15, 18). As elevation occurs, the junction between the surface and neural ectoderm on each side bends as the <u>incipient</u> neural folds. Elevation seems to occur more vigorously in the future midbrain than in more cranial or caudal areas. This notion is suggested by the consistent presence of a large bulge in the apical side of each

Fig. 10. Stereopair of a midsagittal slice through a chick embryo at stage 5. NP, neural plate; arrow, primitive pit. Bar = 100 μm.

Fig. 11. Stereopair of a midsagittal slice through a chick embryo at stage 7. NP, neural plate; arrow, primitive pit; arrowhead, early head fold. Bar = 200 μm.

Fig. 12. Stereopair of a midsagittal slice through a chick embryo at stage 7. Arrow, anterior intestinal portal; arrowhead, head fold; asterisk, apical side of the lateral wall of the neural groove at the future midbrain region; star, foregut. Bar = 100 μm.

Fig. 13. Stereopair of a midsagittal slice through a chick embryo at stage 8-. Arrow, bilateral furrow associated with convergence of the neural folds (neural fold = area just above furrow); arrowhead, head fold; asterisk, apical side of the lateral wall of the neural groove at the future midbrain region. Bar = 200 μm.

Fig. 14. Transverse slice through the neural groove (NG) (future midbrain) of a chick embryo at stage 7. The slice is viewed from its caudal side so that the foreground is more caudal than the background. AIP, anterior intestinal portal; F, foregut; M, mesoderm; asterisks, lateral body folds; stars, incipient neural folds. Bar = 100 μm.

Fig. 15. Transverse slice through the neural groove (NG) (future midbrain) of a chick embryo at stage 8 (four-somite stage). The slice is viewed from its caudal side so that the foreground is more caudal than the background. B, blastoderm; F, foregut; M, mesoderm; SCP, subcephalic pocket; SE, surface ectoderm; stars, incipient neural folds. Bar = 75 μm.

Fig. 16. Transverse cryofracture through the neural groove (NG) (future hindbrain) of a chick embryo at stage 9 (seven-somite stage). The slice is viewed from its caudal side so that the foreground is more caudal than the background. AIP, anterior intestinal portal; C, coelom; EN, endoderm; M, mesoderm; N, notochord; NE, neural ectoderm; SE, surface ectoderm; SO, somatic mesoderm; SP, splanchnic mesoderm; arrow, apposition of surface and neural ectoderm of neural fold; arrowheads, bilateral furrows associated with convergence; asterisk, surface ectodermal trough associated with the neural fold. Bar = 75 μm.

Fig. 17. Transverse cryofracture through the caudal end of the neural groove (NG) (future spinal cord) of a chick embryo at stage 10. M, mesoderm; N, notochord; NE, neural ectoderm; SE, surface ectoderm; arrowheads, bilateral furrows associated with convergence; arrows, apposition of surface and neural ectoderm of neural folds; asterisk, surface ectodermal trough associated with the neural fold. Bar = 25 μm.

Morphogenetic Processes in the Epiblast

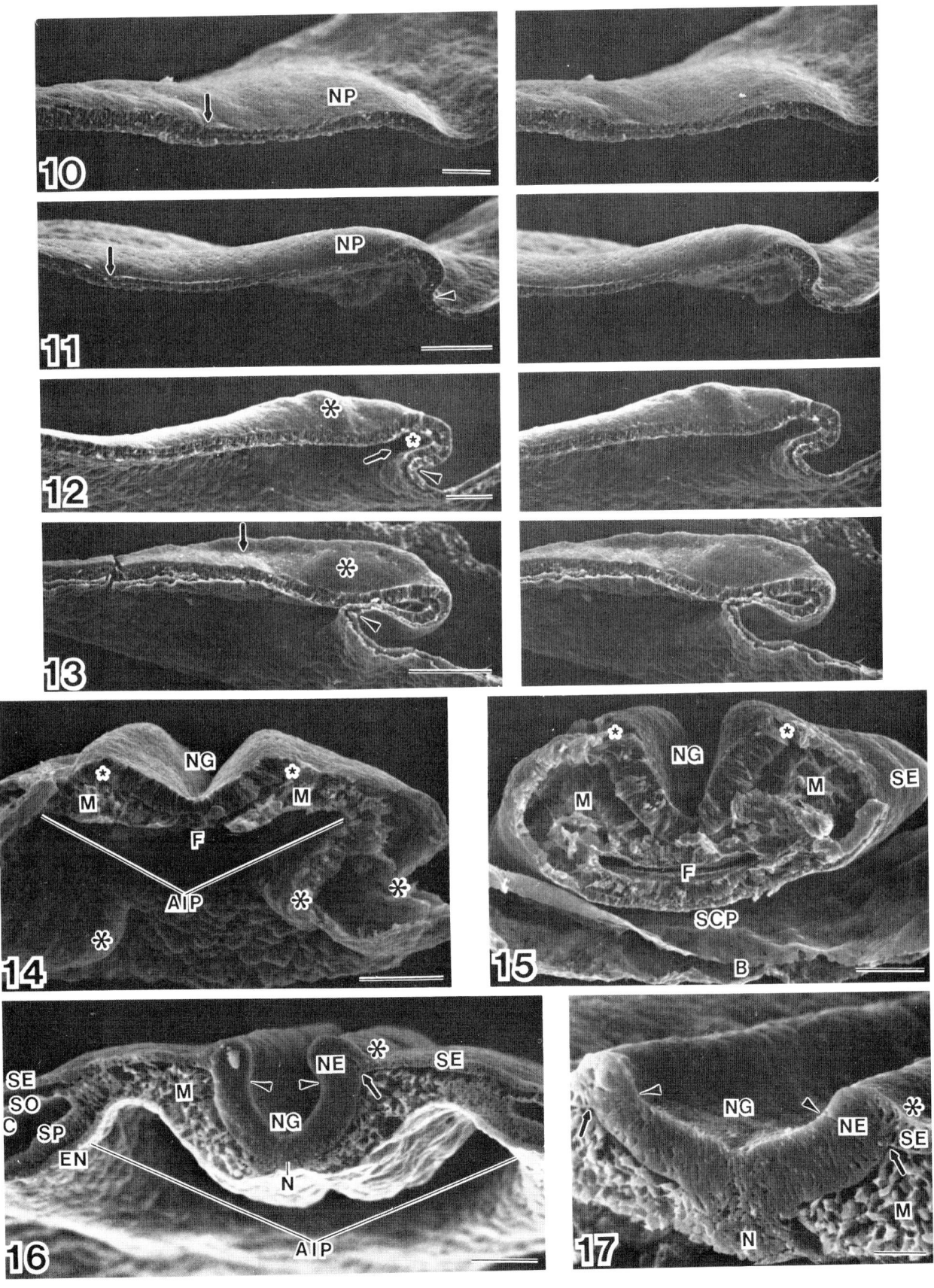

lateral wall of the neural groove in the future midbrain region (Figs. 12, 13).

Convergence is associated with the formation of bilateral furrows or depressions on the apical sides of the dorsolateral walls of the neural groove (Figs. 16, 17). These furrows are always associated with the appearance of the definitive neural folds. The definitive neural folds are bilaminar, consisting of an inner (or medial) layer of neural ectoderm and an outer (or lateral) layer of surface ectoderm (Figs. 16, 17). The two layers of each definitive neural fold seem to be firmly attached to one another. A shallow surface ectodermal trough flanks each definitive neural fold (arrows, Figs. 9, 21; asterisks, Figs. 16, 17). In brain regions, and throughout most of the length of the spinal cord, convergence begins only after elevation has occurred (Fig. 16). However, in caudal regions of the spinal cord, the definitive neural folds, surface ectodermal troughs and bilateral furrows form prior to elevation (Fig. 17). Regardless of when convergence begins, its function is to bring the contralateral neural folds into apposition across the midline, allowing fusion to occur.

Examination of the neural tube shortly after convergence has occurred, reveals that the shapes of the various regions of the <u>incipient</u> neural tube are largely determined by the amount of convergence occurring in that region. Thus, in brain regions, where extensive convergence occurs, the lumen of the neural tube is broadened mediolaterally (Fig. 19). Throughout most of the length of the spinal cord (except at its cranial and caudal ends) little convergence occurs. Hence, the lumen is almost non-existent when convergence is absent or minimal, and the lumen of the neural tube at these levels is said to be occluded (Fig. 20). This occlusion aids in the rapid expansion of the brain during subsequent development by localizing and confining neural tube fluid (which generates hydrostatic pressure in enlarging brain regions to literally inflate the brain) to cranial areas.

Fusion of Epithelial Sheets

Fusion of epithelial sheets involves at least two processes: adhesion and cell intermingling. In some cases, such as the fusion of palatal shelves to form the secondary palate, epithelial cells also die, allowing the subjacent mesenchymal cells to freely intermingle. When cell death does not occur during fusion (in the case of the neural folds, for example), epithelial cells rearrange themselves so that mesenchymal cells can intermingle.

Fusion occurs during the morphogenesis of many different rudiments. I will describe here only fusion of the neural folds as an example. In chick embryos, closure of the neural groove begins in the midbrain region and then progresses both cranially and caudally (Figs. 21, 22). Prior to apposition of the neural folds, the surfaces of these structures remain smooth exhibiting few surface alterations. In contrast, the neural folds of rodent embryos exhibit numerous, elaborate cell processes prior to their apposition. These processes consist principally of large blebs and complex areas of ruffling (depending upon the species), which form precisely at the junction

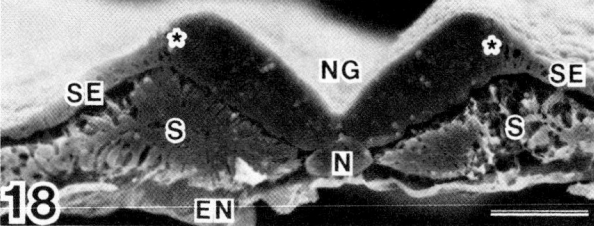

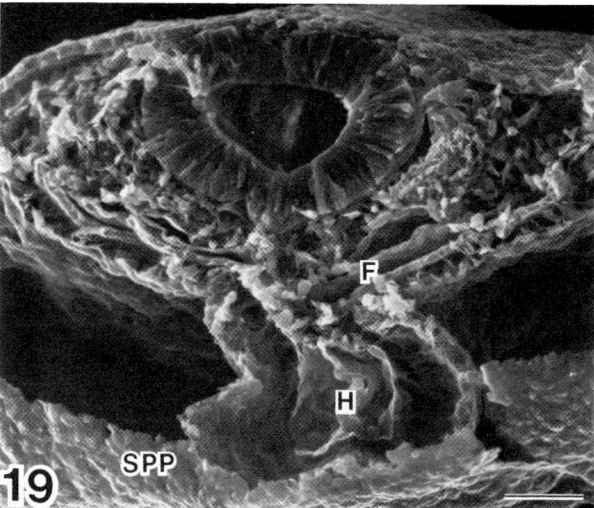

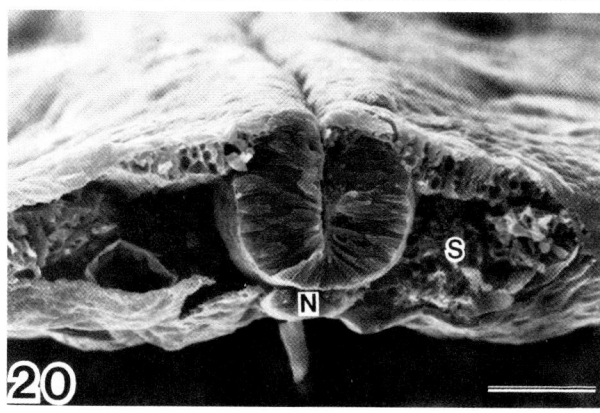

Fig. 18. Transverse cryofracture through the middle of the neural groove (NG) (future spinal cord) of a chick embryo at stage 8+ (five-somite stage). EN, endoderm; N, notochord; S, somite; SE, surface ectoderm; stars, incipient neural folds. Bar = 50 μm.

Fig. 19. Transverse slice through the incipient neural tube (future midbrain) of a chick embryo at stage 9. F, foregut; H, heart; SPP, splanchnopleure. Bar = 50 μm.

Fig. 20. Transverse slice through the incipient neural tube (future spinal cord) of a chick embryo at stage 10. On the left side of the micrograph the somite was removed during slicing. N, notochord; S, somite. Bar = 50 μm.

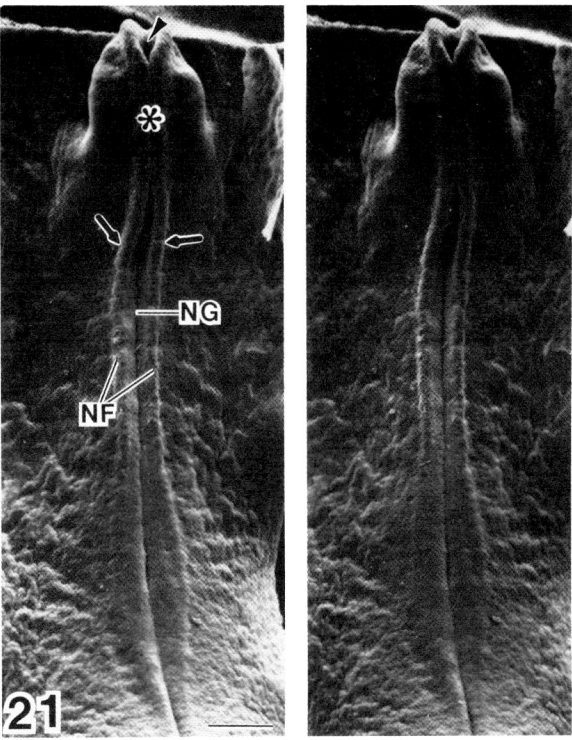
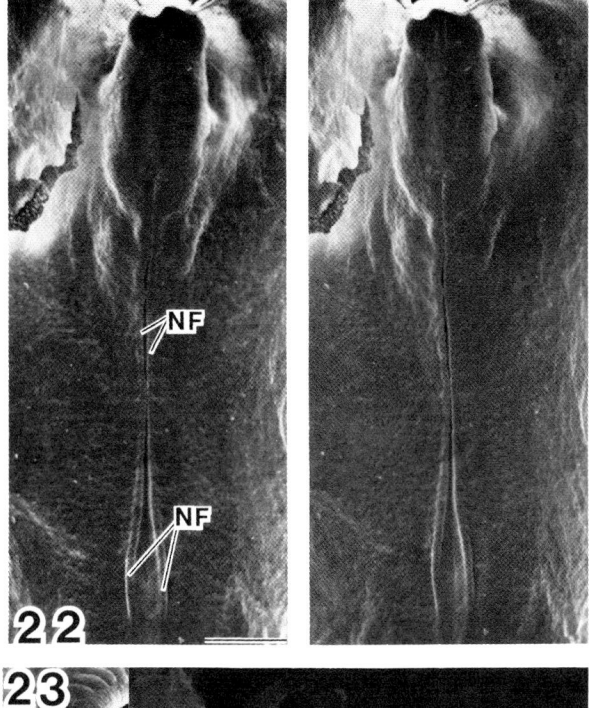
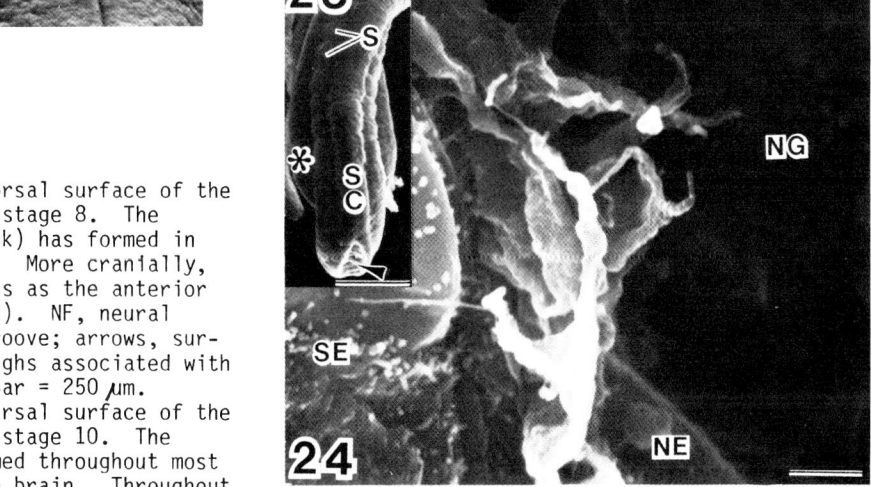

Fig. 21. Stereopair of the dorsal surface of the chick blastoderm at stage 8. The neural tube (asterisk) has formed in the midbrain region. More cranially, the neural tube opens as the anterior neuropore (arrowhead). NF, neural folds; NG, neural groove; arrows, surface ectodermal troughs associated with the neural folds. Bar = 250 μm.

Fig. 22. Stereopair of the dorsal surface of the chick blastoderm at stage 10. The neural tube has formed throughout most of the length of the brain. Throughout most of the length of the spinal cord the neural folds (NF) lie in close proximity to one another. However, at the caudal end of the spinal cord, the folds flare laterad, indicating the region of the posterior neuropore. Part of the blastoderm folded over the cranial tip of the head during processing for microscopy. Bar = 400 μm.

Fig. 23. Dorsal surface of the spinal cord region of a mouse embryo at 9.5 days of gestation. S, somites; SC, spinal cord; arrowhead, posterior neuropore; asterisk, hindleg bud. Bar = 300 μm.

Fig. 24. Enlargement of a neural fold flanking the posterior neuropore from the embryo shown in Fig. 23. A "ruffle" is present at the interface between surface (SE) and neural (NE) ectoderm. NG, neural groove. Bar = 2 μm.

between the surface and neural ectodermal components of each neural fold (Figs. 23, 24). It is reasonable to suggest that these processes might function in guiding the contralateral folds together, but direct evidence for this notion is lacking at present. If it is true that these processes do function in guidance, however, it is unclear why rodents would need some sort of guidance system, whereas chicks would not. In any event <u>after</u> apposition of the neural folds in both avian and mammalian embryos many surface alterations appear. Such alterations include the formation of broad, flattened processes (lamellipodia), as well as long slender processes (filopodia). These processes seem to help cells to migrate across the gap between the apposed folds, causing "healing" and cell intermingling in this area.

After neural folds are brought into apposition, they must be held in place long enough for cell intermingling to occur. Probably multiple factors are involved in maintaining this apposition. Recent evidence suggests that perhaps one factor involved in this process is some sort of a cell surface "glue". This notion is based on the fact that fusing areas are often coated with thick layers of extracellular materials. In the case of the neural folds, the amount of cell surface material coating the lips of the folds (i.e., the areas that become apposed) is greater than that of surrounding areas. Usually, cell surface materials need to be preserved in special ways--that is, extra components need to be added to fixatives to prevent the loss of cell surface materials. However, extracellular material seems to be so abundant in the area of the closing anterior neuropore of chick embryos that it remains intact, even when no special precautions are taken to preserve it (Figs. 25, 26). Alternatively, because of the great depth and narrow width of the closing anterior neuropore, extracellular material may persist in this area not due to its abundance, but due to the decreased likelihood of washing it away during processing.

## Cavitation of Epithelial Cords

Cavitation is a process that occurs during the transformation of the morulae of mammalian embryos into blastocysts, and during the development of a number of organ systems (e.g., formation of the mesonephric ducts and tubules, recanalization of various regions of the gut). I will consider here formation and cavitation of the medullary cord, the rudiment of the caudal portion of the spinal cord.

The medullary cord is derived from the tail bud. The latter consists of an epithelial "bag" containing a core of mesenchymal cells (Figs. 27, 28), and is derived from the cranial portion of the primitive streak (principally, the primitive knot). Recall that the primitive streak is a midline thickening of epiblast. Hence, the tail bud, and the medullary cord as well, are derived directly from epiblast, without the prior formation of distinct germ layers.

The tail bud forms near the time that the posterior neuropore closes (approximately stage 13). With closure of the posterior neuropore, the caudal part of the spinal cord terminates in the tail bud (Fig. 27). Cranial to the tail bud, the germ layers have formed their typical derivatives: ectoderm--surface ectoderm and spinal cord; mesoderm--notochord and segmental plate mesoderm (i.e., mesoderm that forms somites); endoderm--endoderm that will be folded by the tail and lateral body folds to form gut (Fig. 29).

Midway between the levels shown in Figs. 28 and 29 is an area containing the so-called overlap zone (Figs. 30, 31). This zone is merely a region of structural overlap between the caudal end of the neural plate and cranial end of the tail bud. In the overlap zone, the dorsal part of the neural tube is derived from the neural plate, and the ventral part, from the tail bud (medullary cord). Thus, in the overlap zone, the developing neural tube contains a dorsal lumen, formed by closure of the posterior neuropore, and multiple ventral

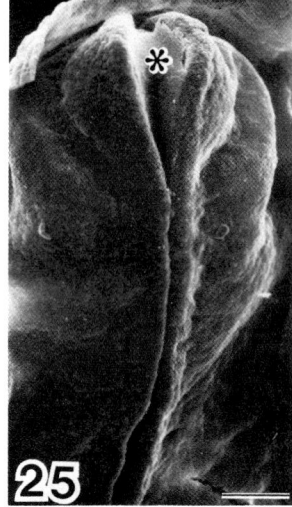
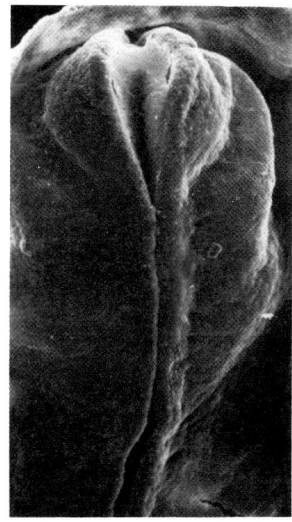
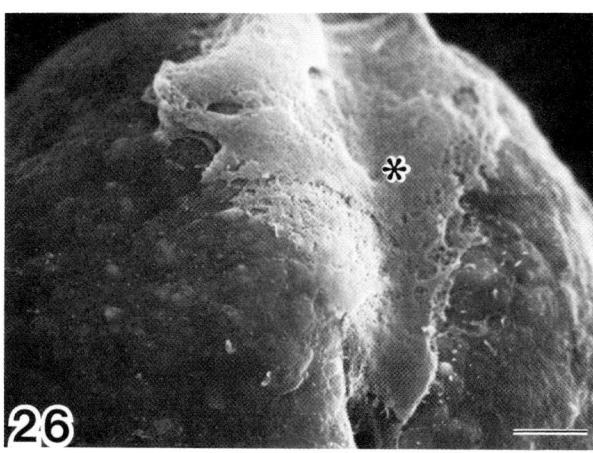

Fig. 25. Stereopair of the dorsal surface of the head of a chick embryo at stage 9. Extracellular material (asterisk) fills the neural groove in the area of the closing anterior neuropore. Bar = 150 μm.

Fig. 26. Dorsal view of the closing anterior neuropore in a chick embryo at stage 9. Extracellular material (asterisk) seems to have splashed over the surfaces of the fusing neural folds. Bar = 20 μm.

lumina, formed by cavitation. All lumina eventually coalesce to form a single lumen.

Formation of the neural tube in caudal regions of chick embryos occurs in four stages (Fig. 32): formation of the medullary cord from the tail bud, differentiation of the medullary cord, cavitation of the cord and coalescence of all lumina into a single, central cavity. Formation of the medullary cord occurs as mesenchymal cells of the tail bud cluster together as an epithelial cord. In other words, formation of the medullary cord involves a mesenchymal-epithelial transformation. Differentiation of this cord involves the elongation of its peripheral cells. The elongated

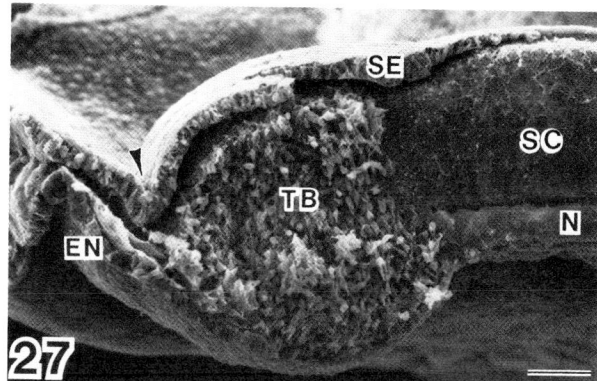

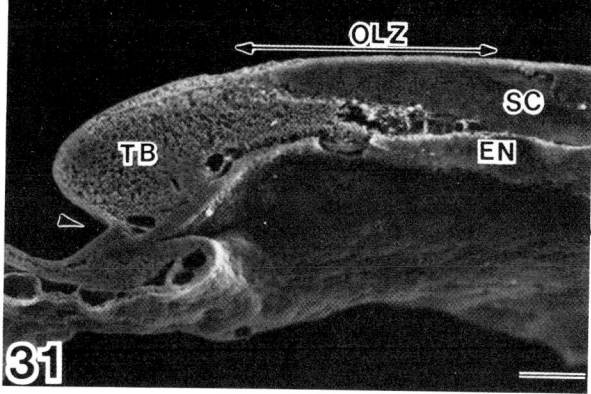

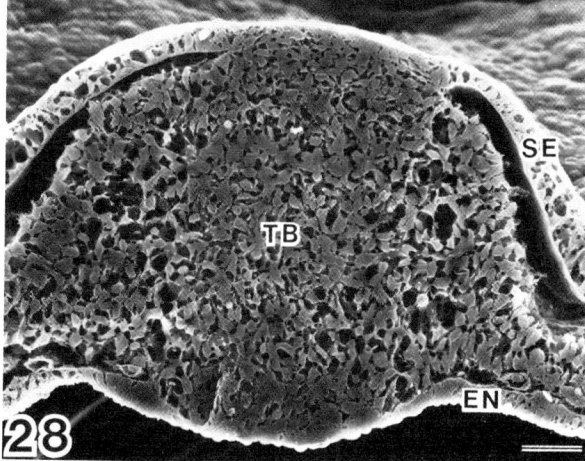

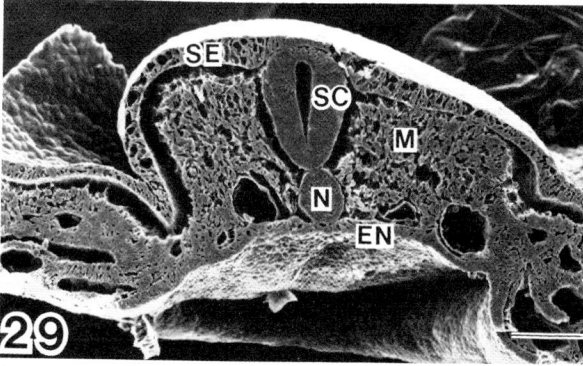

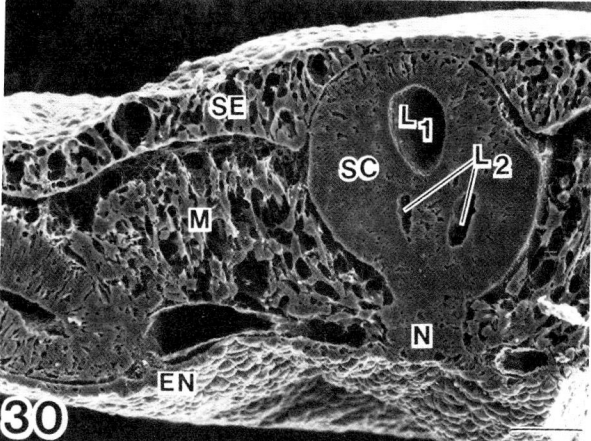

Fig. 27. Parasagittal slice through the caudal region of a chick embryo at stage 15. After closure of the posterior neuropore, the caudal end of the spinal cord (SC) terminates in the tail bud (TB). EN, endoderm; N, notochord; SE, surface ectoderm; arrowhead, tail fold of the body. Bar = 50 μm.

Fig. 28. Transverse cryofracture through the tail bud (TB) of a chick embryo at stage 15. EN, endoderm; SE, surface ectoderm. Bar = 20 μm.

Fig. 29. Transverse cryofracture through the portion of the spinal cord (SC) just cranial to the overlap zone of a chick embryo at stage 15. EN, endoderm; M, mesoderm; N, notochord; SE, surface ectoderm. Bar = 50 μm.

Fig. 30. Transverse cryofracture through the portion of the developing spinal cord (SC) located in the overlap zone of a chick embryo at stage 15. EN, endoderm; $L_1$, lumen formed from the closed neural groove; $L_2$, lumina formed by cavitation; M, mesoderm; N, notochord; SE, surface ectoderm. Bar = 25 μm.

Fig. 31. Midsagittal cryofracture through the caudal end of a chick embryo at stage 15. EN, endoderm; OLZ, overlap zone; SC, spinal cord; TB, tail bud; arrowhead, tail fold of the body. Bar = 100 μm.

peripheral cells collectively form a primitive pseudostratified epithelium whose basal side corresponds to the basal side of the future neural tube and whose apical side is associated with a central cluster of typical mesenchymal cells. Isolated lumina then appear at the interface between peripheral and central cells, and, eventually, these lumina coalesce into a single cavity. The central cells are removed during coalescence. How they are removed is unknown. Apparently, the central cells either migrate caudally, where they in turn form more caudal levels of the medullary cord, or peripherally, becoming intercalated with the peripheral cells to form the walls of the neural tube.

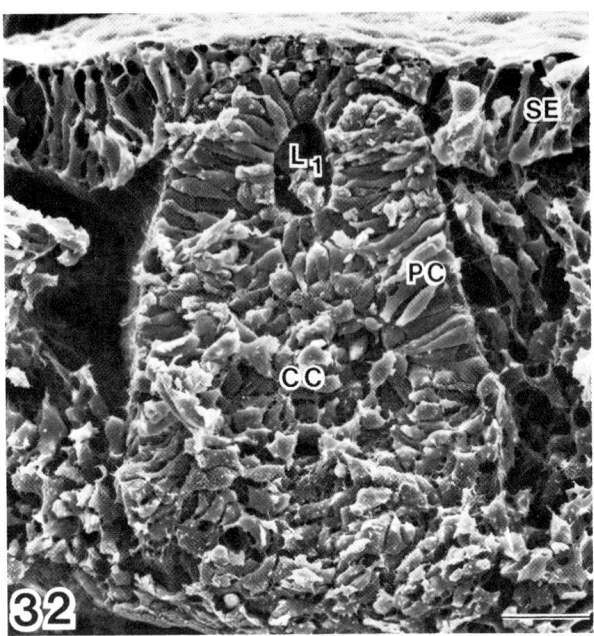

Fig. 32. Transverse slice through the overlap zone of a chick embryo at stage 15. CC, central cells; $L_1$, lumen formed from the closed neural groove; PC, peripheral cells; SE, surface ectoderm. Bar = 25 μm.

### Dispersal of Epithelial Sheets During Cell Migration

An example of cell migration has already been given: the movement of prospective endodermal and mesodermal cells through the primitive streak. Ectodermal cells also undergo migration. The intermingling of cells during fusion of the neural folds involves some, albeit highly limited, cell migration. More extensive migration occurs during the formation and migration of neural crest cells. Like the other examples of cell migration discussed above, formation and migration of the neural crest involves an epithelial-mesenchymal transformation: epithelial cells derived from the neural folds leave the epithelium and disperse as mesenchymal neural crest cells. In mammalian embryos, neural crest cells form precociously as the neural folds are elevating and converging toward the dorsal midline. In chick embryos, formation and migration of neural crest cells is delayed. Hence, neural crest cells originate from the roof of the neural tube in chick embryos rather than from the neural folds (Figs. 33-35). Migration of neural crest cells occurs roughly perpendicular to the craniocaudal axis of the neural tube (Figs. 33, 34). Neural crest cells of the head region migrate laterad between the surface ectoderm and mesenchyme (mesoderm). In the trunk region, neural crest cells migrate as two cell streams: one stream between the surface ectoderm and dorsal (dermomyotome) portions of the somites, the other, between the medial portions of the somites (principally sclerotome) and spinal cord (Fig. 33). Apparently, neural crest cells use both cellular (i.e., the basal side of the surface ectoderm, the dorsal and medial surfaces of the somites, the lateral surfaces of the spinal cord) and extracellular (i.e., basal laminae and extracellular matrix secreted by the surface ectoderm, neural tube, somites, and neural crest cells themselves) substrata for this migration. Neural crest cells assume a variety of irregular shapes during their migration (Fig. 35). Characteristically, these cells are elongated roughly in the direction of their migration, and display numerous, fine filopodia. It is usually impossible to differentiate between the leading and the trailing ends of individual neural crest cells when these cells are examined without reference to their origin or destination, because these two ends have similar morphologies.

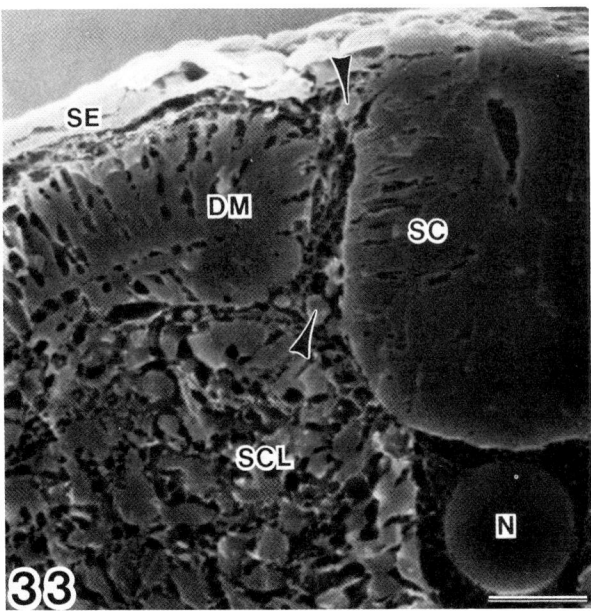

Fig. 33. Transverse cryofracture through the spinal cord (SC) of a chick embryo at stage 14 (22-somite stage). DM, dermomyotome region of somite; N, notochord; SCL, sclerotome region of somite; SE, surface ectoderm; arrowheads, neural crest cells. Bar = 20 μm.

Fig. 34. Stereopair of the dorsal surface of the trunk region (surface ectoderm largely removed) of a chick embryo at stage 20 (40- to 43-somite stage). Neural crest cells (arrowheads) are migrating laterad from the roof of the neural tube (asterisk). S, somite; SE, surface ectoderm. Bar = 100 μm.

Fig. 35. Stereopair enlargement of migrating neural crest cells from the embryo shown in Fig. 33. Asterisk, roof of neural tube. Bar = 5 μm.

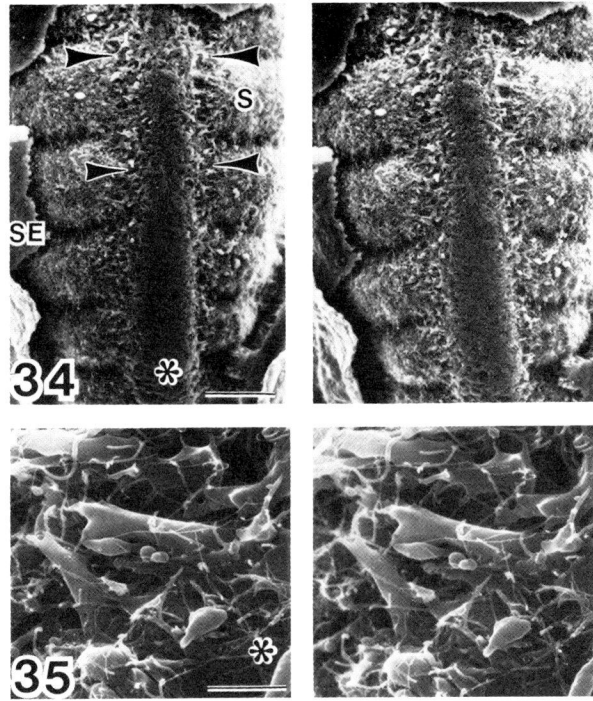

## Summary

Scanning electron microscopy has been used to describe the early morphogenesis of the epiblast. Morphogenetic processes have been discussed individually for simplicity sake, almost as if they occur in isolation. However, it is important to keep in mind that morphogenesis is a highly coordinated phenomenon with multiple morphogenetic events occurring in concert. Because of this fact, morphogenesis of the embryo is a complicated process to analyze, but also because of this fact, there is a great deal of excitement and pure enjoyment in trying to decipher the "way of the embryo." Scanning electron microscopy has helped us immeasurably in this task, and it is likely to play an even more important role in future studies combining the use of scanning electron microscopy and experimental techniques.

## Acknowledgements

I wish to acknowledge the superb technical assistance of Nancy B. Chandler. Dr. Marilyn Fisher provided valuable editorial comments. Karen Evans and Cynthia Shay prepared the typescript. Figs. 1-4, 8, 9, 18 and 21 were modified from Watterson, RL and Schoenwolf, GC (1984). Laboratory Studies of Chick, Pig, and Frog Embryos, 5th edition, Burgess Publishing Company, Minneapolis, Minnesota; Figs. 14 and 15, from Schoenwolf, GC (1982). On the morphogenesis of the early rudiments of the developing central nervous system, Scanning Electron Microsc. 1982; I:289-308; and Figs. 27, 28 and 30 from Schoenwolf, GC (1978). An SEM study of posterior spinal cord development in the chick embryo, Scanning Electron Microsc. 1978; II:739-746. Supported by Grants Nos. HD15231 and NS18112 from the National Institutes of Health.

## Literature Cited

Glücksmann A. (1951). Cell deaths in normal vertebrate ontogeny. Biol. Rev. 26, 59-86.
Hamburger V, Hamilton, HL. (1951). A series of normal stages in the development of the chick embryo. J. Morphol. 88, 49-92.
Lundmark C, Shih J, Tibbetts P, Keller R. (1983). Amphibian gastrulation as seen by scanning electron microscopy. Scanning Electron Microsc. 1983, accepted for publication.
Saunders JW, Jr. (1966). Death in embryonic systems. Science 154, 604-612.
Saunders JW, Jr, Fallon, JF. (1967). Cell death in embryonic systems, in: Major Problems in Developmental Biology, M. Locke (ed), Academic Press, NY, 289-314.
Watterson RL, Schoenwolf GC, Sweeney, RM. (1979). Laboratory Studies of Chick, Pig, and Frog Embryos, 5th edition, Burgess Publishing Company, Minneapolis, Minnesota. p. 5.

## Suggestions for Further Reading

### Gastrulation in Birds
England MA, Wakely J. (1977). Scanning electron microscopy of the development of the mesoderm layer in chick embryos. Anat. Embryol. 150, 291-300.
England MA, Wakely J, Cowper SV. (1978). Scanning electron microscopy of the late primitive streak and head process of the chick embryo. Scanning Electron Microsc. 1978; II: 103-110.
Granholm NH, Baker JR. (1970). Cytoplasmic microtubules and the mechanism of avian gastrulation. Develop. Biol. 23, 563-584.

### Blastoderm Expansion
Chernoff EAG, Overton J. (1977). Scanning electron microscopy of chick epiblast expansion on the vitelline membrane. Develop. Biol. 57, 33-46.
Chernoff EAG, Overton J. (1979). Organization of the migrating chick epiblast edge: Attachment sites, cytoskeleton, and early developmental changes. Develop. Biol. 72, 291-307.
Mayer BW, Jr, Packard DS, Jr. (1978). A study of the expansion of the chick area vasculosa. Develop. Biol. 63, 335-351.

### Mesodermal Patterning and Subdivisions
Bancroft M, Bellairs R. (1976). The development of the notochord in the chick embryo, studied by scanning and transmission electron microscopy. J. Embryol. Exp. Morphol. 35, 383-401.
Langman J, Nelson GR. (1968). A radioautographic study of the development of the somite in the chick embryo. J. Embryol. Exp. Morphol. 19, 217-226.
Meier S. (1979). Development of the chick embryo mesoblast. Develop. Biol. 73, 25-45.

Meier S. (1980). Development of the chick embryo mesoblast: Pronephros, lateral plate, and early vasculature. J. Embryol. Exp. Morphol. 55, 291-306.

Meier S. (1981). Development of the chick embryo mesoblast: Morphogenesis of the prechordal plate and cranial segments. Develop. Biol. 83, 49-61.

Meier S, Jacobson AG. (1982). Experimental studies of the origin and expression of metameric pattern in the chick embryo. J. Exp. Zool. 219, 217-232.

Meier S, Tam PPL. (1982). Metameric pattern development in the embryonic axis of the mouse. I. Differentiation of the cranial segments. Differentiation 21, 95-108.

Tam PPL, Meier S, Jacobson AG. (1982). Differentiation of the metameric pattern in the embryonic axis of the mouse. II. Somitomeric organization of the presomitic mesoderm. Differentiation 21, 109-122.

Youn BW, Malacinski GM. (1981). Somitogenesis in the amphibian Xenopus laevis: Scanning electronic microscopic analysis of intrasomitic cellular arrangements during somite rotation. J. Embryol. Exp. Morphol. 64, 23-43.

Youn BW, Malacinski GM. (1981). Comparative analysis of amphibian somite morphogenesis: Cell rearrangement patterns during rosette formation and myoblast fusion. J. Embryol. Exp. Morphol. 66, 1-26.

Placodes, Pits and Vesicles

Bancroft M, Bellairs R. (1977). Placodes of the chick embryo studied by SEM. Anat. Embryol. 151, 97-108.

Burk D, Sadler TW, Langman J. (1979). Distribution of surface coat material on nasal folds of mouse embryos as demonstrated by Concanavalin A binding. Anat. Rec. 193, 185-196.

Gaare JD, Langman J. (1980). Fusion of nasal swellings in the mouse embryo: DNA synthesis and histological features. Anat. Embryol. 159, 85-99.

Meier S. (1978). Development of the embryonic chick otic placode. I. Light microscopic analysis. Anat. Rec. 191, 447-458.

Meier S. (1978). Development of the embryonic chick otic placode. II. Electron microscopic analysis. Anat. Rec. 191, 459-478.

Smuts MS. (1977). Concanavalin A binding to the epithelial surface of the developing mouse olfactory placode. Anat. Rec. 188, 29-38.

Smuts MS. (1981). Rapid nasal pit formation in mouse embryos stimulated by ATP-containing medium. J. Exp. Zool. 216, 409-414.

Waterman RE, Meller SM. (1973). Nasal pit formation in the hamster: A transmission and scanning electron microscopic study. Develop. Biol. 34, 255-266.

Body Folds

Miller SA. (1982). Differential proliferation in morphogenesis of lateral body folds. J. Exp. Zool. 221, 205-211.

Neural Plate, Neural Groove, Neural Folds and Neural Tube

Burnside B. (1971). Microtubules and microfilaments in newt neurulation. Develop. Biol. 26, 416-441.

Burnside MB, Jacobson AG. (1968). Analysis of morphogenetic movements in the neural plate of the newt Taricha torosa. Develop. Biol. 18, 537-552.

Jacobson AG, Gordon R. (1976). Changes in the shape of the developing vertebrate nervous system analyzed experimentally, mathematically and by computer simulation. J. Exp. Zool. 197, 191-246.

Karfunkel P. (1971). The role of microtubules and microfilaments neurulation in Xenopus. Develop. Biol. 25, 30-56.

Karfunkel P. (1974). The mechanisms of neural tube formation. Internat'l. Rev. Cytol. 38, 245-271.

Schoenwolf GC. (1979). Observations on closure of the neuropores in the chick embryo. Amer. J. Anat. 155, 445-466.

Schoenwolf GC. (1982). On the morphogenesis of the early rudiments of the developing central nervous system. Scanning Electron Microsc. 1982; I: 289-308.

Schroeder TE. (1970). Neurulation in Xenopus laevis. An analysis and model based upon light and electron microscopy. J. Embryol. Exp. Morphol. 23, 427-462.

Waterman RE. (1976). Topographical changes along the neural fold associated with neurulation in the hamster and mouse. Amer. J. Anat. 146, 151-172.

Fusion of Palatal Shelves

Moriarty TM, Weinstein S, Gibson RD. (1963). The development in vitro and in vivo of fusion of the palatal processes of rat embryos. J. Embryol. Exp. Morphol. 11, 605-619.

Waterman RE, Meller SM. (1974). Alterations in the epithelial surface of human palatal shelves prior to and during fusion: A scanning electron microscopic study. Anat. Rec. 180, 111-136.

Waterman RE, Ross LM, Meller SM. (1973). Alterations in the epithelial surface of A/Jax mouse palatal shelves prior to and during palatal fusion: A scanning electron microscopic study. Anat. Rec. 176, 361-376.

Cavitation During Formation of the Neural Tube

Schoenwolf GC. (1977). Tail (end) bud contributions to the posterior region of the chick embryo. J. Exp. Zool. 201, 227-246.

Schoenwolf GC, Delongo J. (1980). Ultrastructure of secondary neurulation in the chick embryo. Amer. J. Anat. 158, 43-63.

Neural Crest

Anderson CB, Meier S. (1981). The influence of the metameric pattern in the mesoderm on migration of cranial neural crest cells in the chick embryo. Develop. Biol. 85, 385-402.

Bancroft M, Bellairs R. (1976). The neural crest cells of the trunk region of the chick embryo studied by SEM and TEM. Zoon 4, 73-85.

Bolender DL, Seliger WG, Markwald RR. (1980). A histochemical analysis of polyanionic compounds found in the extracellular matrix encountered by migrating cephalic neural crest cells. Anat. Rec. 196, 401-412.

LeDouarin NM, Smith J, Teillet M-A, LeLievre CS, Ziller C. (1980). The neural crest and its developmental analysis in avian embryo

chimaeras. Trends in Neurosc. 3, 39-42.
Nichols DH. (1981). Neural crest formation in the head of the mouse embryo as observed using a new histological technique. J. Embryol. Exp. Morphol. 64, 105-120.
Tosney KW. (1978). The early migration of neural crest cells in the trunk region of the avian embryo: An electron microscopic study. Develop. Biol. 62, 317-333.
Verwoerd CDA, van Oostrom CG. (1979). Cephalic neural crest and placodes. Adv. Anat. Embryol. Cell Biol. 58, 1-75.
Weston JA. (1963). A radioautographic analysis of the migration and localization of trunk neural crest cells in the chick. Develop. Biol. 6, 279-310.

## Discussion with Reviewers

M.E. England: The cryofractures are beautifully produced specimens. Would you explain this process for those not familiar with it?
Author: Ethanolic cryofracturing is a procedure developed by W.J. Humphreys and co-workers (Humphreys WJ, Spurlock BO, Johnson JS. (1974). Critical point drying of ethanol-infiltrated, cryofractured biological specimens for scanning electron microscopy. Scanning Electron Microsc., 1974: 275-282). Briefly, in this procedure, tissues are fixed with an aldehyde mixture, postfixed with osmium tetroxide and dehydrated with ethanol. After complete dehydration, tissues are transferred with a pipette to tiny "sausages," fashioned from Parafilm. One end of each sausage is crimped shut with a hemostat, the sausage is filled with absolute ethanol and the other end of the sausage is crimped shut, so that each sausage contains a small piece of tissue surrounded by absolute ethanol. Sausages are then plunged into liquid nitrogen and fractured by placing a razor blade (held with a hemostat and cooled in liquid nitrogen) over the approximate area of the desired fracture plane and hitting the blade with a hammer. Finally, fractured pieces are removed from the liquid nitrogen with forceps, placed in absolute ethanol at room temperature and critical-point-dried from liquid carbon dioxide. It must be emphasized that it is extremely difficult to obtain cryofractures at precisely the desired level and orientation when processing very small, complex embryos. Thus, dozens of cryofractures must be prepared to obtain a few that show desired structures.

S. Meier: What is the evidence that the proamnion represents the future oral membrane? Isn't much of this area colonized by mesoderm that becomes cardiogenic?
Author: Apparently, only part of the proamnion forms the oral membrane. The evidence for this origin is based solely on observations. The cardiogenic mesoderm is generally considered to be located cranial to the proamnion prior to formation of the head fold. I do not know to what extent the proamnion becomes colonized by this mesoderm at later stages.

R.E. Waterman: Could the surface ectodermal troughs that flank the definitive neural folds be fixation artifacts?
Author: It is possible that these troughs are fixation artifacts. However, the fact that they consistently appear precisely at the time at which the surface and neural ectodermal layers of each neural fold become coherent, strongly suggests that they are real depressions. In other words, surface ectodermal troughs do not flank the incipient neural folds even though these structures have elevated. Surface ectodermal troughs only flank the definitive neural folds.

Additional discussion with reviewers of the paper "Development of the Eye of the Chick Embryo" by S. R. Hilfer continued from page 144.

D.C. Beebe: Many of the complex movements described in this chapter are obviously coordinated in space and time. Examples include the simultaneous invagination of optic cup and lens vesicle, the orderly formation of successive layers of retinal neurons, and the sequential differentiation of lens and corneal epithelium from nearly the same population of cells. To what extent is this precise localization and timing of events the reflection of tight intracellular control of gene expression? Alternatively, does this coordination depend primarily upon a "cascade" of cellular interactions in which each event is triggered by the preceding event and then serves as the stimulus for the next series of happenings? Stated in another way, can the precise spatial and temporal order of ocular development be understood best by discovering the mechanisms of gene regulation or by studying the mechanisms of cellular interaction?

Author: With the information presently available, it is not possible to even suggest an answer to the first two questions. It seems obvious that any understanding of developmental processes must come from a synthesis of information on gene regulation and cellular interaction.

D.C. Beebe: In what basic ways is the laminar formation and organization of the retina similar to and/or different from the development and organization of the brain and spinal cord?

Author: There is a great deal of similarity in the formation of the layered structure of the neural retina and parts of the brain, such as the optic tectum and cerebellum. The spinal cord does not become laminar and has a simpler organization.

D.C. Beebe: Do the cell-specific proteins of the lens, cornea, retina, and other ocular tissues appear before, during or after the morphogenesis of these structures?

Author: In those tissues which have been investigated, cell-specific proteins appear during morphogenesis. A good example is crystallin synthesis in the lens.

ON THE MORPHOGENESIS OF THE EARLY RUDIMENTS OF THE DEVELOPING CENTRAL NERVOUS SYSTEM

G. C. Schoenwolf*

Department of Anatomy, University of Utah, School of Medicine,
Salt Lake City, Utah 84132

(Paper received January 18 1982, Final manuscript received August 16 1982)

## ABSTRACT

Neurulation consists of a complex series of events that result in formation of the neural tube, the rudiment of the entire adult central nervous system. This article discusses primary neurulation, a process that occurs in three major stages: formation of the neural plate; bending of the neural plate, with formation of the neural groove and neural folds; and fusion of the neural folds, with formation of the roof of the neural tube and overlying surface epithelium. In addition, new information is presented on the morphogenesis of the chick neuroepithelium during bending of the neural plate. These data show that bending of the neural plate occurs in two distinct steps at all craniocaudal levels: elevation of the lateral parts of the neural plate and the incipient neural folds, and formation and convergence of the definitive neural folds. Elevation is associated with the formation of a midline, ventral locus of bending, and convergence, with the formation of bilateral loci of bending on the future luminal sides of the dorsolateral walls of the neural groove. Several mechanisms possibly involved in the stages and steps of primary neurulation are discussed.

KEY WORDS: Mesenchyme, Neural Crest, Neural Folds, Neural Plate, Neural Tube, Neural Groove, Neuroepithelium, Neurulation, Notochord, Somites

*Address for correspondence:
Department of Anatomy
University of Utah, School of Medicine
Salt Lake City, Utah 84132
Phone No. (801) 581-6453

## INTRODUCTION

Neurulation consists of a series of developmental events that culminate in formation of the neural tube, the rudiment of the entire adult central nervous system. The neural tube forms in two phases in avian and mammalian embryos (Holmdahl, 1925a,b; Schumacher, 1927; Ikeda, 1930; Bolli, 1966; Criley, 1969; Jelínek et al., 1969; Klika and Jelínek, 1969; Lemire, 1969; Dryden, 1973; 1980a,b; Hughes and Freeman, 1974; Lemire et al., 1975; Schoenwolf, 1977; 1978a,b; Schoenwolf and DeLongo, 1980). The brain and a large portion of the spinal cord form during the first phase or primary neurulation. During this phase the ectodermal neural plate "rolls up into a neural tube." The caudal portion of the spinal cord forms during the second phase or secondary neurulation. This phase involves the canalization of the medullary cord, a solid mass of cells derived from the tail bud. The neural tube forms exclusively during a single phase of neurulation in other vertebrate embryos. Thus, in amphibian embryos, the entire neural tube forms during primary neurulation (e.g., Rugh, 1951), whereas in certain fish embryos the neural tube develops wholly during secondary neurulation (Ishii, 1967; Miyayama and Fujimoto, 1977).

I will discuss only primary neurulation in this review article (secondary neurulation has been described and discussed by Schoenwolf and DeLongo, 1980), emphasizing the chick embryo as an important model system. My approach will be to spotlight those stages of primary neurulation that have been studied most thoroughly with electron microscopy. In addition, I will present new data on the morphogenesis of the chick neuroepithelium during the transformation of the neural plate into the neural tube. These latter data suggest that the neural tube forms by the interaction of a series of distinct morphogenetic events; that is, apparently no one mechanism alone causes this process. It is fortunate that neurulation probably does occur in a series of well-defined stages or steps, since this affords us the opportunity to analyze these events experimentally in a piecemeal, but comprehensible, fashion.

## MATERIALS AND METHODS

Several dozen White Leghorn chicken eggs were set in forced-draft, humidified incubators at 38°C until embryos reached stages 5-11 (Hamburger and Hamilton, 1951). Eggs were opened into bowls containing warm saline (0.9% sodium chloride solution), and blastoderms were removed from the yolk, transferred to plastic dishes, and fixed at room temperature for 2 hours with 2% glutaraldehyde, 2% paraformaldehyde in 0.1M cacodylate buffer (pH 7.2) (Karnovsky, 1965). Blastoderms were subsequently washed with buffer, postfixed for 1 hour with osmium tetroxide (1% in 0.1M cacodylate buffer, pH 7.2) and dehydrated with ethanol. Embryos were trimmed from extraembryonic areas, transected at desired levels, or left intact, and processed for either light microscopy (LM) or scanning electron microscopy (SEM).

Embryos processed for LM were transferred to propylene oxide and embedded in Epon/Araldite (Kushida, 1971). Sections were cut at 1 μm with glass knives, mounted on slides and stained with 1% toluidine blue in 1% sodium borate (Richardson et al., 1960).

Embryos processed for SEM were critical-point-dried from liquid $CO_2$ in a Samdri-780 apparatus. Dried specimens were mounted on aluminum stubs with double-stick Scotch (3M) tape and silver conductive paint, coated with gold/palladium in a Hummer VI sputter coater with pulse control and examined with a JEOL JSM-35 scanning electron microscope at 25 kV.

## THE STAGES AND STEPS OF NEURULATION

Primary neurulation occurs in three major stages: formation of the neural plate; bending of the neural plate, with formation of the neural groove and neural folds; and fusion of the neural folds, with formation of the roof of the neural tube and overlying surface epithelium. Additional processes closely related to the latter stages of primary neurulation are the formation and emigration of the neural crest. Only a few brief comments will be included here about these processes since they are discussed elsewhere in this publication (Nichols, 1982). After I have discussed what morphological events happen during the stages of primary neurulation, I will discuss experiments that have attempted to provide insight on how these stages occur. Although our understanding of neurulation is far from complete, it is important to discuss precisely what is (and what is not) known about this process so as to provide impetus and direction for future research.

### Formation of the Neural Plate

Induction: The neural plate forms during the first stage of primary neurulation as the outer, ectodermal layer of the blastoderm thickens in response to induction (reviewed by Saxén, 1980). This thickening is due exclusively to cell elongation, rather than to formation of a stratified epithelium by processes such as mitosis and cell aggregation--the neuroepithelium remains a pseudostratified epithelium until well after its transformation into the neural tube. Several morphological features characterize the onset of induction of the neural plate in chick embryos (England, 1973; 1974; England and Cowper, 1975; 1976). Nuclei of prospective neuroepithelial cells located just cranial to the primitive knot in early stage 5 embryos become aligned in the same apicobasal (dorsoventral) position within the epithelium, forming a "band" that extends across the epithelium in transverse sections. Extreme basal aspects of ectodermal cells in areas undergoing induction exhibit "clear" cytoplasmic regions lacking ribosomes and other organelles, whereas slightly more apical areas exhibit numerous polysomes. Finally, mesodermal cells display numerous filopodia in areas undergoing induction. Each filopodium either splays over the basal aspects of several (usually 4-5) ectodermal cells, or contacts another mesodermal cell (each mesodermal cell contacts 4-6 other mesodermal cells by its filopodia). It has been suggested that the interconnection of cells in these areas might synchronize induction (England and Cowper, 1976). This attractive suggestion may be true; however, it is possible that these filopodia are principally locomotive or skeletal appendages, particularly since gastrulation is also occurring at this time (reviewed by Watterson et al., 1979).

Possible Mechanisms of Cell Elongation: Several studies have attempted to determine the mechanism of cell elongation during formation of the neural plate. It is generally believed that elongation is caused by microtubules, or, at least, that the presence of intact microtubules is required for elongation. Likewise, it is believed that microtubules also maintain the elongated configurations of neuroepithelial cells. These beliefs are based on two types of observations: (1) elongated cells typically contain paraxial microtubules (i.e., microtubules oriented parallel to the direction of elongation) (e.g., amphibian neural plate: Waddington and Perry, 1966; Baker and Schroeder, 1967; Schroeder, 1970; Burnside, 1971; 1973; Karfunkel, 1971; avian neural plate: Lyser, 1968; Messier, 1969; Handel and Roth, 1971; Karfunkel, 1972; Nagele and Lee, 1979; 1980a); mammalian neural plate: Herman and Kauffman, 1966; Wilson and Finta, 1980a,b; medullary cord cells: Schoenwolf and DeLongo, 1980; blastoporal cells: Perry and Waddington, 1966; flask cells of the primitive streak: Granholm and Baker, 1970; lens placode cells: Byers and Porter, 1964; Pearce and Zwaan, 1970); and (2) disruption of these microtubules with colchicine, its analogs or similar compounds results in formation of "colchicine cells" or "figures"--large spherical cells which lack nuclear membranes and display chromosomes arrayed on the metaphase plate (e.g., amphibian neural plate: Karfunkel, 1971; Burnside, 1973; Löfberg and Jacobson, 1974; avian neural plate: Handel and Roth, 1971; Karfunkel, 1972; mammalian neural plate: Ferm, 1963; O'Shea, 1981; lens epithelial cells: Beebe et al., 1979).

Although observations and experiments suggest that microtubules might function in cell elongation during formation of the neural plate, and possibly in the maintenance of the elongated configuration of these cells, the data are far from conclusive. One major fault of previous experiments is that agents used to disrupt microtubules may have other unrelated effects. For example, colchicine prevents the elongation of lens epithelial cells, not by disrupting microtubules per se, but by blocking normal increases in cell volume (Beebe et al., 1979). Low concentrations of colchicine block cell elongation even though microtubules are not disrupted (Beebe et al., 1979). Nocodazole, an antitumor drug that prevents the polymerization of microtubules (DeBrabander and Borgers, 1975; DeBrabander et al., 1976; DeBrabander and Wanson, 1976; Hoebeke et al., 1976), and binds to the colchicine binding site of tubulin (Hoebeke et al., 1976), does not prevent lens epithelial cells from elongating normally, even though microtubules are absent in treated cells (Beebe et al., 1979). A second major fault of previous experiments is that it is impossible to interpret their results definitively. Cells treated with colchicine (and other agents that disrupt microtubules) arrest in metaphase (i.e., the stage at which cells are normally spherical), forming typical "colchicine figures". Furthermore, neuroepithelial cells enter metaphase frequently, since all cells of the neuroepithelium of young embryos are mitotically active and cell generation time is relatively short (i.e., 5-8 hours in the chick neuroepithelium; Fujita, 1962; Langman et al., 1966). Thus, the presence of spherical cells in the neuroepithelium after treatment with colchicine could be due to the: (1) inhibition of elongation, (2) collapse of previously elongated cells or (3) arresting of cells in metaphase. All three interpretations are plausible, but the last one has been largely ignored. Therefore, at present it is possible to conclude only that microtubules might be involved in the elongation of neuroepithelial cells, and in the maintenance of their elongated configurations, but novel experiments are needed to test this possibility directly.

Morphological Features of the Neural Plate: The apical (outer) surface of the ectodermal layer of the embryo becomes subdivided morphologically into surface and neural ectodermal components as the neural plate forms (Figs. 1-3; Backhouse, 1974; Portch and Barson, 1974; Bancroft and Bellairs, 1975; Waterman, 1975; 1976; Nagele and Lee, 1979). Prospective epidermal cells (i.e., surface ectodermal cells) have a cobblestone appearance, and are delineated clearly from one another by short microfolds and microvilli (Figs. 1, 2). Additional microvilli are scattered sparsely over these cells, and each cell displays typically a single, somewhat centrally located cilium. Prospective neuroepithelial cells (i.e., neural ectodermal cells) have a much different appearance (Figs. 1, 3). Cell apices are much smaller than those of surface ectodermal cells, and cell boundaries are often less well defined. Short microvilli and a single cilium project from each neuroepithelial cell.

The apical surfaces of both the surface and neural ectoderm display unusual structures termed "beaded threads" (Figs. 2-4; birds: Backhouse, 1974; Bancroft and Bellairs, 1974; 1975; Bellairs and Bancroft, 1975; Nagele and Lee, 1979; 1980b; mammals: Waterman, 1976; 1979). These structures are far more numerous on the neural ectoderm than on the surface ectoderm, and are also present on the endodermal layer of the early

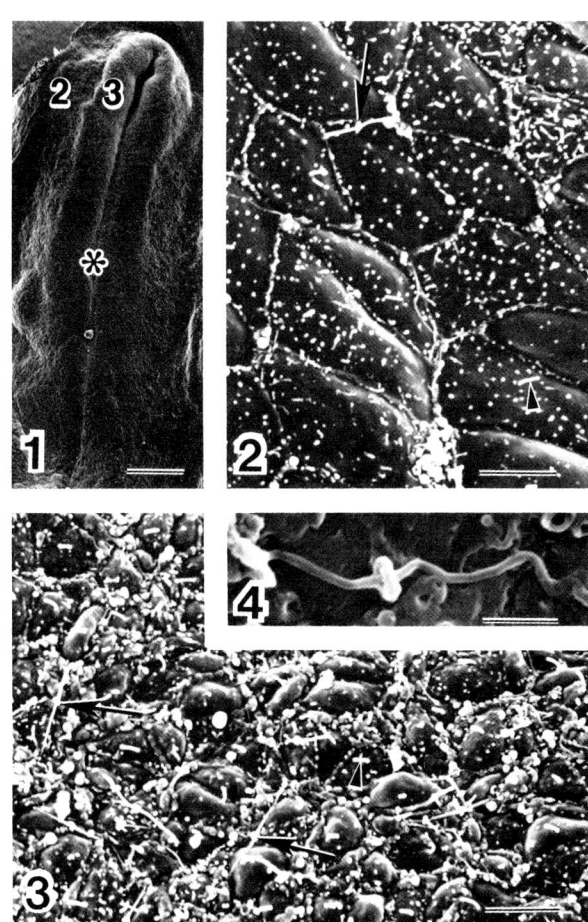

Fig. 1. Dorsal surface of a stage 7 embryo. Asterisk, cranial end of the primitive streak; 2, area of surface ectoderm enlarged in Figure 2; 3, area of neural ectoderm enlarged in Figure 3. Bar = 200 μm.

Fig. 2. Area of surface ectoderm from Figure 1. Arrow, "beaded thread"; arrowhead, cilium. Bar = 5 μm.

Fig. 3. Area of neural ectoderm from Figure 1. Arrows, "beaded threads"; arrowhead, cilium. Bar = 5 μm.

Fig. 4. Enlargement of a "beaded thread" from the neural plate of a stage 6 embryo. Bar = 1 μm.

embryo (Schoenwolf, unpublished observations) and on the ectoderm of the otic placode (Meier, 1978). A "beaded thread" consists of a filopodial-like process bearing a prominent dilatation about midway along its length (Fig. 4). Presumably, "beaded threads" are telophase bridges, since these structures interconnect two cells, contain microtubules and have the characteristic structure of a midbody in the area of the dilatation (Bellairs and Bancroft, 1975; Nagele and Lee, 1979). If these structures are telophase bridges then considerable rearrangement of daughter cells must take place in the neural ectoderm because "beaded threads" frequently span several intervening neuroepithelial cells (Fig. 3; Backhouse, 1974; Bancroft and Bellairs, 1974; 1975; Bellairs and Bancroft, 1975; Waterman, 1976; 1979; Nagele and Lee, 1979; 1980b).

The neuroepithelium is a typical pseudostratified layer throughout the stages of neurulation. Although it is often stated in general articles and textbooks that cells of the neural plate are shaped like columns (and those of the walls of the neural groove and tube, like wedges) SEM of slices through the neuroepithelium demonstrates that this notion is an oversimplification. SEM of such slices from chick embryos reveal that neuroepithelial cells can have at least three different configurations. Most neuroepithelial cells are spindle shaped (Fig. 5: S), containing a bulbous waist that tapers both apically and basally. Others are flask, bottle or wedge shaped (Fig. 5: F), tapering usually apically, but sometimes basally instead. The final type of cell is spherical, often present in pairs and is always positioned near the apex of the neuroepithelium (Fig. 6). These latter cells appear sequestered near the apical side of the epithelium, but they usually exhibit several slender filopodia that project from the cell body basally for various distances, sometimes reaching completely across the epithelium (Fig. 6; Waterman, 1979; García-Porrero and Ojeda, 1981). Apparently, these filopodia are either persisting remnants of cell attachments to the basal aspect of the epithelium, which stretched as these cells rounded up and entered the M phase of the cell cycle, or are new structures that pioneer re-extension of the daughter cells after mitosis is completed.

From the shape of a particular neuroepithelial cell it is possible to ascertain the approximate position of its nucleus, since the most bulbous portion of a flask-shaped or spindle-shaped cell contains the nucleus (this fact is evident when two types of measurements are compared: the greatest diameter of a particular cell from a scanning electron micrograph, and the diameter of a nucleus from a light micrograph of sectioned material--neuroepithelial cell nuclei are too large to be accommodated, except in the bulbous region; Schoenwolf, unpublished observations). Thus, spindle-shaped cells have centrally located nuclei, and wedge-shaped cells have apically or basally located nuclei; spherical cells are in the M phase of the cell cycle and, therefore, lack nuclei. The position of the nucleus is determined by the phase of the cell cycle since interkinetic nuclear migration occurs in the neuroepithelium (Sauer, 1935; Martin and Langman, 1965; Watterson, 1965). Nuclei migrate toward the lumen during early prophase; lie adjacent to the lumen during late prophase, metaphase, anaphase and early telophase; and migrate away from the lumen during late telophase (Watterson, 1965). Because the position of the nucleus apparently affects the shape of cells markedly, it is reasonable to suggest that the phase of the cell cycle might have more effect on cellular morphology than might other events, such as the possible constriction of apical bands of microfilaments. The latter event has been postulated to play a major role in the bending of the neural plate by transforming neuroepithelial cells from columns into wedges (see below).

Bending of the Neural Plate

Bending or folding of epithelial sheets is a common morphogenetic process that occurs during the early formation of the rudiments of a number of organ systems besides the central nervous system (e.g., thyroid rudiment: Shain et al., 1972; Hilfer, 1973; Hilfer et al., 1977; lens vesicle: Wrenn and Wessells, 1969; Zwaan and Hendrix, 1973; submandibular gland: Spooner and Wessells, 1970; 1972; Spooner, 1973a,b; pancreatic rudiment: Wessells and Evans, 1968). Two types of structures form during bending of the neural plate: the neural groove, a gutter-like space; and the neural folds, double-layered structures located at the lateral margins of the neural plate, each of which consists of an inner layer of neural ectoderm and an outer layer of surface ectoderm. Bending of the neural plate serves both to elevate the neural folds as well as to approximate these structures in the dorsal midline.

A major purpose of this article is to show that bending of the neural plate occurs in two distinct steps in chick embryos: elevation of the lateral parts of the neural plate and the incipient neural folds, and formation and convergence of the definitive neural folds. I have studied bending of the neural plate at all craniocaudal levels of this structure by LM, in literally hundreds of plastic sections (too much distortion occurs in paraffin sections to study bending accurately), and by SEM, in dozens of slices cut transversely through embryos. This is the first comprehensive investigation of the morphogenesis of the neuroepithelium that deals with regional differences in this process, and is also the first to provide evidence that bending occurs in two steps. I will first describe bending of the neural plate of the future brain, and then of the future spinal cord. Finally, I will discuss possible mechanisms involved in bending of the neural plate.

Bending of the Neural Plate of the Future Brain: The early neural plate (i.e., at stages 5-6) exhibits a distinct bilateral symmetry, since a well-defined sulcus extends longitudinally down the midline (Fig. 7). The notochord lies beneath this sulcus in the future mid- and hindbrain regions, but it is absent more cranially where scattered cells of the prechordal plate mesoderm occupy the midline. The cells of the neural plate are much taller than the more lateral surface ectodermal cells.

Elevation of the neural plate begins at stages 6-7 (Fig. 8). The neural groove, in the future mid- and hindbrain regions, appears as a "V"-shaped space as elevation occurs and the neural folds begin to form. Elevation of the plate occurs in a gradient fashion--lateral aspects of the plate elevate the most, the medial aspect remains fixed, and intermediate areas of the plate elevate less than lateral areas--and appears similar to the closing of a hinge. In the latter process there is a central pin around which two flat plates rotate. The neural plate appears to bend in this fashion, with the shorter cells in the midline forming a locus for bending that allows the two, relatively flat halves of the neural plate to rotate upward. Because the lateral halves of the neural plate have relatively straight profiles in transverse slices, the neural groove is "V" shaped during the period of elevation (Figs. 8, 9).

The morphology of the neural groove is somewhat different during elevation in the forebrain region than more caudally (Fig. 10). The midline part of the neural groove is "V" shaped, but the neuroepithelium buckles laterally, such that the apical surface of the epithelium on each side forms a broad, convex curve. During the next step of bending (convergence) the lateral portions of the neuroepithelium are inverted, allowing the folds to come into apposition in the dorsal midline. Considerable forces must be generated presumably to effect this process.

The second step in bending of the neural plate is the convergence of the neural folds. If bending of the neural plate occurred strictly in a hinge-like fashion, then the neurocele (the lumen of the neural tube) would be obliterated as the lateral walls of the neural groove came into apposition along their entire extent. This problem is overcome by the formation of two additional loci of bending that are associated with the convergence of the neural folds. These areas of bending are indicated by shallow furrows on the future luminal sides of the dorsolateral walls of the neural groove (Fig. 11). With formation of these bilateral furrows the neural groove changes from "V" shaped to diamond shaped (Fig. 11), and convergence of the neural folds ensues.

The onset of convergence is associated with the formation of the definitive neural folds. The definitive folds are bilaminar structures consisting of an inner (medial) layer of neural ectoderm covered by surface ectoderm (Figs. 11-13). The bilateral furrows associated with convergence always form after close apposition is established between the two layers of each neural fold (cf. Figs. 12, 13; only apposition has been established in Fig. 11, whereas apposition has been established and convergence has begun in Fig. 13). Furthermore, the bilateral furrows always form at the precise level at which the surface and neural ectodermal layers of the neural folds diverge from one another (Fig. 13).

Bending of the Neural Plate of the Future Spinal Cord: Bending of the neural plate in the region of the future spinal cord occurs in two steps, as in brain regions, but some differences exist between these two areas. The early neural plate of the future spinal cord is bilaterally symmetric, as is the case more cranially, with a shallow median sulcus separating the right and

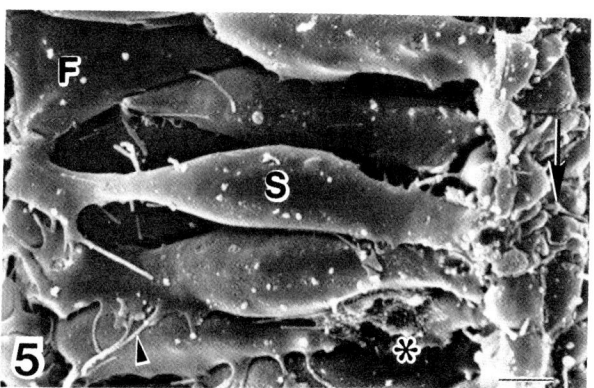

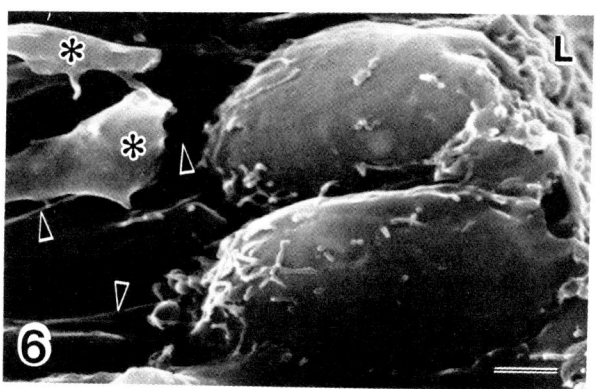

Fig. 5. Exposed surface of the neuroepithelium from the lateral wall of the neural groove (future spinal cord) of a stage 8 embryo. Flask-shaped (F) and spindle-shaped (S) cells are evident. Arrow, "beaded thread" on the apical surface of the epithelium; arrowhead, filopodia emanating from one cell and spreading over the surface of an adjacent one; asterisk, "nest" that presumably contained a cell in the M phase of the cycle prior to transection. Bar = 2 μm.

Fig. 6. Exposed surface of the apical part of the neuroepithelium from the lateral wall of the neural tube (future spinal cord) of a stage 9 embryo. Slender filopodia (arrowheads) extend basally from the cell bodies of the newly formed daughter cells. Asterisks, apical processes of neuroepithelial cells dislodged during transection; L, lumen of the incipient neural tube. Bar = 2 μm.

left halves of the plate (Fig. 14). An intimate spatial relationship exists between the flattened notochord and the supranotochordal cells of the neural plate. The latter cells appear to be pressed against the notochord and are about one-half as tall as the more lateral neuroepithelial cells overlying the segmental plates (Figs. 14, 15). This difference in cell height would presumably facilitate the elevation of the lateral areas of the neural plate at subsequent stages, by providing a locus of bending.

The neural folds in the future spinal cord region elevate during stages 8-10. It is obvious that during these stages the dorsal surface of the notochord is firmly attached to the ventral surface of the midline portion of the neural plate. This is evidenced by three facts: (1) The supranotochordal region is deformed as the incipient neural folds elevate, resulting in the appearance of a series of well-defined pits in this area (Figs. 15, 16). Elevation of the folds presumably exerts tension on the supranotochordal cells (which are apparently tacked down to the notochord), causing stretching (i.e., formation of the pits), particularly at the interfaces between the supranotochordal and more lateral plate cells. (2) During transection, separation often occurs between the neural plate and segmental plates, but rarely between the neural plate and notochord. This latter structure is often left suspended from the midline portion of the neural plate, in complete isolation from other mesodermal cells (Fig. 17). (3) After mild trypsinization (0.5% "1:250" trypsin + 0.2% EDTA in saline), the neuroepithelium separates readily from adjacent organ rudiments, except the notochord. Usually the notochord must be removed with forceps (Fisher and Schoenwolf, unpublished studies).

The basal aspect of the neural plate is covered with a basal lamina prior to the onset of bending (Bellairs, 1959; England, 1974; England and Cowper, 1976). During elevation of the lateral margins of the neural plate, discontinuities appear in the basal lamina of transected specimens in the areas of the incipient neural folds (Fig. 17: arrowhead). Whether these discontinuities represent former sites of attachment between the segmental plates and overlying ectoderm, or are associated with the formation of the neural folds, is unknown.

The neural folds are well defined by the time the lateral margins of the neural plate have elevated about halfway up, and convergence begins shortly after this time. Convergence of the folds is associated with the formation of bilateral furrows and a diamond-shaped neural groove, as in cranial regions.

In caudal regions of the developing spinal cord, the cranial end of the primitive streak is overlapped by the floor of the neural groove (Figs. 19, 20; Schoenwolf, 1979a). Although the floor of the neural groove is much thicker than in more cranial regions of the forming spinal cord (cf. Figs. 18, 19), the floor still acts as a locus for bending during elevation, giving the neural groove its characteristic "V" and diamond shapes (Figs. 19, 20).

The neural folds are highly developed in the region of the closing posterior neuropore: the entire dorsal halves of the lateral walls of the neural groove form the inner portions of the neural folds (Fig. 20). The surface ectoderm just lateral to the neural folds in this region forms wedge-like structures that extend ventrad about halfway down the lateral wall of the neural groove (Fig. 20: asterisk). These ectodermal wedges or "spurs" are characteristic features of caudal areas (Dryden, 1973; Schoenwolf, 1979a).

Bilateral furrows form precociously in caudal regions of stage 9-10 embryos (i.e., in the region of the sinus rhomboidalis, the area of continuity between the neural groove and primitive groove), appearing in most embryos prior to the onset of elevation. Less elevation of the neural folds is required in caudal regions because of the presence of an overlap zone (i.e., the neural folds are already elevated considerably due to the fact that the neural groove rests on top of the forming tail bud; Schoenwolf, 1979a).

Possible Mechanisms of Bending of the Neural Plate: Bending of the neural plate is a complex process that has been studied experimentally for about 50 years (for reviews see Curtis, 1967; Karfunkel, 1974; M. Jacobson, 1978; Lemire et al., 1978). Many different mechanisms have been hypothesized to be involved in this process. For the sake of simplicity these possible mechanisms can be categorized into two groups: intrinsic factors and extrinsic factors. Intrinsic factors consist of changes restricted to the neural plate—changes that would generate forces sufficient to elevate the neural folds and pull these structures medially. Such factors include the possible constriction of apical bands of microfilaments in neuroepithelial cells (for a review see especially Karfunkel, 1974), differential rates of mitosis (Derrick, 1937; Bragg, 1938; Burt, 1943; Davis, 1944; Gillette, 1944; Jelínek and Friebova, 1966) or cell elongation (Schroeder, 1970) and increase in the cohesiveness of neuroepithelial cells (Brown et al., 1941; Karfunkel et al., 1978). Extrinsic factors consist of forces originating outside of the neural plate—forces that would be sufficient to elevate the neural folds and displace these structures medially. Such factors include the possible mediad migration of surface epithelial cells (Bragg, 1938; Gillette, 1944; Schroeder, 1970) or mesodermal cells (Jacobson and Löfberg, 1969), elongation of mesodermal cells underlying the elevating neural plate (Schroeder, 1970), stretching of the neuroepithelium craniocaudally during elongation of the embryo (Jacobson and Gordon, 1976a,b; Gordon and Jacobson, 1978; A. Jacobson, 1978; 1980; 1981) and increases in the volume of extracellular matrix and in the number of mesenchymal cells underlying the elevating portions of the neural plate (Morriss and Solursh, 1978a,b; Schoenwolf and Fisher, 1982). Extrinsic factors could either pull the neural folds medially (in the case of forces generated by elongation of the embryo) or push these structures toward the dorsal midline (all other extrinsic forces). It is likely that bending of the neural plate is caused not by merely one factor--especially when this process occurs in two distinct steps--but rather that multiple factors play important roles in this process.

I will discuss below three plausible mechanisms that evidence currently suggests might be the major ones involved in bending of the neural plate: constriction of apical bands of microfilaments, stretching of the neuroepithelium craniocaudally, and inflation of the extracellular matrix. Other mechanisms might also be involved, but at the present time these three mechanisms seem to me to be the most likely ones.

The hypothesis that bending of the neural plate is caused by constriction of apical bands of microfilaments is based on three types of observations and experiments: (1) the apices of neuroepithelial cells contain circumferential bands of microfilaments (amphibians: Baker and Schroeder, 1967; Schroeder, 1970; Burnside, 1971; Karfunkel, 1971; birds: Karfunkel, 1972; Schroeder, 1973; Camatini and Ranzi, 1976; Nagele and Lee, 1980a; mammals: Freeman, 1972; Morriss and New, 1979; Wilson and Finta, 1980a,b); (2) microfilaments in neuroepithelial cells appear to be capable of contraction: (a) during bending of the neural plate, the apical bands of microfilaments thicken, their circumferences reduce

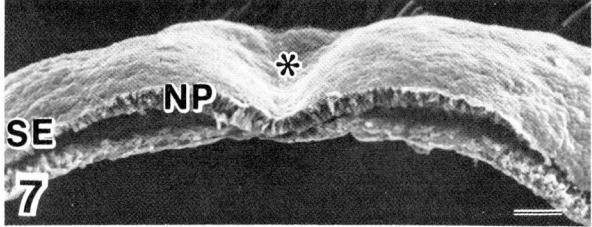

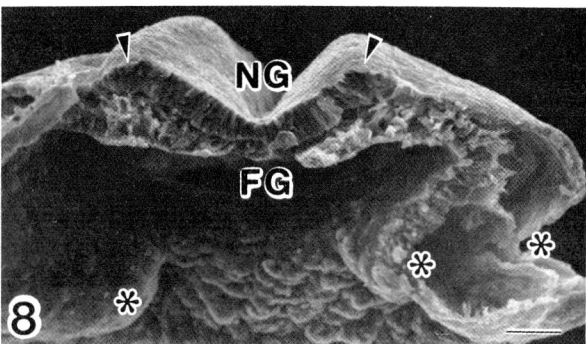

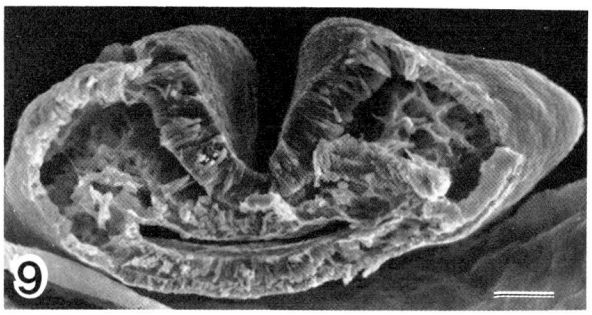

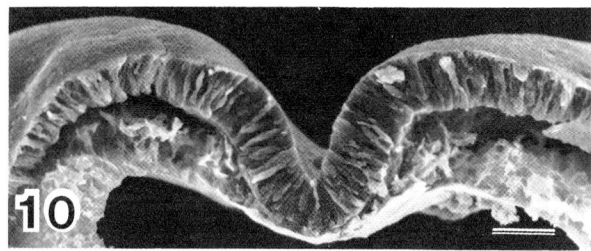

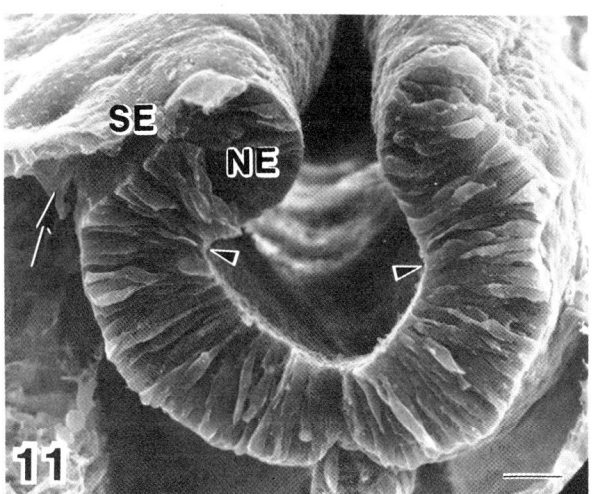

Fig. 7. Transverse slice through the neural plate (future midbrain) of a stage 6 embryo. Asterisk, midline sulcus separating the two halves of the neural plate; NP, neural plate; SE, surface ectoderm. Bar = 50 μm.

Fig. 8. Transverse slice through the neural groove (future midbrain) of a stage 7 embryo. Arrowheads, incipient neural folds; asterisks, portions of body folds; FG, foregut; NG neural groove. Bar = 50 μm.

Fig. 9. Transverse slice through the neural groove (future midbrain) of a stage 8 embryo. Bar = 50 μm.

Fig. 10. Transverse slice through the neural groove (future forebrain) of a stage 7 embryo. A few prechordal plate mesodermal cells are intercalated between the midline neural ectoderm and endoderm. Note that the lateral walls of the neural groove flare laterad, away from the midline. Bar = 40 μm.

Fig. 11. Transverse slice through the neural groove (future hindbrain) of a stage 9 embryo. The two layers of the neural folds (the surface, SE, and neural, NE, ectoderm) are present on the left side of the micrograph; the surface ectoderm has been removed on the right. Convergence of the neural folds has begun, resulting in formation of distinct bilateral furrows (arrowheads) on the apical surfaces of the dorsolateral walls of the neural groove. Arrow, distortion of the surface ectoderm created during slicing. Bar = 20 μm.

proportionately and the total length of microfilaments per cell remains relatively constant, suggesting a possible "sliding filament" action (Burnside, 1971); (b) apical microfilaments bind heavy meromyosin, suggesting that they have actin-like properties, and apices of neuroepithelial cells contain myosin (Nagele and Lee, 1978; 1980a); (c) actin apparently becomes redistributed, from basally to apically, in the cells of the cranial neuroepithelium of mouse embryos at the time that the neural folds change from a biconvex to a biconcave morphology (Sadler et al., 1982); (d) papaverine--a smooth muscle relaxant that inhibits the release of bound calcium (Imai and Takeda, 1967)--prevents bending of the neural plate (Moran and Rice, 1976); (e) ionophore A23187--an antibiotic that promotes calcium transport and release (Reed and Lardy, 1972)--accelerates bending of the neural plate (Moran and Rice, 1976; Lee et al., 1977); and (f) calcium, which might serve as a regulator of microfilament contraction in neuroepithelial cells, is located in coated vesicles in these cells (Nagele et al., 1981), and is released during neurulation (Moran, 1976); and (3) disrupting microfilaments with various cytochalasin congeners or vinblastine sulfate inhibits bending of the neural plate (amphibians: Karfunkel, 1971; Burnside, 1972; 1973; birds: Karfunkel, 1972; Linville and Shepard, 1972; Messier and Auclair, 1974; Lee and Kalmus, 1976; mammals: Wiley, 1980; Morris-Kay, 1981; O'Shea, 1981).

Although apical bands of microfilaments appear to be involved in bending of the neural plate, there are two major weaknesses in the hypothesis that bending is caused by changes in the shapes of neuroepithelial cells as generated by the contraction of microfilaments. First, it has never been proven that microfilaments within apices of neuroepithelial cells actively constrict during bending of the neural plate (although this is an assumption that is stated frequently as a fact in the literature, even appearing in the title of a paper by Nagele and Lee, 1980a, where no evidence is presented that it occurs). It is obvious that in amphibians many neuroepithelial cells become wedge shaped during bending of the neural plate (Schroeder, 1970; Burnside, 1971; Karfunkel, 1971), but it is unknown whether: (a) cells become wedge shaped actively (due to the contraction of their apical microfilaments, or some other mechanism) or passively (because the plate bends, deforming the apices of these cells), (b) enough cells become wedge shaped (assuming the change in shape is active) to cause bending of the neural plate, or (c) such changes in the shapes of neuroepithelial cells are unique to amphibians (i.e., such changes are not obvious in other vertebrates). In an elegant series of studies by A. Jacobson and co-workers (Burnside and Jacobson, 1968; Burnside, 1971; Jacobson and Gordon, 1976a,b) it has been amply demonstrated that the surface area of the neural plate of Taricha torosa embryos shrinks during stages 13-15 (i.e., stages prior to the elevation of the neural folds), and that different areas of the neural plate shrink to different extents. As shrinkage occurs, corresponding changes occur in cell heights, such that the volume of the neural plate remains constant (i.e., there is an inverse relationship between apical surface area and cell height). Shrinkage of the neural plate appears to be an autonomous process, since neural plates that are isolated from adjacent cells and cultured at stage 13 still undergo essentially normal shrinkage. Furthermore, when neuroepithelial cells from one area of the neural plate are transplanted to another area that normally undergoes a different amount of shrinkage, cells shrink according to their original positions, ignoring the fact that they are in new sites. These observations strongly suggest that extrinsic factors have no role in the early shaping of the amphibian neural plate. It must be emphasized that similar analyses have not been done during stages of bending of the neural plate, so the possible roles of shrinkage in this process have not been evaluated. Furthermore, the mechanism of shrinkage is unclear. Reduction of the surface area of the neural plate could be due to either apical constriction of microfilaments (as

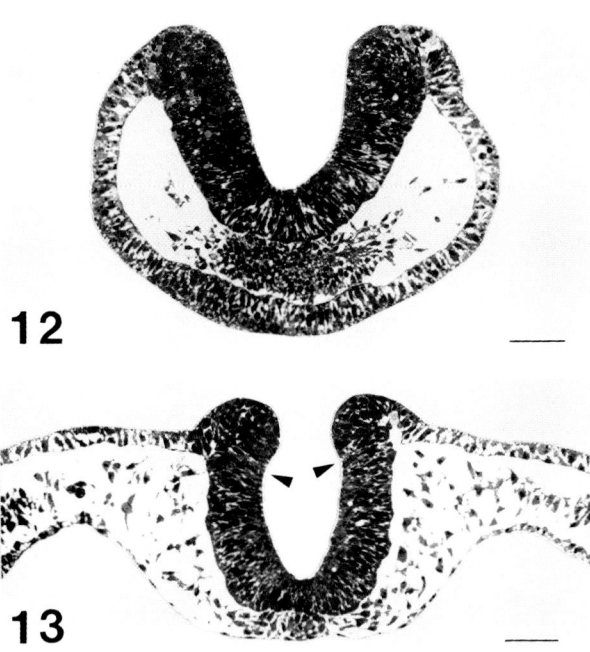

Fig. 12. Transverse section through the neural groove (future forebrain) of a stage 8 embryo. The surface and neural ectodermal layers of the neural folds are apposed closely to one another. Elevation of the neural folds is completed, but convergence has not yet begun. Bar = 50 μm.

Fig. 13. Transverse section through the neural groove (future hindbrain) of a stage 9 embryo. Convergence has begun. Arrowheads, bilateral furrows. Bar = 60 μm.

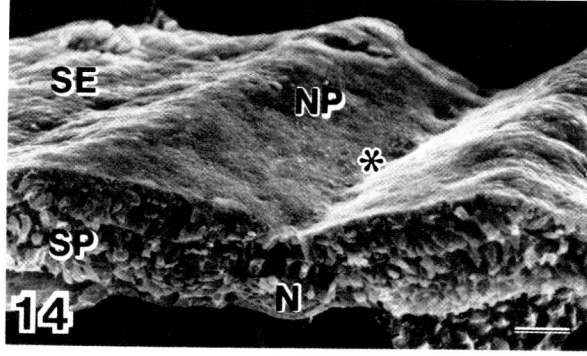
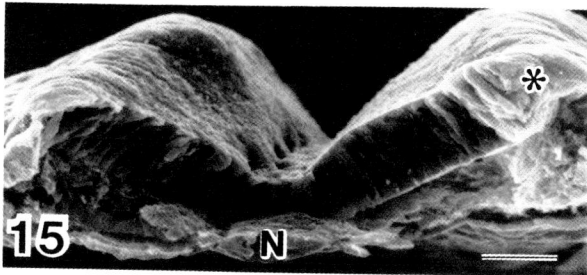
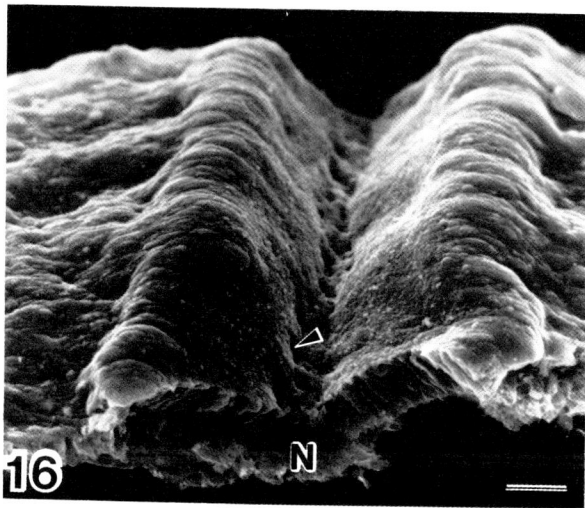
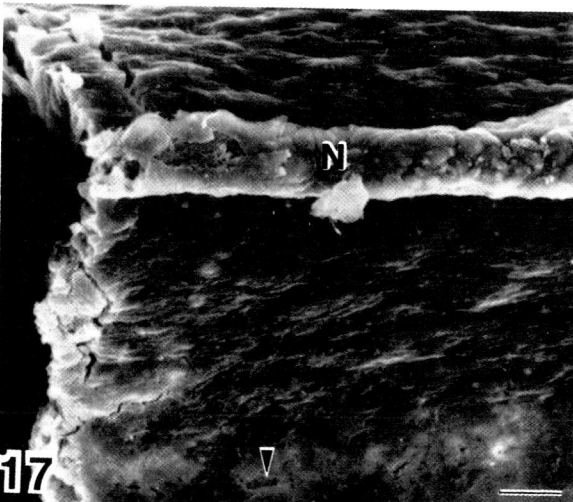
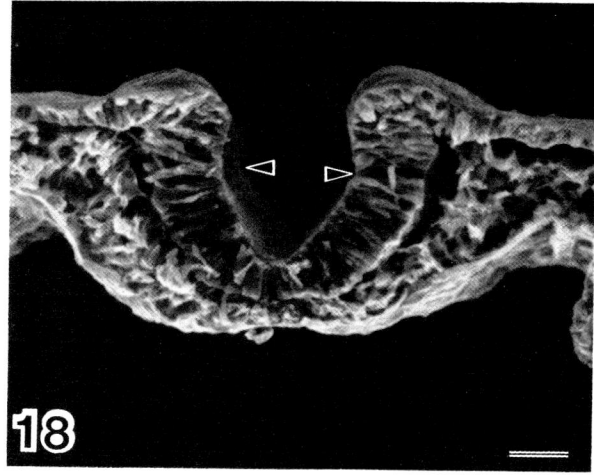

Fig. 14. Transverse slice through the neural plate (future spinal cord) of a stage-8 embryo. Asterisk, midline sulcus separating the two halves of the neural plate; N, notochord; NP, neural plate; SE, surface ectoderm; SP segmental plate. Bar = 30 µm.

Fig. 15. Transverse slice through the neural groove (future spinal cord) of a stage 8 embryo. The neural folds (asterisk) are beginning to form. N, notochord. Bar = 30 µm.

Fig. 16. Same specimen as that shown in Figure 14 tilted to reveal the craniocaudal extent of a series of irregular pits (arrowhead) associated with the supranotochordal cells of the neuroepithelium. N, notochord. Bar = 30 µm.

Fig. 17. Transverse slice through the neural groove (future spinal cord) of a stage 8 embryo. The neural plate separated from the underlying cells during transection, except for those of the notochord (N). The slice is viewed from the basal (ventral) aspect of the neural plate and notochord. The basal aspect of the neural plate is covered by a basal lamina (wisps of this material run horizontally across the plate and obscure the basal ends of the neuroepithelial cells). Periodic discontinuities are present in the basal lamina in the region of the incipient neural fold (arrowhead). Bar = 30 µm.

Fig. 18. Transverse slice through the neural groove (future spinal cord) of a stage 9 embryo. Convergence has begun, as indicated by the presence of bilateral furrows (arrowheads). The neural folds are well formed. Bar = 30 µm.

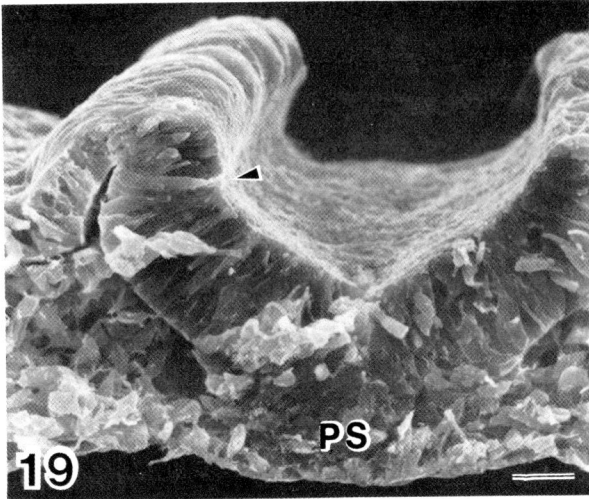

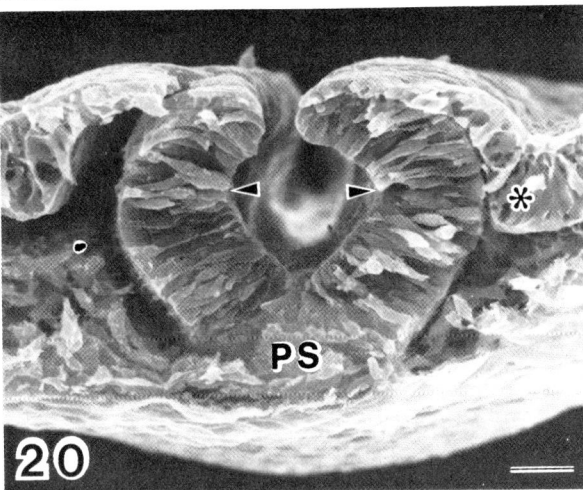

Fig. 19. Transverse slice through the neural groove (future spinal cord) of a stage 10 embryo. Bilateral furrows (arrowhead) indicate that convergence is under way. The primitive streak (PS) occupies the ventral midline. Bar = 20 μm.

Fig. 20. Transverse slice through the neural groove (future spinal cord) of a stage 11 embryo. Convergence is almost completed. On the left side of the micrograph the surface and neural ectodermal layers of the neural fold were separated during transection. Arrowheads, bilateral furrows; asterisk, wedge-like area of surface ectoderm; PS, primitive streak. Bar = 20 μm.

The second major fault of the hypothesis that bending is the result of microfilament-mediated, apical constriction is that it is unclear whether cytochalasins inhibit bending of the neural plate by disrupting microfilaments per se, because these drugs have a multitude of other effects (e.g., cytochalasin B inhibits protein, glycoprotein and mucopolysaccharide synthesis; glucose metabolism; nucleotide and sugar transport across the plasmalemma; etc.; for reviews see Burnside and Manasek, 1972; Holtzer and Sanger, 1972; Pollard and Weihing, 1974). Further complicating the matter is the fact that cytochalasin B curbs interkinetic nuclear migration (Messier and Auclair, 1974; Lee and Kalmus, 1976). Because considerable changes occur in cell shape during this latter process, inhibition of bending may be due to this side effect, rather than to the prevention of microfilament-mediated apical constriction. Also in neurulating hamster embryos dysraphic defects still result even when embryos are treated with concentrations of cytochalasin that do not disrupt microfilaments (Wiley, 1980). It was concluded in this latter study, as in certain previous studies (Tannenbaum et al., 1977a,b; Stagno and Low, 1978), that the principal effect of cytochalasin is on the plasmalemma.

The observations of the present study on the morphogenesis of the chick neuroepithelium suggest that bending of the neural plate does not occur uniformly along its entire width. The neural groove is "V" shaped during elevation, with a single locus of bending in the ventral midline. This suggests that if bending is a result of apical constriction of neuroepithelial cells, then this constriction must be localized initially to the supranotochordal region. If all of the cells of the neural groove constricted apically then the neural groove would be "U" shaped rather than "V" shaped (i.e., the apical sides of the lateral walls of the neural groove would form concave curves in transverse sections if apical constriction occurs uniformly, rather than straight lines or convex curves, as shown by the present study). Likewise, two additional loci of bending appear as the neural folds begin their convergence. This again suggests that apical constriction is localized to precise areas of the wall of the neural groove, rather than being widespread. Further studies are necessary to compare, and perhaps quantify, possible differences in the shapes, cytoskeletal structures, etc., of neuroepithelial cells located in loci of bending and in adjacent areas of the neuroepithelium.

Finally, it should be mentioned that microfilaments might have a vital role in bending of the neural plate that is not considered usually: these structures might have an important skeletal function. Apical ends of neuroepithelial cells are interconnected by junctions originally called "terminal bars" (Sauer, 1935). Transmission electron microscopy of thin sections and freeze-fracture replicas have revealed that three types of intercellular junctions are present: gap and tight junctions and desmosomes (Burnside, 1971; 1973; Revel et al., 1973; Decker and Friend,

is usually assumed) or to cell elongation (since the increase in the heights of groups of neuroepithelial cells is proportionate to the amount of shrinkage of their surface areas).

1974; Bellairs et al., 1975; Revel and Brown, 1976; Schoenwolf and Kelley, 1980; Chamberlain and Scales, 1981; Decker, 1981). Microfilaments appear to insert into these junctions (e.g., Nagele and Lee, 1980a), and, apparently, they have a role in holding cells together since cells dissociate from epithelial layers when early embryos are treated with cytochalasin B (Stagno and Low, 1978). Furthermore, treatment of neurulating chick embryos with cytochalasin B disrupts apical intercellular junctions, presumably stopping interkinetic nuclear migration in the process (Messier and Auclair, 1977). Thus, a principal function of microfilaments during neurulation might be to stabilize the apical surface of the neuroepithelium, aiding in formation of the neural groove.

The hypothesis that bending of the neural plate is caused by stretching of the neuroepithelium craniocaudally has been suggested by A. Jacobson and co-workers. Computer modeling studies show that it is possible to generate the characteristic "keyhole" shape of the amphibian neural plate by programming in both a reduction in surface area (or an increase in cell height) and craniocaudal elongation, demonstrating that their contention is at least plausible (Jacobson and Gordon, 1976a,b; Gordon and Jacobson, 1978; A. Jacobson, 1978; 1980; 1981). Furthermore, during closure of the neural groove in amphibian embryos, the rate of craniocaudal elongation of the midline of the embryo increases tenfold from previous stages, and then decreases rapidly after closure is completed (Gordon and Jacobson, 1978; A. Jacobson, 1978; 1980). Likewise, in stage 9 chick embryos, the portion of the embryo containing the closing neural tube exhibits a 29% increase in length over a 4 hour period during which measurements were made, whereas more cranial and caudal areas showed, respectively, only a 3-4% and a 7% increase in length (Jacobson, 1980). These observations have led Jacobson (1980) to conclude that craniocaudal "elongation is largely responsible for plate shaping and may be the dominant factor in buckling the plate out of the plane and folding it into a tube." Although it is clear that elongation occurs, how this process is brought about is unknown. Some studies have suggested that elongation requires the presence of the notochord (Jacobson and Gordon, 1976a), whereas others (Malacinski and Youn, 1981; Youn and Malacinski, 1981) have shown that elongation (as well as morphogenesis of the neural plate and formation of the neural tube) is normal in "notochord-defective" embryos. At present it is obvious that a direct relationship exists between stretching of the neuroepithelium and closure of the neural groove, but new experiments are needed to determine how this elongation occurs and whether this intriguing relationship is causal or merely coincidental.

The final hypothesis about bending of the neural plate that I will discuss is that bending is caused by inflation of the extracellular matrix underlying the neural plate. During formation and bending of the neural plate, the areas subjacent to the neuroepithelium, neural folds and prospective epidermis contain head mesenchymal or somitic cells (Figs. 7-10, 12-15, 18-20) embedded in an extracellular matrix rich in glycosaminoglycans (birds: Pratt et al., 1975; Solursh, 1976; Fisher and Solursh, 1977; mammals: Solursh and Morriss, 1977; Morriss and Solursh, 1978a,b) and glycoproteins, such as fibronectin (Waterman and Balian, 1980). It is reasonable to suggest that increase in the number of mesenchymal cells underlying the neural plate, or elongation of mesodermal cells during somitogenesis (Lipton and Jacobson, 1974), might assist in bending of the neural plate, but little has been done to directly test this notion. Similarly, it is reasonable to suggest that changes in the extracellular matrix might generate forces sufficient to assist in bending of the neural plate. In our laboratory we have begun to test the possible roles of the extracellular matrix in this process in chick embryos. Because hyaluronic acid (HA) is a major component of the extracellular matrix, constituting at least 84% of the glycosaminoglycans synthesized by early chick embryos (Solursh, 1976), and since HA is associated with a high degree of hydration (which would presumably cause it to swell shortly after it was secreted into the extracellular compartment) (Laurent, 1970) we have degraded HA in ovo, by injecting Streptomyces hyaluronidase subblastodermically (Schoenwolf and Fisher, 1982). Neurulation in treated embryos has been examined so far only in the region of the future spinal cord, due to technical reasons. In this area, neural tube defects (i.e., "spina bifida") are present in 60-94% of the embryos (depending on the particular enzyme batch) following degradation of HA. Furthermore, neural folds elevate normally in defective regions, but convergence, with formation of bilateral furrows, does not occur. Elevated neural folds in many cases not only fail to converge but diverge instead, actually flaring laterally. Control experiments have ruled out the possibilities that neural tube defects are caused by a non-specific protease contaminant of S. hyaluronidase, or by the digestion products of HA. Our observations suggest that in the region of the forming spinal cord the extracellular matrix plays little or no role in the elevation of the neural folds, but plays a major role in the convergence of these structures. It must be emphasized that all drugs have side effects (as discussed above)--S. hyaluronidase is certainly no exception--and it remains to be determined whether convergence is blocked specifically due to the absence of HA, disruption of the integrity of other components of the extracellular matrix, or merely some unrelated side effect. However, two other studies appear to provide light on this subject. In one (Copp and Wilson, 1981) it was demonstrated that differences exist in certain constituents of the extracellular matrix (i.e., HA and sulfated glycosaminoglycans) of normal and mutant (loop-tail) embryos, implying a role for extracellular materials in closure of the neural tube. In a second study (Morriss-Kay, 1981), it was shown that treatment of mouse embryos with β-D-xyloside (a substance that inhibits proteoglycan synthesis) causes neural tube defects. Thus, it appears that glycosaminoglycans and proteoglycans might have roles in formation of the neural tube, at least in avian and mammalian embryos, but what those specific roles might be is unclear at present.

I have attempted in the preceding section to discuss the major mechanisms that seem to be involved in bending of the neural plate. It is evident that much remains to be learned about this process. For each possible mechanism of bending there are enough faults with the experimental designs, or insufficient experiments, to make interpretation of the available data difficult at best. At the present time it is possible to conclude only that the above three mechanisms appear to be associated with bending of the neural plate. Other mechanisms are possibly involved also. Further difficulties result from the possibility that bending might occur by different mechanisms in different species (or that one mechanism might be more active than another in a particular organism)--perhaps it is not valid to extrapolate data from frogs to chicks to mice--and bending might occur differently at various craniocaudal levels of the same organism. Finally, the developing central nervous system is highly sensitive to a wide variety of teratogens, making it difficult to differentiate between the principal effect of a particular experimental treatment, and unrelated side effects. Only after these formidable problems are overcome will we be able to state with confidence what mechanisms cause bending of the neural plate.

Fusion of the Neural Folds

Fusion of the neural folds results ultimately in the formation of the roof of the neural tube and an intact overlying layer of surface epithelium. In addition the formation and early migration of the neural crest occur during this stage in chick embryos.

Fusion of the neural folds has been studied with both SEM and transmission electron microscopy in amphibians (Tarin, 1971; Rice and Moran, 1977; Mak, 1978), birds (Portch and Barson, 1974; Revel, 1974; Bancroft and Bellairs, 1975; Revel and Brown, 1976; Santander and Cuadrado, 1976; Schoenwolf, 1978a; 1979b, Silver and Kerns, 1978) and mammals (Waterman, 1975; 1976; 1979; Goodman and Waterman, 1976; Geelen and Langman, 1979). The specific mechanisms involved in the approximation and fusion of the neural folds are unknown. In mouse and hamster embryos, cellular protrusions (i.e., filopodia, blebs and ruffles) might function in cellular recognition and serve to guide the contralateral folds together (Waterman, 1975; 1976). But similar protrusions are still present in embryos treated in utero with excess Vitamin A (Goodman and Waterman, 1976)--embryos in which the neural folds fail to approximate and fuse--and in anencephalic embryos of the Oel strain (Waterman, 1979), suggesting that such protrusions might not actually function in closure of the neural groove. These protrusions might reflect instead changes involved in the liberation of the neural crest from the depths of the neural folds, or might even be associated with the formation of the definitive neural folds.

In chick embryos, protrusions are present only rarely on the surfaces of the neural folds prior to their approximation (Fig. 21; Schoenwolf, 1978a; 1979b; Silver and Kerns, 1978; Nagele and Lee, 1979). However, in a study by Bancroft and Bellairs (1975) cellular "threads" were sometimes observed spanning the gap between the contralateral neural folds prior to their approximation, suggesting that these protrusions might function actively in closure of the neural groove by towing the folds together. But it seems more likely that these "threads" are artifacts generated as previously approximated, and partially fused, folds are separated from one another during processing (Schoenwolf, 1979b). In any event, approximation of the neural folds is followed by extensive cell rearrangements, and numerous filopodia and blebs form on the dorsal surface of the fusing neural folds (Fig. 22; Gouda, 1974; Revel, 1974; Bancroft and Bellairs, 1975; Revel and Brown, 1976; Santander and Cuadrado, 1976; Silver and Kerns, 1978; Schoenwolf, 1979b).

The neural folds, as stated previously, consist of an inner layer of neural ectoderm and an outer layer of surface ectoderm. Several features of the neural folds have been described in the chick embryo (Schoenwolf, 1979b). The surface ectodermal component of each neural fold reflects over the crest of this structure, merging with the neural ectoderm lining the neural groove. Each neural fold contains a superficial (upper) half, consisting of neural ectoderm covered by surface ectoderm, and a deep (lower) half, composed entirely of neural ectoderm (Schoenwolf, 1979b: his Figs. 8, 17, 22). Initial contact between the neural folds usually occurs near the junction between these halves in cranial and trunk regions, but is restricted principally to surface ectoderm in the region of the closing posterior neuropore. Subsequent fusion of the folds at all craniocaudal levels involves both ectodermal layers (Fig. 23), but is completed first in the surface ectodermal layer (Fig. 24). Similar features occur during fusion of the neural folds of mammalian embryos (Geelen and Langman, 1977).

Extracellular materials form thick coats on the crests of the neural folds during closure of the neural groove in amphibians (Moran and Rice, 1975; Rice and Moran, 1977; Mak, 1978), birds (Lee, Sheffield et al., 1976, Lee et al., 1978; Silver and Kerns, 1978) and mammals (Sadler, 1978). These materials may serve as temporary adhesives, binding the leading edges of the neural folds together as fusion occurs. This view has been tested by experiments in which either concanavalin A or trypsin is applied to neurulating embryos. Both of these substances inhibit neurulation, but their modes of action are unclear. The authors of these studies suggest that neurulation is inhibited because the normal distribution of extracellular materials on the neural folds is presumably disrupted, preventing the folds from adhering across the midline (Lee, 1976; Lee, Nagele and Kalmus, 1976; Lee, Sheffield et al., 1976; Lee et al., 1978). However, other defects are also present, such as the inhibition of interkinetic nuclear migration (Lee, 1976) and gross deformation of the embryo (see especially Lee et al., 1978: their Fig. 17), which makes these experiments difficult to interpret.

Following contact of the neural folds in the dorsal midline, focal intercellular junctions (tight or gap?) form between cells of the apposed folds in birds (Santander and Cuadrado, 1976; Schoenwolf, 1979b), and mammals (Geelen and Langman, 1979). These junctions presumably act in holding the neural folds together, or perhaps they function in cell-to-cell communication. Studies using freeze fracture techniques are needed to identify the types of intercellular junctions present in fusing areas, so that some insight into their possible roles in fusion might be obtained.

In chick embryos, neural crest cells form and begin their migration as fusion of the neural folds occurs (Fig. 24). In cranial regions of mouse embryos, the neural crest form and begin migrating precociously, as the neural folds are elevating (Nichols, 1981). Thus, depending on the species (and the precise craniocaudal level), formation of the neural crest can begin either prior to or after formation of the neural tube. In the chick embryo, neural crest cells migrate

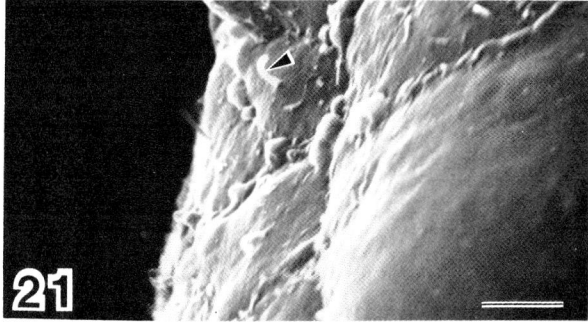

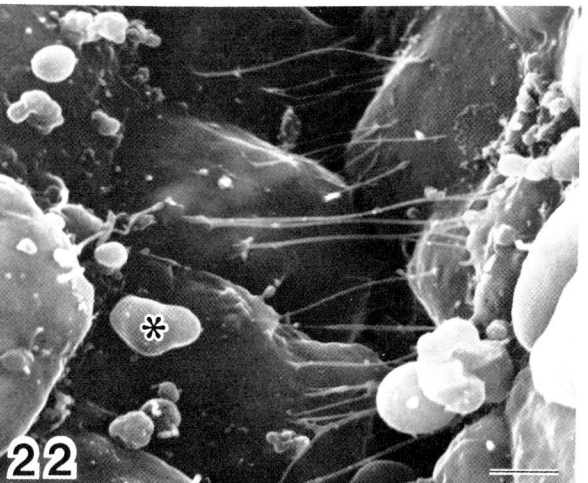

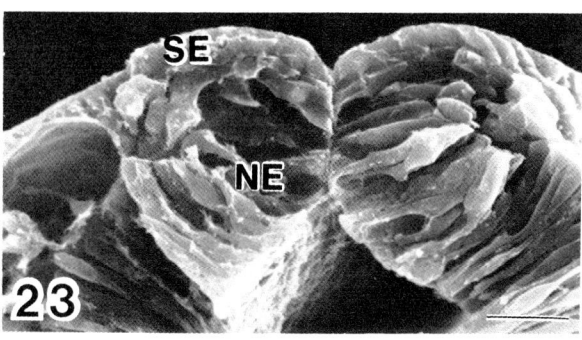

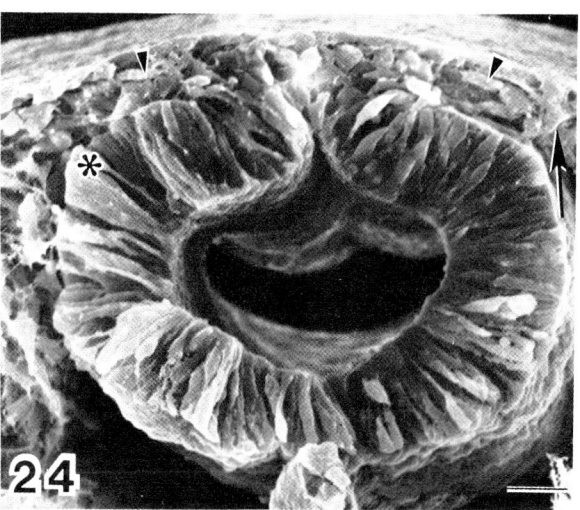

Fig. 21. Right neural fold from the future hindbrain of a stage 8 embryo (apposition of the folds was imminent). Arrowhead, cilium. Bar = 2 μm.

Fig. 22. Dorsal view of the fusing neural folds in the future hindbrain region of a stage 9 embryo. The cells of the two folds are in contact across the midline and many filopodia span the area undergoing fusion. Small bleb-like structures (asterisk) are prevalent. Bar = 2 μm.

Fig. 23. Transverse slice through the dorsal part of the incipient neural tube (future hindbrain) of a stage 9 embryo. Contact has been established across the midline between the two neural folds, with the surface (SE) and neural (NE) ectodermal layers of each fold participating in this contact. Bar = 20 μm.

Fig. 24. Transverse slice through the neural tube (future mid- or hindbrain) of a stage 10 embryo. Neural crest cells (arrowheads) are migrating from the roof of the neural tube. Arrow, persisting connection of the surface and neural ectodermal layers of the neural fold; asterisk, presumed site of a previous connection between the surface and neural ectodermal layers of the neural fold. Bar = 20 μm.

laterad from the roof of the neural tube, between the surface and neural ectodermal components of the fusing neural folds (Fig. 24). During the earlier convergence of the neural folds these two layers were closely apposed and interconnected in each fold, as described above. Presumably, connections between the two layers of each fold must be severed, so that these layers separate from one another, forming a space through which migration of the neural crest can occur. Neural crest cells might play an active role in the separation of these two layers, since separation occurs as the neural crest form. Further studies will be necessary to test this notion, and, if it is true, to determine how the neural crest cells bring separation about.

## CONCLUSIONS

As discussed in this report, primary neurulation occurs in three stages: the formation of the neural plate, the bending of the neural plate and the fusion of the neural folds. Much remains to be learned about how each of these events occurs. Particularly: What are the mechanisms of cell elongation during formation of the neural plate? How do neuroepithelial cells maintain their elongated configurations? What factors cause elevation and convergence during bending of the neural plate? What directs (organizes) cell rearrangement during fusion of the neural folds? These questions remain as challenges for future research. Research that will hopefully not only answer these fascinating and important questions, but will also provide insight on why neurulation frequently goes awry, resulting in serious congenital anomalies.

## DEDICATION

Dedicated to Professor Ray L. Watterson on the occasion of his retirement after over forty years of altruistic service in teaching, administration and research.

## ACKNOWLEDGMENTS

The superb technical assistance of Nancy B. Chandler is acknowledged gratefully. I am most thankful for many stimulating discussions with Dr. Marilyn Fisher throughout the course of this investigation, and her editorial comments during the preparation of the manuscript. The secretarial assistance of Arlene P. Mullins was invaluable. Supported by Grant Nos. HD15231 and NS18112 from the National Institutes of Health.

## REFERENCES

Backhouse M. (1974). Observations on the development of the early chick embryo, Scanning Electron Microsc. 1974:525-532.

Baker PC, Schroeder TE. (1967). Cytoplasmic filaments and morphogenetic movement in the amphibian neural tube. Develop. Biol. 15, 432-450.

Bancroft M, Bellairs R. (1974). The onset of differentiation in the epiblast of the chick blastoderm (SEM and TEM). Cell Tiss. Res. 155, 399-418.

Bancroft M, Bellairs R. (1975). Differentiation of the neural plate and neural tube in the young chick embryo. Anat. Embryol. 147, 309-335.

Beebe DC, Feagans DE, Blanchette-Mackie EJ, Nau ME. (1979). Lens epithelial cell elongation in the absence of microtubules: Evidence for a new effect of colchicine. Science 206, 836-838.

Bellairs R. (1959). The development of the nervous system in chick embryos, studied by electron microscopy. J. Embryol. Exp. Morphol. 7, 94-115.

Bellairs R, Bancroft M. (1975). Midbodies and beaded threads. Amer. J. Anat. 143, 393-398.

Bellairs R, Breathnach AS, Gross M. (1975). Freeze-fracture replications of junctional complexes in unincubated and incubated chick embryos. Cell Tiss. Res. 162, 235-252.

Bolli VP. (1966). Sekundäre Lumen bildungen in Neuralrohr und Ruckenmark Menschlicher Embryonen. Acta Anat. 64, 48-81.

Bragg AN. (1938). The organization of the early embryo of Bufo cognatus as revealed especially by the mitotic index. Z. Zellforsch. Mikros. Anat. 28, 154-178.

Brown MG, Hamburger V, Schmitt FO. (1941). Density studies on amphibian embryos with special reference to the mechanism of organizer action. J. Exp. Zool. 88, 353-372.

Burnside B. (1971). Microtubules and microfilaments in newt neurulation. Develop. Biol. 26, 416-441.

Burnside B. (1972). Experimental induction of microfilament formation and contraction. J. Cell Biol. 55, 33a. (abstract).

Burnside B. (1973). Microtubules and microfilaments in amphibian neurulation. Amer. Zool. 13, 989-1006.

Burnside MB, Jacobson AG. (1968). Analysis of morphogenetic movements in the neural plate of the newt Taricha torosa. Develop Biol. 18, 537-552.

Burnside B, Manasek FJ. (1972). Cytochalasin B: Problems in interpretating its effects on cells. Develop. Biol. 27, 443-444.

Burt AS. (1943). Neurulation in mechanically and chemically inhibited Amblystoma. Biol. Bull. 85, 103-115.

Byers B, Porter KR. (1964). Oriented microtubules in elongating cells of the developing lens rudiment after induction. Proc. Nat. Acad. Sci. (USA). 52, 1091-1099.

Camatini M, Ranzi S. (1976). Ultrastructural analysis of the morphogenesis of the neural tube, optic vesicle and optic cup in chick embryo. Acta Embryol. Exp. 1, 81-113.

Chamberlain JG, Scales DL. (1981). Gap junctions in freeze fracture replicas of rat embryo brains. Cell Biol. Internat'l Rep. 5, 219-220.

Copp, SN, Wilson DB. (1981). Cranial glycosaminoglycans in early embryos of the looptail (Lp) mutant mouse. J. Craniofacial Genet. and Develop. Biol. 1, 253-260.

Criley BB. (1969). Analysis of the embryonic sources and mechanisms of development of posterior levels of chick neural tubes. J. Morphol. 128, 465-501.

Curtis, ASG. (1967). The Cell Surface: Its Molecular Role in Morphogenesis, Academic Press, New York, 306-311.

Davis, JO. (1944). Photochemical analysis of neural tube formation. Biol. Bull. 87, 73-95.

DeBrabander M, Borgers M. (1975). The formation of annulated lamellae induced by the disintegration of microtubules. J. Cell Sci. 19, 331-340.

DeBrabander MJ, Van de Veire RML, Aerts FEM, Borgers M, Janssen PAJ. (1976). The effects of Methyl [5-(2-Thienylcarbonyl)-1H-benzimidazol-2-yl] carbamate, (R 17934; NSC 238159), a new synthetic antitumoral drug interfering with microtubules, on mammalian cells cultured in vitro. Cancer Res. 36, 905-916.

DeBrabander M, Wanson JC. (1976). The role of microtubules in the reassociation of isolated hepatocytes from adult rats. Internat'l Cong. Cell Biol. 1976, 87a. (abstract).

Decker, RS. (1981). Disassembly of the zonula occludens during amphibian neurulation. Develop. Biol. 81, 12-22.

Decker RS, Friend DS. (1974). Assembly of gap junctions during amphibian neurulation. J. Cell Biol. 62, 32-47.

Derrick, GE. (1937). An analysis of the early development of the chick by means of the mitotic index. J. Morphol. 61, 257-284.

Dryden R. (1973). Spontaneous and Experimentally-Induced Spina Bifida in Chick Embryos, Ph.D. Thesis, University of Birmingham, England.

Dryden RJ. (1980a). Duplication of the spinal cord: A discussion of the possible embryogenesis of diplomyelia. Develop. Med. Child Neurol. 22, 234-243.

Dryden RJ. (1980b). Spina bifida in chick embryos: Ultrastructure of open neural defects in the transitional region between primary and secondary modes of neural tube formation, in: Advances in the Study of Birth Defects, Vol. 4, T.V.N. Persaud (ed), MTP Press Ltd., Lancaster, 75-100.

England MA. (1973). The occurrence of a band of nuclei in primary neural induction in the chick embryo. Experentia 29, 1267-1268.

England MA. (1974). Cytoplasmic changes in primary neural induction. Experentia 30, 808-809.

England MA, Cowper SV. (1975). Primary neural induction as studied by scanning electron microscopy. Experentia 31, 1449-1451.

England MA, Cowper SV. (1976). A transmission and scanning electron microscope study of primary neural induction. Experentia 32, 1578-1580.

Ferm VH. (1963). Colchicine teratogenesis in hamster embryos. Proc. Soc. Exp. Biol. Med. 112, 775-778.

Fisher M, Solursh M. (1977). Glycosaminoglycan localization and role in maintenance of tissue spaces in the early chick embryo. J. Embryol. Exp. Morphol. 42, 195-207.

Freeman (1972). Surface modifications of neural epithelial cells during formation of the neural tube in rat embryo. J. Embryol. Exp. Morphol. 28, 437-448.

Fujita S. (1962). Kinetics of cellular proliferation. Exp. Cell Res. 28, 52-60.

García-Porrero JA, Ojeda JL. (1981). A stereoscan analysis of cell surface characteristics during the interkinetic nuclear migration in normal and colchicine-treated developing chick retina. Experentia 37, 181-182.

Geelen JAG, Langman J. (1977). Closure of the neural tube in the cephalic region of the mouse embryo. Anat. Rec. 189, 625-640.

Geelen JAG, Langman J. (1979). Ultrastructural observations on closure of the neural tube in mouse. Anat. Embryol. 156, 73-88.

Gillette R. (1944). Cell number and cell size in the ectoderm during neurulation (Amblystoma maculatum). J. Exp. Zool. 96, 201-222.

Goodman P, Waterman RE. (1976). SEM observations on abnormal neural tube closure. Clin. Res. 24, A131. (abstract).

Gordon R, Jacobson AG. (1978). The shaping of tissues in embryos. Sci. Amer. 238, 106-113.

Gouda JG. (1974). Closure of the neural tube in relation to the developing somites in the chick embryo (Gallus gallus domesticus). J. Anat. 118, 360-361.

Granholm NH, Baker JR. (1970). Cytoplasmic microtubules and the mechanism of avian gastrulation. Develop. Biol. 23, 563-584.

Hamburger V, Hamilton HL. (1951). A series of normal stages in the development of the chick embryo. J. Morphol. 88, 49-92.

Handel MA, Roth LE. (1971). Cell shape and morphology of the neural tube: Implications for microtubule function. Develop. Biol. 25, 78-95.

Herman L, Kauffman SL. (1966). The fine structure of the embryonic mouse neural tube with special reference to cytoplasmic microtubules. Develop. Biol. 13, 145-162.

Hilfer SR. (1973). Extracellular and intracellular correlates of organ initiation in the embryonic chick thyroid. Amer. Zool. 13, 1023-1038.

Hilfer SR, Palmatier BY, Fithian EM. (1977). Precocious evagination of the embryonic chick thyroid in ATP-containing medium. J. Embryol. Exp. Morphol. 42, 163-175.

Hoebeke J, Van Nijen G, DeBrabander M. (1976). Interaction of oncodazole (R 17934), a new antitumoral drug, with rat brain tubulin. Biochem. Biophys. Res. Commun. 69, 319-324.

Holmdahl DE. (1925a). Die erste Entwicklung des Körpers bei den Vögeln und Säugetieren, inkl. dem Menschen, besonders mit Rücksicht auf die Bildung des Rückenmarks, des Zöloms und der entodermalen Kloake nebst einem Exkurs über die Entstehung der Spina Bifida in der Lumbosakralregion. I. Gegenbaurs Morphol. Jahrb. 54, 333-384.

Holmdahl DE. (1925b). Die erste Entwicklung des Körpers bei den Vögeln und Säugetieren, inkl. dem Menschen, besonders mit Rücksicht auf die Bildung des Rückenmarks, des Zöloms und der entodermalen Kloake nebst einem Exkurs über die Entstehung der Spina Bifida in der Lumbosakralregion. II-V. Gegenbaurs Morphol. Jahrb. 55, 112-208.

Holtzer H, Sanger JW. (1972). Cytochalasin-B: Microfilaments, cell movement and what else? Develop. Biol. 27, 444-446.

Hughes AF, Freeman RB. (1974). Comparative remarks on the development of the tail cord among higher vertebrates. J. Embryol. Exp. Morphol. 32, 355-363.

Ikeda Y. (1930). Beitrage zur normalen und abnormalen Entwicklungsgeschichte des caudalen Abschnittes des Ruckenmarks bei menschlichen Embryonen. Z. Anat. Entw. Gesch. 92, 380-491.

Imai S, Takeda K. (1967). Effects of vasodilation on the isolated Taenia coli of the guinea pig. Nature 213, 509-511.

Ishii K. (1967). Morphogenesis of the brain in medaka, Oryzias latipes. I. Observations on morphogenesis. Sci. Rep. Tohoku Univ. Ser. IV (Biol). 33, 97-104.

Jacobson AG. (1978). Some forces that shape the nervous system. Zoon 6, 13-21.

Jacobson AG. (1980). Computer modeling of morphogenesis. Amer. Zool. 20, 669-677.

Jacobson AG. (1981). Morphogenesis of the neural plate and tube, in: Morphogenesis and Pattern Formation, Connelly et al. (eds), Raven Press, New York, 233-263.

Jacobson AG, Gordon R. (1976a). Changes in the shape of the developing vertebrate nervous system analyzed experimentally, mathematically and by computer simulation. J. Exp. Zool. 197, 191-246.

Jacobson AG, Gordon R. (1976b). Nature and origin of patterns of changes in cell shape in embryos. J. Supramol. Struct. 5, 371-380.

Jacobson C-O, Löfberg J. (1969). Mesoderm movements in the amphibian neurula. Zool. Bdr. Uppsal. 38, 233-239.

Jacobson M. (1978). Developmental Neurobiology, 2nd ed., Plenum Press, New York, 16-20.

Jelínek R, Friebova Z. (1966). Influence of mitotic activity on neurulation movements. Nature (Lond.). 209, 822-823.

Jelínek R, Seichert V, Klika E. (1969). Mechanism of morphogenesis of caudal neural tube in the chick embryo. Folia Morphol (Praha). 17, 355-367.

Karfunkel P. (1971). The role of microtubules and microfilaments in neurulation in Xenopus. Develop. Biol. 25, 30-56.

Karfunkel P. (1972). The activity of microtubules and microfilaments in neurulation in the chick. J. Exp. Zool. 181, 289-302.

Karfunkel P. (1974). The mechanisms of neural tube formation. Internat'l Rev. Cytol. 38, 245-271.

Karfunkel P, Hoffman M, Phillips M, Black J. (1978). Changes in cell adhesiveness in neurulation and optic cup formation. Zoon 6, 23-31.

Karnovsky MJ. (1965). A formaldehyde-glutaraldehyde fixative of high osmolality for use in electron microscopy. J. Cell Biol. 27, 137a-138a. (abstract).

Klika E, Jelínek R. (1969). The structure of the end and tail bud of the chick embryo. Folia Morphol. (Praha). 17, 29-40.

Kushida H. (1971). A new method for embedding with Epon 812. J. Electron Microsc. 20, 206-207.

Langman J, Guerrant RL, Freeman BG. (1966). Behavior of neuro-epithelial cells during closure of the neural tube. J. Comp. Neurol. 127, 399-412.

Laurent TC. (1970). Structure of hyaluronic acid, in: Chemistry and Molecular Biology of the Intercellular Matrix, E.A. Balazs (ed), Academic Press, New York, 703-732.

Lee H-Y. (1976). Inhibition of neurulation and interkinetic nuclear migration by Concanavalin A in explanted early chick embryos. Develop. Biol. 48, 392-399.

Lee H-Y, Kalmus GW. (1976). Effects of cytochalasin B on the morphogenesis of explanted early chick embryos. Growth 40, 153-162.

Lee H-Y, Nagele RG, Kalmus GW. (1976). Further studies on neural tube defects caused by Concanavalin A in early chick embryos. Experentia 32, 1050-1052.

Lee H, Nagele R, Karasanyi N. (1977). Inhibition of neural tube closure by ionophore A 23187 in chick embryos. Experentia 34, 518-520.

Lee H-Y, Sheffield JB, Nagele RG. (1978). The role of extracellular material in chick neurulation. II. Surface morphology of neuroepithelial cells during neural fold fusion. J. Exp. Zool. 204, 137-154.

Lee H-Y, Sheffield JB, Nagele RG, Kalmus GW. (1976). The role of extracellular material in chick neurulation. I. Effects of Concanavalin A. J. Exp. Zool. 198, 261-266.

Lemire RJ. (1969). Variations in development of the caudal neural tube in human embryos (Horizons XIV-XXI). Teratology 2, 361-370.

Lemire RJ, Beckwith JB, Warkany J. (1978). Anencephaly, Raven Press, New York, 119-122.

Lemire RJ, Loeser JD, Leech RW, Alvord EC. (1975). Normal and Abnormal Development of the Human Nervous System, Harper and Row, Hagerstown, Maryland, 71-83.

Linville GP, Shepard TH. (1972). Neural tube closure defects caused by cytochalasin B. Nature New Biol. 236, 246-247.

Lipton BH, Jacobson AG. (1974). Analysis of normal somite development. Develop. Biol. 38, 73-90.

Löfberg J, Jacobson C-O. (1974). Effects of vinblastine sulphate, colchicine, and guanosine phosphate on cell morphogenesis during amphibian neurulation. Zoon 2, 85-98.

Lyser KM. (1968). Early differentiation of motor neuroblasts in chick embryo as studied by electron microscopy. II. Microtubules and neurofilaments. Develop. Biol. 17, 117-142.

Mak LL. (1978). Ultrastructural studies of amphibian neural fold fusion. Develop. Biol. 65, 435-446.

Malacinski GM, Youn BW. (1981). Neural plate

morphogenesis and axial stretching in "notochord-defective" Xenopus laevis embryos. Develop. Biol. 88, 352-357.
Martin A, Langman J. (1965). The development of the spinal cord examined by autoradiography. J. Embryol. Exp. Morphol. 14, 25-35.
Meier S. (1978). Development of the embryonic chick otic placode. II. Electron microscopic analysis. Anat. Rec. 191, 459-478.
Messier P-E. (1969). Effects of β-mercaptoethanol on the fine structure of the neural plate cells of the chick embryo. J. Embryol. Exp. Morphol. 21, 309-329.
Messier P-E, Auclair C. (1974). Effects of cytochalasin B on interkinetic nuclear migration in the chick embryo. Develop. Biol. 36, 218-223.
Messier P-E, Auclair C. (1977). Alteration of apical junctions and inhibition of interkinetic nuclear migration by cytochalasin B and trypsin. Acta Embryol. Exp. 3, 341-356.
Miyayama Y, Fujimoto T. (1977). Fine morphological study of neural tube formation in the teleost, Oryzias latipes. Okajimas Fol. Anat. Jap. 54, 97-120.
Moran DJ (1976). A scanning electron microscopic and flame spectrometry study on the role of $Ca^{2+}$ in amphibian neurulation using papaverine inhibition and ionophore induction of morphogenetic movement. J. Exp. Zool. 198, 409-416.
Moran D, Rice RW. (1975). An ultrastructural examination of the role of cell membrane surface coat material during neurulation. J. Cell Biol. 64, 172-181.
Moran D, Rice RW. (1976). Action of papaverine and ionophore A 23187 on neurulation. Nature 261, 496-499.
Morriss GM, New DAT. (1979). Effect of oxygen concentration on morphogenesis of cranial neural folds and neural crest in cultured rat embryos. J. Embryol. Exp. Morphol. 54, 17-35.
Morriss GM, Solursh M. (1978a). The role of primary mesenchyme in normal and abnormal morphogenesis of mammalian neural folds. Zoon 6, 33-38.
Morriss GM, Solursh M. (1978b). Regional differences in mesenchymal cell morphology and glycosaminoglycans in early neural-fold stage rat embryos. J. Embryol. Exp. Morphol. 46, 37-52.
Morriss-Kay GM. (1981). Growth and development of pattern in the cranial neural epithelium of rat embryos during neurulation. J. Embryol. Exp. Morphol. 65 (Suppl), 225-241.
Nagele RG, Lee H-Y. (1978). Motility-related proteins in developing neuroepithelial cells in the chick. Amer. Zool. 18, 608.
Nagele RG, Lee H-Y. (1979). Ultrastructural changes in cells associated with interkinetic nuclear migration in the developing chick neuroepithelium. J. Exp. Zool. 210, 89-106.
Nagele RG, Lee H-Y. (1980a). Studies on the mechanism of neurulation in the chick: Microfilament-mediated changes in cell shape during uplifting of neural folds. J. Exp. Zool. 213, 391-398.

Nagele RG, Lee H. (1980b). A transmission and scanning electron microscopic study of cytoplasmic threads of dividing neuroepithelial cells in early chick embryos. Experentia 36, 338-340.
Nagele RG, Pietrolungo JF, Lee H. (1981). Studies on the mechanisms of neurulation in the chick: The intracellular distribution of $Ca^{++}$. Experentia 37, 304-306.
Nichols DH. (1981). Neural crest formation in the head of the mouse embryo as observed using a new histological technique. J. Embryol. Exp. Morphol. 64, 105-120.
Nichols DH. (1982). The extracellular matrix in neural crest formation and migration, Scanning Electron Microsc. 1982; in press.
O'Shea S. (1981). The cytoskeleton in neurulation: Role of cations, in: Progress in Anatomy, Vol. I, R.J. Harrison (ed), Cambridge Univ. Press, Great Britain, 35-60.
Pearce TL, Zwaan J. (1970). A light and electron microscopic study of cell behavior and microtubules in the embryonic chicken lens using Colcemid. J. Embryol. Exp. Morphol. 23, 491-507.
Perry MM, Waddington CH. (1966). Ultrastructure of the blastopore cells in the newt. J. Embryol. Exp. Morphol. 15, 317-330.
Pollard TD, Weihing RR. (1974). Actin and myosin and cell movement, in: CRC Critical Reviews in Biochemistry, Vol. 2, Cleveland, Ohio, 1-65.
Portch PA, Barson AJ. (1974). Scanning electron microscopy of neurulation in the chick. J. Anat. 117, 341-350.
Pratt RM, Larsen MA, Johnston MC. (1975). Migration of cranial neural crest cells in a cell-free hyaluronate-rich matrix. Develop. Biol. 44, 298-305.
Reed PW, Lardy A. (1972). A 23187: A divalent cation ionophore. J. Biol. Chem. 247, 6970-6977.
Revel J-P. (1974). Scanning electron microscope studies of cell surface morphology and labeling, in situ and in vitro, Scanning Electron Microsc, 1974, I: 541-548.
Revel J-P, Brown SS. (1976). Cell junctions in development, with particular reference to the neural tube. Cold Springs Harbor Symp. Quant. Biol. 40, 443-455.
Revel J-P, Yip P, Chang LL. (1973). Cell junctions in the early chick embryo. A freeze etch study. Develop. Biol. 35, 302-317.
Rice RW, Moran DJ. (1977). A scanning electron microscopic and x-ray microanalytic study of cell surface material during amphibian neurulation. J. Exp. Zool. 201, 471-478.
Richardson KC, Jarett L, Finke EH. (1960). Embedding in epoxy resins for ultrathin sectioning in electron microscopy. Stain Technol. 35, 313-323.
Rugh R. (1951). The Frog: Its Reproduction and Development, McGraw-Hill, New York, 123-138.
Sadler TW. (1978). Distribution of surface coat material on fusing neural folds of mouse embryos during neurulation. Anat. Rec. 191, 345-350.

Sadler TW, Greenberg D, Coughlin P, Lessard JL. (1982). Actin distribution patterns in the mouse neural tube during neurulation. Science 215, 172-174.

Santander RG, Cuadrado GM. (1976). Ultrastructure of the neural canal closure in the chicken embryo. Acta Anat. 95, 368-383.

Sauer FC. (1935). The cellular structure of the neural tube. J. Comp. Neurol. 63, 13-23.

Saxén L. (1980). Neural induction: Past, present, and future, in: Current Topics in Developmental Biology, Vol. 15, Chapter 12, Academic Press Inc., New York, 409-418.

Schoenwolf GC. (1977). Tail (end) bud contributions to the posterior region of the chick embryo. J. Exp. Zool. 201, 227-246.

Schoenwolf GC. (1978a). An SEM study of posterior spinal cord development in the chick embryo, Scanning Electron Microsc. 1978; II: 739-746.

Schoenwolf GC. (1978b). Effects of complete tail bud extirpation on early development of the posterior region of the chick embryo. Anat. Rec. 192, 289-296.

Schoenwolf GC. (1979a). Histological and ultrastructural observations of tail bud formation in the chick embryo. Anat. Rec. 193, 131-148.

Schoenwolf GC. (1979b). Observations on closure of the neuropores in the chick embryo. Amer. J. Anat. 155, 445-466.

Schoenwolf GC, DeLongo J. (1980). Ultrastructure of secondary neurulation in the chick embryo. Amer. J. Anat. 158, 43-63.

Schoenwolf GC, Fisher M. (1982). Analysis of the effects of Streptomyces hyaluronidase on formation of the neural tube. J. Embryol. Exp. Morphol., in press.

Schoenwolf GC, Kelley RO. (1980). Characterization of intercellular junctions in the caudal portion of the developing neural tube of the chick embryo. Amer. J. Anat. 158, 29-41.

Schroeder TE. (1970). Neurulation in Xenopus laevis. An analysis and model based upon light and electron microscopy. J. Embryol. Exp. Morphol. 23, 427-462.

Schroeder TE. (1973). Cell constriction: Contractile role of microfilaments in division and development. Amer. Zool. 13, 949-960.

Schumacher S. (1927). Über die sogenannte Vervielfachung des Medullarrohres (bzw. des Canalis centralis) bei Embryonen. Z. Mikrosk.-Anat. Forsch. 10, 75-109.

Shain WG, Hilfer SR, Fonte VG. (1972). Early organogenesis of the embryonic chick thyroid. I. Morphology and biochemistry. Develop. Biol. 28, 202-218.

Silver MH, Kerns JM. (1978). Ultrastructure of neural fold fusion in chick embryos, Scanning Electron Microsc. 1978; II: 209-215.

Solursh M. (1976). Glycosaminoglycan synthesis in the chick gastrula. Develop. Biol. 50, 525-530.

Solursh M, Morriss GM. (1977). Glycosaminoglycan synthesis in rat embryos during the formation of the primary mesenchyme and neural folds. Develop. Biol. 57, 75-86.

Spooner BS. (1973a). Morphogenesis of vertebrate organs, in: Concepts of Development, J.W. Lash, J.R. Whittaker (eds), Sinnauer Assoc. Inc., Stanford, Connecticut, 213-240.

Spooner BS. (1973b). Microfilaments, cell shape changes, and morphogenesis of salivary epithelium. Amer. Zool. 13, 1007-1022.

Spooner BS, Wessells NK. (1970). Effects of cytochalasin B upon microfilaments involved in morphogenesis of salivary epithelium. Proc. Nat. Acad. Sci. (USA). 66, 360-364.

Spooner BS, Wessells NK. (1972). An analysis of salivary gland morphogenesis: Role of cytoplasmic microfilaments and microtubules. Develop. Biol. 27, 38-54.

Stagno PA, Low FN. (1978). Effects of cytochalasin B on fine structure of organized endodermal cells of the early chick embryo. Amer. J. Anat. 151, 159-172.

Tannenbaum J, Tanenbaum SW, Godman GC. (1977a). The binding sites of cytochalasin D. I. Evidence that they may be peripheral membrane proteins. J. Cell Physiol. 91, 225-238.

Tannenbaum J, Tanenbaum SW, Godman GC. (1977b). The binding sites of cytochalasin D. II. Their relationship to hexose transport and to cytochalasin B. J. Cell Physiol. 91, 239-248.

Tarin D. (1971). Scanning electron microscopical studies of the embryonic surface during gastrulation and neurulation in Xenopus laevis. J. Anat. 109, 535-547.

Waddington CH, Perry MM. (1966). A note on the mechanism of cell deformation in the neural folds of the amphibia. Exp. Cell Res. 41, 691-693.

Waterman RE. (1975). SEM observations of surface alterations associated with neural tube closure in the mouse and hamster. Anat. Rec. 183, 95-98.

Waterman RE. (1976). Topographical changes along the neural fold associated with neurulation in the hamster and mouse. Amer. J. Anat. 146, 151-172.

Waterman RE. (1979). Scanning electron microscope studies of central nervous system development. Birth defects: Original Article Series 15, 55-77.

Waterman RE, Balian G. (1980). Indirect immunoflourescent staining of fibronectin associated with the floor of the foregut during formation and rupture of the oral membrane in the chick embryo. Anat. Rec. 198, 619-635.

Watterson RL. (1965). Structure and mitotic behavior of the early neural tube, in: Organogenesis, R.L. DeHaan, H. Ursprung (eds), Holt, Rinehart and Winston, New York, 129-159.

Watterson RL, Schoenwolf GC, Sweeney RM (1979). Laboratory Studies of Chick, Pig, and Frog Embryos, 4th ed., Burgess Publishing Company, Minneapolis, Minnesota, 15-20.

Wessells NK, Evans J. (1968). Ultrastructural studies of early morphogenesis and cytodifferentiation in the embryonic mammalian pancreas. Develop. Biol. 17, 413-446.

Wiley MJ. (1980). The effects of cytochalasins on the ultrastructure of neurulating hamster embryos in vivo. Teratology 22, 59-69.

Wilson DB, Finta LA. (1980a). Fine structure of the lumbosacral neural folds in the mouse

embryo. J. Embryol. Exp. Morphol. 55, 279-290.
Wilson DB, Finta LA. (1980b). Early development of the brain and spinal cord in dysraphic mice: A transmission electron microscopic study. J. Comp. Neurol. 190, 363-371.
Wrenn JT, Wessells NK. (1969). An ultrastructural study of lens invagination in the mouse. J. Exp. Zool. 171, 359-368.
Youn BW, Malacinski GM. (1981). Axial structure development in ultra-violet-irradiated (notochord-defective) amphibian embryos. Develop. Biol. 83, 339-352.
Zwaan J, Hendrix RW. (1973). Changes in cell and organ shape during early development of the ocular lens. Amer. Zool. 13, 1039-1049.

## DISCUSSION WITH REVIEWERS

T.W. Sadler: With respect to the formation of the bilateral furrows during convergence of the neural folds, what forces presumably mediate bending in this particular area? Are we again left with the old stand-by of microfilaments or are other forces involved? In this respect, we have observed a region of physiological necrosis that consistently occurs in the area of the furrows in cranial (not caudal) regions of mouse embryos. Could cell death play a role in this process and in other areas of bending of the neural tube? Have you observed similar necrotic areas in the chick?

Author: At present I do not know what forces generate the formation of the bilateral furrows and convergence. Constriction of apical microfilaments might be involved, or the furrows might form passively, as neural folds are pushed (or migrate) toward the midline. Experiments are being planned (and others are under way) to differentiate among these possibilities.

It is well known that cell death is involved in the shaping of a number of organ rudiments and it might be involved in the formation of the bilateral furrows in some organisms. We have not observed necrotic areas in the chick, such as you describe for cranial regions of the mouse.

T.W. Sadler: In a recent article (Sadler et al., 1982) we have demonstrated (using antiactin antibodies) what appears to be an increased concentration of actin in basal areas of mouse neuroepithelial cells during cranial fold elevation followed by a relocation of the actin to apical cell areas at the commencement of convergence. Is it possible that the initial biconvex structure of the forebrain is produced at least in part by constriction of basal filaments?

Author: Certainly.

A.G. Jacobson: I suspect that because of fixation methods used by the authors (and others), that some aspects of neurulation that are described may be artifacts. We (Lipton and Jacobson, 1974) gave evidence that early chick embryo stages are distorted by the usual fixation methods that remove the yolk before fixation. The "V" shape of the plate is likely to be such an artifact. We suggested that these early stages must be fixed in situ on the yolk to avoid such artifacts. Our Figure 20C, D compares neural plates in the rhomboidal sinus of stage 12 embryos. When fixed after removal from the yolk (whether Bouin's or glutaraldehyde/osmium is the fixative) the plate has assumed a "V" shape. If fixed on the yolk the plate is quite flat except for the lateral edges that are already turned 90° upward.

There are probably tensions in the plate that are normally countered by tensions from the spreading blastoderm, the latter being relieved when the embryo is cut off the yolk before fixation. The tensions in the plate could then create the "V" shape before fixative is applied.

I believe the 90° bends at the lateral edges of the plate, present when the plate is otherwise flat, must be the same bends described in this review as the bilateral furrows associated with convergence. In my plastic thick sections of material fixed in situ, these bilateral folds rise with the plate as it bends. The position of the bilateral folds exactly corresponds to the extent of adhesion between the basal surfaces of plate and epidermis.

Please comment on the above statements.

Author: According to Lipton and Jacobson's "Materials and Methods" section, eggs were opened into a dry finger bowl, and a glass ring (15 mm diameter by 5 mm in height) was placed on the yolk, making a well around the embryo. Embryos were fixed with glutaraldehyde for 15 minutes, and then the primary fixative was removed and replaced with osmium tetroxide. Embryos were subsequently removed from the yolk and placed in fresh osmium for a total period of 45 minutes.

Lipton and Jacobson's technique was not used for several reasons. First, it is my contention that their technique might distort the neural groove by pulling this structure laterally. According to their technique, eggs are opened into a dry finger bowl. This causes the yolk to flatten, presumably stretching the embryo. Thus, it is possible that the elevated neural folds might be pulled laterally, causing the neural plate to assume a more flattened configuration. In addition, in their technique a glass ring is placed around the embryo. Such rings (as fabricated according to their specifications) have an average weight of 0.71 grams--a formidable weight compared to that of an early embryo. Placing this weight around the embryo might stretch the embryo, distorting the neural groove and flattening the neural plate. Second, in our hands Lipton and Jacobson's technique is messy. Subvitelline fluid usually coagulates on the surface of embryos fixed on the yolk (the vitelline membrane is removed prior to fixation during our procedure), yielding most unattractive SEM images. Third, when embryos are fixed on the yolk, the primary fixation time must be relatively short (for their technique to be convenient), raising the possibility of inadequate primary fixation. Finally, glutaraldehyde should be washed out of tissues prior to osmication to prevent the precipitation of osmium (Meek, G.A., 1976, "Practical Electron Microscopy for Biologists", Second Ed., John Wiley & Sons, New York, pp. 21). This is not possible when embryos are fixed on the yolk.

It must be emphasized that the study by Lipton and Jacobson did not give evidence that early chick embryos are distorted by the usual fixation methods that remove the embryo from the yolk before fixation. In this study the authors failed to compare (see "Materials and Methods" and legends for Figure 20), just the effects of fixation before or after removal of the embryo from the yolk. Note that Figures 20A and C (the sections that show distortion) are 5 μm paraffin sections from embryos fixed with Bouin's solution after the embryos were removed from the yolk, whereas Figures 20B and D (the sections that show a flat neural plate) are 1 μm plastic sections from embryos fixed with glutaraldehyde/osmium tetroxide before the embryos were removed from the yolk. In our laboratory we have found that embedding with paraffin causes the characteristic distortions of the embryo described by Lipton and Jacobson. This statement is based on the examination of scores of paraffin and plastic sections of early chick embryos. In addition the shape of the neural groove is frequently distorted, such that it is more typically "U" shaped rather than "V" shaped, and the close approximation between the surface and neural ectodermal layers of each neural fold is often disrupted. Thus, it is clear from our data that the differences shown by Lipton and Jacobson in Figure 20 are due principally to distortion caused by embedding in paraffin, and not to fixation on or off the yolk.

Bilateral furrows form prior to elevation of the neural folds in only one area of the embryo, the sinus rhomboidalis. This area is unusual because of the presence of an overlap zone. In addition to being restricted in location, such precocious furrows can be found only in stage 9-10 embryos. I agree with Lipton and Jacobson's observations on this point. However, in Jacobson's comment above he states: "In my plastic thick sections of material fixed in situ, these bilateral folds rise with the plate as it bends." It must be emphasized that this is true only for the region of the sinus rhomboidalis. More cranially, bilateral furrows clearly do not rise with the plate as it bends, as is evident in Lipton and Jacoboson's plastic sections of embryos fixed on the yolk (their Figs. 6-8).

A.G. Jacobson: I agree with the author that mesenchyme and matrix may assist in causing neurulation, but question whether they are essential. Isolated newt neural plates will roll up into tubes despite the absence of mesenchyme and matrix (Jacobson and Gordon, 1976a). Please comment.

H.-Y. Lee and R.G. Nagele: An isolated presumptive neural plate, when cultured under various experimental conditions, is known to undergo neurulation. How would the author account for this fact if the extracellular matrix or somite mesodermal cells play important roles in neurulation?

Author: It is true that when a newt neural plate is isolated and cultured it rolls up into a structure resembling a neural tube, but this might be an artifact of culturing an epithelial sheet. Support for this possibility comes from the fact that neural plates isolated at stage 13 immediately roll up into a tube, unless they are prevented from doing so by covering them with beaded coverslips (Jacobson and Gordon, 1976a). Bending of the neural plate does not normally occur in vivo until stage 15 (approximately 12.5 hours after stage 13; Burnside and Jacobson, 1968). Furthermore, transverse slices of stage 25 neural plates either roll up into tubes, or remain flattened (Jacobson, 1981). When they do roll up into tubes they roll up in the wrong direction (i.e. with their apical ends outward and their basal ends inward). Further experiments are needed to resolve these difficulties before it can be concluded that isolated neural plates undergo neurulation.

R. Bellairs: I agree that the notochord probably is firmly attached to the supranotochordal cells of the neural plate, but think that pits in this area may be artifacts. Is this possible?

Author: Yes. Unfortunately, it is never truly possible to differentiate between artifacts and subtle surface modifications. However, the consistent presence of these pits during elevation of the neural folds suggests that they are probably real.

STRUCTURAL ANALYSIS OF EXTRACELLULAR MATRIX PRIOR TO THE MIGRATION OF CEPHALIC NEURAL CREST CELLS

*D. L. Bolender; W. G. Seliger; R. R. Markwald and P. R. Brauer

Department of Anatomy
Texas Tech University Health Sciences Center
Lubbock, Texas 79430

## Abstract

Cephalic neural crest cells enter cell free areas containing abundant extracellular matrix (ECM). Previous histochemical studies have identified both sulfated and non-sulfated glycosaminoglycans within this matrix. In the present study, ultrastructural examination of the ECM demonstrated an anastomosing network of pleomorphic, cetyl pyridinium chloride-dependent strands within cell free spaces and in association with the basement membrane of the surface ectoderm. Thin section analysis revealed that the strands consisted of three components: (1) 3-5nm filament meshwork; (2) electron dense amorphous material and (3) 30nm granules. In contrast, the ECM associated with the basement membrane consisted principally of a continuum of electron dense, amorphous material. The molecular ordering of ECM within crest cell pathways was compared to the well-characterized, hyaluronate-rich, premigratory matrix of cardiac jelly.

KEY WORDS: Cephalic neural crest, extracellular matrix, glycosaminoglycans, collagen

*Current address:
Department of Anatomy
Medical College of Wisconsin
Milwaukee, WI 53226
Phone no.: (414) 257-8474

## Introduction

The neural crest consists of a transient population of cells originating from the dorsolateral margin of the neural folds. They migrate extensively throughout the embryo supplying pigmented and neuronal cells [Weston, 1970; Le Douarin, 1976]. That portion of the crest cells derived from the cranial neural folds exhibits even greater phenotypic expression, contributing significantly to the mesenchyme-derived structures within the forming head and neck region including bone, cartilage, glandular stroma and other connective tissue components [Johnston, 1966; Noden, 1978]. As the cephalic neural crest cells leave the neural tube, they enter a relatively cell-free space (CFS) between the surface ectoderm, neural tube, and adjacent mesodermally derived (MD) mesenchymal cells [Pratt et al., 1975; Bolender et al., 1980a]. Subsequent accumulation of MD mesenchymal cells reduces the CFS to laterally positioned channels, one beneath the surface ectoderm and the other external to the neural tube [Bolender et al., 1980a]. A rich fibrillogranular extracellular matrix (ECM) is present within the CFS and consists of glycosaminoglycans (GAG), collagen and possibly glycoprotein [Pratt et al., 1975, Bolender et al., 1980a,b]. Autoradiographic [Pratt et al., 1975] and histochemical [Fisher and Solursh, 1977; Bolender et al., 1980a] studies have demonstrated the presence of sulfated and non-sulfated GAG components within the CFS.

The neural crest system seems ideally suited for studies involving the effects of the extracellular environment on development since abundant evidence indicates that both the differentiation and migratory patterns of neural crest cells are determined by environmental cues. The differentiation of crest-derived adrenergic or cholinergic neurons seems dependent on the type of mesenchyme encountered [Teillet et al., 1978], whereas cartilage differentiation in certain areas of the head region results from interaction between the cephalic crest cells and pharyngeal endoderm [Holtfreter, 1968; Epperlein, 1974]. Using heterotopic transplants between different regions of cephalic crest cell outgrowth (for example, mesencephalic region to metencephalic), Noden [1975] has demonstrated that the transplanted crest cells mimic the migratory pattern normally followed by cells of the host tissue. Such studies provide prima facie evidence that, at least for cephalic neural crest cells, morphogenetic events such as migration and differentiation are determined not by the cells, but by their microenvironment.

The significance of the matrical identifications of GAG and collagen components in the pathway of "seeded" neural crest cells is that these extracellular macromolecules have been shown to possess stimulatory and feedback properties [see Manasek, 1975, for a review]. For example, corneal differentiation and enhanced production of its matrix is markedly promoted by GAG and collagen [Meier and Hay, 1974]. Toole and Gross [1971] have shown that hyaluronate promotes the early differentiation and migration of blastemal cells, but its turnover is required for their overt in situ differentiation. A similar situation occurs during migration of endocardial cushion cells where the initial wave of "pioneering" cells encounters a hyaluronate-rich matrix which is subsequently converted to one in which the predominant GAG is sulfated [Markwald et al., 1978, 1979]. Direct action of hyaluronate on cell behavior has been demonstrated in vitro where its addition to the medium enhances migratory capacity of endocardial cushion mesenchymal cells [Bernanke and Markwald, 1979]. In addition, hyaluronate has been shown to regulate cell proliferation [Cohn et al., 1977] and the maintenance of morphogenetic structure [Bernfield et al., 1973; Banerjee et al., 1977]. Sulfated GAG also has been shown to have positive feedback on matrix synthesis [Huang, 1974].

Hyaluronate seems to be the principal GAG synthesized by early chick [Solursh, 1976] and rat [Solursh and Morriss, 1977] embryos analyzed biochemically. As stated previously, hyaluronate has already been identified histochemically in the pathway of trunk [Derby, 1978; Pintar, 1978] and cephalic [Bolender et al., 1980a] neural crest cells, in the heart [Markwald and Adams-Smith, 1972; Markwald et al., 1978] and several other areas of the early embryo [Kvist and Finnegan, 1970]. Ultrastructurally, hyaluronate has been characterized as a randomly arranged network of the 3-5nm filaments [Markwald et al., 1978, 1979; Solursh et al., 1979; Singley and Solursh, 1980]. Previous ultrastructural studies involving neural crest cells have concentrated on cell morphology during migration, with little or no emphasis being placed on the organization of the ECM encountered by the crest cells [Johnston and Listgarten, 1972; Bancroft and Bellairs, 1976; Tosney, 1978]. The purpose of this report is to identify the molecular ordering of the matrical components within the CFS which will be encountered by the initial wave of migrating cephalic neural crest cells. These matrical identifications will be compared with areas designated as premigratory matrix of the endocardial cushion where ultrastructural identification of matrix molecular ordering has been thoroughly characterized [Markwald et al., 1978, 1979].

## Methods and Materials

Fertilized chicken eggs (White Leghorn) were incubated in a forced-air incubator at 38°C and 60% humidity. Embryos 24-30 hours of age (for neural crest studies) corresponding to stage 8 [Hamburger and Hamilton, 1951] and 55 hours, stage 15 (for hearts) were removed from the surface of the yolk using the filter paper ring method of Low [1967] and placed in warm Tyrode solution for removal of extraembryonic tissues. Embryos and excised hearts were fixed at room temperature in 3% glutaraldehyde with or without the addition of 0.5% cetyl pyridinium chloride (CPCl) which allows for retention of water soluble GAG compounds. For histochemical localization of GAG, embryos and isolated hearts were stained en bloc with colloidal iron (CI) at pH 2.5 or 1.7 and embedded in methacrylate [see Bolender et al., 1980a]. Tissue for ultrastructural studies was further fixed in 1% osmium tetroxide and prepared for routine TEM or SEM examination. Material for SEM studies was microdissected in 100% ethanol prior to critical point drying. Following initial fixation, selected embryos or hearts were treated with either testicular hyaluronidase (0.5mg/ml, Sigma Type I) or Streptomyces hyaluronidase (10TRU/ml, Calbiochem) in 0.1M NaK phosphate buffer (pH 5.5) for two hours at 37°C with agitation. Streptomyces hyaluronidase specifically removes hyaluronate, whereas testicular hyaluronidase removes hyaluronate, chondroitin and chondroitin sulfate A and C [Yamada, 1973; Derby and Pintar, 1978]. Control embryos were simultaneously incubated in the buffer without enzyme.

## Results

Prior to the initial migratory wave of cephalic neural crest cells, a CFS was located beneath the surface ectoderm and lateral to the closing neural folds (figure 1). Scattered MD mesenchymal cells were present within portions of the CFS lateral to the neural tube. Previous studies [Bolender et al., 1980a] have shown these areas to contain a fine filamentous ECM (figure 1).

### Matrix Associated With Basement Membrane

The surface ectoderm (especially the lateral, non-neural portions) rested on a thick basement membrane (figure 2) which reacted positively to CI staining at pH 2.5 and 1.7 suggesting a content of both sulfated and non-sulfated GAG. Examination of this surface with SEM (figure 3) revealed a loose stroma of anastomosing, fibrillar strands often associated with spherical inclusions, typical of that seen in previous reports on the ultrastructure of these spaces by Tosney [1978] and Lofberg et al. [1978]. Thin section analysis demonstrated a conventional basement membrane profile of a lamina lucida sandwiched between the basal surface of the plasma membrane and a dense amorphous layer often associated with collagen fibrils (figure 4.) In embryos placed in fixative containing 0.5% CPCl (figure 5), the fibrillar stroma had become less distinct due to a coating of pleomorphic mat-like material. Conventional basement membrane profiles were no longer apparent but were obscured by a continuum of anastomosing, electron dense, amorphous material intimately associated with the basal surface of the ectoderm (figure 6). Though CI staining indicated the presence of sulfated and non-sulfated GAG, treatment of embryos (en bloc after fixation) with either Streptomyces or testicular hyaluronidase did not remove the amorphous material, though electron density was reduced to some extent (figure 7). These results suggested that the basement membrane associated GAG was tightly bound to another component (represented by the amorphous material) such as glycoproteins, known to be components of basement membranes.

Matrix Within the CFS

Fine filamentous strands extended from the surface ectoderm basement membrane (figure 1 and 2) spanning the CFS and contacting underlying MD mesenchymal cells (in lateral areas). The filaments stained positively with CI at pH 2.5 while at pH 1.7 the dye was bound only as scattered, punctate deposits along the length of the filaments. Previous polyanionic histochemical studies suggested that the filaments consisted primarily of hyaluronate [Bolender et al., 1980a]. When viewed with the SEM, the filaments appeared as CPCl-dependent pleomorphic strands suspended from the basal surface of the ectoderm and extending variable distances into the CFS (figure 8). These strands seemed studded along their length with granular material (figure 8). Correlative TEM (figure 9) demonstrated that these strands consisted of a closely knit meshwork of 3-5nm filaments containing randomly positioned accumulations of amorphous material.

Also associated with the filaments were 30nm granules (figure 10). Treatment of embryos with Streptomyces or testicular hyaluronidase removed this material though some of the amorphous material remained as scattered deposits or globular condensations associated with MD mesenchymal cells or basement membrane matrix (figure 11). Strands of matrix containing 3-5nm filaments were also associated with scattered MD mesenchymal cells. In addition, mesenchymal cells were associated with bead-like patches of amorphous appearing matrix often connected by non-beaded filaments.

Comparison With Cardiac Jelly "Premigratory Matrix"

Prior to the seeding of the endocardial cushion mesenchymal cells from the activated endocardium, the cardiac jelly of the AV-canal contains an abundance of acellular, filamentous premigratory matrix (figure 12). This material stains positively with CI at pH 2.5 and is Streptomyces hyaluronidase labile indicating a significant hyaluronate content [Markwald et al., 1978, 1979]. The same areas viewed with the SEM exhibited an extensive continuum of CPCl-dependent, anastomosing, pleomorphic strands coating a loose microfibrillar stroma (figure 13). Thin section analysis by TEM revealed three components within the premigratory matrix: (1) an intertwined network of 3-5nm filaments; (2) electron dense amorphous material and (3) 30nm granules (figure 14). Treatment of embryos by Streptomyces hyaluronidase caused disruption of premigratory matrix and removal of the 3-5nm filaments with subsequent aggregation of 30nm granules and amorphous material. Further coalescence of granules and amorphous matrix was engendered after incubation with testicular hyaluronidase [Markwald et al., 1978].

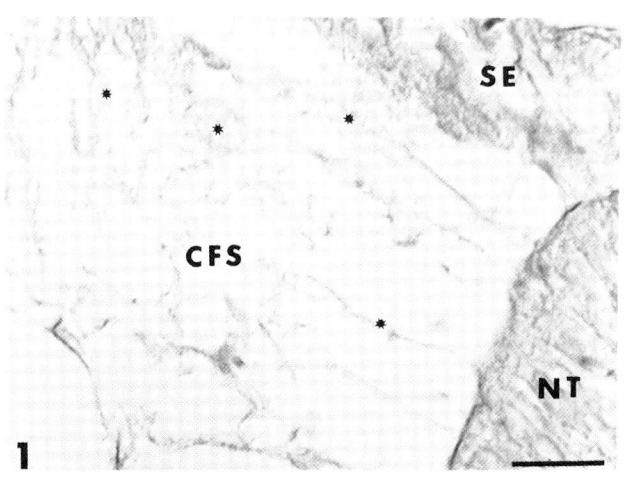

Fig. 1. A large cell free space (CFS) is present in stage 8 embryos (mesencephalic level) between the neural tube (NT) and the surface ectoderm (SE). Filamentous extracellular matrix (asterisks) is present within the CFS and in contact with its boundaries. Nomarski contrast enhanced optics. Bar = 50μm.

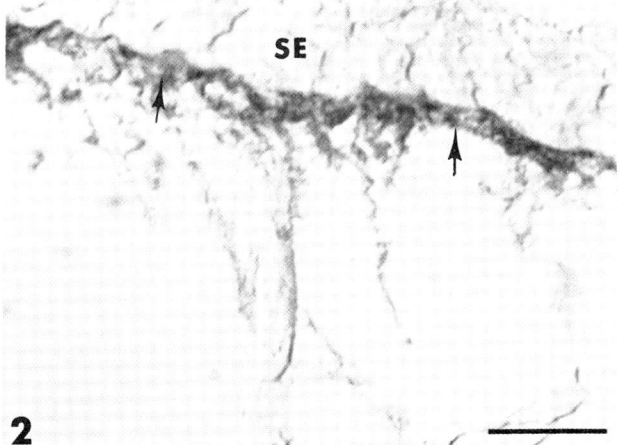

Fig. 2. The surface ectoderm (SE) rests upon thick, amorphous basement membrane (arrows) which stains positively with colloidal iron at pH 2.5. Note the strands of extracellular matrix suspended from the basement membrane which is contacted by scattered mesenchymal cells. Nomarski contrast enhanced optics. Fixation - glutaraldehyde/CPCl. Bar = 50μm.

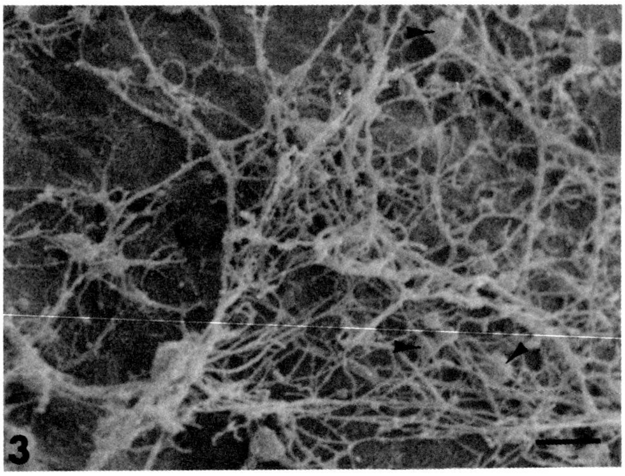

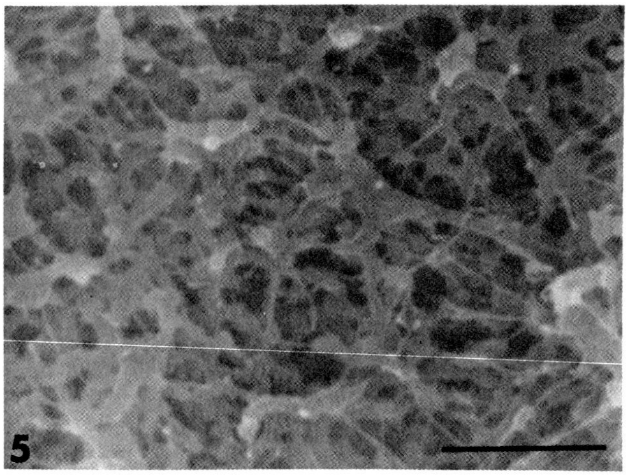

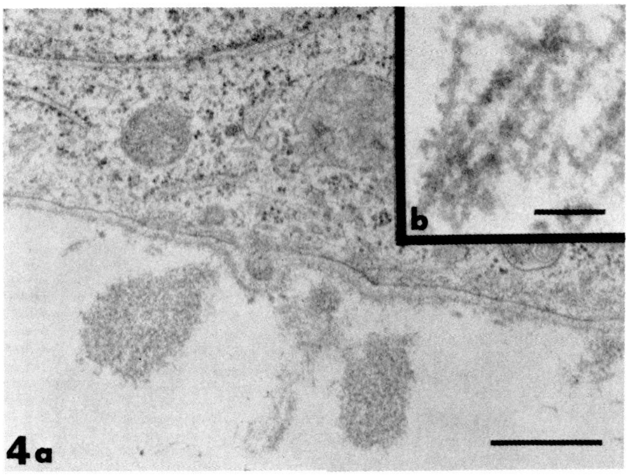

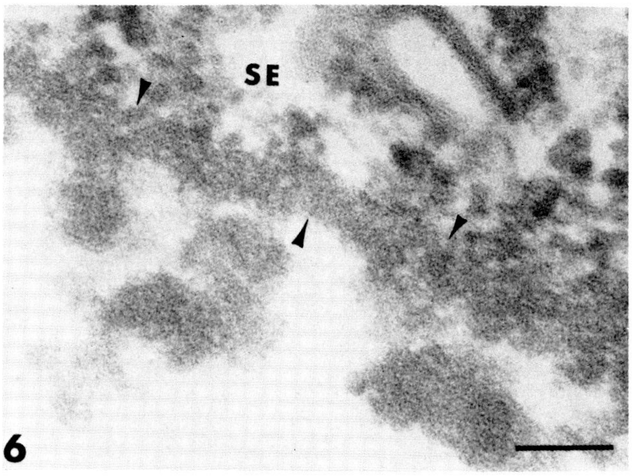

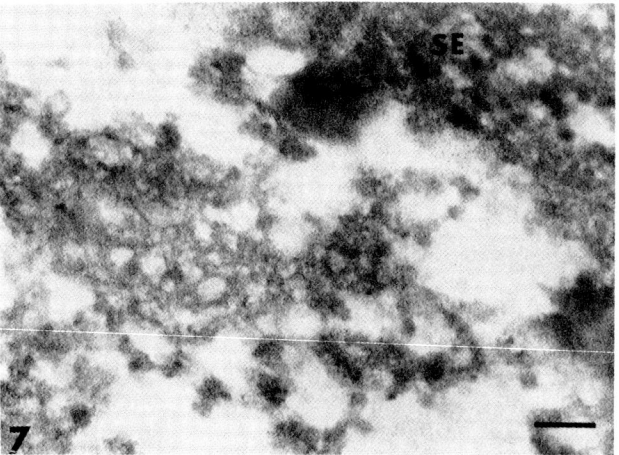

Fig. 3. A loose meshwork of anastomosing fibrils covers the undersurface of the surface ectoderm basement membrane in embryos placed in fixative without addition of CPCl. Numerous spherical inclusions (arrowheads) resembling the interstitial bodies seen by Low [1966] are present within the fibril network. Bar = 1μm.

Fig. 4. (a) In embryos fixed without CPCl, a conventional basement membrane profile is revealed with the TEM. Note interstitial bodies associated with the basement membrane. Bar = 0.5μm. (b) Collagen fibrils are often associated with the basement membrane. Bar = 0.1μm.

Fig. 7. Treatment of embryos with Streptomyces hyaluronidase does not remove basement membrane associated matrix though electron density seems reduced. Surface ectoderm (SE). Bar = 0.1μm.

Fig. 5. A mat-like coating obscures details of the fibrillar network in embryos placed in fixative with CPCl (compare to figure 3). Bar = 1μm.

Fig. 6. A TEM view of the basement membrane area of the surface ectoderm (SE) from an embryo fixed with glutaraldehyde/CPCl. The conventional basement membrane profile seen in figure 4 is no longer apparent but is replaced by a thick amorphous band (between arrows). Pleomorphic strands of this material extend into the CFS. Bar = 1μm.

Cytochemical and biochemical studies have indicated that the 3-5nm filaments are hyaluronate and that the granules contain sulfated GAG [Markwald et al., 1978].

The ECM of premigratory cardiac jelly and that contained within the CFS encountered by cephalic neural crest cells, though similar in composition, exhibited differences which could subsequently effect the migratory patterns of their tenant mesenchyme. The hyaluronate-rich ECM of the AV-canal, visualized as filaments or strands at the LM level, were oriented parallel to the pathway eventually taken by seeded cushion mesenchymal cells. In the CFS, however, hyaluronate-rich strand orientation traversed the migratory area perpendicular to the plane of expected crest cell migration - a direction more favorable to movement of the primary MD mesenchyme. During subsequent stages 9-10, prospective cephalic neural crest cells initiate their migration by proceeding internally into the CFS beneath the surface ectoderm adjacent to the neural tube [Johnston, 1966; Noden, 1978]. Further migration varies regionally depending on the environment encountered [Noden, 1975]. While both cardiac jelly and the CFS contained 3-5nm filamentous networks, 30nm granules and electron dense amorphous material, the CFS seemed to

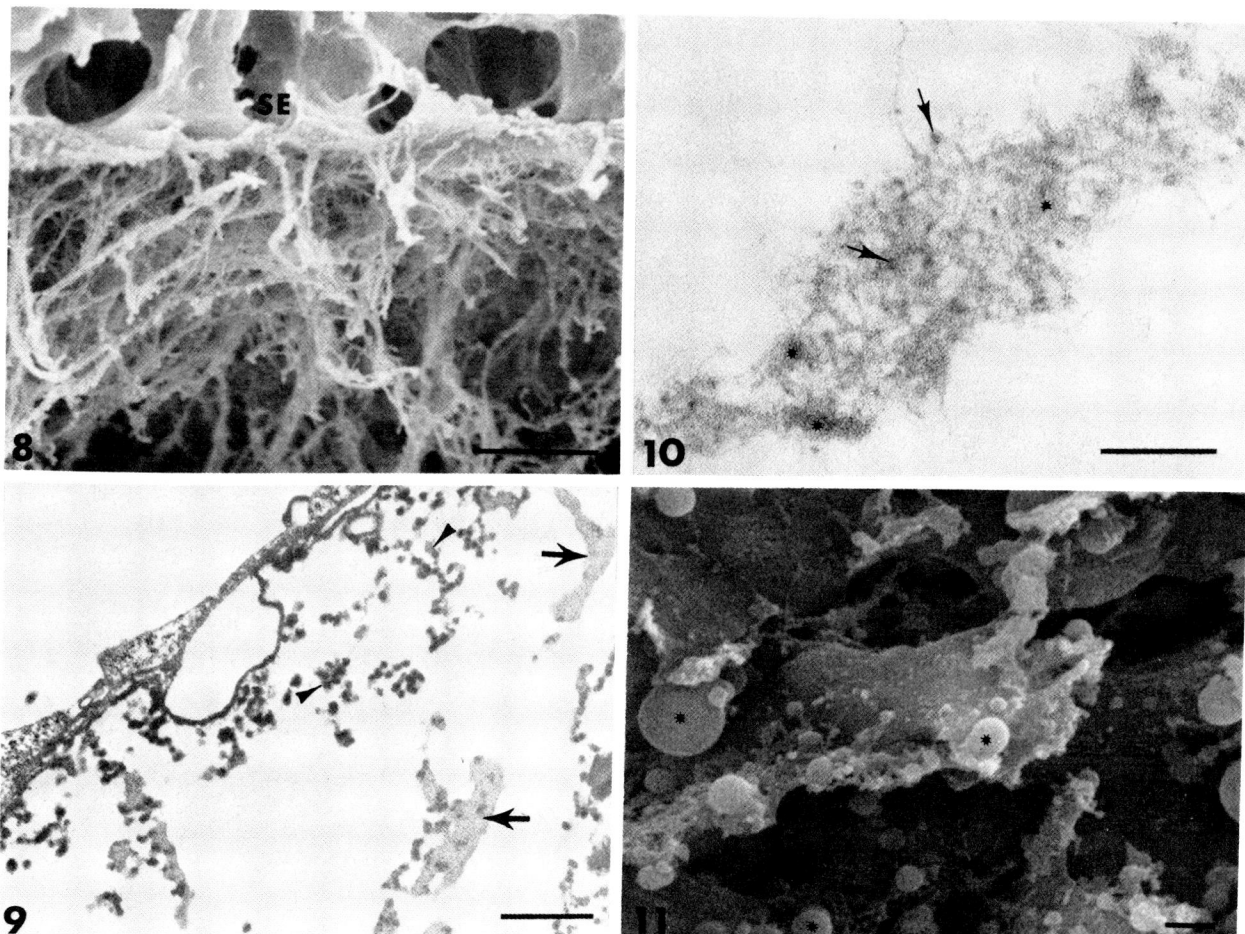

Fig. 8. A cross-sectional view of the surface ectoderm (SE) of the cephalic region of a stage 8 embryo showing CPC1 dependent fibrillogranular strands extending variable lengths into the CFS. Bar = 5μm.

Fig. 9. A TEM view of an area similar to that shown in figure 8. Clumps or pleomorphic strands of amorphous material (arrowheads) are intimately associated with the basement membrane of the surface ectoderm. The strands seen extending into the CFS (figure 8) appear as a network of 3-5nm filaments containing randomly scattered patches of amorphous material (arrows). Bar = 1μm.

Fig. 10. A high power view of the matrix within the CFS. Embedded within the 3-5nm filament meshwork are electron dense amorphous deposits (asterisks) and occasional 30nm granules (arrows). Bar = 0.1μm.

Fig. 11. An SEM view of a stage 8 embryo treated with Streptomyces hyaluronidase for 2 hr. Remnants of the previously demonstrated structured ECM now appear as scattered clumps or spherical bodies (asterisk) coating MD mesenchymal cells and the basal surface of the ectoderm. Bar = 1μm.

have a higher content of the electron dense material (especially beneath the basement membrane) with fewer, visible 30nm granules.

## Discussion

The GAG component of the ECM (especially hyaluronate) has been suggested as a mediator for a variety of morphogenetic events. Timely increases in the amount of hyaluronate in the ECM were seen in corneal stroma [Toole, 1976] during primary mesenchyme mobilization [Solursh, 1976] and prior to formation of the endocardial cushions [Markwald et al., 1978]. Other systems linked to high hyaluronate content include limb development and regeneration [Toole and Gross, 1971; Toole et al., 1977] and epithelial branching [Bernfield et al., 1973]. Biochemical studies have shown hyaluronate to be the predominant GAG in early stages of both chick and rat embryos [Solursh, 1976; Solursh and Morriss, 1977]. It is, in fact, the major product of many embryonic epithelia [Solursh et al., 1979]. In the present study, significant amounts of hyaluronate-rich matrix were located ultrastructurally directly within the pathway of the initial migratory wave of cephalic neural crest cells confirming previous isotope incorporation [Pratt et al., 1975] and histochemical studies [Fisher and Solursh, 1977; Bolender et al., 1980a] of this region.

With such a ubiquitous distribution throughout the embryo, the question arises as to the specific role for GAG (in particular hyaluronate) during morphogenesis. To date, several possible functions for GAG during morphogenesis have been postulated. Expansion of the ECM prior to extensive cell migration has been temporally linked to increases in the hyaluronate component of the corneal stroma [Toole, 1976], neural folds [Morriss and Solursh, 1978], endocardial cushions [Markwald and Adams-Smith, 1972; Markwald et al., 1978] and primitive streak [Solursh, 1976]. These migratory spaces, especially those

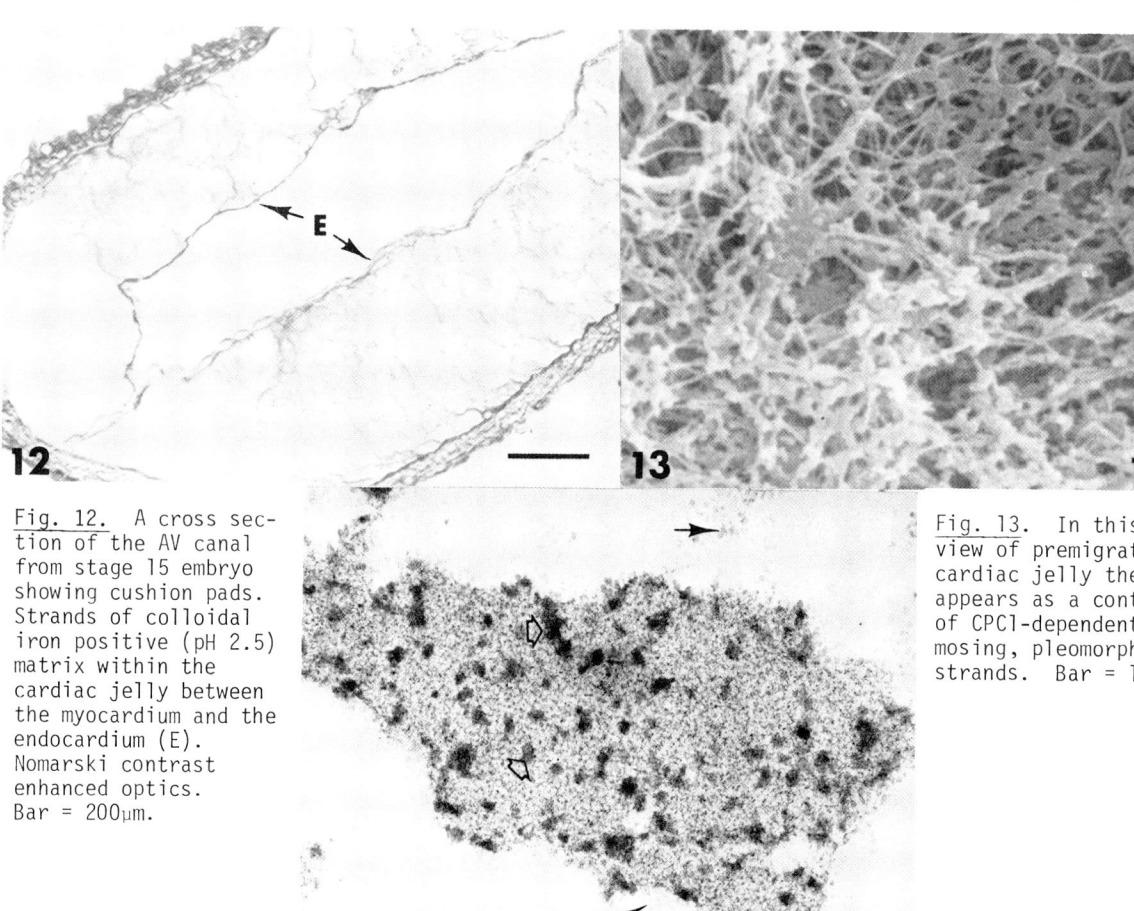

Fig. 12. A cross section of the AV canal from stage 15 embryo showing cushion pads. Strands of colloidal iron positive (pH 2.5) matrix within the cardiac jelly between the myocardium and the endocardium (E). Nomarski contrast enhanced optics. Bar = 200μm.

Fig. 13. In this SEM view of premigratory cardiac jelly the ECM appears as a continuum of CPCl-dependent anastomosing, pleomorphic strands. Bar = 1μm.

Fig. 14. TEM examination of the ECM of the AV canal reveals an intertwined network of 3-5nm filaments containing electron dense amorphous deposits (open arrows) and associated 30nm granules (solid arrows). Bar = 0.5μm.

between adjacent mesenchymal cells, were eliminated in embryos treated in vivo with Streptomyces hyaluronidase [Fisher and Solursh, 1977] which putatively degrades hyaluronate specifically [Derby and Pintar, 1978]. Hyaluronate-rich matrical strands also were shown to separate previously opposed surfaces of the neural ectoderm and neural fold prior to neural crest cell migration [Bolender et al., 1980a]. Results of the above studies suggest that hyaluronate may have enhanced cell migration because of its physical ability to become highly hydrated and expand forming migratory spaces [Laurent, 1970].

In addition to its biophysical properties related to providing spaces for migrating cells, the presence of a hyaluronate-rich matrix has been suggested as a direct migratory stimulus [Toole, 1976; Markwald et al., 1978, 1979]. Crest-derived, corneal fibroblasts [Johnston et al., 1974] invaded the corneal stroma immediately after increases in hyaluronate content [Toole and Trelstad, 1971; Trelstad et al., 1974]. Endocardial cushion mesenchymal cells delaminated and developed numerous motility-like appendages and became motile when exposed to a cardiac jelly rich in hyaluronate [Markwald et al., 1979]. Similar cells exposed to cardiac jelly treated with 6-diazo-5-oxo-L-norcleucine (a potent inhibitor of GAG synthesis) appeared rounded, exhibited few if any motility-like processes and seemed to aggregate rather than populate the cushion pad as do cells exposed to unaltered matrix [Markwald and Bernanke, 1979]. Studies on both systems, though demonstrating a definite link between cell migration and increases in extracellular hyaluronate provided no proof for its direct action on migrating cells. However, addition of hyaluronate to endocardial cushion cells cultured on hydrated collagen lattices resulted in enhanced migratory capacity, penetration rate and increased cell perimeter providing the first definitive proof that an ECM moeity (hyaluronate) has a direct, positive affect on cell migration [Bernanke and Markwald, 1979].

Unlike cushion cells which delaminate and migrate as individual cells, the initial wave of cephalic neural crest cells appeared as a sheet of cells having numerous lateral contacts [Johnston, 1966; Weston, 1970]. Rather than penetrate matrical lattices, crest cells migrate instead beneath the surface ectoderm basement membrane [Noden, 1975; Bolender et al., 1980a]. The present study has demonstrated that this basement membrane is structurally interfaced with ECM containing sulfated and non-sulfated GAG tightly bound to an electron dense component. Though further investigation is needed to identify this electron dense component, one possible identification of this material is that it is non-collagenous, glycoprotein - a known component of embryonic basement membranes [Hay, 1978]. Both fibronectin and laminin (non-collagenous glycoproteins) have been identified as components of basement membranes associated with several migratory systems and have been suggested as mediators of attachment and aggregation during cell migration [Culp et al., 1978;

Ekblom et al., 1980]. The presence of sulfated GAG within this basement membrane was of interest since heparin sulfate has been shown to alter the configuration of fibronectin [Jilek and Hormann, 1979], rendering it more favorable for cell movement. Such interactions have been suggested as mediating attachment and detachment of migrating cells [Culp et al., 1979]. The presence of hyaluronate in the ECM might interfere with the binding reactions involved in these transformations thus inhibiting cell movement or imparting differential migratory capabilities to specific groups of cells. In fact, studies have demonstrated the presence of laminin and fibronectin (both glycoproteins) within the basement membrane of the surface ectoderm [Leivo et al., 1980] where they would be available to migrating crest cells.

Cell systems exhibiting extensive migratory capacity such as the neural crest most likely rely on abundant cell-cell interactions between a variety of cell surfaces. Alteration of the extent to which cells interact with each other may result from passage through a matrix with a high content of hyaluronate [Nichols et al., 1977]. Within the trunk region, portions of the crest cell population (i.e., melanoblasts and sympathoblasts) temporarily stopped migrating when contained within a hyaluronate-rich matrix [Derby, 1978] while, in adjacent areas depleted of this material, cells differentiated into spinal ganglia. Thus, hyaluronate may retard or mask expression of cells or groups of cells until the appearance of proper morphogenetic cues - for example, the binding of fibronectin for cell attachment. Studies have suggested that, though some neural crest cells can produce fibronectin in vitro, most crest cells [Loring et al., 1977; Newgreen and Thiery, 1980] and other invasive cell populations (e.g., malignant cells) [Black, 1980] are deficient in fibronectin production. The presence of hyaluronate may block cell surface interaction with external fibronectin interfering with normal cell motility. Toole et al. [1972] suggest that certain cells can release a hyaluronidase which may remove the masking effect of hyaluronate allowing expression of the differentiated state. Recently Orkin and Toole [1980] have characterized two types of hyaluronidase existing within embryonic cells.

In summary, neural crest cells like many other migratory cell systems within the embryo encounter a matrix rich in hyaluronate. We have demonstrated this to be a Streptomyces hyaluronidase sensitive network of 3-5nm filaments organized into fibrillogranular strands suspended from the basement membrane of the surface ectoderm. This hyaluronate material is often interfaced with collagen fibrils and sulfated GAG becoming more amorphous and electron dense with proximity to the basement membrane. Such a matrix may have a variety of effects on determining crest cell behavior and differentiation. However, information for specific phenotypic expression such as cartilage differentiation is probably derived from gene-translated products such as glycoproteins. Further study utilizing biochemical and cytochemical technique should enable us to identify the interactions involved and the molecular ordering present between the various components of the ECM to be encountered by cephalic neural crest cells.

## Acknowledgments

The authors express their appreciation to Mrs. Makiko Hartman for her expert technical assistance and to Ms. Valinda Bradshaw for typing the manuscript. This project was supported by NIDR Grant DE04603.

## References

Bancroft M, Bellairs R. The neural crest cells of the trunk region of the chick embryo studies by SEM and TEM. Zoon. 4, 1976, 73-85

Banerjee S D, Cohn R H, Bernfield M R. Basal lamina of embryonic salivary epithelia. Production by the epithelium and role in maintaining lobular morphology. J. Cell Biol. 73, 1977, 445-463.

Bernanke D H, Markwald R R. Effects of hyaluronic acid on cardial cushion tissue cells in collagen matrix cultures. Tx. Rept. Biol. Med. 39, 1979, 271-285.

Bernfield M F, Cohn R H, Banerjee S D. Glycosaminoglycans and epithelial organ formation. Amer. Zool. 179, 1973, 167-182.

Black P R. Shedding from the cell surface of normal and cancer cells. Adv. Can. Res. 32, 1980, 75-200.

Bolender D L, Seliger W G, Markwald R R. A histochemical analysis of polyanionic compounds found in the extracellular matrix encountered by cephalic neural crest cells. Anat. Rec. 196, 1980a, 401-412.

Bolender D L, Seliger W G, Markwald R R, Brauer P R. Extracellular matrix modifications prior to cephalic neural crest cell migration; In: Proceedings of the International symposium on current research trends in prenatal craniofacial development. Elsevier North Holland, Amsterdam, 1980b, in press.

Cohn R H, Banerjee S D, Bernfield M R. Basal lamina of embryonic salivary epithelia. Nature of glycosaminoglycan and organization of extracellular materials. J. Cell Biol. 73, 1977, 464-478.

Culp L A, Rollins B J, Buniel J, Hitri S. Two functionally distinct pools of glycosaminoglycan in the substrate adhesion site of murine cells. J. Cell Biol. 79, 1978, 788-801.

Derby M A. Analysis of glycosaminoglycans within the extracellular environments encountered by migrating neural crest cells. Develop. Biol. 66, 1978, 321-336.

Derby M A, Pintar J E. The histochemical specificity of *Streptomyces* hyaluronidase and chondroitinase ABC. Histochem. J. 10, 1978, 529-547.

Ekblom P, Alitato K, Vaheri A, Timpl R, Saxen L. Induction of a basement membrane glycoprotein in embryonic kidney: Possible role of laminin in morphogenesis. Proc. Nat. Acad. Sci. USA, 77, 1980, 485-489.

Epperlein, H H. The ectomesenchymal-endodermal interaction-system (EEIS) of trituris alpestris in tissue culture. 1. Observations on attachment, migration and differentiation of neural crest cells. Differentiation 2, 1974, 151-168.

Fisher M, Solursh M. Glycosaminoglycan localization and role in maintenance of tissue spaces in the early chick embryo. J. Embryol. Exp. Morph. 42, 1977, 195-207.

Hamburger V, Hamilton H L. A series of normal stages in the development of the chick embryo. J. Morphol. 88, 1951, 49-92.

Hay E D. Role of basement membranes in development and differentiation; In: Biology and Chemistry of basement membranes, N. A. Kefalides (ed.). Academic Press, New York, 1978, 119-136.

Holtfreter J. Mesenchyme and epithelia in inductive and morphogenetic processes; In: Epithelial-mesenchymal interactions, R. Fleischmajer and R. Billingham (eds.). Williams and Wilkins, Baltimore, 1968, 1-30.

Huang D. Effect of extracellular chondroitin sulfate on cultured chondrocytes. J. Cell. Biol. 62, 1974, 881-886.

Jilek F, Hormann H. Fibronectin (Cold-Insoluble Globulin), VI Influence of heparin and hyaluronic acid on the binding of native collagen $^{(1,2)}$. Hoppe-Seyler's Z. Physiol. Chem. 360, 1979, 597-603.

Johnston M C. A radioautographic study of the migration and fate of cranial neural crest cells in the chick embryo. Anat. Rec. 156, 1966, 143-156.

Johnston M C, Listgarten M A. Observations on the migration, interaction, and early differentiation of orofacial tissues; In: Developmental aspects of oral biology, H. C. Slavkin and L. A. Bavetta (eds.). Academic Press, New York, 1972, 53-80.

Johnston M C, Bhakdinaronk A, Reid Y C. An expanded role of the neural crest in oral and pharyngeal development; In: Fourth symposium on oral sensation and perception: Development in the fetus and infant, J. F. Bosma (ed.). USGPO (National Institutes of Health), Bethesda, 1974, 37-52.

Kvist T N, Finnegan C V. The distribution of glycosaminoglycans in the axial region of the developing chick embryo. I. Histochemical analysis. J. Exp. Zool. 175, 1970, 221-240.

Laurent T C. Structure of hyaluronic acid; In: Chemistry and molecular biology of the intercellular matrix, Vol. 2, E. A. Balazs (ed.). Academic Press, New York, 1970, 703-732.

Le Douarin N M. Cell migration in early vertebrate development studied in interspecific chimeras. Ciba Foundation Symposium 40, 1976, 71-76.

Leivo I, Vaheri A, Timpl R, Wastinovaara J. Appearance and distribution of collagens and laminin in the early mouse embryo. Develop. Biol. 76, 1980, 100-114.

Lofberg J, Ahlfors K. Extracellular matrix organization and early neural crest cell migration in the axolotl embryo. Zoon. 6, 1978, 87-101.

Loring J, Erickson C, Weston J A. Surface proteins of neural crest, crest-derived and somite cells in vitro. J. Cell Biol. 75, 1977, 71a.

Low F N. Interstitial bodies in the early chick embryo. Am. J. Anat. 128, 1966, 45-56.

Low F N. Developing boundary (basement) membranes in the chick embryo. Anat. Rec. 159, 1967, 231-238.

Manasek R J. The extracellular matrix: A dynamic component of the developing embryo. Curr. Topics Develop. Biol. 10, 1975, 35-102.

Markwald R R, Adams-Smith W N. Distribution of mucosubstances in the developing rat heart. J. Histochem. Cytochem. 20, 1972, 896-907.

Markwald R R, Fitzharris T P, Bank H, Bernanke D H. Structural analyses on the matrical organization of glycosaminoglycans in developing endocardial cushions. Develop. Biol. 62, 1978, 292-316.

Markwald R R, Fitzharris T P, Bolender D L, Bernanke D H. Structural analysis of cell: matrix association during the morphogenesis of atrioventricular cushion tissue. Develop. Biol. 69, 1979, 634-654.

Markwald R R, Bernanke D H. Structural analysis of 6-diazo-5-oxo-L-norleucine (DON) effects upon early cushion tissue morphogenesis; In: Perspectives in cardiovascular research, T. Pexieder (ed.). Raven Press, New York, 1980, 237-251.

Meier S, Hay E D. Stimulation of extracellular matrix synthesis in the developing cornea by glycosaminoglycans. Proc. Nat. Acad. Sci. USA 71, 1974, 2310-2313.

Morriss G M, Solursh M. Regional differences in mesenchymal cell morphology and glycosaminoglycans in early neural-fold stage rat embryos. J. Embryol. Exp. Morph. 46, 1978, 37-52.

Newgreen D, Thiery J P. Fibronectin in early avian embryos: Synthesis and distribution along migration pathways of neural crest cells. Cell Tiss. Res. 211, 1980, 269-291.

Nichols D H, Kaplan R A, Weston J A. Melanogenesis in cultures of peripheral nervous tissue. II. Environmental factors determining the fate of pigment forming cells. Develop. Biol. 60, 1977, 226-237.

Noden D M. An analysis of the migratory behavior of avian cephalic neural crest cells. Develop. Biol. 42, 1975, 106-130.

Noden D M. Interactions directing the migration and cytodifferentiation of avian neural crest cells; In: Specificity of embryological interactions, D. R. Garrod (ed.). Chapman and Hall, London, 1978, 5-49.

Orkin R W, Toole B P. Chick embryo fibroblasts produce two forms of hyaluronidase. J. Cell Biol. 85, 1980, 248-257.

Pintar J E. Distribution and synthesis of glycosaminoglycans during quail neural crest morphogenesis. Develop. Biol. 67, 1978, 444-464.

Pratt R M, Larsen M A, Johnston M C. Migration of cranial neural crest cells in a cell-free hyaluronate-rich matrix. Develop. Biol. 44, 1975, 298-305.

Singley C T, Solursh M. The use of tannic acid for the ultrastructural visualization of hyaluronic acid. Histochemistry 65, 1980, 93-102.

Solursh M. Glycosaminoglycan synthesis in the chick gastrula. Develop. Biol. 50, 1976, 525-530.

Solursh M, Morriss G M. Glycosaminoglycan synthesis in rat embryos during the formation of the primary mesenchyme and neural folds. Develop. Biol. 57, 1977, 75-86.

Solursh M, Fisher M, Singley C T. The synthesis of hyaluronic acid by ectoderm during early organogenesis in the chick embryo. Differentiation 14, 1979, 77-85.

Teillet M A, Cochard P, Le Douarin N M. Relative roles of the mesenchymal tissues and of the complex neural tube-notochord on the expression of adrenergic metabolism in neural crest cells. Zoon 6, 1978, 115-122.

Toole B P. Morphogenetic role of glycosaminoglycans (acid mucopolysaccharides) in brain and other tissues; In: Neuronal recognition, S. H. Barondes (ed.). Plenum Press, New York, 1976, 275-329.

Toole B P, Gross J. The extracellular matrix of the regenerating newt limb: Synthesis and removal of hyaluronate prior to differentiation. Develop. Biol. 25, 1971, 57-74.

Toole B P, Trelstad R L. Hyaluronate production and removal during corneal development in the chick. Develop. Biol. 26, 1971, 28-35.

Toole B P, Jackson G, Gross J. Hyaluronate in morphogenesis: Inhibition of chondrogenesis in vitro. Proc. Nat. Acad. Sci. USA 69, 1972, 1384-1386.

Toole B P, Okayama M, Orkin R W, Yoshimura M, Muto M, Kaji A. Developmental roles of hyaluronate and chondroitin sulfate proteoglycans; In: Cell and tissue interactions, J. W. Lash and M. M. Burger (eds.). Raven Press, New York, 1977, 139-154.

Tosney K W. The early migration of neural crest cells in the trunk region of the avian embryo: An electron microscopic study. Develop. Biol. 62, 1978, 317-333.

Trelstad R L, Hayashi K, Toole B P. Epithelial collagens and glycosaminoglycans in the embryonic cornea. Macromolecular order and morphogenesis in the basement membrane. J. Cell Biol. 62, 1974, 815-830.

Weston J A. The migration and differentiation of neural crest cells. Advan. Morphogen. 8, 1970, 41-114.

Yamada K. The effect of digestion with Streptomyces hyaluronidase upon certain histochemical reactions of hyaluronic acid-containing tissues. J. Histochem. Cytochem. 21, 1973, 794-803.

## Discussion With Reviewers

Reviewer I: I am struck by the similarities between the two systems described here as regards their matrical organization and the migration of mesenchymal cells. Could these matrical relationships reflect universally found interactions between cells and their ECM, required in such developing systems?

Authors: The existence of generalized matrical relationships is a tempting speculation especially in regard to cell migration. As discussed in the manuscript, striking similarities in matrical organization in conjunction with mesenchymal cell migration not only occur in the neural crest and endocardial cushion systems, but also in corneal fibroblast migration and limb formation. In each system, high levels of hyaluronate are combined with other GAG, a collagen network and the presence of other non-collagenous glycoproteins (especially fibronectin). However, all of the above components can be localized throughout the embryo. Thus, if generalized or universal microenvironments are present within the embryo, their ubiquitous distribution would limit their instructional capabilities to general information such as initiation or stimulation of cell motility rather than provide a template for overt differentiation.

Reviewer I: What significant differences could there be in the mesenchyme-like neural crest cells which prevent them from taking advantage of the CFS matrix utilized by the MD mesenchymal cells mesenchyme for migration and restrict their movement to the areas subjacent to the basement membrane of the ectoderm and neural tube?

Authors: We doubt that the neural crest cells themselves possess any unique characteristics which determine their migratory pathway. Instead, their migratory behavior seems to be determined by their environment, shown by the heterotropic transplantation experiments of Noden (see manuscript for references). In the case of the cephalic crest cells, we believe that the mesenchymal cells (derived from mesoderm) may play a role in altering or programming the environment in such a manner that the crest cells are directed (at least in certain areas of the head region) to the subectodermal pathway.

Reviewer I: Recent research has implicated the non-collagenous glycoprotein component laminin, found in several basement membranes of developing systems as having a role in aggregation of mesenchymal-type cells and thus influencing their movement. Do you have any evidence for laminin playing a part in the restriction of movement of neural crest cells along the areas adjacent to the basement membranes of the ectoderm and the neural tube?

Authors: We have no evidence for such a role for neural crest cells at present. Work is in progress to determine the distribution of laminin and other basement membrane markers and record any temporal or spacial variation which parallels neural crest cell migration.

Reviewer II: CPCl is an anionic detergent and is detrimental to cell preservation, as shown in your figure 6, 7 and 9 (see also Tosney, 1978). Is it possible that, during fixation in the presence of CPCl, material is extracted from the cells and deposited along matrix strands, where it gives rise to or alters the appearance of pleomorphic materials seen in SEM?

To what extent do you feel that CPCl treatment preserves the in situ relationships among ECM components? For instance, can CPCl disrupt ionic interactions among ECM elements, allowing material to be precipitated randomly upon the matrical scaffolding of the cell-free spaces? Also, could you comment on the relative merits of using CPCl and tannic acid (c.f., Singley and Solursh, 1980. Histochem. 65:93-102) to determine the macromolecular ordering among hyaluronate and other ECM components?

H. Ris (1980 "38th Ann. Proc. Electra Microscopy Soc. Amer." p. 812-813) has recently shown that, under common critical point drying circumstances, material may be precipitated upon the internal filaments systems in cells, giving an anamalous rounded or pleomorphic appearance to the internal lattice. In your opinion, could this anomaly also affect the SEM visualization of ECM strands?
[Note: The authors feel that the three questions listed above dealing with matrix preservation can be answered collectively in the following response.]

Authors: Most embryonic tissues are highly aqueous and no doubt the polyanionic GAG moieties within ECM interact with other matrix components (e.g., collagen and glycoprotein) in such an aqueous environment. Thus, in order to retain these substances when studying the composition of ECM, primary fixation must be carried out in either non-aqueous solutions or in fixatives containing additives to retain water-soluble compounds. Even with these procedures some shrinkage, collapse or other artifactual changes will occur. In our studies of the composition of ECM, CPCl has provided consistent preservation of polyanionic matrix components, especially GAG. Though fixation with CPCl diminishes the quality of cell morphology when compared to fixation with glutaraldehyde alone or glutaraldehyde/tannic acid, cellular integrity is still maintained. If in fact cellular material was being extracted and deposited in extracellular spaces, as suggested by Reviewer II, such cells should appear vacuolated or possibly shrunken and exhibit significant discontinuity in the cell membrane. No such modifications occurred in any of the cells observed in the present study nor was any material precipitated on the external surface of the ectoderm. In addition, treatment of sections with polycationic markers results in labeled matrix but not labeled cells. If such pleomorphic material was extracted from cells then label would be present over both cytoplasm and matrix. Further proof of the integrity of CPC1 preserved matrix can be seen in the following series of micrographs which compare the ECM of the cardiac jelly fixed with glutaraldehyde/0.5% CPC1 (A); freeze-substituted, unfixed hearts (B); and freeze-dried, unfixed hearts (C). Regardless of treatment, ECM exhibited characteristics of the hyaluronic-rich premigratory matrix similar to those described in the manuscript-specifically, pleomorphic mat-like material coating a loose network of filaments. For details on methodology and results of the above techniques see Markwald et al., SEM 1981, this volume, and Kitten et al., Anat. Rec., 1981, 199:143A.

Critical point drying also seems to have little or no effect on the integrity of ECM. Striking similarities between preservation of matrix components using methods employing critical point drying (glutaraldehyde/CPC1, freeze substitution) and freeze-dried, unfixed matrix (which does not utilize critical point drying) would suggest that no artifactual deposits occur on the fibrillar network as a result of this procedure.

In our hands the consistency of matrix preservation obtained with CPC1 cannot be duplicated by addition of tannic acid to the fixative. Though more ECM material is retained by addition of tannic acid than in fixation with glutaraldehyde alone, especially beneath the basement membrane, the majority of the hyaluronic-rich matrix observed after fixation with glutaraldehyde/CPC1 is not preserved by the addition of tannic acid to fixative. The ECM of the chick wing bud discussed by Singley and Solursh [1980] is localized within a network of intercellular spaces whereas the CFS is a larger, more accessible area. It is possible that the tannic acid-matrix complex shown by Singley and Solursh was sufficiently bonded to allow physical trapping within the smaller intercellular spaces, but not sufficient to prevent extraction as observed in our studies involving the more open CFS.

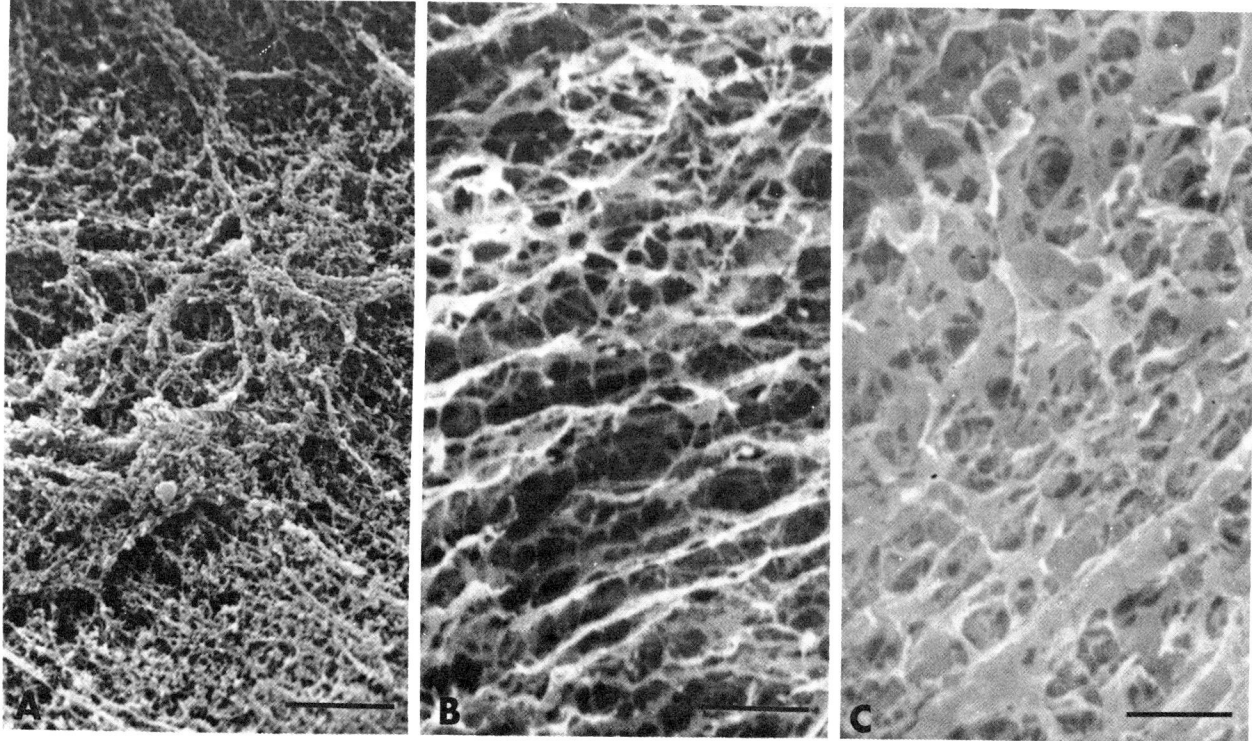

Discussion Figures:
A. Premigratory cardiac jelly from a 72 hr embryo fixed with 3% glutaraldehyde/0.5% CPC1. Compare with figures 5 and 13 in the paper. A coating of pleomorphic material covers a network of intertwined fibrils. Bar = 5 μm.
B. Premigratory matrix from a similar staged embryo placed in 20% glycerol (cryoprotectant), plunged into liquid nitrogen cooled freon, and processed for freeze substitution with acetone. Note similarities of unfixed matrix to that preserved with CPC1 fixative (A). Bar = 5 μm.
C. Cardiac jelly from a 72 hr embryo which was freeze dried and examined with SEM without subsequent critical point drying. Matrix preservation is identical with that seen in figures A and B. Bar = 5 μm.

Reviewer I - David H. Bernanke
Reviewer II - Kathryn W. Tosney
Reviewer III - Gary C. Schoenwolf

Reviewer II: You appear to describe mainly the mesencephalic neural crest pathway throughout your paper. At this level, crest cells migrate en masse between the ectoderm and mesodermal mesenchyme. At other levels, however, this pattern is not the case: at more anterior levels, the crest cells migrate between the ectoderm and optic epithelium, while at more posterior cephalic levels, the ectodermal bias is less or is lost [Noden, 1975]. First, can your observations be generalized to all cephalic levels? Second, did you note any difference in the distribution or arrangement of ECM components that correlates with different pathways? For instance, does the reported differential distribution of granules under the ectoderm also occur in the metencephalon, where many crest cells do not migrate subectodermally?

Authors: The arrangement of the ECM as characterized in our study is present within the CFS at all levels in the cephalic region of the stage 8 chick embryo. However, a regional analysis of ECM utilizing later stages (10-14), including those stages described by Noden [1975], is in progress. It appears that at all cephalic levels a significant number of crest cells are available for ectodermal interaction whether it be for a majority of their migration (mesencephalic or prosencephalic) or only for portions of it (metencephalic). Though three distinct pathways for cephalic crest cells have been demonstrated [Noden, 1975], all require various degrees of cell migration. Thus, changes in matrix components influencing cell migration present in the various pathways may not be as striking as the distinctive migratory patterns observed in the different cephalic levels.

Reviewer III: Have you exposed living embryos in ovo to Streptomyces hyaluronidase to determine if more matrical material is removed when tissues are treated prior to fixation? If not, would you expect to find any differences following this protocol?

Authors: We have not exposed embryos to enzyme treatment prior to fixation. The interpretation of such results would depend on the role of hyaluronate within the matrix and its interaction and binding properties with other components. If it functions as an organizing or binding component interacting with other matrix moieties then enzyme treatment prior to fixation might remove all matrix. In contrast, if the CFS is dependent on hydration properties of hyaluronate to maintain its integrity, pretreatment with Streptomyces hyaluronidase would cause collapse of the CFS as suggested by Fisher and Solursh [1976]. Such a condition might obscure the relationships between the remaining ECM moieties making it very difficult to determine their molecular ordering. This problem does not occur when hyaluronate is enzymatically removed from the ECM after fixation in situ.

Editor: Please provide additional details about your fixative.
Author: For fixation we used a 0.065M Cacodylate buffer (pH 7.2).

# DEVELOPMENT OF THE EYE OF THE CHICK EMBRYO

S. R. Hilfer

Department of Biology, Temple University, 12th and Norris Streets
Philadelphia, PA 19122

Phone No. (215) 787-8863

(Paper received January 28 1983, Complete manuscript received June 30 1983)

## Abstract

Vertebrate eye development begins with the formation of the optic vesicles as outgrowths of the forebrain. These initial pouches grow laterally and can be subdivided into optic stalk and optic vesicle. The axis of growth then shifts to produce optic vesicles that enlarge dorsally to lie alongside the expanding diencephalon. Concomitant invagination of the optic vesicles and the overlying ectoderm produces the optic cup and lens. During later stages, the lens detaches from the surfaces ectoderm and the optic cup forms the neural retina and the pigmented epithelium. Experimental analysis of eye development has revealed an intimate relationship between invagination of the lens and optic cup. The primordia of the lens and neural retina become adherent, as a result of changes in the extracellular matrix, before invagination commences. Interference with matrix synthesis causes abnormal development of the optic cup, and subsequent abnormalities of the lens. The forces that control invagination are under investigation. Lens formation may result from internal contractile forces as well as from forces exerted by surrounding cells. Characteristic changes in cell shape and cytoplasmic organization occur during invagination of the neural retinal primordium. These and the effects of inhibitory drugs suggest the involvement at least in part of a contractile mechanism during optic cup formation.

Key words: organogenesis, eye development, lens development, retinal development, optic vesicle, optic cup, extracellular matrix, microfilaments, calcium dependency, cell shape changes.

## Introduction

The eye is one of the first organs to form in the developing embryo. Development of the optic primordium is very similar in all vertebrate embryos (see O'Rahilly, 1975 for development in humans). The chick embryo has been a favorite model for studying eye development because of the early appearance and large size of its optic primordia. This discussion is based upon development in the chick embryo from stage 9 to stage 44, with emphasis on the earlier stages. Staging is based upon the illustrations of Hamburger and Hamilton (see Hamilton, 1952). Three aspects of eye development will be examined. Formation of the optic vesicle and invagination to form optic cup and lens vesicle are described first. Cellular organization is covered next, followed by a brief discussion of experiments related to control of early organogenesis.

### Formation of the Optic Vesicle

The eye primordia are first recognizable at stage 9 (approximately 31 hours of incubation) as lateral outgrowths of the forebrain (Fig. 1a). In this illustration the head was cut parallel to the surface of the blastoderm, exposing the neural tube cavity. Initially, the optic primordia are almost hemispherical but within a few hours (stage 10) enlargement of the distal portion results in the formation of larger terminal optic vesicles and narrower optic stalks (Figs. 1b, 2a). The primordium is shown in frontal section of the head in Fig. 1b and in surface view with the ectoderm peeled away (Fig. 2a). The virtually symmetrical optic primordium becomes asymmetric at stage 11 by dorsal expansion of the optic vesicle (Figs. 1c; 2b). There is little additional lateral growth of the optic primordium after stage 11; however, the head becomes wider as a result of expansion of the brain cavity so that the optic vesicles become farther apart. Continued dorsal growth during stage 12 (Figs. 1d; 2c) and stage 13 (Figs. 1e, f; 2d) results in a flattened vesicle that lies parallel to the surface of the dorsally enlarging brain. The two eyes of an embryo need not develop in synchrony. For example, one eye often is equivalent to stage 13 while the other is entering the next developmental stage when it forms a lens primordium (Fig. 1g).

### Invagination of the Optic Vesicle

Toward the end of stage 13, the optic vesicle attains its maximum height (Fig. 3a) and the overlying ectoderm forms the lens primordium, or placode (see below). Coincidental folding of the lens placode and the outer half of the optic vesicle (the retinal disc) produces the lens pit and optic cup. The optic cup consists of two layers, the lateral retinal disc and the medial future retinal pigmented epithelium (RPE) of the eye. Folding begins in the region of apposition of the lens placode and retinal disc (Fig. 3b). As invagination progresses during stage 14, the center of the retinal disc approaches the surface of the future RPE (Fig. 3c). With time, the space between the inner and outer walls of the optic vesicle (the optocoel) gradually is eliminated as the two layers become apposed over their entire surface. Most of this space disappears during stage 15 (Fig. 3d).

Folding of the optic vesicle wall is not uniform. It must be remembered that the optic vesicle is connected to the brain ventrally by a stalk. The central invagination extends into the stalk region during the early stages of folding (Fig. 4a). With progressive invagination of the retinal disc, a deep groove is formed in the ventral margin of the optic cup which extends into the ventral surface of the optic stalk (Fig. 4b). This groove is called the choroid fissure. It is the pathway for the blood vessels that eventually will vascularize the eye cavity.

### Formation of the Lens Vesicle

Late in stage 13, the lens placode protrudes from the surface of the embryo (Figs. 1g; 3a). When the lens placode begins to invaginate early in stage 14, a shallow depression is formed (Figs. 3b; 5a). During stage 14, invagination proceeds more rapidly in the dorsal part of the lens placode. As a result, the lens pit is eccentric, with the trough-shaped opening pointing ventrally (Fig. 5g) toward the end of stage 14. During stage 15 the opening (pore) becomes round as the ventral margin of the lens becomes delineated (Fig. 5c) and the lens pit becomes a vesicle (Fig. 5h). The lens pore becomes smaller during stage 15 as the attachment to the ectoderm continues to narrow (Figs. 5c, d). Finally, the attachment to the ectoderm is marked by a small depression and the lens has become a closed vesicle (Figs. 5e, i). The ectoderm covering the lens becomes smooth but the lens remains connected to the ectoderm by a short stalk (Fig. 5f). This connection becomes lost during the next few stages. The cornea is formed from the ectoderm that covers the invaginating lens. The shape of the optic cup also changes as the lens pore closes and the lens detaches from the ectoderm. Late in stage 14, the lens protrudes from the optic cup (Figs. 3c; 5b). During stage 15 the optic cup grows laterally to surround the lens

---

**Fig. 1** Sections through the eye regions of chick embryos from stages 9 to 13. Bar = 50 μm. ANP = anterior neuropore, D = dorsal, Ect = ectoderm, Mes = mesenchyme, OS = optic stalk, OV = optic vesicle, V = Ventral.

**a.** Anterior portion of a stage 9 chick embryo, cut through the frontal plane to expose the neural tube. The optic vesicles form hemispherical bulges on either side of the forebrain. The anterior neuropore marks the opening where the neural folds have not yet fused. Mesenchyme has migrated to the ventral part of the body only in the region posterior to the optic primordia.

**b.** Left optic primordium of a stage 10 embryo. The embryo was cut through the optic region on a plane perpendicular to the long axis of the brain. The optic vesicle is of slightly greater diameter than the optic stalk. A few mesenchymal cells lie ventral to the optic stalk but the optic vesicle is in direct contact with the surface ectoderm. The lighter protrusions from the apical surface (arrow) of this and succeeding figures are thought to represent dying cells.

**c.** Right optic primordium at stage 11. The optic vesicle has enlarged dorsally but the optic stalk is approximately the same diameter as at stage 10. The mesenchyme has been pulled away from the medial surface of the optic vesicle, to expose the fibrillar attachments (arrow) and the pitted surface (arrowhead) of the epithelium.

**d.** Left optic primordium at stage 12. The optic vesicle has elongated in a dorsal direction, but the other dimensions of the primordium are similar to those at earlier stages. Note the fibrils between the mesenchyme and back of the optic vesicle (arrow).

**e.** Right optic primordium early in stage 13, cut toward the back of the primordium. Dorsal expansion has resulted in a larger optic vesicle.

**f.** Left optic primordium late in stage 13.

**g.** Low magnification of the head at the end of stage 13. The optic vesicle has reached its maximal height. The primordium on the left is at a slightly later stage than on the right; the ectoderm on the left has formed a lens placode.

**Fig. 2** Heads of chick embryos that have been partially stripped of the ectoderm to expose the surfaces of the optic primordia. Bar = 50 μm. ANP = anterior neuropore, D = dorsal, Ect = ectoderm, Mes = mesoderm, OS = optic stalk, OV = optic vesicle, V = ventral.

**a.** Front view of a stage 10 head. The optic vesicles project laterally. The anterior neuropore is a slit.

**b.** View from above and to the side of a stage 11 head. The optic vesicle has a slight upward turn.

**c.** Front view of the left side of a stage 12 head. The optic vesicle clearly protrudes above the surface of the optic stalk.

**d.** Front view of the left side of a stage 13 head. The first points of fusion between optic vesicle and ectodermal basal laminae are present as torn shreds (arrowheads). The X on the optic stalk is a contaminant.

Development of the Eye

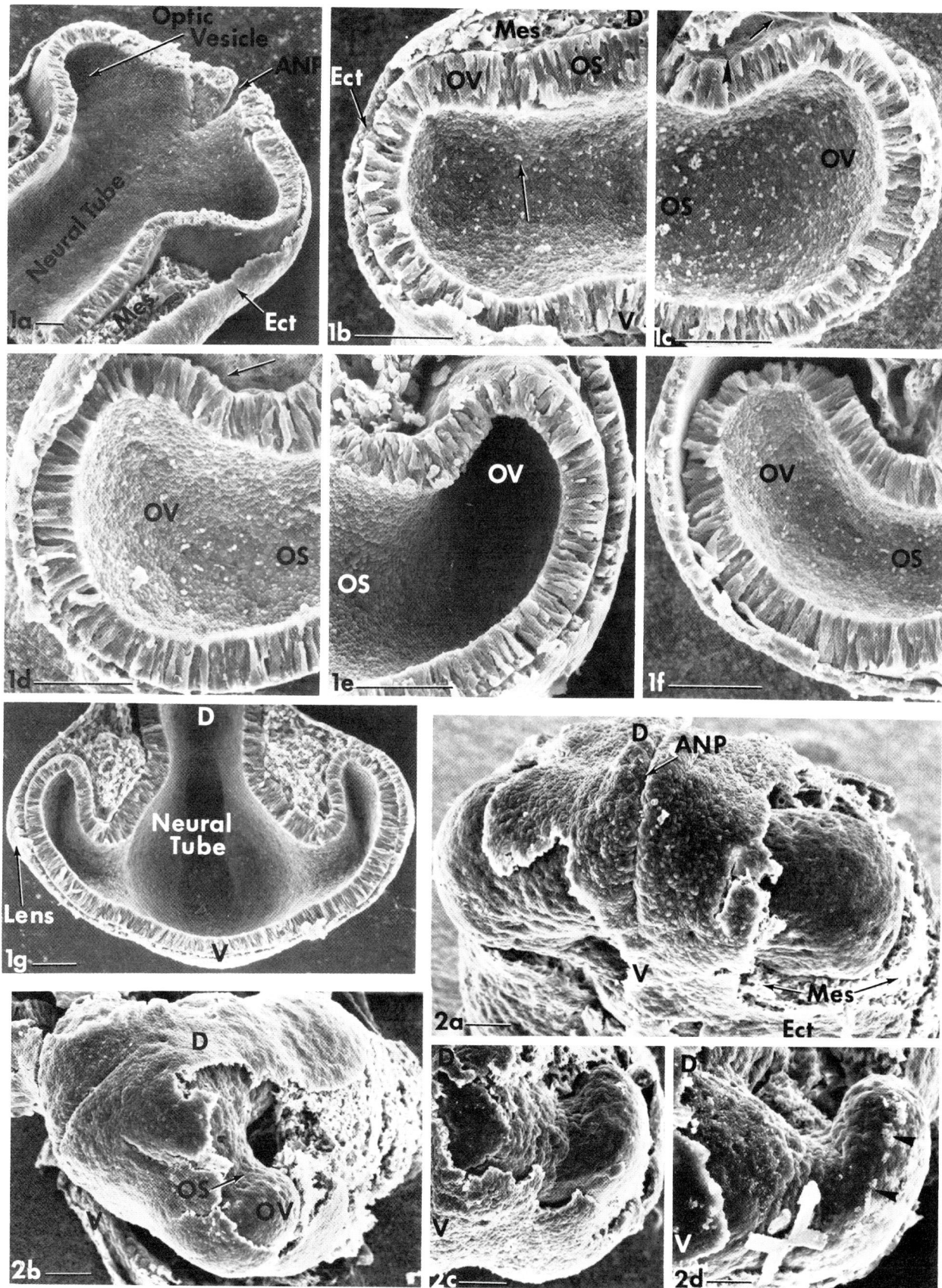

vesicle (Figs. 3d; 5b; then 5c). During stage 16 the surface of the optic cup becomes separated from the surface of the lens as a result of enlargement of the optic cup (Figs. 5e, f). The space that is formed is called the vitreous cavity and is filled with fluid.

### Cellular Changes During Optic Cup and Lens Vesicle Formation

The cells of the optic vesicle and optic stalk comprise a pseudostratified columnar epithelium from stage 9 to stage 13 (Fig. 6a). Embryonic epithelial layers commonly are pseudostratified. As is characteristic of such epithelial layers, the cells of the optic vesicle have tortuous shapes and stretch from the apical (optocoel) to basal (surface covered by basal lamina) surface of the cell layer. The widest part of the cell contains the nucleus and this portion lies apically, basally, or in an intermediate position in adjacent cells. This distribution of the nuclei gives the epithelium a stratified appearance, thus the name pseudostratified. The position of the nucleus in pseudostratified epithelia has been shown to depend upon the stage of the mitotic cycle that the cell is in; basal during DNA synthesis, apical during mitosis, and intermediate between these two time points in the cell cycle ($G_1$ and $G_2$). The change in position of the nuclei is called interkinetic migration. There are minor differences in the heights of the cells in different regions of the optic outgrowth, with those of the stalk tending to be taller than those of the optic vesicle. The presumptive lens epithelium, in contrast, consists of two layers; an outer flattened periderm and a basal layer of loosely joined cuboidal to low columnar cells (Fig. 6a). During lens formation these cuboidal cells elongate to approximately twice their previous height (Fig. 6b). The cells of the retinal disc and the ventral optic stalk also increase in height to become approximately one and one half times as tall as previously (Fig. 6c). Elongation, or palisading, begins in the center of the retinal disc and spreads to the margins during stage 14, while invagination is occurring. Little change in cell height occurs in the future pigmented layer until stage 15, when the cells of this layer become shorter (Fig. 3d; 5h) as the optic cup surrounds the invaginating lens. This difference in cell height of the future neural and pigmented layers persists during the enlargement of the vitreous cavity over the next few stages (Figs. 5e, f).

The cell apices are relatively smooth (Figs. 1; 3; 7), and show a minimum of the blebbing which is common in other organs during morphogenesis. The scattered protrusions that are seen on these surfaces at all stages between 10 and 14 (Figs. 1 & 7) have been ascribed to extrusion of dying cells from surfaces. A large amount of cell death has been reported in both the developing lens and optic primordium (Schook, 1980; Silver, 1981). The activity appears to diminish in the retinal disc at stage 15 (Fig. 7b). The sizes of cell apices are difficult to measure in scanning electron micrographs, but during invagination the cells of the margin between the future neural and pigmented layers assume different shapes from those in the center (Fig. 7a). The cause of this change is more obvious in transmission electron micrographs. From stage 10 to stage 13 all of the cells possess bundles of microfilaments that circle the cell apices at the level of the junctional complexes (Figs. 8a, b). When invagination commences, organized filament bundles no longer are found in the cells toward the center of the retinal disc (Figs. 9a, b). The cells at the margin, in contrast, have thicker filament bundles and the apices are flattened (Fig. 9c) along the direction of the fold marking the boundary between the future neural and pigmented layers (Fig. 7a).

During later development, the retinal disc becomes the complex, multilayered retina. The pseudostratified cells begin to divide so as to produce two and then several layers; interkinetic migration of the nuclei ceases. By stage 23 (4 days incubation) layering of the retina is clearly visible (Fig. 10a). As the layers become more defined, the retina becomes thicker so that it is twice the height at stage 28 (6 days, Fig. 10b) than it was at stage 23. Cell differentiation begins at the inner (vitreal) surface and progresses outward toward the pigmented layer. Cell differentiation is preceded by withdrawal of the cells from the mitotic cycle. The first specialized cells to form are ganglion cells which send their axons through spaces between cells in the optic stalk to synapse with neurons in the optic tectum of the midbrain. The first ganglion cells appear during the fourth day, in the central part of

**Fig. 3** Formation of the optic cup. Bar = 10 μm. CF = choroid fissure, Ect = ectoderm, L = lens, LP = Lens placode, LV = lens vesicle, OS = optic stalk, OV = optic vesicle, RD = retinal disc, RPE = presumptive retinal pigmented epithelium.
**a.** Left eye primordium at the end of stage 13. The ectoderm has formed a lens placode and the adjacent cells of the optic vesicle have palisaded to form the retinal disc.
**b.** Right eye primordium early in stage 14. The lens placode and retinal disc have started to invaginate and the optocoel is flattened.
**c.** Late in stage 14 the retinal disc approaches the future retinal pigmented epithelium and the optocoel is reduced to a slit between the two cell layers. The lens also is cup-shaped with a distinct lens pit. Note the sheets of extracellular material at the margins of the lens (arrowheads) and the strands between the retinal disc and lens (arrow).
**d.** By stage 15 the lens has formed a vesicle and the retinal disc is apposed to the future pigmented layer of the optic cup. The primordium was cleaved through the choroid fissure. Pits are visible in the surface of the lens vesicle and the future pigmented layer (arrowheads). CPC precipitable material (arrows) is present between the lens and ectoderm, between lens and retinal disc and behind the optic cup.

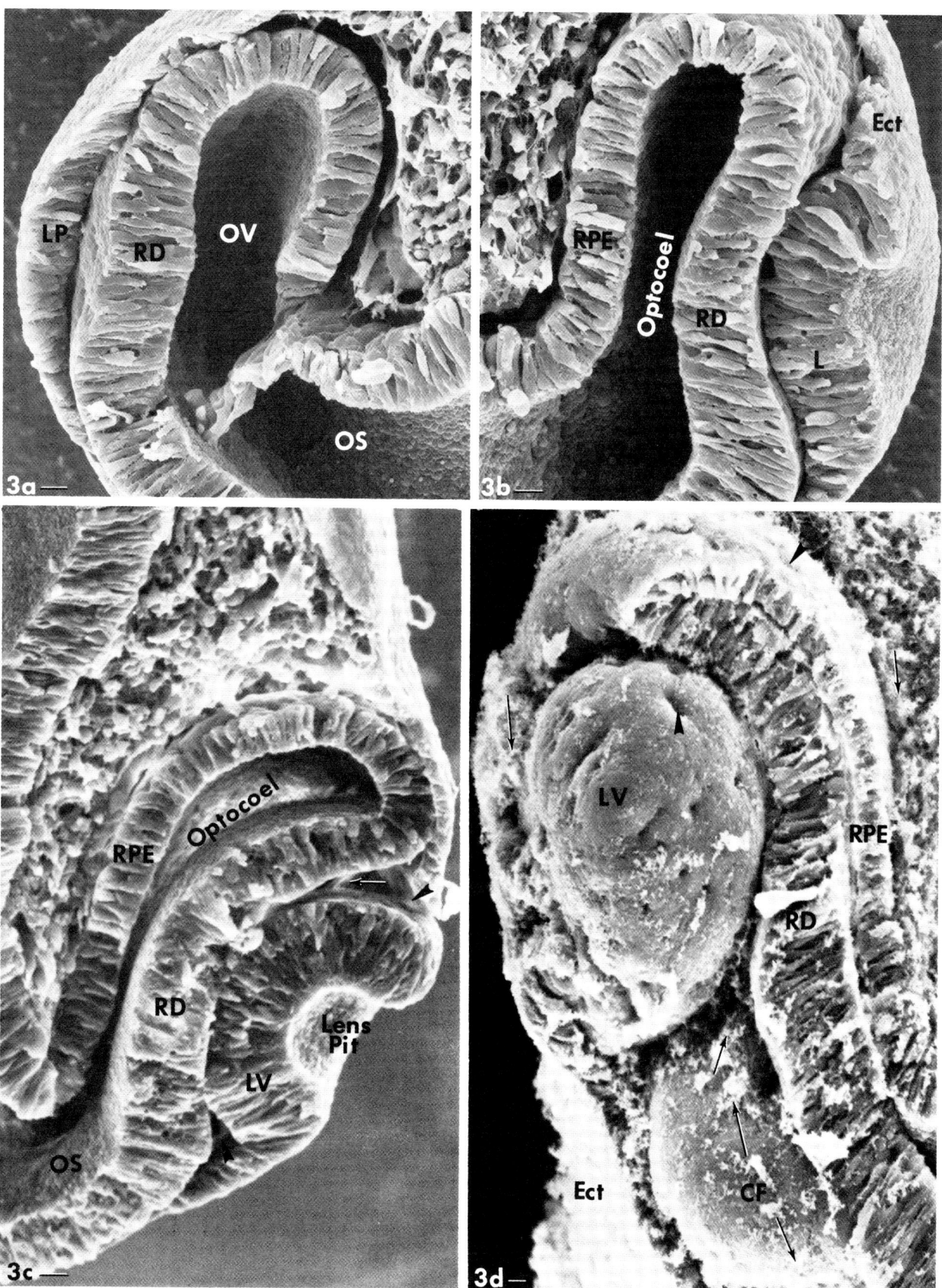

the retinal disc (the fundus), near the choroid fissure. Differentiation of ganglion cells progresses toward the periphery of the retinal disc; growth of the retina occurs at the periphery until after hatching and new ganglion cells continually are formed in this region. The neuroblasts that are the precursors of the ganglion cells form the inner layer of cells at stage 23 (Fig. 10a). By stage 29, ganglion cell bodies and axons have formed in this region adjacent to the basal lamina that delimits the vitreous cavity (Fig. 10b). By stage 35 (9 days) a ganglion cell layer is well established (Fig. 10c) It is likely that the surface characteristics of the cells change during this time (Sheffield & Lynch, 1981).

External to the ganglion cell layer, the dendrites of ganglion cells form a nucleus-free band called the inner plexiform layer. Processes of the developing amacrine and bipolar

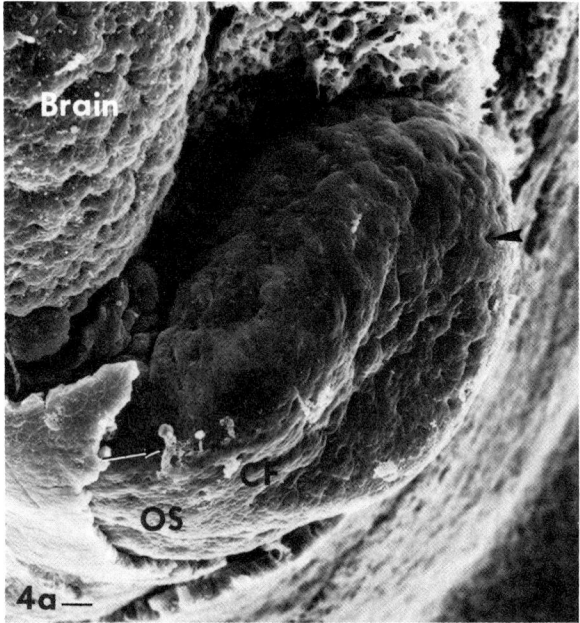

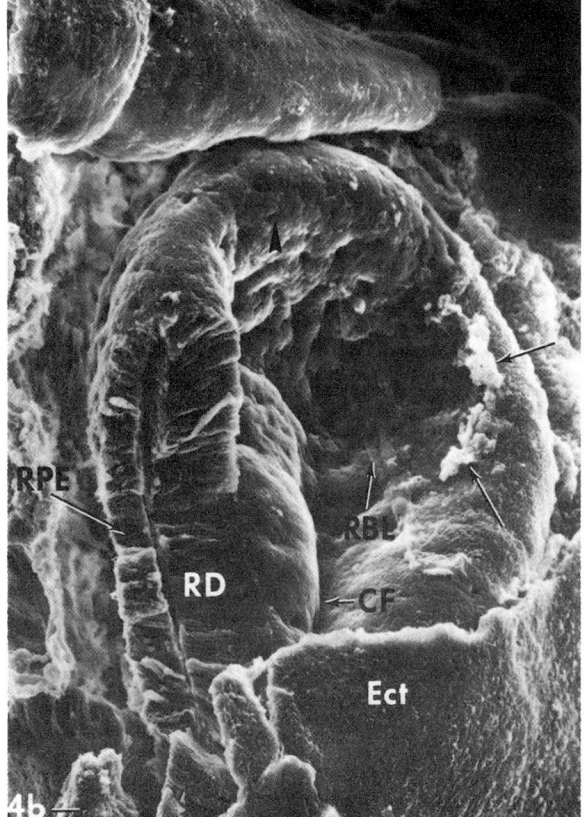

**Fig. 4** Formation of the choroid fissure (CF) Bar = 10 μm. Ect = ectoderm, OS = optic stalk, RBL = retinal basal lamina, RD = retinal disc, RPE = future retinal pigmented epithelium.
**a.** The ectoderm has been stripped from the surface of an early stage 14 optic primordium. Invagination has commenced in the center of the retinal disc, and extends ventrally as the beginning of the choroid fissure. Remnants of the fused basement lamellae remain at the edge of the optic stalk (arrow) and pits are visible in the retinal disc surface (arrowhead).
**b.** Late in stage 14, removal of the ectoderm and lens exposed the pitted (arrowhead) basal surface of the retinal disc. The margin of the retinal disc is marked by shreds of torn fused basement lamella (arrows). The edge of the torn retinal basal lamina is marked. The choroid fissure forms a deep groove from the ventral margin of the optic cup into the ventral optic stalk.

**Fig. 5** Formation of the lens. Bar = 50 μm. CF = choroid fissure, Ect = ectoderm, OC = optic cup, OS = optic stalk, PC = presumptive cornea, RPE = presumptive retinal pigmented epithelium, RD = retinal disc. a,b, and c are surface views of embryos cleaved through the head at the level of the eye primordia. d,e, and f are cut to expose the basal surface of the lens primordia. g, h, and i are cut through the lens primordia.
**a.** Early in stage 14 the lens placode has a shallow central depression and protrudes from the surface of the embryo.
**b.** Late in stage 14, the lens is cup-shaped and the surface is relatively smooth. Note the CPC precipitate in the lens pit, neural tube cavity, and optocoel (arrows).
**c.** Early in stage 15, the lens is rounder and the lens pore smaller. The specimen was compressed slightly during cleaving.
**d.** Right eye of an early stage 15 embryo cleaved to expose the surface of the intact optic cup. The cup-shaped lens fits into the cavity of the optic cup. Anchoring of the lens to the optic cup is visible in the choroid fissure (arrow). The band of fused basement lamella is marked by patches at the edge of the optic cup and on the optic stalk (arrowheads).
**e.** Left eye of a late stage 15 embryo cleaved through the frontal plane, dorsal to the optic stalk. This is the inverted top half of the embryo. The lens pore is reduced to a narrow opening and a lens stalk is forming. The

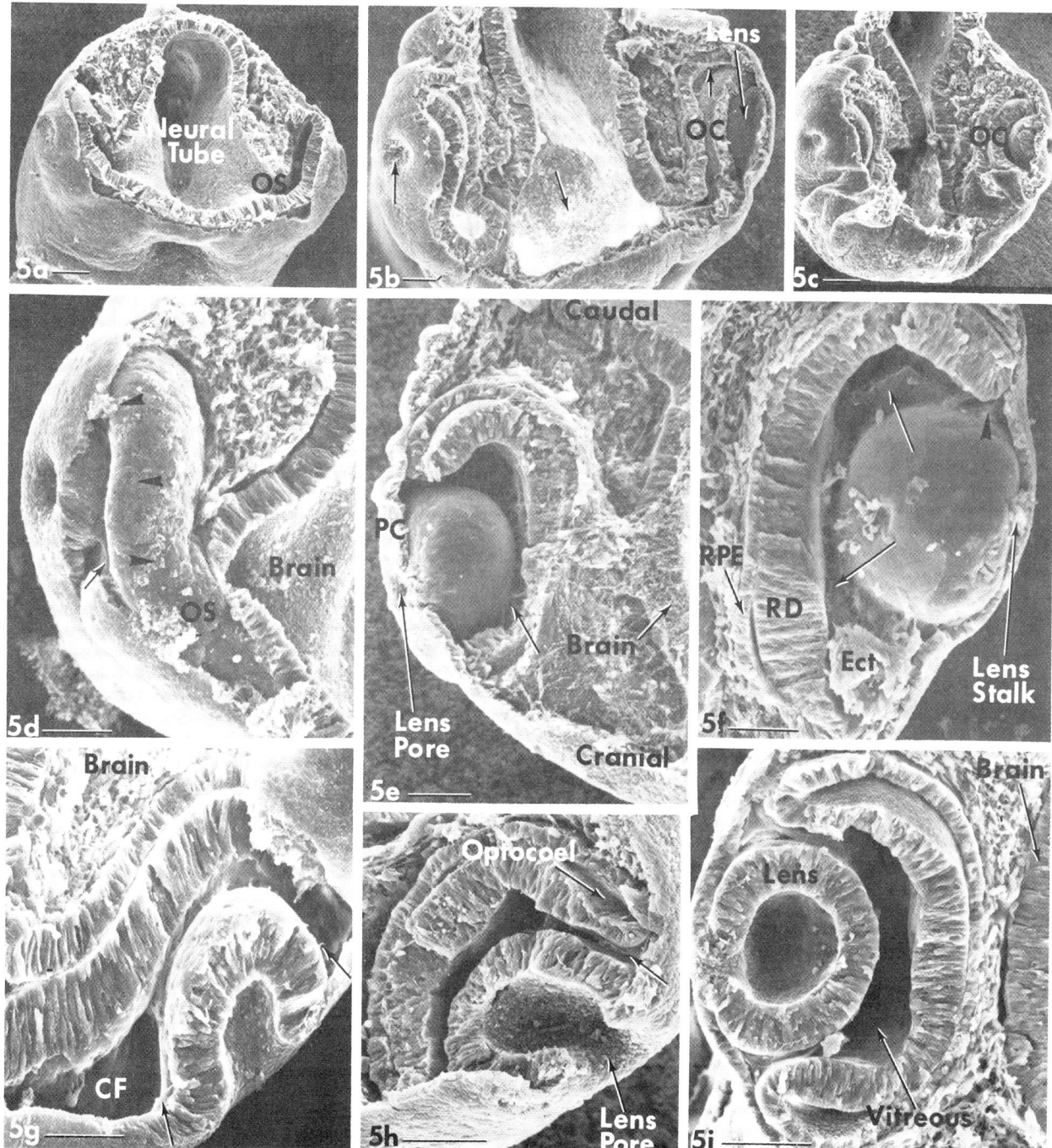

ectoderm covering the lens will form the cornea. Note the attachments between the lens and retinal disc (arrow) and the CPC precipitate in the future vitreous cavity.

**f.** Right eye of a stage 16 embryo. A piece of ectoderm has fallen into the choroid fissure. The lens is held to the ectoderm by a short lens stalk. The future retinal pigmented epithelial cells are shorter than at earlier stages. The lens is anchored to the central retinal disc by several sheet-like membranes of fine fibrils (arrows) as well as at the margins of the cup (arrowhead)

**g.** Right eye late in stage 14. The trough-shaped lens pit opens ventrally. Note the thickened lamellae dorsal and ventral to the lens primordium (arrows).

**h.** Right eye at stage 15. The lens opening has reversed its orientation as the ventral wall of the lens pore closes. The basal laminae and fused basement lamella are exposed at the dorsal margin of the lens (arrow).

**i.** Lens vesicle at stage 16. The lens vesicle is still attached to the surface ectoderm by a stalk having a small pore. Lens is attached to the optic cup only at its margins.

cell populations penetrate this layer as the cells differentiate in the cell layer external to the inner plexiform layer. Later, another layer of fibers, the outer plexiform layer, containing processes of the horizontal, bipolar and photoreceptor cells appears just beneath the most superficial cell layer. The outermost layer of the neural retina forms the photoreceptors, the rods and cones. By stage 44 (18 days) most of these cell types can be distinguished (Fig. 10d). The glial (support cell) component, the Mueller cells, also can be identified. In the central retina there are many more cones than rods. The receptor portions of both rods and cones, the outer segments, are beginning to form as extensions of the bulbous inner segments.

The structure of the pigmented layer of the retina also changes as it matures. The low pseudostratified columnar epithelium of the stage 16 embryo (e.g., Figs. 6b, 7b) becomes low columnar by stage 23 (Fig. 10a). By stage 29 the pigmented cell layer has flattened and is so tightly apposed to the neural retina that the demarcation line is indistinct (Fig. 10b). Later, the pigment cells form processes that interdigitate with the photoreceptor inner and outer segments (Fig. 10d). The external or choroid coat of the eye is formed from mesenchyme cells at least in part of neural crest origin. This layer contains blood vessels that lie at the basal surface of the RPE (Figs. 10b, c).

### Extracellular Matrix

Extracellular macromolecules are found on all cell surfaces. They exist as coats on the apical surface of cell sheets exposed to a lumen, on the lateral surfaces between adjacent cells and at the cell bases. The apical and basal granular material can be visualized by precipitating agents such as cetylpyridinium chloride (CPC). A dense coating of extracellular material first appears at the luminal surface of the optic vesicle and lens placode as the two layers begin to invaginate (Figs. 5b, 11a). The precipitate becomes more heavily concentrated toward the margin of the retinal disc late in stage 14 (Fig. 11b) and forms a thin layer between pigmented surfaces by stage 15, while precipitated material begins to appear in the vitreous cavity (Figs. 3d; 11c). The material is highly acidic and has a particulate appearance when precipitated with CPC (Fig. 11a).

Extracellular material at the cell base is present in the form of a basal lamina and granular and fibrillar material that is external to this layer. Many of the extracellular macromolecules are glycoconjugates; they consist of proteins with attached polysaccharide chains. Glycoproteins, with short chains and proteoglycans with long chains are part of the amorphous material at the cell membrane and in the basal lamina. Proteoglycans comprise much of the amorphous material external to the basal lamina. The fibrillar material within the basal lamina and external to it consists primarily of collagen, also classified as a glycoprotein because it contains short sugar chains. The optic vesicle and overlying ectoderm possess basal laminae from the earliest stages of the eye development. The space between these two epithelia is devoid of mesenchymal cells until late in development, well after the lens detaches from the surface ectoderm. It has been shown that the extracellular matrix at these early stages is synthesized and secreted by the epithelial layers. Fine fibrils connect the two extracellular sheets (Figs. 1b, e, f; 12). Before invagination begins, the basal laminae fuse to make a thicker basement lamella at the margin of the lens placode and retinal disc while the central region remains connected by fine strands of material (Fig. 6b). The basal laminae at the margin of the retinal disc and lens primordium are easier to visualize during invagination (Figs. 5d, g, h; 7b; 11b; 13) than at earlier stages. When the lens is removed, the basal lamina of the retinal disc often is partly torn away (Fig. 4b). The fused laminae at the margin of the retinal disc remain as torn shreds (Fig. 4a, b; 5d; 11b). As invagination proceeds and the optic cup and lens pit are formed, the central attachments become more pronounced and form delicate sheets during stages 15 and 16 (Figs. 5e, f). The basal lamina of the central lens cells also becomes thicker and denser (Figs. 7a; 11b; 13) and the basal cell surfaces are visible only when this layer is torn (Fig. 13). The lens covering becomes supplemented by an external layer of fibrillar collagen; this thickened layer is called the capsule. When thin sections are made through a late stage 14 embryo parallel to the lens surface, the relationship between the lens and retinal disc can be examined by transmission electron microscopy (Fig. 14). Because of the plane of section, the neural retina surrounds the lens and the space between is the precursor of the vitreous cavity. The indented surfaces of both retina and lens (Fig. 14a) correspond to the pits that are seen in scanning electron micrographs of the retinal (Figs. 4a, b; 11c) and lens (Figs. 3d; 5g; 11c) basal surfaces. The retinal and lens surfaces are covered with a basal lamina (Fig. 14a). An additional layer, the future lens capsule, surrounds the lens. This layer is rich in collagen fibrils and granular material (Fig. 14b). The basal lamina of the retinal disc covers the neural retina throughout later development (Fig. 10).

The future RPE also is covered by a basal lamina from the earliest stages of its development. Adjacent mesenchyme cells are anchored to this extracellular layer by fine fibrils (Figs. 1c, d), similar to those found between the retinal disc and overlying ectoderm. This association between the mesenchyme and future pigmented layer persists throughout optic cup formation (Figs. 5d-f; 6c; 7a; 11b). In addition to fibrils, the mesenchyme-pigmented cell interface contains CPC-precipitable material (Figs. 3d; 11b, c) that probably is proteoglycan. The presence of CPC-precipitable material between the lens and the neural retina was mentioned earlier. The mesenchymal cells

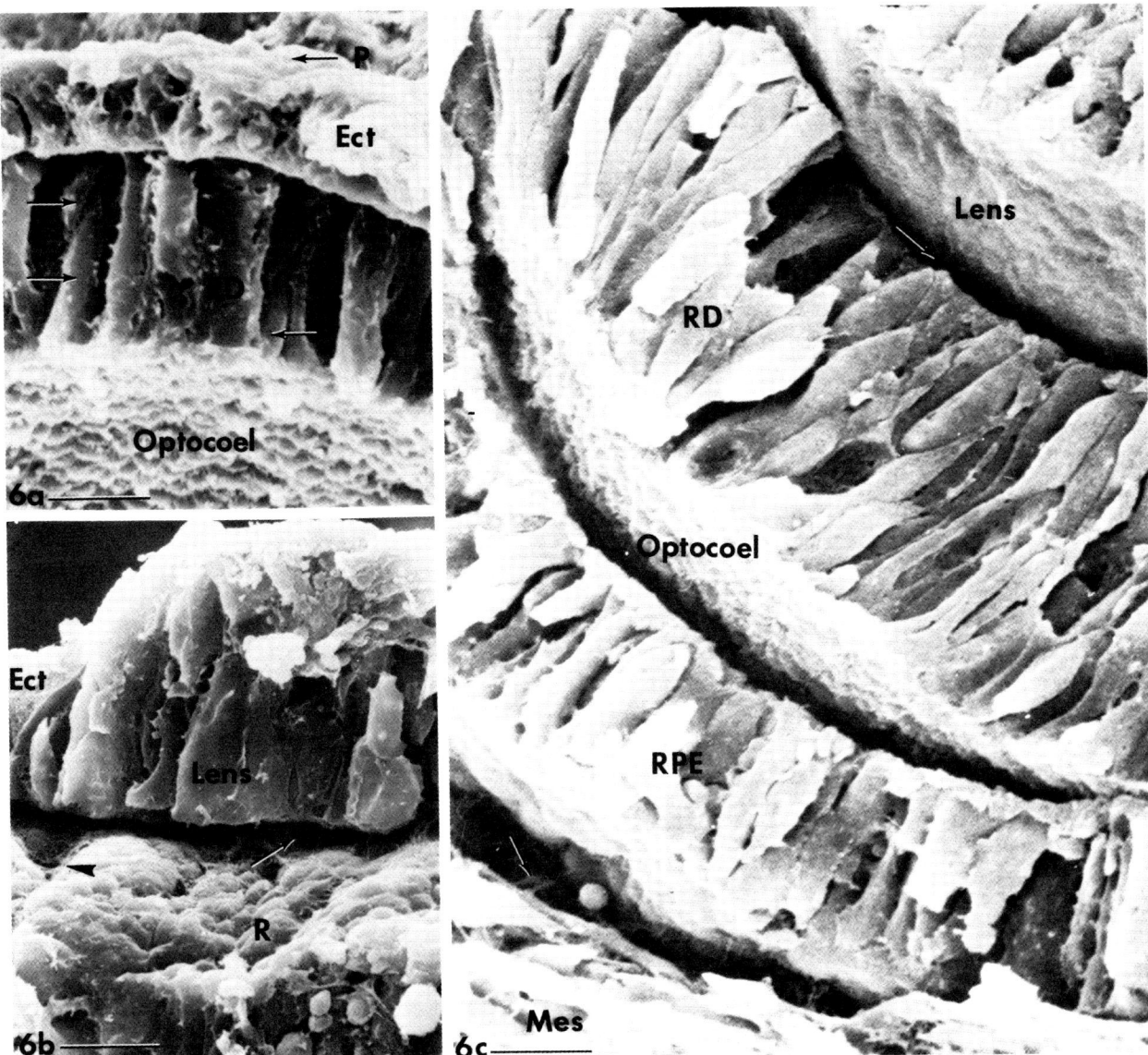

**Fig. 6** Cellular shape changes. Bar = 10μm, Ect = ectoderm, Mes = mesoderm, P = periderm layer of ectoderm, R = basal surface of retinal disc, RD = retinal disc, RPE = future retinal pigmented epithelium.
**a.** Small portion of the lateral optic vesicle wall and adjacent ectoderm. The ectoderm consists of a low columnar basal layer and a flattened surface peridermal layer. The retinal disc cells are pseudostratified with the nuclei at various levels (arrows) within the cell layer.
**b.** Small portion of the lens placode and adjacent ectoderm late in stage 13. The lens cells have doubled in height compared with the adjacent ectoderm and form a pseudostratified epithelium. The basal surface of the retinal disc is pitted (arrowhead). Fine fibrils interconnect the lens placode and retinal disc (arrow).
**c.** Small portion of the optic cup early in stage 15. The cells of the retinal disc have elongated and are twisted about each other. The cells of the future retinal pigmented epithelium have not changed in height. Fine fibers connect the lens to the retinal disc and the mesenchyme to the future pigmented layer (arrows).

adjacent to the future pigment cell layer eventually form the choroid coat (Fig. 10, see above).

### Experimental studies on Early Eye Development

Much of the work on induction of the optic primordium and lens has been done on amphibian embryos. Spemann (1938) and his coworkers were responsible for shaping much of the current thought on induction of the eye. The region of the chick blastoderm that has eye-forming potency has been investigated by mapping experiments and by transplantation. It has been shown that the eye region is determined by the head process stage (Coulombre, 1965; Hilfer et al., 1981).

It has been suggested that there is a close relationship between invagination of the optic vesicle and the lens placode (see Schook, 1978). Under certain conditions, however, these two processes are separable. Optic vesicles can be stimulated to invaginate at stage 10, well before the normal stage, by treatment with adenosine triphosphate (ATP) under conditions that allow the nucleotide to enter cells. Optic cups form in a matter of minutes after such treatment. Treatment with an ionophore which selectively increases the intracellular calcium concentration also causes precocious invagination of the optic vesicle. On the other hand, drugs that prevent the entry of calcium into cells and drugs that inhibit the calcium-regulatory protein, calmodulin, prevent optic cup formation. The response of the optic vesicle to ATP and agonists and antagonists of calcium transport as well as the changes in microfilament organization during invagination (Fig. 9), make it tempting to speculate that contractile proteins are involved in optic cup formation.

Extracellular macromolecules also seem to be involved in optic cup formation. Tunicamycin is a drug that specifically inhibits the glycosylation of glycoproteins. When heads of stage 12 embryos are incubated overnight with tunicamycin, optic cup formation is inhibited while relatively normal but small lens vesicles are formed (Fig. 15). Under these conditions, the apical surface of the retinal disc contains little CPC-precipitable material. The synthesis of glycoconjugates can be monitored by measuring the incorporation of radioactive glucosamine into precipitable material. Normal heads incorporate radioactive glucosamine at a slow rate until invagination begins; the rate then increases significantly. Heads incubated with tunicamycin incorporate radioactive glucosamine at the slower initial rate of the normal heads but do not exhibit an increased rate of incorporation. Tunicamycin may act at the basal surface of the retinal disc as well as at the apex to inhibit invagination. Not only is the CPC-precipitable material absent from the optocoel of tunicamycin-treated embryos, but also defects occur in the attachment between the lens epithelium and the retinal disc. Thus, glycoconjugates that appear at the onset of invagination may be involved in optic cup formation, but their role has not been identified.

Much of the experimental work on lens development has been done on later stages. However, it is known that the optic cup produces a factor which induces lens formation in the overlying ectoderm as well as in ectoderm from other regions of the embryo (van Doorenmaalen, et al., 1982). As mentioned above, lens invagination can occur under conditions that prevent optic cup formation, such as after treatment with tunicamycin (Hilfer, et al., 1981). The cells of the lens placode contain bundles of microfilaments that circle the apices in the appropriate position to be involved in invagination. Adhesion of the lens epithelium and optic vesicle at the margin of the lens placode and the retinal disc acts to confine the

---

**Figs. 7, 8, and 9.** Changes in cellular shape. BLL = basal lamina of lens, BLR = basal lamina of retinal disc, Ect = ectoderm, J = junctional complex, Mes = mesenchyme, MF = microfilament bundle, RPE = future retinal pigmented epithelium, RD = retinal disc.

**Fig. 7** Scanning electron micrographs of optic cups. Bar = 10 μm.
**a.** Stage 14 optic cup and lens, viewed from behind. The dense basal lamina of the lens is torn at the margin where it was fused with the basal lamina of the optic cup (arrow). The cell apices are distorted at the margin of the retinal disc and future retinal pigmented epithelium (within box). The basal surface of the future retinal pigmented epithelium is pitted (arrowhead).
**b.** Dorsal part of the optic cup and lens at stage 15. The retinal disc and lens basal laminae have been exposed as well as the point of fusion (arrow). The retinal disc apical surface is virtually free of the protruding cell fragments seen at earlier stages. The mesenchyme is closely apposed to the basal surface of the future retinal pigmented epithelium.

**Fig. 8** Transmission electron micrographs of cell apices from a stage 12 future retinal disc region. Bar = 1 μm.
**a.** Section parallel to the cell surface through the level of the junctional complexes. Microfilament bundles tend to circle the cell apices (arrows).
**b.** Section perpendicular to the apical cell surface. The microfilament bundles lie at the level of the junctional complexes.

**Fig. 9** Transmission electron micrographs of cell apices from the retinal disc at stage 14. Bar = 1 μm.
**a.** Section parallel to the surface at the center of the retinal disc, at the level of the junctional complexes. Microfilaments form only a few, small bundles.
**b.** Section perpendicular to the surface of the central retinal disc cells. Few microfilaments are oriented parallel to each other.
**c.** Section through the junction of the retinal disc and future retinal pigmented epithelium, similar to the region boxed in Fig. 7a. The section passes from the optocoel in the lower left corner, through the apical region to the level of the junctional complexes at the top of the figure. The cell apices are distorted at the level of the junctional complexes. Microfilament bundles are pronounced at this level (arrows).

# Development of the Eye

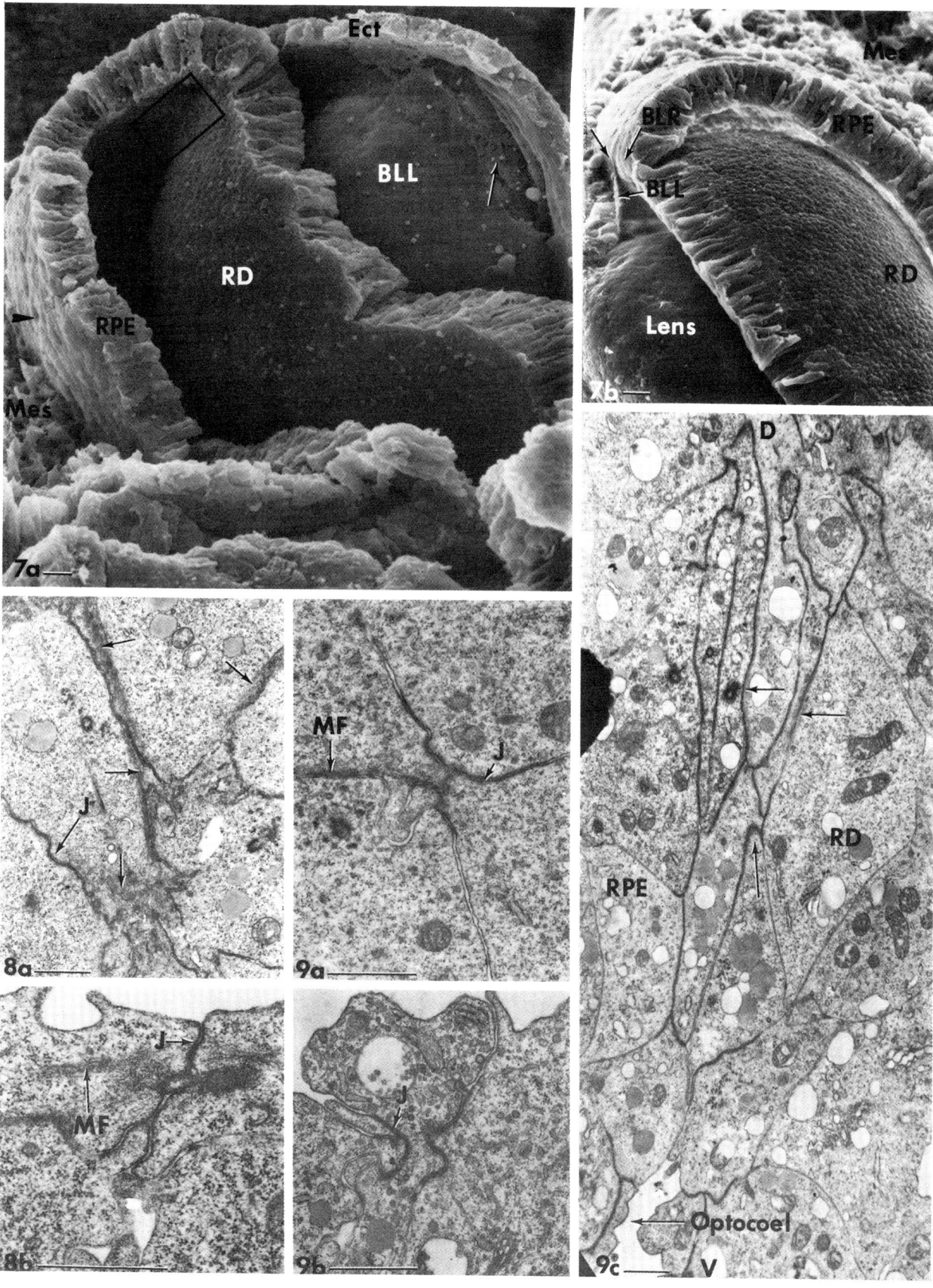

lens-forming region. Studies of division patterns, increases in cell number, and changes in cell height have resulted in the proposal that lens invagination could be a passive response to changes in lens cell packing rather than by an active process, such as contraction of cell apices (Zwaan & Hendrix, 1973). The lens does form in embryos that are treated with ionophore but not after treatment with ATP. Drugs that prevent calcium transport or inactivate calmodulin inhibit lens formation just as they inhibit optic cup formation. Thus, lens invagination must be at least in part motivated by other than by externally applied forces. Because of the attachments between the lens epithelium and retinal disc, it is difficult to assess how much of the motive force of one layer causes the other to invaginate.

Lens cells undergo cytodifferentiation soon after detachment of the lens vesicle from the surface ectoderm. The cells of the inner hemisphere of the lens vesicle elongate to form the first fiber cells. The cells make a series of specific proteins, the lens crystallins. These appear first in the outer cells that will comprise the epithelium of the mature lens. Crystallins appear later in the first fiber cells. The cells at the margin between the epithelium and the fibers are responsible for the formation of more fiber cells and the growth of the lens (Beebe, et al., 1981).

Growth of the eye is controlled by interaction between the lens and the optic cup (Coulombre, 1965). If the optic vesicle is prevented from reaching the overlying ectoderm either through developmental error or through experimental interference, a lens placode will not form and the optic cup that forms will be imperfect and small, or microophthalmic. Removal of a lens also results in reduction in the size of the optic cup. Transplantation of a lens from another species or placement of two lenses in the position occupied by the normally single lens will result in regulation of the transplant to the proper size for the optic cup. Developmental abnormalities or experimental manipulations that permit fluid to leak from the vitreous cavity, thus reducing the pressure in the cavity will result in reduction in the size of both the lens and the optic cup.

## Conclusion

The three-dimensional image produced by the scanning electron microscope is exceptionally useful when applied to the study of developing organs. Formation of the eye involves a complex set of coordinated shape changes in the epithelial sheets that produce the optic cup and lens. These changes involve not only folding of the sheets into vesicles, but also distortions of the cell surfaces and changes in the organization of the intervening extracellular matrix. Although each of these changes has been described from light and transmission electron microscopic investigations, the true interrelationships among the tissues become more recognizable as a result of the three-dimensional images. Examples of these interrelationships are 1) the fit of the curved surfaces of the retinal disc and lens epithelium, 2) the grooved and pitted basal surfaces of these two layers and their relationship to the attachments of extracellular fibers, and 3) the changes in cell height at earlier stages and in layering of the retinal disc at later stages of development. An understanding of organ formation requires a grasp not only of three-dimensional organization but also of the fourth dimension of time. The images produced by the scanning electron microscope serve as a bridge to achieving that understanding.

## Acknowledgements

I wish to express my sincere appreciation to Dr. Jyh-jia W. Yang for the use of a number of her scanning electron micrographs, that represent part of her thesis research. I also greatly appreciate the four scanning electron micrographs supplied by Dr. Joel B. Sheffield, and his help in writing the section on later retinal development. This work would not have been possible without the highly competent

---

Fig. 10 Development of the retina. Bar = 10 μm. Axon = ganglion cell axon layer, BL = basal lamina of the retinal disc. BV = blood vessel, Gang = ganglion cell layer, IPL = inner plexiform layer, IS = inner segment of photoreceptor, Muller = Muller cell process, OPL = outer plexiform layer, OS = outer segment of photoreceptor, RPE = developing retinal pigmented epithelium.
**a.** Section of the developing retina at stage 23 (4 days). The pigmented epithelial cells are low columnar. The neural retina consists of several layers of cells. Those adjacent to the basal lamina are the neuroblasts that will form ganglion cells.
**b.** Stage 29 (6 days). The pigmented epithelial layer is closely apposed to the neural retina. Blood vessels lie at the choroid-pigmented epithelial interface. Layering of the neural retina is more pronounced than at earlier stages. The ganglion cell bodies form a distinct layer and their axons begin to form a separate layer at the inner (basal) surface.
**c.** Stage 35 (9 days). The pigmented epithelial cells have flattened. The ganglion cell layer is distinct and their axons form a full layer. The inner plexiform layer has appeared. Mueller cells can be identified as processes that span the ganglion cell and axon layers.
**d.** Stage 44 (18 days). By a few days before hatching, most of the structure of the mature retina has appeared. The ganglion cell axons form bundles at the basal surface. The inner plexiform layer is a thick tangle of nerve cell processes and the outer plexiform layer is visible. Processes of the pigmented epithelial cells surround the photoreceptor endings. The photoreceptor inner segments have developed and the outer segments are forming. (Courtesy of Dr. J.B. Sheffield and C.W. Taylor.)

# Development of the Eye

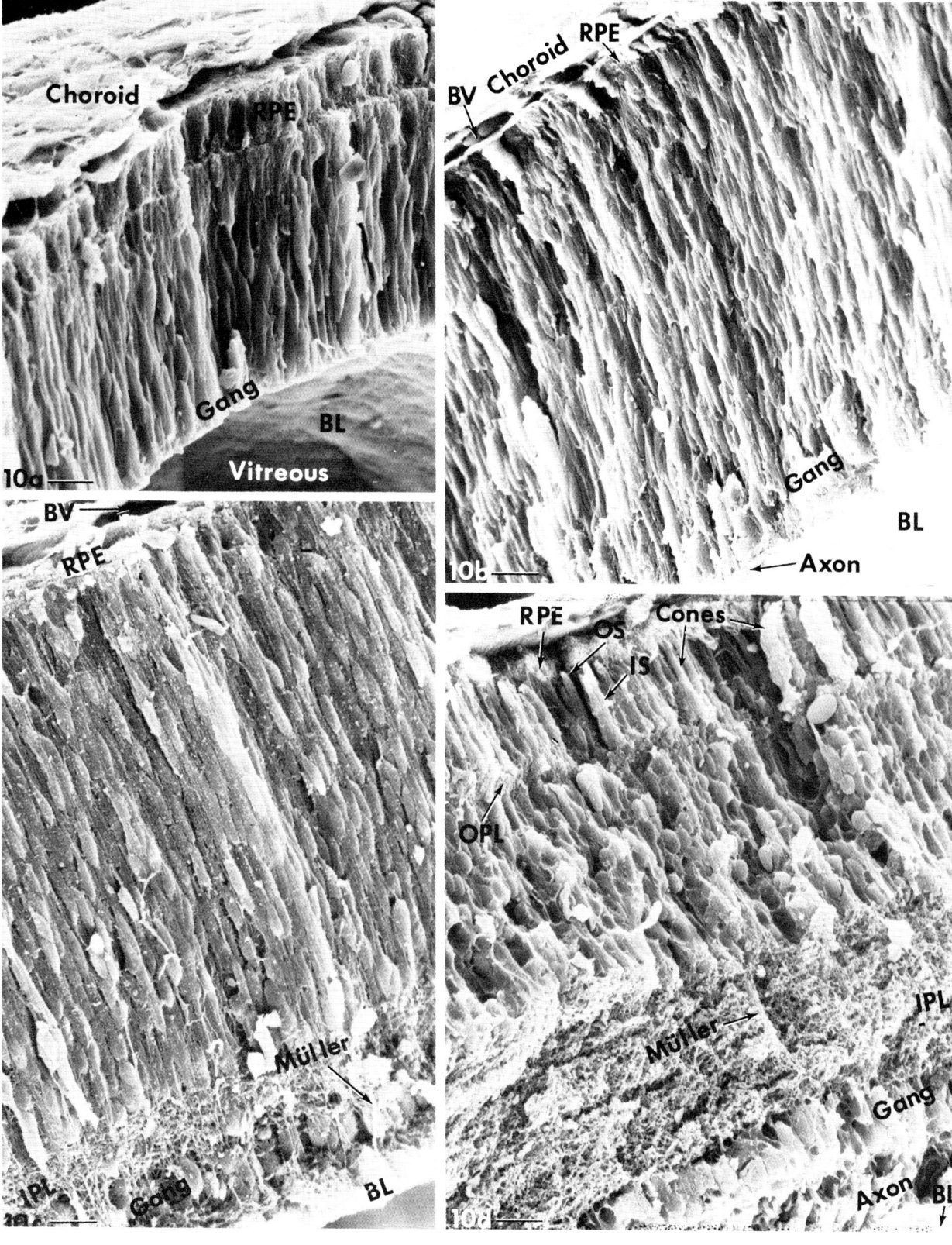

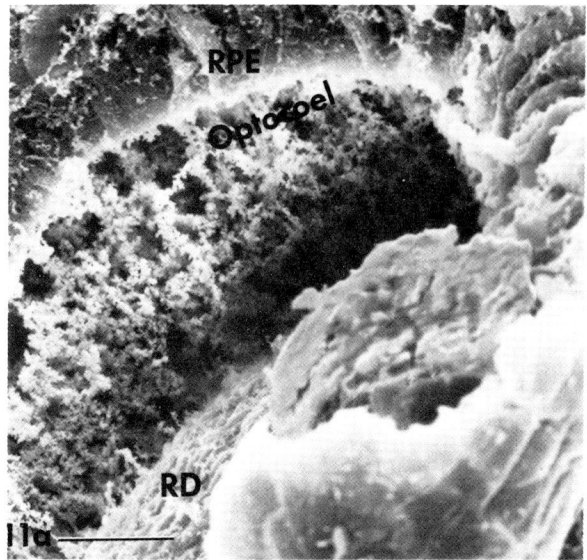
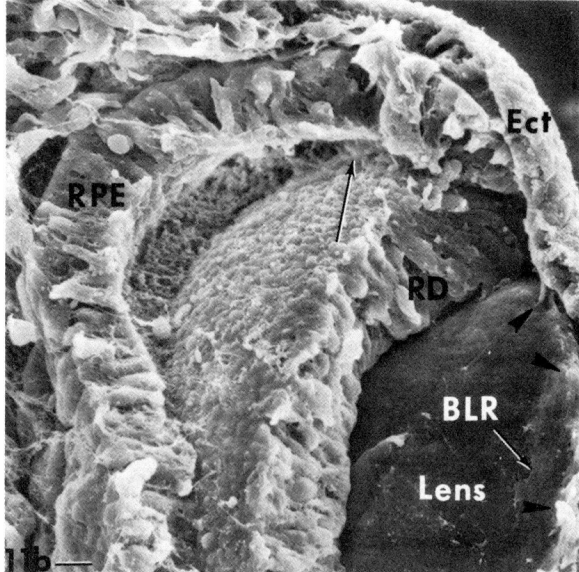
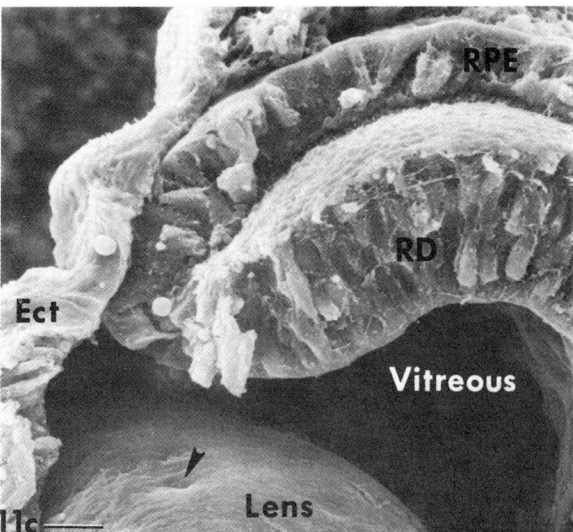

**Fig. 11** CPC-precipitable material of the optic cup. Bar = 10 µm, BLR = retinal disc basal lamina, Ect = ectoderm, RD = retinal disc, RPE = future retinal pigmented epithelium.
**a.** Dorsal portion of an optic cup at stage 14. The optocoel is filled with a granular and fine fibrillar CPC precipitate.
**b.** Higher magnification of a part of Figure 5b, showing the dorsal region of the optic cup and lens late in stage 14. A heavy amorphous precipitate is present at the junction of the retinal disc and future pigmented layer (arrow). Some precipitate is seen as fine strands at the interface between the mesenchyme and future pigmented layer. The fused basement lamella is visible at the edge of the lens (arrowheads). A fragment of the retinal basal lamina remained attached to the sample when the outer portion of the optic cup was removed.
**c.** Higher magnification of a part of Figure 5e, showing the dorsal part of the eye primordium at stage 15. CPC-precipitable material is present in the vitreous cavity, on the apical surface of the retinal disc, and on the basal surface of the future retinal pigmented epithelium. Pits are visible in the dense covering over the lens surface (arrowhead) and the basal surface of the retinal disc.

**Fig. 12** Optic vesicle of a stage 13 embryo. Bar = 10 µm. Ect = ectoderm. The embryo was cut in the horizontal plane through the optic stalk and the embryo tilted to examine the basal surface of the optic vesicle and the ectoderm. Note the fine, interconnecting fibrils.

**Fig. 13** Right eye primordium of a late stage 14 embryo. Bar = 20 µm. BLL = lens basal lamina, BLR = retinal disc basal lamina, RD = retinal disc, RPE = future retinal pigmented epithelium. Half of the optic cup was removed leaving behind the retinal and ectodermal basal laminae at the top of the figure. The point of fusion of the two intact layers is marked by an arrowhead and the torn edge by an arrow. The basal lamina of the lens primordium was torn to reveal the basal cell surfaces.

**Fig. 14** Transmission electron micrographs of the matrix between the lens and optic cup at stage 14. BLL = lens basal lamina, BLR = retinal basal lamina, LC = future lens capsule, RD = retinal disc.
**a.** Bar = 1 µm. The section was cut parallel to the surface of the optic cup. A section was selected through the back of the lens primordium, where it was surrounded by the retinal disc. Both the retinal disc and lens surfaces are covered by basal laminae. The surfaces of the lens (arrowhead) and retinal disc (arrows) are indented. Fibrils in cross and oblique section are clustered in and around the pits. The lens surface is covered with a dense meshwork of fibrillar and granular material, the precursor of the lens capsule.
**b.** High magnification of the future lens capsule in a region similar to the box in Figure 14a. Bar = 100 nm. The fibrils having crossbanding suggestive of collagen (arrows) probably belong to the vitreous.

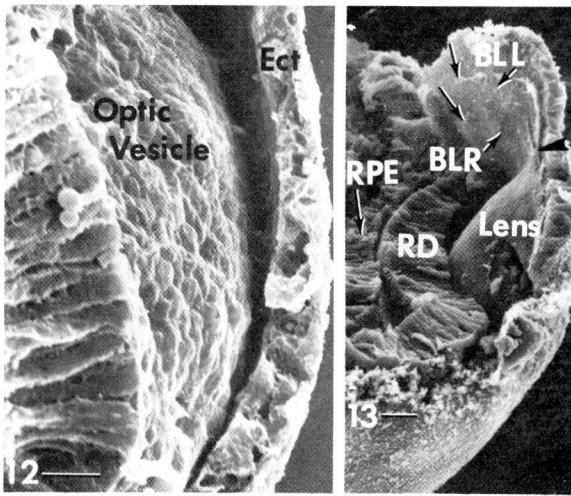
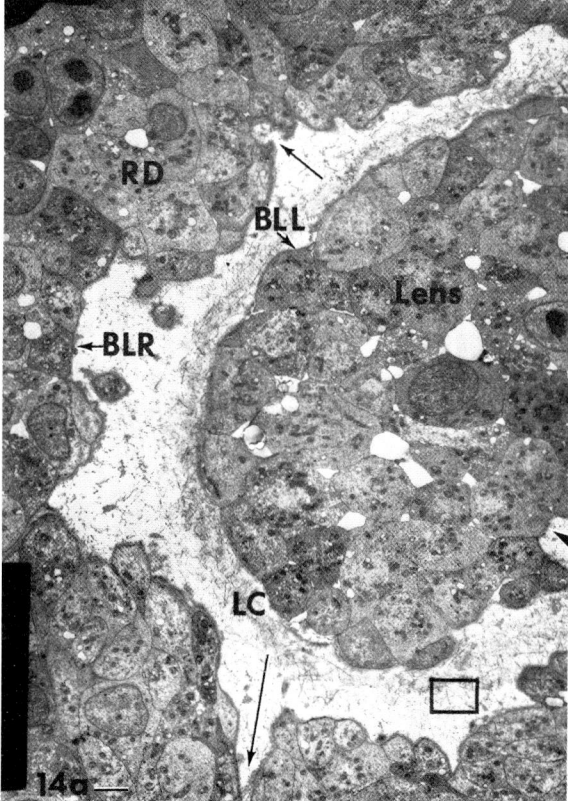
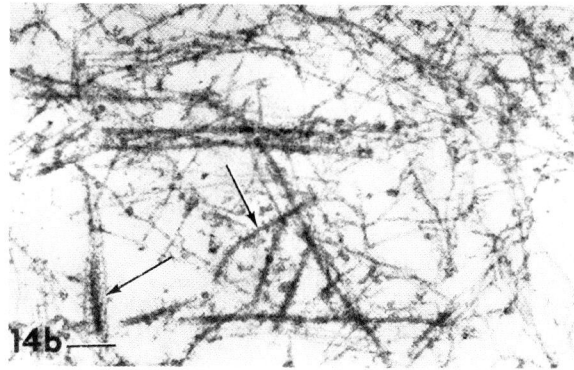
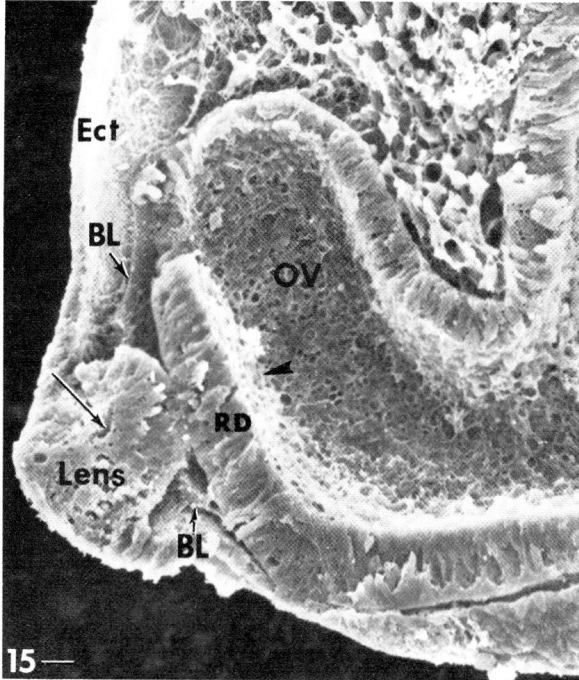

**Fig. 15** Effects of tunicamycin. Bar = 10 μm. BL = basal lamina, Ect = ectoderm, OV optic vesicle, RD = retinal disc. A stage 12 head after 24 hours in organ culture. The lens has formed a vesicle with a small cavity (arrow). The optic vesicle shows no sign of invagination. There is virtually no CPC-precipitate on the apical cell surfaces, which appear distorted (arrowhead). However, CPC precipitate is present in the neural tube cavity. The basal lamina appears either to have stretched or to have detached at the margins of the retinal disc.

assistance of Joyce W. Brown and Michael Czeredarczuk. My thanks also go to Jo-Ann Felder for her skills with the word processor. The original research reported here was supported in part by DHHS research grant EYO1934, a Fight For Sight grant-in-aid, Fight For Sight, Inc., New York City, Biomedical Research Support Grant RR07115, and a Temple University grant-in-aid.

### References

Beebe DC, Johnson MC, Feagans DE, and Compart PJ. (1981). The mechanism of cell elongation during lens fiber cell differentation, in: Ocular Size and Shape: Regulation During Development, S.R. Hilfer and J.B. Sheffield (eds.); Springer-Verlag, NY, 79-98.

Coulombre AJ. (1965). The eye, in: Organogenesis, R.L. DeHaan and H. Ursprung (eds.). Holt, Rinehart, and Winston, NY 219-251.

Hamilton HL. (1952). Lillie's Development of the chick, Holt, NY, 78-91.

Hilfer SR, Brady RC, and Yang, J-jY. (1981). Intracellular and extracellular changes during early ocular development in the chick embryo, in: Ocular Size and Shape: Regulation During Development, S.R. Hilfer and J.B. Sheffield (eds.) Springer-Verlag, NY, 47-78.

O'Rahilly R. (1975). The prenatal development of the human eye, Exp. Eye Res. 21, 93-113.

Schook P. (1978). A review of data on cell actions and cell interactions during morphogenesis of the embryonic eye, Acta Morphol. Neerl-Scand., 16, 267-286.

Schook P. (1980). A spatial analysis of the localization of cell division and cell death in relationship with the morphologenesis of the chick optic cup, Acta Morphol. Neerl-Scand. 18, 213-229.

Sheffield JB and Lynch M. (1981). Cell Surface differentiation in the embryonic chick retina, in: Ocular Size and Shape: Regulation During Development, S.R. Hilfer and J.B. Sheffield, (eds.), Springer-Verlag, NY, 99-122.

Silver J. (1981). The role of cell death and related phenomena during formation of the optic pathway, in: Ocular Size and Shape: Regulation During Development, S.R. Hilfer and J.B. Sheffield, (eds.) Springer-Verlag, NY, 1-24.

Spemann H. (1938). Embryonic Development and Induction, Yale Univ. Press, CT, 40-96.

van Doorenmaalen WJ, van der Starre H, Janssen PT and van der Starre-van Bekkum M. (1982). Molecular biology of lens induction, in: Cellular Communication During Ocular Development, J.B. Sheffield and S.R. Hilfer (eds.), Springer-Verlag, NY 97-109.

Zwaan, J. and R.W. Hendrix (1973). Changes in cell and organ shape during early development of the ocular lens, Amer. Zool. 13, 1039-1049.

## Discussion with Reviewers

D.J. Moran: In view of the dependence of optic cup formation on the extracellular matrix, would you speculate on how these materials may function in optic cup morphogenesis?

Author: Extracellular matrix molecules accumulate at the apical surface of the retinal disc within the optocoel, and at the interface between the basal surfaces of retinal disc and lens placode. The material within the optocoel appears as invagination commences and is highly acidic. The time of appearance is suggestive of a role in invagination and these macromolecules may act to sequester calcium ions that are needed to initiate invagination. Movement of calcium ions into the retinal disc cells seems to be involved in the initiation of invagination. Alternatively, the material may be involved at a later stage in optic cup formation. It may act to stabilize the retinal disc against the future RPE until the RPE cells can assume this function.

The matrix between the retinal disc and lens placode also appears to play a role in invagination. The major role may be mechanical; attachment of the lens placode to the optic vesicle so that invagination of both occurs in coordinated fashion. Other more subtle changes in extracellular macromolecules may occur in this region. For instance, initiation of invagination may depend upon a change in the glycoconjugates of the basal cell membranes. Measurements of changes in glycosylation patterns and heparan sulfate proteoglycan content in developing organs is a very active area of current research.

G.C. Schoenwolf: The experiments with ATP, ionophore, etc. seem to suggest an active role for microfilaments in formation of the optic cups, but not the lens vesicles. How do you reconcile these experiments with the paucity of microfilaments shown in most areas of the optic vesicle undergoing invagination?

Author: Actually, the distribution of microfilaments during invagination of the optic vesicle is consistent with the interpretation of a contractile event at the margin of the retinal disc, where maximal bending of the cell sheet occurs. Spreading of cells apices at the center of the retinal disc also may be dependent upon microfilaments. Either calcium-dependent relaxation of an actin-myosin interaction or depolymerization of actin filaments is consistent with the loss of microfilament bundles in these cells during invagination. A role for microfilaments in lens invagination has not been totally eliminated, since lens vesicles form in a calcium-dependent manner in the model system.

G.C. Schoenwolf: When precocious invagination of the optic cups is induced with ATP, are there any differences in morphology (i.e., the shape of the invaginating cup; the distributions or sizes of microfilament bundles) as compared to normal invagination with simultaneous invagination of the lens placode?

Author: The preparations are difficult to fix for electron microscopy after these treatments. However, preliminary results suggest that the same changes in shape of the optic vesicle and microfilament organization occur in this in vitro system as in vivo.

---

For additional discussion see page 96.

# STEREO SCANNING ELECTRON MICROSCOPY OF THE CRYSTALLINE LENS

J.R. Kuszak*, M.S. Macsai and J.L. Rae

Departments of Pathology, Physiology and Ophthalmology,
Rush Medical College, Chicago, Illinois 60612

(Paper received March 14 1983, Complete manuscript received August 4 1983)

## Abstract

We have used an improved protocol to prepare human, human neonatal, rat and frog lenses for examination by stereo scanning electron microscopy. In this manner, complete and accurate images of the changes in lens cell shape, size and surface complexity are revealed as they differentiate and develop from cuboidal epithelial cells into elongate fiber cells. This method also shows that the apical ends of elongating fibers are variably expanded as they interface with the overlying lens epithelium. Apical ends are most expanded as they contact pre-germinative zone epithelial cells and least enlarged as they contact transitional zone cells. By examining the interlocking devices on opposed fibers in frog, rat and human lenses we determined that there are standard types and interlocking patterns in all lens species. Finally, stereo SEM reveals that the ridges previously reported on aged human nuclear fibers are also seen on human neonatal cortical fibers and that these ridges may actually be interlocked villous or fingerlike projections.

KEY WORDS: Crystalline Lens, Stereo Scanning Electron Microscopy, Terminal Differentiation, Interlocking Devices, Lens Epithelium.

*Address for correspondence:
Rush Medical College
Department of Pathology
600 S. Paulina
Chicago, IL 60612    Phone No. (312) 942-6789

## Introduction

The function of the crystalline lens is to collect divergent rays of light and transform them into a single focused ray to be refracted onto the retina. To perform this role, the lens must remain transparent as rays of light are refracted through the membrane, cytoplasm and extracellular spaces of this multicellular organ. Troekel (1962) and Duke-Elder and Abrams (1970) have suggested that the structure of lens cells are a unique adaptation for lens function. If so, then understanding lens function requires knowledge of lens cell morphology. However, several of the unique structural features of the lens makes this an extremely difficult organ to examine morphologically.

The anterior surface of the lens is covered by a monolayer cuboidal epithelium which contains cells in every stage of the cell cycle (Harding et al., 1971). The bulk of the lens is composed of fiber-like cells derived from a region of this epithelium known as the germinative zone. Cells in this zone, overlying the lens equator, are induced to terminally differentiate and elongate into lens fibers. These newly formed fibers are added onto the periphery of the existing lens mass. Lens fiber cells are hexagonally shaped, elongate crescents. The fiber cells are aligned in columns radiating from the core of the lens. Each radial cell column (RCC) consists of progressively older fibers extending from the oldest region of the lens, the lens nucleus, through the middle region of the lens, the lens cortex, and finally to the periphery of the lens, the bow region. Since the lens grows continuously in this manner, fiber cells become progressively more internalized with age, and are retained permanently within the core of the lens.

A lens measuring 3mm in diameter has superficial cortical fibers as long as 7.5 cm in length. The older more internalized fibers are as short as 1.5 cm in length. The ends of all fibers curve in opposite directions away from their axis as they approach and interlock at the lens sutures. Therefore fibers are actually S-shaped. The extreme length and S-like curvature of fibers makes it improbable if not impossible to capture a representative fiber cell in a single thick section for examination by light microscopy (LM). It is even more difficult to serially reconstruct fibers for examination at the ultrastructural level by transmission electron microscopy (TEM). In addition, the large amount of trimming required for TEM precludes direct examination and comparison of lens cell morphology in a single lens. A direct comparison of the morphology of cells throughout a lens is necessary because the lens is not composed of uniformly shaped cells. The shapes and surface morphology of lens cells vary as a function of their state of differentiation and age (Kuwabara, 1975; Kuszak et al., 1980; and Kuszak and Rae, 1982).

Scanning electron microscopy (SEM) permits examination of whole cells and even large surface areas comprising numerous cells within a single organ. Therefore, it is ideally suited for examining the complex morphology of the lens. Kuwabara (1970) was the first to examine the lens by SEM. He noted that artifactual cracks developed within lenses prepared for SEM during the process of critical point drying (CPD). By enlarging these cracks, he demonstrated that large segments of cortical fiber cell surfaces could be exposed for examination. Unfortunately, the cracks afforded only minimal exposure of the surfaces of nuclear fibers. Furthermore, these cracks failed to expose the surfaces of either epithelial or differentiating cells.

Recently, we have found that the quality of lens specimens prepared for SEM depends in part on the osmolality of the fixative-buffer mixture (Kuszak and Rae, 1982). Lenses prepared for SEM in an isotonic fixative-buffer mixture do not crack during CPD. However, these lenses can be separated by dry fracture between radial cell columns (RCCs) to reveal essentially every lens cell in an antero-posterior plane. Therefore, it is now possible to examine and compare directly, within a single lens, the shapes, surface morphology and structural organization of all lens cells.

The contrast mechanism described for SEM and that which operates in producing an optical image for the human eye are comparable. The TV raster images produced by the SEM provide near perfect reproduction of the microscaled surfaces of SEM specimens. SEM stereomicroscopy provides accurate three dimensional imaging of step or height differences in cell surface morphology as well as an appreciation of cell shape that is unparalleled by any other morphologic technique. Therefore, we have used SEM stereomicroscopy to examine the morphology and structural organization of cells in the crystalline lens. In this report, we will present stereomicrographs of representative central, pre-germinative, germinative and transitional zone lens epithelial cells. We will also demonstrate the development of interlocking devices and the basic interdigitating pattern between representative cortical and nuclear fiber cells in amphibian (frog) and mammalian (human and rat) lenses.

## Materials and Methods

Adult lenses from small Rana pipiens (northern variety) frogs and Sprague-Dawley rats were used in this study. Human lenses used in this study were obtained from the Illinois Eye Bank. Immediately following sacrifice the eyes were removed from the head of rats and frogs and placed into appropriate Ringer solution. After removing the posterior portion of the globe, the lenses still attached by the zonules were immersed in 10 ml of an appropriate fixative-buffer mixture. The determination of appropriate fixative-buffer mixtures for rat and human lenses and the protocol for preparing these lenses for SEM were as described previously for frog lenses by Kuszak and Rae (1982). Specimens were examined in a JEOL JSM 35c scanning electron microscope at 15kV. Stereopairs were taken with a Polaroid camera system at F11 by tilting the specimen stage $\pm 6°$ to the 0° plane of RCCs.

## Results

### Lens epithelium and elongating fibers

Lens epithelial cell shape and surface morphology vary in the central, pregerminative, germinative and transitional zones. In general, the cells increase in size while decreasing in surface complexity as one moves from the central to transitional zones.

Central zone epithelial cells are shown in figure 1. These cells have amorphous cell bodies that correlate well with the marked lateral infoldings of lens epithelial cells shown by TEM (Kuwabara, 1975; and Maisel et al., 1981). Pregerminative zone cells are shown in figure 2. The cell bodies of these cells are slightly less irregular in shape and are slightly taller than central zone cells. The germinative zone cells shown in figure 3 are low columnar cells that have begun the process of terminal differentiation and are assuming the hexagonal shape of fibers. The transitional zone cells shown in figure 4 are the largest lens cells prior to fiber elongation. They are hexagonal in shape with four approximately equal narrow faces and two approximately equal broad faces. The broad faces of these cells become oriented parallel to the lens surface as they rotate in the bow region during the initiation of elongation. Examination of the bow region suggests that the four narrow faces and two broad faces of transitional zone cells are the forerunners of the four narrow and two broad faces of fiber cells (fig. 5). The apical and basal surfaces of the transitional zone cells appear to be maintained as the anterior and posterior ends of elongating fibers. As fibers elongate, these ends are the surfaces that interface with the overlying epithelium (anterior ends) and the capsule (posterior ends).

In general, the thickness and width of fully elongated fibers is of the order middle region > posterior end > anterior end. The opposite is

Fig. 1 Central epithelial cells. Arrowheads indicate the expanded apical end of an apposed underlying fiber. Note the underlying fibers are devoid of interlocking devices and are thinnest as they pass beneath this region of the epithelium.

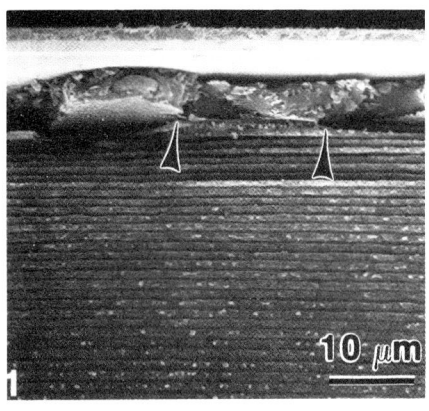 

Fig. 2 Pre-germinative zone cells. Arrowheads indicate the expanded apical end of an apposed underlying fiber. Note the thickened underlying fibers feature interlocking devices.

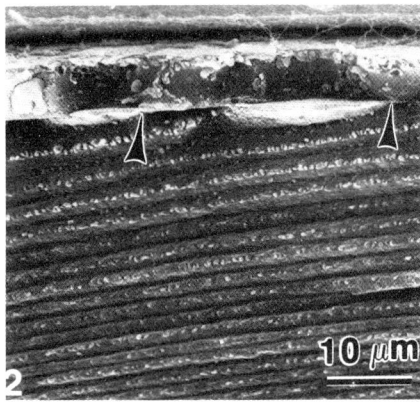 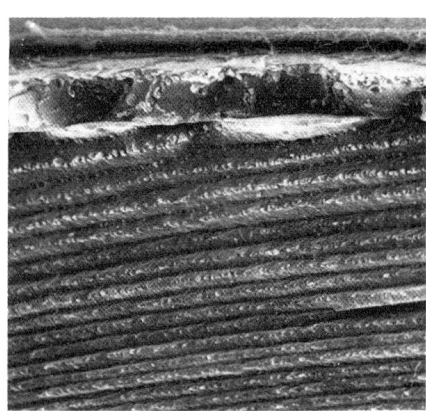

Fig. 3 Germinative zone cells. Arrowheads indicate the expanded apical ends of two apposed underlying fibers. Note the underlying fibers feature interlocking devices and are thickest as they pass beneath this region of the epithelium.

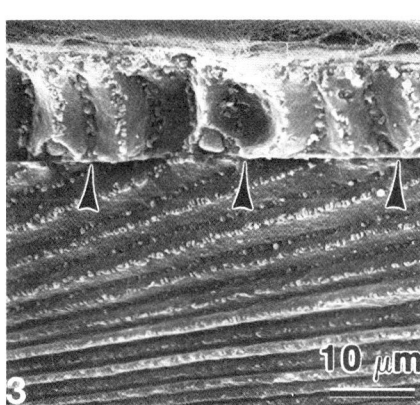 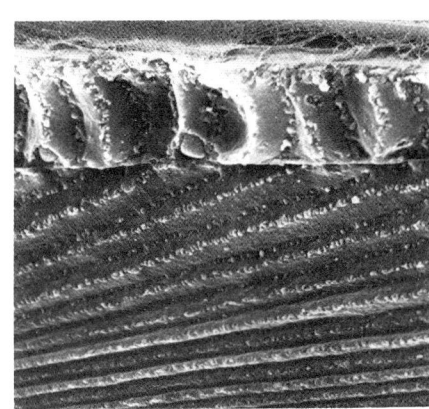

Fig. 4 Transitional zone cells. Arrowheads indicate the expanded apical ends of three apposed underlying fibers. Note the underlying fibers feature interlocking devices and are moderately thickened.

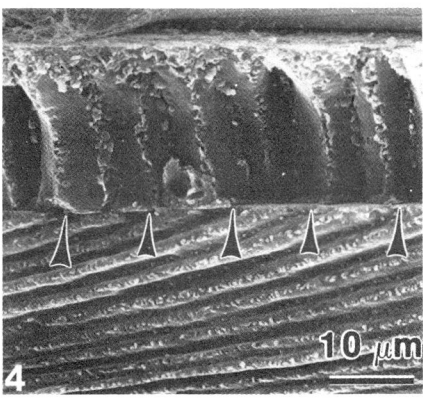 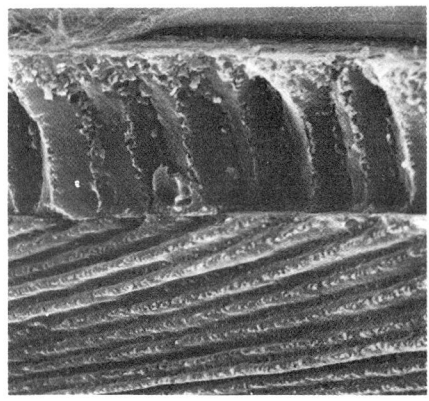

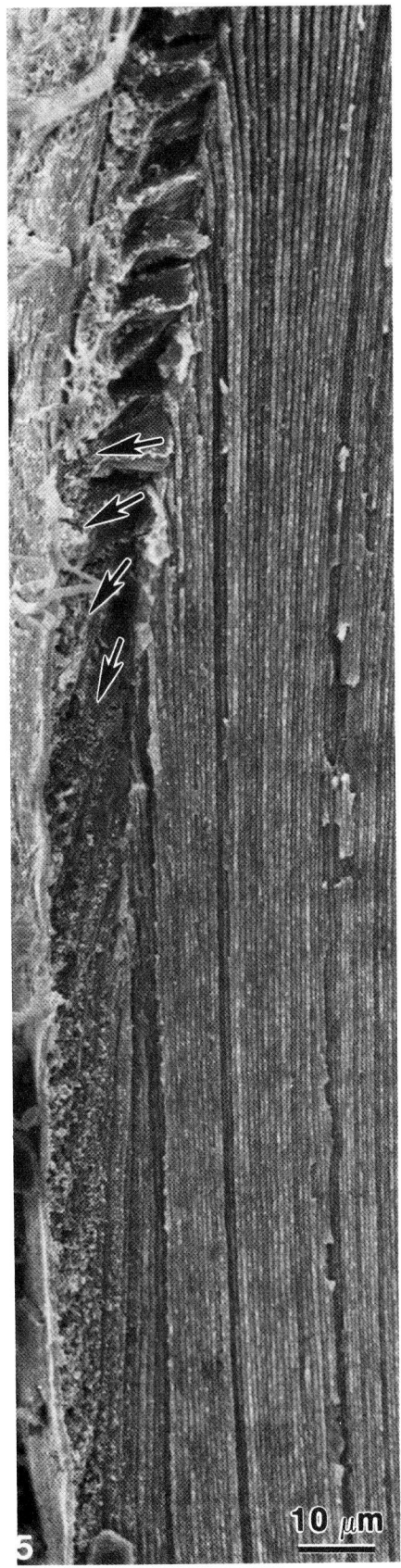 

Fig. 5 The bow region. Note the rotation of the transitional zone cells as they begin to elongate and are apposed onto the existing lens mass (arrows). The elongating fibers are very thin and have not yet begun to develop interlocking devices.

true of elongating fibers. The thickness and width of these cells is of the order anterior end > posterior end > middle region. As a result of increased thickness and width, the anterior ends of elongating fibers are flared expansions. The size of these flared ends varies in accordance with the region of the lens epithelium that they contact as they elongate toward the anterior suture. The size of the anterior expansions which contact the overlying epithelium is of the order pre-germinative zone > germinative zone > central zone > transitional zone. Since the area of the anterior expansion is always greater than the surface area of the overlying epithelial cells, more than one epithelial cell contacts a single anterior expansion. On an average, five to seven pre-germinative zone (fig. 2), four or five germinative zone (fig. 3), two or three central zone (fig. 1) and two transitional zone cells (fig. 4) contact a single underlying anterior expansion.

The surface morphology of elongating fibers varies along the length of the fiber. The anterior ends of elongating fibers have interlocking devices similar to those found on mature fibers, as they contact the transitional, germinative and pre-germinative zones. As these ends contact the central zone cells, these devices are eliminated (figs. 1-5). Interlocking devices are not found along the lengths of elongating fibers. The posterior ends of elongating fibers also vary in size as this end elongates along the capsule towards the posterior suture. The posterior ends are largest at the onset of fiber elongation and then they become progressively smaller as these ends approach the posterior suture.

Interlocking devices of fully elongated fibers

The development of interlocking devices occurs within fifty fibers following the completion of fiber elongation in the adult frog lens (fig. 6). This process appears to begin earlier in the adult rat lens (fig. 7). There are

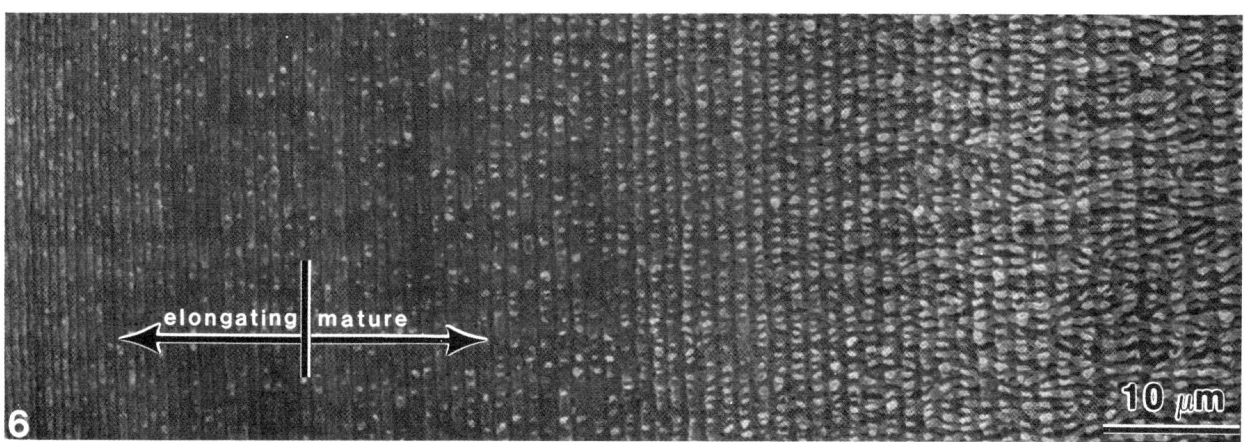

Fig. 6 The development of interlocking devices in the adult frog lens begins after the completion of fiber elongation. This process is completed within the first fifty fully elongated fibers.

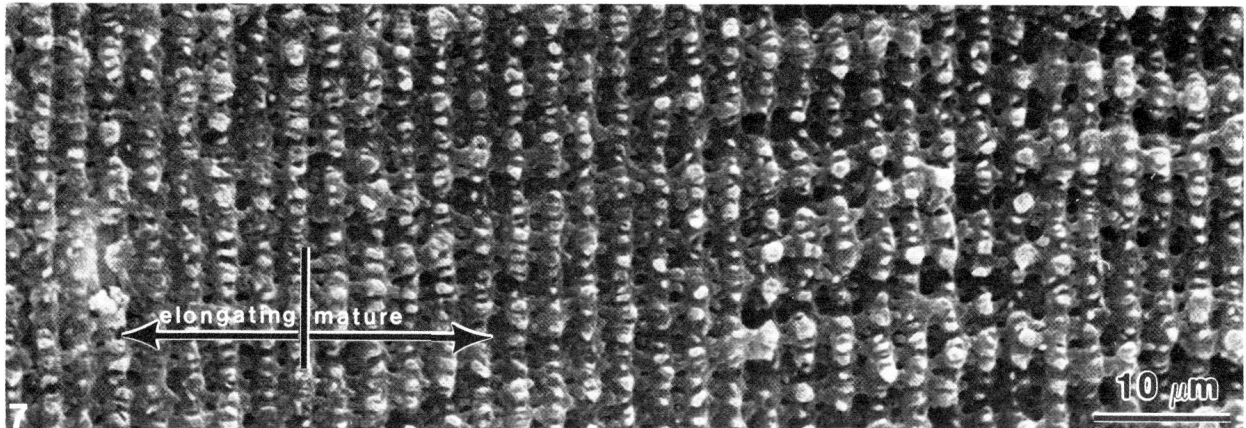

Fig. 7 The development of interlocking devices in the adult rat lens begins prior to the completion of fiber elongation and proceeds more gradually than in the frog lens.

standard types of interlocking devices and a standard pattern of interdigitation between fibers in all the lenses we examined. Ball-like devices, that consist of a head atop a short stalk or neck, are formed at the angle of narrow faces. The heads of these devices fit into complementary shaped sockets formed at the angle of narrow and broad faces in adjacent growth rings. Flap-like devices are formed at the angle of narrow and broad faces. These devices fit into complementary shaped, shallow imprints formed on the narrow faces of fibers in adjacent growth rings. The balls and flaps and their complementary sockets and imprints are repeated in an alternating pattern along the length of fibers (fig. 8). Narrow faces of representative cortical fibers from frog, rat, neonatal and adult human lenses, are shown in figures 8, 9, 12 and 13 respectively. Although the size and shapes of the interlocking devices vary slightly from species to species, the basic interdigitating pattern is the same.

The broad faces of fully elongated fibers in frog and rat lenses have smooth membrane with a few small ball-like devices that fit into complementary shaped sockets in fibers of apposed growth rings. Representative broad faces of fibers in frog and rat lenses are shown in figs. 10 and 11 respectively. Variable sized raised plaques are seen on the broad faces of frog lens fibers. These plaques are similar to the gap junction plaques seen on the broad faces of chick lens fibers (Kuszak et al., 1978; and Kuszak et al., 1980). The plaques seen on frog lens fibers are similar in size and density to gap junctions observed on these surfaces by freeze fracture analysis (unpublished observations). Plaques of this type have not been reported on the broad faces of rat lens fibers. Ultrastructural studies (freeze fracture) suggest that the size and number of gap junctions conjoining the broad faces of rat lens fibers are less than those observed in both frog and chicken (Goodenough, 1979). Goodenough (1979) has reported that the density of intramembrane particles (IMPs) is very high in rat lens fibers, making identification of gap junctions on these surfaces even by freeze fracture more difficult. The high density of IMPs, the low number and the small size of gap junctions on rat lens fibers may explain why these intercellular membrane specializations are unresolvable by SEM in this lens.

The surface morphology of human lens fibers is markedly different than the other lenses we have examined (figs. 12 and 13). While the types of interlocking devices along these fibers and their pattern of interdigitation are comparable, the membrane surfaces are convoluted or undulating. High magnification stereomicrographs of these surface suggest that finger or villous-like processes may interlock into the convolutions or undulations. An earlier SEM study of chick lens revealed similar fingerlike devices on the broad faces of the posterior ends of fibers (Kuszak et al., 1980). The broad faces of nuclear fiber membrane in frog and rat lens also have a convoluted appearance (fig. 14). Kuszak and Rae (1982) have shown that the convoluted appearance of frog nuclear fibers is age-related. However, our examination of human neonatal lenses reveals a similar membrane appearance in the very young superficial cortical fibers.

## Discussion

With stereo SEM it is possible to demonstrate unequivocally that the crystalline lens is not composed of uniformly shaped cells. The shapes and surface morphology of lens epithelial cells vary as a function of their stage in the cell cycle. Elongating fiber shape and surface morphology vary as these cells lengthen into mature fibers. Mature fiber cell shape and surface morphology vary as a function of their age. Despite this lack of cellular uniformity. stereo SEM demonstrates that the cells of the crystalline lens are organized into a highly ordered structure.

The variable size of contact between the apical ends of elongating fibers and the apical surfaces of lens epithelial cell may be significant. The central epithelium has been shown to be electrically coupled to the underlying elongating fibers (Rae and Kuszak, 1983). This study has revealed that the size of the apical ends of elongating fibers is variable as they contact lens epithelial cells in different stages of the cell cycle. The size of the apical ends of elongating fibers is greatest as they pass beneath pre-germinative zone cells. Gap junctions have been implicated in the control of development (Loewenstein, 1979). It would be interesting to determine if the number or size of gap junctions between lens epithelial cells and the leading ends of elongating fibers varies as the elongating fiber ends contact epithelial cells in different stages of the cell cycle. Furthermore, since the ends of elongating fibers are constantly moving at a slow rate toward the anterior suture the communication pathways between these ends and the overlying lens epithelial cells must continuously be formed and reformed.

The interlocking devices are generally considered to maintain the ordered alignment of fibers (Dickson and Crock, 1975; Kuwabara, 1975; Harding et al., 1976; and Kuszak et al., 1980). This stereo SEM study reveals that there are only slight differences between the types and patterns of interlocking devices in lenses from a variety of species. The lenses of these species have markedly different accommodative power. Further studies will be necessary to determine if the interlocking devices play any role during accommodation.

The surface morphology of the human lens appears to be considerably different than the other lenses we examined. Kuwabara (1975) and Dickson and Crock (1972, 1975) reported that the membrane of monkey and human lens becomes

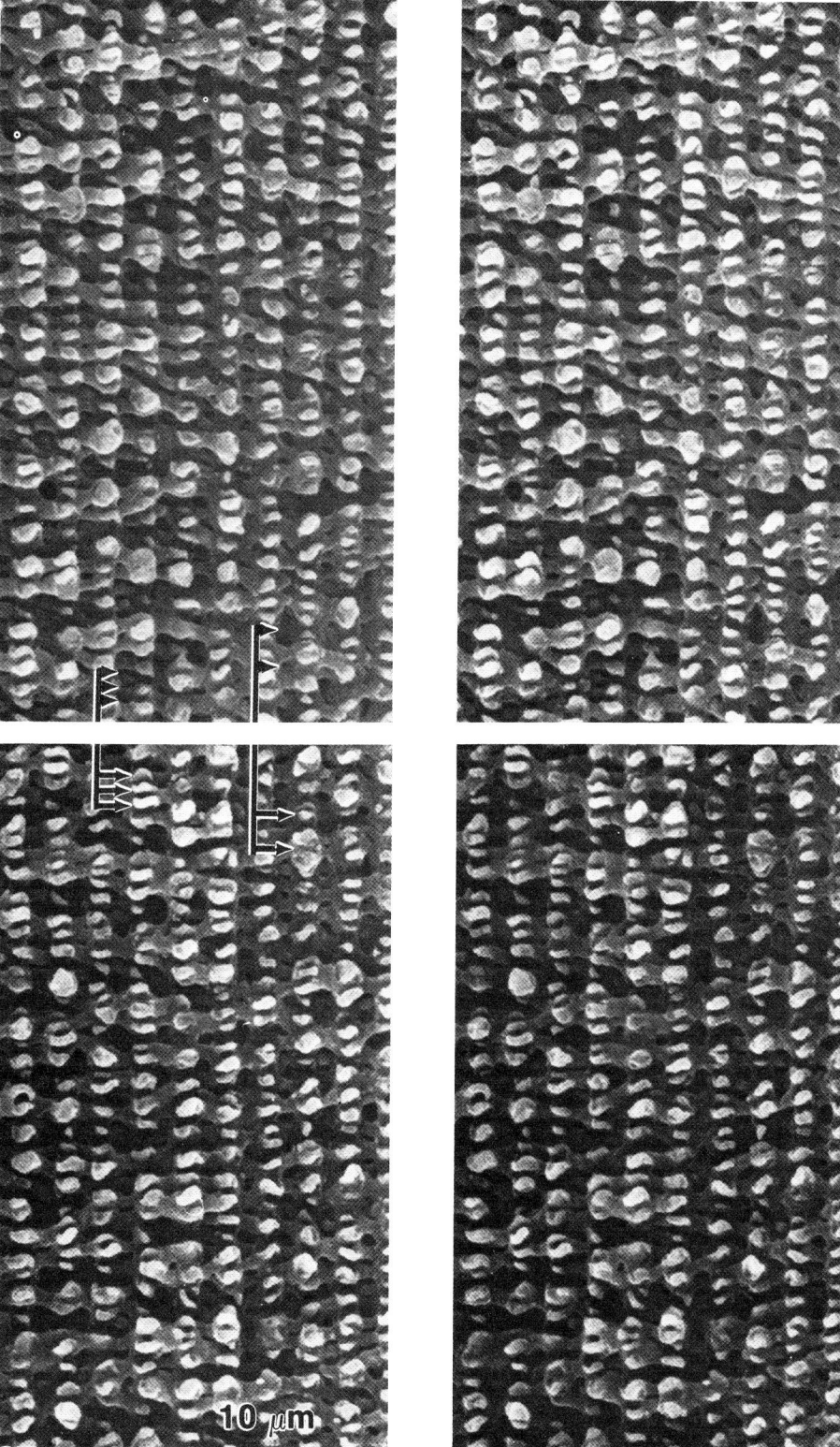

Fig. 8 Apposed cortical fibers from adult frog lens. Note the exact complementarity of the ball-like interlocking devices and their sockets (double headed arrows).

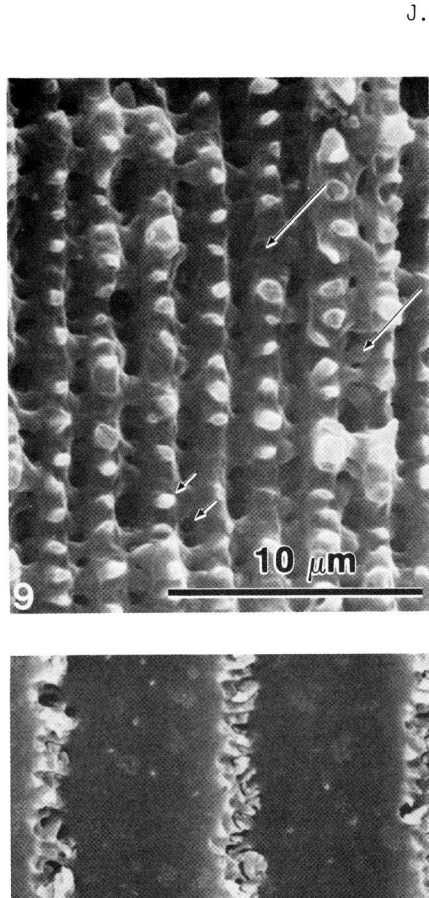

Fig. 9 Cortical fibers from adult rat lens. Note the interlocking devices and their pattern of interdigitation is comparable to the frog lens. Balls and sockets (short arrows), flaps and imprints (long arrows).

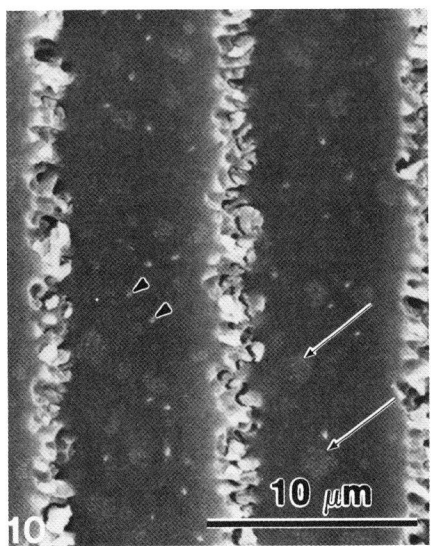

Fig. 10 The long sides of cortical fibers from adult frog lens feature raised plaques which are presumably gap junctions (arrows) and small ball-like interlocking devices (arrowheads).

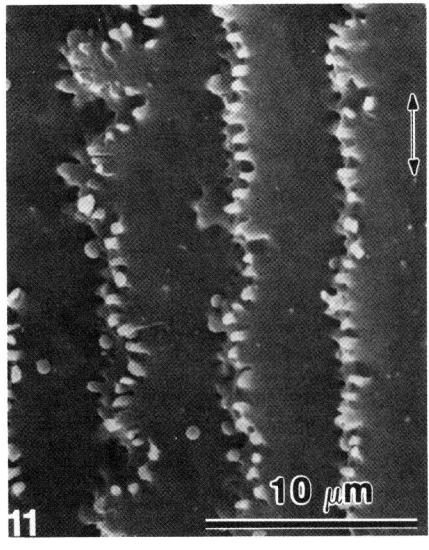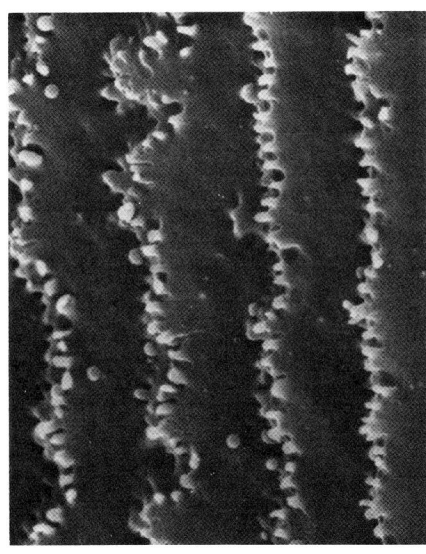

Fig. 11 The long sides of cortical fibers from adult rat lens feature smooth membrane with small ball-like interlocking devices and complementary shaped sockets (double arrows).

Fig. 12 Cortical fibers from adult human lens. The types of interlocking devices and their interdigitating pattern are comparable to the frog and rat lens. However, the membrane surface is convoluted or furrowed.

Fig. 13 Cortical fibers from human neonatal lens.

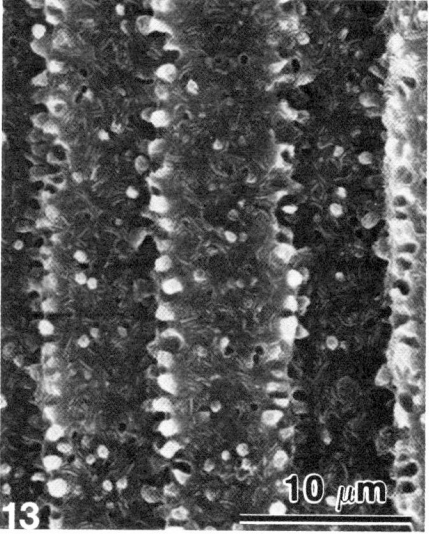 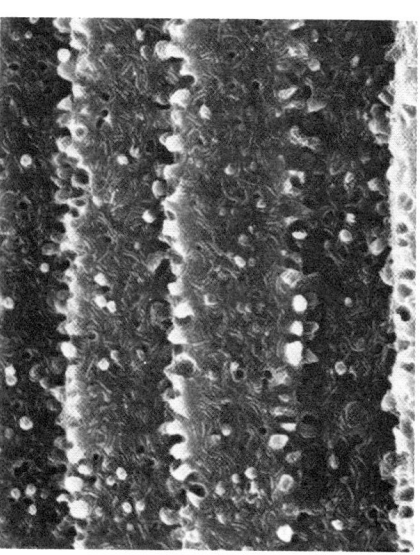

Fig. 14 The long sides of nuclear fibers from adult frog lens. The membrane has been transformed from smooth to coarse and convoluted or furrowed as a result of age.

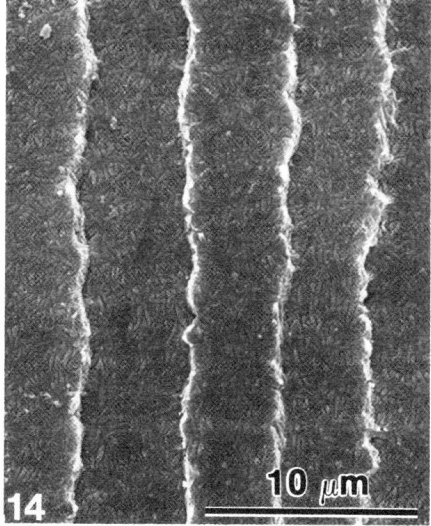 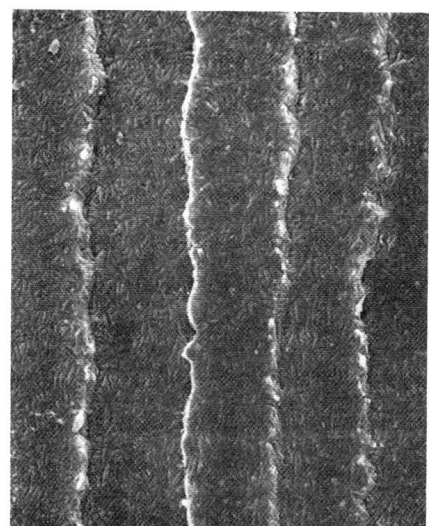

convoluted as fibers age. Zamphigi et al. (1982) have recently shown by the ultrastructural technique of freeze-fracture, that the membrane of bovine fibers is also convoluted. Our study demonstrates that the superficial cortical fiber membrane of human neonatal lenses is also convoluted, suggesting that this type of membrane is not necessarily a consequence of advanced age. However, we are not certain that convoluted fiber membrane is not an artifact of preparation. Lenses of experimental animals are removed and placed in fixative within seconds after sacrifice. It is impossible to obtain and fix human lenses so rapidly following death. The possibility that the convoluted membrane of human lenses represents autolytic changes that occur between death and fixation cannot be discounted. Furthermore, if the convoluted membrane of human and bovine lenses is not an artifact of preparation, it brings into question the role of gap junctions in higher mammalian lenses. Convoluted membrane is inconsistent with gap junctions because these specialized intercellular contacts require planar regions of membrane to be conjoined across a narrowed extracellular space. Gap junctions are believed to be of the utmost importance in lens physiology by providing pathways for rapid exchange of nutrients between lens fibers (Bloemendal, 1972; Benedetti et al., 1976; Kuszak et al, 1978; Goodenough, 1979; Mathias et al., 1979). Zampighi et al. (1982) have suggested that the convoluted membrane of the bovine lens is in fact the non-gap junctional membrane in the bovine lens. They describe this membrane as a junctional membrane designed to eliminate extracellular space. Further ultrastructural studies of membranes in different species will be necessary to determine the contribution of membrane and gap junctions to lens function.

## Acknowledgements

This work supported by the William B. and Rhoda Wyeth Brinton Memorial Foundation, the Louise C. Norton Trust, the Regenstein Foundation and NIH grant no. EY 03282 to J. L. Rae. Ms. Macsai was a post-sophomore fellow in the Post Sophomore Fellowship Program of the Department of Pathology at Rush Medical College during the course of this study. The authors wish to thank Mr. William Leonard for expert technical assistance and Ms. Geri Byrd for typing the manuscript.

## References

Benedetti EL, Dunia I, Bentzel CJ, Vermorken AJM, Kibbelaar M and Bloemendal H. (1976). A portrait of plasma membrane specializations in eye lens epithelium and fibers. Biochimica et Biophysica Acta 457, 353-384.

Bloemendal H, Zweers A, Vermorken F, Dunia I and Benedetti EL. (1972). The plasma membranes of eye lens fibers. Biochemical and structural characterization. Cell Differ. 1, 91-106.

Dickson DH and Crock GW. (1972). Interlocking patterns of primate lens fibers. Invest. Ophthal. 13, 809-815.

Dickson DH and Crock GW. (1975). Fine structure of primate lens fibers, in: Cataract and Abnormalities of the Lens, J.G. Bellows (ed.), Grune and Stratton, Inc., New York, 49-58.

Duke-Elder S and Abrams D. (1970). The refraction of light, in: Systems of Ophthalmology, S. Duke-Elder (ed.), C.V. Mosley Co., St. Louis, Vol. V, 74-78.

Goodenough DA. (1979). Lens gap junctions: A structural hypothesis for non-regulated low-resistance intercellular pathways. Invest. Ophthal. 18, 1104-1122.

Harding CV, Reddan JR, Unakar NJ and Bagchi M. (1971). The control of cell division in the ocular lens. Int. Rev. Cytol. 31, 215-300.

Harding CV, Susan S and Murphy H. (1976). Scanning electron microscopy of the adult rabbit lens. Ophthal. Res. 8 (6), 443-455.

Kuszak JR, Maisel H and Harding CV. (1978). Gap junctions of chick lens fiber cells. Exp. Eye Res. 27, 495-498.

Kuszak JR, Alcala JR and Maisel H. (1980). The surface morphology of embryonic and adult chick lens fiber cells. Am. J. Anat. 159, 395-410.

Kuszak JR and Rae JL. (1982). Scanning electron microscopy of the frog lens. Exp. Eye Res. 35, 499-519.

Kuwabara T. (1970). Surface structure of the eye tissue. Scanning Electron Microsc. 1970; 185-192.

Kuwabara T. (1975). The maturation of the lens cell: A morphological study. Exp. Eye Res. 20 427-443.

Loewenstein WR. (1979). Junctional intercellular communication and the control of growth. Biochimica et Biophysica Acta. 560, 1-65.

Maisel H, Harding CV, Alcala JR, Kuszak JR and Bradley R. (1981). The Morphology of the Lens, in: Molecular and Cellular Biology of the Eye Lens, H. Bloemendal (ed.), John Wiley and Sons Inc., New York, 49-84.

Mathias RT, Rae JL and Eisenberg RS. (1979). Electrical properties of structural components of the crystalline lens. Biophys. J. 25, 181-201.

Rae JL and Kuszak JR. (1983). The electrical coupling of epithelium and fibers in the frog lens. Exp. Eye Res. 36, 317-326.

Troekel S. (1962). The physical basis for transparency of the crystalline lens. Invest. Ophthal. 1 (4), 443-501.

Zampighi G, Simon SA, Robertson JD, McIntosh TJ and Costello MJ. (1982). On the structural organization of isolated bovine lens fibers junctions. J. Cell Biol. 93, 175-189.

## Discussion with Reviewers

S.R. Hilfer: Would you care to speculate about the relationship of the end bulges of the elongating fibers, the specialized contacts that are made with the epithelium and the control of maturation?
Authors: This is the first morphological study that has shown a variation in the size and complexity of the apical ends of elongating fibers as they pass beneath lens epithelial cells in different stages of the cell cycle. At this time there is no evidence that this structural diversity is functionally related to the control of fiber maturation.

B.W. Streeten: Is there evidence that metabolite entry into the lens is through pathways between the lens epithelium and the underlying fibers?
Authors: Rae and Kuszak (1983) have shown that the lens epithelium is electrically coupled to the underlying fiber mass. In addition, Goodenough et al. (1980) have presented evidence that is compatible with metabolic cooperation between the lens epithelium and the underlying lens fibers.

S.R. Hilfer: What induces the initiation of lens fiber formation?
Authors: It is not known what induces lens fiber formation throughout the lifetime of the organism. Primary fiber cell differentiation appears to be induced by a substance present in the posterior part of the eye. Coulombre and Coulombre (1963) demonstrated that rotation of developing lenses in 5 day old chick embryos resulted in anterior vesicle cells being induced to become fibers and the cessation of normal primary fiber formation. Recently, McAvoy (1980) has shown that embryonic rat lens epithelia cocultured in vivo with neural retina will form fiber-like cells. Beebe, Feagans and Jebens (1980) have isolated a 60,000 molecular weight glycoprotein from chick vitreous humor which they have labeled lentropin. This factor causes elongation and increased synthesis of specific crystallines in cultured lens epithelial cells from 6-day old chick embryos.

N.J. Unakar: Could you briefly describe the previously used method (Kuszak and Rae, 1982) for preparing lenses?
B.W. Streeten: Could you give specific directions for the preparation of lenses? What was the fixative?
Authors: For frog lens we had previously determined that 2.5% glutaraldehyde prepared in 0.05 M sodium cacodylate was essentially isotonic for this animal (Kuszak and Rae, 1982). The osmolarity of rat and human tissue is not the same as frog. Consequently, different concentrations of fixative buffer mixture are required. For the rat we determined that 2.5% glutaraldehyde in 0.07 M sodium cacodylate was appropriate. For the human lens we used 2.5% glutaraldehyde in 0.12 M sodium cacodylate. We arrived at this mixture for the human lens quite empirically. This mixture is actually hypersomotic. The inaccessibility of human lenses immediately following death prevents us from determining what is the proper fixative-buffer mixture for this lens.

B.W. Streeten: Did the authors investigate if the convoluted membrane of the human nuclear lens fibers could be altered by using different concentrations of fixative-buffer mixtures?
Authors: Again, we are unable to address this question because we have not been able to obtain sufficient numbers of fresh human lenses.

B.W. Streeten: Are there any differences between the convoluted surfaces of adult human nuclear fibers as compared with the convoluted surfaces of human neonatal cortical fibers?
Authors: We observed neither qualitative nor quantitative differences in the convoluted membranes of these lenses.

T. Kuwabara: Can the authors offer any "tips" as to how they prepare their specimens?
Authors: The following steps in our protocol are different from those generally described for preparing the lens for SEM. First, we fix our lenses in an isotonic fixative buffer mixture at room temperature. We extensively wash the specimens free of glutaraldehyde prior to osmication and free of osmium tetroxide prior to dehydration. Third, we dehydrate through ascending concentrations of alcohol (30,50,70,95 ethanol) to 100% ethanol overnight. We then critically point dry our lenses in Freon. Immediately following this procedure we secure the specimens to stubs using silver paste under a dissecting microscope. As soon as the adhesive is dry we sputter coat the specimens in vacuo with gold. Handling of the specimens under the dissecting microscope is done with number five "Biologie" EM forceps with care being taken never to touch the surfaces to be scanned.

## Discussion with Reviewers References

Beebe DC, Feagans DE and Jebens HAH. (1980). Lentropin: A factor in vitreous humor which promotes lens fiber cell differentiation. Proc. Natl. Acad. Sci. USA 77, 490-493.

Coulombre JL and Coulombre AJ. (1963). Lens development: Fiber elongation and lens orientation. Science 142, 1489-1490.

Goodenough DA, Dick II JSB and Lyons JE. (1980). Lens metabolic cooperation: A study of mouse lens transport and permeability visualized with free substitution and autoradiography and electron microscopy. J. Cell Biol. 86, 576-589.

McAvoy JW. (1980). Induction of lens fiber differentiation by neural retina. Invest. Ophthalmol. 19(suppl), 115.

## TISSUE INTERACTIONS DURING AXIAL STRUCTURE PATTERN FORMATION IN AMPHIBIA

G. M. Malacinski[1], B. W. Youn[2], and A. Jurand[3]

Department of Biology Indiana University Bloomington, IN 47405[1], Department of Biology Princeton University, Princeton, NJ 08540[2], Department of Genetics, University of Edinburgh Edinburgh, U.K.[3]

### Abstract

Tissue interactions have traditionally been assigned important roles in establishing the pattern of amphibian axial structure morphogenesis. Those interactions have been postulated to generate the patterns of neural fold morphogenesis, neural tube formation, and somite development. A review of axial structure development together with a brief discussion of the classical viewpoint, is presented. A re-examination of axis formation has recently been carried out with the SEM. Embryos which displayed major defects in notochord development, ranging from diminished length to complete obliteration, were produced by irradiating fertile eggs prior to first cleavage. A comparative SEM analysis of normal and "notochord defective" embryos revealed that, contrary to previous reports, the notochord is apparently a dispensable component of the developing axial structure system. Lastly, TEM examination of the notochord defective embryos allowed some insight into the ultrastructural alterations which occur in the notochord and neural tube cells of irradiated embryos. Additional information about the structure of the notochord, and the cellular mechanics of somitogenesis emerged from those studies.

Key words: neural tube; anuran embryogenesis; notochord; Xenopus embryos; axial structures; amphibian morphogenesis; UV irradiation; primary induction; somitogenesis; tissue interactions

### Introductory Remarks

Axial structure development in vertebrates involves the establishment of the pattern of morphogenesis of several tissue types, including the neural tube, notochord, and somites. Axis formation has been studied extensively in the embryos of several organisms, including the amphibia, chicken, and mammal. Some of the earliest (eg. Lehmann, 1929), as well as some of the most extensive (eg. Hamilton, 1969; Jacobson and Gordon, 1976) analyses have, however, employed amphibian embryos as experimental material. Consequently, a rather complete picture of the descriptive aspects of pattern formation during amphibian axis development is available (reviewed by Karfunkel, 1974). Comparisons of the data accumulated with the amphibian embryo to the information derived from the study of the chick embryo (eg. Jurand, 1962; Bancroft and Bellairs, 1976; Meier, 1979) suggest that generally similar patterns of axial structure morphogenesis are employed by relatively diverse organisms.

The axial structure organs have not been "determination" and/or "differentiation" of the relevant tissues prior to the presumed tissue interaction itself. Nevertheless, several types of interactions have been postulated to be involved in generating the pattern of axial structure morphogenesis. Some of those interactions are summarized in Table 1, and will be reviewed below.

### Review of The Presumed Roles of Tissue Interactions in Axis Formation

Establishment of the amphibian primary embryonic axis is probably the most dynamic morphogenetic event which follows gastrulation. Of late, it has been most extensively studied in the anuran amphibian, Xenopus laevis.

Segregation of the pre-notochordal mesoderm cells can be observed as early as the embryonic stage at which the blastopore is not quite completely closed (Youn et al., 1980). The notochord develops as a distinct, highly organized group of cells which eventually become separated from the paraxial mesoderm. Elongation of the notochord takes place, and the distal regions of the neural plate elevate. The elevated borders of the neural plate move towards the dorsal midline of the embryo, where they converge, and fuse, to form the neural tube. Differentiation of the notochord occurs, and by the time the neural folds have begun to fuse it can easily be recognized as a morphologically unique, rod-like structure.

The presomitic cells can be distinguished from the lateral and midline mesodermal cells even prior to the completion of the closure of the blastopore. The surfaces and shapes of those presomitic cells are different from the surrounding cells (Youn et al.; 1980). During further embryogenesis the somitic mesoderm segments into blocks of tissue (somite files), which later undergo a change in orientation vis a vis the dorsal midline. Eventually the majority of the somite cells which lie parallel to the notochord differentiate into myoblasts which make up the blocks of the mature muscle tissue.

Classically, tissue interactions between the emerging mesoderm, neural ectoderm, notochord, and somitic mesoderm have been viewed to play very important roles in establishing the pattern of axial structure morphogenesis. Several of those presumed roles of tissue interactions are summarized in Table 1. As the information contained in that table reveals, the notochord has--traditionally--been considered to play a key role in axis formation. No doubt its large size, prominent location along the dorsal midline, and sequence of development, account for the diverse functions which have been ascribed to it. Whether the notochord does indeed have a "a causative" or "direct" influence on the pattern of development of the tissues which surround it (eg. neural tube, somites) has not been previously established by direct experimentation. Microsurgical techniques have been employed. For example, notochordectomy of the urodele embryo has been attempted (Kitchen, 1949; Jacobson and Gordon, 1976). Likewise, the chemical treatment with lithium chloride of the whole embryo has been carried out (Lehmann, 1935). Those relatively large scale perturbations, present the possibility that various morphological side effects or indirect consequences could enter into the interpretation of the data. In fact, it has recently been argued that the role ascribed to tissue interactions in axis formation has historically been exaggerated. (Hay and Meier, 1978).

In an attempt to learn more about the role various tissue interactions play in axial structure development, our laboratory has recently been engaged in a detailed and comprehensive analysis of several aspects of the pattern of Xenopus axis morphogenesis. Irradiation of the uncleaved egg with low doses of ultraviolet light irradiation (UV) generates embryos which display several abnormal features of axial structure development. Some of those embryos lack a complete notochord, and exhibit altered somite pattern formation. SEM and TEM

Table 1. Summary of several of the presumed roles of tissue interactions in amphibian axis formation

| developmental stage | aspect | type of tissue interaction | reference |
|---|---|---|---|
| early embryogenesis (gastrulation) | "determination" of the prospective neural ectoderm | ectoderm-mesoderm (i.e. primary embryonic induction) | Spemann and Mangold, 1924; reviews in Nakamura and Toivonen, 1978 |
| neurulation | "induction" of the neural plate | notochord interacts with the overlying ectoderm | Nieuwkoop and Nigtevecht, 1954 |
| neurulation | neural plate shaping and cell movement patterns | presence of the notochord beneath the neural epithelium | Jacobson and Gordon, 1976 |
| neurulation | neural tube organization | notochord "shapes" neural tube | Holtfreter and Hamburger, 1955 |
| somitogenesis | enhancement of growth and segmentation | notochord/somite contact | Takaya, 1956; 1961 |
| somitogenesis | muscle development | notochord stimulates muscle differentiation | Yamada, 1939 |
| further development | axial stretching | notochord/neural tube/ somite contact | Kitchen, 1949; Lehman, 1935 |

analyses of those embryos has led to several novel observations on (1) the role of the notochord in neural tissue morphogenesis; (2) the ultrastructure of axial organs; and (3) the cellular mechanics of somite pattern formation. Those observations, and the new interpretations concerning tissue interactions they permit, will be described in the following sections of this manuscript.

## Materials and Methods

*Source of embryos*: Xenopus laevis eggs were artificially inseminated and chemically dejellied using previously published methods (eg. Youn and Malacinski, 1980). Irradiation of uncleaved eggs was performed with a germicidal lamp (MP-R2, Ultraviolet Products Co.) which emits light exclusively at 254 nm. A dose of approximately 10,000 ergs/cm$^2$ provided the irradiated embryos which were employed in these studies. After irradiation embryos were permitted to develop to the appropriate stage (Nieuwkoop and Faber staging series, 1967) and were then collected and fixed for subsequent electron microscopy (Fig. 1).

*Electron Microscopy*: For SEM observations embryos were fixed in 2% glutaraldehyde in 0.1 M cacodylate-HCl buffer (pH 7.6) and washed in the same buffer for 24 hrs (Keller and Schoenwolf, 1977). The epidermis, as well as the neural tube, was removed from some embryos by dissection with a fine steel knife and forceps. Dissected embryos were dehydrated in increasing concentrations of ethanol and then critical point dried with liquid $CO_2$. Specimens were mounted on aluminum stubs with conducting silver paint, and then coated with gold-palladium (60:40) in a Denton 503 vacuum evaporator. The specimens were examined with an ETEC Autoscan U-1 SEM and photographed on Polaroid type 55 positive/negative film.

For TEM observations embryos were fixed overnight at 4°C with 2.5% glutaraldehyde in 0.1M cacodylate buffer (pH 7.6). After two rinses with buffer, the specimens were post-fixed with 1% osmium tetroxide. Dehydration with ethanol was followed by the use of an epoxy resin for embedding (Jurand and Ireland, 1965). Ultrathin sections were mounted on collodion and carbon coated grids. They were treated with lead citrate (Reynolds, 1963) and 2.5% uranyl acetate. Specimens were examined with a Philips model 300 TEM.

## Novel Observations on Neural Tube Morphogenesis in "notochord-defective" Embryos

Irradiated embryos were permitted to develop to approximately stage 22-24. From a group of irradiated embryos which displayed a broad range of UV effects, those which appeared to be normal (e.g. Fig. 2a) or which exhibited only minor abnormalities in their anterior axial structures (eg. Fig. 2; b,c) were collected. Acephalic embryos (eg. Fig. 2d) were discarded. After fixation, the epidermis was peeled off. Measurements (length, width) of the neural tube were made, and those embryos which displayed a normal sized neural tube were examined further (Fig. 1).

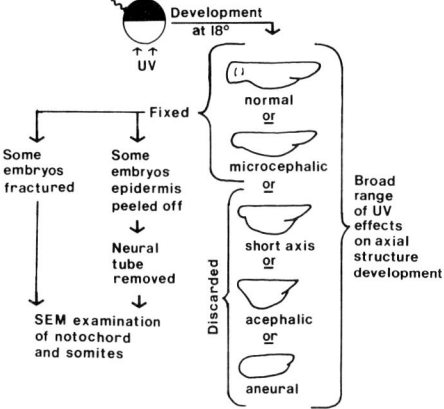

Figure 1. Procedure for producing and examining embryos which display defects in axial structure morphogenesis. Figure 2 contains photographs of the anterior axial (head) defects.

The neural tube was dissected away, and the embryos were prepared for SEM analysis. Figure 3 contains low power views of whole embryos and Fig. 4 contains higher power views of structural details. Among the irradiated embryos various notochord morphologies were observed. Some embryos contained a completely normal notochord. Those notochords could not be distinguished from the notochord of unirradiated, control embryos (Figs. 3a and 4a). Other embryos contained a notochord which was normal in length, but somewhat diminished in width, especially in the anterior region of the embryo (Fig. 3b and 4b). Some embryos exhibited incomplete notochord formation. In those embryos the notochord was shortened in length, and usually diminished in width. The extent of anterior shortening varied from embryo to embryo. The notochord develops in a progressive fashion, from the posterior to the anterior direction (Youn et al., 1980). Figure 5 contains a summary of the elongation sequence for the Xenopus notochord. Most likely, therefore, aberrations in notochord development could be expected to be most prominent in the anterior region, as shown in Fig. 3b. Embryos which contained a partial notochord, as well as those embryos which completely lacked a notochord (see below), displayed fusion of the somite files along the dorsal midline where the notochord was absent.

The entire notochord was missing along the whole length of some embryos (Fig. 3c). The somite files could be observed to have fused beneath the neural tube along the entire length of the dorsal midline. Figure 4c contains higher magnification views of that region. As was mentioned previously, for the dissections described in Figs. 3 and 4, embryos which displayed a normal neural tube were selected for detailed examination. Figure 6 shows a cross-sectional view of a stage 22-24 notochordless embryo. The intact neural tube can be observed, as well as the region along the midline where the somites fused.

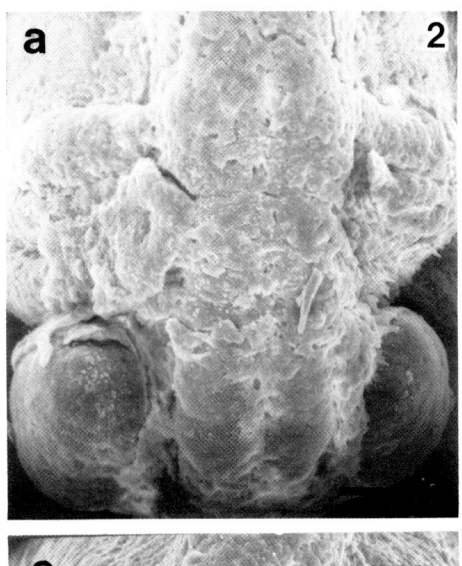
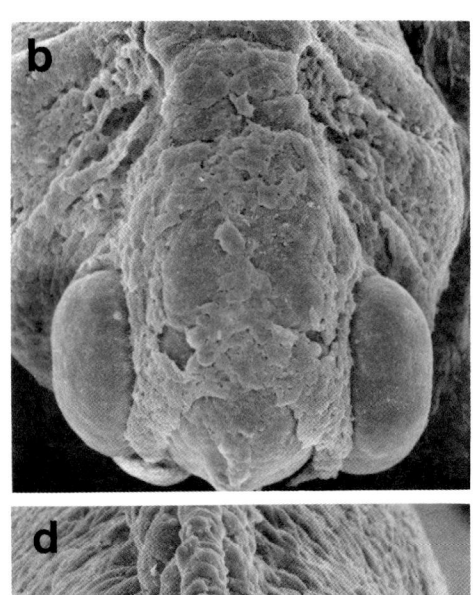
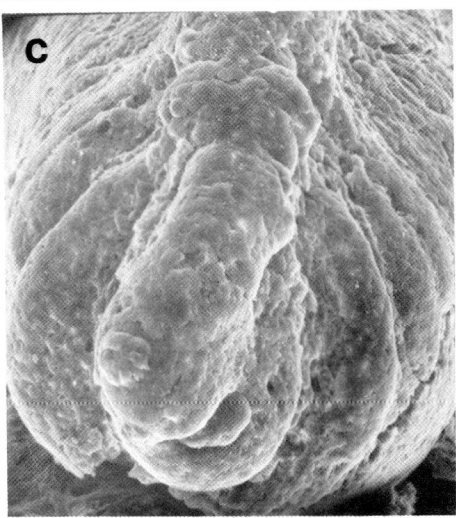
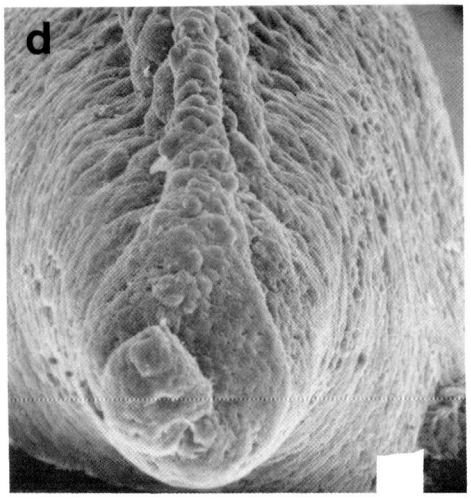

Figure 2. Dorsal-frontal view of intact (epidermis not peeled off) irradiated embryos at stage 22-24. (a) control, unirradiated embryo; (b) UV'd slightly microcephalic embryo; (c) microcephalic embryo; with short axis; (d) acephalic embryo. Bar equals 0.1 mm.

Figure 3. Dissection (epidermis peeled off and neural tube removed) of stage 22-24 irradiated embryos revealed a broad range of notochord morphologies. (a) normal notochord; (b) partial notochord (notochord absent in the anterior half of the embryo); (c) notochord completely absent. Anterior end of embryo is at top of photo. Bar equals 0.5 mm.

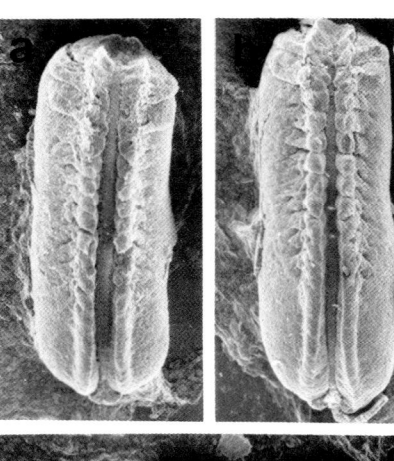
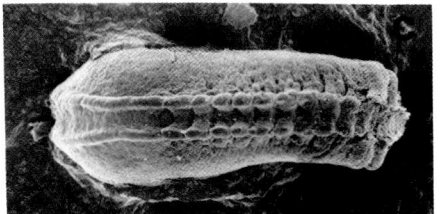

In order to determine whether (a) the notochord never formed properly in notochord defective embryos, or (b) the notochord developed at early stages but later regressed, irradiated embryos were examined at earlier stages. Several stage 15-17 embryos were dissected. The results clearly demonstrated that the notochord was not present at the earlier developmental stages during which the notochord normally completes its posterior-anterior elongation. Figure 7 contains photographs which illustrate this point. The same types of abnormalities in notochord morphology described in Fig. 3 can be observed in the earlier stage embryos shown in Fig. 7.

The main conclusion which can be drawn from these observations is that the neural tube can develop in the absence of a notochord (Youn and Malacinski, 1981a). The notochord apparently does not play an indispensable role in the development of the gross morphological features of the neural tube.

The results of previous studies on the role of the notochord in axial structure pattern formation were interpreted to mean the notochord plays an important role in directing the cell movements of neural plate morphogenesis (Jacobson and Gordon, 1976), and axial stretching (Kitchen,

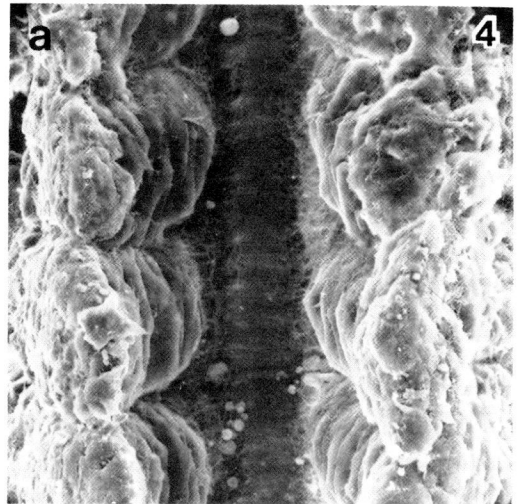
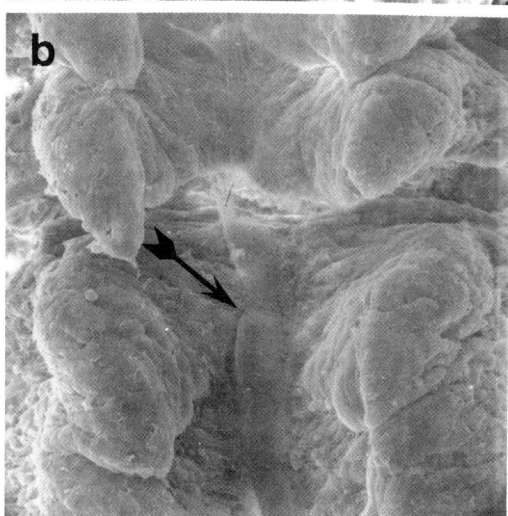
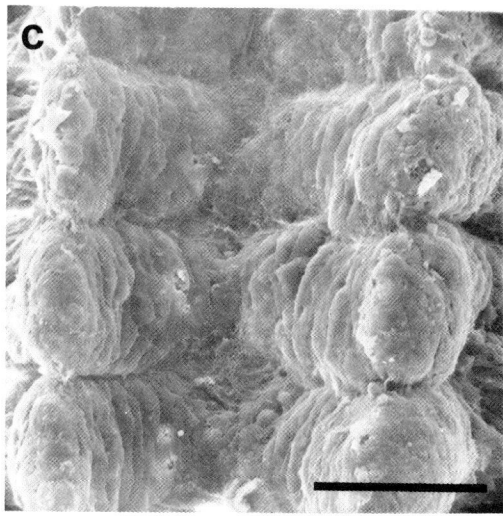

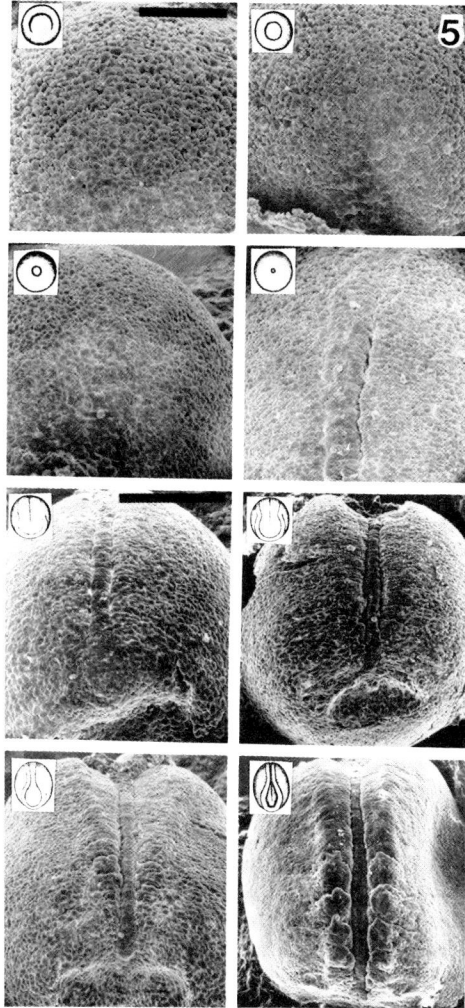

Figure 5. Progressive development of the notochord in the posterior to anterior direction. Epidermis removed to reveal development of the mesoderm. Inserts indicate embryonic stage of development. Bar represents 0.2 mm for top 4 photos and 0.4 mm for bottom 4 photos.

Figure 4. Higher magnification view of the region between the somite files which is normally occupied by the notochord. For the preparation of these samples the epidermis and neural tube were removed, as in Fig. 3. (a) normal notochord; (b) partial notochord (arrow points to anterior most terminus of notochord); (c) notochord completely absent. Specimens oriented with anterior end at top of photo. Bar equals 0.1 mm.

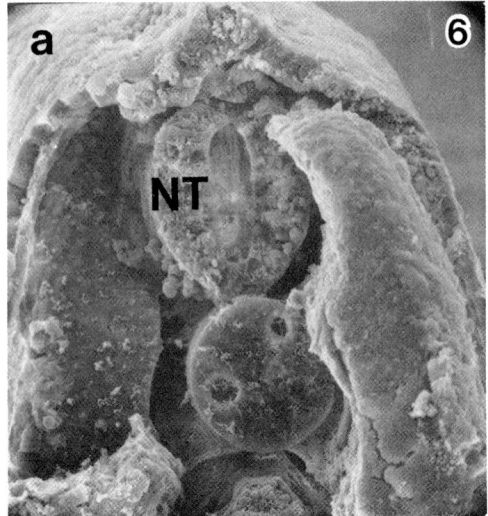

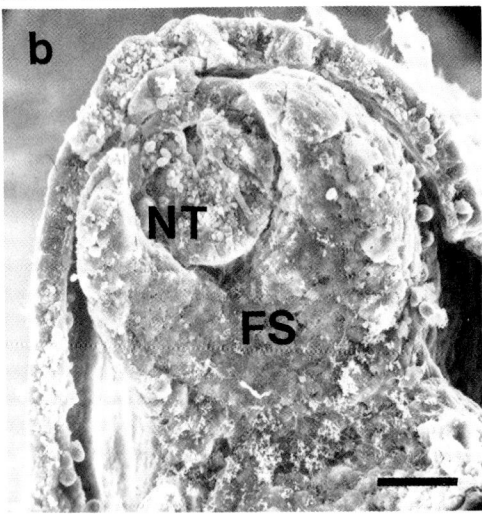

Figure 6. Cross-sectional view of a normal (a) and notochordless (b) embryo Neural tube (NT) and region of fusion of the somites (FS) are shown. Bar equals 0.05 mm.

1949). Those previous studies employed, however, direct microsurgery on normal embryos. Such operations probably generated non-specific side effects on the pattern of axial structure morphogenesis. Ultraviolet irradiation, when employed at relatively low doses (Youn and Malacinski, 1980) has, however, proved to have relatively specific effects on notochord formation in Xenopus embryos (Youn and Malacinski, 1981a).

From the above observations, at least two important directions for future experimentation emerge. First, it would be interesting to trace the fate of the presumptive notochord cells in notochordless embryos. Using recently developed technologies for studying cell lineages (eg. Weisblat et al., 1978), it should be possible to determine whether the presumptive notochord cells degenerate get lost, or differentiate into other structures (eg. somites). Second, it would be useful to examine other amphibian embryos with the UV technology. It would be worthwhile to

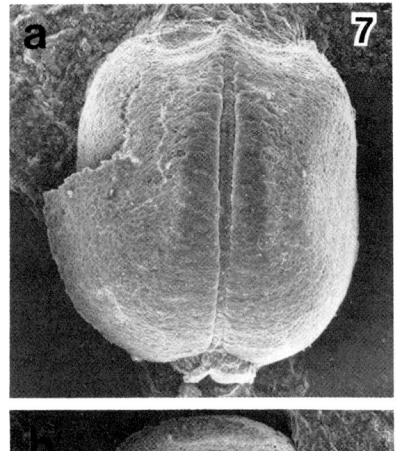

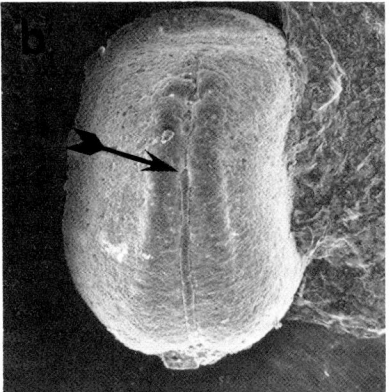

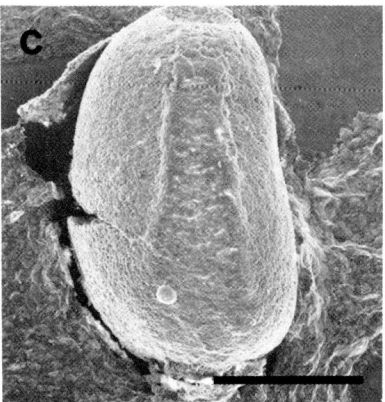

Figure 7. Notochord defects displayed by stage 15-17 embryos from which the epidermis was removed. (a) complete notochord; (b) partial notochord (arrow points to anterior most terminus of notochord); (c) notochord completely absent. Specimens oriented with anterior end at top of photo. Bar equals 2.5 mm.

establish whether other embryos, including those of urodele species which may employ alternative developmental pathways for the origin of the notochordal mesoderm (Lovtrup, 1975), can dispense with the notochord.

### Recent Observations on the Ultrastructure of the Neural Tube and Notochord in Irradiated Embryos

Although the gross morphologies observed with the SEM of the neural tube and "reduced size" notochord appeared normal in irradiated embryos, more detailed analyses were carried out. The rationale behind this experimental approach is relatively straightforward: Since UV generates distinct and reproducible perturbations in axial structure development, perhaps further study with the TEM would lead to insights into the differentiation program of the neural tube and notochord.

One feature of neural tube cells in UV'd embryos which was observed with the TEM is the abnormal morphology of some of the mitochondria. They were often vacuolated and occasionally bent in shape. The diameters of these swollen mitochondria were frequently increased. Figure 8 contains photographs of mitochondria in both control (unirradiated) and UV'd neural tube cells.

Similar observations concerning swollen mitochondria have been previously reported by other authors. Ikenishi et al. (1974) and Smith and Williams (1979), for example, reported on the presence of vacuolated and swollen mitochondria in irradiated amphibian embryos. It is entirely possible that altered mitochondria result from direct UV hits on the population of egg mitochondria which normally reside very close to the surface of the egg. Egg mitochondria have been demonstrated to generate the mitochondria of primoridal germ cells (Dawid and Blackler, 1972). It is, therefore, reasonable to speculate that neural tube mitochondria may also be derived from mitochondria of the egg. The altered neural tube mitochondria might have developed from egg mitochondria which sustained a direct UV hit.

It is also possible that the abnormal mitochondria may have originated from more indirect causes. The neural tube cells of irradiated embryos contained an increase number of lysosomes in their cytoplasia. Control neural tube cells almost never contained lysosomes. Fewer than one lysosome per 50 cells was observed. Neural tube cells of irradiated embryos, however, displayed 1-2 lysosomes per ultrathin section. In addition to the vesicles which exhibited features typical of lysosomes, somewhat larger electron-dense vesicles were also observed. These larger vesicles often appeared to contain mitochondrial remnants such as degenerated cristae. Those "autophagic vesicles" (Dean, 1977) occasionally were observed to have fused with mitochondria (Fig. 9). The meaning of the presence of increased numbers of lysosomes is not yet understood. Perhaps the neural tube of irradiated embryos begins to atrophy. In as much as normal organ and tissue shaping processes involve cell death and destruction (Saunders, 1966), the increase in the number of lysosomes may reflect unbalanced organ development. The neural tube of irradiated embryos is frequently somewhat narrower in diameter in its anterior end than in control embryos (Fig. 2). It is conceivable that UV irradiation leads in an indirect way to the unregulated proliferation of lysosomes, which lead to cell destruction. Further experimentation might profitably be directed towards a TEM analysis of neural tube cells from much later stage irradiated embryos. Alternatively, the functional significance of the altered mitochondria and increased lysosome content might be directly tested by grafting experiments. Segments of the neural tube of irradiated embryos could be grafted, in a reciprocal fashion, into the neural tube area of normal embryos. The developmental pattern which results might give some indication as to whether those altered cell organelles ultimately affect neural tube function.

The ultrastructure of the notochordal sheath is altered in irradiated embryos which display an abnormal notochord. The sheath normally consists of a basal lamina which rests on the notochord and serves to cover its entire surface. Layered upon the basal lamina are several (approx. 4-8) sheets of collagen fibers. Those sheets are approximately 15 nm thick, and are spaced approximately 15 nm apart. A zone of loosely organized fibers is attached to the outermost layer of collagen fibers (Fig. 10a). These features are very distinct in the later stage embryos. (eg. stage 36).

The notochordal sheath of an irradiated embryo almost completely lacks the organized layer of collagen fibers. The zone of loosely organized fibers was, however, present. It was directly associated with the basal lamina of the notochord (Fig. 10b). Future analyses should perhaps be directed towards an examination of the ontogeny of the highly organized collagen fibers in normal notochord development. It is possible that normal notochordal sheath development passes through a stage which resembles that of the irradiated sheath shown in Fig. 10b. Development of the notochordal sheath may actually be arrested, rather than merely altered, in irradiated embryos.

### Results of Analyses of Somite Pattern Formation in Notochordless Embryos.

As the information contained in Figs. 3, 4, and 6 indicates, the patterning of the somite segmentation process does not require the presence of the notochord. The more classical point of view, which emphasized the role of the notochord in somite development (Table 1), is therefore no longer tenable. Not only are the cellular mechanics of segmentation normal in "notochord-defective" embryos, but the somite counts are also normal in embryos irradiated with low doses of UV.

Previous studies on somitogenesis in normal Xenopus embryos led to the concept that whole segmented somites rotate through 90° (Hamilton, 1969). Initially, the paraxial mesoderm cells elongate perpendicularly to the notochord. Then segmentation occurs. Finally, each somite block was postulated to rotate through 90° with its medial edge moving forward. As a result of that presumed "block rotation", individual somites develop into bundles of spindle-shaped cells (myoblasts) which lie parallel to the notochord. The evidence from light microscopy studies quite clearly favored the model that the whole somite block rotates as a unit through 90° (Hamilton, 1969).

Observations on "notochord-defective" embryos which displayed fused somites and apparently normal myoblasts aligned parallel to

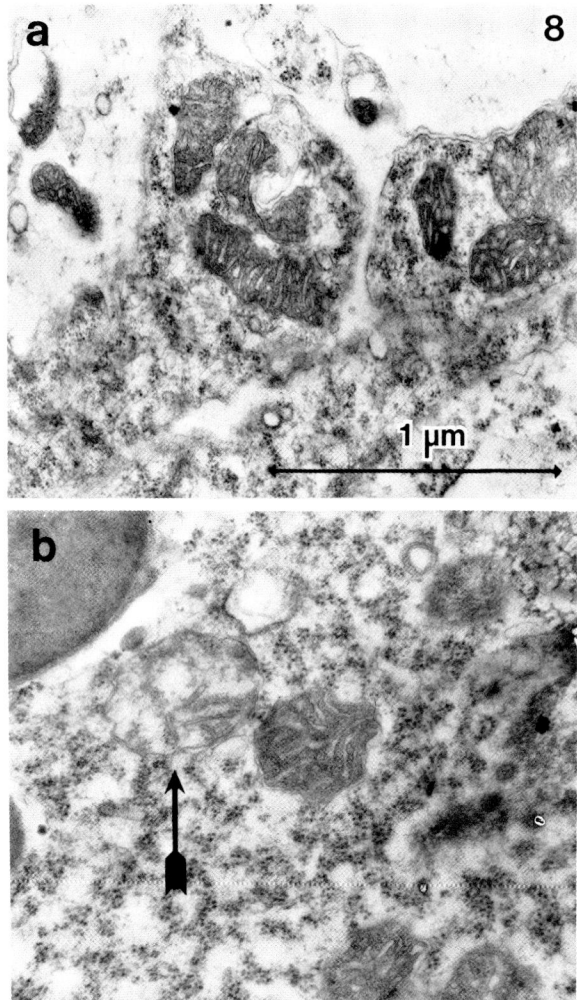

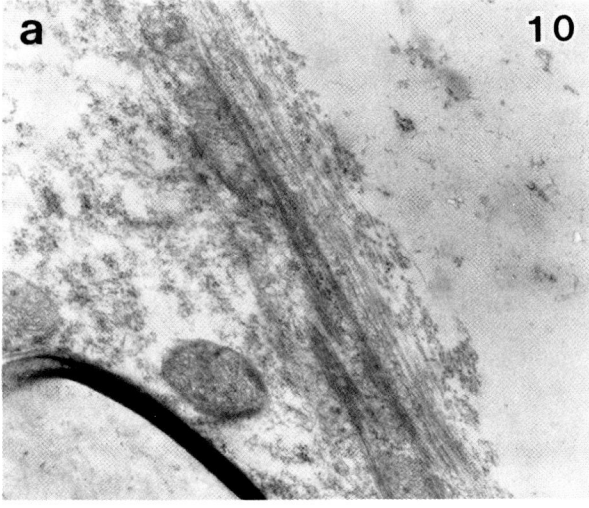

Figure 8. Transmission electron micrograph which shows normal mitochondrial size and morphology of neural tube cells (a). Swollen, more vacuolated mitochondria of neural tube cells of irradiated stage 32 embryo are shown (arrow) in (b).

Figures 8, 9, and 10 are at same magnification.

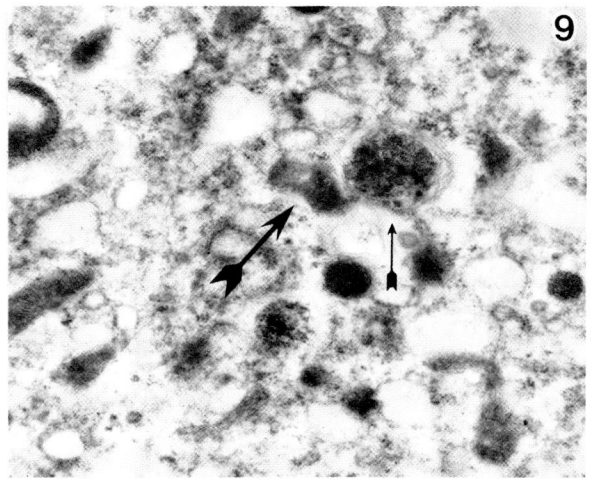

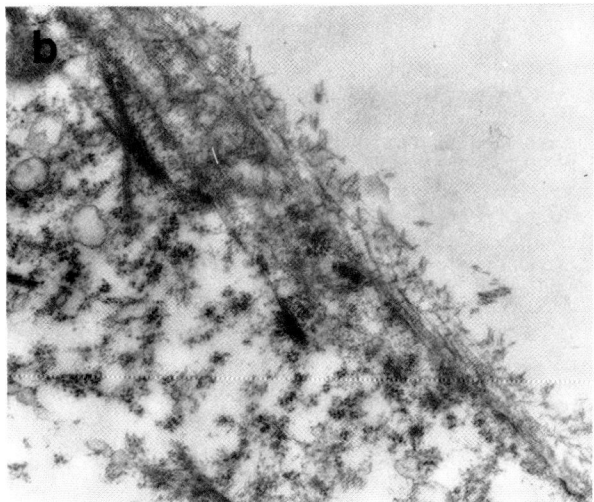

Figure 10. TEM view of the notochordal sheath of stage 36 embryos. (a) the parallel layers of collagen fibers and the outermost zone of loosely organized matrix of a control (unirradiated) embryo. (b) irradiated notochordal sheath lacks layers of collagen fibers, but does contain zone of loosely organized fibers.

the notochord led to a critical examination of the cellular mechanics of somite rotation. Fusion of the somites which occurs in irradiated embryos should prevent the block rotation of individual somites. That is not, however, the case. A detailed SEM study of individual somite cell arrangements during somite rotation generated the conclusion that individual cell rearrangements, rather than block rotation of a whole somite, accounts for the rotation process (Youn and Malacinski, 1981b). The basis of that recent study was the SEM examination of a series of longitudinal fractures through the upper,

Figure 9. TEM view of a neural tube cell from an irradiated embryo which shows the fusion of a lysosome (large arrow) with a mitochondrion (small arrow).

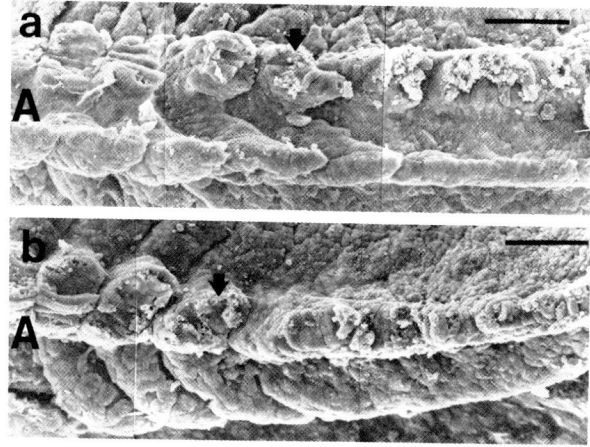

Figure 11. Arrangements of cells in ths fused somites of a stage 22-24 irradiated embryo. (a) dorsal view of cells in the fractured upper region of the fused somite. Cells can be observed to be undergoing normal rotation. (b) ventral view of cells in the fractured region of the somite below the fusion site. The embryo was photographed in the inverted configuration, so the orientation of the cells is the opposite of (a). A = anterior. Arrows indicate most posterior segmented somite. Bar equals 0.1 mm.

middle (fusing site), and lower (beneath the fusing site) regions of fused somites. Figure 6 illustrates the type of axial structure morphology which was subjected to this kind of examination. In Fig. 11a a fracture through the upper region of the somites of an irradiated embryo is displayed. The intrasomitic cellular arrangement of the prospective myotomal cells is completely normal. Likewise, examination of the lower region of a fused somite (Fig. 11b) revealed that a normal cell rotation process was underway. The perpendicular arrangement of cells in the unsegmented mesoderm, the oblique orientation of the turning cells in the most posterior somite, and the parallel arrangement of cells in the more anterior somites, were observed. These observations support the conclusion that the rearrangement of individual cells within the somite leads to the myotomal cells changing position to lie parallel to the long axis of the embryo.

The arrangement of the cells in the fusion region is complex. Cells in that region are oriented in several directions. In some instances the cells are parallel to the main axis, while in other cases they display the perpendicular orientation. Those cells in the fusion region are genuine myoblasts. TEM observations provide a direct confirmation of that fact. Figure 12 illustrates the state of cyto-differentiation of those cells. The extent of differentiation of those myoblasts may actually be somewhat less advanced than normal somite cells. The arrangement of the myofibrils in the fusion area is much less orderly (less parallel) than in the cells in control (unirradiated) somites.

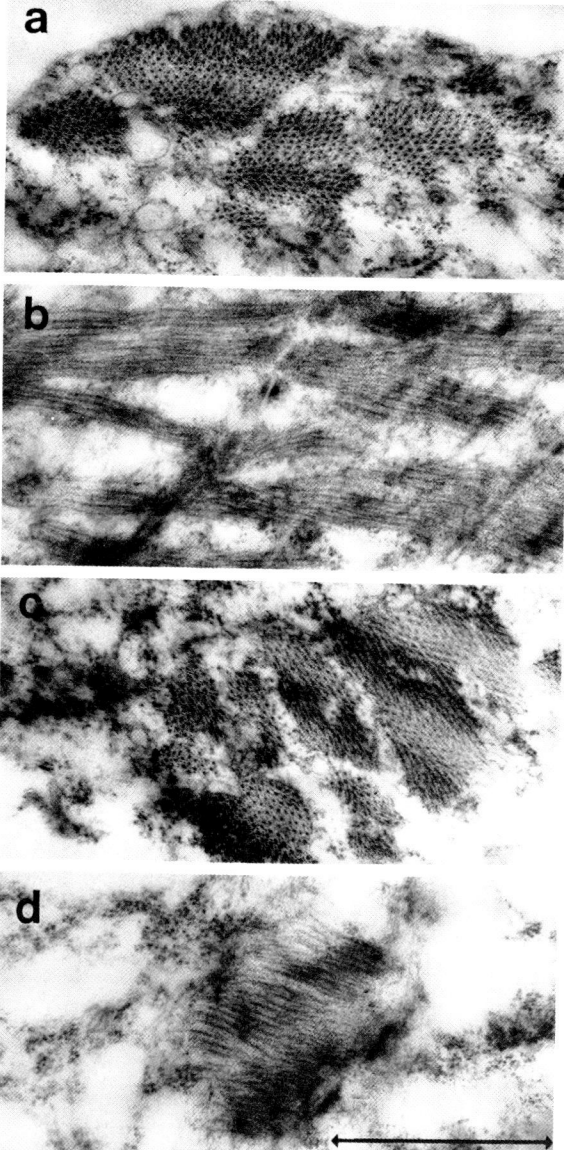

Figure 12. TEM photo (bar equals 1 µm) of the ultrastructure of a normal somite cell (a,b), and a cell in the fused region of a somite (c,d). The myofibrils can be seen in cross-section in (a) and (c), and in a longitudinal array in (b) and (d).

Further analyses--with the TEM--should perhaps be directed towards establishing the mechanism of the fusion process. The study of the manner in which the somites fuse across the midline might prove interesting. The analysis of the role the cell surface, as well as the role microtubules and microfilaments play in the fusion process may lead to further interesting insights into cell and tissue interactions during axial structure morphogenesis. These observations on the identity of the cells in the fusion area should also provoke further discussion of the fate of the presumptive notochord cells in notochordless embryos.

## Concluding Remarks

The ability to produce "notochord-defective" embryos with ultraviolet irradiation has provided an important opportunity for a re-examination of the role tissue interactions "do or don't" play during axial structure pattern formation. As a brief review of the more classical points of view (Table 1) indicated, the prominent position and relatively large size of the amphibian notochord probably account for the plethora of roles assigned to it. Many of those assignments have been re-evaluated during the past few years by the authors of this report. As this review article has emphasized, several of the roles previously assigned to the notochord should now be considered irrelevant.

## Acknowledgement

The authors gratefully acknowledge financial support provided by NSF grant number PCM 80-06343.

## References

Bancroft M, Bellairs R. (1976). The development of the notochord in the chick embryo, studied by scanning and transmission electron microscopy. J. Embryol. exp. Morph. 35: 383-401.

Dawid IB, Blackler AW (1972). Maternal and cytoplasmic inheritance of mitochondrial DNA in Xenopus. Develop. Biol. 29: 152-161.

Dean RT. (1977). Lysosomes. Edward Arnold Publishers, Ltd. London.

Hamilton L. (1969). The formation of somites in Xenopus. J. Embryol. exp. Morph. 22: 253-264.

Hay ED, Meier S. (1978). Tissue interaction in development. In "Textbook of Oral Biology" (J. Shaw et al., eds.), pp. 3-23. W.B. Saunders, Philadelphia.

Hirose G, Jacobson M. (1979). Clonal organization of the central nervous system of the frog. I. Clones stemming from individual blastomeres of the 16-cell and earlier stages. Develop. Biol. 71: 191-202.

Holtfreter J, Hamburger V. (1955). Amphibians. In "Analysis of Development" (B.H. Willier, P.A. Weiss and V. Hamburger, eds.), pp. 230-296. W.B. Saunders Co., Philadelphia.

Ikenishi K, Kotani M, Tanabe K. (1974). Ultrastructural changes associated with UV irradiation in the "germinal plasm" of Xenopus laevis. Develop. Biol. 36: 155-168.

Jacobson AG, Gordon R. (1976). Changes in the shape of the developing vertebrate nervous system analyzed experimentally, mathematically and by computer simulation. J. Exp. Zool. 197: 191-246.

Jacobson A. (1962). The development of the notochord in chick embryos. J. Embryol. exp. Morph. 10: 602-621.

Jurand A, Ireland, NJ. (1965). A slow rotary shaker for embedding in viscous media. Stain Technol. 40: 233-234.

Jurand A. (1962). The development of the notochord in chick embryos. J. Embryol. exp. Morph. 10: 602-621.

Karfunkel P. (1974). The mechanisms of neural tube formation. Intl. Rev. Cytol. 38: 245-271.

Keller RE, Schoenwolf GC. (1977). An SEM study of cellular morphology, contact and arrangement as related to gastrulation in Xenopus laevis. Wilhelm Roux's Arch. 182: 165-186.

Kitchen JC. (1949). The effects of notochordectomy in Ambystoma mexicanum. J. Exp. Zool. 112: 393-415.

Lehmann FE. (1929). Entwicklungsstorungen in der Medullaran von Triton, erzeugt durch Unterlagerungsdefekte. Wilhelm Roux's Arch. Entwicklungsmech. Organismen 108: 243-282.

Lehmann FE. (1935). Die Entwicklung von Ruckenmark, Spinalganglien und Wirbelanlagen in Chordalosen Korperregionen von Triton larven. Rev. Suisse de Zool. 42: 405-415.

Lovtrup S. 1975. Fate maps and gastrulation in amphibia -A critique of current views. Canad. J. Zool. 53: 473-479.

Meier S. (1979). Development of the chick embryo mesoblast: Formation of the embryonic axis and establishment of metameric pattern. Develop. Biol. 73: 25-45.

Nakamura O, Toivonen S. (1978). Organizer - A Milestone of a Half- Century From Spemann. Elsevier/North-Holland Biomedical Press, New York.

Nieuwkoop PD, Nigtevecht GV. (1954). Neural activation and transformation in explants of competent ectoderm under the influence of fragments of anterior notochord in urodeles. J. Embryol. exp. Morph. 2: 175-193.

Nieuwkoop PD, Faber J. (1967). Normal Table of Xenopus laevis (Daudin). 2nd ed. North-Holland Publishing Co., Amsterdam.

Reynolds ES. (1963). The use of lead citrate at high pH as an electron-opaque stain in electron microscopy. J. Cell. Biol. 176: 208-213.

Saunders JW Jr. (1966). Death in embryonic systems. Science 154: 604-616.

Smith LO, Williams M. (1979). Germinal plasm and germ cell determinants in anuran amphibians. Symp. British Soc. Develop. Biol. 4: 167-197.

Spemann H, Mangold H. (1924). Uber Induktion von Embryonalanlagr durch Implantation artfremder Organisatoren. Wilhelm Roux Archiv 120: 384-706.

Takaya H. (1956). Notochordal influence upon the differentiation and segmentation of muscle tissue. Annot. Zool. Jap. 29: 133-138.

Takaya H. (1961). Significance of the notochord for the differentiation and growth of the embryonic trunk in amphibia. Embryologia 6: 123-134.

Weisblat DA, Sawyer RT, Stent GS. (1978). Cell lineage analysis by intracellular injection of a trace enzyme. Science 202: 1295-1298.

Yamada T. (1939). Uber bedeutungsfremde Selbstdifferenzierung der prasumptiven Ruckenmuskulatur des Molchkeims bei Isolation. Okajimas Fol. Anat. Jap. 18: 565-568.

Youn BW, Malacinski GM. (1980). Action spectrum for ultraviolet irradiation inactivation of a cytoplasmic component(s) required for neural induction in amphibian egg. J. Exp. Zool. 211: 369-377.

Youn BW, Keller RE, Malacinski GM. (1980). An atlas of notochord and somite morphogenesis in several anuran and urodelean amphibians. J. Embryol. exp. Morph. 59: 223-247.

Youn BW, Malacinski GM. (1981a). Axial structure development in UV-irradiated (notochord-defective) amphibian embryos. Develop. Biol. 83 (April): in press.

Youn BW, Malacinski GM. (1981b). Somitogenesis in the amphibian Xenopus laevis: Scanning electron microscopic analysis of the pattern of intrasomitic cellular arrangements during somite rotation. J. Embryol. exp. Morph. in press.

## Discussion with Reviewers

A.G. Jacobson: Is there any possibility that in UV'd embryos the prospective notochord cells differentiated instead into prospective somites, and thereby managed to induce the central nervous system?
Authors: That is a distinct possibility. We have not yet attempted to mark prospective notochord cells and follow their fate in UV'd embryos. Should some of them actually differentiate into somite cells which have neural induction capacity, our main conclusion about a structually intact notochord being a dispensable component would, however, stand.

A.G. Jacobson: Don't your observations on the shape of the neural tube in UV'd embryos (eg. Fig. 6 b) support, rather than contradict, Holtfreter and Hamburger's earlier (1955) contention that the notochord is required for normal neural tube cell arrangements?
Authors: Figure 6b does indeed display a neural tube with an abnormally small neural canal, and with an extra thick ventral region. That is a typical observation. We have, however, examined dozens of similar fractured embryos and find--in approximately 10% of the notochordless ones--neural tubes which look completely normal. We believe, therefore, that although the internal structure of the neural tube is usually somewhat irregular, occasionally normal neural tubes do form in notochordless embryos.

J. Fallon: UV irradiation appears to be a non-specific insult, and subject, therefore, to some of the same criticisms the authors leveled at previous work on this subject. Would the authors please clearly state their rationale for the present studies?
Authors: Our rationale is straightforward: Low doses of UV generate a class of embryos which display specific and reproducible effects on notochord and anterior axial structure development. These embryos provide the positive result--the neural tube, somites, and other axial structures undergo morphogenesis in the absence of a notochord. Those observations are different from previous attempts at either chemical or microsurgical notochordectomy. Those previous observations often yielded negative data. That is, neural morphogenesis, somitogenesis, and axial elongation did not proceed normally in notochordless embryos. This important distinction between the present data and previous data indicates that UV can indeed be employed as a relatively specific probe. As is illustrated in Fig. 1, UV usually generates a wide spectrum of developmental abnormalities. Embryos with those severe, relative non-specific effects were not employed in these studies.

J. Fallon: Could the notochord have been present at an earlier developmental stage? How is it known that those very embryos which lacked a notochord at an earlier stage would have developed a normal neural tube at a later stage?
Authors: That question raises an important point. Needless to say, it is not possible to examine the extent of notochord development at earlier stages in the same embryos which are subjected to SEM analysis later. We have, however, carried out an extensive SEM analysis at various earlier developmental stages. The details are included in Youn and Malacinski, 1981a. We found no evidence at the earlier stages for formation/degeneration of the notochord. In some classes of embryos (eg. severe microcephaly), 100% of the embryos are notochordless at all stages, and yet they developed a neural tube, somites, etc.

J. Fallon: What is the value of speculation concerning the injured neural tube mitochondria having been derived from the oocyte?
Authors: The cells with the abnormal mitochondria (as well as increased numbers of lysosomes) arose approximately 2 1/2 days after the egg was irradiated. Several (between 14 and 16) cell divisions intervened between the egg and tailbud stage. The abnormal mitochondria could represent either (1) cytoplasmic segregation during embryogenesis of egg mitochondria of the vegetal hemisphere which sustained direct UV hits, or (2) abnormal daughter mitochondria which are the progeny of the damaged egg mitochondria. Both of these possibilities are of interest in view of current notions of mitochondrial biogenesis in early amphibian embryogenesis.

R. Flickinger: Could ultraviolet irradiation inhibit primarily notochord differentiation because of a higher content of RNA in these cells?
Authors: Our previous work on the action spectrum of the UV target (Youn and Malacinski, 1980) yielded data which are more consistent with the notion that the target is a protein, rather than a nucleic acid. As well, we have been unable to achieve photoreversal of the UV effects, which might be expected of a nucleic acid target.

D.L. Stocum: What is the possibility that presumptive notochord cells (i.e. undifferentiated or even undetermined midline mesoderm cells) might induce neural tube formation (as opposed to determination of neural ectoderm) before being lost, destroyed, or differentiating along some other path?
Authors: If the presumptive notochord cells of notochordless embryos actually do move into position along the dorsal midline, without later differentiating into an organized and visible notochord, there is a good possibility those cells might exert their normal inductive influence on neural tube formation. Cell-marking experiments which trace the fate of presumptive notochord cells in irradiated embryos could possibly resolve this very important issue. The crux of this matter concerns the cell migration

behavior of the presumptive notochord cells. We have not yet accumulated any data which deal with this issue.

D.L. Stocum: If the notochord of presumptive notochord cells are not necessary for neural tube formation, are the somites necessary?
Authors: It would be fascinating to talk about "the role of somite-driven tissue interactions" in axial structure morphogenesis. However, we know of no method (except for microsurgery--which has some limitations) that could be employed on whole embryos to inhibit somitogenesis.

D.L Stocum: Are altered mitochondria and lysosomes observed in other tissues of irradiated embryos besides neural tube? Altered mitochondria and increased numbers of lysosomes might be due to a general effect of early irradiation, in which case they would be observed in all tissues, or due to a faulty tissue interaction, in which case they might be restricted to the neural tube.
Authors: Somite cells in irradiated embryos display a normal complement of cytoplasmic organelles. Other cells or tissues have not been examined. This issue relates to an earlier point made by one of the reviewers. We speculated then that cytoplasmic segregation may account for the accumulation of damaged egg vegetal hemisphere mitochondria in the neural tube. If, in fact, only neural tube cells contain altered organelles, this matter would be worth pursuing.

B.M. Carlson: How do the authors regard the functional significance of the amphibian notochord, in view of their current findings?
Authors: During early development the notochord probably functions, somewhat indirectly, as a "partition" along the dorsal midline. Its main function is to contribute to "pattern formation". Blocks of somites from right and left sides are prevented from fusing, and the shape of the spinal column is anticipated by the presence of the notochord. In later embryogenesis the notochord probably serves a more important structural role in stabilizing dorsal organs, and in spinal column development. Direct "inductive" roles for the notochord in early tissue interactions are ruled out by the present findings.

B.M. Carlson: If the notochord is perceived of as less than critical to the individual embryo, what selection pressures would lead to its preservation through vertebrate evolution?
Authors: As mentioned above, the notochord can be considered an important component of the axial structure system. If it actually does play a role early in "pattern formation" and later in stabilizing the somites and neural tube, it should be considered indispensable to the organism. Through evolution it might be modified, but not eliminated.

# CELL MOVEMENT AND CONTRACTION IN SOMITE DEVELOPMENT

Ellen A. G. Chernoff

Case Western Reserve University, Department of Developmental
Genetics and Anatomy, 2119 Abington Road, Cleveland, OH 44106
Phone no.: (216) 368-2388

(Paper received January 19 1984, Completed manuscript received January 15 1985)

## Abstract

During somite formation the segmental plate mesoderm, lying on either side of the axial organs, reorganizes into roughly spherical pairs of epithelial structures. This segmentation process includes changes in cell shape and position, cell-cell and cell-substratum adhesive properties and accumulation of extracellular matrix material which proceed down the anterior-posterior axis. Later in somite development the sclerotome region "disperses", migrating around the spinal cord where it produces the cartilage model of the vertebral column. Experimental manipulation of segmentation and sclerotome dispersal with drugs affecting microfilaments, microtubules and calcium-dependent contraction suggest that cells migrate into position, elongate, and undergo apical contraction as part of the segmentation process. This process of calcium-dependent, possibly calmodulin-mediated, contraction can be both stimulated precociously and inhibited, showing similarities with contractile morphogenetic events in epithelial organ systems such as eye and thyroid. Similar experiments with drugs affecting contractile microfilaments demonstrate that active cell movement, along with extracellular matrix production, is involved in sclerotome dispersal.

KEY WORDS: organogenesis, segmentation, segmental plate mesoderm, somite formation, sclerotome dispersal, calcium-dependent contraction, cell shape changes, microfilaments, microtubules, extracellular matrix

## Introduction

Segmentation, in vertebrate embryos, is the process of forming a series of paired, transitory "vesicular" structures called somites from a primary mesodermal tissue (segmental plate) found on either side of the early axial structures (notochord, neural plate/tube). Somite formation and development has been described as a transition in tissue organization from mesenchymal to epithelial and back to mesenchymal form. A collection of loosely associated (mesenchymal) cells in the segment plate reorganizes to closely apposed cells that form the epithelial somite. Later in somite development there is a return to mesenchymal organization as the ventromedial wall of the somite vesicle breaks down and the sclerotome portion of the somite "disperses". The sclerotome is the cartilage-forming region of the somite. Somitic cartilage eventually surrounds the spinal cord and establishes the framework for the vertebral column. The dorsal portion of the somite forms the dermamyotome, which eventually gives rise to the dermis of the skin and to the skeletal muscles. These reorganizations of tissue structure involve changes in cellular adhesivity, specialized cell junctions, extracellular matrix (ECM) material, and cytoskeletal organization. My own studies of somitogenesis have involved establishing the existence of a calcium-dependent contraction event in segmentation and examination of the contribution of cell movement to somite formation and to sclerotome dispersal. In this paper I will describe my own work in relation to other studies of segmental plate organization, segmentation, and sclerotome dispersal and try to integrate the various observations on morphogenetic forces underlying somitogenesis using examples from my own scanning electron microscopy (SEM) studies. All of the micrographs I will show are of stage 14-15 chick embryos (staging as in Hamburger and Hamilton, 1951), which occurs at approximately 2.5 days of incubation. There are 22 pairs of somites at stage 14, and the segmental plate mesoderm is in the trunk region (between the prospective limb bud areas).

## Segmental Plate Organization

The early events in formation of the segmental plate mesoderm include shearing of the primary mesoderm from anterior to posterior (cranial to caudal) by the regression of Hensen's node (avian) and formation of the notochord, association of the tissue with neural epithelium, and condensation toward the midline (in anterior regions - Lipton and Jacobson, 1974a). Accompanying changes in cell shape and orientation have been described in detail by Bellairs (1979). Somite formation proceeds from anterior to posterior and the segmental plate mesoderm lies on either side of the axial structures extending from the last somite caudally to Hensen's node (avian) or to the anterior end of the primitive streak (mammalian). (Fig. 1). The dorsal and ventral surfaces of the segmental plate are covered by ectoderm and endoderm, respectively (Figs. 2-4).

Figure 5 shows the dorsal side of normal segmental plate with the ectoderm dissected away after fixation. Cells appear loosely associated and contact each other with numerous cell processes. On the ventral side (Fig. 6), with the endoderm removed, cells lie in a planar arrangement. In more anterior portions of the segmental plate, more ECM is seen on both dorsal and ventral surfaces (Fig. 7a,b,c,) and more ECM accumulates with time. In sagittal fracture, (Fig. 8) segmental plate appears as a loosely organized mesenchymal tissue. The cells are stellate and slightly flattened along the dorsal-ventral axis. The tissue appears the same in cross-fracture. The segmental mesoderm is attached, initially, to the neural tube (Fig. 9) and the overlying ectoderm (Bellairs & Portch, 1977) by fine cell processes. The ventro-medial portion of the segmental plate is connected with the notochord ECM by matrix fibers (Fig. 10) composed of collagen and glycosaminoglycan (Lash & Vasan, 1978).

A variety of experiments have suggested that there is a pre-patterning of the segmental plate mesoderm prior to the actual segmentation event. When separated from the neural tube and notochord surgically, 10 to 12 somites (in chick and quail) form at one time in the segmental plate, after a lag, instead of in sequence. (Packard and Jacobson, 1976; Packard, 1980). Meier (1979) demonstrated a morphological basis for the prepatterning by showing 10 to 12 areas of circular organization on the surfaces of the segmental plate. These areas, termed somitomeres, have now been shown to correlate with the location of the somites (Packard and Meier, 1983). Somitomeres are difficult to see and require stereo SEM imaging. The forces that produce this pre-patterning remain under investigation. The involvement of Hensen's node, the axial structures, ectoderm and endoderm have all been considered (Bellairs and Portch, 1977; Meier and Jacobson, 1982; Packard and Meier, 1983;

FIGURE 1. Control stage 14 chick embryo in ring culture (see Chernoff and Lash, 1981). Explanted with surrounding extraembryonic membranes grown ventral side up over L-15 culture medium. Shown from heart level to posterior end. OA= omphalomesenteric artery. Arrows mark last fully formed somite (som), notochord (noto, dark structure), and neural tube (NT, bright structure). Light micrograph. Bar=100μm

FIGURE 2. This figure and Fig. 5 are a set showing cross-fracture through the nascent somite pair (nasc. som.) from a stage 14-15 embryo. Both halves are shown to give an impression of the three dimensional arrangement of cells in the forming somite. end=endoderm, ect=ectoderm, NT=neural tube, no=notochord. Bar=100μm

FIGURE 3. This half of the fracture pair retains more of the ECM that surrounds the somites and forms the perinotochordal sheath. ECM=extracellular matrix. Bar=100μm

FIGURE 4. Dorsal surface of cultured trunk segment like that used in figures 2 and 3 (see Chernoff and Hilfer, 1982) shows close packed polygonal cells of the covering ectoderm(ect). Bar=100μm

FIGURE 5. Control segmental plate mesoderm, dorsal surface (ectoderm dissected off following fixation). Fine cell processes and some ECM fibers are visible. Bar=10μm

FIGURE 6. Control segmental plate, ventral side (endoderm off). Cells lie in a more planar arrangement on this surface. Bar=10μm

Christ et al., 1972, 1974; Sandor and Fazakas-Todea, 1980).

## Segmentation

The segmental plate mesoderm lying on either side of the notochord forms somites by a process of segmentation that proceeds from the anterior to the posterior end of the embryo. A number of morphogenetic forces are involved in the transition from segmental plate mesoderm to epithelial somite. Progressive changes in cellular adhesivity appear to occur along the embryonic axis as somites form (Bellairs and Portch, 1977; Bellairs, 1979). Cell-cell and cell-substratum adhesivity increase from segmental plate to somite stages (Bellairs et al., 1978, 1980). Changes in cell junctions accompany this process (Trelstad et al., 1967; Lipton and Jacobson, 1974a; Solursh et al., 1979; Bellairs, 1979). The role of extracellular matrix material (ECM) in somitogenesis has been an active area of investigation. One ECM component, fibronectin, seems to stimulate segmentation in vitro (Cheney et al., 1980) presumably by affecting cell adhesion. Perinotochordal sheath fibers may 'stabilize' newly formed somites (Lipton and Jacobson, 1974a,b). It has been suggested that collagen fibers of the segmental plate and surrounding tissues anchor nascent somite cells and aid in elongation (Bellairs, 1979) (Figures

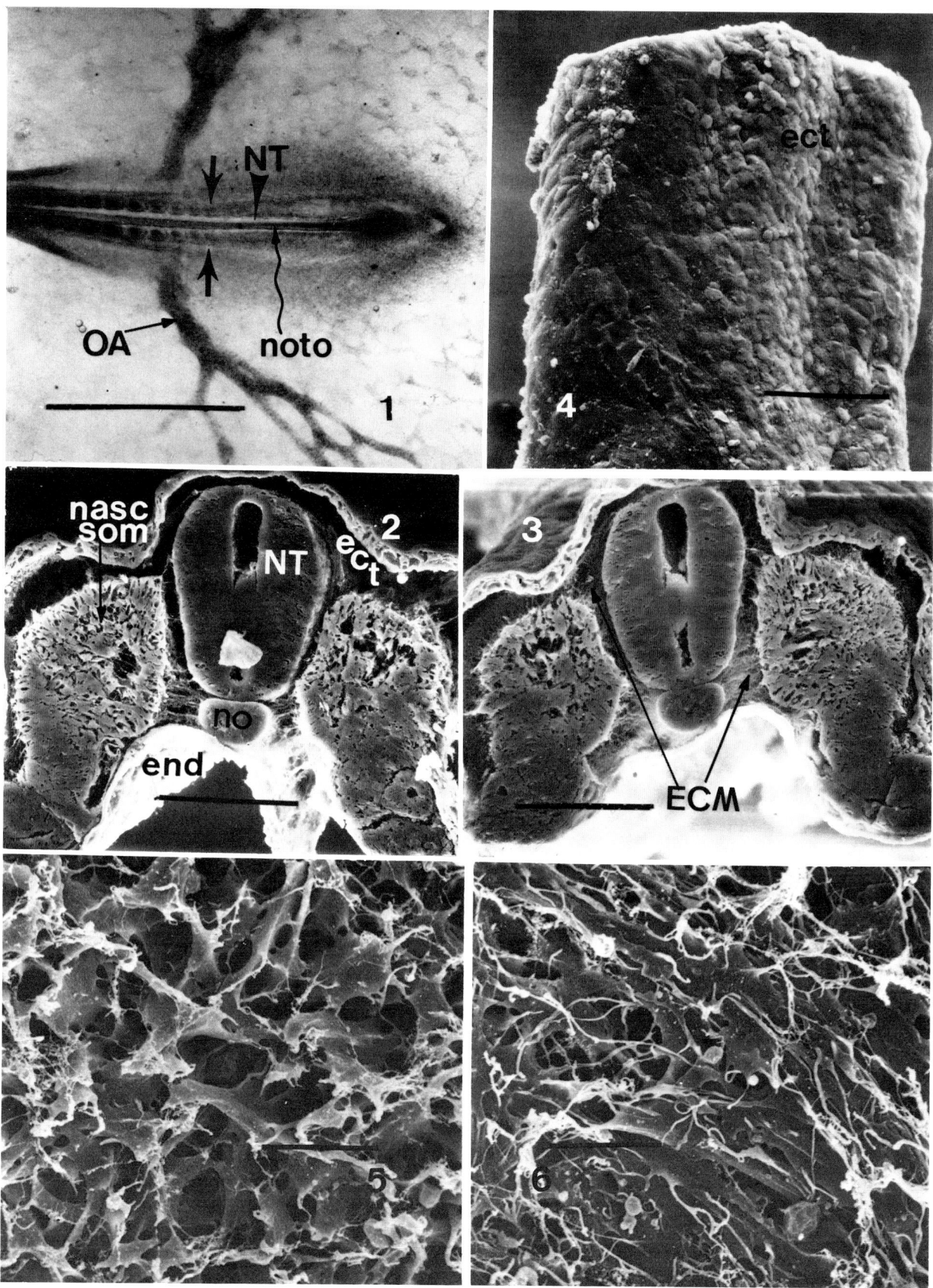

2,3,10,11). However, these cell-matrix interrelationships alone cannot account for the major changes in cell organization during somitogenesis.

Changes in cell shape during segmentation have been described in a variety of contexts (Williams, 1910; Lipton and Jacobson, 1974a; Bellairs, 1979; Meier, 1979). Of principal interest here is the elongation of cells and apical constriction during formation of the epithelial somite (i.e., a contractile event). The existence of a contractile event in segmentation is suggested by two lines of experimental evidence. The changes in cellular morphology during segmentation are consistent with contraction. The detailed studies of Bellairs (1979) describe the changing appearance of the surface of the segmenting mesoderm. Segmentation seems to begin on the dorsal side of the segmental plate with elongation of cells in that region (Platt, 1889; Lipton and Jacobson, 1974a; Bellairs, 1979). The cells of the somite epithelium are elongated and tend to be narrower toward the lumen (Figs. 12,13). The position of the nucleus in a cell of the somite epithelium varies with the stage of the cell cycle (Fig. 13). Nuclei move toward the apical end of the cell (center of the somite) in preparation for cell division (Langman and Nelson, 1968; Bellairs, 1979). The elongated somite cells contain many microtubules whereas few are reported in segmental plate cells (Bellairs, 1979). Microtubules in segmental plate cells are probably just not as highly oriented as those in somites. The epithelial somite appears to be under tension through attachment to neural tissue, ectoderm, endoderm and the dorsal aorta (Bellairs, 1979; Lipton and Jacobson, 1974a). The presence of microtubules and tension forces are thought to contribute to somite cell elongation. This elongation and narrowing of the cell apices is reminiscent of morphological changes that occur in epithelial systems in which contraction has been found to play a role (Wessells et al., 1971; Burnside, 1973; Schroeder, 1973). A second line of thought has grown out of studies on early stages of somitogenesis and observations on segmentation in the absence of the axial structures. As stated in the previous section, Meier (1979) described somitomeres, circular areas of orientation visible on the dorsal and ventral mesoblast surfaces. These appear during condensation of the paraxial mesoblast as the segmental plate forms. They represent a stage in the process that results in segmentation. In Meier's stereo SEM study, somitomeres seem to indicate the boundaries on the segmental plate of 10 or 12 prospective somites. This is consistent with the observation (Lipton and Jacobson, 1974b; Packard and Jacobson, 1976; Packard, 1978) that 10-12 somites will form simultaneously from segmental plate separated from the axial structures. This simultaneous formation has been cited as evidence for a contraction event in somitogenesis (Meier, 1979). It should be

FIGURE 7. This series of three micrographs shows accumulation of ECM with time. Sequence is posterior to anterior. (a) shows dorsal surface of segmental plate as in figure 5. (b) shows more anterior region of segmental plate in same stage 15 embryo with covering of ECM fibers. (c) shows dorsal surface of 2 fully-formed somites at the level of the omphalomesenteric artery. The somites are densely covered with ECM. Arrows show neural crest cells migrating across the somites. Bar=10μm

FIGURE 8. Sagittal fracture of normal segmental plate mesoderm in culture 4 hours. Most cells are slightly flattened along the dorso-ventral axis (dorsal surface is at top of figure) and stellate in profile. Bar=10μm

FIGURE 9. Fine cell processes connect the segmental plate (stage 14, ectoderm off) with the neural tube (NT). Bar=10μm

FIGURE 10. On the ventral surface (stage 15, endoderm off) ECM surrounds the notochord (Noto) and ECM fibers connect the notochord and the segmental plate mesoderm. Bar=10μm

FIGURE 11. Higher magnification view of cross-fractured nascent somite shown in figures 2 and 3. Cells are accumulating into an epithelial arrangement (Epith) at the somite periphery and elongating. Cells seem to be accumulating at the somite core. The tissue is surrounded by ECM. NT=neural tube. Bar=10μm

noted that the response is simultaneous and not instantaneous segmentation; segmentation took 14-17 hr in the posterior (Lipton and Jacobson, 1974b) and approximately 10 hr in the anterior region (Packard and Jacobson, 1976). The situation may not be one of contraction upon release of tension or physical restraint through anchoring to adjacent tissues but, instead, may reflect release from axial control of sequential segmentation.

In my own work, the existence of a calcium-dependent process in somite formation was investigated in the embryonic chicken using drugs and conditions that inhibit shape changes in other systems (Hilfer et al., 1981; Brady and Hilfer, 1982). Calcium-dependence of segmentation was investigated by culturing embryonic trunk explants with and without $Ca^{++}$ in the medium and by treatment with calcium antagonists and agonists. Calcium activation of non-muscle contractile systems is thought to reside with calcium dependent regulatory proteins, such as calmodulin (Cheung, 1980; Klee et al., 1980). The participation of calmodulin in segmentation was investigated by using calmodulin antagonists in culture (Brady and Hilfer, 1982). The effects of the calmodulin antagonists were compared with those of cytochalasin D (CD), which affects contractile microfilaments by another mechanism (Brown and Spudich, 1979; Lin, et al., 1980; MacLean-Fletcher and Pollard, 1980). A calcium agonist, the $Ca^{++}$ ionophore A23187 was used to stimulate contraction of somitic tissue to observe precocious somite

Somite Development

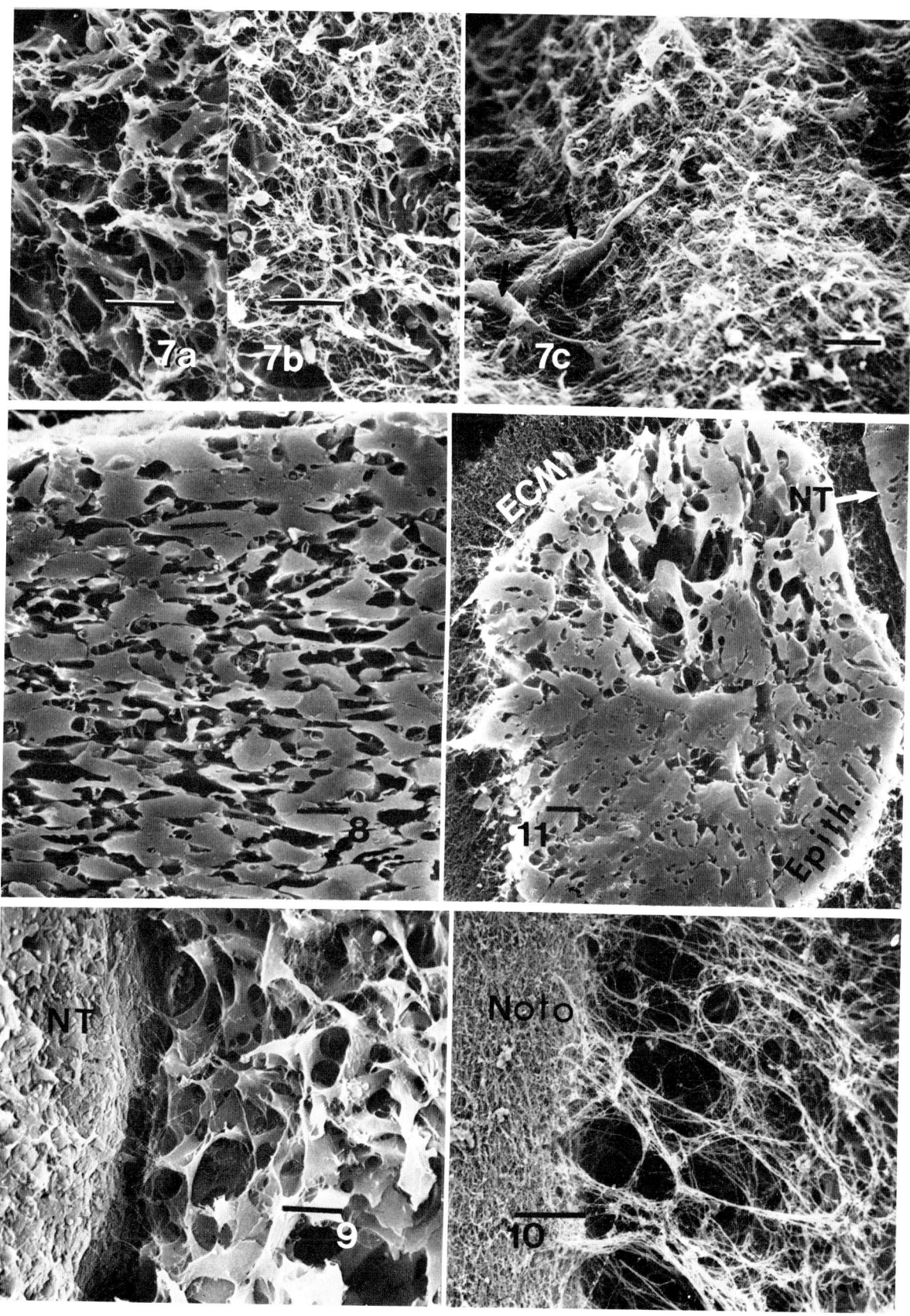

formation. The role of microtubules in somite cell shape changes was explored with nocodazole, an inhibitor of microtubule polymerization. Scanning electron microscopy of frozen, fractured (Humphreys, et al., 1974) somites and segmental plate tissue was used to analyze the contributions of cell movement, elongation, and apical constriction to somitogenesis. The results are consistent with the involvement of a calcium-dependent process in somite formation (Chernoff and Hilfer, 1982).

The responses of segmental plate and somites to the drugs used in this study emphasize that the segmentation process has a number of components. $Ca^{++}$ antagonists (verapamil, papaverine) (Chernoff and Hilfer, 1982), calmodulin antagonists (trifluoperazine (TFP)), cytochalasin (CD), and nocodazole all reversibly arrest somite formation at the stage of drug treatment (Chernoff and Hilfer, 1982; Figs. 14-16). The importance of cell movement in segmentation is illustrated by the prevention of accumulation of segmental plate cells into the somite periphery. (Compare drug-treated segmental plates; Fig. 14-16; and drug-treated nascent somites; Figs. 17,18, with normal tissue at the same time of incubation (Figs. 11,12).

TFP and CD prevented the apical constriction necessary to form the epithelial somite (compare Figs. 19, 20 with Fig. 12). The effects of nocodazole show that segmental plate cell morphology and somite epithelial cell elongation are heavily dependent upon cytoplasmic microtubules (Figs. 18,21). Microtubules have been implicated in directionality of cell movement in other systems (e.g., Gail and Boone, 1971). The $Ca^{++}$ ionophore A23187 has a very rapid, specific effect on somitogenesis. The nascent somite pair of a trunk segment in culture rapidly separates from the segmental plate, directly implicating apical contraction (Chernoff and Hilfer, 1982). Segmental plate tissue merely condenses in the presence of $Ca^{++}$ ionophore.

## Sclerotome Dispersal

Later in somite development the sclerotome "disperses". Some cell junctions are lost and ECM is elaborated (Fisher and Solursh, 1977; Solursh et al., 1979). The sclerotome then expands into the space around the notochord, a process that has generally been considered to mark the onset of active sclerotome migration (Trelstad et al, 1967; Hay, 1968; Ebendal, 1977). Both ECM production and active cell movement are important in sclerotome migration. ECM can play different roles in a tissue at different stages in development. The ECM may stabilize the sclerotome in early development (Lipton and Jacobson, 1974a,b). It may later serve as a substrate during migration into the perinotochordal area (Ebendal, 1977). Later still, this matrix is a stimulator of somite chondrogenesis (Lash and Vasan, 1978). Solursh, et al., (1979) showed that treatment

FIGURE 12. Cross-fracture of epithelial (fully formed) somite. Typical elongated epithelial cells (epith) with narrowed apices surround core cells (core). Bar=10μm

FIGURE 13. A higher magnification view of somite epithelium shows the varying levels of nuclei within the single cell layer (arrowheads). Location of the nucleus is cell-cycle-dependent. Bar=10μm

FIGURE 14. Sagittal fracture of segmental plate treated with TFP for 4 hrs. Cells have rounded up and tissue is more condensed than in control (figure 8). During time of drug treatment control explants have formed 1-2 new somite pairs. Bar=10μm

FIGURE 15. Dorsal surface of segmental plate mesoderm (ectoderm off) treated with CD for 5 hours. Cells have rounded up. ECM fibers cover the surface (compare with normal tissue, figures 5 and 7). Bar=1μm

FIGURE 16. Cross-fracture of segmental plate treated with nocodazole. Cells have completely lost their stellate shape and tissue is highly condensed. NT=neural tube. Bar=10μm

FIGURE 17. TFP-treated nascent somite viewed in cross-section. Compare with figure 11. Cells are rounded up and distinction between accumulating epithelial cells and core cells is largely lost. Bar=10μm

---

of sclerotome-forming somites with glycosaminoglycan lyases, (enzymes which digest important ECM components) caused the collapse of sclerotome and halted dispersal. They suggested that sclerotome cells are passively pushed apart by the accumulating ECM. However, sclerotome cells assume a filopodial, lamellipodial morphology as they disperse and move toward the notochord (Hay, 1968; Ebendal, 1977). This morphology is consistent with that of actively moving cells in other tissues of the embryonic chick (Trelstad et al., 1967; Ebendal, 1976; Chernoff and Overton, 1977; Bard et al., 1975; Nelson and Revel, 1975; Ho and Shimada, 1978). Treatment of somites that have formed distinct dermatome and sclerotome with CD (Figs. 22,23) shows sclerotome dispersal to be arrested. If active cell movement was not involved in sclerotome dispersal, then the sclerotome cells should have been passively pushed apart by ECM formation in the presence of CD, since CD did not affect ECM production (Chernoff and Lash, 1981). Active cell movement is also consistent with the increase in cell-substratum adhesivity during somite development (Bellairs and Portch, 1977; Bellairs, et al., 1978). It is likely that cell-substrate adhesivity continues to increase as ECM accumulates. Bellairs et al. (1980) have shown that sclerotome tissue does spread more in vitro than do nascent somites.

## Summary

Since the somites form in a temporal sequence along the embryonic axis, different events take place at a given time at different

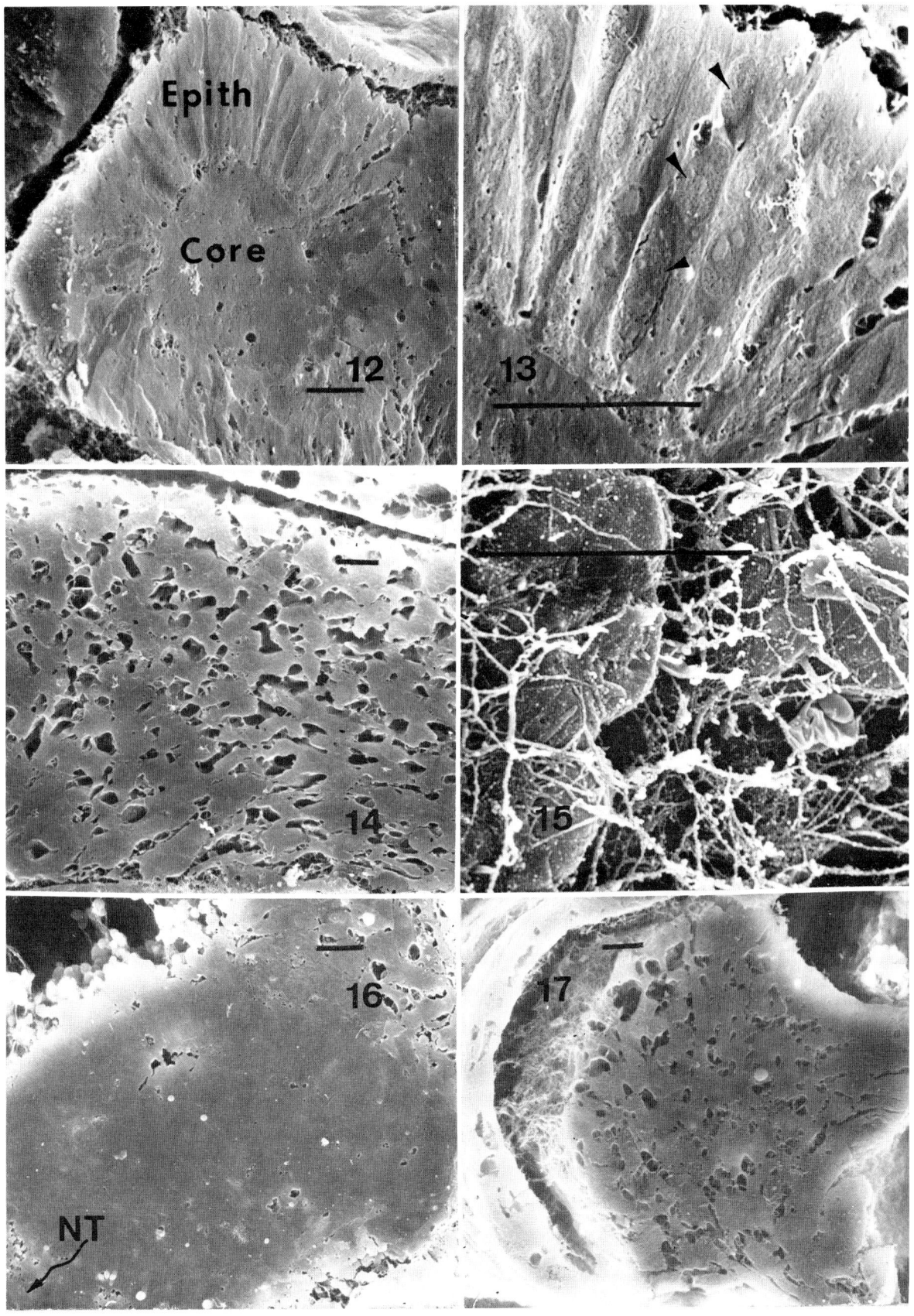

levels of the embryo. Moreover, within each somite the events do not occur in a strictly coordinated fashion; some somite epithelial cells may be undergoing apical constriction while others are still migrating into position. In this regard the somite system is unlike the epithelial systems in which cells are organized as sheets at the start of the morphogenetic event. Invagination or branching in epithelial systems is a coordinated process of cell elongation and apical constriction. It does appear that somite segmentation shares underlying mechanisms with epithelial systems such as thyroid eye, and neural tube, including $Ca^{++}$-dependent (possibly calmodulin-mediated) apical constriction and microtubules dependent cell elongation. The process of sclerotome dispersal involves active cell migration through ECM-filled spaces and the elaboration of ECM within the sclerotome itself.

## Acknowledgements

The studies on cell movement and contraction in somite formation were performed in the laboratory of Dr. S. Robert Hilfer, Department of Biology, Temple University, Philadelphia, PA 19122. The study of sclerotome dispersal was performed in the laboratory of Dr. James W. Lash, Department of Anatomy, University of Pennsylvania School of Medicine, Philadelphia, PA. 19104, and completed in the laboratory of Dr. Hilfer.

## References

Bard, JBL., Hay, E., Meller, SM. (1975). Formation of the endothelium of the avian cornea: A study of cell movement in vivo. Develop. Biol. 42, 334-361.

Bellairs, R. (1979). The mechanism of somite segmentation in the chick embryo. J. Embryol. Exp. Morphol. 51, 227-243.

Bellairs, R., Curtis, ASG., Sanders, EJ. (1978). Cell adhesiveness and embryonic differentiation. J. Embryol. Exp. Morphol. 46, 207-213.

Bellairs, R., Portch, PA. (1977). Somite formation in the chick embryo. In "Vertebrate Limb and Somite Morphogenesis: British Society for Developmental Biology, Symposium 3" (DA. Ede, JR. Hinchliffe, M. Balls, eds.), pp. 449-463. Cambridge Univ. Press, Cambridge.

Bellairs, R., Sanders, EJ., Portch, PA (1980). Behavioral properties of chick somitic mesoderm and lateral plate when explanted in vitro. J. Embryol. Exp. Morphol. 56, 41-58.

Brady, RC. Hilfer, SR. (1982). Optic cup formation: a calcium-regulated process. Proc. Natl. Acad. Sci., USA, 79, 5587-5591.

Brown, SS., Spudich, JA. (1979). Cytochalasin inhibits the rate of elongation of actin filament fragments. J. Cell Biol. 83, 657-662.

Burnside, B. (1973). Microtubules and microfilaments in amphibian neurulation. Amer. Zool. 13, 989-1006.

Cheney, CM., Seitz, AW., Lash, JW. (1980) Fibronectin, cell adhesion and somite formation. J. Cell Biol. 87, 94a.

Chernoff, EAG. Hilfer, SR. (1982). Calcium dependence and contraction in somite formation. Tissue and Cell 14, 435-449.

Chernoff, EAG. Lash, JW. (1981). Cell movement in somite formation and development in the chick: inhibition of segmentation. Develop. Biol., 87, 212-219.

Chernoff, EAG., Overton, J. (1977). Scanning electron microscopy of chick epiblast expansion on the vitelline membrane. Develop. Biol. 57, 33-46.

Cheung, WY. (1980). Calmodulin plays a pivotal role in cellular regulation. Science 207, 19-27.

---

FIGURE 18. Cross-fractured nascent somite treated with nocodazole. Elongation of peripheral cells is lost (Epith). NT=neural tube. Bar=10μm

FIGURE 19. TFP-treated somite that has fractured slightly obliquely. Cells rounded. Some cells appear elongated, but apical constriction is lost. (compare with figure 12) NT=neural tube, epith=epithelium. Bar=10μm

FIGURE 20. Cross-fracture of epithelial somite from CD-treated explant. The epithelial cells (epith) are rounded up and apical narrowing is lost. (compare figure 12). Bar=10μm

FIGURE 21. Epithelial somite treated with nocodazole. Epithelial cells have rounded compared with controls, but apical constriction can still be seen in some regions of the periphery. NT=neural tube. Bar=10μm

FIGURE 22. Cross section of CD-treated stage 15 embryo, somite number 20, in culture 8 hours. The forming dermamyotome (derm) and sclerotome (scl) are indicated. ECM is visible in the somitocoel (arrowhead). NT=neural tube, noto=notochord. Bar=10μm

FIGURE 23. Cross-fracture of control, somite number 20 from explant of stage 15 embryo in culture for 8 hours. Only small somitocoel remains (arrowhead). Dermamyotome (derm) formation is more advanced than in the drug-treated somite (see figure 22) and sclerotome dispersal has started. NT=neural tube, noto=notochord. Bar=10μm

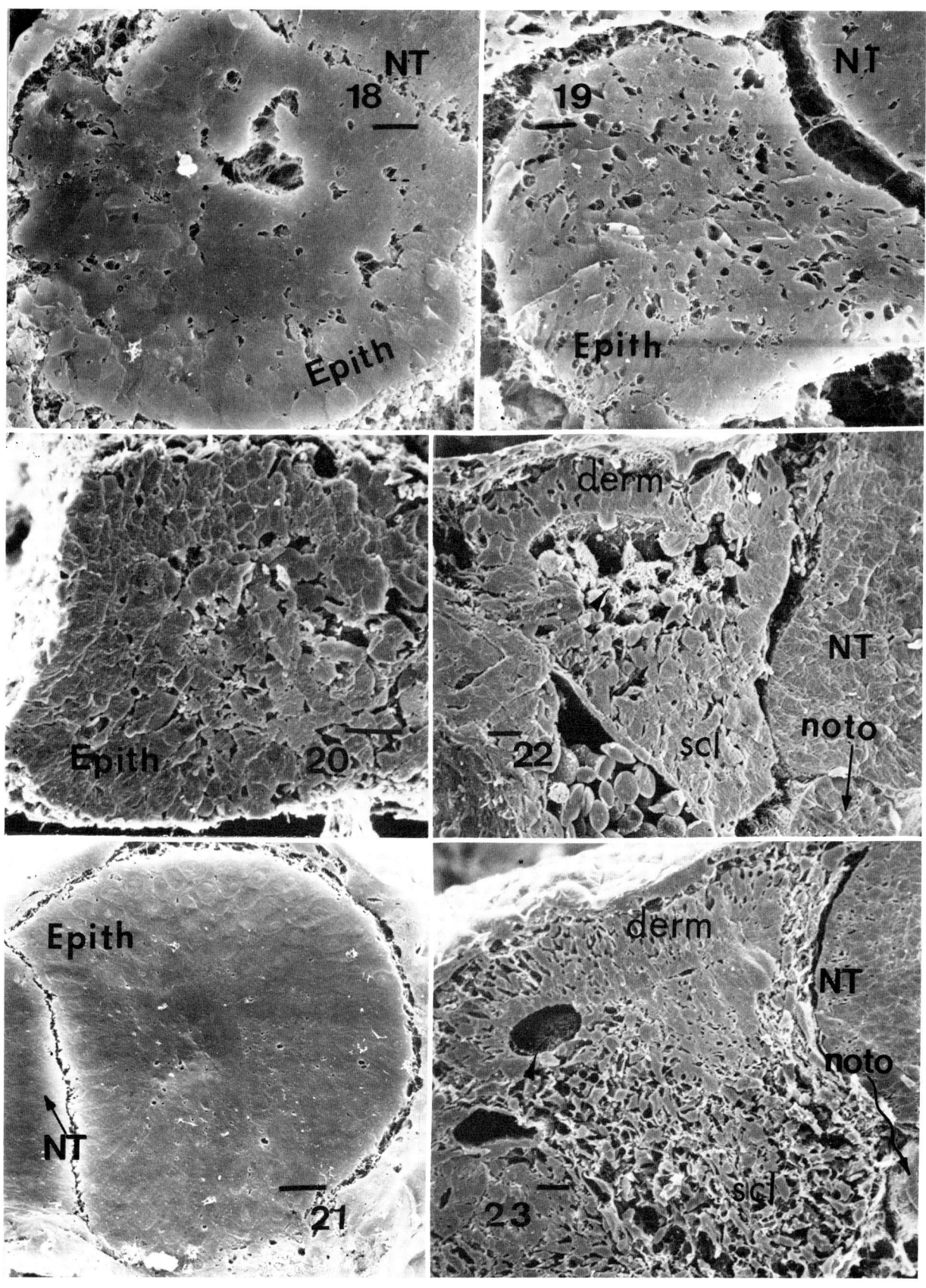

Christ, B., Jacob, HJ., Jacob, M. (1972). Experimental analysis of somitogenesis in the chick embryo. 138, 82-97.

Christ, B., Jacob, HJ., Jacob, M. (1974). Somitogenesis in the chick embryo. Determination of Segmentation. Verh. Anat. Ges., 68, 573-579.

Ebendal, T. (1976). Migratory mesoblast cells in the young chick embryo examined by scanning electron microscopy. Zoon 4, 101-108.

Ebendal, T. (1977). Extracellular matrix fibrils and cell contacts in the chick embryo: Possible role in orientation of cell migration and axon extension. Cell Tissue Res. 175, 439-458.

Fisher, M., Solursh, M. (1977). Glycosaminoglycan localization and the role in maintenance of tissue spaces in the early chick embryo. J. Embryol. Exp. Morphol. 42, 195-207.

Gail, MH. Boone, CW. (1971). Effect of colcemid on fibroblast motility. Expl. Cell Res., 65, 221-227.

Hamburger, V., Hamilton, HL. (1951). A series of normal stages in the development of the chick. J. Morphol. 88, 49-92.

Hay, ED. (1968). Organization and fine structure of epithelium and mesenchyme in the developing chick embryo. In "Epithelial-Mesenchymal Interactions: 18th Hahnemann Symposium" (R. Fleishmajer,RE. Billingham, eds.), pp. 31-55. Williams & Wilkins, Baltimore, MD.

Hilfer, SR., Brady, RC. Yang, JW. (1981). Intracellular and extracellular changes during early ocular development in the chick embryo. In Ocular Size and Shape: Regulation During Development (eds. SR. Hilfer, JB. Sheffield), pp. 47-78. Springer-Verlag, New York.

Ho, E., Shimada, Y. (1978). Formation of the epicardium studied with the scanning electron microscope. Develop. Biol. 66, 579-585.

Humphreys, WJ., Spurlock, BO., Johnson, J. (1974). Critical point drying of ethanol-infiltrated, cryofractured biological specimen. Scanning Electron Microsc. 1974: 276-282.

Klee, CB., Crouch, TH., Richman, PG. (1980). Calmodulin. Annu. Rev. Biochem. 49, 489-515.

Langman, J. Nelson, GR. (1968). A radioautographic study of the development of the somite in the chick. J. Embryol. Exp. Morphol., 19, 217-226.

Lash, JW., Vasan, NS. (1978). Somite chondrogenesis in vitro stimulation by exogenous extracellular matrix components. Develop. Biol. 66, 151-171.

Lin, DC., Tobin, KD., Grumet, M., Lin, S. (1980). Cytochalasins inhibit nuclei-induced actin polymerization by blocking filament elongation. J. Cell Biol. 84, 455-460.

Lipton, BH., Jacobson, AG. (1974a). Analysis of normal somite development. Develop. Biol. 38, 73-79.

Lipton, BH., Jacobson, AG. (1974b). Experimental analysis of the mechanisms of somite morphogenesis. Develop. Biol. 38, 91-103.

MacLean-Fletcher, S., Pollard, TD. (1980). Mechanism of action of cytochalasin B on actin. Cell 20, 329-341.

Meier, S. (1979). Development of the chick mesoblast: Formation of the embryonic axis and establishment of the metameric pattern. Develop. Biol. 73, 25-45.

Meier, S., Jacobson, AC. (1982). Experimental studies of the origin and expression of metameric pattern in the chick embryo. J. Exp. Zool. 219, 217-232.

Nelson, GA., Revel, JP. (1975). Scanning electron microscopic study of cell movement in the corneal endothelium of the avian embryo. Develop. Biol. 42, 315-333.

Packard, Jr, DS. (1978). Chick somite determination: the role of factors in young somites and the segmental plate. J. Exp. Zool., 203, 295-306.

Packard, Jr, DS. (1980). Somitogenesis in cultured embryos of the Japanese quail, Coturnix coturnix, japonica. Amer. J. Anat. 158, 83-91.

Packard, Jr. DS. Jacobson, AG. (1976). The influence of axial structures on chick somite formation. Develop. Biol. 53, 36-48.

Packard, Jr, DS., Meier, S. (1983) An experimental study of the somitomeric organization of the avian segmental plate. Develop. Biol. 97, 191-202.

Platt, JB. (1889). Studies on the primitive axial segmentation of the chick. Harvard Univ. Museum Comp. Zool. Bull., 17, 171-190.

Sandor, S., Fazakas-Todea, I. (1980). Researches on the formation of axial organs in the chick embryo X. Further investigations on the role of ecto- and endoderm in somitogenesis. Rev. Roum. Morphol. Embryol. 26, 29-32.

Schroeder, TE. (1973). Cell constriction: contractile role of microfilaments in division and development. Am. Zool., 13, 949-960.

Solursh, M., Fisher, M., Meier, S., Singley, CT. (1979). The role of extracellular matrix in the formation of the sclerotome. J. Embryol. Exp. Morphol. 54, 75-98.

Trelstad, RL., Hay, ED., and Revel, JP. (1967). Cell contact during early morphogenesis in the chick embryo. Develop. Biol. 16, 78-106.

Wessells, NK., Spooner, BS., Ash, JF., Bradley, MO., Luduena, MA., Taylor, EL., Wrenn, JT., Yamada, KM. (1971). Microfilaments in cellular and developmental processes: Contractile microfilament machinery of many cell types is reversibly inhibited by cytochalasin B. Science 171, 135-143.

Williams, LW. (1910). The somites of the chick. Amer. J. Anat. 11, 55-100.

## Discussion with Reviewers

S.R. Hilfer: It has been suggested that oriented collagen fibers direct the migration of sclerotome cells toward the notochord and spinal cord. Do you find evidence for such a phenomenon?

Author: The extracellular matrix fibers (collagen with hyaluronic acid and chondroitin sulfate proteoglycan) direct sclerotome migration in the sense that the cells use the fibers as their physical support for migration. The perinotochordal matrix is a dense meshwork and except for a general extension of the fibrous mass between the notochord and somite, there was little indication of radial orientation in my own samples or the micrographs from other studies. In fact, sclerotome cell processes are seen on fine fibers oriented in any direction as the cells move in the general direction of the notochord. It should be noted that movement toward the notochord is only the first stage in sclerotome migration. As the process continues the sclerotome surrounds the spinal cord, and the matrix around the developing spinal cord (neural tube) is a dense mat of randomly-oriented fibers.

S. R. Hilfer: What is the current opinion of how spinal cord and notochord induce somite formation?

Author: The role of neural tube and notochord in somitogenesis is still a matter of active investigation. A number of studies have suggested that the shearing of the primary mesoderm by the regression of Hensen's node (notochord formation) is a key step in triggering the start of somite formation (Lipton and Jacobson, 1974a, b; Bellairs, 1979). Transplanted nodes or primitive streaks can induce supernumerary somites (summarized in discussion section of Packard and Meier, 1983). The implication is that nodal regression is associated with somitomere formation. Nothing is yet known about the nature of the interaction.

Contact with notochord or neural tube may be important in other aspects of somite formation. Continued contact of newly formed anterior somites with the axial structures is required for stability of the somites. The more posterior somites form is a "more advanced" stage than the anterior somites with regard to morphology, stability when separated from the embryonic axis, and chondrogenesis in vitro. This could result from the relatively longer exposure of posterior somitomeres to the axial structures or their extracellular matrix. It should be pointed out that the overlying ectoderm and endoderm and their matrix material may also interact with forming somites and could be involved in the same processes.

Additional discussion with reviewers of the paper "Selected Views of Early Heart Development by Scanning Electron Microscopy" by D.A. Hay, R.R. Markwald and T.P. Fitzharris continued from page 190.

T. Pexieder: What is the role of cardiac jelly in looping?
Authors: Presently, the exact role of cardiac jelly in looping is unknown.

T. Pexieder: Please relate your bulbus cordis to the terms conus and truncus which are frequently used in clinical cardiology.
Authors: The bulbus cordis is generally considered to consist of three parts: A) The proximal portion becomes part of the right ventricle, distal to the ventricular trabeculations. B) The middle portion is the conus, which encompasses the semilunar valves and the roof (outlet) of the right ventricle. C) The distal, elongated portion is the truncus arteriosus, which becomes the root and proximal part of the aorta and the pulmonary artery.

T. Pexieder: What evidence do you have for AP septal cells' contribution to the membranous portion of the interventricular septum?
Authors: This is the traditional viewpoint as proffered by the works of Van Mierop and Patten. (Patten BM. (1968). in: Human Embryology, 3rd ed., McGraw-Hill Book Co., New York, 171-173.) However, this doesn't mean that this is a totally correct view, but the purpose of this manuscript is that of illustrating introductory cardiac morphogenetic events, not discussing the mechanisms of development.

D.E. Morse: Do you feel that the IA septum orchestrates the spatial growth of the cushions?
Authors: It does appear that the IA septum may influence the direction of atrioventricular cushion growth, but it most likely does not initiate nor control the rate of growth.

D.E. Morse: Are there any congenital anomalies in which the endocardial cushions meet and fuse when a defective septum primum is present?
Authors: Yes there are. However, it should be emphasized that atrial septal defects (ASD) most frequently occur in the attenuated portion of the interatrial septum (IAS), without involving the thick, basal portion with which the AV cushions fuse. In fact, there is an interatrial communication known as the "ostium primum type of ASD" that does not correspond in location to the embryonic ostium primum, but to that of the AV septum, which includes both the basal IAS as well as the AV cushions. Moreover, endocardial cushions defects (ECD) may also be associated with the more typical ASD's, while other ECD's exist in which the atrial septum is normally developed and complete.

G.C. Schoenwolf: What is the evidence that the coronary arteries arise from the epicardium, rather than by sprouting from the aortic arches or aortic sac?
Authors: The studies by Manasek (1971) and by Viragh and Challice (1981) indicate that the earliest recognizable coronary vessels are located adjacent to the epicardium and only later on are found in continuity with the aorta. (Manasek FJ. (1971). The ultrastructure of embryonic myocardial blood vessels. Devel. Biol. 26, 42-54). (Viragh S, Challice CE. (1981). The origin of the epicardium and the embryonic myocardial circulation in the mouse. Anat. Rec. 201, 157-168).

SELECTED VIEWS OF EARLY HEART DEVELOPMENT BY SCANNING ELECTRON MICROSCOPY

Don A. Hay[1]*, Roger R. Markwald[2] and Timothy P. Fitzharris[3]

[1]Department of Biology, Stephen F. Austin State University, Nacogdoches, TX 75962. [2]Department of Anatomy, School of Medicine, Texas Tech University, Lubbock, TX 79430. [3]Department of Anatomy, Medical University of South Carolina, Charleston, SC 29403

(Paper received May 21 1983, Completed manuscript received August 18 1984)

## Abstract

This tutorial on cardiac development is designed to acquaint the novice student of embryology with the key events that occur during cardiogenesis. Each of the following events is depicted through a series of scanning electron micrographs which convey the spatial relationships between minute, yet essential structures: fusion of paired heart tubes; looping; partitioning of the common atrium, the atrioventricular canal, the primitive ventricle and the outflow tract. The cellular and biochemical mechanisms responsible for these events cannot be adequately determined nor illustrated by SEM and therefore will not be considered in detail herein.

KEY WORDS: Heart development, Fusion of heart tubes, Cardiac septation, Outflow tract.

*Address for correspondence:
Don A. Hay, Department of Biology, Box 13003, Stephen F. Austin State University, Nacogdoches, Texas 75962.    Phone no.: (409) 569-3601.

## Introduction

Heart development involves a series of complex, interdependent morphogenetic events including cell migration, epithelial fusion and cell shape changes (Manasek, et al., 1972; Manasek, 1976). In the chick and mammals these processes transform a pair of primitive heart tubes into a four-chambered structure having two sets of valves which separate pulmonary blood flow from systemic flow. Although not the first organ to form, it is the first to acquire function. Perhaps due to its morphogenetic complexity and early functional status, heart malformations are among the most frequent of birth defects (Pexieder, 1980).

### Fusion of Paired Heart Tubes and Looping

Heart development is closely linked to the development of the coelom or body cavity. Following the segregation of the blastomeres into the three basic germ layers, the mesoderm lying at the periphery of the embryonic shield (termed lateral plate mesoderm) "splits" (i.e., hollows out) to form a cavity within the embryo lined by mesoderm. That portion of the "split" mesoderm which contacts the endoderm of the foregut is termed splanchnic mesoderm. In the anterior end of the embryo, the foregut endoderm influences the associated splanchnic mesoderm to differentiate into cardiac tissues (Jacobsen and Duncan, 1968; Lemanski, et al., 1979). The rapid growth of the nerve tube in comparison to the rest of the embryo brings the heart into the position shown in Figure 1.

If the splanchnic mesoderm is removed as in Figure 2, the cardiac tissues are revealed as a pair of hollow tubes. These are separated from each other by the developing foregut and associated splanchnic mesoderm. In slightly younger embryos (Figures 3,4), the tubes are more widely separated with each tube consisting of two epithelia separated by an acellular expanse called "cardiac jelly". Thus each heart tube is actually a tube within a tube. The inner tube of epithelium is termed the endothelium or endocardium (Pexieder, 1981); the outer tube is the myocardium. The intervening cardiac jelly is secreted by the myocardium and contains a variety of extracellular macromolecules including collagen, hyaluronate, and glycoproteins such as fibronectin (Johnson et al., 1974; Hurle et al., 1980; Icardo and Manasek, 1983).

Figure 1. Ventral view of a 27 hr chick embryo having 8 somites (an early Hamburger-Hamilton stage 9). The developing heart region (asterisks) is covered by splanchnic mesoderm. Arrowhead denotes anterior end of the closed neural tube (N). Caudal to the anterior intestinal portal (arrow) the paired dorsal aortae (DA) are seen on either side of the neural tube (N). Bar = 300 μm.

Figure 2. Transverse section between the two asterisks of the same embryo shown in Figure 1. The paired heart tubes (asterisks lie in their lumens) can be seen beneath the foregut endoderm (FGE). The arrowheads denote the columnar epithelium of the ventral gut endoderm which is in close contact with the developing endocardium (E) and myocardium (M). These endodermal cells have been shown to produce glycoproteins which may initiate differentiation of splanchnic mesoderm into cardiac tissue. N (neural tube), S (somites). Bar = 75 μm.

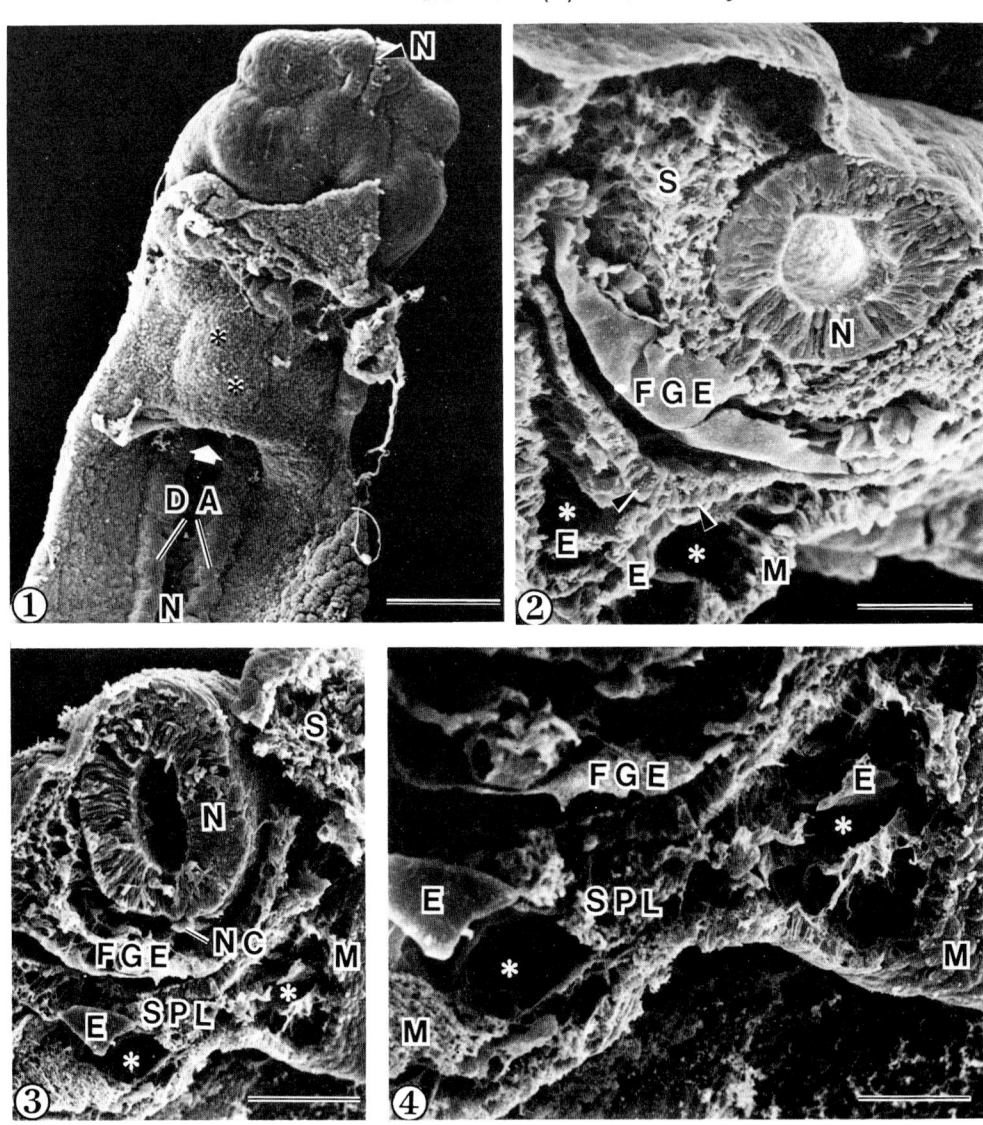

Figure 3. Transverse cut through the forming heart region of an embryo slightly younger than that of Figure 1 (approximately 25 hr or a Hamburger-Hamilton stage 8). Note the paired heart tubes (asterisks) are more widely separated from each other by splanchnic mesoderm (SPL) associated with the foregut endoderm (FGE). E (endocardium), M (myocardium), S (somite), NC (notochord). Bar = 75 μm.

Figure 4. Higher magnification of the previous figure. Same captions as before. Bar = 30 μm.

The fate of the paired heart tubes is to fuse into a single, straight tube termed the primary or simple tubular heart (Figure 5). The tube-within-a-tube arrangement is preserved after fusion of the two endothelia (Figure 6). The hydration characteristics of the cardiac jelly add turgor to the wall of the heart but additionally the macromolecules of cardiac jelly may function by their alignment or chemical composition to permissively or actively coordinate subsequent morphogenetic events (Manasek, 1976; Runyan and Markwald, 1983). One such event is "looping". During looping, the heart bends upon itself to form a U-shaped organ that foreshadows the adult pattern and establishes the fundamental spatial relationship upon which all future development hinges (Manasek, 1976). Looping is a genetic event which occurs even if the heart is removed from the embryo and seems to result from an interaction of cardiac

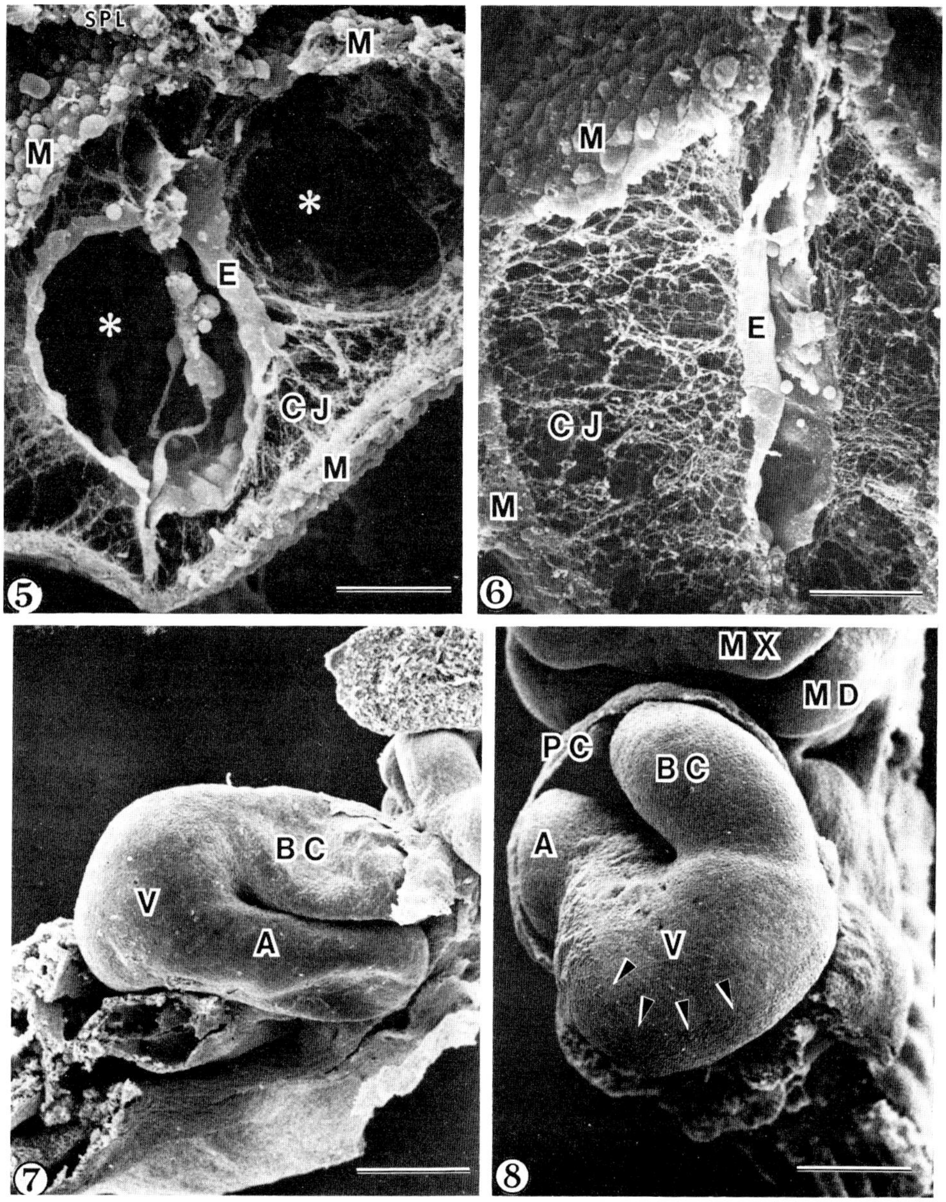

SEM in Cardiac Development

Figure 5. Late Hamburger-Hamilton stage 9 embryo (30 hr). Fusion of the two heart tubes (asterisks) has begun. Splanchnic mesoderm (SPL) at the top of the micrograph contributes to the transitory dorsal mesocardium. Note alignment of the fibers in the cardiac jelly (CJ). The fusion of the myocardial tubes occurs prior to and independent of endocardial fusion. The absence of a complete endocardium (E) in the right heart tube is probably an artifact of processing. M (myocardium). Bar = 35 μm.

Figure 6. The primary or simple tubular heart formed after fusion is complete (i.e., approximately 33 hr or a Hamburger and Hamilton stage 10). Embryonic hematopoietic cells are seen in the lumen lined by endothelial-like endocardium (E). Extracellular macromolecules of the cardiac jelly (CJ) extend from the myocardium (M) to the endocardium. Bar = 25 μm.

Figure 7. A Hamburger-Hamilton stage 15 embryo of 55 hr dissected to reveal the "looped" heart. Looping occurs between the primitive ventricle (V) and the bulbus cordis (BC) creating a crease termed the bulboventricular sulcus. A (atrium). Note: The negatives for figures 7 and 8 were inadvertently inverted during printing and subsequently lost. The correct orientation would have the atrium and the bulbus cordis reversed, i.e., the atrium to the embryo's left, the bulbus cordis to its right. Bar = 250 μm.

Figure 8. A Hamburger-Hamilton stage 18+ embryo of 72 hr dissected to reveal the heart after looping is complete. Note the transected pericardial cavity (PC) derived from splanchnic mesoderm. The arrowheads denote the epicardial sheet partially extended across the surface of the ventricle (V). MX (maxillary process), MD (Mandibular process), A (atrium), V (ventricle), BC (bulbus cordis). Bar = 300 μm.

jelly components and the myocardium (Manasek and Monroe, 1972; Manasek et al., 1972). The bend or "crease" forms between the future ventricle and the outflow tract or bulbus cordis (Figure 7). Regional dilations and constrictions accompany or follow the looping process serving to demarcate the primitive chambers (Figure 8). Close inspection of the ventricular surface (Figure 8) will reveal a third epithelial tissue, the epicardium, which appears initially as a sheet of migratory cells that extend across the myocardium (Ho and Shimada, 1978). Like the endocardium and myocardium the epicardium is a derivative of splanchnic mesoderm which will subsequently give origin to the connective tissue and coronary blood vessels of the heart.

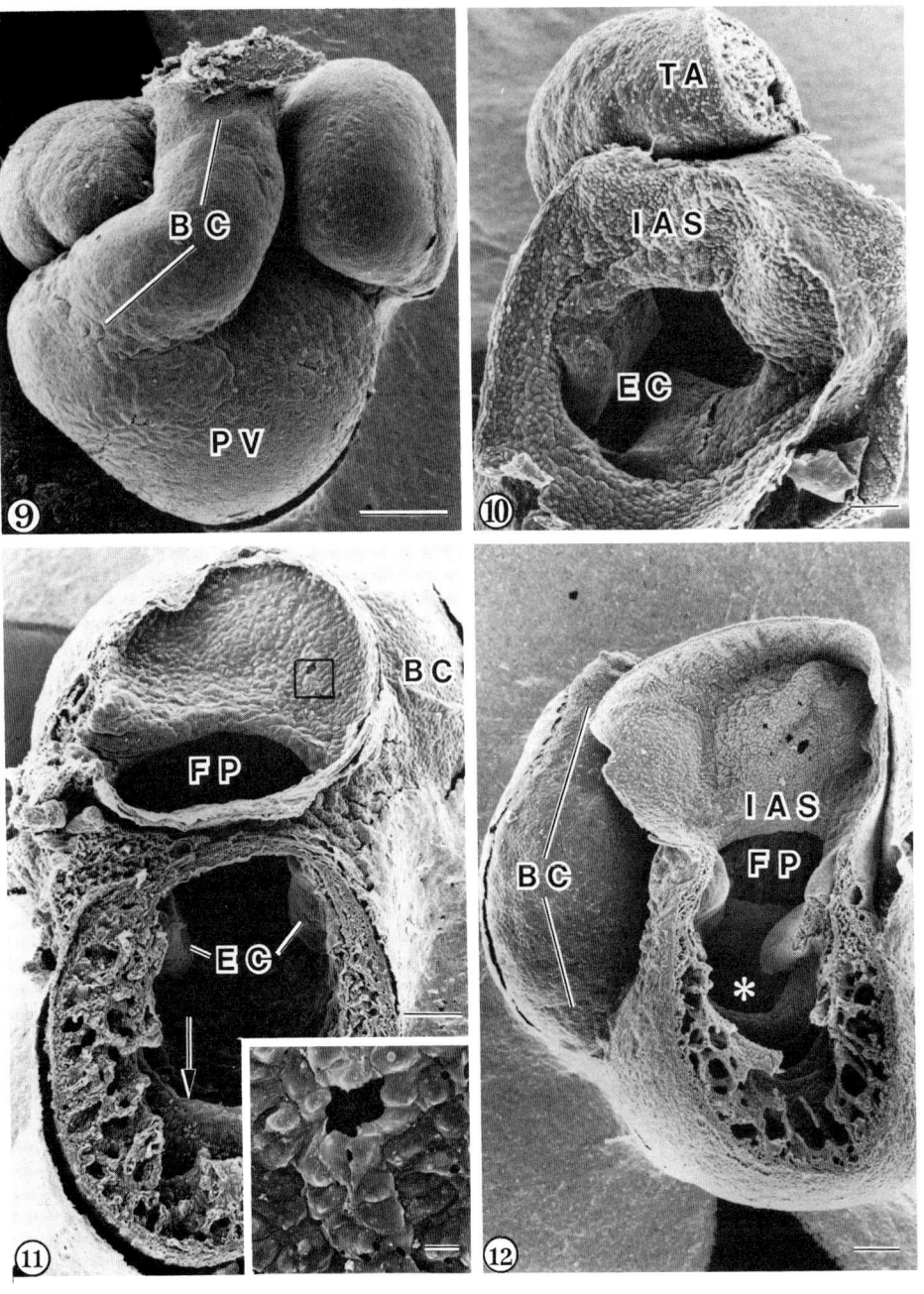

Figure 9. Embryonic chick heart, stage 25, 4 1/2 days' incubation. Demarcation of the primitive ventricle (PV) into right and left chambers by an interventricular sulcus is not yet apparent. The outflow tract, or bulbus cordis (BC), may be seen anterior to the right and left atria. Bar = 100 μm.

Figure 10. Embryonic heart, left supero-lateral view of the AV canal, stage 24, 4 days. The interatrial septum (IAS) arches above the endocardial cushions (EC), but is continuous with the base of each cushion. The opening formed by the interatrial septum and the AV cushions is called the foramen primum. The truncus arteriosus (TA), which is the distal segment of the bulbus cordis, will form the roots and proximal portions of the aorta and pulmonary artery. Bar = 100 μm.

Figure 11. Embryonic heart, right lateral view, stage 24, 4 1/2 days. The foramen primum (FP) diminishes in size as interatrial septal growth continues towards the AV cushions (EC). The early muscular IV septum (arrow) emerges as the most prominent of several concave ridges. Bar = 100 μm. Inset: Enlargement of boxed area. Tiny perforations coalesce in the interatrial septum to form the foramina secunda. Bulbus cordis (BC). Bar = 10 μm.

Figure 12. Left lateral view stage 25, 4 3/4 days. The interatrial septum (IAS) exhibits multiple foramina. The ridge visible behind the foramen primum (FP) is the lateral wall of the right AV canal. The opening (*) beneath the AV cushions is the IV foramen primum. Note the trabeculated ventricular wall. BC (bulbus cordis). Bar = 100 μm.

## Partitioning of Common Atrium, AV Canal and Primitive Ventricle

Blood enters the U-shaped tube through a saclike confluence of veins called the sinus venosus and exits through a series of vessels called aortic arches. Between the sinus venosus and aortic arches, the walls of the heart begin to expand laterally in two distinct regions: that portion immediately adjacent to the sinus venosus forms an atrium, while the region at the base of the "U" forms the primitive ventricle (Figure 9). The distal part of the tube does not undergo extensive expansion, but becomes the definitive outflow tract, termed the bulbus cordis. The lumen of each chamber gradually becomes subdivided by muscular and/or membranous partitions called septa.

Figure 13. Interior view of the AV canal (*) and roof of right ventricle, Stage 25, 5 days. Note: The same view in a human embryonic heart would be considered a frontal view. Arrow indicates direction of blood flow from the right ventricle into the bulbus cordis (BC). Leading edge of interatrial septum (to right of asterisk) may be seen between the endocardial cushions, while the prominent bulboventricular fold (to left of asterisk) approaches the right side of the AV cushions. Bar = 100 μm.

Figure 14. Left lateral view, stage 25+, 5 1/4 days. The interatrial septum (IAS) has reached the AV canal, thus closing the foramen primum. However, blood still enters the left atrium through the numerous perforations in the septum (foramina secunda). BC (bulbus cordis). Bar = 100 μm.

Figure 15. Right lateral view, stage 25+, 5 1/4 days. Venous blood enters the right atrium from the sinus venosus (dotted arrow). The interventricular foramen primum (*) is surrounded by the AV cushions above and the concave crest of the muscular IV septum below. Blood from the right ventricle enters the bulbus cordis (curved black arrow) and passes distally to the aortic arches, where the aorticopulmonary septum (B/W arrow) divides the outflow tract. Bar = 100 μm.

Figure 16. Left lateral view, stage 26, 5 1/2 days. Fat arrows reveal line of demarcation between left atrial wall and interatrial septum. Septal strands have thickened, reflecting a trend that continues through hatching. The interatrial septum is completely joined to the partially fused AV cushions (clear arrow). Muscular IV septum (B/W arrow) is aligned with the right tubercle of the dorsal AV cushion. BC (bulbus cordis). Bar = 100 μm.

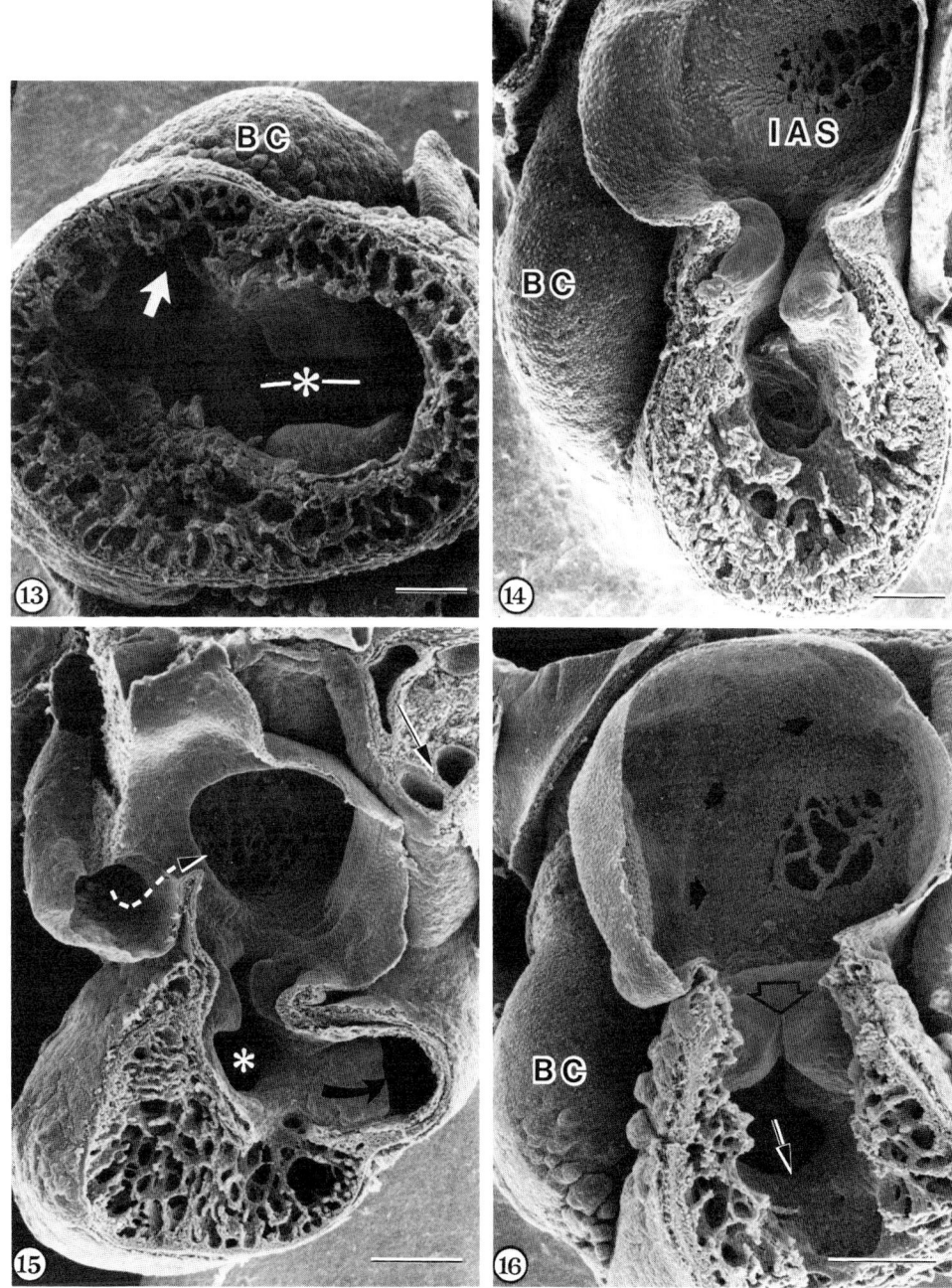

Partitioning or septation occurs as somewhat distinct, yet concomitant processes in four different regions of the heart: the atrium, the ventricle, the bulbus cordis (outflow tract) and the junction of the atrium and ventricle (AV canal). The interatrial (IA), interventricular (IV) and bulbar septa all converge at the AV canal where they eventually join their AV canal counterpart. For purposes of clarity, septation of each region will be considered individually, although the interrelationship of septa necessitates a certain degree of duplication.

Atrium

Early in the third day of embryonic development in the chick (beginning of the fifth week in the human), the IA septum (septum primum) first appears as a narrow ridge in the postero-superior wall of the atrium. As the septum grows toward the AV canal, its concave leading edge becomes continuous anteriorly and posteriorly with the bases of the respective gelatinous, block-like structures called AV endocardial cushions (Figure 10). The opening circumscribed by the IA septum and the AV cushions is termed the foramen (or ostium) primum, but it is temporary, diminishing in size and finally disappearing when the free edge of the septum fuses with the AV cushions (Figures 11, 12, 14, 15, 16). Prior to completion of the septum primum, tiny perforations (Figure 11) form, coalesce and enlarge (Figs. 12, 14, 15), resulting collectively in a foramen secundum. The formation of the foramen secundum is of extreme functional importance, since this enables blood entering the right atrium from the sinus venosus (Figure 15) to flow into the left atrium, thus maintaining the distribution of blood to both sides of the heart. As chick cardiac development continues, the strands separating perforations progressively thicken, obliterating some of the smaller holes (Figure 16). At hatching, adjacent strands fuse to one another, thereby occluding the remaining foramina and establishing intact, separate flow patterns in right and left sides of the heart. For a more thorough description of the chick interatrial septum formation see Hendrix and Morse (1977), and Morse and Hendrix (1980). Development of the interatrial septum in the human differs significantly from that in the chick with regard to the mechanism by which the foramen secundum is closed. For a discussion of human interatrial septation, see Van Mierop (1969 and 1979) and Langman (1981).

AV Canal

As the primitive atrium and ventricle begin to expand, that portion of the tubular heart between the two chambers (interconnecting them) remains relatively constricted and is therefore called the atrioventricular (AV) canal. At 3 1/2 days' incubation in the chick (the end of the fourth week in humans), gelatinous pads of mesenchyme called AV endocardial cushions emerge from the anterior and posterior walls of the AV canal (Figures 10 and 11). During the next 2 1/2 days the opposing cushions grow toward one another until they fuse (Figures 12-18), thereby dividing the AV canal into right and left orifices. While the endocardial cushions expand into the lumen, the basal portion of the IA septum acquires continuity with the cushions soon after their appearance (Figure 10) by progressively fusing with the superior aspect of both cushions, thus bridging the AV canal (Figures 12, 13, 14, 15). The intimate relationship of the basal IA septum with the AV cushions (Figures 15 and 16) in effect aligns the opposing cushions, virtually ensuring that they will meet one another despite the constant movement of the heart. When the fusion of the IA septum to the AV cushions is complete (during the fifth day in chicks, fifth week in humans), the foramen primum is obliterated and the faces of apposing cushions begin to fuse (Figure 16). Such fusion occurs sequentially from the atrium toward the ventricle (Figures 16 and 17). The resulting arched structure that spans the canal is called the septum intermedium (Figure 18). The left portion of the septum intermedium eventually becomes remodeled to contribute to the anterior cusp of the mitral (bicuspid) valve while part of the right side of the septum becomes incorporated into the membranous interventricular septum and serves as the posterior wall of the aortic vestibule. For a more detailed explanation of AV cushion development and fusion, see Hay and Low (1972), Hay (1978) and Van Mierop (1979).

Ventricle

Unlike the active processes of septation that occur in the atrium and the AV canal, ventricular septation is more accurately considered a passive process. Instead of actively growing toward the AV canal, as does the free edge of the interatrial septum (septum primum), the crest of the muscular interventricular (IV) septum maintains a relatively constant distance from the AV cushions (150-200 $\mu$m) while the base of the septum elongates. This is accomplished by the lateral expansion and trabeculation of the right and left ventricular myocardium away from the septum, thereby increasing the length and diameter (and therefore volume) of the chambers.

During the third and fourth day of embryonic growth, several crescent shaped ridges appear in the primitive ventricle (Figure 12). One of the ridges, aligned beneath the right edge of the AV cushions, becomes more prominent that the other (due to the apposition of adjacent trabeculae), and becomes the muscular IV septum (Figures 11 and 12) (Harh and Paul, 1975). The crest of the septum becomes infiltrated with endocardial cushion mesenchyme and extracellular matrix similar to that found in the AV canals. The opening circumscribed by the crest (Figure 12) and the bulboventricular fold (Figure 13), through which blood from the primitive left ventricle flows into the developing right ventricle, is termed the primary IV foramen. As the heart develops, the primary IV foramen persists (Figures 15-18) and gradually enlarges, becoming continuous with the root of the aorta (aortic vestibule; Figure 13) due to the migration and contribution of the proximal portion of the aorticopulmonary septum to the membranous IV septum. (Pexieder, 1978; Van Mierop, 1979).

Figure 17. Left lateral view, stage 27, 5 3/4 days, greater right atrial pressure forces interatrial strands into left atrium. The sinus venosus (SV) extends across the posterior wall of the left atrium, but empties only into the right atrium (see Figure 15). The septum intermedium (SI) resulting from the fusion of the AV cushions arches above the IV foramen primum (*). Bar = 100 μm.

Figure 18. Left lateral view, stage 28, 6 days. Plane of section is such that the muscular IV septum (IVS) is presented in profile and in cross-section, revealing both ventricular chambers (RV = right ventricle). Note the continuity of fused cushions (SI) with the muscular interventricular septum. BC (bulbus cordis). Bar = 100 μm.

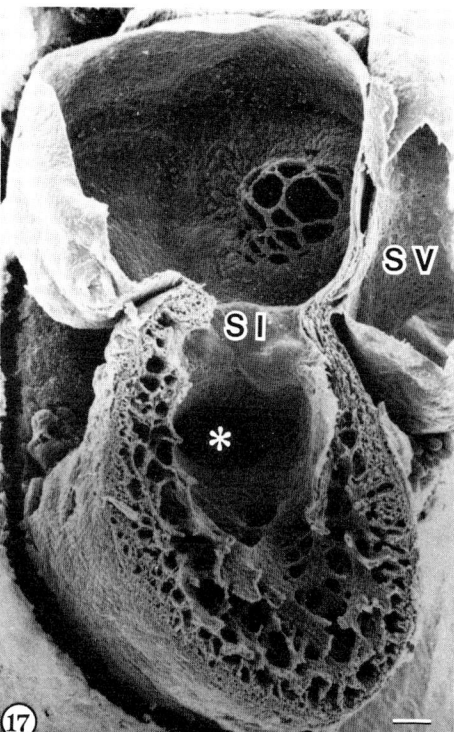
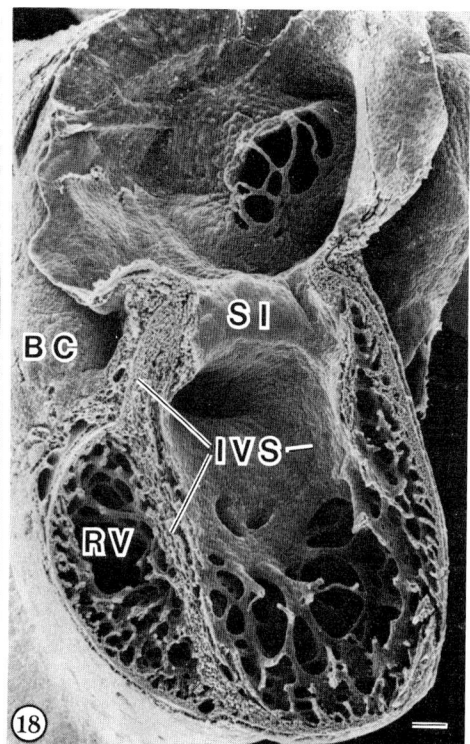

The AV endocardial cushions contribute to the anterior leaflet of the mitral valve, but the bulk of the cusps of both AV valves form from myocardium by the undercutting and remodeling of the muscular IV septum as well as the lateral myocardial walls.

### Division of the Outflow Tract

As the heart is becoming partitioned internally, the outflow tract (OT) which connects the primitive ventricle to the aortic arches is also preparing to undergo its morphogenetic movements which will establish definitive pulmonary and systemic circuits in the adult. This hollow tube (Fig. 19) will undergo a longitudinal

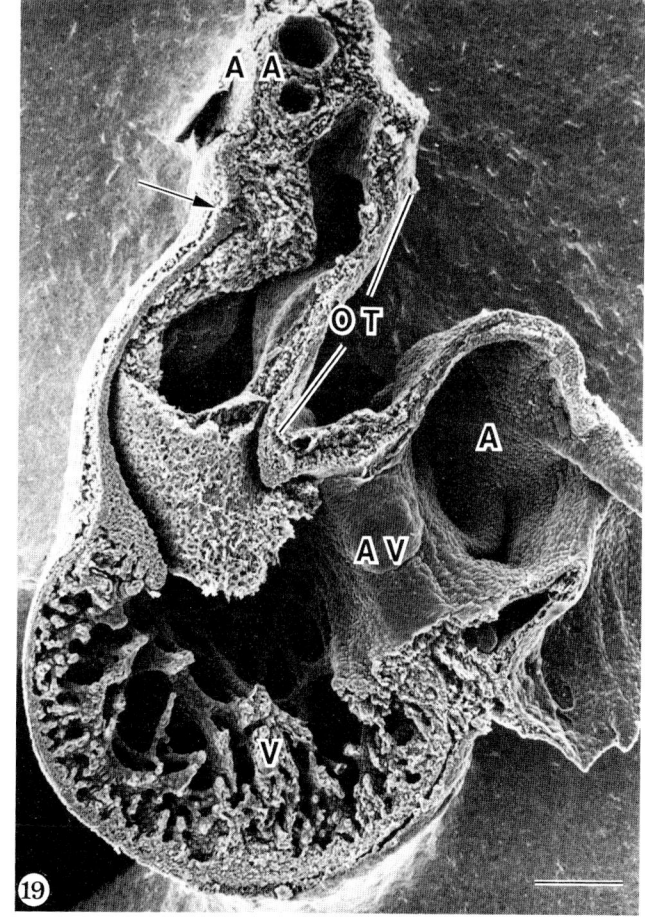

Figure 19. A low power survey photograph of an embryonic chick heart (stage 26, 5 days) demonstrates the relative positions of the atrium (A), ventricle (V), lateral wall of atrioventricular canal (AV), outflow tract (OT), and aortic arch (AA) regions. The arrow indicates the limit of the myocardial sheath which surrounds the outflow tract at this stage of development. The conotruncal ridges are easily seen in the lumen of the outflow tract. Bar = 100 μm.

division into two major vessels: the aorta and pulmonary artery which will connect with the left and right ventricles respectively (Kramer, 1942; Jaffe, 1967). This seemingly straightforward division is complicated by a 180° repositioning of the vessels which is accomplished by a spiral set of ridges comprised of cushion tissue and a specialized subpopulation of this tissue called the aortico-pulmonary (AP) septum. Adjacent to the aortic arch region, a proliferation of cushion tissue occurs which presages the formation of the spiral ridges which change position (i.e., rotate around an axis) as they progress down the outflow tract (OT) towards the ventricle (Figs. 20, 21, 22). In favorable orientations, a condensed population of mesenchyme is seen (Fig. 23) which in 3-dimensional array would resemble a boomerang tucked in the groove of the division of the major vessels with the arms directed toward the myocardial sleeve which ensheaths the outflow tract at this stage of development (Thompson and Fitzharris, 1979a, 1979b). Without implicating a cause/effect relationship, the following events occur simultaneously during proper division of the OT: the sheath of myocardium which was adjacent to the aortic arches will retract the entire length of the OT. Concurrently the

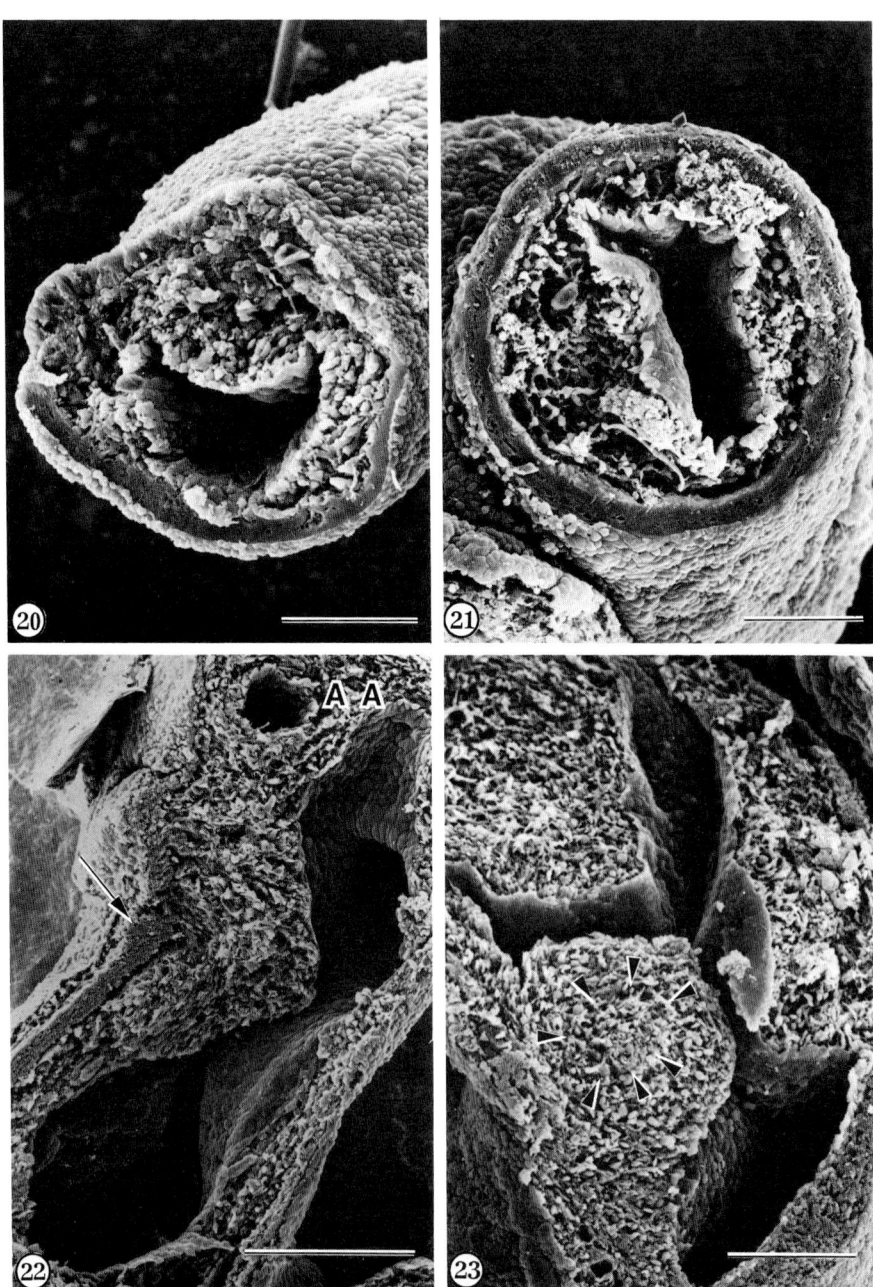

Figures 20 and 21. Cross-sections of similarly staged embryos adjacent to the aortic arch region and at the midway point respectively. Note that the lumen is not concentric with the outer myocardial sheath, but rather has a distinctive ridge which clearly shifts position along the length of the outflow tract. Bar = 100 μm.

Figure 22. A higher magnification of Fig. 19 which shows the relationship between the two major lumens and the crests of the spiral ridges which will begin fusing at the point of the myocardial sheath (arrow). The fusion will extend slightly downstream towards the aortic arches (AA) and eventually traverse the entire upstream position of the outflow tract. Bar = 100 μm.

Figure 23. A medium power cross section of the outflow tract which is about halfway through the septation process. Note the two ridges which are in the process of fusing as well as the region of condensed mesenchyme which constitutes the AP septum (B/W arrows encircle area). Bar = 100 μm.

Figure 24. The final consequence of division of the outflow tract in the aortic arch region is the establishment of the major vessels which feed the lungs and systemic circulation. Bar = 100 μm.

Figure 25. The final consequence of division of the outflow tract in the ventricular region is the establishment of pulmonary and systemic circuits via the pulmonary artery and aorta respectively. The exits for these vessels are guarded by the semilunar valves (SV) which derive their physical support from the AP septum (AP) and adjacent fibrous rings. Bar = 100 μm.

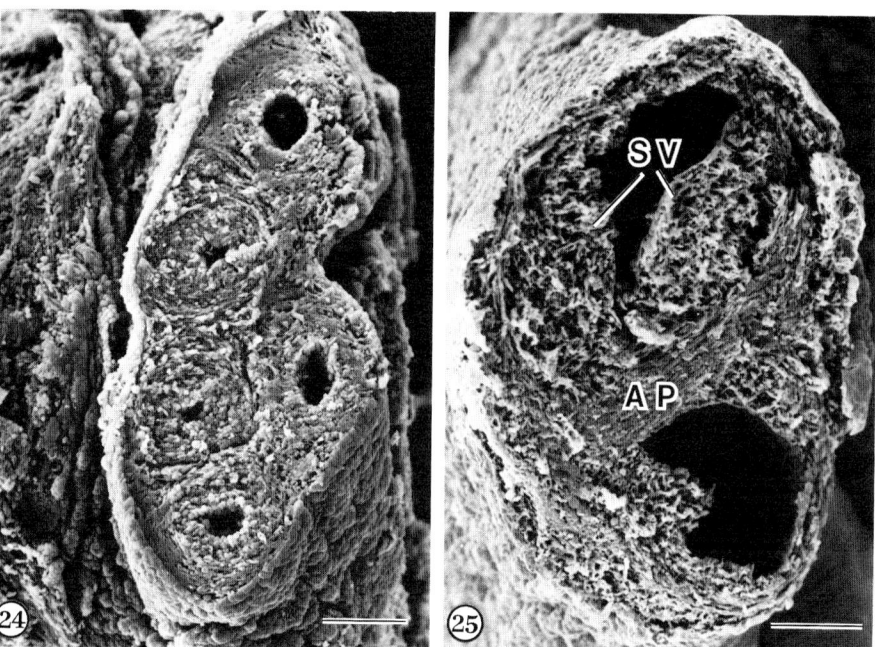

spiral ridges will fuse along their crests creating a bifurcation which will create the aorta and pulmonary artery. The AP septum traverses the length of the truncus arteriosus in conjunction with the retraction of the myocardium and also undergoes a 180° rotation so that the "arms" of the AP septum change their relative position internally. This complete morphogenetic movement leaves the 5 major vessels intact at the arches (Fig. 24) and completely divides the OT at the mouth of the ventricle into aorta and pulmonary artery (Fig. 25). A portion of the AP septum remains as the conus ligament between these vessels and provides the internal architectural framework for the cardiac skeleton which will support the semilunar valves that guard these exit sites. A smaller portion of the AP septum will penetrate into the ventricular region and contribute cells to the membranous portion of the interventricular septum (see section on AV canal). It is clear from this analysis that any interference with this process can lead to a congenital heart defect, such as transposition of the great vessels in which the AP septum forms but the relative 180° shift of tissue does not occur and the vessels are connected to the wrong left and right ventricular chambers. For additional reading, see de la Cruz, et al., (1977), Pexieder (1978) and Rychter (1978).

## Acknowledgements

The authors wish to thank Ms. Alexis P. Sage and Maki Harman for valuable technical assistance and Mr. Bret Bannon and Ms. Anita Cummings for typing the manuscript. This work was supported in part by an American Heart Association - Florida Affiliate Grant-In-Aid (AG128) from the Heartland Field Area, to Dr. Hay and by an NIH grant (HL19136) to Dr. Markwald.

## References

de la Cruz MV, Sanchez C, Arteaga MM and Arguello C. (1977). Experimental study of the development of the truncus and the conus in the chick embryo. J. Anat. 123, 661-686.

Harh JY, Paul MH. (1975). Experimental cardiac morphogenesis. I. Development of the ventricular septum in the chick. J. Embryol. Exp. Morph. 33, 13-28.

Hay DA. (1978). Development and fusion of the endocardial cushions, in: Morphogenesis and Malformation of the Cardiovascular System. G.C. Rosenquist and D. Bergsma (eds.) Alan R. Liss, Inc. NY, Birth Defects: Original Article Series, XIV, 69-90.

Hay DA, Low FN. (1972). The fusion of dorsal and ventral endocardial cushions in the embryonic chick hearts: A study in fine structure. Am. J. Anat. 133, 1-24.

Hendrix MJC, Morse DE. (1977). Atrial septation. I. Scanning electron microscopy in the chick. Devel. Biol. 57, 345-363.

Ho E, Shimada Y. (1978). Formation of the epicardium studied with the scanning electron microscope. Devel. Biol. 66, 579-585.

Hurle JM, Icardo JM and Ojeda JL. (1980). Compositional and structural heterogeneity of the cardiac jelly of the chick embryo tubular heart: A TEM, SEM and histochemical study. J. Embryol. Exp. Morph. 56, 211-223.

Icardo JM, Manasek FJ. (1983). Fibronectin distribution during early chick embryo heart development. Devel. Biol. 95, 19-30.

Jacobsen AG, Duncan TT. (1968). Heart induction in salamanders. J. Exp. Zool. 167, 365-375.

Jaffe OC. (1967). The development of the arterial outflow tract in the chick embryo heart. Anat. Rec. 158, 35-42.

Johnson RC, Manasek FJ, Vinson WC and Seyer JM. (1974). The biochemical and ultrastructural demonstration of collagen during early heart development. Devel. Biol. 36, 252-271.

Kramer TC. (1942). The partitioning of the truncus and the formation of the membranous portion of the interventricular septum in the human heart. Am. J. Anat. 71, 343-370.

Langman J. (1981). Medical Embryology, 4th ed. Williams and Wilkins, Co., MD. 157-183.

Lemanski LF, Paulson DJ and Hill CS. (1979). Normal anterior endoderm corrects the heart defect in cardiac mutant salamanders (Ambystoma mexicanum) Science 204, 860-862.

Manasek FJ. (1976). Heart development: Interactions in cardiac morphogenesis. in: The Cell Surface in Animal Embryogenesis and Development. G. Post and G.L. Nicholson (eds.), Elsevier-North Holland, Amsterdam, 545-598.

Manasek FJ and Monroe RG. (1972). Early cardiac morphogenesis is independent of function. Devel. Biol. 27, 584-588.

Manasek FJ, Burnside MB and Waterman RE. (1972). Myocardial cell shape changes as a mechanism of embryonic heart looping. Devel. Biol. 29, 349-371.

Morse DE, Hendrix MJC. (1980). Atrial septation II. Formation of the foramina secunda in the chick. Devel. Biol. 78, 25-35.

Pexieder T. (1978). Development of the outflow tract of the embryonic heart. in: Morphogenesis and Malformation of the Cardiovascular System, G.C. Rosenquist and D. Bergsma (eds.), Alan R. Liss, Inc. NY, Birth Defects: Original Article Series, XIV, 29-68.

Pexieder T. (1980). Cellular abnormalities leading to congenital heart disease. in: Pediatric Cardiology. Vol 4. M.J. Godman (ed.), Churchill Livingstone, London, 24-32.

Pexieder T. (1981). Prenatal development of the endocardium: a review. Scanning Electron Microscopy 1981;II:223-253.

Runyan RB, Markwald RR. (1983). Invasion of mesenchyme into three-dimensional collagen gels: A regional and temporal analysis of interaction in embryonic heart tissue. Devel. Biol. 95, 108-114.

Rychter Z. (1978). Analysis of relations between aortic arches and aorticopulmonary septum. in: Morphogenesis and Malformation of the Cardiovascular System, G.C. Rosenquist and D. Bergsma (eds.), Alan R. Liss, Inc. NY, Birth Defects: Original Article Series, 14(7), 443-448.

Thompson RP, Fitzharris TP. (1979a). Morphogenesis of the truncus arteriosus of the chick embryo and heart: The formation and migration of mesenchymal tissue. Am. J. Anat. 154, 545-556.

Thompson RP, Fitzharris TP. (1979b). Morphogenesis of the truncus arteriosus of the chick embryo heart: Tissue reorganization during septation. Am. J. Anat. 156, 251-264.

Van Mierop LHS. (1969). Embryology of the heart. in: The CIBA Collection of Medical Illustrations, The Heart. F.H. Netter, (ed.), Pharmaceutical Co., Summit NJ, 5, 112-130.

Van Mierop LHS. (1979). Morphological development of the heart. in: Handbook of Physiology - The Cardiovascular System I. Williams and Wilkins, Co. Baltimore, Maryland, 1-28.

### Discussion with Reviewers

T. Pexieder: Is the myocardium the only source of cardiac jelly?
Authors: No. Although the myocardium appears to be the major source of cardiac jelly, cells of the ventral surface of the foregut can and probably do secrete into the cardiac jelly space.

T. Pexieder: Please explain what you mean by "...looping is a genetic event..."?
Authors: Manasek and Monroe (1972) demonstrated that embryonic hearts exposed to high potassium concentrations still underwent looping despite the lack of blood flow. Thus, a hemodynamic factor does not appear to be necessary. Rather the differentiation of the myocardium may be influential (Manasek, et al., 1978) (Manasek FJ, Kulikowski RR and Fitzpatrick L.(1978)). Cytodifferentiation: A causal antecedent of looping? in: Morphogenesis and Malformation of the Cardiovascular System, G.C. Rosenquist and D. Bergsma (eds), Alan R. Liss, Inc. NY, Birth Defects: Original Article Series, 161-178.

---

For additional discussion see page 180.

PRENATAL DEVELOPMENT OF THE ENDOCARDIUM: A REVIEW

Tomas Pexieder

Institute of Histology and Embryology
University of Lausanne
Rue du Bugnon 9
CH - 1011 Lausanne CHUV
Switzerland

## Abstract

The chronology of SEM studies of the embryonic endocardium is followed in this review by discussion of species, stages and localizations studied. In reviewing the methodology of SEM studies of the embryonic endocardium, particular weight is given to standard methods which can be applied to all species of interest. Two main aspects are more deeply analysed: the perfusion fixation and the effects of the osmolarity of the fixative vehicle. Using these standardized techniques, the embryonic endocardium of chick, mouse, dog, human and, to a lesser extent, rat hearts are described in SEM. All species investigated presented microvilli, ruffles, filopodia, cytosegresomes, intercellular openings and phagocytes. Marginal folds, lamellipodia, dividing cells and incomplete endocardium could be observed in some species only. Each of these microappendages is discussed in relationship to observations of other authors on four levels - embryonic endocardium, adult endocardium, embryonic endothelium and adult endothelium.

The general tendency in differentiation of the embryonic endocardium results in a progressive loss of the majority of the microappendages mentioned.

Contrary to a relative absence of interspecific differences in endocardial morphology as seen in SEM, there is a strong variation of this morphology relating to the intracardiac localization of the endocardial cells.

The discovery of autolytic postmortem changes in the material from pregnancies terminated by prostaglandins leads to the recommendation that the further use of this source of embryonic and fetal material be discouraged.

Finally, the modifications of the morphology of embryonic endocardial cells under the effects of cytochalasin B, altered hemodynamics, and the hereditary congenital heart defects of the Keeshond strain of dogs are discussed, using the above-mentioned principles of four levels.

KEY WORDS: Embryology, Heart, Endocardium, Species differences, Fixation, Vehicle osmolarity, Prostaglandin, Hemodynamics, Animal, Human.

## Introduction

The heart is one of the first organs to function in the developing embryo. In the human, for example, its contractions start at the 21st day of gestation, when the woman frequently does not yet recognize that she is pregnant. The very first components of the embryonic heart to appear are its endocardial lining and its myocardium. The myocardium is the source of the contractile force essential for its pumping action and, according to some authors (Manasek, 1976; Manasek et al., 1978) also the driving force of the initial phases of the heart organogenesis. The endocardium represents the interface with the pumped fluid - the blood - and as such is relevant to the hemodynamic aspects of cardiac development (Rychter and Lemez, 1961; Rychter, 1962). As malformations of the heart (i.e. congenital heart disease) occur in the human population with an incidence of 1%, it is of importance to investigate the early prenatal phases of normal and abnormal cardiac morphogenesis. Whereas the embryonic myocardium has been extensively studied during the last two decades, mainly by TEM (Manasek, 1968, 1970, 1979), investigations on the embryonic endocardium can be more effectively carried out by SEM. In addition, the embryonic epicardium has recently been studied (Shimada and Ho, 1980).

The prenatal ontogenesis of the endocardium as seen in SEM will be discussed under the following headings:
I. Chronology of the SEM studies of the embryonic endocardium; II. Species investigated; III. Methodology of SEM studies of the embryonic endocardium; IV. Essential features of the endocardial morphology; V. Local variations of the surface morphology; VI. Experimental and pathologic alterations of the embryonic endocardial morphology.

### I. Chronology of SEM studies of the embryonic endocardium

The first studies by SEM of the embryonic endocardium were performed on the rat by Markwald et al. (1975) (see also Markwald and Fitzharris, 1974). In the same year, following the availability of critical point drying equipment, the SEM morphology of a normal chick embryo endocardium was presented (Pexieder, 1975, 1976). At the same time it was possible to present a first account of an experimental modification of the embryonic endocardium (Pexieder, 1976a). Also Johnson (1976) described the ventricular mural endocardium of the chick embryo. Hendrix (1977) and Hendrix and Morse (1977) described the SEM ultrastructure of atrial septation in human fetus and chick embryos, and a report on the SEM morphology of dividing cells and phagocytes in the endocardium of chick embryos was published (Pexieder, 1977) together with further details on experimental modification of the endocardium (Pexieder, 1977a).

In 1978 Los and Langemeijer-van Eijndthoven (1978) presented SEM observations on the AV cushions in mouse embryonic hearts, and concomitantly developments in preparation technique, and interspecific and topographic comparisons on the SEM morphology of the chick and mouse embryonic endocardium were made (Pexieder, 1978). Hay (1978) discussed the morphology of the endocardium on the chick embryonic atrioventricular (AV) cushions, and Morse (1978) extended his observations on the endocardium during formation of the foramina secunda. Some SEM features of the endocardium on the fusing AV cushions were reported by Los (1978), and further work on the experimental hemodynamic alterations of the embryonic endocardial morphology appeared (Pexieder, 1978a).

The ultrastructure of the endocardium as seen by SEM on the chick embryo semilunar valves was described by Hurle in 1979, and information on the endocardial morphology was presented by Steding et al. (1979). The cardiac jelly colonization was analysed experimentally by Bolender and Markwald (1979) in chick embryonic hearts.

Pexieder, (1980), presented correlated SEM and TEM observations on some of the endocardial ultrastructural features in the chick and mouse embryos.

SEM observations on the cytology of fusion of the AV cushions in the chick and mouse embryonic hearts was investigated by Los (1981). In the same monograph various other aspects of the embryonic endocardial SEM ultrastructure (Pexieder, 1981) and its interpretation (Fitzharris, 1981) in terms of cell physiology were presented.

## II. Species, stages and localizations studied

Chick has been the species most frequently studied (Pexieder, 1975, 1976; Hendrix and Morse, 1977; Los, 1978; Los and Langemeijer-Eijndthoven, 1978; Morse, 1978; Pexieder, 1978; Hurle, 1979; Steding et al., 1979; Bolender and Markwald, 1979; Pexieder, 1980; Los, 1981). Generally the developmental period studied has ranged from 2ed (embryonic day) to 21 ed, with most of the studies concentrated between 4ed and 7ed, i.e. the main period of heart organogenesis. The greatest interest has focused on the atrioventricular cushions (Los, 1978; Los and Langemeijer-Eijndthoven, 1978; Hay, 1978; Bolender and Markwald, 1979; Los, 1981) and the conotruncus (bulbus) (Pexieder, 1975, 1976, 1978; Steding et al., 1979; Pexieder, 1980). Less frequently studied were the atria (Hendrix and Morse, 1977; Morse, 1978; Pexieder, 1980) and the semilunar valves (Hurle, 1979).

The next species studied was the mouse (Pexieder, 1977; Los and Langemeijer-Eijndthoven, 1978; Los, 1978; Pexieder, 1978; Los, 1981). The group of Los was interested in the fusing AV cushions, and thus studied days 13 to 15, whereas the author's group was interested in the conotruncus (bulbus) and covered days 9 to 19.

Rat embryonic hearts were investigated between 10ed and 13.5ed by Markwald et al. (1975) and by Markwald and Fitzharris (1974). Recently the author's laboratory has investigated the conotruncus (bulbus) of some rat embryos on the 15ed from the laboratory of Prof. W. Scott at The Children's Hospital Research Foundation in Cincinnati (Ohio). These observations will be discussed later on in this review, together with studies on the conotruncus (bulbus) region of the embryonic hearts of normal (Beagle) and malformed (Keeshond) dog embryos at periods ranging from stage XV to stage XXII. These dogs have been raised by Dr. D. F. Patterson at the Genetics Section of the School of Veterinary Medicine of the University of Pennsylvania at Philadelphia.

There is a single report on human embryonic heart relating to its atrial septum at 4 months of gestation (Hendrix, 1977). Our observations on the conotruncus (bulbus) from 80 human embryos and fetuses ranging in age between 4 and 14 weeks of gestation (since last menstrual period) will later be presented. This material was obtained from the legal interruptions of pregnancy performed at the Department of Obstetrics and Gynecology, University of Lausanne. Further observations on SEM morphology of aortic endothelium in 5, 6, 7 and 8 months old human fetuses have been reported by Fujimoto et al. (1975) and Matonoha and Zechmeister (1978).

In our chick and mouse studies reported here, a sampling interval of 8 hours and a minimal sample size of 5 embryos at each stage were investigated. In the observations on rat embryos the sample size was 10: 5 embryos on each stage were studied also in the dog, whereas in the human embryos and fetuses, the minimal sample size was 3.

## III. Methodology of the SEM studies of the embryonic endocardium

The general methodological problems of the SEM investigations of embryonic material have been reviewed recently by Waterman (1980), and those of the circulatory system by Hollweg and Buss (1980). Thus only the features specific to the study of the embryonic endocardium will be discussed here. There are two principal constraints: the first is the very small size of embryonic hearts and the second is the hidden location of the endocardium. Because of the former the majority of authors use immersion fixation (Markwald et al., 1975; Hendrix, 1977; Hendrix and Morse, 1977; Hurle, 1979; Steding et al., 1979; Bolender and Markwald, 1979), rather than perfusion fixation (Pexieder, 1975, 1976, 1976a, 1976b, 1977, 1977a, 1978, 1978a, 1980, 1981). The latter constraint imposes the necessity of disclosing the endocardial surface by various kinds of microdissection. As the remaining part of the preparatory procedures does not vary substantially among the different laboratories, the details of the author's set up will be presented, followed by a critique of some particular aspects, such as osmolarity of the fixative.

Since 1969 a technique of microperfusion of the embryonic heart has been used for fixation, employing micropipets (Fig. 1) as used by Rychter (Prague, Czechoslovakia) in his studies of the topography of laminar blood streams (Rychter and Lemež, 1961). (Cf. also Abrunhosa, 1972.) The chick and mouse embryos are dissected free from their embryonic membranes and placed lying on their backs in dishes with Ringer solution of appropriate osmolarity. Under the dissecting microscope the tip of the micropipet is then introduced into the apex of the left ventricle. The rat and dog embryos were similarity perfused at the Cincinnati and Philadelphia laboratories. The intact human embryos from terminations of pregnancy (using special forceps) were immediately perfused in the operating theatre, using the same technique.

All embryos, whatever the species, were perfused with 2% glutaraldehyde (Serva) and 1% formaldehyde in 0.05M cacodylate buffer, the osmolarity being adjusted to species-dependent isotonicity. The main purpose of this perfusion was to clear the heart of blood, to maintain the ventricular cavities inflated, and to ensure immediate contact between the endocardium and fixative. The design of the procedure is presented in Fig. 2. The perfusion fixation was followed by immersion in the same fixative for 30 min., following which the embryos were rinsed with 0.05M isotonic cacodylate buffer for 1 hour. Postfixation in 1% $OsO_4$ (in the previously mentioned buffer) was then performed for $1\frac{1}{2}$ hour, and the embryos then stored in buffer before the internal relief of the heart was exposed by microdissection (under a dissecting microscope) using a technique described previously (Pexieder, 1978). At this stage macrophotographs were taken for the organ level studies. The preparation continued with ethanol dehydration; 20 minutes in each of 70%, 90% and two 100%. The intermediate fluid used for critical point drying was Freon 113. The specimens were processed through 25, 50, 75, and two 100% mixtures of Freon 113 and ethanol, each time for 20 minutes. Critical point drying was performed in a Bomar SPC 900 EX apparatus, using Freon 12 as transitional medium. The dried specimens were sputtered with 300 nm of gold using a S150 Edwards Sputter coater. The samples were examined in a Jeol Co. JSM-35 scanning electron microscope at 25 kV.

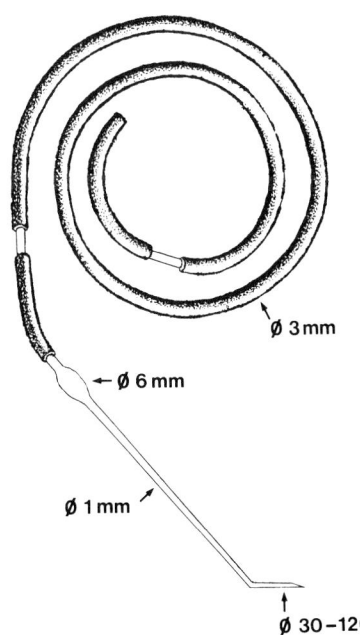

Fig. 1. The micropipette and its mouthpiece used for perfusion fixation of embryos.
(Ø = diameter)

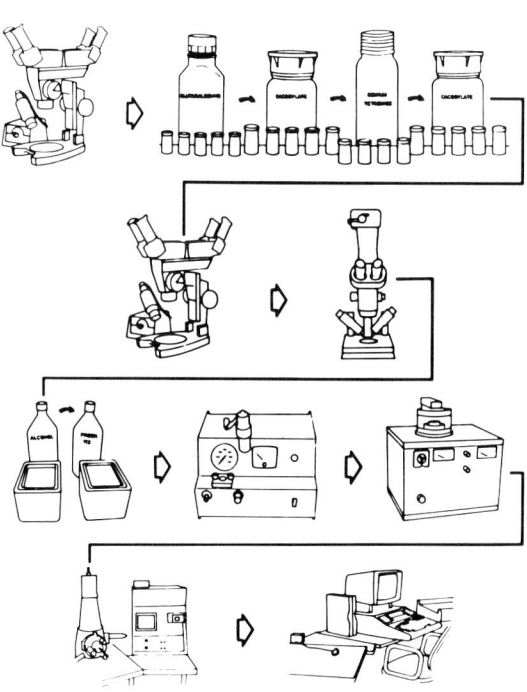

Fig. 2. The procedure used for preparation of the embryonic hearts for SEM.
For details see text.

Two aspects of the fixation procedure deserve comment. The first concerns the use of formaldehyde as compared with paraformaldehyde and the second concerns the tonicity of the fixative.

In the early stages of TEM studies of the chick embryo heart in the author's laboratory, comparative studies using both paraformaldehyde and technical formaldehyde stabilised with $CaCO_3$ were performed. The micrographs showed much better structure preservation with the latter. It was thought possible that the effect might be due to differing amounts of calcium ions, which are known for their membrane protection effects during fixation (Busson-Mabillot, 1971). The colorimetric dosage of $Ca^{++}$ using methylthymol blue showed that, whereas 25% glutaraldehyde contains 0.01 meq/l $Ca^{++}$, paraformaldehyde 0.04 meq/l and cacodylate buffer 0.14 meq/l, the 30% technical formaldehyde has a $Ca^{++}$ concentration of 0.32 meq/l producing a 0.49 meq/l $Ca^{++}$ concentration in the glutaraldehyde-formaldehyde fixative as used.

It is well known that fixative tonicity is of importance. Although the majority of authors prefer isotonic adjustment of the fixative (as reviewed recently by Schiff and Gennaro, 1979), some authors have discussed the osmotic effects of the glutaraldehyde (Mathieu et al., 1978) but consider the total osmolarity as being of greater importance. The studies in the author's laboratory indicate the desirability of controlling the osmolarity of all the various formulas used in specimen preparation before employing them. It has been found, for instance, that "mammalian Ringer" (Millonig, 1976) has an effective osmolarity of 410.

It has been shown by Litke and Low (1977) that the embryonic tissues are particularly sensitive to deviations from isotonicity of the fixative vehicle. This is easily understood when the high water content of the embryo is considered together with the developmental changes in the isotonic point. For example, in the chick embryo, the osmotic pressure of the embryo is 248 mOsm/l on the 2nd ed, 251 on 3rd ed, 255 on 4th ed, 264 on 5th ed, 272 on 6th ed and 276 on 7th ed (Romanoff, 1967). Using TEM (Chavaz and Pexieder, 1976), the osmotic reactivity of the embryonic cardiac tissues was studied. Chick embryo hearts were perfused on the 4ed with 2% glutaraldehyde and 1% formaldehyde in 0.05M cacodylate buffer. The tonicity of the buffer (75 mOsm) was modified by the addition of NaCl to obtain 210, 230, 250, 270, 290, 330 mOsm/l. Three embryos were sampled for each buffer osmolarity. Ultrathin sections cut in the region of the distal ventral cushion were studied. On three micrographs per embryonic heart, each at 4,000x, the number of mitochondria with normal, hypotonic, and hypertonic morphology were counted and expressed in percentages. Using the arctangent transformation the figures were submitted to Student t-test. The percentages of hypertonic and hypotonic mitochondria were significantly different for each vehicle osmolarity. The same was true for the percentage of normal mitochondria, with the exception of 270 versus 290 mOsm where the difference was not significant. Fig. 3 represents a set of polynomial regressions describing the relationship between the vehicle osmolarity and the morphology of mitochondria in the endocardium. The proportion of normal mitochondria is highest at 270 mOsm. With decreasing osmolarity the percentage of hypotonic mitochondria increases, as does the percentage of hypertonic mitochondria when the osmolarity of the fixative is increased beyond the isotonic region.

Using a similar experimental design, the changes in the density of microvilli (number of microvilli per unit of endocardial cell surface) were studied (Pexieder, 1976b). 4ed chick embryos (3 per given osmolarity) were perfused with 2% glutaraldehyde and 1% formaldehyde in 0.05M cacodylate buffer, the osmolarity being as indicated above. After microdissection and preparation for SEM, 8 micrographs per embryonic heart were taken of the proximal left bulbar cushion at X 5,900 magnification. On the micrographs the surface of individual endocardial cells was measured using a planimeter and the number of microvilli on each endocardial cell was counted. The results are shown in Fig. 4. The density of microvilli decreased significantly between 100 and 150 mOsm/l (t-test, $\alpha \leq 0.05$). It did not change significantly during the increase from 250 to 270 mOsm/l

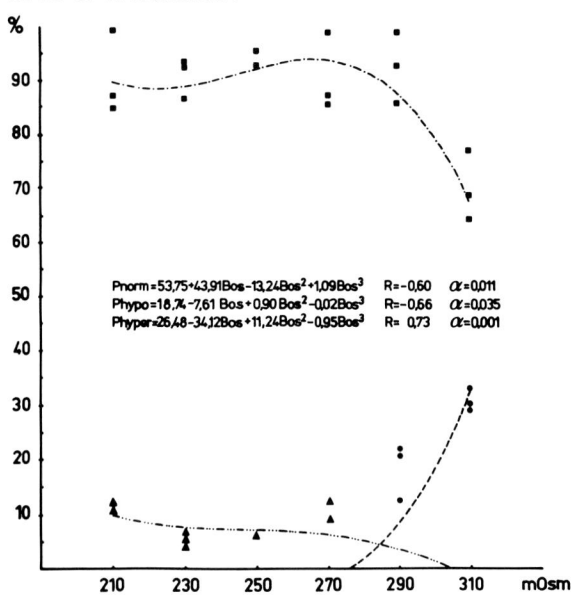

Fig. 3. TEM appearance of the chick embryonic endocardial mitochondria expressed as a function of the buffer osmolarity.
Normal mitochondria = ■ and -.-.,
hypotonically altered mitochondria = ▲ and -...-,
hypertonically altered mitochondria = ● and ---.

(t-test, $\alpha \leq 0.01$). The highest density was obtained also at 290 mOsm/l. Further increase in osmolarity resulted in a highly significant decrease in the microvilli density (t-test, $\alpha \leq 0.0005$). The osmotic fragility of microvilli has been described also by Arborgh et al. (1976) from their experiments on cells in tissue culture. It is suggested that the changes in the density of microvilli can be interpreted as follows: in hypotonic buffer, cells are swelling, and the increased cell volume introduces an increase in cell surface thereby decreasing the density of microvilli, both by virtue of the distention of the cell membrane and because of the eventual use of microvilli as spare material to cope with the increase in cell membrane surface (Erickson and Trinkhaus, 1976; Lee and Chien, 1979). As the osmolarity reaches the isotonic point the cells reach their physiological conditions which correspond to the highest density of microvilli. At that point also the cell to cell variations, as reflected by the standard deviations, are largest. It seems logical that at isotonic conditions the individual cells will most easily express their individuality. Thus hypertonic conditions would harm the cell and decrease its number of microvilli. Similar observations were made by Naef et al. (1978) in synaptic vesicles.

How can these differing views on the relative importance of the vehicle versus glutaraldehyde osmolarity during fixation be reconciled? The study of Mathieu et al. (1978) suffers two major drawbacks. The first one is that they obtain various buffer osmolarities by changing their molar concentration. The changing buffer concentration can alter chemically the reaction pattern of the cells investigated. A better method would be to use the same buffer concentration but to vary its osmolarity by some osmotically active ions, such as $Na^{++}$. The chemical nature of the buffer was shown to be of importance by Squier et al. (1976). They observed that phosphate buffer and saline with identical osmolarity had different effects on cell volume. The second drawback of the study of Mathieu et al. is that the glutaraldehyde osmolarity cannot be changed without changing also the glutaraldehyde concentration, and therefore also its effect on stabilization of membranes, which necessarily results in changing osmotic reactivity of the cell. When buffer osmolarity is kept constant, the variations of total osmolarity are essentially an expression of changing glutaraldehyde concentration. When aqueous solutions of glutaraldehyde are used (Arborgh et al., 1976), the so-called "osmotic effect" of glutaraldehyde disappears. There may be other reasons why the present results differ from those of Mathieu et al., e.g., the combined use of glutaraldehyde and formaldehyde, and also the presence of calcium ions in the fixative. Other considerations entering the analysis of this complex problem have been discussed recently by Kaufman (1980). Perhaps Squier et al. (1976) summarize best this problem: "Thus the osmolarity of buffer and fixative are to a certain extent additive, and both need to be considered in formulating a fixative; however the glutaraldehyde probably does not exert its full osmotic effect, as indicated by its measured osmolarity, since its action as a fixative limits the responsiveness of the cells".

## V. Essential features of the endocardial morphology

Up to the present, only Ojeda and Hurle (1975), Johnson (1976) and more recently Steding et al. (1980), have published some SEM observations on the constitution of the early embryonic heart endocardial lining. Investigations in the author's laboratory on the chick embryo heart development (starting with 2ed 8h) have shown that in heart organogenesis the endocardial lining morphology, as seen in SEM, is remarkably constant (Pexieder, 1978). The comparison of 5 different species, prepared and examined under strictly the same techniques, has shown that there are almost no species differences in the endocardial morphology. Its essential features are: endocardial cells

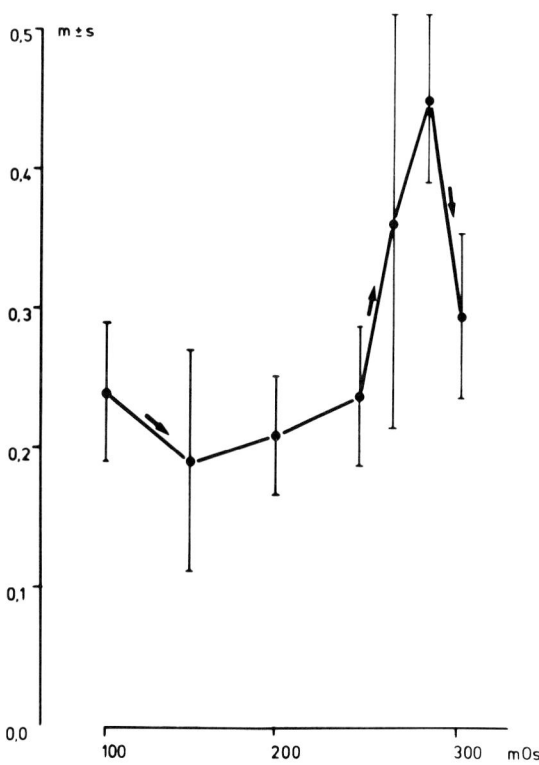

Fig. 4. Variation in surface density of microvilli (number of microvilli per square micrometer) on chick embryonic endocardial cells in relationship to changing buffer osmolarity. Each point is based on at least 20 cells.

with varying numbers of microvilli, marginal folds, ruffles, lamellipodia, filopodia, cytosegresomes, intercellular clefts, incomplete endocardium and phagocytes. These different cellular forms have specific locations in the embryonic heart that change according to the stage of development. Almost all of these forms are present in the embryonic heart from the very beginning of its development.

The various features (illustrated in Figures 5 to 15) will now be discussed individually.

A. Endocardium without (or with rare) microvilli

This kind of endocardial cell has been observed in all 5 species investigated (Figs 5a, 5b, 6a, 6b, 7a, 15B).

Many of the endocardial cells seen in the chick embryo outflow tract were flat squamous-like cells with some microvilli (Pexieder, 1975). Their size was 8.5 x 12.5 μm. The diameter of microvilli was 0.2 - 1 μm and their length 1 - 1.5 μm. Under high magnification fine pores (diameter 0.1 μm) could be seen in the cell membrane. They were interpreted as invaginations of pinocytotic vesicles (Pexieder, 1976). Variations in morphology of endocardial cells covering chick atrioventricular cushions, as well as the presence of irregularly distributed microvilli and microplicae, were described by Hay (1978). Polygonal flattened cells without microvilli were observed by Hurle (1979) on the ventricular surface of chick embryonic semilunar valves. Between 2ed 16h and 3ed Bolender and Markwald (1979) found a second type of endocardial cell overlapping or overgrowing its neighbour in regions where a seeding of the cardiac jelly takes place. No overlapping of endocardial cells was seen by Los and Langemeijer-Eijndthoven (1978) in mouse atrioventricular cushions. Reduced number of microvilli on endocardial cells in flattened areas of the rat embryonic heart were found by Markwald et al. (1975). In comparison, the adult endocardium seems to be characterized by almost complete lack of microvilli (Buss et al., 1973; Harasaki et al., 1975) and the presence of nuclear bulges (Wheeler et al., 1972, 1973; Peine, 1974; Harasaki et al., 1975; Missirlis and Armeniades, 1977; Sarphie and Allen, 1977; Sarphie and Hawkins, 1980). Nuclear bulges were mentioned in the embryonic endocardium in the rat (Markwald et al., 1975) and in the human (Hendrix, 1977), but have been considered by others (Buss et al., 1973; Peine and Low, 1975) to be the consequence of myocardial contraction and/or drying and dehydration.

Even if the nuclear bulges persisted in distended blood vessels (Clark and Glagov, 1976), they were interpreted as the consequence of cellular contraction due to injury by Jørgensen and Svendsen (1978); only few endothelial cells could exhibit small microvilli (Buss and Hollweg, 1977) or a modest number of small blebs (Schaub et al., 1980).

B. Endocardium with large numbers of microvilli

Again this kind of morphology was found in all species investigated (Figs 5c, 5d, 6c, 6d, 7b, 15H). There were usually 20 - 40 microvilli per endocardial cell (Pexieder, 1976). Hay (1978) has described microvilli interdigitation in fusing chick embryo atrioventricular cushions. Abundant microvilli were also described by Hurle (1979) on the arterial pole of semilunar valves in the chick embryo heart. Profuse arrays of endothelial projections, probably microvilli, have been observed in the aorta, the pulmonary artery and the ductus arteriosus of human fetuses (Fujimoto et al., 1975).

In the adult heart, villiform membrane projections were described by Maguire (1972) in the mouse and by Sarphie and Allen (1977) in the dog. More densely packed microvilli were found on aortic semilunar valves (Buss et al., 1973) and atrioventricular valves (Edanaga, 1975; Harasaki et al., 1975). Sarphie (1980) has found more microappendages in the dog atrium than in ventricles.

Microvilli or finger-like projections were seen, as well on the endothelium of blood vessels (Smith et al., 1971), but they seem to be more rare than on the endocardium (Buss and Hollweg, 1977; Chi and Schwartz, 1978; Buss et al., 1979).

C. Endocardium with marginal folds

Marginal folds were observed less frequently than microvilli (Figs 7c, 11a, 15A, 15D, 15H). They were most easily found in the chick embryo, some have been observed also in the human embryonic heart, but they were much less frequent or even missing in hearts of mouse, rat and dog embryos.

Long tongue-like cytoplasmic protrusions corresponding to the rim of endothelial cells were reported by Los and Langemeijer-Eijndthoven (1978) and Hay (1978) in chick embryo atrioventricular cushions, increasing in frequency and size during the fusion of these cushions (Los, 1978). Even though Markwald et al. (1975) described the accumulation of microvilli at sites of cell contact in 10ed rat embryonic heart, perhaps as precursors of marginal folds seen on 11ed, Los (1981)

---

Fig. 5. a) Endocardial cells with rare microvilli. Note the presence of a mitosis (left arrow) and marginal folds (right arrow). All scale bars in this figure = 10 μm. Proximal left bulbar cushion. Chick embryo, 7ed.
b) Endocardial cells with rare microvilli. Aortic infundibulum. Mouse embryo, 12ed.
c) Endocardial cells with large numbers of microvilli. Distal ventral bulbar cushion. Chick embryo, 5ed 8h.
d) Endocardial cells with large numbers of microvilli. Bulboventricular border. Mouse embryo, 15ed.
e) Ruffles (arrows) associated with endocardial cell limits. Distal ventral bulbar cushion. Chick embryo, 4ed.
Reproduced by the permission of Futura Publ. Co., N.Y., from Pexieder (1980).
f) Ruffles (arrows) associated with endocardial cell limits. Interventricular foramen. Mouse embryo, 14ed.

# Embryonic and Foetal Endocardium

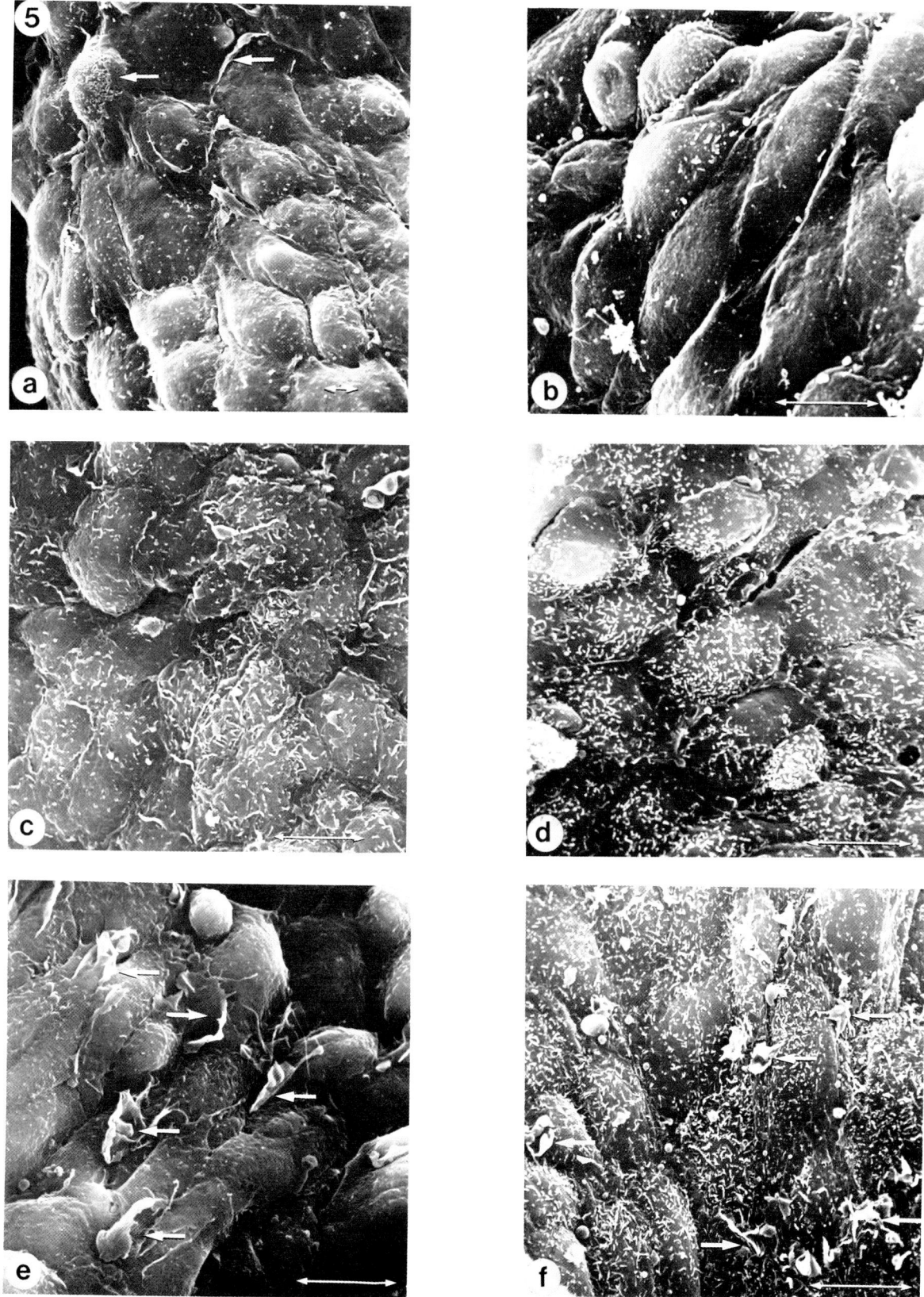

Fig. 5 - See Facing page for Legends.

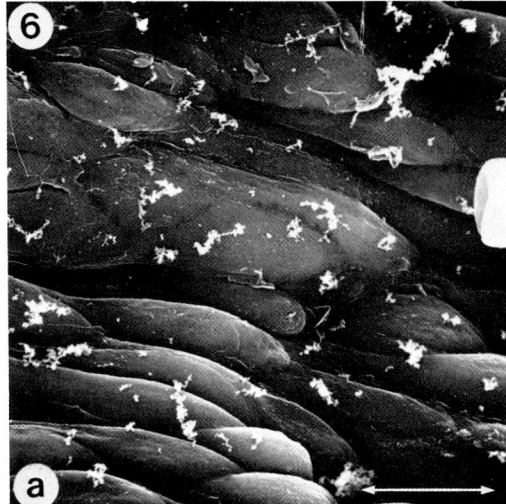

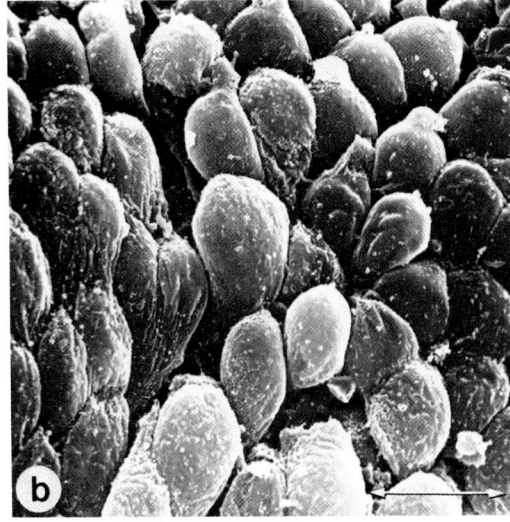

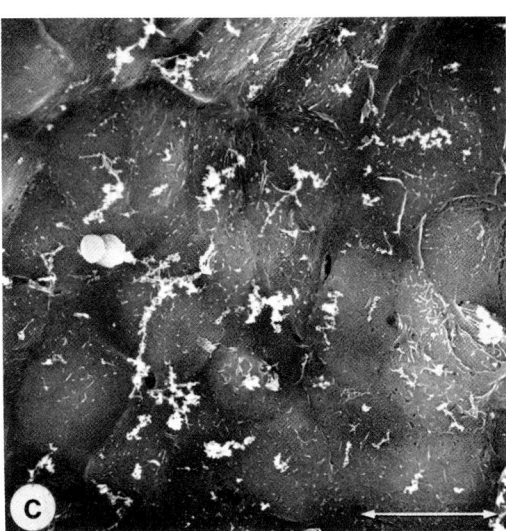

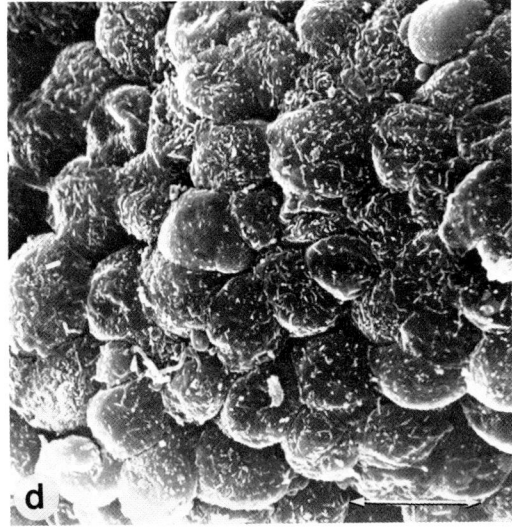

Fig. 6. a) Endocardial cells with rare microvilli. Semilunar valves. Rat embryo, 15ed.
All scale bars in this figure = 10 μm.
b) Endocardial cells with rare microvilli. Atrioventricular cushions. Dog embryo, stage XVI.
c) Endocardial cells with more frequent microvilli. Distal ventral bulbar ridge. Rat embryo, 15ed.
d) Endocardial cells with large numbers of microvilli. Atrioventricular cushions. Dog embryo, stage XVI.
e) Ruffles (arrows) associated with endocardial cell limits. Bulboventricular border. Dog embryo, stage XVI.

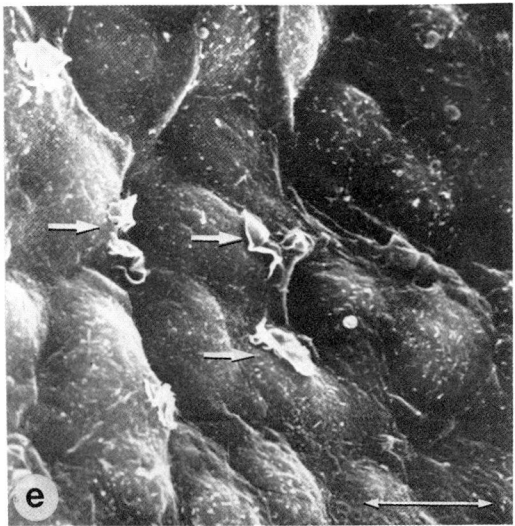

observed that the rat embryonic endocardium displays even less protrusion than the mouse. Marginal microvilli and plasmalemma folds were seen by Morse (1978) on the muscular part of the interventricular septum in the chick embryo. Johnson (1976) and later also Bolender and Markwald (1979) have described them on the ventricular mural endocardium of young chick embryos. In 1979, Hurle noted marginal folds on the ventricular face of semilunar valves of chick embryos before hatching.

The adult endocardium presented marginal folds in atria of dogs and rats (Peine, 1974) and monkeys (Harasaki et al., 1975). Marginal folds have been also reported in the left ventricle and conus arteriosus of rats and rabbits (Edanaga, 1975), as well as on the human mitral valve (Lim and Boughner, 1977). Harasaki et al. (1975) observed that marginal folds and microvilli are easily lost during preparation. This may explain why Peine and Low (1975) stated that there is a considerable disagreement in the literature concerning the presence or absence of microappendages of the plasmalemma in both TEM and SEM. The marginal folds should certainly not be confused with the fine-line-like appearance of interendothelial cell junctions (Wheeler et al., 1972) or so called "Grenzleisten" (Buss et al., 1973) described especially after pretreatment with $AgNO_3$ (Garbarsch and Christensen, 1970).

In the endothelium of blood vessels, marginal folds can occur in dog pulmonary artery (Smith et al., 1971), rabbit aorta (Edanaga, 1974; Thomsen and Kjeldsen, 1978), and in rat aorta (Chi and Schwartz, 1978).

D. Ruffles

The ruffles deriving from the cell membrane have been observed in all species examined except the rat (Figs 5e, 5f, 6e, 8a, 10b, 15D, 15G, 15K), but it is reasonable to expect that the examination of more developmental stages might well reveal them there also. They have been described as very thin cellular protrusions of the size 4.4 x 0.4 μm (Pexieder, 1976) reminiscent of the structures frequently seen in SEM micrographs of cells in culture (Collins et al., 1980). Ruffling was also reported by Johnson (1976). It is not clear whether they are identical with the microplicae defined as shell-shaped cell processes by Hay (1978). Pseudopodial processes with ruffled membranes were reported in the chick embryonic endocardium by Shimada and Ho (1980). As TEM studies have shown, the ruffles could possibly originate from the endocardium itself or could be the ruffles of subendocardial (most frequently cushion tissue) mesenchymal cells (Pexieder, 1980).

In the adult mouse heart, Maguire (1972) observed cusp shaped membrane projections in the flatter expansive areas of the endocardium. Cell

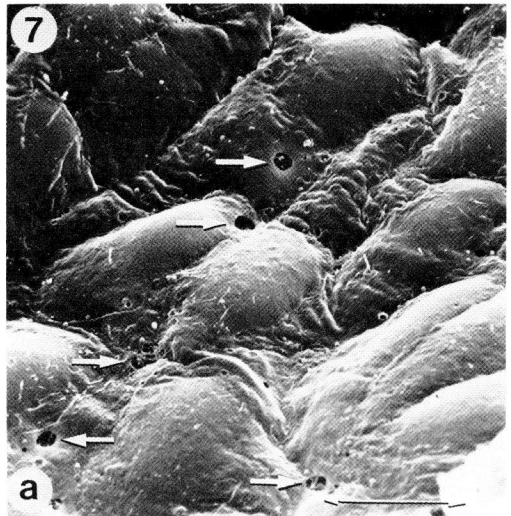

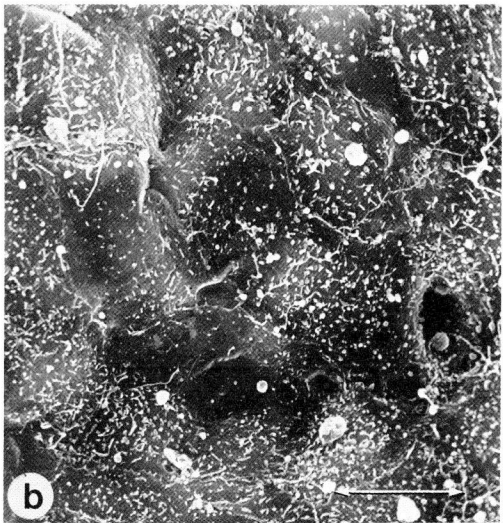

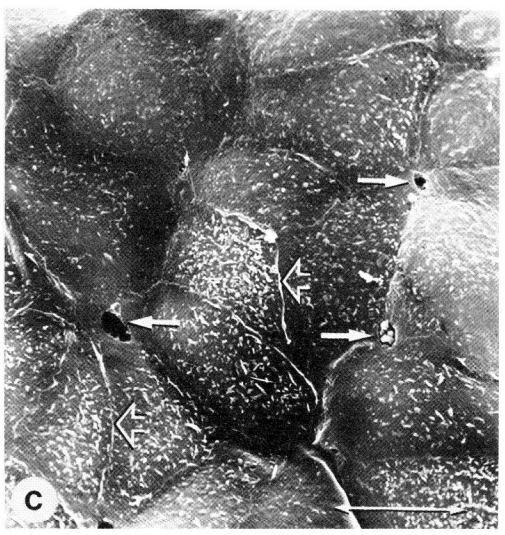

Fig. 7. a) Endocardial cells with rare microvilli. Note the presence of intercellular openings (arrows).
All scale bars in this figure = 10 μm.
Bulboventricular border. Human embryo, 7 weeks of gestation.
b) Endocardial cells with large numbers of microvilli. Bulboventricular border. Human embryo, $6\frac{1}{2}$ weeks of gestation.
c) Marginal fold (arrowhead) and intercellular openings (arrows) in the endocardium. Atrioventricular cushions. Human embryo, $6\frac{1}{2}$ weeks of gestation.

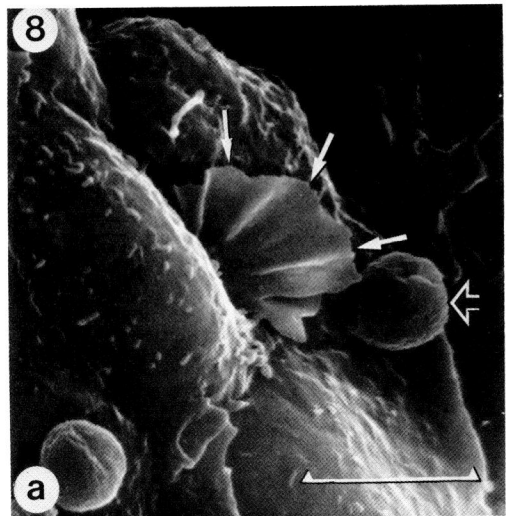
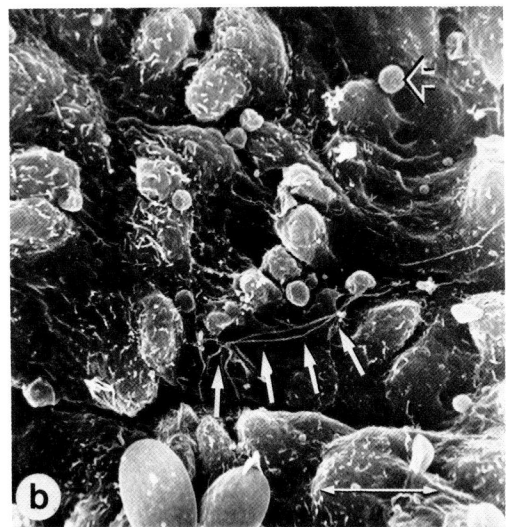
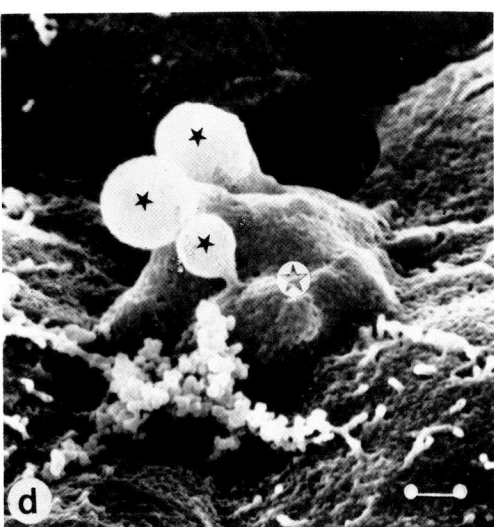

margin ruffling was described by Sarphie and Allen (1977) in dog hearts.

In the present studies, saline or buffer wash was not used prior to fixation, and thus the observed ruffles are certainly not the artefact reported by Peine and Low (1975), nor that due to immersion or the pressureless perfusion fixation of blood vessels as described by Clark and Glagov (1976).

Recently there has seemed to be some confusion in the interpretation of the functional importance of the ruffles. Fitzharris (1981) has suggested that ruffling has nothing to do with movement. He proposed replacing the term ruffle by microplica (infolding or outpocketing of adjacent cells), which increase the adhesiveness between the cells. Los (1981) even interpreted ruffling as a consequence of failure of the cell to form new adhesions, considering ruffling as specific for the fusion of atrioventricular cushions, and explaining its relative scarcity in mouse and rat embryos by a more firm attachment of the endocardium to the cushion tissue mesenchyme. Figs 5e, 5f, 6e, 8a, 10b and 15D represent supplementary evidence for the contention that ruffling is not specific for the regions of fusion only (Pexieder, 1981b).

It is believed from the evidence of the

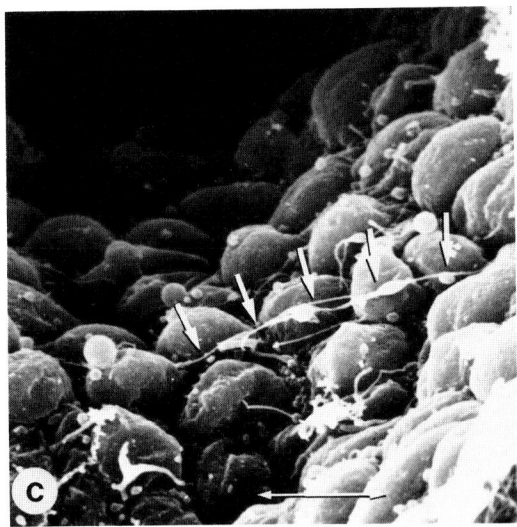

Fig. 8. a) Ruffles (lamellipodia) (arrows) and cytosegresomes (arrowhead) in the endocardium. Distal ventral bulbar cushion. Chick embryo, 3ed 16h. Scale bar = 5 μm.
b) Filopodia (arrows) on the endocardium. Cytosegresome (arrowhead). Bulboventricular border. Chick embryo, 3ed. Scale bar = 10 μm.
c) Filopodia (arrows) on the endocardium. Aortopulmonary septum. Mouse embryo, 12ed. Scale bar = 10 μm.
d) Endocardial cellular defecation of primary (white asterisk) and secondary (black asterisks) cytosegresomes. Aortic infundibulum. Chick embryo, 7ed. Scale bar = 1 μm.

author's laboratory that ruffles are related to the motility of cells (cf Abercrombie, 1961) and that this has been demonstrated convincingly in vitro on cultured cells (Revel, 1974; Shay and Walker, 1980). That endocardial cells can move has been demonstrated by Bolender and Markwald (1979) and also by Fitzharris (1981a) in studies

Fig. 9. a) Lamellipodium between endocardial cells. Atrioventricular cushions. Dog embryo, stage XVII. Scale bar = 1 µm.
b) Filopodium (arrows) extending over the endocardium. Distal ventral bulbus ridge. Rat embryo, 15ed. Scale bar = 10 µm.
c) Endocardial filopodia (arrows). Right atrioventricular orifice. Dog embryo, stage XVI. Scale bar = 10 µm.
d) Cytosegresome defecation from an endocardial cell. TEM. Distal ventral bulbar cushion. Chick embryo, 4ed. Scale bar = 5 µm.
e) Several cytosegresomes (asterisks) in the neighborhood of a dividing cell (arrow). Right atrioventricular orifice. Dog embryo, stage XVIII. Scale bar = 10 µm.

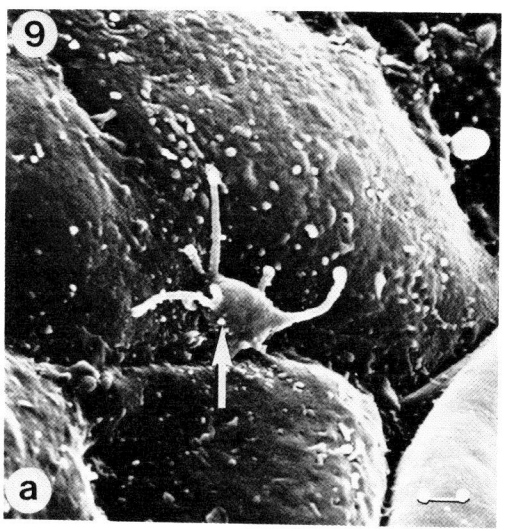

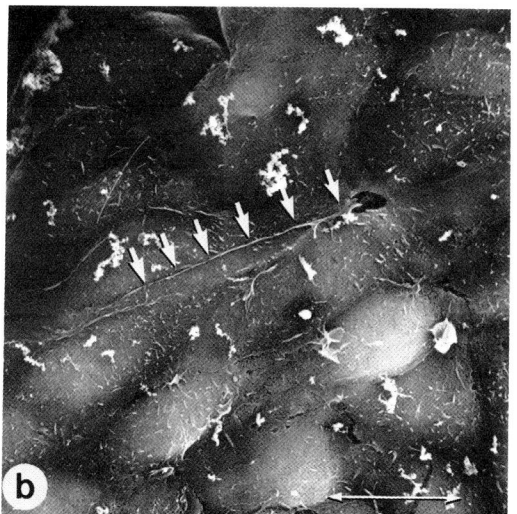

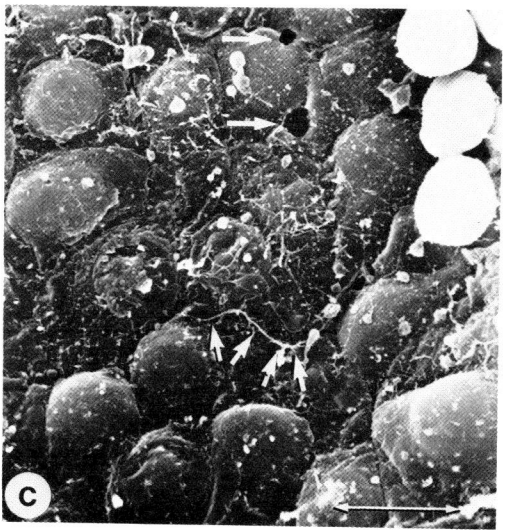

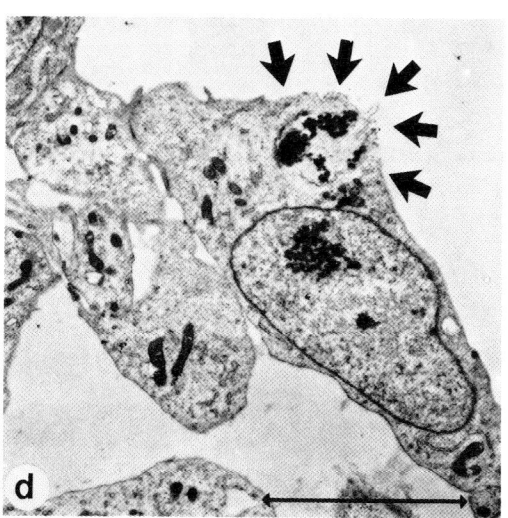

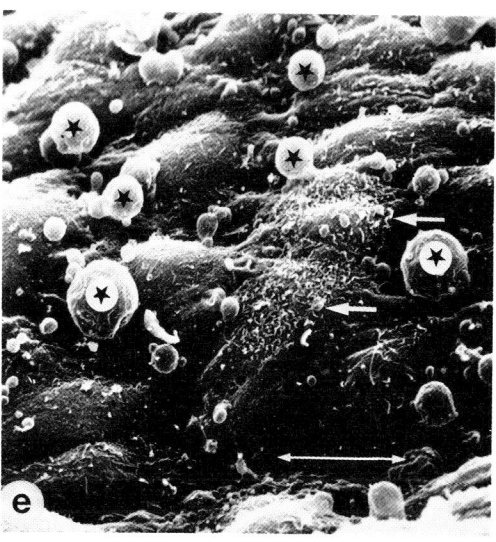

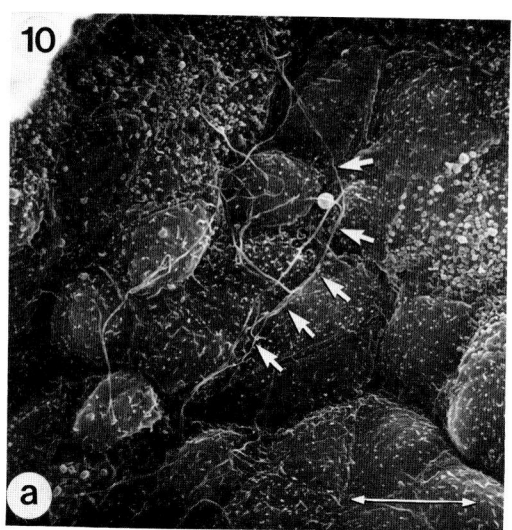
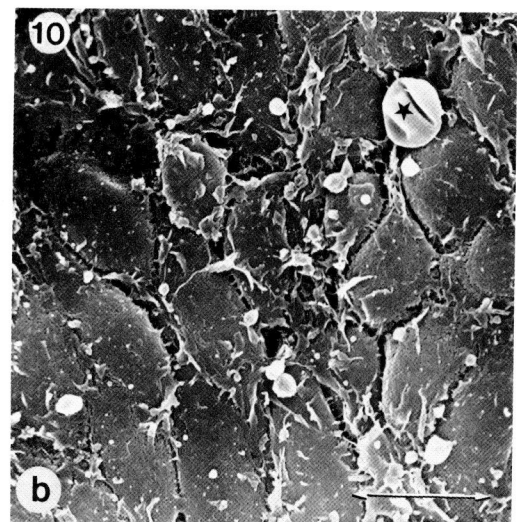

Fig. 10 (above), Fig. 11 (below). See facing page for legend.

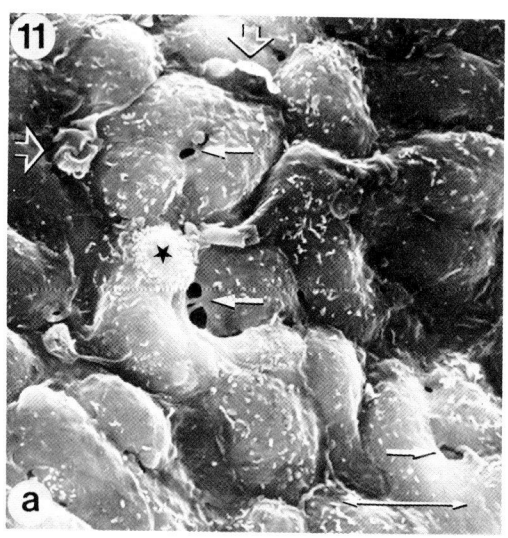
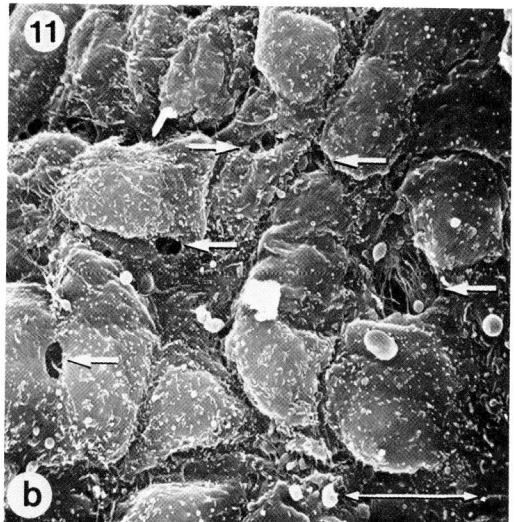
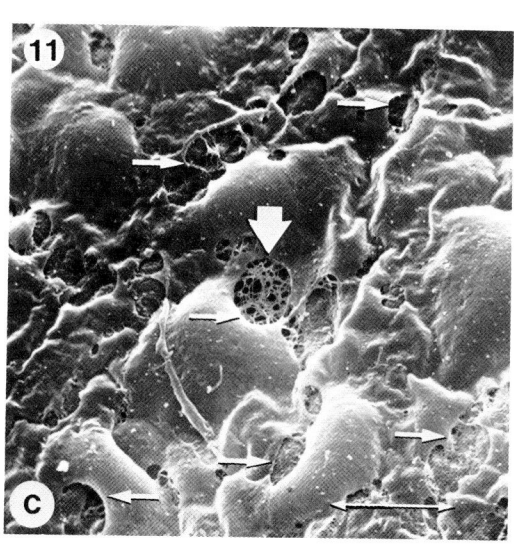
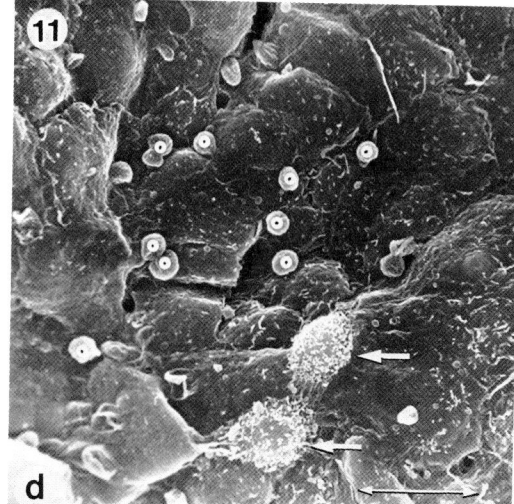

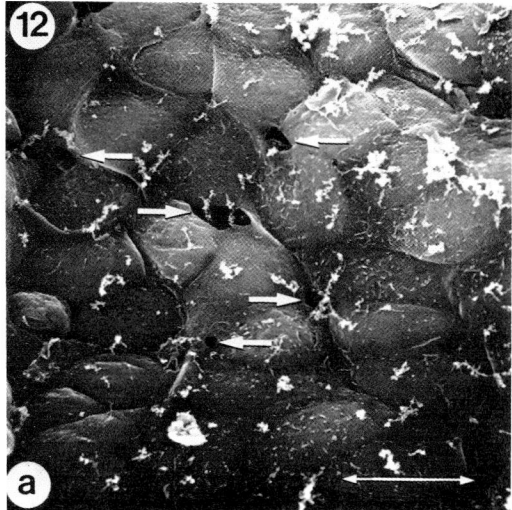
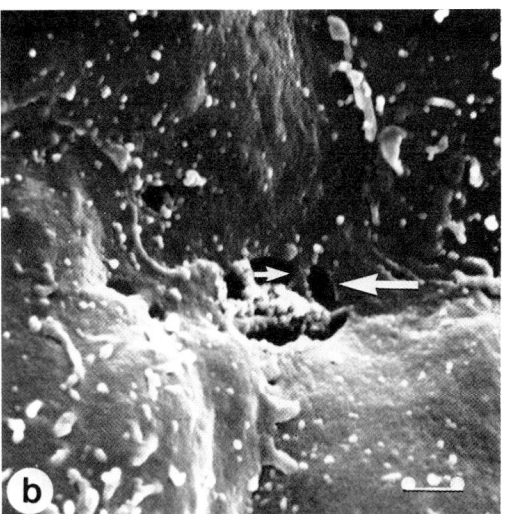
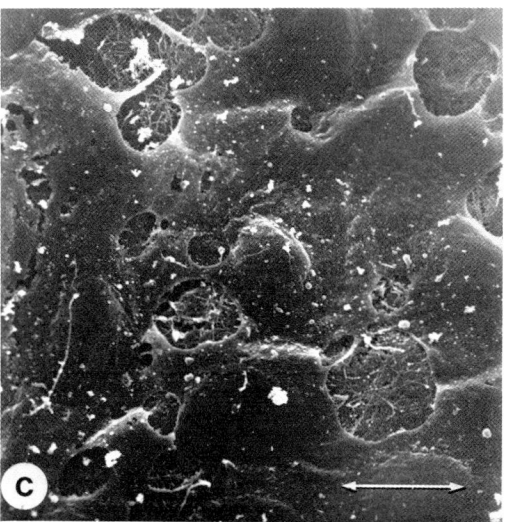
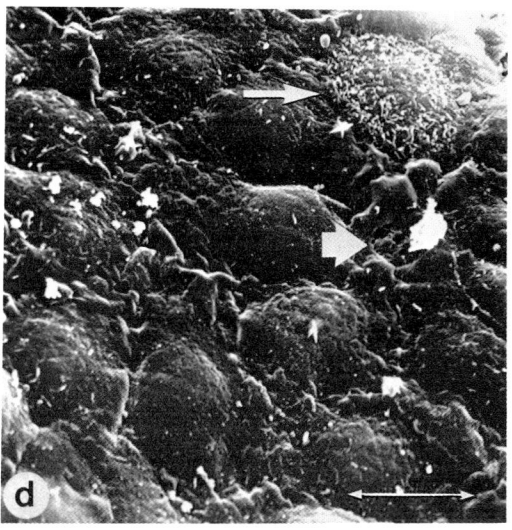

Fig. 12. a) Intercellular openings (arrows) between endocardial cells. Distal ventral bulbar ridge. Rat embryo, 15ed. Scale bar = 10 μm.
b) Intercellular opening (arrow) partially filled with fine mesh of cytoplasmic extensions (arrowhead). Atrioventricular cushions. Dog embryo, stage XVII. Scale bar = 1 μm.
c) Incomplete coverage of the subendocardium (arrows) by the endocardial cells. Left atrium. Dog embryo, stage XVIII. Scale bar = 10 μm.
d) Endocardial cell in mitosis (arrow). Note also the intercellular opening (arrowhead). Right atrioventricular orifice. Dog embryo, stage XVIII. Scale bar = 10 μm.

Fig. 10. a) Filopodia (arrows) overlying several endocardial cells. Pulmonary infundibulum. Human embryo, 8½ weeks of gestation. Scale bar = 10 μm.
b) Cytosegresome (asterisk). Note the presence of numerous ruffles associated with cell limits. Pulmonary infundibulum. Human embryo, 10 weeks of gestation. Scale bar = 10 μm.

Fig. 11. a) Openings (arrows) between endocardial cells. Arrowheads indicate marginal folds. Proximal right bulbar cushion. Chick embryo, 3ed 16h. All scale bars in this figure = 10 μm.
b) Intercellular openings (arrows). Distal ventral bulbar ridge. Mouse embryo, 13ed 8h.
c) Endocardial cells incompletely covering the subendocardium (arrows) the fibrillar nature of which is seen at the arrowhead. Right atrium. Chick embryo, 5ed 8h.
d) Dividing endocardial cells in telophase (arrows). Note the numerous cytosegresomes (black points). Proximal right bulbar cushion. Reproduced by the permission of the National Foundation March of Dimes, N.Y., from Pexieder (1978). Chick embryo, 3ed.

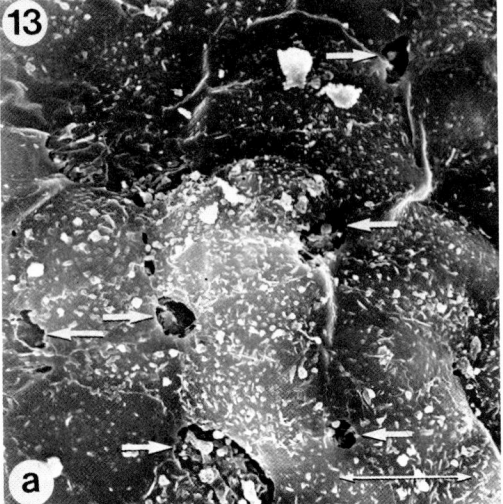

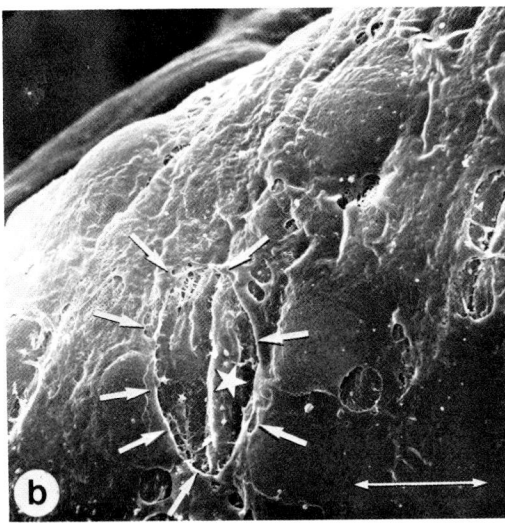

Fig. 13. a) Intercellular openings (arrows) between endocardial cells. Human embryo, 6½ weeks of gestation. Atrioventricular cushions.
All scale bars in this figure = 10 µm.
b) Incomplete coverage (arrows) of the myocardial fibres (asterisk) by the endocardial cells. Right atrium. Human embryo, 8½ weeks of gestation.

of the cardiac jelly colonization by cells originating in endocardium.

E. Lamellipodia

Isolated cases of lamellipodia have been observed in the chick embryonic heart (Fig. 8a). It is difficult to classify the structure seen in the heart of a dog embryo (Fig. 9) which might also have been called a nascent filopodium. It does not seem conclusively demonstrated whether these structures should be considered more as particular forms of ruffles. Similar formations may be seen in Fig. 1b of Los (1981) who interprets them as retracting, partially adhering, endocardial cells, in fusing atrioventricular cushions. It seems that Buss et al. (1973) observed some lamellipodia in their SEM studies of blood vessels. It appears that there is confusion in the terminology used to describe the various types of microappendages found on endocardial and endothelial cells.

F. Filopodia

Filopodia of varying lengths, sometimes crossing several endocardial cells, were found in ventricular part of the embryonic heart of all species investigated (Figs 8b, 8c, 9b, 9c, 10a).

Filopodia were recorded also by Johnson (1974) in his TEM study of ventricular endocardium in the chick embryo. Morse (1978) reported numerous long cytoplasmic processes extending across several flat endocardial cells.

Hair-like projections of cytoplasm, 3 µm in diameter and 150 - 160 µm in length, were described on the endothelium of rabbit aorta by Tokunaga et al. (1973) and by Edanaga (1974).

G. Cytosegresomes

Cytosegresomes (Trump and Ericsson, 1965) and their intraluminal passage have been observed in all species studied (Figs 8a, 8d, 9d, 10b, 11d, 15B), with the exception of mouse and rat embryos. As the presence of cytosegresomes is more or less proportional to the incidence of physiological cell death in and beneath the endocardium, the absence of these in the mouse and rat correlates well with the lesser extent of cell death in the outflow tract of these two species (Pexieder, 1975a).

The exocytosis of cytosegresomes at the borders of bulbar (conotruncal) cushions in chick embryo (Pexieder, 1975) is seen as vesicles of varying size from 1.2 to 3.6 µm (Pexieder, 1976). An increase in their number coincides with the merging of proximal (conus) and distal (truncus) bulbar (conotruncal) cushions occurring on 5ed in the chick embryo. TEM proof of their existence and ultrastructure has been published (Pexieder, 1980). These cytosegresomes should not be confused with membrane blisters reported by Shelton and

Fig. 14. a) Phagocytes present on the endocardium. Asterisks label some of the phagosomes. Bulboventricular border. Chick embryo, 4ed 16h.
Scale bar = 10 µm.
b) Phagocyte on the endocardium. Asterisks identify some of the phagosomes. Proximal right bulbar cushion. Dog embryo, stage XVI.
Scale bar = 5 µm.
c) Phagocytes traversing the endocardium (asterisks). Right atrium. Mouse embryo, 13ed.
Scale bar = 1 µm.
d) Phagocytes (asterisks) penetrating through the intercellular openings of the endocardium into the heart lumen. Interventricular foramen. Human embryo, 6½ weeks of gestation. Scale bar = 10 µm.
e) Phagocyte (asterisk) on the endocardium. Distal ventral bulbar ridge. Rat embryo, 15ed. Scale bar = 10 µm.
f) Phagocyte spreading on the endocardium. Note the presence of a pseudopod (arrows) and phagosomes (asterisks). Right atrioventricular orifice. Dog embryo, stage XVIII. Scale bar = 5 µm.

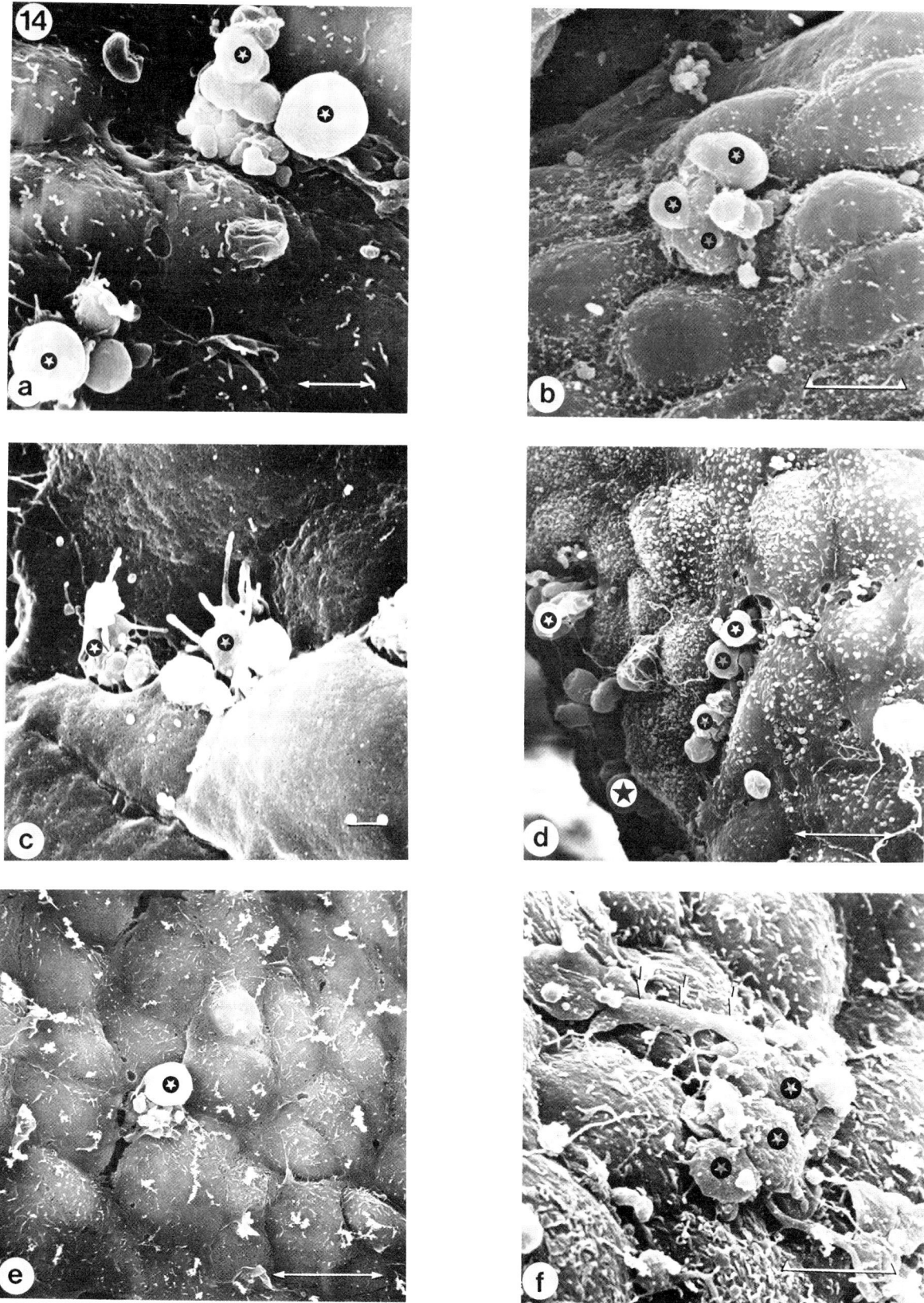

Fig. 14. See facing page for legend.

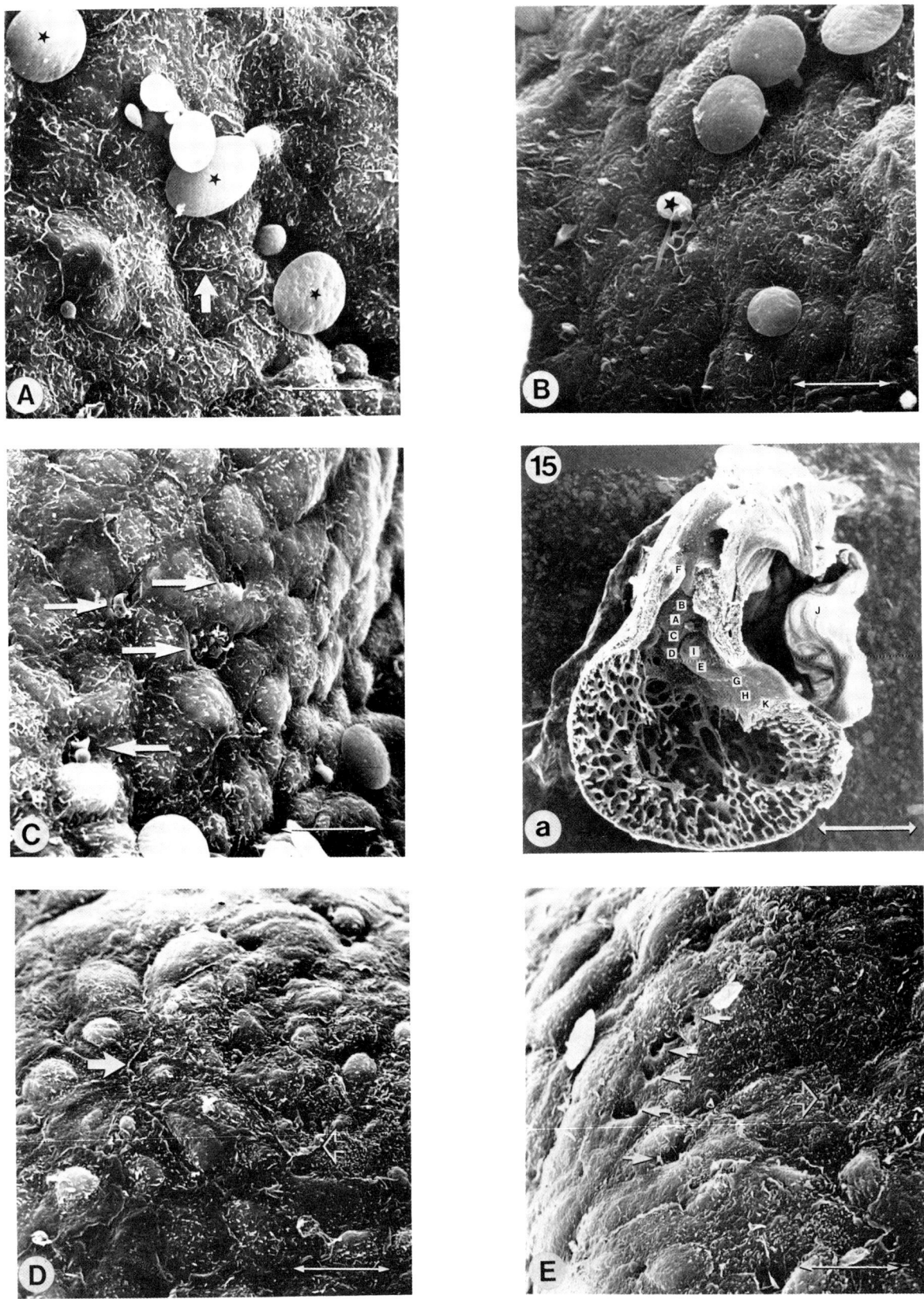

Fig. 15 (continued on next two pages). See facing page for legend.

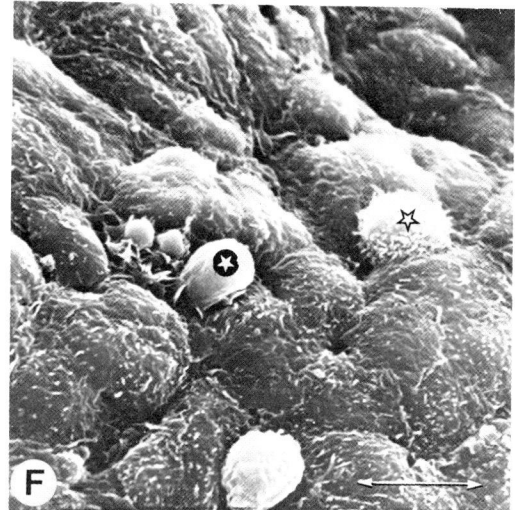
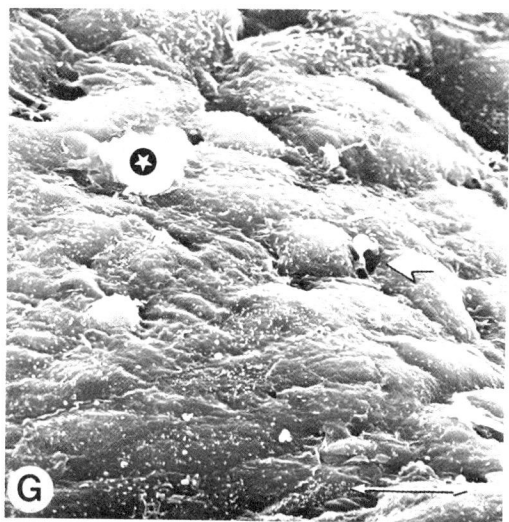
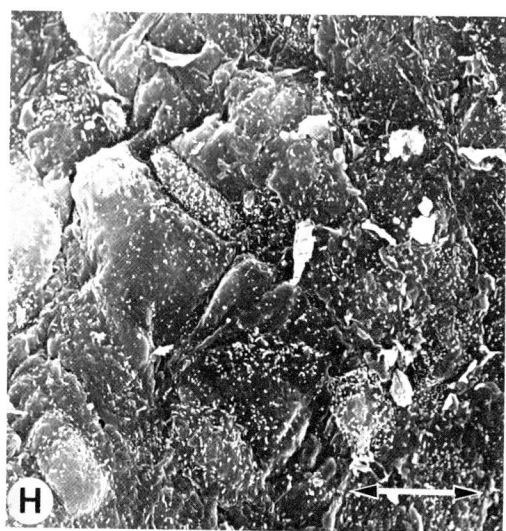

See next page for Fig. 15 I, J & K.

Fig. 15. Local variations of the endocardial morphology inside the embryonic heart.
a) Right half of the right ventricle showing the localizations where the detailed micrographs (A-J) have been taken. Chick embryo, 5ed 8h. Scale bar = 400 μm.
A. Endocardial cells with marginal folds (arrowhead) on the distal ventral bulbar cushion. Note the shape preservation of the erythrocytes. They are biconvex in birds.
In this and following pictures, the scale bar = 10 μm.
B. Cytosegresome (asterisk) and smaller density of microvilli at the top of the distal ventral bulbar cushion.
C. Intercellular openings (arrows) at the bottom of the distal ventral bulbar cushion.
D. Marginal folds (arrow) and ruffles (arrowheads) on the bottom of the distal ventral bulbar cushion.
E. Intercellular openings (arrows) and ruffles (arrowhead) at the upper portion of the proximal right bulbar cushion.
F. Phagocyte (round asterisk) penetrating the endocardium of the pulmonary artery semilunar valve anlage through an intercellular opening. Note the presence of a mitotic cell (asterisk).
G. Phagocyte (asterisk) and ruffles in an intercellular opening (arrowhead) on the endocardium covering the upper border of the right AV canal.
H. Endocardial cells of varying size, with large numbers of microvilli and some marginal folds from the center of the right AV canal.
I. Phagocytes (asterisks) on the endocardium overlying the top of the proximal right bulbar cushion.
J. Incomplete endocardium in the posterior wall of the right atrium. Arrowheads indicate the defects in which sometimes the fibrillar structure of the subendocardium (1) and/or myocardial fibres (2) can be seen.
K. Many ruffles (arrowheads) and intercellular openings (arrows) on the endocardium situated at the bottom of the right AV canal.

Mowczko (1978) after glutaraldehyde fixation followed by osmium tetroxide postfixation. The essential differences are the size (the blisters have a diameter of 0.2 - 0.6 μm only), the content (the blisters are wholly empty, whereas the cytosegresomes contain cytoplasm and cell organelles) and, finally, the localization (the blisters are disposed along cell limits, whereas the cytosegresomes are to be found at random over all of the cell surface).

Cytosegresomes can be found also in illustrations from papers of other authors, even if they are not commented upon as such, e.g., Fig.20 from Markwald et al. (1975); a "rounded cell" in Fig. 12 from Morse (1978) is also a cytosegresome. Also the "blebs" described by Hurle (1979) on the arterial pole of chick embryo semilunar valves are probably cytosegresomes.

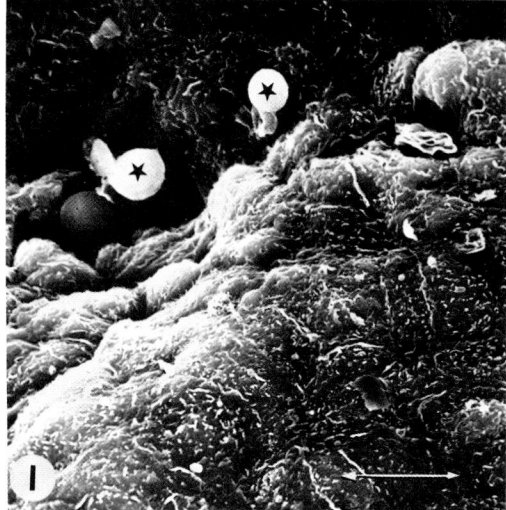
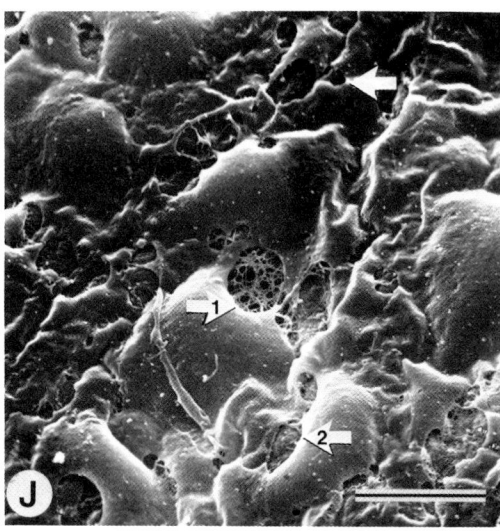
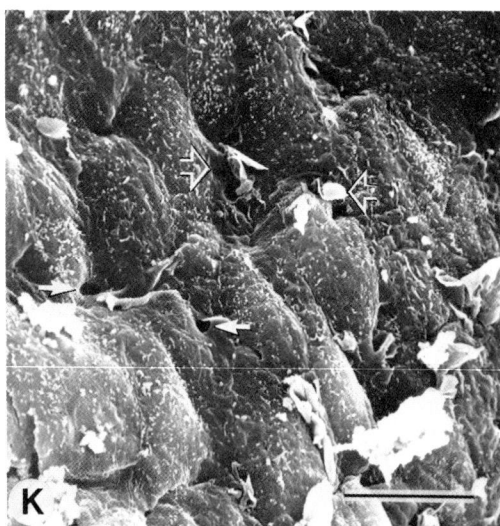

Fig. 15 (cont'd). See previous page for legend.

## H. Intercellular openings

It was a surprise to find that the general idea of a "watertight endocardium" obtained from LM studies is shown not to be valid examined in SEM. Intercellular clefts are a frequent feature of the embryonic endocardium, and these were observed in all 5 species studied (Figs 7a, 7c, 11a, 11b, 12a, 12b, 13a, 14d, 15C, 15E, 15F, 15G, 15K).

In 1975 "dehiscences" at the cell limits communicating with the cardiac jelly were described (Pexieder, 1975). Their usual size was 3 x 2 μm (Pexieder, 1976). They have been seen also in TEM studies (Pexieder, 1980).

Markwald et al. (1975) described intercellular gaps or pores of 0.5 - 3 μm in size in the endocardium of atrioventricular cushions of 11.5 ed rat embryos. These openings were reported in ventricular mural endocardium by Johnson (1976). Inter- and intracellular fenestrations were seen at the margins of foramina secunda in the chick embryo by Hendrix and Morse (1977). Local endocardial interruptions with protruding mesenchymal cells were reported by Los and Langemeijer-Eijndthoven (1978) on the atrioventricular cushions of the mouse embryonic heart (Los, 1978). Large gaps or pits between the cells covering the semilunar valves of chick embryo great vessels were identified by Hurle (1979). On the other hand, the intercellular clefts seen in an illustration from Steding et al. (1979) must be considered fixation and drying artefacts. Large intercellular spaces were found between 3ed 8h and 3ed 13h in mural endocardium of chick embryonic hearts by Bolender and Markwald (1979). These authors consider such a formation as consistent with endocardial motility. According to their opinion, the intercellular spaces are the consequence of atrioventricular endocardial cells pulling free from the surface epithelium (endocardium) and migrating into cardiac jelly. They suggest that these pores can allow rapid hydration of the cardiac jelly matrix and a rapid entrance of inorganic ions. The diffusible cations could then be immobilized by interactions with polyanionic glucosaminoglycans. The observations of Pexieder (1976, 1980) and Los (1978) show that the intercellular pores allow for the passage from the lumen into the subendocardium, as mentioned above, but also in a reverse sense, from the subendocardium (cushion tissue) into the cardiac lumen. Los (1981) stated that there are fewer intercellular pores in the chick embryos than in the mouse and rat embryos. This may hold true for the fusing AV cushions, but this could not be confirmed in material for the outflow tract, where the mouse embryonic heart showed the least frequency of these pores.

In the adult mouse heart intra- and intercellular pores were reported by Maguire (1972), but their existence was denied in the rat (Noack et al., 1973). They were also observed in the dog aortic valve endothelium by Hammon et al. (1974).

Intercellular openings have not been reported in SEM studies of typical blood vessels.

### I. Incomplete endocardium

A most particular feature has been observed in the atria of chick, dog and human embryos (Figs 11c, 12c, 13b, 15J), corresponding to large "defects" in the endocardial lining having the size of one or more endocardial cells. In such areas the subendocardium or even the cardiac myocytes are largely and directly exposed to the blood stream. Studies of mouse and rat embryonic hearts in the author's laboratory are not sufficiently complete to allow for a categorical denial of the existence of these surface differentiations in the two rodent species mentioned. When describing this unique morphology for the first time (Pexieder, 1980), it was suggested that it might have been related to the part of the atria derived from the sinus venosus. Similar observations in investigations of embryonic and adult endocardium or endothelium by other authors do not appear to have been made.

### J. Mitosis

Despite the rather high rate of DNA synthesis in the embryonic endocardium (Pexieder, 1978; Paschoud and Pexieder, 1981), the observation of a dividing endocardial cell in SEM is very rare (Figs 9d, 11d, 12d, 15F). This is easily understood if one considers as Bolender and Markwald (1979) did, that the mitoses are most frequently oriented perpendicular to the endocardial surface, resulting in mitotic transposition of endocardial daughter cells into cardiac jelly. Sarphie (1980) has indicated that as endocardial cells approach division, they engage in elaboration of microvilli and blebs. If the plane of division is not lying in the plane of endocardium as in Fig. 9d, then the mitotic cells can be identified only as single cells with very high numbers of microvilli. SEM pictures of individual dividing cells have been published (Pexieder, 1977, 1978, 1980). The great difficulty of presenting adequate SEM micrographs of this endocardial feature in rat embryos is certainly due to an insufficient amount of material investigated, whereas this feature seems really to be more rare or morphologically less easily identifiable in the mouse. Hurle (1979) described abundant mitoses on the arterial face of chick embryo semilunar valves, but did not present a corresponding SEM micrograph. Dividing cells were never reported in SEM studies of normal adult endocardium and/or endothelium.

### K. Macrophages

A typical feature of the embryonic endocardium is the presence of phagocytes or macrophages. They could be observed in chick, mouse, rat, dog and human embryonic hearts (Fig. 14). Some of them have been described in previous publications (Pexieder, 1975, 1976, 1977). In 1980 TEM evidence of dead and dying cells and macrophages lying under and on the embryonic endocardium was presented (Pexieder, 1980). Los and Langemeijer-Eijndthoven (1978) found the macrophages on interrupted spots of endocardium in fusing mouse atrioventricular cushion. These phagocytes are most probably related to the phenomena of physiological cell death accompanying the heart morphogenesis. This topic has been extensively discussed in reviews by the present author (Pexieder, 1975a, 1981a). Macrophages on normal healthy adult endocardium and endothelium were never reported in SEM. Wheeler et al. (1972) have seen macrophages characterized by numerous ruffles and bulbous surfaces on allograft human valves obtained 4 months after transplantation.

### VI. Local variations in surface morphology inside the embryonic heart

In the heart of chick embryos, as well as (at reduced scale) in those of mouse embryos, specific investigations relating the endothelial cell morphology, described in section V, to the topography of the internal relief of the developing heart have been undertaken. These investigations consisted in meander-like scanning of the surface at the microdissected halves of the embryonic heart at primary magnification X 1.100. Features of interest were photographed and the localization of the particular field noted on a previously prepared overview picture of the specimen (primary magnification X 16). Usually some 40 to 50 micrographs were obtained per microdissected half, such as is presented in Fig. 15a. This approach was published for the first time in 1978 (Pexieder, 1978). Micrographs forming Fig. 15 of the present review represent a selection from the photographs obtained in that particular specimen, as an example.

This kind of investigation permits the following conclusions concerning the chick embryo heart on 5ed 8h: there are both more and less active zones of the endocardium, where under "active" we understand the presence of specialized microappendages discussed in the preceding section. The active zones correspond generally to structures involved in the morphogenesis of the heart, particularly to its septation, e.g., bulbar (conotruncal) and atrioventricular cushions. The less active zones, e.g., mural and ventricular endocardium, are characterized by varying sizes of endocardial cells and varying density of microvilli. These aspects will need a quantitative approach for further analysis.

Concerning the bulbar (conotruncal) cushions, study of Fig. 15 allows general statements: whereas on the top of the cushions the cells display fewer microvilli but frequently phagocytes and cytosegresomes, the upper and lower borders show the presence of intercellular clefts, marginal folds and ruffles. Similar distributions are observed at and around the atrioventricular canal. Using this method, it is possible to follow the developmental trends of a given region of the embryonic heart in terms of its endocardial morphology, stage by stage. This kind of observation was made on the chick embryo, but a full account would be beyond the scope of this review. Two structures will be taken as examples - the proximal bulbar cushions and the semilunar valves.

The formation of proximal bulbar (conus) cushions occurs in the chick embryo at 3ed. At that time the endocardium is characterized by the presence of marginal folds and cytosegresomes, on 3ed 16h fewer cytosegresomes and fewer microvilli occurring than at 3ed. Many ruffling edges can be observed on the distal part of proximal (conus) bulbar cushions. On 4ed 8h intercellular clefts, some of them giant ones with traversing phagocytes are seen on the proximal (conus) bulbar cushions. On 5ed 8h, intercellular clefts open in high numbers on the proximal part of the mentioned cushions. 6ed is marked by the merging of proximal (conus) and distal (truncus) cushions. This process is accompanied by the massive presence of cytosegresomes.

First signs of development of semilunar valves from the distal (truncus) bulbar cushions (Pexieder, 1978) can be observed in the chick embryo on 5ed 8h. At this stage the endocardium of the valves is characterised by numerous cytosegresomes and otherwise flattened cells. On 6ed intercellular clefts with protruding ruffles appear just beneath the valvular lip, whereas the lower parts of the valve display marginal folds. Detachment of a cytosegresome transformed subsequently into a filopodium is seen at 6ed 16h. At that time, distal portions of the semilunar valves show flattened cells with remnants of marginal folds. The marginal folds and ruffles disappear almost completely on 7ed, with few intercellular clefts remaining on the proximal side of the valves. A proximodistal gradient of microvillous density does exist at that stage. Even if an interpretation in terms of cellular physiology and/or morphogenetic action is at present rather difficult, two general issues governing the morphology of endocardial cells should be considered. They are: the functional state of a cell and hemodynamics. It is well known from the tissue culture work in vitro that the kind, amount and localization of various microappendages are dependent upon the functional and metabolic state of a cell (Bell and Revel, 1980); on the other hand, as will be shown in greater detail in the next section, modifications of embryonic hemodynamics can produce changes in the morphology of the endocardial cell. The final state, as demonstrated by SEM, is an integration of the functional state of a cell with the various parameters of the hemodynamic action of the blood stream. Both are localization and developmental stage dependent.

## VII. Pathologic and experimental modifications of the embryonic and fetal endocardial morphology as seen in SEM

SEM studies of the embryonic and/or fetal endocardium under pathologic or experimental conditions are almost non-existent. In this section four conditions will be discussed: the morphology of the human fetal endocardium from the prostaglandin-induced second trimester abortions, the embryonic endocardium in a hereditary congenital heart disease, the effects of limiting endocardial mobility by cytochalasin B, and the effects of altering embryonic hemodynamics.

### A. Prostaglandins

Today prostaglandins $F_{2\alpha}$ and $E_{2\alpha}$ ($PGF_{2\alpha}$ and $PGE_{2\alpha}$) are widely used for terminations of pregnancy after 12 weeks of gestation. The material from such abortions may, theoretically, be considered for various studies. Under the conditions used in the present study, the interruption was started by an intravaginal priming dose of 7.5 mg of $PGF_{2\alpha}$ in the form of a gel preceding by 14 hrs the first paraamniotic application of another 7.5 mg. This treatment was repeated in 2-3 hour intervals until the expulsion of the fetus (usually 8 to 15 hours from the first application) (De Grandi, 1979). At that time the fetus (already presenting signs of insufficient circulation) was perfusion fixed as described in chapter II.

In SEM the endocardium of these fetuses presented varying degrees of abnormalities. The very first abnormal sign was the swelling of microvilli and appearance of a multitude of blisters (Fig. 16d). This was followed by alterations of the endocardial cell membrane resulting in its "sieve-like" appearance and the beginning retraction of cell borders (Fig. 16a). The cellular retraction continued, giving the cells bulging aspects and creating artificial intercellular dehiscences (Fig. 16b) communicating with the subendocardium. The next step in this process was the endocardial cell denudation (Fig. 16c). The initial changes were amplified by the preparative procedures (dehydration, drying), resulting in the formation of bizarre structures (Fig. 16e).

Considering the mode of action of $PGF_{2\alpha}$, it is difficult to conceive that the described changes are the effects of its direct action on the endocardium. The $PGF_{2\alpha}$ produces contractions of the uterine musculature. These contractions kill the fetus by, among other factors, interrupting the placental circulation, which results in anoxia. Finally the death of the fetus results in its expulsion. This sequence of events explains the 8 to 15 hrs interval separating the first $PGF_{2\alpha}$ treatment and the effective fetus dismissal. The large variability of this interval also explains the wide variety of cellular changes. Consequently these changes must be considered as signs of cell injury predominantly autolytic and postmortem in nature.

In the adult human endocardium, prolonged anoxia resulted in coarse granular appearance of cell surfaces (Katsumoto and Watanabe, 1972). An experimental infarction in the dog was followed after 20 minutes by an irreversible injury resulting in separation of adjacent cells along their boundaries followed by the shedding of some endocardial cells (Gavin et al., 1973; Johnson et al., 1979). In human hearts suffering infectious endocarditis, moderate postmortem autolysis was characterized by considerable shrinkage of individual cells, retracting cell junctions and small

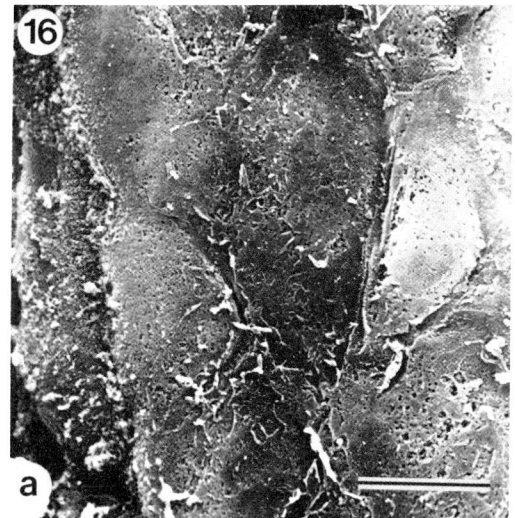
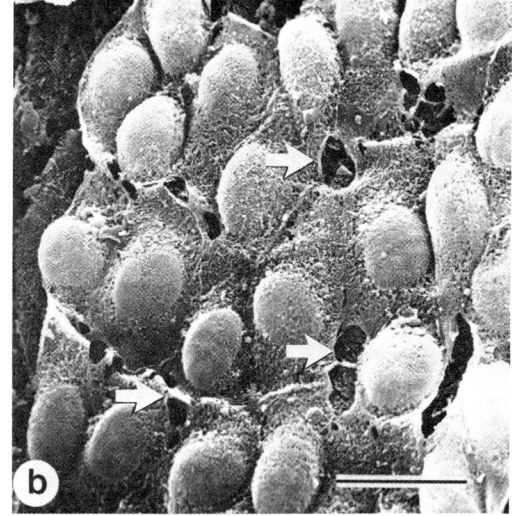

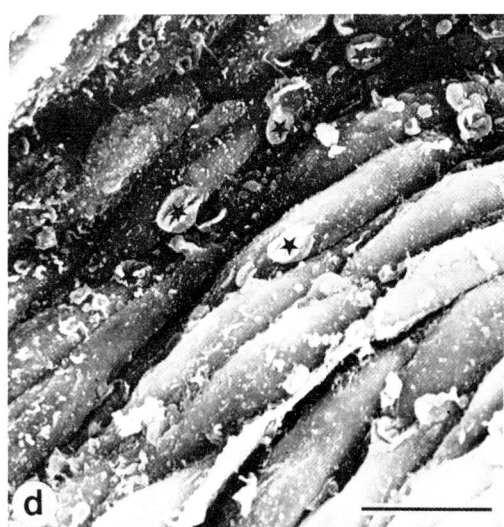
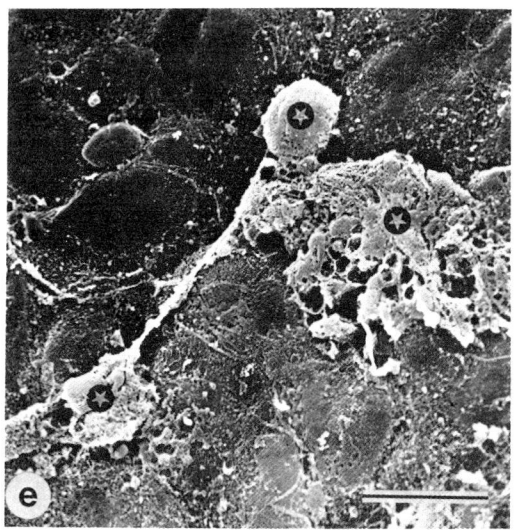

Fig. 16. Abnormal changes of the endocardium from prostaglandin-induced human pregnancy terminations.
All scale bars in these figures = 10 μm.
a) Erosions of the endocardial cell membranes. Pulmonary infundibulum. Human fetus, 12 weeks of gestation.
b) Bulging of endocardial cell nuclei together with the retraction of the cytoplasm resulting in creation of giant artificial intercellular holes (arrows). Atrioventricular orifice. Human fetus, 16 weeks of gestation.
c) Complete denudation of the subendocardium. Some contracted endocardial cells (asterisk) are still sticking to the fibrillar subendocardial meshwork. Left ventricle. Human fetus, 16 weeks of gestation.
d) Blebbing (asterisks) of the endocardial cells and edematous degeneration of the microvilli. Ductus arteriosus. Human fetus, 12 weeks of gestation.
e) Autolysing phagocytes on the endocardium (asterisks). Pulmonary infundibulum. Human fetus, 12 weeks of gestation.

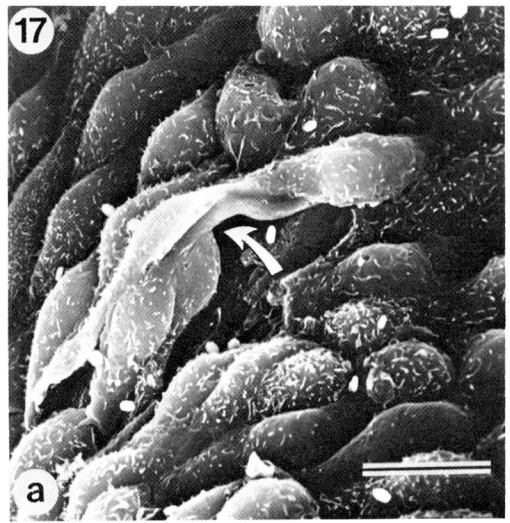
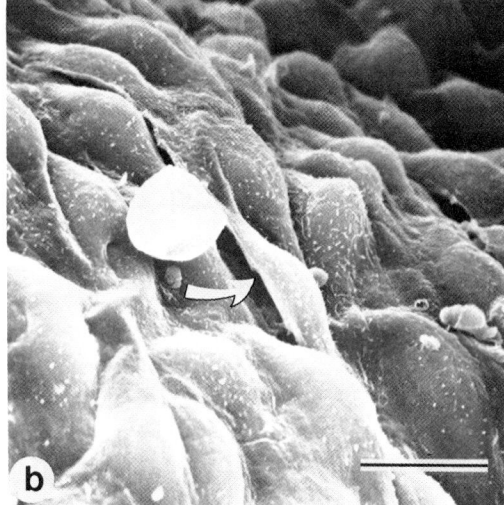
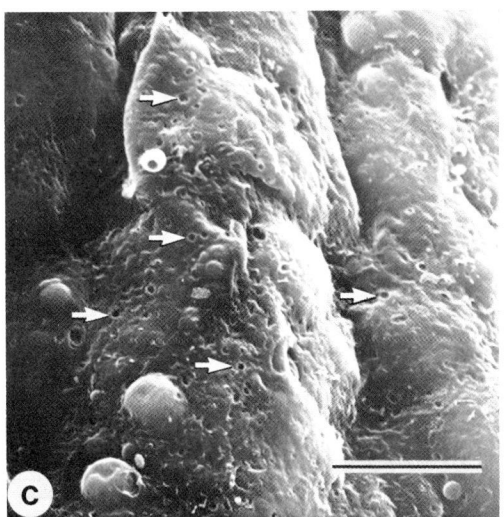
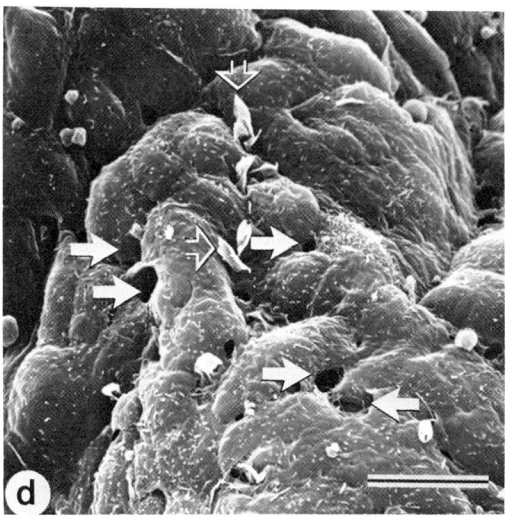

Fig. 17. Experimental modifications of the endocardial morphology.

a & b) Double layered endocardium (arrows) from hearts suffering hereditary congenital heart disease. Ventricular septal defect and proximal right bulbar cushion. Keeshond dog embryo, stage XVI. Scale bar = 10 µm.

c) Small surface defects ("craters") from hearts with hemodynamically-induced congenital heart disease (arrows). Proximal left bulbar cushion. Chick embryo. 4ed 16h. Scale bar = 5 µm.

d) Empty and widened intercellular openings (arrows) from hearts with hemodynamically-induced congenital heart disease. Note the presence of some ruffles (arrowheads). Proximal left bulbar cushion. Chick embryo, 5ed. Scale bar = 5 µm. Reproduced by the permission of Association des Anatomistes, Nancy, from Pexieder (1976a).

holes in the plasma membrane (Wright and Kirschner, 1979).

Many of the features observed in prostaglandin-aborted fetuses have been described previously on the endothelium of blood vessels under different experimental conditions as agent - or as etiology - unspecific reaction of the cell to injury. For instance edematous microvilli were reported in hypovitaminosis C (Weber et al., 1970) or in blood vessels rinsed by saline before perfusion fixation (Edanaga, 1974). Formation of blebs occurred after ischemia (Sunaga et al., 1973), in autogenous venous grafts (Crissman et al., 1976), following hypovitaminosis A (Bayer et al., 1976), bad fixation (Buss and Hollweg, 1977), and surgical stress (Stewart et al., 1978). The appearance of cell membrane discontinuities was described after the exposure of human umbilical vein to endotoxin lipid A (Chen et al., 1978) and in monkeys' aorta after atherogenous diet (Taylor et al., 1978). The retraction of cell borders was seen under various circumstances, such as after adrenaline or cholesterol administration (Shimamoto, 1974), in veinous autografts (Ramos et al., 1976), hypovitaminosis A (Bayer et al., 1976), in organ-cultured vessels (Buck, 1977), after free fatty acids injection (Sedar et al., 1978), after atherogenous diet (Taylor et al., 1978) and after surgical trauma (Stewart et al., 1973). Even more common are observations on cellular retraction and nuclear bulging such as after cholesterol diet (Shimamoto et al., 1969; Svendsen, 1979), hypoxia (Kjeldsen and Thomsen, 1975), endotoxin treatment (Reidy and Bowyer, 1977), clamp ischemia (Fonkalsrud et al., 1977), or free fatty acid application (Sedar et al., 1978). In the next degree of injury, the endothelial cells remained attached only at poles (Jørgensen and Svendsen, 1978) before nuclear extrusion (Fonkalsrud et al., 1977; Sedar et al., 1978) and final denudation (Reichle et al., 1973; Stewart et al., 1973; Ramos et al., 1976; Guidoin et al., 1978; Wechezak and Mansfield, 1979; Pagnanelli et al., 1980). The present observations demonstrate that the human fetal material from the prostaglandin-induced terminations of pregnancy is unsuitable for ultrastructural studies. Also its use in other studies as tissue culture or biochemical analysis should be discouraged because it represents essentially dead, autolysing tissues.

B. Hereditary congenital heart disease

The main purpose of the studies on the dog was to investigate the cellular pathogenesis of heart malformations which are hereditary, with multifactorial inheritance/etiology, in the Keeshond strain (Patterson et al., 1974; Van Mierop et al., 1977; Van Mierop and Patterson, 1978, 1980). Until now, the only qualitative alteration seen has been the doubling of the endocardial lining of the bulbar (conotruncus) cushions in the abnormal heart. This was seen first in the light microscope on serial sections and, later on (Figs 17a, b), SEM specimens from the same strain. This feature was absent in control hearts of Beagle embryos. At this moment no precise explanation is forthcoming. It occurs at the time the cushion cells are already formed and, therefore, it is unclear whether or not it can be related to overgrowth of the endocardium preceding the cushion cells seeding reported by Bolender and Markwald (1979). Studies are in progress for further elucidation of the cellular changes underlying the abnormal heart development in this animal model (Pexieder, 1980).

C. Cytochalasin B

Cytochalasin B was used by Bolender and Markwald (1979) to test their hypothesis on endocardial cell mobility as a parameter of endocardial cushion tissue cell derivation. The chick embryos were treated at 2ed 7h with 5 µg of cytochalasin B and examined on 3ed, 3ed 8h, 3ed 16h and 4ed. The cytochalasin B treatment, which disrupts microfilaments, did not present hypertrophy of the endocardium and the reorientation of the endocardial cells observed in controls. Absent in treated embryos were the overlapping and overgrowth of cells. On the cardiac jelly surface of these cells less motile processes were seen. Consequently, the AV endocardium (where cushion tissue cell formation normally occurs) resembled the nonprogenitive mural endocardium (with normally no cushion tissue cell formation). Bolender and Markwald (1979) interpret their experiment as if the decreased endocardial motility resulted in a disruption but not a total arrest of cushion seeding. The full importance of this experiment can be understood only in the context of the whole work done by Markwald (Markwald et al., 1977, 1978, 1979; Markwald and Bernanke, 1981) and Fitzharris (Fitzharris et al., 1979, 1980; Kinsella and Fitzharris, 1980) and cannot be treated in detail within the framework of the present review.

D. Effects of hemodynamic factors

The experimental or pathologic modifications of the embryonic circulation can alter the normal development of the heart and result in congenital malformations in this organ (for more details on this particular background, see Pexieder, 1978a). The author's laboratory has been interested in the cellular and subcellular aspects of interaction of the embryonic endocardium with the blood stream.

In 1976 a series of experiments were performed in which both the right and left sixth aortic arches (future pulmonary arteries) were clamped on 4ed in the chick embryo, resulting in the production of a ventricular septal defect (Pexieder, 1976a). The embryonic hearts were studied 8, 16, 24 and 32 hours after the intervention using the techniques described in sections III and V of this review. It was observed that 8 hours after the operation, ruffles reported in section IV disappeared completely, whereas giant intercellular openings occurred on the top of the proximal left bulbar cushion (Fig. 17d). Very small crater-like formations formed at the bottom of the same cushion. Numerous phagocytes were seen traversing the endocardium 16 hours after the intervention, and "craters" (Fig. 17c) formed on the top of distal ventral bulbar cushions. Eight hours later (24 hrs since the intervention), cytosegresomes, described also in section IV, were missing on the convex side of the bulbus. These appeared after a delay of eight hours, 32 hours after the clamping procedure, were smaller and less frequent. In an extension of this work (Pexieder, 1977a, 1979), attempts were made to correlate these ultrastructural changes and their topography at the organ level with changes of hemodynamic parameters, such as pressure and flow, and it has been proposed that there is an association of giant intercellular openings and craters with increased blood flow on one hand and of transit of phagocytes with decreased blood flow on the other. One of the most convincing documents is the picture, published in 1978 (Pexieder, 1978a) of an endocardial cell, 32 hours after the clamping procedure, literally torn out in the direction of the blood stream.

No similar observations could be found in the literature on SEM morphology of the adult endocardium. The changes in cellular lining of bladders developed as a component of a left ventricle assist device and described by Wechezak and Mansfield (1979), are in part due to blood flow (cell polarization) but in part also the consequence of its wall motion (cell retraction from the substrate). Various craters, crater-like formations of crateriform lacunes, were frequently reported in the endothelium of blood vessels under various experimental conditions, as in hypovitaminosis C (Weber et al., 1970), ischemia (Sunaga et al., 1973; Kawamura et al., 1973; Nelson et al., 1975, 1976), cholesterol and adrenaline treatment (Shimamoto, 1974), at branching points (Buss and Hollweg, 1977) and in low pressure perfusion fixation (Lee and Chien, 1979). Even if these structures are considered by Nelson et al. (1975) as another nonspecific reaction of the endothelial cell to injury, they differ from the formations observed in the author's laboratory basically in their size. The craters in adult endothelial cells are usually 7 to 10 µm in diameter, whereas our "craters" were the size of a fraction of a µm.

## Conclusions

It has been shown that perfusion fixation of the embryonic heart is feasible and represents the best practice for the SEM study of the embryonic endocardium. The osmotic fragility of the endocardial lining of the prenatal heart requires that the tonicity of the fixative vehicle be carefully controlled.

The chick, mouse, rat, dog, and human embryos could be fixed and otherwise prepared using strictly the same methods so that the subsequent SEM studies were quite comparable. Despite the wide divergences of the heart development at the organ level in these species, the underlying cellular elements, at least insofar as the endocardium is concerned, are fairly uniform. Almost all species presented more or less microvilli, marginal folds, lamellipodia, filopodia, cytosegresomes, intercellular openings, mitotic cells and phagocytes on their embryonic endocardium. More efforts will be needed to devise a reasonable and unified nomenclature or terminology for these microappendages of the endocardial cells. Even more numerous undertakings are necessary in studies which will allow the interpretation of what the presence of these structures in the embryonic heart really means in terms of cellular function, as well as in terms of heart morphogenesis. Studies in the author's laboratory have shown that the use of material from prostaglandin-induced abortions should be avoided. Even if the embryonic endocardium has been shown to react to some chemical, genetic and hemodynamic factors in a basically similar way, as the adult endocardium and endothelium (with the exception of the platelet, mononuclear and leukocyte participation) vast horizons remain for further exploration.

## Acknowledgements

The author's original work reported in this review was supported by grants no. 3.465.0.75, 3.162.0.77 and 3.673.0.80 from the Swiss National Science Foundation.

We are thankful to Drs D.F. Patterson (Philadelphia) and J.W. Scott, Jr. (Cincinnati) for making the dog and rat embryos available and for the performing perfusion fixations.

We acknowledge the precious contribution of Dr. P. Janeček from the Department of Obstetrics and Gynecology (Lausanne) in providing intact human embryos.

Our thanks are due to Miss M. Vuillemin (B. Sci.) from my Department for help with the study of some of the mouse embryos. The skillful photographs of Miss C. Dumauthioz, as well as no less effective secretarial help of Mrs E. Warwood and Mrs M. Devolz are appreciated.

## References

Abercrombie, M. The bases of the locomotory behaviour of fibroblasts. Exp. Cell Res., suppl. $\underline{8}$, 1961, 188-198.

Abrunhosa, R. In vivo microperfusion fixation of embryos for ultrastructural studies. J. Ultrastruct. Res. $\underline{38}$, 1972, 188-212.

Arborgh, B., Bell, F., Brunk, U. and Collins, V. P. The osmotic effect of glutaraldehyde during fixation. A transmission electron microscopy, scanning electron microscopy and cytochemical study. J. Ultrastruct. Res. $\underline{56}$, 1976, 339-350.

Bayer, R.C., Smith, M.A. and Ringer, R.K. The influence of vitamin A deficiency on aortic endothelium of Japanese quail. Artery $\underline{2}$, 1976, 423-430.

Bell, P.B. and Revel, J.-P. Scanning electron microscope application to cells and tissues in culture. In: Biomedical research applications of scanning electron microscopy, vol. 2, G.M. Hodges and R.C. Hallowes (eds). Academic Press, London, 1980, 1-64.

Bolender, D.L. and Markwald, R.R. Epithelial-mesenchymal transformation in chick atrioventricular cushion morphogenesis. Scanning Electron Microsc. 1979/III, 313-322.

Buck, R.C. Organ cultures of rat aorta: a scanning and transmission electron microscopic study. Exp. Mol. Pathol. $\underline{26}$, 1977, 260-277.

Buss, H., Dahm, H.H. and Lindenfelser, R. Die Oberfläche des Endocards des Rattenherzens. Rasterelektronenmikroskopische Untersuchungen. Beitr. Path. $\underline{148}$, 1973, 340-359.

Buss, H. and Hollweg, H.G. Scanning electron microscopy of blood vessels : A review.

Scanning Electron Microsc. 1977/II, 467-476.

Buss, H., Schneider, J. and Hollweg, H.G. Endothelial surface of large veins of rabbit - Scanning electron microscopic observations. Path. Res. Pract. 165, 1979, 392-410.

Busson-Mabillot, S. Influence de la fixation chimique sur les ultrastructures. I.Etude sur les organites du follicule ovarien d'un poisson téléostéen. J. Micr. 12, 1971, 317-348.

Chavaz, P. and Pexieder, T. Quelle osmolarité choisir pour la fixation en microscopie électronique? Acta anat. (Basel) 95, 1976, 142-143.

Chen, S., Barnhart, M.I., Gemski, P.M. and Alving, C.R. Impact of lipid from endotoxin on endothelium. Scanning Electron Microsc. 1978/II, 357-366.

Chi, E.V. and Schwartz, S.M. Surface replicas of aortic endothelium. Scanning Electron Microsc. 1978/II, 479-484.

Clark, J.M. and Glagov, S. Luminal surface of distended arteries by scanning electron microscopy: Eliminating configurational and technical artefacts. Brit. J. exp. Path. 57, 1976, 129-135.

Collins, V.P., Brunk, U., Frederiksson, B.-A. and Westermark, B. Transmission and scanning electron microscopy of whole glioma cells cultured in vitro. Scanning Electron Microsc. 1980/II, 223-230.

Crissman, R.S., Ross, J.N. and Dosick, S.M. Scanning electron microscopy of endothelial damage induced by autogenous vein graft techniques. Anat. Rec. 184, 1976, 384.

De Grandi, P. Interruption de grossesse. Mode d'usage des prostaglandines dans le service de gynécologie-obstétrique du CHUV, Lausanne. Rev. méd. Suisse rom. 99, 1979, 665-670.

Edanaga, M. A scanning electron microscope study on the endothelium of the vessels. I.Fine structure of the endothelial surface of aorta and other arteries in normal rabbits. Arch. histol. jap. 37, 1974, 1-14.

Edanaga, M. A scanning electron microscope study on the endothelium of vessels. II.Fine surface structure of the endocardium in normal rabbits and rats. Arch. histol. jap. 37, 1975, 301-312.

Erickson, C.A. and Trinkhaus, J.P. Microvilli and blebs as sources of reserve surface membrane during cell spreading. Exp. Cell Res. 99, 1976, 375-384.

Fitzharris, T.P. Discussion. In: Mechanisms of cardiac morphogenesis and teratogenesis. T.Pexieder (ed.). Raven Press, New York, 1981, 278-279.

Fitzharris, T.P. Endocardial shape change in the truncus during cushion tissue formation. In: Mechanisms of cardiac morphogenesis and teratogenesis, T. Pexieder (ed.). Raven Press, New York, 1981a, 227-235.

Fitzharris, T.P., Markwald, R.R. and Dunn, B.E. Effects of beta-aminopropionitrile fumarate (BAPN) on early heart development. J. Molec. Cell. Cardiol. 12, 1980, 553-578.

Fitzharris, T.P., Thompson, R.P. and Markwald, R.R. Matrical ordering in the morphogenesis of tunica media. Texas Rep. Biol. Med. 39, 1979, 287-304.

Fonkalsrud, E.W., Sanchez, M., Zerubavel, R. and Mahoney, A. Serial changes in arterial endothelium following ischemia and perfusion. Surgery 81, 1977, 527-534.

Fujimoto, S., Yamamoto, K. and Takeshige, Y. Electron microscopy of endothelial microvilli of large arteries. Anat. Rec. 183, 1975, 259-266.

Garbarsch, C. and Christensen, B. Scanning electron microscopy of aortic endothelial cell boundaries after staining with silver nitrate. Angiologica 7, 1970, 365-373.

Gavin, J.B., Wheeler, E.E. and Herdson, P.B. Scanning electron microscopy of the endocardial endothelium overlying early myocardial infarcts. Pathology 5, 1973, 145-148.

Guidoin, R., Martin, L., Levaillant, P., Gosselin, C., Domurado, D., Marois, M., Awad, J. and Blais, P. Endothelial lesions associated with vascular clamping. Surface micropathology by scanning electron microscopy. Biomat. Med. Devices Art. Org. 6, 1978, 179-198.

Hammon, J.W. Jr., O'Sullivan, M.J., Oury, J. and Fosburg, R.G. Allograft cardiac valves: a view through the scanning electron microscope. J. thorac. cardiovasc. Surg. 68, 1974, 352.

Harasaki, H., Suzuki, I., Tanaka, J., Hanano, H. and Torisu, M. Ultrastructure research of the endocardial endothelium. Arch. histol. jap. 38, 1975, 71-84.

Hay, D.A. Development and fusion of the endocardial cushions. In: Morphogenesis and malformation of the cardiovascular system. G.C.Rosenquist and D. Bergsma (eds). Birth Defects: Orig. Art. Ser. XIV, No 7. Alan R. Liss, New York, 1978, 69-90.

Hendrix, M.J.C. Scanning electron microscopy of atrial septation in the human fetus. Anat. Rec. 187, 1977, 602.

Hendrix, M.J.C. and Morse, D.E. Atrial septation. I. Scanning electron microscopy in the chick. Develop. Biol. 57, 1977, 345-363.

Hollweg, H.G. and Buss, H. Problems with preparation of blood vessels for scanning electron microscopy. A critical review. Scanning 3, 1980, 3-14.

Hurle, J.M. Scanning and light microscope studies of the development of the chick embryo semilunar heart valves. Anat. Embryol. 157, 1979, 69-89.

Johnson, R.C. Ultrastructure of developing ventricular endocardial endothelium in the avian embryo. Amer. Zool. 14, 1974, 1301.

Johnson, R.C. A scanning electron microscopic study of the developing chick mural endocardium. Anat. Rec. 184, 1976, 438.

Johnson, R.C., Crissman, R.S. and Didio, L.J.A. Endocardial alterations in myocardial infarction. Lab. Invest. 40, 1979, 183-193.

Jørgensen, L. and Svendsen, E. Endothelial cell injury. In: International conference on atherosclerosis, L.A. Carlson, R. Paoletti and G. Weber (eds). Raven Press, New York, 1978, 561-566.

Katsumoto, K. and Watanabe, S. Endocardial surface changes observed with a scanning electron microscope and phospholipid change during extracorporeal circulation in man. Jap. Circulat. J. 36, 1972, 523-528.

Kaufmann, P. Der osmotische Effekt der Fixation auf die Placentastruktur. Anat. Anz. 148, Erg.H., 1980, 351-352.

Kawamura, J., Sunaga, T., Mulhern, H. and Nelson, E. Ischemia of the common carotid artery in rabbits. Scanning and transmission electron microscopic study of the luminal surface. Scanning Electron Microsc. 1973, 465-472.

Kinsella, M.G. and Fitzharris, T.P. Origin of cushion tissue in the developing chick heart : Cinematographic recordings of in situ formation. Science 207 (4437), 1980, 1359-1360.

Kjeldsen, K. and Thomsen, H.K. The effect of hypoxia on the fine structure of the aortic intima in rabbits. Lab. Invest. 33, 1975, 533-543.

Lee, M.M.L. and Chien, S. Morphologic effects of pressure changes on canine carotid artery endothelium as observed by SEM. Anat. Rec. 194, 1979, 1-14.

Lim, K.O. and Boughner, D.R. Scanning electron microscopical study of human mitral valve chordae tendineae. Arch. Path. Lab. Med. 101, 1977, 236-239.

Litke, L.L. and Low, F.N. Fixative tonicity for scanning electron microscopy of delicate chick embryos. Amer. J. Anat. 148, 1977, 121-129.

Los, J.A. Cardiac septation and development of the aorta, pulmonary trunk, and pulmonary veins: previous work in the light of recent observations In: Morphogenesis and malformation of the cardiovascular system, G.C. Rosenquist and D. Bergsma (eds). Birth Defects: Orig. Art. Ser. XIV, No 7. Alan R. Liss, New York, 1978, 109-138.

Los, J.A. Introduction (cell surface). In: Mechanisms of cardiac morphogenesis and teratogenesis, T. Pexieder (ed.). Raven Press, New York, 1981, 255-266.

Los, J.A. and Langemeijer-van Eijndthoven, E. The fusion of the atrioventricular endocardial cushions in the heart of the chick and the mouse embryo. Acta morph. neerl. scand. 16, 1978, 138-139.

Maguire, K.F. Scanning electron microscopy of mouse endocardium and venae cordis minimae foramina. Z. Anat. EntwGesch. 139, 1972, 107-114.

Manasek, F.J. Embryonic development of the heart. I. A light and electron microscopic study of myocardial development in the early chick embryo. J. Morph. 125, 1968, 329-366.

Manasek, F.J. Histogenesis of the embryonic myocardium. Amer. J. Cardiol. 25, 1970, 149-168.

Manasek, F.J. Heart development: Interactions involved in cardiac morphogenesis. In: Cell surface reviews, vol. 1: The cell surface in animal embryogenesis and development, G. Poste and G.L. Nicolson (eds). North Holland Publ. Co, Amsterdam New York - Oxford, 1976, 545-597.

Manasek, F.J. Organization, interactions, and environment of heart cells during myocardial ontogeny. In: Handbook of physiology, section 2: The cardiovascular system, vol. 1: The Heart, R.M. Berne (ed.). Amer. Physiol. Soc., Bethesda, 1979, 29-42.

Manasek, F.J., Kulikowski, R.R. and Fitzpatrick, L. Cytodifferentiation: a causal antecedent of looping? In: Morphogenesis and malformation of the cardiovascular system, G.C. Rosenquist and D. Bergsma (eds). Birth Defects: Orig. Art. Ser. XIV, No 7. Alan R. Liss, New York, 1978, 161-178.

Markwald, R.R. and Bernanke, D.H. Structural analysis of 6-diazo-5-oxo-L-norleucine effects upon early cushion tissue morphogenesis. In: Mechanisms of cardiac morphogenesis and teratogenesis, T. Pexieder (ed.). Raven Press, New York, 1981, 237-251.

Markwald, R. and Fitzharris, T.P. Early endocardial development in the rat. Anat. Rec. 178, 1974, 411-412.

Markwald, R.R., Fitzharris, T.P. and Adams Smith, W.N. Structural analysis of endocardial differentiation. Develop. Biol. 42, 1975, 160-180.

Markwald, R.R., Fitzharris, T.P., Bank, H. and Bernanke, D.H. Structural analyses on the matrical organization of glycosaminoglycans in developing endocardial cushions. Develop. Biol. 62, 1978, 292-316.

Markwald, R.R., Fitzharris, T.P., Bolender, D.L. and Bernanke, D.H. Structural analysis of cell: matrix association during the morphogenesis of atrioventricular cushion tissue. Develop. Biol. 69, 1979, 634-654.

Markwald, R.R., Fitzharris, T.P. and Manasek, F.J. Structural development of endocardial cushions. Amer. J. Anat. 148, 1977, 85-121.

Mathieu, O., Claassen, H. and Weibel, E. Differential effect of glutaraldehyde and buffer osmolarity on cell dimensions: a study on lung tissue.

J. Ultrastruct. Res. 63, 1978, 20-34.

Matonoha, P. and Zechmeister, A. Scanning electron microscopic observation of intimal surface of normal and atherosclerotic arteries. Acta morph. Acad. Sci. Hung. 26, 1978, 173-184.

Millonig, G. Laboratory manual of biological electron microscopy. Mario Saviolo Inc., Vercelli 1976, 66p.

Missirlis, Y.F. and Armeniades, C.D. Ultrastructure of the human aortic valve. Acta anat. 98, 1977, 199-206.

Morse, D.E. Scanning electron microscopy of the developing septa in the chick heart. In: Morphogenesis and malformation of the cardiovascular system, G.C. Rosenquist and D. Bergsma (eds). Birth Defects: Orig. Art Ser. XIV, No 7. Alan R. Liss, New York, 1978, 91-107.

Naef, W., Munz, K. and Waser, P.G. Morphometric analyses of the electric organ of Torpedo: The influence of different fixative modes on the vesicle diameter. Histochemistry 58, 1978, 193-201.

Nelson, E., Gertz, S.D., Rennels, M.L., Forber, M.S. and Kawamura, J. Scanning and transmission electron microscopic studies of arterial endothelium following experimental vascular occlusion. In: The cerebral vessel wall, J. Cervos-Navarro, E. Betz, F. Matakas and R. Wullenweber (eds). Raven Press, New York, 1976, 33-40.

Nelson, E., Sunaga, T., Shimamoto, T., Kawamura, J., Rennels, M.L. and Hebel, R. Ischemic carotid endothelium. Arch. Path. 99, 1975, 125-131.

Noack, W., Schweichel, J.U. and Lunkenheimer, P. Elektronenmikroskopische und Rastermikroskopische Untersuchungen zur Morphologie der Sinusoide im Herzen der Ratte. Z. Anat. EntwGesch. 141, 1973, 171-178.

Ojeda, J.L. and Hurle, J.M. Cell death during the formation of tubular heart of the chick embryo. J. Embryol. exp. Morph. 33, 1975, 523-534.

Pagnanelli, D.M., Pair, T.G., Rittoli, H.V. and Kobrine, A.I. Scanning electron micrographic study of vascular lesions caused by microvascular needles and suture. J. Neurosurg. 53, 1980, 32-36.

Paschoud, N. and Pexieder, T. Patterns of proliferation during the organogenetic phase of heart development. In: Mechanisms of cardiac morphogenesis and teratogenesis, T. Pexieder (ed.). Raven Press, New York, 1981, 73-88.

Patterson, D.F., Pyle, R.L. and Van Mierop, L. Hereditary defects of the conotruncal septum in Keeshond dogs: Pathologic and genetic studies. Amer. J. Cardiol. 34, 1974, 187-205.

Peine, C.J. Scanning electron microscopy of selected surfaces of the dog and rat heart. Anat. Rec. 178, 1974, 436.

Peine, C.J. and Low, F.N. Scanning electron microscopy of cardiac endothelium of the dog. Amer. J. Anat. 142, 1975, 137-158.

Pexieder, T. SEM investigations on physiological cell death in the chick embryo heart. Experientia 31, 1975, 745.

Pexieder, T. Cell death in the morphogenesis and teratogenesis of the heart. Adv. Anat. Embryol. Cell Biol. 51 (3), 1975a, 1-100.

Pexieder, T. Rasterelektronenmikroskopische Beobachtungen der Oberfläche der Herzbulbuswülste der Hühnerembryonen. Anat. Anz. 140, Erg.H., 1976, 747-754.

Pexieder, T. Effets de l'hémodynamique sur la morphologie de l'endocarde embryonnaire. Bull. Ass. Anat. 60, 1976a, 399-406.

Pexieder, T. The role of buffer osmolarity in fixation for SEM and TEM. Experientia 32, 1976b, 806-807.

Pexieder, T. Pathogénie des malformations cardiaques. Bull. Fond. Suisse Cardiol. 8, 1977, 42-48.

Pexieder, T. SEM observations of the embryonic endocardium under normal and experimental hemodynamic conditions. Bibl. Anat. 15, 1977a, 531-534.

Pexieder, T. Development of the outflow tract of the embryonic heart. In: Morphogenesis and malformation of the cardiovascular system, G.C. Rosenquist and D. Bergsma (eds). Birth Defects: Orig. Art. Ser. XIV, No 7. Alan R. Liss, New York, 1978, 29-68.

Pexieder, T. Discussion of the topics: Effects of modifying the embryonic circulation. In: Morphogenesis and malformation of the cardiovascular system, G.C. Rosenquist and D. Bergsma (eds). Birth Defects: Orig. Art Ser. XIV, No 7. Alan R. Liss, New York, 1978a, 449-455.

Pexieder, T. Mechanisms of teratogenesis in hemodynamically induced ventricular septal defect. In: Advances in the detection of congenital malformation, E.B. Van Julsingha, J.M. Tesh and G.M. Fara (eds). Eur. Teratology Soc., Wethersfield, 1979, 264-268.

Pexieder, T. Cellular mechanisms underlying the normal and abnormal development of the heart. In: Etiology and morphogenesis of congenital heart disease, R. Van Praagh and A. Takao (eds). Futura Publ. Co., New York, 1980, 127-153.

Pexieder, T. Mechanisms of cardiac morphogenesis and teratogenesis. Perspectives in Cardiovascular Research, vol. 5. Raven Press, New York, 1981, 512p.

Pexieder, T. Introduction (Cell death). In: Mechanisms of cardiac morphogenesis and teratogenesis, T. Pexieder (ed.). Raven Press, New York, 1981a, 93-99.

Pexieder, T. Discussion. In: Mechanisms of cardiac morphogenesis and teratogenesis,

T. Pexieder (ed.). Raven Press, New York, 1981b, 278.

Ramos, J.R., Berger, K., Mansfield, P.B. and Sauvage, L.R. Histologic fate and endothelial changes of distended and nondistended vein grafts. Ann. Surg. 183, 1976, 205-228.

Reichle, F.A., Stewart, G.J. and Essa, N. A transmission and scanning electron microscopic study of luminal surfaces in Dacron and autogenous vein bypasses in man and dog. Surgery 74, 1973, 945-960.

Reidy, M.A. and Bowyer, D.E. Scanning electron microscopy: morphology of aortic endothelium following injury by endotoxin and during subsequent repair. Atherosclerosis 26, 1977, 319-329.

Revel, J.P. Scanning electron microscope studies of cell surface morphology and labeling, in situ and in vitro. Scanning Electron Microsc. 1974, 541-548.

Romanoff, A.L. Biochemistry of the avian embryo. J. Wiley & Sons, New York, 1967, 398p.

Rychter, Z. Experimental morphology of the aortic arches and the heart loop in chick embryo. Adv. Morphogenes. 2, 1962, 333-371.

Rychter, Z. and Lemež, L. Vascular system of the chick embryo. VIII.On the relationship between the experimentally produced left-sided arcus aortae to the right ventricle. Čs. Morfol. 9, 1961, 55-68.

Sarphie, T.G. Pleomorphic surface features of mammalian endocardium: Fine structure of canine bicuspid valves. J. Mol. Cell. Cardiol. 12, 1980, 241-256.

Sarphie, T.G. and Allen, D.J. Scanning electron microscopy of dog endocardium. Anat. Rec. 187, 1977, 705.

Sarphie, T.G. and Hawkins, W.E. Electron microscopy of mammalian endocardium covering AV valves. Anat. Anz. 148, Erg.-H., 1980, 567.

Schaub, R.G., Rawlings, C.A. and Stewart, G. Scanning electron microscopy of canine pulmonary arteries and veins. Amer. J. vet. Res. 41, 1980, 1441-1446.

Schiff, R.I. and Gennaro, J.F. Jr. The role of the buffer in the fixation of biological specimens for transmission and scanning electron microscopy. Scanning 2, 1979, 135-148.

Sedar, A.-W., Silver, M.J., Kocsis, J.J. and Smith, J.B. Fatty acids and the initial events of endothelial damage seen by scanning and transmission electron microscopy. Atherosclerosis 30, 1978, 273-284.

Shay, J.W. and Walker, C. Introduction to cells in culture as studied by SEM. Scanning Electron Microsc. 1980/II, 171-178.

Shelton, E. and Mowczko, W.E. Membrane blisters: A fixation artifact - A study in fixation for scanning electron microscopy. Scanning 1, 1978, 166-173.

Shimada, Y. and Ho, E. Scanning electron microscopy of the embryonic chick heart: Formation of the epicardium and surface structure of the four heterotypic cells that constitute the embryonic heart. In: Etiology and morphogenesis of congenital heart disease, R. Van Praagh and A. Takao (eds). Futura Publ. Co., Mount Kisco, New York, 1980, 63-80.

Shimamoto, T. Injury and repair in arterial tissue. Contraction and blebbing of endothelial cells in atherogenesis and thrombogenesis and abnormal secretion of sex hormone in Takayasu's disease. Angiology 25, 1974, 682-718.

Shimamoto, T., Yamashita, Y., Numano, F., Sunaga, T. The endothelial cell damages of pre-atheromatous and atheromatous lesions observed by scanning electron microscope. Proc. Jap. Acad. 45, 1969, 761-766.

Smith, U., Ryan, J.W., Michie, D.D. and Smith, D.S. Endothelial projections as revealed by scanning electron microscopy. Science 173, 1971, 925-927.

Squier, C.A., Hart, J.S. and Churchland, A. Changes in red blood cell volume on fixation in glutaraldehyde solutions. Histochem. 48, 1976, 7-16.

Steding, G., Seidl, W., Kluth, D. and Schulze, M. Die Entstehung des Endocards. Untersuchungen an Hühnerembryonen. Anat. Anz. 148, Erg.-H., 1980, 365-367.

Steding, G., Seidl, W. and Schwartz, P. Die Entwicklung der Scheidewände in der Ausflussbahn des Herzens. Präparationen an Hühnerembryonen. Anat. Anz. 146, Erg.-H., 1979, 565-576.

Stewart, G.J., Ritchie, W.G.M. and Lynch, P.R. A scanning and transmission electron microscopic study of canine jugular veins. Scanning Electron Microsc. 1973,473-480.

Stewart, G.J., Stern, H.R. and Schaub, R.G. Endothelial alterations, deposition of blood elements and increased accumulation of $^{131}$I-albumin in canine jugular veins following abdominal surgery. Thromb. Res. 12, 1978, 555-563.

Sunaga, T., Shimamoto, T. and Nelson, E. Correlated scanning and transmission electron microscopy of arterial endothelium. Scanning Electron Microsc. 1973,459-464.

Svendsen, E. Focal endothelial cell injury in rabbit aorta, aggravation of injury by 2 days of cholesterol feeding. Acta path. microbiol. scand. 87, 1979, 123-130.

Taylor, K., Glagov, S., Lamberti, J., Vesselinovitch, D. and Schaffner, T. Surface configuration of early atheromatous lesions in controlled-pressure perfusion-fixed monkey aortas. Scanning Electron Microsc. 1978/II, 449-458.

Thomsen, H.K. and Kjeldsen, K. The surface structure of the thoracic aorta in normal rabbits. Scanning Electron Microsc. 1978/II, 791-796.

Tokunaga, J., Osaka, M. and Fujita, T. Endothelial surface of rabbit aorta as observed by scanning electron microscopy. Arch. histol. jap. 36, 1973, 129-141.

Trump, B.F. and Ericsson, J.L.E. Some ultrastructural and biochemical consequences of cell injury. In: The inflammatory process, B.W. Zweifach, L. Grant and R.T. McCluskey (eds). Academic Press, New York, 1965, 35-120.

Van Mierop, L.H.S. and Patterson, D.F. The pathogenesis and spontaneously occurring anomalies of the ventricular outflow tract in Keeshond dogs: embryologic studies. In: Morphogenesis and malformation of the cardiovascular system, G.C. Rosenquist and D. Bergsma (eds). Birth Defects: Orig. Art. Ser. XIV, No 7. Alan R. Liss, New York 1978, 361-375.

Van Mierop, L.H.S. and Patterson, D.F. Pathogenesis of conotruncal defects and some other cardiovascular anomalies in the Keeshond dog. In: Etiology and morphogenesis of congenital heart disease, R. Van Praagh and A. Takao (eds). Futura Publ. Co., Mount Kisco, N.Y., 1980, 177-193.

Van Mierop, L.H.S., Patterson, D.F. and Schnarr, W.R. Hereditary conotruncal defects in Keeshond dogs - Embryologic studies. Amer. J. Cardiol. 40, 1977, 936-951.

Waterman, R.E. Preparation of embryonic tissues for SEM. Scanning Electron Microsc. 1980/II, 21-44.

Weber, G., Tosi, P. and Cellesi, C. Osservazioni, in microscopia elettronica a scansione, sugli aspetti ortologici dell'endotelio aortico, della vena cava inferiore e dell'endocardio (valvolare e parietale) di Cavia cobaya e sulle modificazioni di essi in corso di scorbuto sperimentale. Arch. "De Vecchi" LVI, 1970, 1-25.

Wechezak, A.R. and Mansfield, P.B. Environmental influence of endothelial surface characteristics. Scanning Electron Microsc. 1979/III, 857-864.

Wheeler, E.E., Gavin, J.B. and Herdson, P.B. A scanning electron microscopy study of human heart valve allografts. Pathology 4, 1972, 185.

Wheeler, E.E., Gavin, J.B. and Herdson, P.B. A study of endocardial endothelium using freeze-drying and scanning electron microscopy. Anat. Rec. 175, 1973, 579-584.

Wright, J.P. and Kirschner, R.H. Scanning electron microscopy of infective endocarditis. Scanning Electron Microsc. 1979/III, 793-800.

## Discussion with Reviewers

F.J. Manasek: In Fig. 3, were the test-wise $\alpha$ levels (or P levels) of the pair-wise T tests adjusted to deal with the problems that are implicit in a posteriori statistical procedures, or did you use some appropriate multiple range test?
Author: After you have raised the question of a posteriori statistical procedures we did submit our data to the analysis of variance and a battery of multiple range tests (Tukey, LSD, Duncan). Both showed significances at $\alpha \leq 0.001$ to 0.032.

F.J. Manasek: Did you perform appropriate tests for heteroscedasticity before performing the polynomial regression in Fig. 3? It appears from the figure that the variances are unequal.
Author: The Bartlett test did not confirm your suspicion of variance heterogeneity.

E.B. Clark: I believe the appropriate statistics for the analysis of these data is a one-way analysis of variance using either Tukey's HSD multiple comparison procedure or some other similar comparison.
Author: Following your recommendation, we did submit our data to the one-way analysis of variance, as well as to Tukey's HSD multiple comparison procedure. The one-way analysis of variance did produce $\alpha \leq 0.001$. Tukey's procedure allowed us to distinguish three significantly ($\alpha \leq 0.05$) different subsets (250 and 310 mOsm; 270 and 310 mOsm; 270 and 290 mOsm).

E.B. Clark: While I agree that density should increase with increasing osmolarity, I cannot account for the sudden decrease in density at an osmolarity of 320 mOsm. If this is as a result of the microvilli damage as you suggest, there should be some morphologic evidence for this damage.
Author: We are sure that the microvilli did not show altered morphology in SEM at high osmolarities and that their area ratio decreased unexpectedly. Actually we can only speculate on an eventual cell damage (e.g. metabolic) by such a hypertonic buffer. This might then lead indirectly to the decrease of their density without visible alterations of their integrity.

D. Morse: It is indicated that endocardial cytosegresomes contain cytoplasm and organelles. Also their population density is relative to the cell death occurring in and subjacent to a given area of endocardium. The author's earlier papers show an intraluminal release of cytosegresomes. What is the physiological importance of these exocytotic vesicles? Since cytosegresomes are a characteristic of the endocardial lining rather than phagocytes, what is their physiological relationship to cell death?
Author: The "exocytotic vesicles" represent the extracellular release of cytosegresomes (cellular defecation) described also in other tissues (e.g.

Rybicka, Virchows Arch. B 28, 1978, 119-133; Munnell and Cork, Amer. J. Path. 98, 1980, 385-394; Stauber et al., Exp. Molec. Path. 34, 1981, 87-93). Any process that will increase the cellular autophagy (e.g. cell injury, extracellular debris, degradation metabolite from dying cells) with subsequent production of cytosegresomes is susceptible of inducing these vesicles. The cellular debris and metabolites from dying cells may stimulate the autophagy in neighbouring cells.

R.R. Markwald: Is there any evidence that macrophages present in the cardiac jelly are actually derived in situ from the endocardium?
Author: We do not believe, and could never observe, that cardiac jelly macrophages are derived in situ from the endocardium. They differentiate from the cushion tissue mesenchymal cells (Pexieder, 1975). Such phagocytes can then migrate through the intercellular openings (Fig. 14d) or be seen lying on the endocardial surface (Fig. 14f).

D. Morse: A common feature of phagocytic cells is a multilobed nucleus. What evidence does the author have to indicate that the structures labeled in Fig. 14a, b, f are really phagosomes, and not nuclear lobes or some other cytoplasmic feature? Along the same line, why is the cell in Fig. 8d termed an endocardial cell, rather than a phagocyte?
Author: The multilobed appearance of phagocytic cells is a TEM and not SEM visible feature. Our diagnosis of phagocytes is based on TEM investigations (Pexieder, 1980), as well as on the papers relating the SEM appearance of phagocytes in various systems (e.g. Carr and Tonner, Scanning Electron Microsc. 1979/III, 637-644; Carr, In: Carr and Daems (eds): The reticuloendothelial system, vol. 1. Plenum Press, New York, 1980, 259-296). Whereas in Fig. 14a, b and f, the phagocytes have the size of 10 to 15 μm and are either penetrating between the endocardial cells or lying on their surface, the cytosegresome in Fig. 8d is an integral part of the endocardial cell and reaches the size of 6 μm only.

C.E. Challice: In our TEM studies we have frequently seen regions where the myocardial cytoplasm is a very thin lamella, such that it would be difficult to assess the actual thickness of cytoplasm, since it is commensurate with the membrane thickness. Can such regions be detected by SEM. If damaged, couldn't they be confused with the holes?
Author: Our TEM studies did not show such a degree of thinning in the endocardium. The lamellae, if existing, should be electronopaque, not allowing the observation of what is behind them, which is not the case (Fig. 15J). In practically all the intercellular openings we could see the elements of underlying tissues (e.g. collagen fibres). Also the phagocytes sometimes penetrating these openings (Fig. 14d) when observed in SEM, as well as in TEM, did not show any sign of interaction with such a hypothetical lamella.

D. Morse: What are the criteria for calling the cells in Fig. 13b "myocardial fibres" ?
Author: These criteria are based on the knowledge of structure of the atrial walls. The latter contains collagen fibres (diameter 20-50 nm), fibroblasts and myocytes (diameter 7-10 μm). The diameter of our structure (7.5 μm) and its length (in part beneath the endocardium) lead us to the term used.

T.P. Fitzharris: If many cells have blebs, what is the significance? How does this relate to development?
Author: The presence of "blebs" is interpreted as a sign of activation and metabolic remodelling, sometimes preceding cell death. Table 1 gives an example of how the developmental changes in morphology of the embryonic endocardium are related to organogenetic events. For instance the merging of proximal and distal bulbar cushions is accompanied by an increase in the number of cytosegresomes.

---

DEVELOPMENTAL CHANGES IN THE MORPHOLOGY OF THE CHICK EMBRYONIC ENDOCARDIUM

Proximal bulbar (conus) cushions

TABLE 1

| Stage | Organogenetic event | Feature of endocardial morphology | Trend |
|---|---|---|---|
| 3 ed | Formation of bulbar cushions | Marginal folds<br>Cytosegresomes | +<br>± |
| 3 ed 16 h | | Microvilli<br>Cytosegresomes<br>Ruffles | ▼<br>▼<br>▲ |
| 4 ed 8 h | | Intercellular openings<br>Macrophages - traversing | +<br>+ |
| 5 ed 8 h | Formation of semilunar valves | Intercellular openings | ▲ |
| 6 ed | Merging of proximal and distal bulbar cushions | Cytosegresomes<br>Marginal folds | ▲<br>+ |
| 6 ed 16 h | | Cytosegresomes<br>Flattened cells | ▼<br>+ |
| 7 ed | End of heart organogenesis | Marginal folds<br>Ruffles<br>Intercellular openings | ▼<br>▼<br>▼ |

For additional discussion see page 248.

# DIFFERENT MODES OF PRONEPHRIC DUCT ORIGIN AMONG VERTEBRATES

Thomas J. Poole[*] and Malcolm S. Steinberg

Department of Biology, Princeton University, Princeton, NJ 08544

(Paper received April 12 1983, Complete manuscript received February 29 1984)

## Abstract

It is possible to distinguish differences in pronephric duct morphogenesis by using scanning electron microscopy to observe the results of blocking, marking and grafting experiments as well as the normal course of development. Here we compare the mode of pronephric duct development in embryos representing three orders of vertebrates: birds (class, Aves; order, Gallus); frogs (class, Amphibia; order, Anura); and salamanders (class, Amphibia; order, Urodela). The axolotl (a urodele) pronephric duct is formed by the caudal extension of a solid stream of cells segregated below somites 2 through 7. During its migration, cells are rearranged so that a short, wide rudiment is extended to form a long, thin one of similar volume. The pronephric duct rudiment of Xenopus laevis (an anuran) shows no evidence of caudal migration. Rather, pronephric duct cells are segregated out in situ by the formation of a fissure which separates them from the lateral plate mesoderm over ten somite widths. The chick pronephric duct forms a part of the intermediate mesoblast that extends by a caudal migration which does not, however, involve extensive cell rearrangements. Instead, cells near the tip of the duct rudiment in the chick proliferate, extending the duct by true growth as well as by active cell locomotion.

KEY WORDS: pronephric duct, morphogenesis, comparative embryology, cell migration, salamander, frog, chick, development.

[*]Address for correspondence:
Thomas J. Poole
Department of Anatomy and Cell Biology
State University of New York
Upstate Medical Ctr.
Syracuse, NY 13210

## Introduction

The development of the pronephric or primary nephric duct represents a critical early phase of vertebrate urogenital morphogenesis. Since the turn of the century, it has been examined in a variety of vertebrates initiating considerable debate over its manner of origin and the means by which it arrives at the cloaca. We have found that a combination of classical embryological surgical and marking techniques and scanning electron microscopy (SEM) allows one to begin to resolve the variation with species of the mode of duct origin and some of its evolutionary implications.

The organization of kidneys in primitive vertebrates such as the hagfish suggests that ancestral vertebrates had a complete set of segmental tubules which connected the coelom at each somite level with a duct or ureter which ran backwards to the cloaca (Goodrich, 1958). Development of these segments proceeded in a cranial to caudal sequence coupled closely to segment differentiation. More advanced vertebrates have evolved more specialized and differentiated tubule-containing organs called the pronephros, mesonephros and metanephros. These are essentially aggregates of uriniferous tubules which form from the anterior, middle and posterior portions of the intermediate mesoderm respectively. The pronephros and pronephric duct develop first, and successive kidneys use the pronephric duct as their main drainage channel when they form. In addition, the pronephric duct serves a morphogenetic function by both inducing mesonephric tubule formation and affecting the structure of the cloaca (Waddington, 1938).

## Embryological Methods

Axolotls (Ambystoma mexicanum) were obtained from spawnings of our colony and the Indiana University Axolotl Colony. African clawed frog (Xenopus laevis) embryos were obtained from matings evoked by injection of human chorionic gonadotropin as described by Gurdon (1967). Chick (Gallus gallus) embryos were incubated at 39°C for 30 to 55 hours.

Surgical operations on amphibian embryos were carried out under aseptic conditions in full strength Steinberg's solution modified by the addition of calcium chloride, to make the final calcium concentration 2mM, and antibiotics (gentamycin, 70 mcg/ml, penicillin, 5 units/ml, streptomycin 5 mcg/ml) using procedures and operating dishes like those described by Jacobson (1967).

Operations were performed on chick embryos explanted onto a semi-solid substrate with the body of the upright embryo suspended over a punched hole filled with the same solution (see Augustine, 1977). Barriers to migration were fragments of eggshell membrane inserted with the aid of tungsten needles and forceps. The cultured chick embryo was incubated at 37°C in a humidified incubator for 10 to 12 hours before fixation.

## Results and Discussion

### Directed Migration and Cell Rearrangement of the Salamander Duct Rudiment

Figure 1 is an SEM photograph of an axolotl embryo with the ectoderm peeled off its right side. The mesoderm is composed at this stage of the segmenting somitic mesoderm, the pronephros and pronephric duct rudiment and lateral mesoderm. The posterior tip of the pronephric duct rudiment is marked by an arrow. If one marks across a portion of the duct rudiment with a vital dye such as Nile blue sulfate (Figure 2), then one may readily demonstrate the caudal extension of the duct rudiment over the 12-15 hours of its development. A short, wide rudiment extends to over double its length to form a long, thin duct rudiment nearing the caudal destination of the cloacal wall.

What is the cellular basis of this extension? Figure 3 is a series of closer views of the duct rudiment several somites behind the anterior end of the pronephros at succesive stages of development. You can count six to eight cells across the diameter in Figure 3a. When the posterior extension is almost complete, a later stage embryo at the same somite level and magnification has a duct rudiment only two or three cells wide (Figure 3d). This demonstrates that the elongation accompanying posterior extension involves extensive cell rearrangement; that is, the cells within the duct have changed neighbors. This rearrangement occurs before the duct has a lumen (i.e., while it is a solid mass). There is also no change in volume and very little change in the number of cells during this morphogenetic event (see Figure 6 in Poole and Steinberg, 1981).

The posterior tip apparently provides the motive force for this cell rearrangement as it displays ultrastructural features characteristic of actively locomoting cells. Figure 4a is a micrograph of the posterior half of a duct rudiment in the progress of migration. The somite and lateral mesoderm is visible and the duct rudiment is extending at the boundary in the direction of the arrow. Figure 4b is a close-up of that duct rudiment's posterior tip. The cells within the rudiment overlap in the manner of fish scales and extend filopodia and lamellipodia on more posterior rudiment cells as well as onto the cells of the somite and lateral mesoderm

which comprise the substrate for migration.

Blockage and transplantation experiments demonstrate that the salamander duct rudiment is capable of streaming across foreign terrain. In particular, if one transplants a rudiment just beginning migration onto the flank of another embryo at the same stage and place the secondary rudiment at the same axial level, then one may observe elongation of the secondary duct dorso-caudally across the lateral mesoderm. Figure 5 shows an example where the secondary duct has streamed across the flank and has almost reached (arrow) the edge of the primary duct. If the graft was placed closer to the host's duct path, the secondary duct would have fused with the primary.

In summary, the salamander pronephric duct rudiment segregates from the dorsal portion of the flank as an ovoid, solid mass of cells which is remodeled into a long, thin strand by active cell locomotion of the posterior tip cells and extensive cell rearrangements.

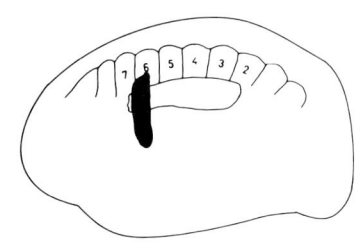

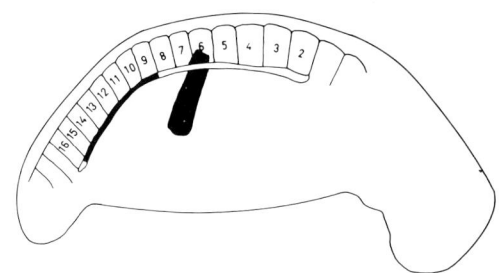

Figure 2. Camera-lucida tracings of a vitally stained axolotl embryo. The distal segment of the pronephric duct stained with Nile blue sulfate at an early tailbud stage has elongated and moved caudad after 12 additional hours of development.

Figure 3. SEM photographs of pronephric duct development at the level of the sixth trunk somite. From stages 22 to 32 (4 hour intervals, a-d), the axolotl pronephric duct thins markedly by cell rearrangement as it elongates. Bar equals 200 μm.

Figure 4. SEM of the posterior tip of an axolotl pronephric duct rudiment. The arrow marks the direction of migration. a) Bar equals 200 μm, b) Bar equals 40 μm.

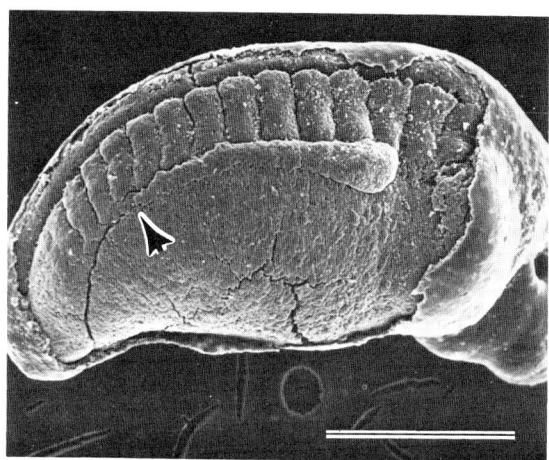

Figure 1. Scanning electron micrograph of an axolotl embryo whose ectoderm was peeled off the right side after fixation. An arrow marks the caudal tip of the pronephric duct rudiment. Bar equals 1 mm.

Captions for Figures 3 and 4 on the facing page.

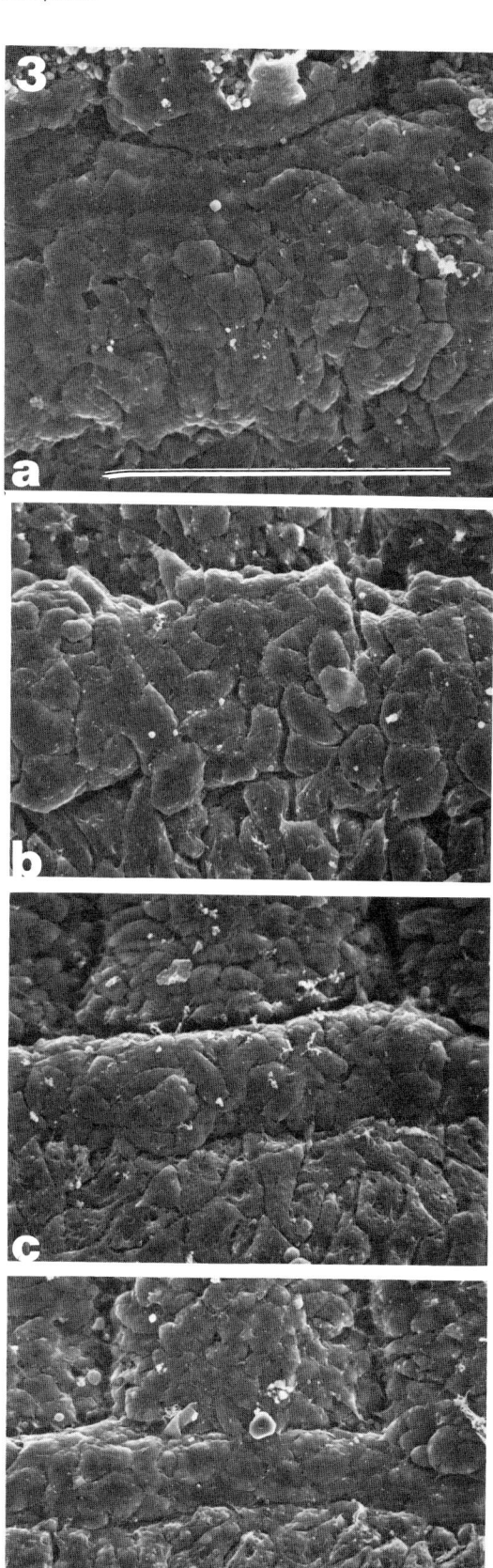

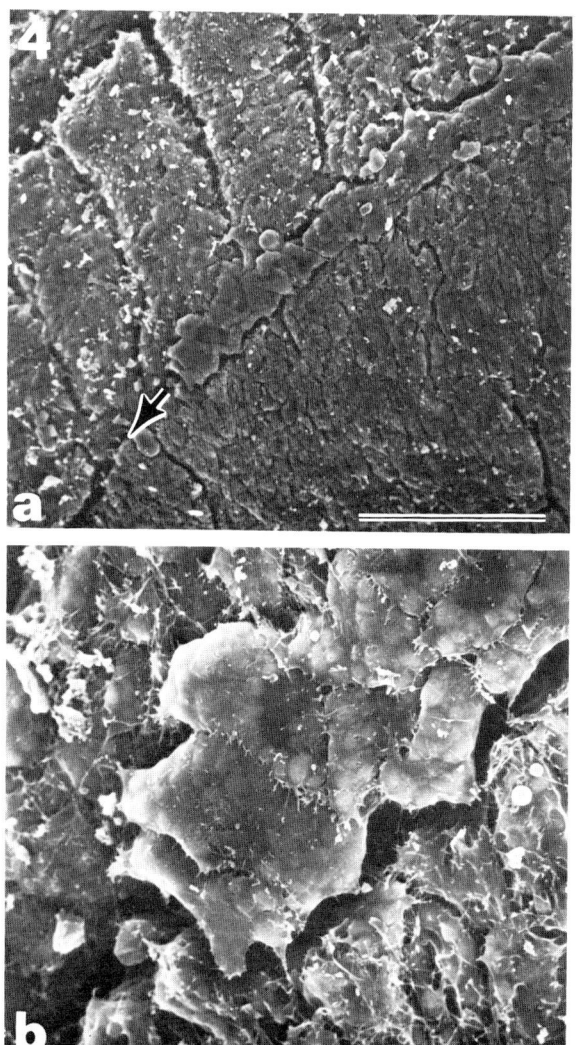

## Segregation of the Frog Duct Rudiment

Figure 6a shows a micrograph of a *Xenopus* embryo at an early stage of duct segregation (the ectoderm of the right side has been peeled away). The fissure separating the presumptive duct mesoderm from the dorsal portion of the lateral mesoderm is well defined at the anterior end (the right side of the photo) and is progressively less well defined, and even, discontinuous as one proceeds caudad. With 18 to 20 hours further development at 25°C, the duct morphology has changed to that shown in Figure 6b. The duct rudiment has not extended any farther but has become more clearly demarcated and slightly narrower due to the rounding of the originally slightly flattened, segregated duct mesoderm into a solid, cylindrical cord. This change is illustrated by comparison of the duct rudiment at the same level at stages 23 and 31. The area of the duct below pronephric somite 5 at these stages is shown in Figure 7. There is no marked change in the number of cells across the diameter; that is, there is no evidence of extensive cell rearrangement as was the case in the axolotl (compare to figure 3).

The apparent mechanism as revealed morphologically has been confirmed by vital dye marking, deletion and transplantation experiments. Marking across prospective duct mesoderm anterior to somite 10 during the stages of duct segregation did not result in streaking or translocation of dye in the duct that one would expect if there was caudal migration. Segregation of duct mesoderm, although normally occurring in an anterior-posterior sequence, is unimpaired by isolation of posterior portions of the rudiment at stage 22 when there is no indication of a fissure at such levels. Figure 8 shows the result of transecting somite, duct and a portion of lateral mesoderm at the level of pronephric somite 5. The duct still clearly extends to pronephric somite 10 by stage 32, the age at which this embryo was fixed.

Is the *Xenopus* duct rudiment mesoderm, which displays no migratory propensity during normal development, capable of migration if placed in the foreign environment of the flank? If a young duct rudiment is transplanted to the flank of a similar stage host, the secondary duct mesoderm displays virtually no capacity for migration. In contrast to the mechanism of duct formation by caudal extension in the axolotl, the pronephric duct in *Xenopus laevis* forms *in situ* along almost its entire length.

## Directed Migration and Growth of the Bird Duct Rudiment

The dorsal aspect of a 15 somite chick embryo is shown in Figure 9 (the surface ectoderm has been removed). The ectoderm, lateral mesoderm, segmented and unsegmented somite mesoderm, dorsal surface of the neural tube and intermediate mesoblast (the rudiment of the pronephric duct and possibly the posterior cardinal vein) are also visible. With the exception of the intermediate mesoblast, the mesoderm is covered by fine extracellular fibers. This difference between the duct rudiment and its substratum can be more clearly seen at higher magnification.

Figure 10 shows the area in the box in Figure 9 at a higher magnification. The surfaces of somite and lateral mesoderm are covered by 50-100 nm collagen fibers. The cells near the tip of the extending rudiment are markedly flattened and elongated in the antero-posterior direction (the axis of migration). A series of micrographs at different axial levels (Figure 11) shows the sequence of changes in the intermediate mesoblast morphology which could also be seen at the same axial level in a series of older stages. Cells within the duct become increasingly closely apposed (Figures 11a,b). The free space is apparently eliminated as the duct becomes compacted, and then extracellular fibers extending from somite and lateral mesoderm partially obscure the duct from surface view (Figure 11c). Eventually the somite and lateral mesoderm make contact (Figure 11d) and totally engulf this more anterior portion of the duct rudiment.

Although they do not immediately suggest a mode of migration, observations of the chick intermediate mesoblast show tip cells sometimes contacting lateral mesoderm, sometimes somite mesoderm and sometimes stretched between the two. This suggests that the flattened stellate cells are capable of some dorso-ventral movement but tend to center themselves at the V-shaped boundary of somite and lateral mesoderm that is their path. The duct does not appear to thin with extension, the tip having a similar appearance in embryos of 14 to 18 somites. Overton (1959) demonstrated a relatively high rate of mitosis in the chick pronephric duct near its tip (although not higher than mitotic rates in the adjacent mesoderm at the same A-P level). A large number of rounded, mitotic cells have not, however, been visible in the SEM.

The onset of migration has been confirmed by inserting fragments of eggshell membrane to block the tip of the rudiment of embryos explanted *in vitro* on a semisolid substratum. We have found that such embryos develop normally for at least 12 hours during the period of development of the duct (2-3 day old embryos). Figure 12 shows a typical consequence of

---

Figure 5. SEM of the right side of an axolotl embryo which received a secondary pronephric duct rudiment. The graft streams dorso-caudally across lateral mesoderm. The arrow marks the distal tip of the secondary duct. Bar equals 0.5 mm.

Figure 6. SEM photographs of two *Xenopus laevis* embryos at an early (a) and late (b) stage of pronephric duct development. The ectoderm was peeled off the right side. Arrows indicate the caudal limit of duct organization. Bar = 0.5 mm.

Figure 7. SEM of the area six somite widths from the anterior end of the pronephros at an early (a) and late (b) stage of development. Bar equals 100μm.

Figure 8. SEM of the result of splitting a *Xenopus* embryo at an early stage. 12 hours later the duct rudiment is visible posterior to the incision (arrow). Bar equals 0.5 mm.

Figure 9. Dorsal view of a 15 somite chick embryo. The ectoderm has been removed over a large area. The left intermediate mesoblast posterior tip is in a box and enlarged in Figure 10. Bar equals 0.5 mm.

Pronephric Duct Development

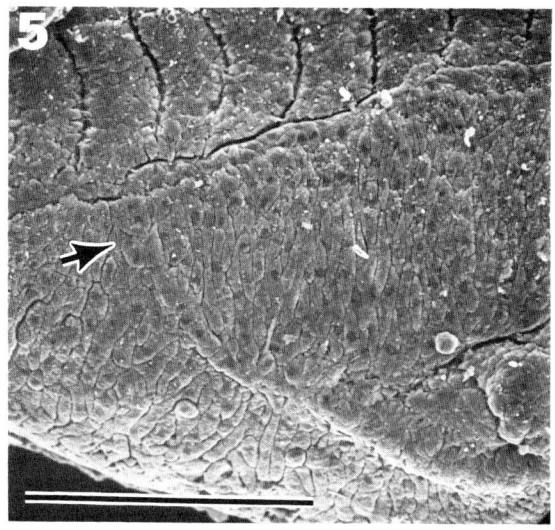

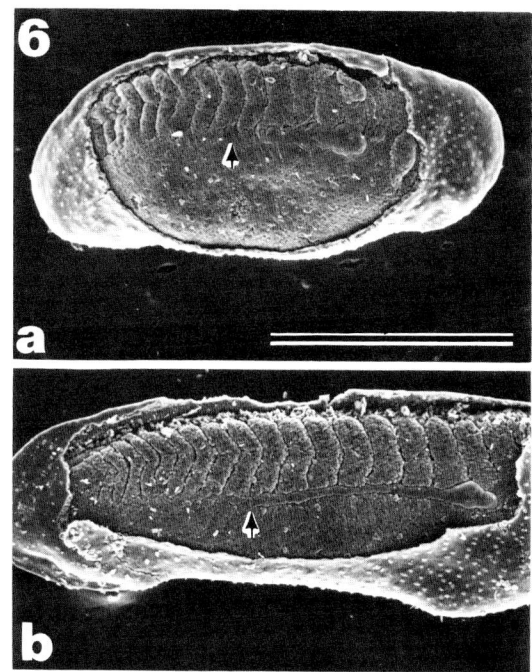

Captions for Figures 5 to 9 are on the facing page.

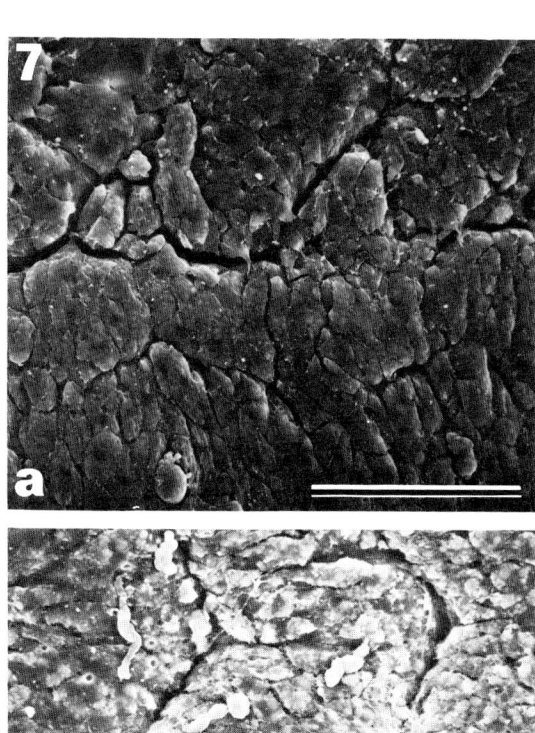

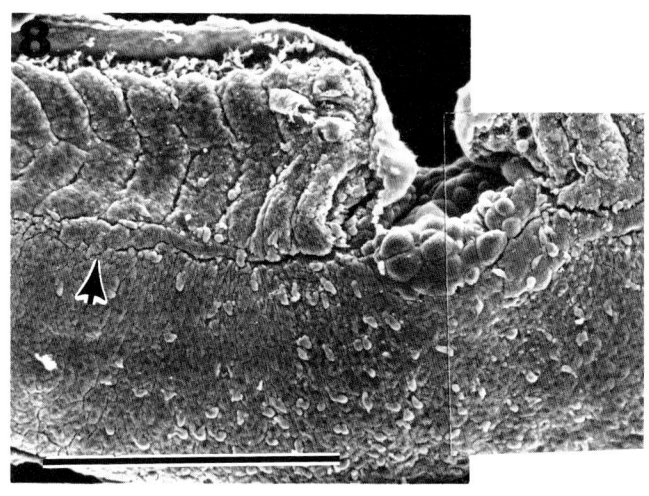

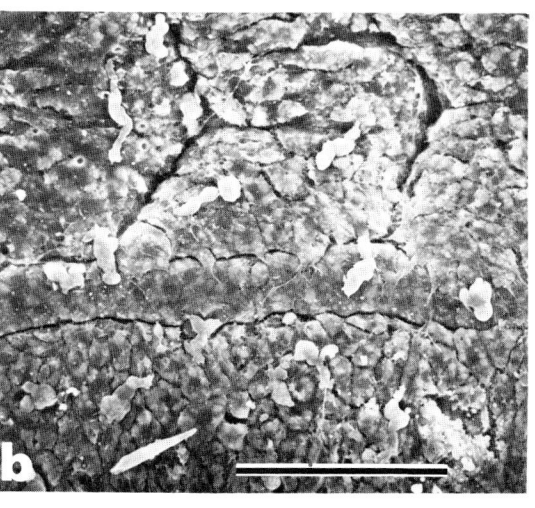

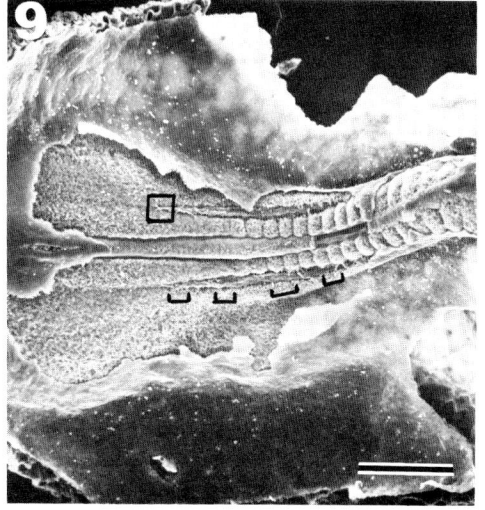

inserting eggshell membrane in an 18-somite chick embryo just behind somite 17. Such an operation, in all 5 cases, had no effect upon the caliber or extent of the duct after 12 hours of further development. In sharp contrast, when an eggshell membrane barrier was inserted into an 18-somite chick embryo split 3 somite widths behind the last fully formed somite (i.e. behind presumptive somite 21), none of 5 cases showed any extension of the duct rudiment on the operated side (Figure 12b).

In summary, the chick pronephric duct forms by caudal extension of the intermediate mesoblast which may also give rise to the posterior cardinal vein. It apparently elongates by active cell locomotion accompanied by cell proliferation in a region near the tip.

## Acknowledgements

This work was supported by research grants PCM76-84588 from the National Science Foundation and CA9167 from the NCI, DHEW. The electron microscopy was performed in departmental facilities supported by the Whitehall Foundation. Figures 2 and 3 are reproduced with permission from J. Embryol. Exp. Morph. 63: 1-16.

## References

Augustine JM. (1977). Mesodermal expansion after arrest of the edge in the area vasculosa of the chick. J. Embryol. Exp. Morph. 41, 175-188.

Goodrich ES. (1958). Studies on the Structure and Development of Vertebrates, Dover Publications, New York, 657-719.

Gurdon JB. (1967). African clawed frogs. in: Methods in Developmental Biology, F.H. Wilt and N.K. Wessells (eds.), T.Y. Crowell Co., New York, 75-84.

Jacobson AG. (1967). Amphibian cell culture, organ culture and tissue dissociation. in: Methods in Developmental Biology, F.H. Wilt and N.K. Wessells (eds.), T.Y. Crowell Co., New York, 531-542.

Overton J. (1959). Mitotic pattern in the chick pronephric duct. J. Embryol. Exp. Morph. 7, 275-280.

Poole TJ, Steinberg MS. (1981). Amphibian pronephric duct morphogenesis: Segregation, cell rearrangement and directed migration of the Ambystoma duct rudiment. J. Embryol. Exp. Morph. 63, 1-16.

Waddington CH. (1938). The morphogenetic function of a vestigial organ in the chick. J. Exp. Biol. 15, 371-376.

## Suggestions for Further Reading

Fox H, Hamilton L. (1964). Origin of the pronephric duct in Xenopus laevis. Arch. Biol. (Liege) 751, 254-251.

Holtfreter, J. (1944). Experimental studies on the development of the pronephros. Rev. Canad. Biol. 3, 220-250.

Meier S. (1980). Development of the chick embryo mesoblast: pronephros, lateral plate and early vasculature. J. Embryol. Exp. Morph. 55, 291-306.

Poole TJ, Steinberg MS. (1977). SEM-aided analysis of morphogenetic movements: development of the amphibian pronephric duct. Scanning Electron Microsc. 1977;II: 43-52.

Poole TJ, Steinberg MS. (1982). Evidence for the guidance of pronephric duct migration by a craniocaudally travelling adhesion gradient. Develop. Biol. 92, 144-158.

## Discussion with Reviewers

P.B. Armstrong: How extensive is the cell rearrangement in the elongating urodele pronephric duct? If a segment of the duct is stained as in Fig.2: top, do the borders of the stained portion remain clearly demarcated, or is there considerable intermingling of stained and unstained cells across the border during the course of elongation?

Authors: The spreading of the stain in the duct seems to be limited to the local exchange of cell neighbors necessary to extend it to the cloaca. For example in Figure 2 the posterior tip of the duct does not become stained but rather the unstained portion becomes longer and thinner as does the stained portion.

P.B. Armstrong: The migration of the urodele pronephric duct resembles spreading of tissue sheets, as during epiboly, save that the duct spreads as a linear cord of cells rather than as a 2-dimensional sheet. An interesting question is the character of the factors that restrain cells craniad from the actively migrating tip from spreading as a sheet dorsally over the somites or ventrally over the lateral mesoderm. The cells appear to be motile (filopods and what appear to be lamellipods are visible in Figure 3, and the rearrangement of cells depicted in Figure 3a-d suggests that these cells are probably locomotory). Cells of the duct can migrate over lateral mesoderm peripheral to the track usually traversed by the duct (Figure 5). What information is available to decide amongst various

---

Figure 10. The intermediate mesoblast posterior tip extends filopodia and flattened lamellipodia onto the collagen covered lateral and somite mesoderm over which it is migrating. Bar equals 50 μm.

Figure 11. A series of SEMs at different anterior-posterior levels of the right mesoblast of the embryo in Figure 9. (a) Cells are loosely packed and stellate-shaped near the posterior end. Further anterior (b-d) the cells become more closely apposed and the somite mesoderm eventually covers a portion of the mesoblast. Bar equals 50 μm.

Figure 12. The results of inserting a fragment of eggshell membrane (EM) either 3 somite-widths anterior to (a) or 3 somite-widths posterior to (b) the extending intermediate mesoblast tip. In (a) there is mesoblast on both sides of the embryo. In (b) the left mesoblast has extended but the right mesoblast is absent (arrow) behind the fragment of shell membrane. Bar equals 100 μm.

Pronephric Duct Development

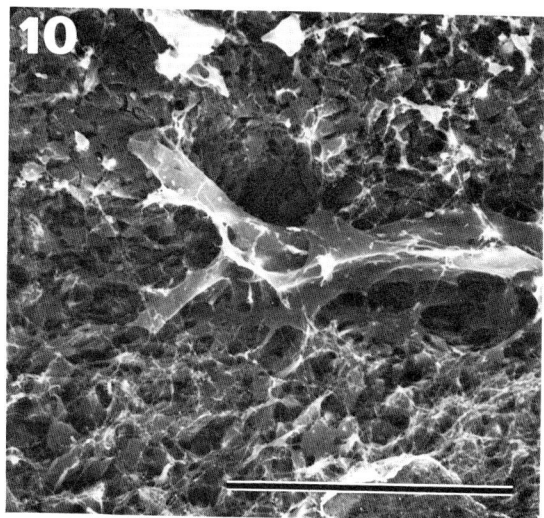

Captions for Figures 10-12 are on the facing page.

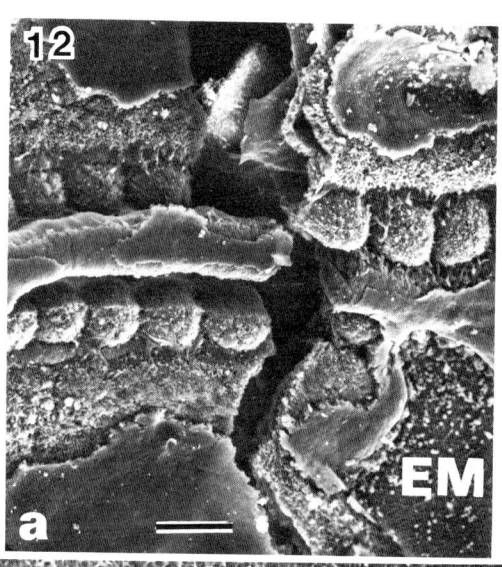

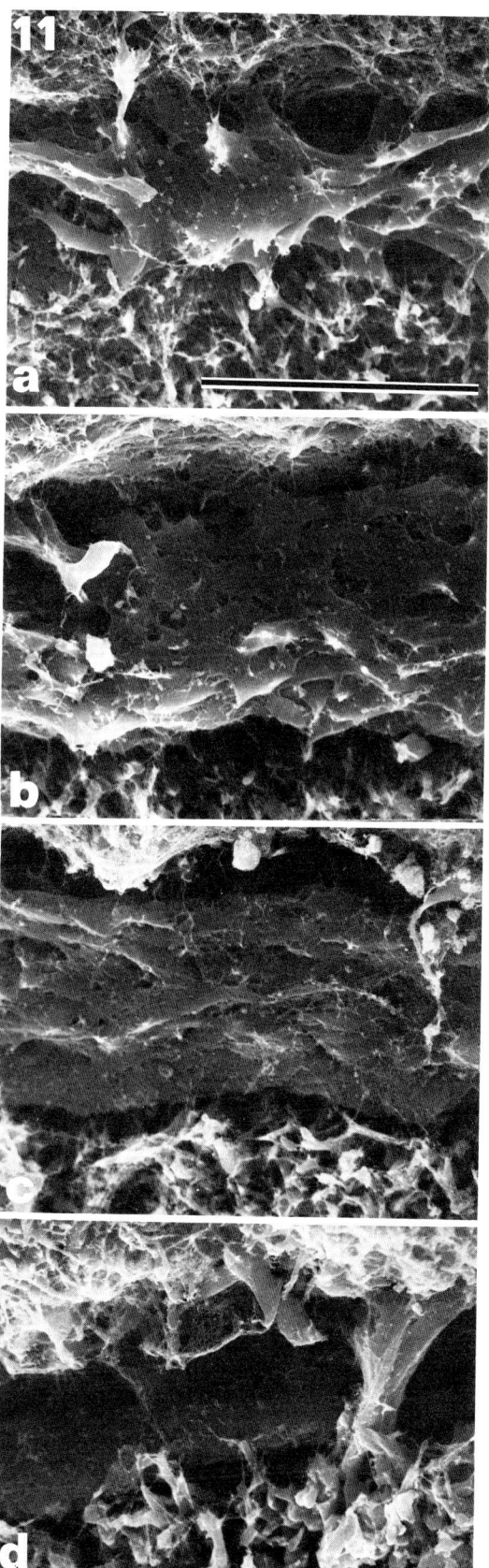

227

possibilities for preventing ventral spreading over the lateral mesoderm of craniad portions of the duct?

Authors: This is a subject we have dealt with in considerable detail elsewhere (Poole and Steinberg, 1982). In brief, we found, from grafting secondary duct rudiments to different anteroposterior levels of the flank, that only a narrow region of the flank to either side of the duct tip will support extensive cell migration. Moreover, this region passes caudad across the flank as a developmentally regulated wave at the same rate as duct extension. It is correct that the more anterior cells of the duct rudiment have migratory propensity. If one grafts the anterior half of the duct rudiment posteriad on the flank then it too is capable of streaming across the flank (Poole and Steinberg, 1982, figure 9). Therefore the constraints on cell migration seem to be a property of the substrate.

S. Meier: Would you say that the pattern and mode of pronephric duct extension in axolotl embryos is more like that of chick than frog? What do you feel are the evolutionary implications of your observations and experiments on these vertebrate embryos?

Authors: It is clear that both chick and axolotl form their pronephric duct by the directed caudal extension of a rudiment segregating more anteriorly. Since those studying pronephric duct formation traditionally distinguished in situ formation from directed caudal migration, it is tempting to say that these two modes of duct morphogenesis are more similar to each other than to the frog which forms its duct in situ. The mechanisms of duct formation in the chick and the axolotl are, however, really strikingly different. The axolotl duct forms by cell rearrangement and the chick duct apparently by cell multiplication although both rudiments display extensive cell migration. It is perhaps most correct to suggest only that higher vertebrates have an increased tendency to develop the pronephric duct precociously in comparison to kidney tube development (Goodrich, 1958) and that these three groups have evolved different developmental devices for that purpose.

APPLICATION OF SCANNING ELECTRON MICROSCOPY TO KIDNEY DEVELOPMENT AND NEPHRON MATURATION

Andrew P. Evan*, Vincent H. Gattone, II, and Philip M. Blomgren

Department of Anatomy, Indiana University School of Medicine,
635 Barnhill Drive, Indianapolis, Indiana 46223

(Paper received February 14 1983, Complete manuscript received January 8 1984)

## Abstract

The present study uses scanning electron microscopy to review mammalian kidney development using the dog model. Canine mesonephros and metanephros are examined, focusing on the structure of the nephron from each and the maturation of the different cell types found in the metanephric nephron. The mesonephros possesses numerous nephrons each with a glomerulus, proximal, distal, and collecting tubule which empty into the mesonephric (Wolffian) duct. The mesonephros does not have cortical and medullary regions. The metanephric nephron not only possesses a glomerulus and the same tubular segments as described for the mesonephric nephron (proximal, distal, and collecting tubules), but also has tubular segments that are found in the medulla, namely the loop of Henle and thick ascending limb. Mammalian metanephric glomerulogenesis and tubulogenesis follows an orderly centrifugal pattern (from inner to outer cortex).

The glomerular visceral epithelium initially has a columnar shape which is modified into a cuboidal sheet possessing numerous slender foot processes. The endothelium begins as a double layer with minimal fenestrae. With maturation, a single cell layer is formed possessing many large fenestrae. The maturation of the proximal tubule is documented by changes in the apical microvilli, and the foldings of the lateral and basal cell surfaces. Initially, the collecting tubule is lined by a primordial cell which differentiates into both the principal and intercalated cells.

KEY WORDS: Pronephros, Mesonephros, Metanephros, Glomerulus, Proximal tubule, Collecting tubule.

*Address for correspondence:
Indiana University, Department of Anatomy
Medical Science Building, Rm. 258
635 Barnhill Drive
Indianapolis, IN 46223  Phone No. (317) 264-8102

## Introduction

The normal development of the mammalian kidney has been of widespread interest because of its complex pattern of maturation. Three separate types of kidneys are formed in the human fetus with only one remaining as the primary urinary organ (the metanephric kidney).

The development of the permanent kidney is particularly fascinating in that the nephrons and microvessels have a specialized organization. While light microscopic and transmission electron microscopic studies have provided details on the events of the genesis of the glomeruli, tubules, and vascular system [5,6,8,10,11,12,15,18,19,21], it is difficult to appreciate the extensive variation in the level of maturation of the nephron. Not only is there a centrifugal pattern of nephron development with its oldest structures in the juxtamedullary zone, but the cells of a single glomerulus may exhibit several stages of maturation.

Scanning electron microscopy would appear to add a valuable technical approach in studying the development of the kidney, since large areas of tissue may be examined in three dimensions. Moreover, with the wide range of magnifications of the SEM, it is possible to examine important details that are occurring at the cellular level. All cell surfaces can be observed by SEM if one employs special preparation techniques such as freeze fracturing and our dissection/acid-hydrolysis protocol.

In the present study, SEM has been used in an effort to demonstrate the usefulness of this technique to study spatial and temporal relationships in the developing kidney. All three types of kidneys, the pronephros, mesonephros, and metanephros, are examined.

## Overview of the Embryology of Mammalian Kidneys

Mammals develop three kidneys in the course of their intrauterine period [1,19,20]. The kidneys are, in order of their appearance, the pronephros, mesonephros, and metanephros (figures 1 and 2). The first two regress in utero in most vertebrates. However, the pronephros remains as the functional kidney in

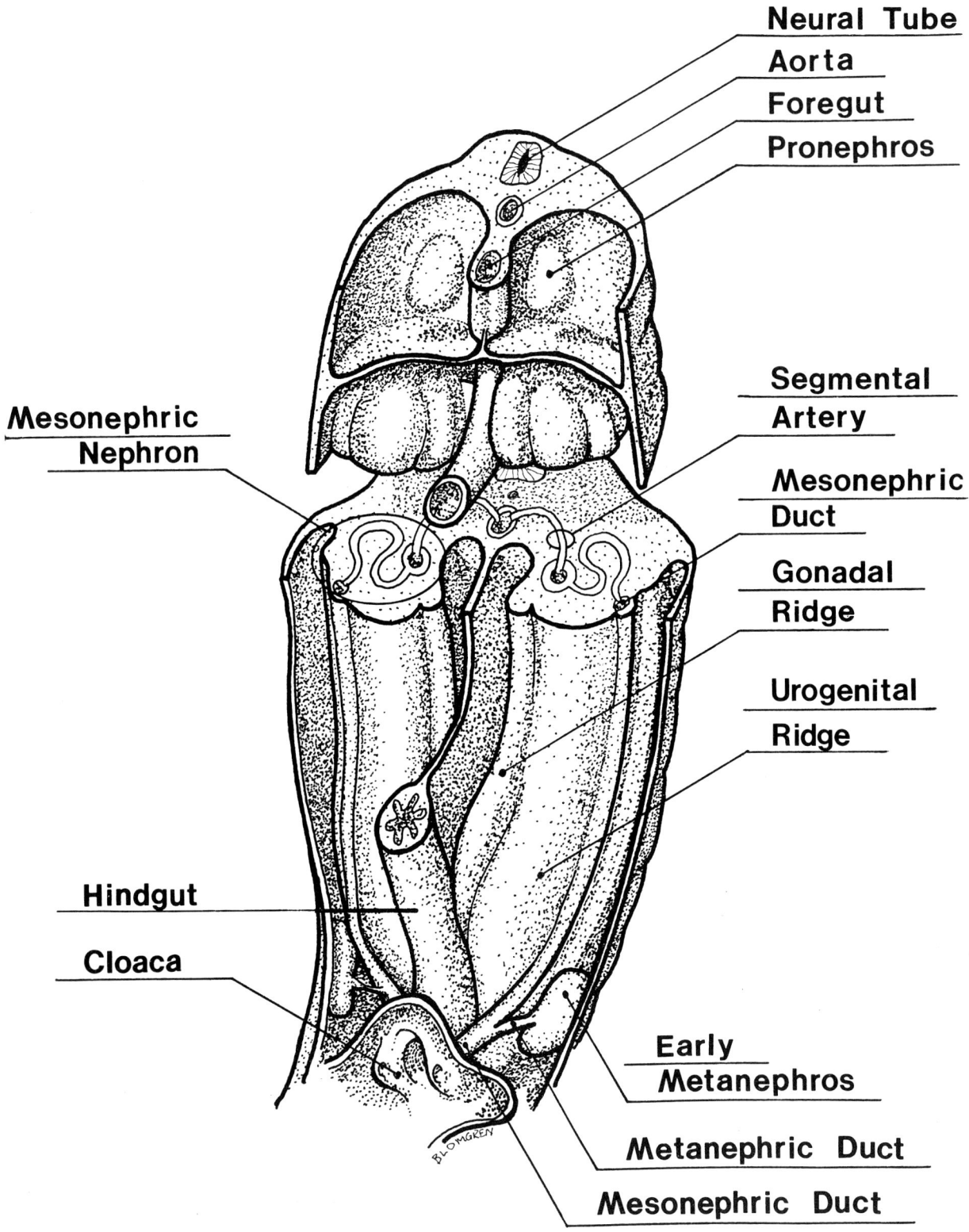

cyclostomes and some teleosts, whereas the mesonephros remains as the functional kidney of most anamniotes.

The pronephros develops at cervical somite levels. The mammalian pronephros consists of pronephric tubules attached to the pronephric duct. This is a non-functional kidney which possesses no glomeruli in mammals and is very short-lived.

The second kidney or mesonephros, which is also a temporary kidney in mammals, possesses glomeruli and is functional. In each mesodermal ridge (the urogenital ridge) along the peritoneal aspect of the dorsal body wall, mesonephric vesicles develop medial to the mesonephric duct (formerly the pronephric duct). These vesicles elongate. One end of each vesicle connects to the mesonephric duct, while the other end is invaginated by capillaries from the dorsal aorta to form the mesonephric glomerulus. Between the glomerulus and mesonephric duct, each vesicle further lengthens to form the tubular components of the mesonephric nephron (figure 2b). Since the mesonephric nephrons develop segmentally, cranial to caudal, within the urogenital ridge, various stages of nephron development and/or regression can be observed at any one time. The mesonephric duct is located along the ventrolateral aspect of the urogenital ridge (figures 3b and 3c) and extends caudally to connect to the cloaca (figure 1).

Just proximal to where the mesonephric duct enters the cloaca, an evagination appears which is termed the metanephric (or ureteric) bud (figures 1 and 2a). This evagination elongates and projects into the mesenchyme of the dorsal wall in the pelvic region of the embryo and becomes surrounded (capped) by condensed mesenchyme. As this dilated diverticulum migrates cranially, it starts to divide to form the major and minor calyces of the metanephros. The metanephros maintains its connection to the mesonephric duct and cloaca via the metanephric duct (which becomes the ureter). Outgrowths

==========

Figure 1. Developing mammalian kidneys.

Schematic representation of the three kidneys that develop in mammals. Keep in mind that all three kidneys are never present at the same time. The pronephros (which normally disappears by the time the embryo reaches the age shown) develops in the cervical region of the embryo. The pronephric duct is renamed the mesonephric duct when the mesonephric nephrons connect to it. Mesonephric nephrons lie in the transverse plane of the embryo and are piled on top of each other forming a ridge running along the dorsal abdominal wall. The nephron consists of a glomerulus and associated tubule. Along the ridge (termed the urogenital ridge) lies the gonadal ridge. The mesonephric duct runs the length of the urogenital ridge and ends caudally in the cloaca. A diverticulum from the mesonephric duct forms the metanephric duct (which becomes the ureter) and induces formation of the metanephros.

from the minor calyces develop and project radially into the condensed mesenchyme, known as metanephric blastema or metanephrogenic tissue (figure 2c). These outgrowths form the collecting duct portion of the nephron, while the metanephric blastema is the origin for the rest of the nephron (glomerular and tubular epithelial components) (figure 3d). The scheme of metanephric nephrogenesis is illustrated in figure 5. Metanephric blastema condensation occurs adjacent to the ampulla of the collecting duct (just under the renal capsule). This condensed blastema differentiates into a vesicle which will form an S-shaped structure. The deeper portion (medullary side) of the S will be involved in the formation of the glomerulus and Bowman's capsule, while the outer (capsular) two-thirds form the tubular components of the nephron. The outer end of the S-shaped vesicle will unite with the arched collecting duct (figure 5), thereby allowing the lumen of the collecting duct to become continuous with the lumen of the renal vesicle. Thereafter, the renal vesicle continues its development with the differentiation and maturation of the components of the nephron (glomerulus and tubular segments). The main components of the nephron are the renal corpuscle, the proximal tubule, the thin limbs, the distal tubule, and the connecting segment [23]. The collecting duct continues to grow radially toward the kidney capsule while initiating the formation of new generations of nephrons. This continual addition of new nephrons in the sub-capsular region creates a centrifugal pattern of nephrogenesis (figures 4 and 5). After the first generation of nephrons (the juxtamedullary nephrons) form their loops of Henle, the medulla of the metanephric kidney can be distinguished (compare figures 2d and 2e). This is due to the elongation of Henle's loops toward the hilum. It is of interest that this first generation of metanephric nephrons will eventually degenerate. Table 1 presents some critical time points in mammalian kidney development for several animal models.

## Structure of the Nephron

Mesonephros. The mesonephric nephron (figure 6) consists of a renal corpuscle (glomerulus and capsule) and various tubular segments (proximal, distal, and collecting tubules). The ultrastructure of the mammalian mesonephric nephron has not been extensively studied. Of special interest are the SEM studies in lamprey [17], rabbit [22], dog [9], and opossum [16]. The description herein is of the mesonephros of the dog. Embryos presented are 29 and 44 days of gestation and were immersion fixed.

The mesonephric glomerulus and tubular segments appear very similar to their metanephric counterpart. The mesonephric glomerulus consists of capillary loops possessing a fenestrated endothelium. The loops are covered on their outer surface by a visceral epithelium (podocytes) characterized by their

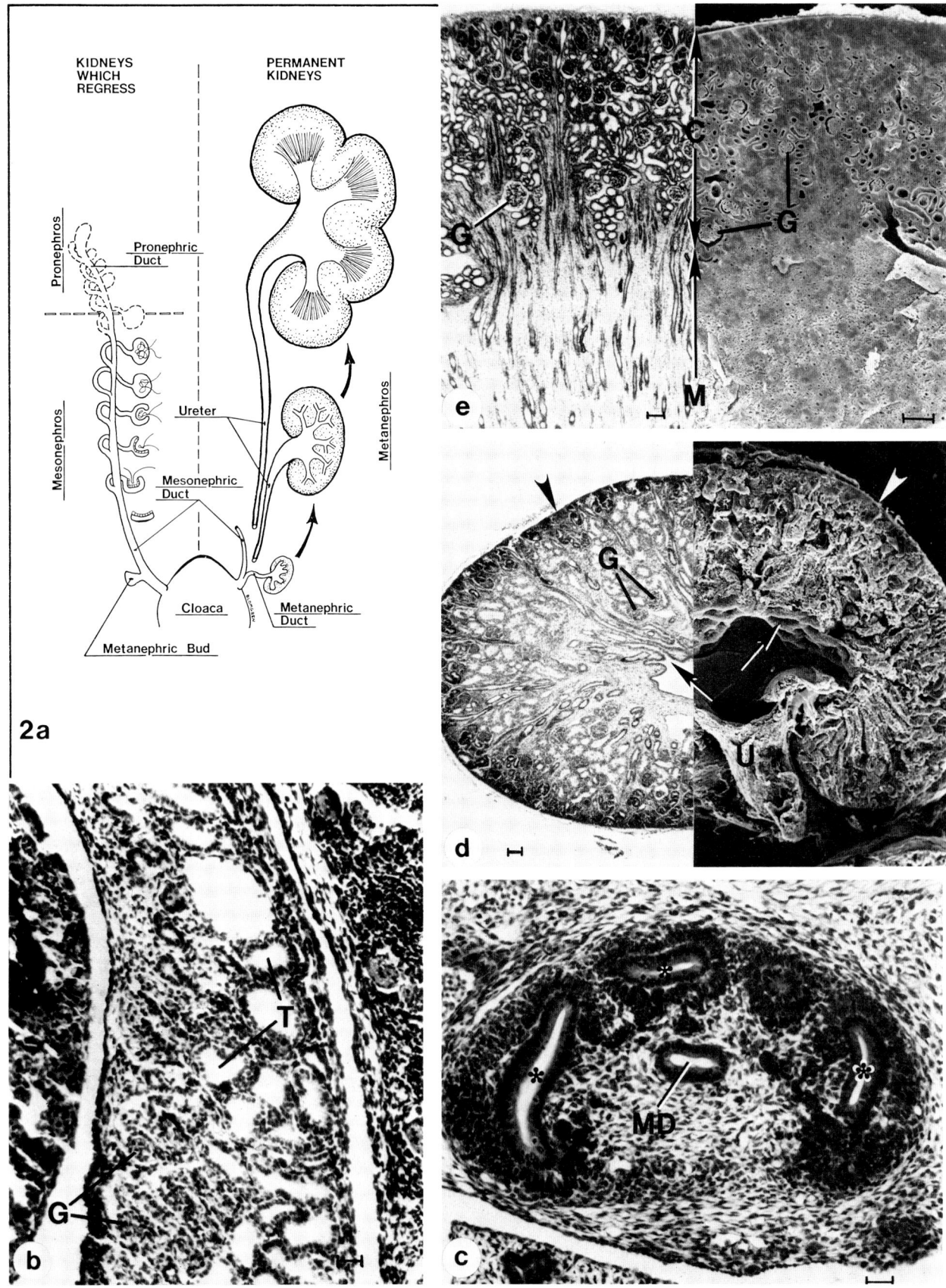

numerous slender foot processes (figures 6a and 6b). The first segment of the mesonephric tubule is the proximal tubule, which drains Bowman's space (figure 6c). The cells of the proximal tubule of most mammals are characterized by lush microvilli on their apical surface. However, in the dog, the initial segment of the proximal tubule possesses cells with a variable amount of microvilli (figure 6c). In more distal portions of the proximal tubule, the apical surface of all cells appear uniform with a single cilium and lush microvilli (figure 6d). The distal tubule of the dog mesonephros is difficult to distinguish from the collecting duct just by surface features. The rabbit, however, has very distinct segmentation of the mesonephric tubule [22]. In this animal, the proximal tubule has an initial and terminal segment, whereas the distal tubule has three segments: a pre-attachment, attachment, and ampulla segment. The rabbit collecting tubule possesses two cell types: principal and intercalated cells. The intercalated cells are similar to those seen in the metanephric kidney and have small folds called microplicae. The mesonephric collecting tubule of the dog seems to contain a single cell type, the principal cell. This cell type is characterized by a single cilium and stubby microvilli (figure 6e). The collecting tubule terminates at the mesonephric (Wolffian) duct by entering perpendicular to the axis of the duct. The apical surfaces of the mesonephric duct cells are cobblestone in appearance, and each possesses a single cilium and many blunt microvilli.

Metanephros. The basic structure of the metanephric nephron [3,7,14,23] is shown schematically in figure 7. Bowman's epithelium and most of the tubular segments develop from the metanephric blastema (non-stippled area in figure 7), while the collecting tubule segments develop from outgrowths of the metanephric duct. The tubular segments consist of the proximal tubule, loop of Henle, distal tubule, and collecting tubule. Each of these tubule segments can be further subdivided.

The metanephric glomerulus consists of three cell types: visceral epithelium, glomerular endothelium, and mesangium. A network of capillary loops called the glomerulus are covered on their outer surface by a visceral epithelium and are lined by a fenestrated endothelium internally (figure 7a). Mesangial cells are seen positioned between capillary loops. The visceral epithelial cells have an elaborate pattern of foot processes which interdigitate with neighboring foot processes (figures 7a and 7b). The proximal tubule of the dog is described as having three [7] or four [3] segments: two portions in the pars convoluta or convoluted region, and one or two portions in the pars rectae or straight region. However,

Figure 2. Scheme of kidney development.

Descriptions will be for figures starting in the upper left and proceeding counterclockwise around the plate.

a. Schematic representation of the development of the three paired kidneys present during the intrauterine life of mammals. The kidneys represented on the left side are the pronephros (without glomeruli) and mesonephros (possessing glomeruli and tubules) which regress in utero. Development of mesonephric nephrons occurs in a cranial to caudal direction with some of the basic stages of development illustrated. A diverticulum which evaginates from the mesonephric duct initiates the formation of the metanephros and is called the metanephric bud. The diverticulum elongates, and the distal portions cause the development of the kidney while the proximal portion remains attached to the mesonephric duct and is called the metanephric duct. The metanephric duct will form the ureter as the metanephros moves in a cranial direction. The right side of the schematic drawing represents development of the metanephros or permanent kidney for mammals. Three stages of metanephros development are represented with the earliest at the bottom (which correlates to figure 2c) and the dog kidney at term at the top (correlating to figure 2e). The intermediate stage of metanephros development correlates with figure 2d.

b. Light micrograph of the mesonephros from a 29-day dog embryo in which a row of glomeruli (G) and their associated tubules (T) can be seen. Bar = 100 μm.

c. Light micrograph of a very early stage of metanephros development showing the metanephric duct (MD) and its branches (*) which are surrounded by metanephric blastema. Bar = 100 μm.

d. Metanephros from a 44-day dog fetus. The left part of the figure is a light micrograph showing the presence of glomeruli (G) and a prominent nephrogenic zone (arrowhead). A collecting tubule entering the kidney hilium (arrow) can also be seen. There is no distinct medulla present yet. The right half of the figure is a scanning micrograph of a comparable kidney with the nephrogenic zone (arrowhead) and ureter (U). The termination points of the collecting tubules (arrow) can also be seen. LM & SEM Bars = 100 μm.

e. Metanephric cortex (C) and medulla (M) from a newborn puppy which possesses a nephrogenic zone in outer cortex and definitive glomeruli (G) in the inner cortex. LM & SEM Bars = 100 μm.

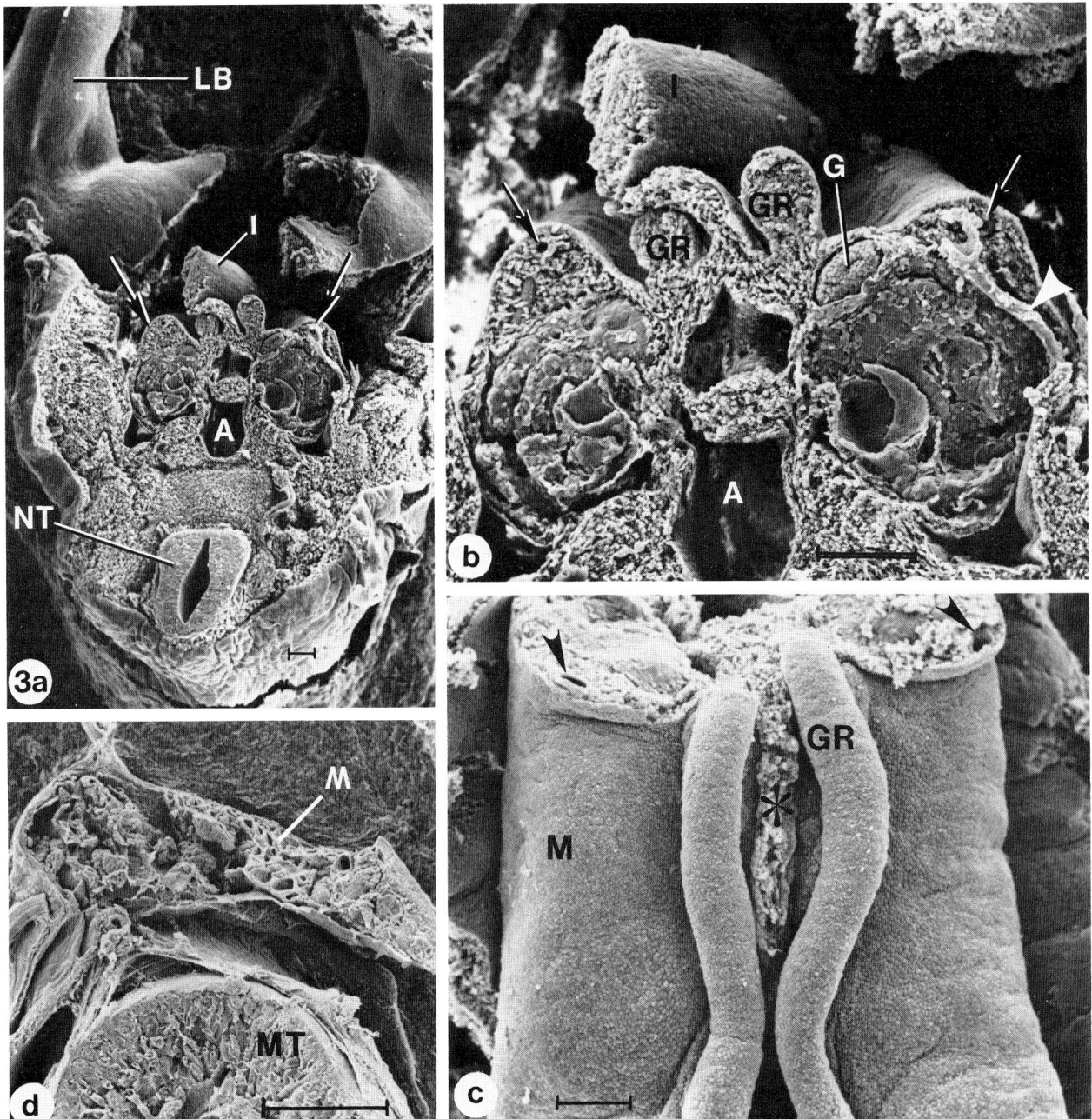

Figure 3. Canine mesonephros.

a. Low magnification scanning micrograph of a transverse section through a 29-day dog embryo. The urogenital ridge (arrows) is relatively large and attached to the dorsal aspect of the coelomic wall. For orientation, note the neural tube (NT), intestines (I), aorta (A), and limb bud (LB). Bar = 100 µm.

b. Enlargement of central part of figure 3a. The medial location of the glomerulus (G) is seen, as well as the ventrolateral location of the mesonephric duct (arrows). Note the collecting tubule (arrowhead) which is going into the right mesonephric duct. Segmental arteries or arterioles from the aorta (A) feed into the glomerular capillary loops. The paired gonadal ridges (GR) are seen near the intestine (I). Bar = 100 µm.

c. Scanning micrograph of the ventral aspect of the dorsal body wall with its urogenital ridge. Note the mesonephros (M) and mesonephric duct (arrowheads), as well as the gonadal ridge (GR). The dorsal wall attachment of the mesentery of the intestine is also seen (*). Bar = 100 µm.

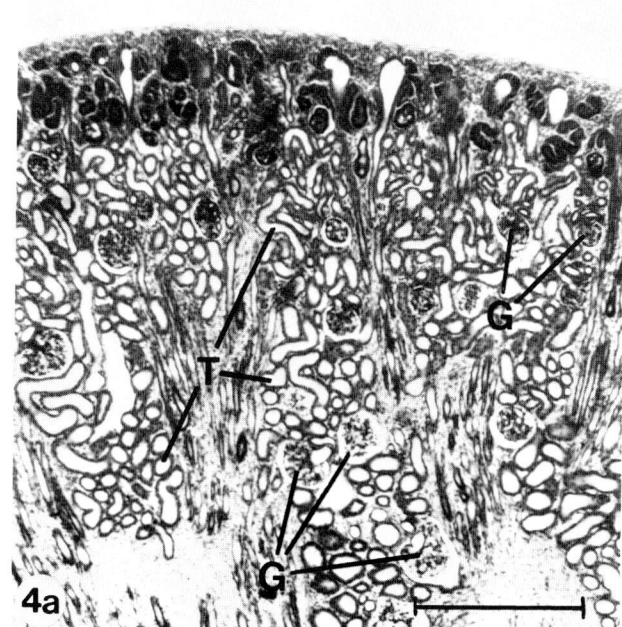
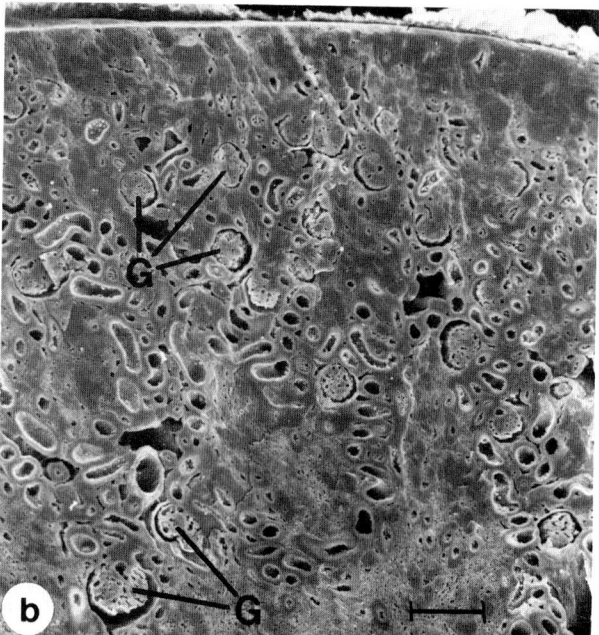

Fig. 3d. Scanning micrograph of mesonephros (M) and metanephros (MT) from a 44-day dog fetus. The metanephros starts to function while the mesonephros is still functional [20]. The left side of the micrograph projects to the midline of the fetus while the top projects cranially. Bar = 100 μm.

Figure 4. Metanephric glomerulogenesis.

The newborn puppy is still initiating new nephrons in the outer cortex (subcapsular). This centrifugal pattern of nephron maturation can be seen by comparing the outer cortex with the inner cortex in these micrographs. Note that in the deeper regions of the cortex (bottom of micrographs), the inner cortical glomeruli (G) have more capillary loops and are bigger. The tubular (T) portions of the nephrons are also more pominent in the inner cortex.

a & b. LM & SEM Bars = 100 μm.

Figure 5. Puppy renal cortex.

Schematic representation of metanephric nephrogenesis. All nephrons are initiated in the subcapsular region with a capping or condensation of metanephric blastema adjacent to the ampulla of the collecting duct. The condensed metanephric blastema forms a small tear-drop shaped vesicle which elongates into an S-shaped vesicle. The inner third of the S-shaped vesicle will develop into the glomerulus and Bowman's capsule, whereas the outer two-thirds will form most of the tubular component of the nephron, forming the proximal

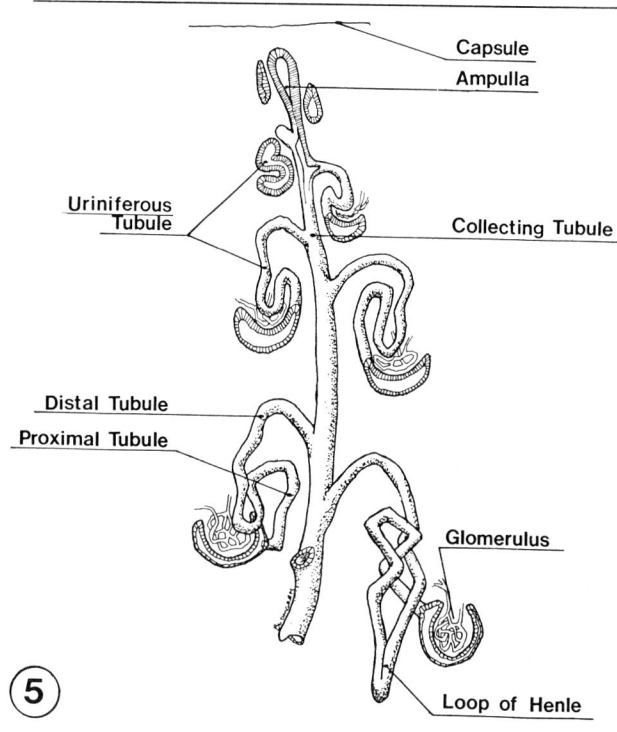

tubule, loop of Henle, and distal tubule. Meanwhile, the outer end of the S connects to the collecting duct which becomes somewhat arched. The subcapsular region of the collecting duct continues in its radial growth, initiating succeeding generations of nephrons as it proceeds. Therefore, the kidney has nephrons at many different stages of maturation with the youngest or least mature nephrons closest to the capsule (see figure 4).

Table 1

| MAMMALIAN KIDNEY DEVELOPMENT (TIME IN DAYS) | | | | | | | |
|---|---|---|---|---|---|---|---|
| APPEARANCE OF: | | | | | | | |
|  | PRONEPHROS | MESONEPHROS | MESONEPH. DUCT ENTERS U.G. SINUS | URETIC BUD | BUD WITH "CAP" | METANEPHROS | GESTATIONAL PERIOD |
| MAN | 22 | 24 | 28 | 28 | 32 | 35-37 | 267 |
| MACAQUE |  |  | 28-29 | 29-30 | 31-32 | 38-39 | 167 |
| GUINEA PIG | 16 | 17 | 19 | 20 | 21 | 23 | 67 |
| RABBIT | 8.5 | 9.0 | 11.5 | 11.5 | 13 | 14 | 32 |
| RAT | 10 | 11.5 | 12 | 12.3 | 12.5 | 12.5 | 22 |
| MOUSE | 8 | 9.5 | 11 | 11 |  | 11 | 19 |
| HAMSTER |  | 8.5 | 9.3 | 9.3 | 9.5 | 10 | 16 |
| CHICK | 1.5 | 2.3 | 3 | 4 | 5 | 6 | 21 |

FROM HOAR AND MONIE [13]
IN: DEVELOPMENTAL TOXICOLOGY, 1981

## Development of the Metanephric Nephron

The kidney, because of its complicated vasculo-tubular relationships, presents a number of interesting problems in its development. While light and transmission electron microscopy studies have provided details on events of the morphogenesis of glomeruli and tubules, it remains difficult with these conventional techniques to appreciate the various stages of maturation of the different nephrons located throughout the kidney cortex. For example, there are differences between the zones of the kidney (outer, middle, and inner), as well as the individual cells of a single glomerulus. A glomerulus may exhibit all stages of maturation. Thus, if one is to understand the functional capabilities of the developing kidney, it becomes necessary to know the state of maturity of all nephrons. The SEM is the most feasible means to approach this problem [5], since large areas of the kidney may be examined in three dimensions. Moreover, with the wide range of magnifications of the SEM, it is possible to examine important details in the development of

Figure 6. Mesonephric nephron.

Descriptions will be for the micrographs starting in the upper left and proceeding in a clockwise direction around the plate.

a & b.  Mesonephric glomerulus (G) which possesses visceral epithelial cells or podocytes (P) with foot processes (arrow). a. Bar = 10 μm. b. Bar = 1 μm.

c.  Initial part of the proximal tubule (PT) (lower left) showing cuboidal cells with a variable amount of microvilli. Bowman's capsule (BC) is to the upper right. Bar = 1 μm.

d.  The cells along most of the length of the proximal tubule (PT) are cuboidal and have a lush carpet of microvilli on their apical surface. Bar = 1 μm.

e.  Cells of the mesonephric collecting tubule (CT) are cuboidal with short microvilli and a single cilium on their apical surface. Bar = 1 μm.

f.  The mesonephric duct (MD) consists of cuboidal cells which have an apical cobblestone appearance. The cells have a domed apical surface with stubby microvilli and a single cilium. LM inset of mesonephric duct. SEM Bar = 10 μm.

g.  Schematic representation of the mesonephric nephron: glomerulus (G), Bowman's capsule (BC), proximal tubule (PT), distal tubule (DT), collecting tubule (CT), and mesonephric duct (MD). Also shown in the drawing is the gonadal ridge (GR) and the aorta (A) which gives branches to the mesonephric glomerulus.

all proximal tubule cells are cuboidal in shape with an apical brush border (i.e., carpet of microvilli) (figure 7c). The loop of Henle consists of a descending thin and an ascending thin segment of varying lengths. They are longer for the inner cortical nephrons than outer ones. The thin segments found in the medulla are composed of a squamous epithelium lined by stubby microvilli and a cilium, depending on the animal model (figure 7d). The distal tubule is considered as having three basic segments: a thick ascending limb (TAL), macula densa, and a convoluted portion. The cells of the thick ascending limb are cuboidal in shape in the medulla and low cuboidal in the cortex (figure 7e). In the cortex, the TAL has a straight segment which is continuous with the macula densa, then a short post-macula densa segment before it joins the distal convoluted tubule. The macula densa is that specialized region of the distal tubule adjacent to the hilus of the glomerulus and is composed of low columnar cells. The distal convoluted tubule is comprised of cuboidal shaped cells lined by stubby microvilli and a single cilium on their apical surface (figure 7f). The cortical collecting tubule is comprised of two cell types: the principal (light) cell and intercalated (dark) cells (figure 7g), both of which are cuboidal. The principal cell has numerous apical stubby microvilli and a single cilium, while the apical surface of the intercalated cell is characterized by numerous microplicae. The collecting tubule of the outer medulla is similar to the cortical collecting tubule. However, in the inner medulla, only principal cells are found and they are covered by numerous small microvilli (figure 7h).

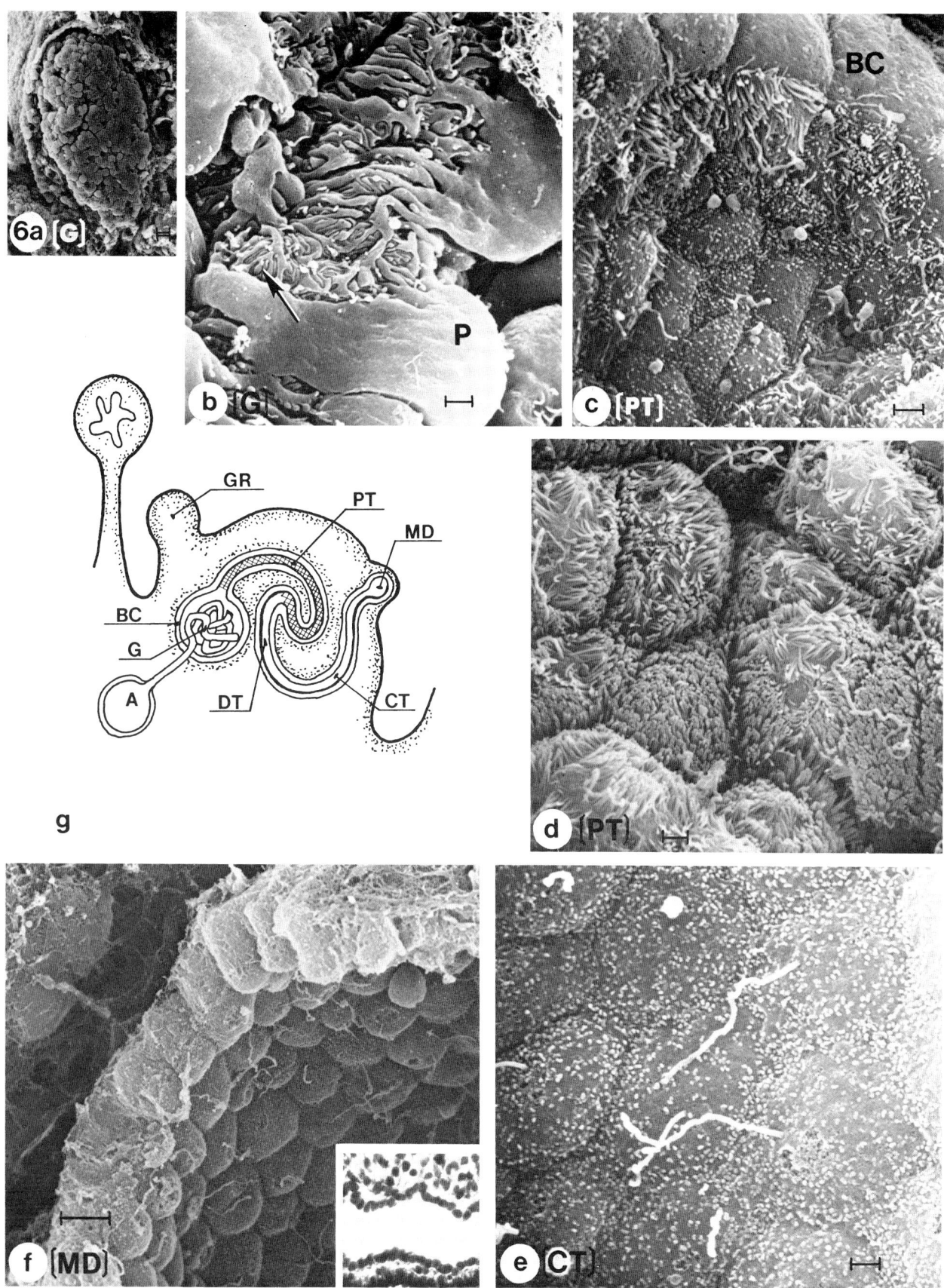

nephron segments. In the following description, the SEM has been employed in an effort to demonstrate the usefulness of this technique to study spatial and temporal relationships in the developing kidney [6,10,11,12,15,21]. Particular emphasis will be placed on the renal corpuscle, proximal tubule, and collecting tubule.

Renal Corpuscle. Nephrogenesis continues to occur within the newborn puppy kidney for two to three weeks. Thus, at two days post-natally, a wide range in glomerular maturation exists. The smallest and most immature glomeruli are at the periphery of the cortex (subcapsular), while the largest and most advanced glomeruli (characterized by extensive capillary expansion and looping) are located adjacent to the medulla (figures 4a and 4b).

Visceral Epithelium. Outer cortical glomeruli can be recognized by the grape-cluster appearance of their visceral epithelium (figure 8a). Microvilli are irregularly distributed over the surface of the densely-packed podocytes (figure 8b). Such podocytes are columnar to cuboidal in shape when observed in profile (figure 9a), and their cell bodies make direct contact with the thin basement membrane in the absence of foot processes. At the same time that the capillaries begin to elongate and expand, one of the first alterations of the podocyte is detected by the formation of a long, broad process that partially enwraps the capillary. No pedicels are observed in these glomeruli.

Continued capillary growth occurs in the mid cortex and is accompanied by a slight flattening of podocytes and their separation from one another (figure 8c). Laterally directed primary processes of irregular shape and length begin to appear (figure 8d), and occasionally pedicel-like structures are seen to arise from these processes. Concurrently, similarly small, thin extensions pass directly from the podocyte cell body to the capillary basement membrane without any consistency in their size or distribution.

The largest glomeruli are found within the inner cortex (figure 8e) and present the most mature appearance, although not like the adult (figure 8g). However, considerable heterogeneity in podocyte development exists. Some podocytes are flattened and spread apart, while others remain rounded and closely packed. Individual capillary loops can also be distinguished. Primary or secondary processes display pedicels which are well organized in some areas while still disorganized in other areas (figure 8f). The adult-like pedicels appear long and slender, whereas the immature foot processes have knob-like swellings (figure 8f). Scanning electron micrographs of cryofractured specimens further illustrate and clarify those features just described for the visceral epithelium of inner cortical glomeruli (figures 9e and 9f). Freeze fracturing of the tissue is accomplished by placing the samples in a sac containing 100% ethanol and then immersing the whole sac into liquid nitrogen. The frozen tissue is then fractured with a hammer and chisel. It has been suggested that maturation of the visceral epithelium occurs in response to the increasing surface area of the glomerular endothelium and/or in response to the passage and interaction of substances transmitted across

Figure 7. Metanephric nephron.

The metanephric nephron with a schematic representation of a typical nephron in the center of the plate. The nephron is composed of a glomerulus (G), proximal tubule (PT), loop of Henle (LH), thick ascending limb (TAL), distal tubule (DT), cortical collecting tubule (CCT), and medullary collecting tubule (MCT).

a. Glomerular capillary showing visceral epithelial (podocyte) process (P) with associated foot processes (arrow), the capillary endothelial cell (EC), and the fenestrated nature of this endothelium (arrowhead). Note the mesangium (M) at the tethered side of the capillary loop. Bar = 1 μm.
b. Low magnification scanning micrograph of a glomerulus (G) with its podocytes (arrows) and capillary loops. Bar = 10 μm.
c. Apical and lateral cell surfaces of proximal tubule cells (PT) with lush apical microvilli (arrowhead) and elaborate lateral surface foldings (arrow). Bar = 1 μm.
d. Thin descending loop of Henle (LH) with its squamous epithelium with cilium and stubby microvilli. Note that the cilium (arrowhead) is located at the perinuclear (bulging) region of the cell. Bar = 1 μm.
e. Medullary thick ascending limb of Henle (TAL) which is part of the pars recta segment of the distal tubule. The cells are cuboidal and possess a cilium and short microvilli on their apical surface. Bar = 1 μm.
f. Distal convoluted tubule (DT) with microvilli and cilium and, from the broken edge of the tubule, the folds formed by its basal and basolateral ridges (arrowhead). Bar = 1 μm.
g. Apical appearance of cortical collecting tubule (CCT) with its two cell types, the principal cell (PC) and intercalated cell (IC); the principal cell is characterized by a cilium and short microvilli, while the intercalated cell has an elaborate pattern of microplicae. Bar = 1 μm.
h. Apical and lateral surfaces of the cells of the medullary collecting tubule (MCT). Only one type of cell is seen. It displays numerous apical and lateral surface microvilli. Bar = 1 μm.

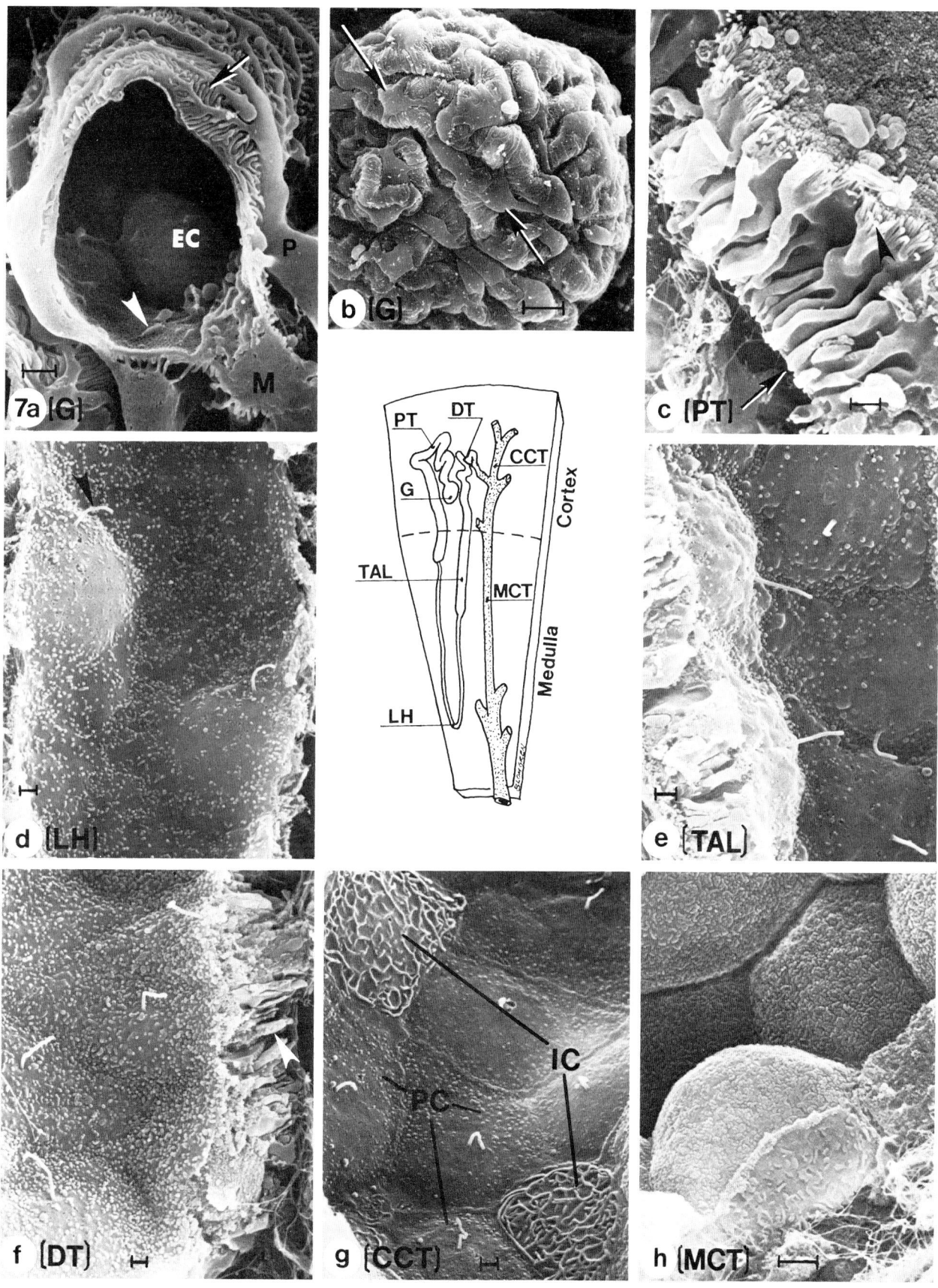

the capillary endothelium. However, recent observations [2] on maturing renal corpuscles growing in culture revealed normal visceral epithelial development even though no glomerular capillaries were present. The visceral epithelium of the adult dog is seen in figures 8g and 8h.

Glomerular Capillary Endothelium. Progressive glomerular modifications reflecting maturation are not limited to the visceral epithelium [11,12,15]. Indeed, the glomerular capillary endothelium undergoes marked changes in structure before it attains the characteristic sieve-like appearance observed in the adult (figure 9h). Subcapsular (outer cortical) glomeruli are approximately half the size of juxtamedullary glomeruli and possess few capillary loops (figure 9a). The endothelium comprising these loops is initially thickened and possesses few fenestrae (figure 9b). Soon, tiny indentations ($27.9 \pm 2.3$ nm), herein termed pinholes, are observed scattered irregularly along the endothelial surface. However, not all outer cortical glomeruli exhibit these features. On occasion, the endothelium has long, attenuated processes that interlace, creating holes that vary greatly in diameter (0.3 - 1.6 $\mu$m). These larger holes appear to represent a second population of fenestrae. Multiple layers of endothelium are also sometimes noted.

In contrast, glomeruli within the mid cortex are larger, due in part to increased capillary looping, and present a heterogeneous appearance (figure 9c). Some portions of capillaries possess both populations of fenestrae, as evidenced by the great variation in their size (0.05 - 1.2 $\mu$m) and shape (figure 9d), separated either by patches of thickened endothelium or by endothelial processes. Other portions exhibit pores that are more consistent in size ($56.9 \pm 2.6$ nm) and shape and thus reflect an increase in maturity.

Juxtamedullary (inner cortical) glomeruli show extensive vascularization and endothelial organization (figures 9e and 9f), but still do not approach those found in the adult. Fenestrae increase in diameter, averaging $74.6 \pm 2.4$ nm, yet continue to be interrupted by thickened regions (figure 9f). Some cells still lack pores or possess pores of variable shape and size. In our studies, examination of glomerular endothelium utilizing the SEM has never revealed the presence of diaphragms spanning the fenestrae, whereas these structures are occasionally observed with TEM. The adult canine glomerulus and endothelium are seen in figures 9g and 9h. Endothelial pores in the adult average $72.3 \pm 4.3$ nm.

Several recent studies [15,21] following the development of the renal corpuscle in the rat have described similar findings to that seen in the puppy [10,11,12].

Proximal Tubule

Proximal tubules found throughout the cortex of the puppy kidney are extremely immature in comparison to the adult. This immaturity is easily recognized when the proximal tubules are microdissected free [6], in that the convoluted and straight portions are greatly limited in their lengths (figure 10a).

The entire length of proximal tubules located in the outer cortex is comprised of short, cuboidal cells with relatively smooth, straight surfaces (figure 10b). Many cells lack lateral ridges and lateral-basal processes, but do exhibit stubby, abbreviated basal villi. However, some variability does exist; i.e., adjacent tubular cells may possess more extensive lateral processes. The intracellular organelles of the outer cortical proximal tubules are limited both in their number and organization. Mitochondria are few in number, appear round, and lack any orientation to the

---

Figure 8. Maturation of the glomerular visceral epithelium.

a & b. Outer cortical glomerulus from a young puppy. The rounded visceral epithelial cells look much like a bunch of grapes. The cells do not appear to have the specialized foot processes, at least at the surfaces exposed to Bowman's space. a. Bar = 10 $\mu$m. b. Bar = 1 $\mu$m.

c & d. Mid-cortical glomerulus from a young puppy. The surface of the glomerulus is still dominated by the epithelial cell bodies; however, there are numerous areas in which podocyte processes and pedicels are seen (arrowhead). The capillary loops are not yet obvious. c. Bar = 10 $\mu$m. d. Bar = 1 $\mu$m.

e & f. Inner-cortical glomerulus from a young puppy. The glomerular capillary loops with their associated epithelium comprise the predominant surface area. Podocyte (visceral epithelium) cell bodies (arrow) can still be appreciated. Note the presence of bulbous processes attached to the podocyte processes (arrowhead). e. Bar = 10 $\mu$m. f. Bar = 1 $\mu$m.

g & h. Adult dog glomerulus with its capillary loops covered by podocytes and their processes. Note that with maturation (figures a, c, e & g), podocyte cell bodies (arrow) become less obvious and capillary loops become more obvious. h. Fully mature podocyte (P) with its processes which terminate in pedicels or foot processes (arrowhead). g. Bar = 10 $\mu$m. h. Bar = 1 $\mu$m

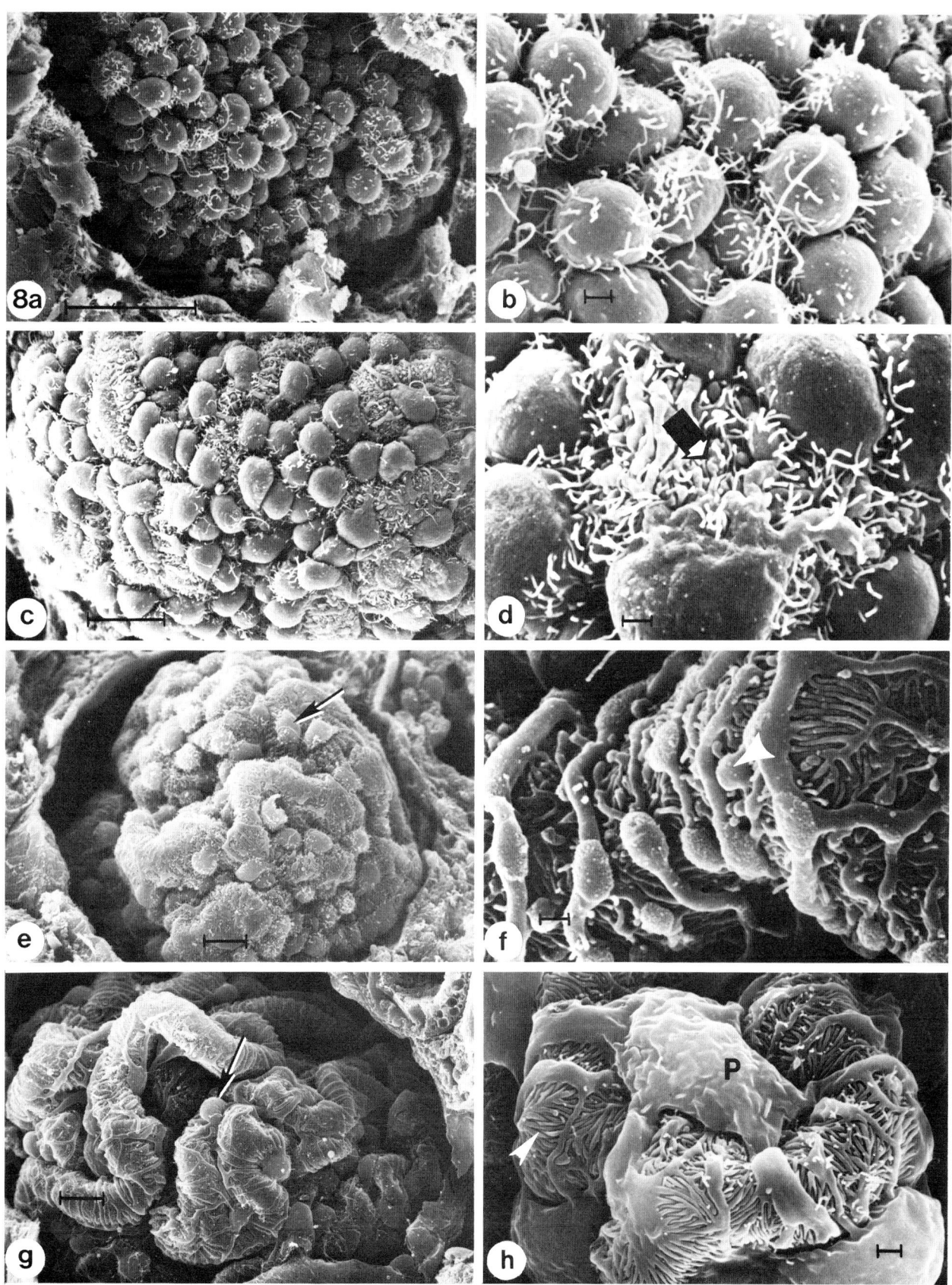

basilar interdigitations. There are relatively few lysosomes, pinocytic vesicles, and cisternae of the rough endoplasmic reticulum. Thus, these data suggest that the outer proximal tubules lack morphological segmentation.

When the convoluted portions of proximal tubules in the mid-cortex are examined, the primary feature noted is heterogeneity of the lateral surface contours. Around the circumference of a single tubule, cells may exhibit only subtle ridges and depressions (figure 10c), while cells elsewhere display numerous branched processes. However, none of the lateral folds show the vertical alignment characteristic of the adult (figure 10e). Distinct, finger-like basal villi (figure 10d) are observed in contrast to blunt processes of the outer cortex. Mitochondria, lysosomes, and pinocytic vesicles are more abundant in mid-cortical tubular cells than in the outer cortical cells. Cells of the straight portion of the proximal tubule are similar in both cortical zones. Segmentation of the mid-cortical proximal tubules is not possible because of the immature state of the tubular cells.

The convoluted proximal tubules of the inner cortex exhibit the most elaborate lateral and basal processes. The lateral membranes are now organized into lateral ridges and baso-lateral processes. Many of the mitochondria are arranged within lateral and basal folds, although not yet clearly oriented perpendicular to the length of the tubule. Phagosomes and vesicles occupy a large portion of the apex of each cell. However, the degree of plasmalemmal and intracellular organellar development still does not approach that found in the adult tubules. Therefore, it is still difficult to define the $S_1$ and $S_2$ segments. The cells of the distal portion of the pars recta are simply cuboidal in shape with varying numbers of basal villi. The lateral surfaces rarely exhibit processes. These cells do not closely resemble those of the adult except that they lack lipid droplets. Therefore, we cannot yet label this portion of the proximal tubule as $S_3$.

The morphology of entire proximal tubules from the outer, middle, and inner zones of the cortex can be further examined after the basement membrane has been removed. When this procedure is accomplished, the entire basal surface of the tubule can be scanned. Observations on the basement membrane-free tubules show the basal surface of all proximal tubules of the young puppy kidney to possess large, irregular processes (figure 10d). These processes do not show the discrete organization found in the adult (figure 10f). Furthermore, the proximal tubule cannot be segmented on the basis of basal surface morphology as is possible in the adult.

By one month, segmentations of the entire inner cortical proximal tubules are complete, while the outer tubules are mature by six weeks after birth.

### Loop of Henle and Distal Tubule

At present, there are no complete SEM studies that have followed maturation of the loop of Henle and distal tubule in the puppy. Several TEM reports have shown limited aspects of the development of these segments [4,18].

### Collecting Tubule

For our consideration of the development of the collecting tubule, we will divide it into five different regions. They are the outer cortex, middle cortex, inner cortex, outer medulla, and inner medulla. In the outer cortex or nephrogenic zone of the one week old puppy, the collecting tubules are like a blind, dilated sack, termed the ampulla. The cells lining this portion of the collecting tubule are columnar to cuboidal in shape, possess smooth lateral surfaces and apical microvilli positioned at the margin of the cell, and a centrally placed cilium (figures 11a, 11b, and 11c). Thus, these cells resemble the principal cells of the adult collecting tubule, whereas the intercalated cells, normally found in the adult, are absent (figure 11f).

As one progresses from outer cortex toward the outer medulla, the collecting tubule contains two cell types which are recognized by their apical surface features (figures 11d and 11e). The principal-like cell is again seen, as well as an intercalated-like cell which shows numerous microvilli and a cilium on the apical

---

Figure 9. Maturation of glomerular endothelium.

a & b. Cryofractured outer cortical glomerulus from young puppy showing a very immature glomerulus. a. Note that there are very few capillary loops (arrowheads) which are widely separated. Bar = 10 μm. b. The capillary endothelium has very few fenestrae (arrowhead). Those that are present are small. Bar = 1 μm.

c & d. Mid cortical glomerulus from young puppy. c. Note the relative increase in the number of capillaries (*) as compared to a. Bar = 10 μm. d. The endothelium has both large (arrow) and small (arrowhead) fenestrae. Bar = 1 μm.

e & f. Inner cortical glomerulus from young puppy. e. Note the numerous capillary loops (*). Bar = 10 μm. f. The capillary endothelium possesses numerous uniformly sized fenestrae. Bar = 1 μm.

g. Adult glomerulus with its many capillary loops (*). Bar = 10 μm.

h. The glomerular endothelium has numerous large fenestrae (arrow). Bar = 1 μm.

# SEM in Kidney Development

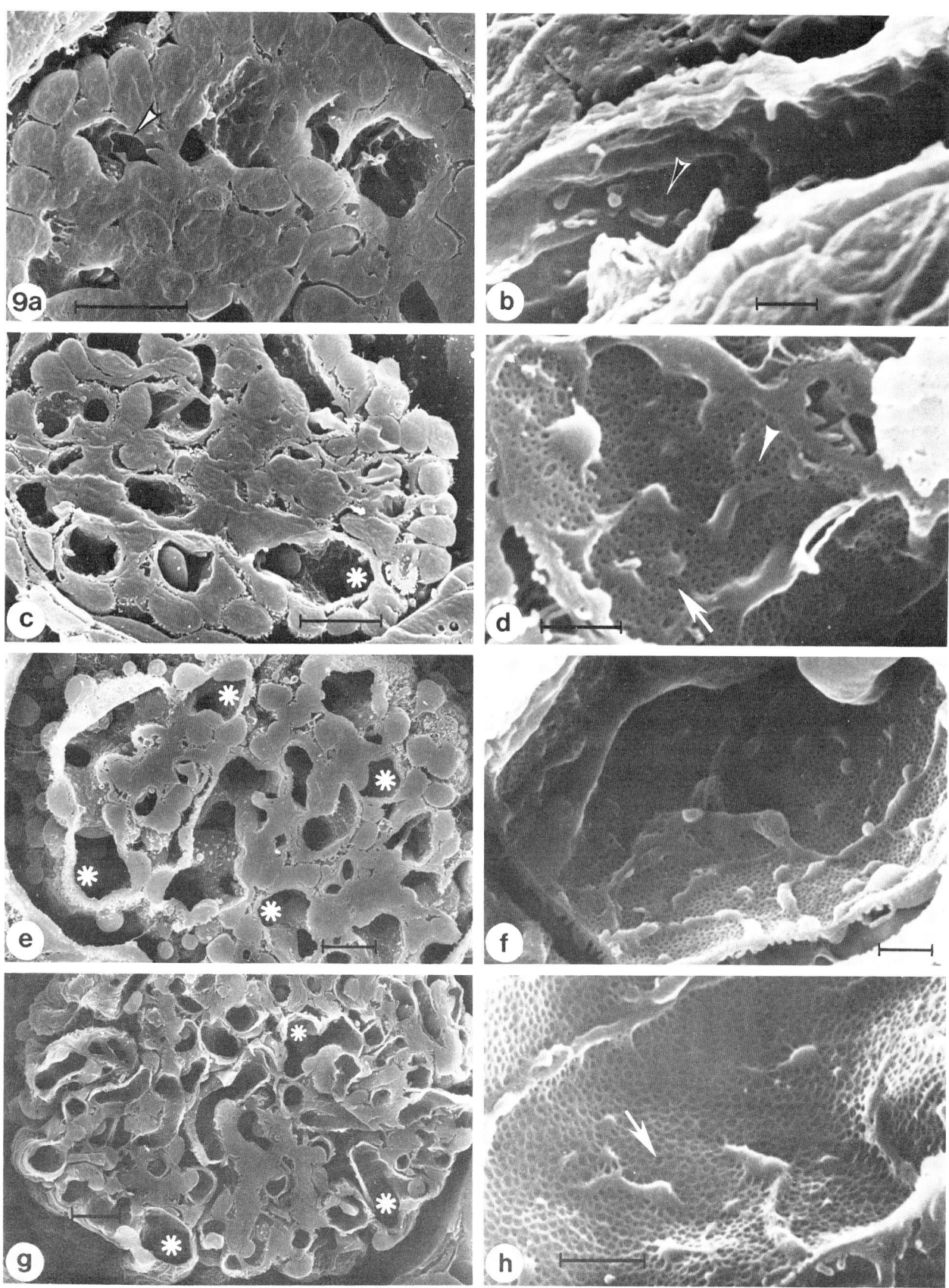

243

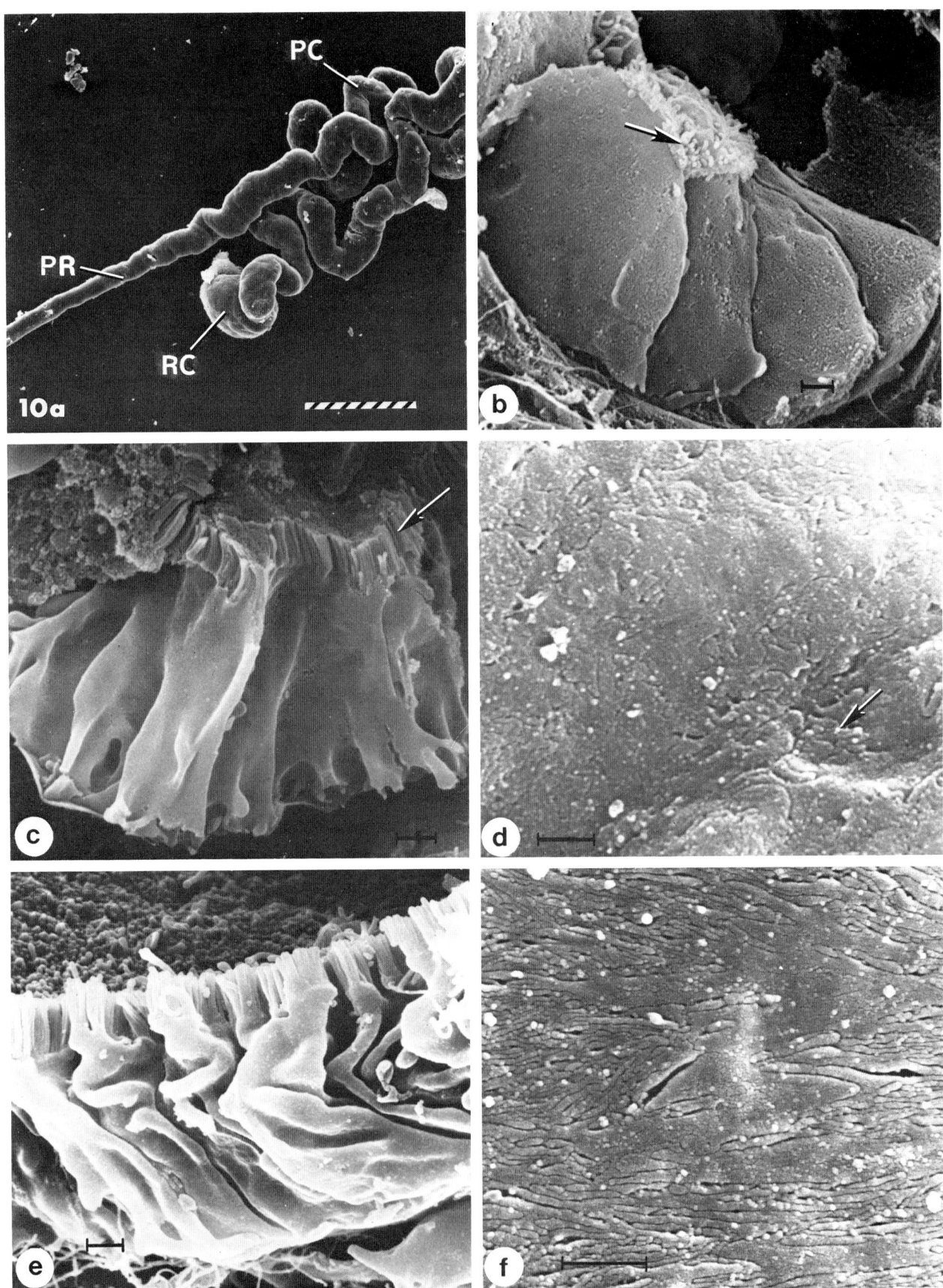

surface. The early intercalated cells must undergo further maturation such that the microvilli are modified into microplicae and the cilium is lost (figure 11d). The percentage of intercalated cells also changes along the length of the neonatal collecting tubule such that the middle cortical collecting tubule possesses 2% intercalated cells, while the inner cortical collecting tubule possesses 5%, and the outer medullary collecting tubule has 6% intercalated cells. It should be noted that the adult collecting tubule possesses approximately 20% intercalated cells along its entire length from outer cortex to outer medulla. As maturation progresses in the puppy, there is a gradual increase in the number of intercalated cells. These data suggest that the principal-like cell is a primordial cell for both the principal and intercalated cells. Clearly, SEM serves as an excellent tool for following differentiation and maturation of the collecting tubule [8].

## Summary

The present study has shown SEM to be a valuable tool in documenting the three-dimensional features of the developing kidney. This is due to the wide range of magnifications and good resolution of the SEM which allows one to observe large areas of tissue. The variations in developmental stages of inner versus outer cortical nephrons are thereby noted, as well as differences in particular cell types. By carefully following the changes in surface specializations of a cell, the level of maturation can be determined; that is (1) the glomerular endothelium gradually forms fenestrae, (2) the glomerular visceral epithelium develops slender foot processes, (3) the proximal tubular cells display apical and basal microvilli and various sized lateral folds, while (4) the dark cells of the collecting tubules present apical microplicae and the light cells, basilar infoldings.

## Acknowledgements

We thank Ms Marcia Powell for the typing of this manuscript. Funding was provided by UPHS grant #R01 HD 13232 and #R01 HD 10214.

## References

1. Arey LB. (1974). Developmental Anatomy. W.B. Saunders Company, Philadelphia. 295-309.
2. Bernstein J, Cheng F, Roszka J. (1981). Glomerular differentiation in metanephric culture. Lab. Invest. 45, 183-190.
3. Bulger RE, Cronin RE, Dobyan DC. (1979). Survey of the morphology of the dog kidney. Anat. Rec. 194, 41-66.
4. Dorup J, Maunsbach AB. (1982). The ultrastructural development of distal nephron segments in the human fetal kidney. Anat. Embryol. 164, 19-41.
5. Evan AP, Dail WG. (1977). Applications of SEM to problems of kidney development, Scanning Electron Microsc. 1977; II:373-380.
6. Evan AP, Gattone VH, Schwartz GJ. (1983) Development of solute transport in rabbit proximal tubule. II. Morphologic segmentation. Am. J. Physiol. 245, F391-F407.
7. Evan AP, Hay DA, Dail WG. (1978). SEM of the proximal tubule of the adult rabbit kidney. Anat. Rec. 191, 397-414.
8. Gattone VH, Evan AP. (1981). The collecting duct of the neonatal rabbit. Anat. Rec. 199, 92A.
9. Gattone VH, Johnson ML, Morse, DE. (1979). A scanning electron microscopic study of developing mesonephric and metanephric renal glomeruli. Micron 10, 201-202.
10. Gattone VH, Johnson ML, Morse DE. (1978). Renal corpuscle development in the dog: A scanning electron microscopic study. Renal Physiol. 1, 338-347.

Figure 10. Maturing proximal tubule.

a. Scanning micrograph of an isolated microdissected metanephric nephron from a 2 day-old puppy. The renal corpuscle RC (Bowman's capsule) can be seen, as well as the proximal convoluted tubule (pars convoluted PC and pars recta PR). Bar = 100 μm.
b. Scanning micrograph of the cells of the proximal portion of the metanephric uriniferous tubule from a developing nephron in the outer cortex from a 2 day-old puppy. Note the flat lateral surface and the scanty microvilli (arrow) on the luminal surface of the cells. Bar = 1 μm.
c. Proximal tubule cell from a mid cortical nephron from a 7 day-old puppy. Note the lush microvilli (arrow) on the apical surface and the basolateral folding of the cell membrane (largely in the lower portion of the cell). Bar = 1 μm.
d. Basal surface of the proximal tubule from a 7 day-old puppy isolated nephron. Note that the basal surface has some interdigitations (arrow) with neighboring cells leading to a somewhat complex basal pattern. Bar = 1 μm.
e. Proximal tubule cell from adult proximal tubule. Note the lush apical microvilli and numerous lateral plates or folds. Bar = 1 μm.
f. The basal surface of the proximal tubule from an adult nephron. Note the elaborate pattern of interdigitating basolateral plates. Bar = 1 μm.

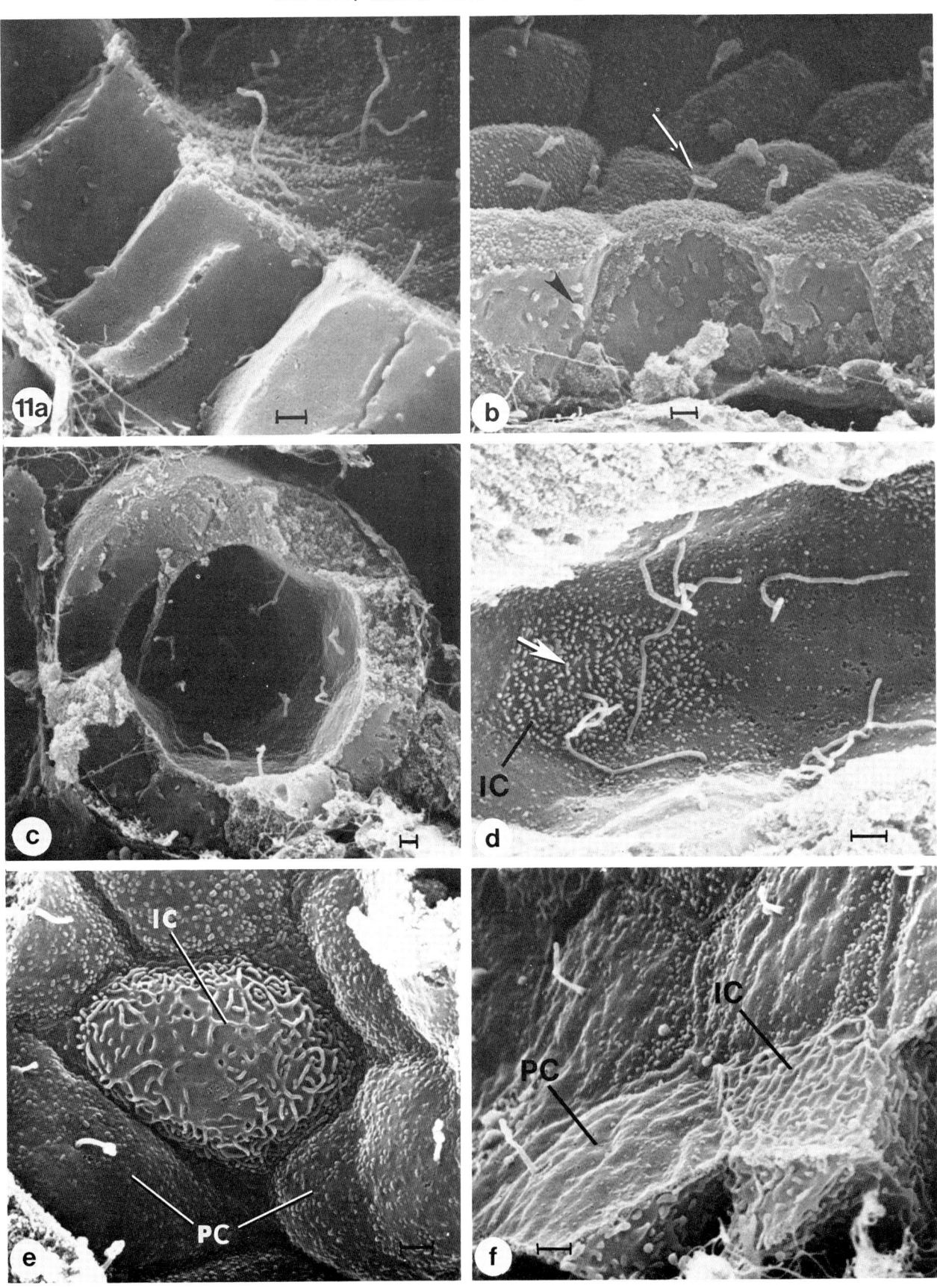

11. Hay DA, Evan AP. (1978). Observations on the development of glomerular capillaries of the puppy kidney, Scanning Electron Microsc. 1978, II:885-892.
12. Hay DA, Evan AP. (1979). Maturation of the glomerular visceral epithelium and capillary endothelium in the puppy kidney. Anat. Rec. 193, 1-22.
13. Hoar RM, Monie IW. (1981). Comparative development of specific organ systems, in: Developmental Toxicology, CA Kimmel and J Buelke-Sam (eds.), Raven Press, NY, 13-33.
14. Kaissling B, Kriz W. (1979). Structural analysis of the rabbit kidney. Adv. Anat. Embryol. and Cell Biol. 56, 1-123.
15. Kazimierczak J. (1980). A study by scanning (SEM) and transmission (TEM) electron microscopy of the glomerular capillaries in the developing rat kidney. Cell Tiss. Res. 212, 241-255.
16. Krause WJ, Cutts JH, Leeson CR. (1979). Morphological observations in the mesonephros in the postnatal opossum, Didelphis virginiana. J. Anat. 129, 377-397.
17. Miyoshi M. (1978). Scanning electron microscopy of the renal corpuscle of the mesonephros in the Lamprey, Entosphenus japonicus Mortens. Cell Tiss. Res. 187, 105-113.
18. Neiss WF. (1982). Histogenesis of the loop of Henle in the rat kidney. Anat. Embryol. 164, 315-330.
19. Oliver J. (1968). Nephrons and Kidneys. Hoeber Medical Division, Harper and Row, NY, 1-112.
20. Potter ED. (1972). Normal and Abnormal Development of the Kidney, Year Book Medical Publishers, Inc., Chicago, 3-79.
21. Spinelli F. (1974). Structure and development of the renal glomerulus as revealed by scanning electron microscopy. Int. Rev. Cytol. 39, 345-379.
22. Tiedemann K, Wettstein R. (1980). The mature mesonephric nephron of the rabbit embryo. I: SEM studies. Cell Tiss. Res. 209, 95-109.
23. Tisher CC (1981). Anatomy of the kidney, in: The Kidney, BM Brenner and FC Rector (eds.), WB Saunders Co., Philadelphia, Chapter 1, 3-75.

Figure 11. Maturing collecting tubule.

a. Apical and lateral surfaces of subcapsular arched collecting ducts from 7-day puppy kidney. Note the single cilium and marginal microvilli and relatively smooth lateral surfaces. Bar = 1 μm.
b. Apical and lateral surfaces of an outer corotical collecting tubule from a puppy. Note that all cells have an apical cilium (arrow) and sparse short microvilli typical of the principal cell. Some lateral surface microvilli (arrowhead) are also appreciated. Bar = 1 μm.
c. Cross-section of an outer cortical collecting tubule with its single ciliated cell type. Note the sparse lateral surface microvilli from the cell surfaces exposed by the break. Bar = 1 μm.
d. Mid cortical collecting tubule from puppy showing a cilium, numerous microvilli, and some small microplicae (arrow) typical of the intercalated cell (IC). Note that the microvilli appear to be lining up to form the microplicae. Bar = 1 μm.
e. Inner cortical - outer medullary region of puppy collecting tubule showing both principal cells (PC) and a mature appearing intercalated cell (IC). Bar = 1 μm.
f. Adult cortical collecting tubule with exposed apical and lateral surfaces of principal (PC) and intercalated cells (IC). Note the numerous lateral surface microvilli on both intercalated and principal cells. Bar = 1 μm.

Editor's Note: All of the reviewers' questions and comments have been appropriately addressed by text changes, hence the paper has no Discussion with Reviewers.

Additional discussion with reviewers of the paper "Prenatal Development of the Endocardium: A Review" by T. Pexieder continued from page 220.

R.R. Markwald: Is there any substantial evidence that there might be two different origins for embryonic endocardium, i.e., could mural endocardium have a separate origin from cushion forming endocardium?
Author: We are not aware of any evidence for dual origin of embryonic endocardium. Another relevant question to be answered in the future is whether the capability to form specific cushions, either atrioventricular or conotruncal, is inherent to specific endocardial regions. If the answer is yes, then how does the endocardium know where to form cushions and where not?

R.R. Markwald: Have any experiments been done transplanting endocardium from one area (e.g. mural) to another (e.g. truncus)?
Author: Despite their importance we are, to the best of our knowledge, not aware of the existence of such experiments.

R. Van Praagh: Did the author learn anything of interest to pediatric cardiologists and cardiac surgeons in his SEM studies of embryonic endocardium? Specifically, did the author gain any insight concerning the morphogenesis of
1) so-called endocardial cushion defects, i.e., complete and incomplete forms of common atrioventricular canal; or
2) membranous ventricular septal defects?
Author: This paper, which is an ultrastructural study at the cellular level, is a by-product of organ level studies related in a more straightforward way to the questions asked by Dr. Van Praagh. These studies are beyond the scope of the present communication and will be published elsewhere. Briefly summarized, the observations on human embryos especially, challenge some established views concerning the developmental changes in the position of the aorta and pulmonary artery, the existence of conotruncal torsions, the conus shift, the reality of the conus absorption and the so-called "transfer" of aorta into the left ventricle. The message this particular paper brings to pediatric cardiologists and cardiac surgeons is that despite the species differences of the heart organogenesis at the organ level, the underlying cellular mechanisms are much alike, in chick embryo as well as in human embryo.

THE DEVELOPMENT OF PHARYNGEAL ENDOCRINE ORGANS IN MOUSE AND CHICK EMBRYOS

S. R. Hilfer* and J. W. Brown

Department of Biology, Temple University
12th and Norris Streets, Philadelphia, PA   19122

(Paper received May 29 1984, Completed manuscript received September 13 1984)

## Abstract

Some interesting differences exist between development of the pharyngeal endocrine organs in mouse and chicken. In both, the thyroid forms as an evagination of the pharyngeal floor in the midline at the level of pharyngeal arch II and moves caudally to the base of the neck. In the chicken, the thyroid divides to form paired organs whereas in the mouse it forms a moustache-shaped bar that later fuses with the parathyroid and ultimobranchial bodies.

The thymi, parathyroids and ultimobranchial bodies form as lateral evaginations of the pharyngeal pouches. The chick forms two pairs of thymi and parathyroids from the third and fourth pouches whereas the mouse forms a single pair of each primordium from the third pouch. Ultimobranchial evaginations form from the caudal wall of the sixth pouch in the chick and from the posterior pharynx in the mouse. In the chick all of these evaginations detach from the pharynx and come to lie along the carotid arteries as separate organs that surround the thyroids. In the mouse the parathyroids and ultimobranchial bodies fuse with the lateral thyroid lobes. The thymi move to the ventral midline and fuse to form a single organ.

Positioning of these organs appears to be related to three developmental events. 1) Outgrowth of the original evaginations and their ventral movements conform to the curved shape of the pharyngeal arches. Ventral growth of the primordia should result in their convergence near the origin of the ventral aortic roots. 2) The thyroid appears to be attached to the common carotid arteries. Lateral growth of the carotids is consistent with the change in thyroid shape. 3) In the mouse the thymi lie ventral to the other organs. Their movement appears to be related to ventral closure of ventral ectoderm in the neck region.

Key words: Thymus, thyroid, parathyroid, ultimobranchial body, embryonic development, pharynx development, neck region development, endocrine gland development.

*Address for correspondence:
For reprints and other information, please contact S.R. Hilfer at above address.
Phone No.: (215) 787-8863.

## Introduction

In the last decade or so, a number of papers have been published on histogenesis of thyroid, parathyroid, ultimobranchial and thymic primordia in birds and mammals. These studies provide modern descriptions, using electron microscopy and histochemistry, of the cytodifferentiation of specific cell types in these organs. With a few exceptions, current understanding of the initial formation of these organs and their movements to the adult locations are based upon studies that were done more than 50 years ago. A few of these older studies were done by reconstruction from serial sections of paraffin-embedded material. Although these studies provide a good picture of pharyngeal development during early embryogenesis, they do not present adequate three-dimensional visualization of organ formation. For this description of pharyngeal derivatives, descriptions from the literature will be supplemented with scanning electron micrographs of Rhode Island red chick embryos and Swiss albino mouse embryos at various stages of development. Chick embryos will be described according to the staging of Hamburger and Hamilton (1951) and mouse embryos according to gestation time (Rugh, 1968; Theiler, 1972). Although the respiratory system also is a pharyngeal derivative, it will not be described here.

Endocrine organs form from the pharynx in all vertebrates. In fact, at least the thyroid has a prevertebrate counterpart; specialized cells of the endostyle in urochordates sequester iodine (Barrington, 1962) and in Amphioxus form iodothyronines similar to vertebrate thyroid hormones (Tong et al., 1962). In all vertebrates the thyroid forms from the floor of the anterior pharynx. Four or more pairs of parathyroid and thymic diverticula form from the pharyngeal pouches in lower vertebrates and postbranchial or ultimobranchial bodies form from the fifth or sixth pouch, or the corresponding region in the higher vertebrates, where these pouches are rudimentary. In all but mammals, thymic primordia form dorsally and parathyroid primordia form ventrally. In mammals the positions are reversed. There is no explanation for this change in origin of the two glands.

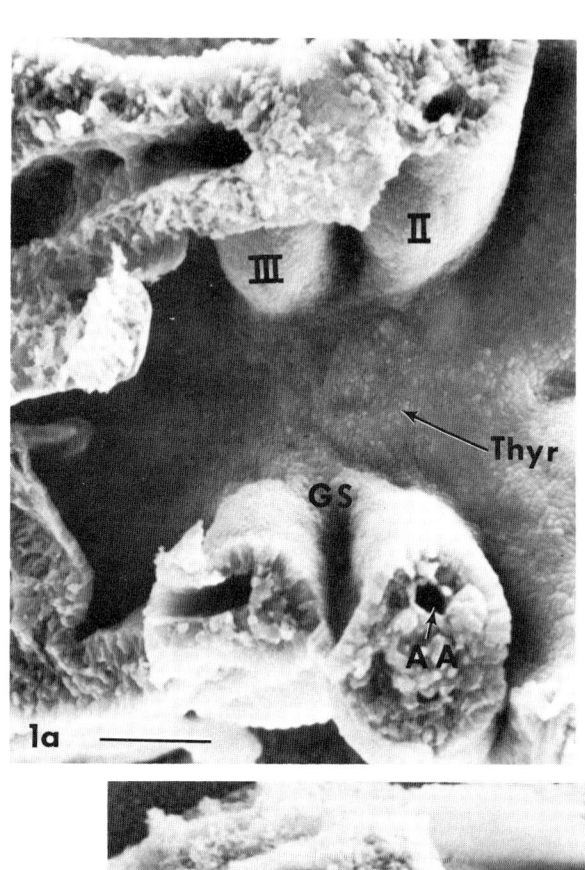
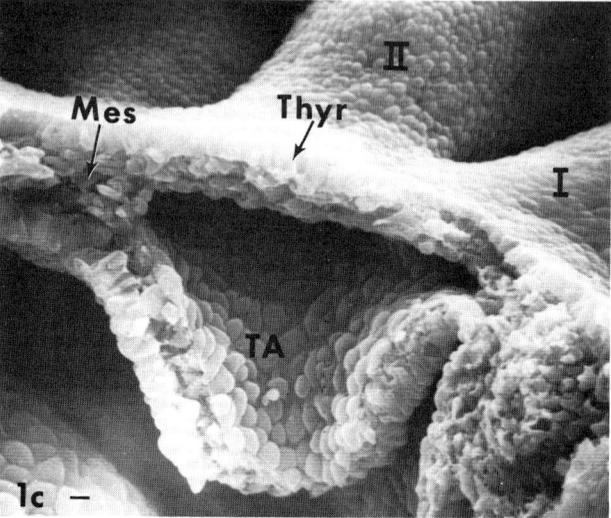
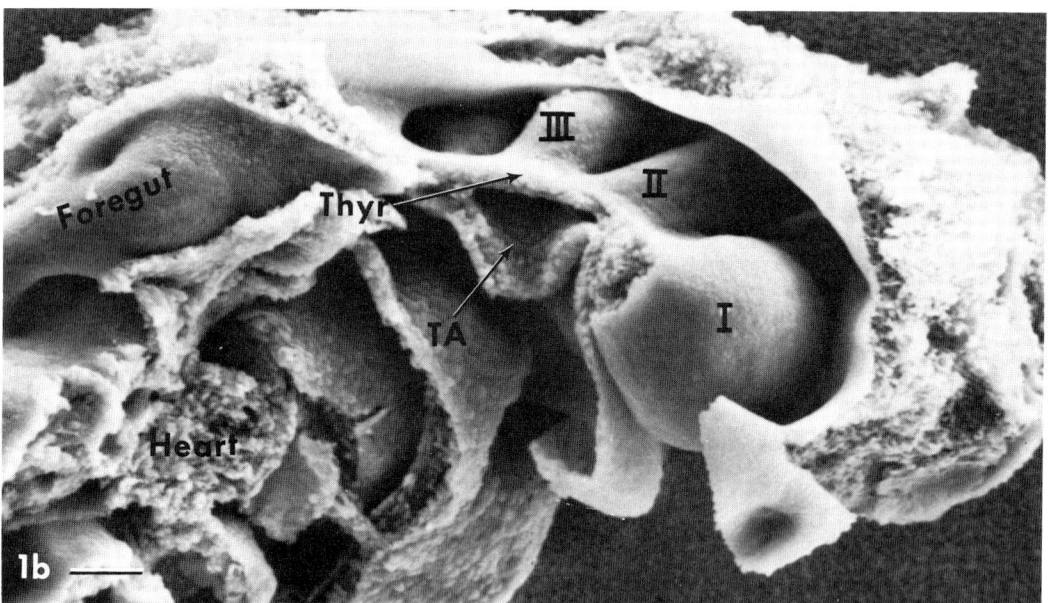

**Figure 1.** Pharyngeal region of mouse embryos at 9 1/2 days of gestation. a. dorsal view of pharyngeal floor. The thyroid placode (Thyr) is a thickening between the second pharyngeal arches (II). The lower jaw (pharyngeal arch I) is at the right side of the figure. The aortic arches (AA) are visible in pharyngeal arches II and III. GS = gill slit. Bar = 100 µm. b. Longitudinal section of another embryo. The embryo was cleaved through the center of the lower jaw (I). The right side of the heart has been cut away, leaving a portion of the truncus arteriosus (TA) and arterial sinus. The cranial end of the cardiac primordium lies in the midline. Bar = 100 µm. c. Enlargement of thyroid region in (b). The thyroid cells are slightly elongated. Mes = mesenchyme. Bar = 10 µm.

## Evagination and Detachment from the Pharynx

### Thyroid

The thyroid is the first of these primordia to form from the pharyngeal endoderm. It has been studied more extensively in the chicken (Hopkins, 1935; Venzke, 1949; Shain, et al., 1972; Hilfer, 1973) than in the mouse (Chardard-Ramboult, 1949; Crisan, 1935). In both species, the thyroid forms as a disc of elongate cells in the ventral midline between the second pharyngeal arches (Fig. 1). The placode forms between the two ventral aortic roots at stage 12 in the chick and late in the ninth day of gestation in the mouse (Fig. 1). A few hours later, the placode has evaginated to form a shallow depression (Fig. 2). When the embryo is cut horizontally through the dorsal pharynx and viewed from above, the thyroid primordium can be recognized by the circular orientation of its cell apices (Figs. 2a & b). In longitudinal section, the cells of the placode are taller and are packed closer together than the neighboring pharyngeal cells (Fig. 2c). Continued evagination produces a pouch surrounded by a slight ridge that protrudes above the pharyngeal surface in the chick (Figs. 3a & b). The pouch is skewed toward the lower jaw and is in intimate contact with the underlying arterial trunks (Fig. 3c). In the mouse, the pouch is more symmetrical and is surrounded by a layer of mesenchyme (Fig. 4).

Detachment of the thyroid from the pharynx begins with narrowing of the pouch and closure of the opening to the pharynx. This process begins at stage 18 in the chick and late in the eleventh day in the mouse. At first a passageway clearly exists between the pharynx and the interior of the pouch (Fig. 5). The opening becomes a narrow tube and finally closes entirely (Fig. 6). While the passageway is still relatively wide, the attachment to the pharynx in the chick becomes narrower than the thyroid vesicle (Fig. 5). In the mouse, the primordium is more of a cup shape (Fig. 4) until the opening closes (Fig. 7). The bulk of the thyroid moves away from the pharyngeal surface but remains attached to it. In the chick, a long stalk is formed (Fig. 8a) and the point of attachment is marked by a depression in the

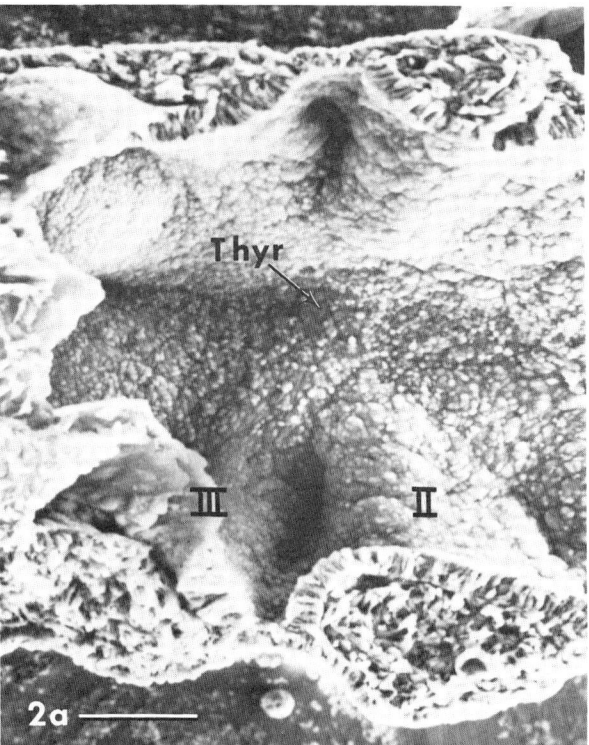

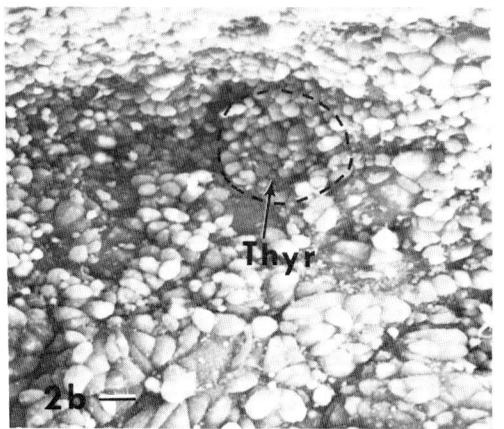

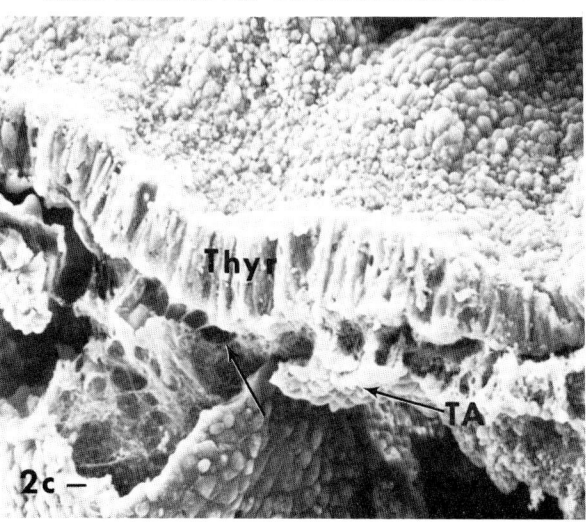

**Figure 2.** Pharyngeal region of chick embryos at stage 14. a. Dorsal view of the pharyngeal region. The thyroid primordium (Thyr) lies between pharyngeal arches II and III and forms a shallow depression. Bar = 100 μm. b. Enlargement of the thyroid region seen in (a). The cell apices are oriented to form concentric rings. Bar = 10 μm. c. Another stage 14 embryo cleaved longitudinally through the thyroid region. The thyroid cells are taller than their neighbors and have elongated relative to the placode stage. A fine meshwork of matrix fibers (arrow) separates the thyroid endoderm from the underlying endothelium of the truncus arteriosus (TA) and ventral aortic roots. Bar = 10 μm.

pharyngeal surface. The thyroid is suspended below the pharynx and becomes surrounded by a mesenchymal capsule (Figs. 8b & c). The stalk attaching the thyroid to the pharynx is shorter in the mouse but otherwise appears similar (Figs. 9a & b). When viewed from below, the thyroid is slightly elongated laterally at this stage in both the mouse (Fig. 9c) and chick.

Other Endocrines

The thymic and parathyroid primordia form as evaginations of the pharyngeal pouches (Verdun 1898; Schreier & Hamilton, 1952; Hammond, 1954). In most amniotes, including the chick, two pairs of diverticula are formed, one pair from the third pouches and another from the fourth pouches (Schreier & Hamilton, 1952). In the mouse, usually only one pair forms from the third pouch. The thymic evaginations form laterally and the parathyroids more medially. The ultimobranchial bodies form as a pair of evaginations from the sixth pouches in birds (Dudley, 1942), or from the posteriolateral pharyngeal wall in mammals (Ishikawa, 1965; Wollman & Hilfer, 1977, 1978). The initial stages of these diverticula can be seen best by observing the surface of the third or fourth pharyngeal pouch of stage 24 chick embryos. When the embryo is cleaved in cross section (Fig. 10a), the pouch can be oriented to allow inspection of the inner surface (Fig. 10b), revealing two evaginations, one dorsal and the other ventral. These are the thymic and parathyroid primordia, respectively. Cleaving the embryo or sectioning through the level of the evaginations results in distortion to the extent that the evaginations become unrecognizable in the chick. They are somewhat easier to locate in 12-day mouse embryos (Fig. 11a) but distortion prevents identification of the specific organ. When the heart and connective tissue are cleaned from the ventral surface of the developing respiratory system, the ultimobranchial bodies and parathyroids can be visualized in mouse embryos (Fig. 11b). In the 12 day mouse embryo or stage 24 (4 day) chick embryo, the thyroid is still a single body suspended from the pharynx (Figs. 8b, 9c, & 11b).

## Migration to Adult Positions

The thyroid forms from the floor of the anterior pharynx at the level of the bifurcation of the ventral aortae. The thymi and parathyroids form within the pharyngeal arches in close proximity to the aortic arches. Much of the later development of these organs is associated with the caudal displacement of the heart, the lateral growth of the aortic vasculature, and closure of the ventral body wall as the amnion undercuts the body. In birds, most of the primordia remain as separate organs, whereas fusion of various components occurs in mammals.

In the chick, the median thyroid vesicle becomes solid and elongates laterally to form a dumbbell shaped structure by stage 25 (Fig. 12). The larger ends curve ventrally to lie along the ventral aortic roots with the thinner isthmus lying closer to the ventral surface of the pharynx. Caudal movement of the heart and major aortic trunks carries the thyroid along so that the thyroid comes to lie between the derivatives of the third and fourth arches. By stage 27, the isthmus has broken to produce left and right thyroids (Fig. 13a) lying against the common carotid arteries (derivatives of the original ventral aortae cranial to the fourth aortic arches). At this stage, the other endocrine organs are deeply embedded in the neck mesenchyme and cannot be viewed by superficial dissection. A longitudinal fracture through one of the thyroids exposes thymus, parathyroid, and ultimobranchial bodies, but not all in the same section. The thymi lie dorsal to the thyroid (Fig. 13b); the ultimobranchial body and parathyroids lie ventral to or at the same level as the thyroid (Fig. 13c).

At later stages of development, the thyroids move farther apart and to the level of the clavicles in the shoulder region. They remain attached to the surfaces of the common carotid arteries close to the origins of the subclavian arteries (Fig. 14a). By eight days of incubation, the thymi have elongated along the surface of the external carotids and form twisted, string-like structures (Figs. 14a & b). The parathyroids are more compact and lie cranial and caudal to the thyroid, also in close proximity to the common carotids. It is reported that the parathyroids may fuse and often only a single parathyroid is seen on one side of the neck. The ultimobranchial bodies are small and are the most caudal of the organs (Fig. 14b). It has been reported that some cells of the ultimobranchial body fuse with the thyroid.

The arrangement of the endocrine organs in the neck of the mouse fetus is markedly different from that of the chick. All of the organs can be seen in ventral view by removal of the superficial layers of connective tissue.

**Figure 3.** Pharyngeal regions of chick embryos at stage 17. a. Diagonal view of the floor of a pharynx showing the pharyngeal arches (Roman numerals) and pouches (Arabic numerals), with the lower jaw (I) on the left. The fourth pouch is incomplete. Aortic arches (AA) are visible in several of the pharyngeal arches. The thyroid (Thyr) is a deep depression in the floor of the pharynx. Bar = 100 µm. b. Enlargement of the thyroid region in (a) after rotation and tilting the specimen. The cells surrounding the thyroid pit form a ridge (arrowheads). The pit is deeper at the caudal end, toward the left of the figure. Bar = 10 µm. c. Another embryo that was cleaved longitudinally close to the midline. The ridge surrounding the thyroid pit is more pronounced caudally than cranially and the evagination is skewed caudally. The thyroid is in close association with the truncus arteriosus (TA) at its juncture with the ventral aortic roots (VAR). Pharyngeal arches II and III are visible. Bar = 10 µm.

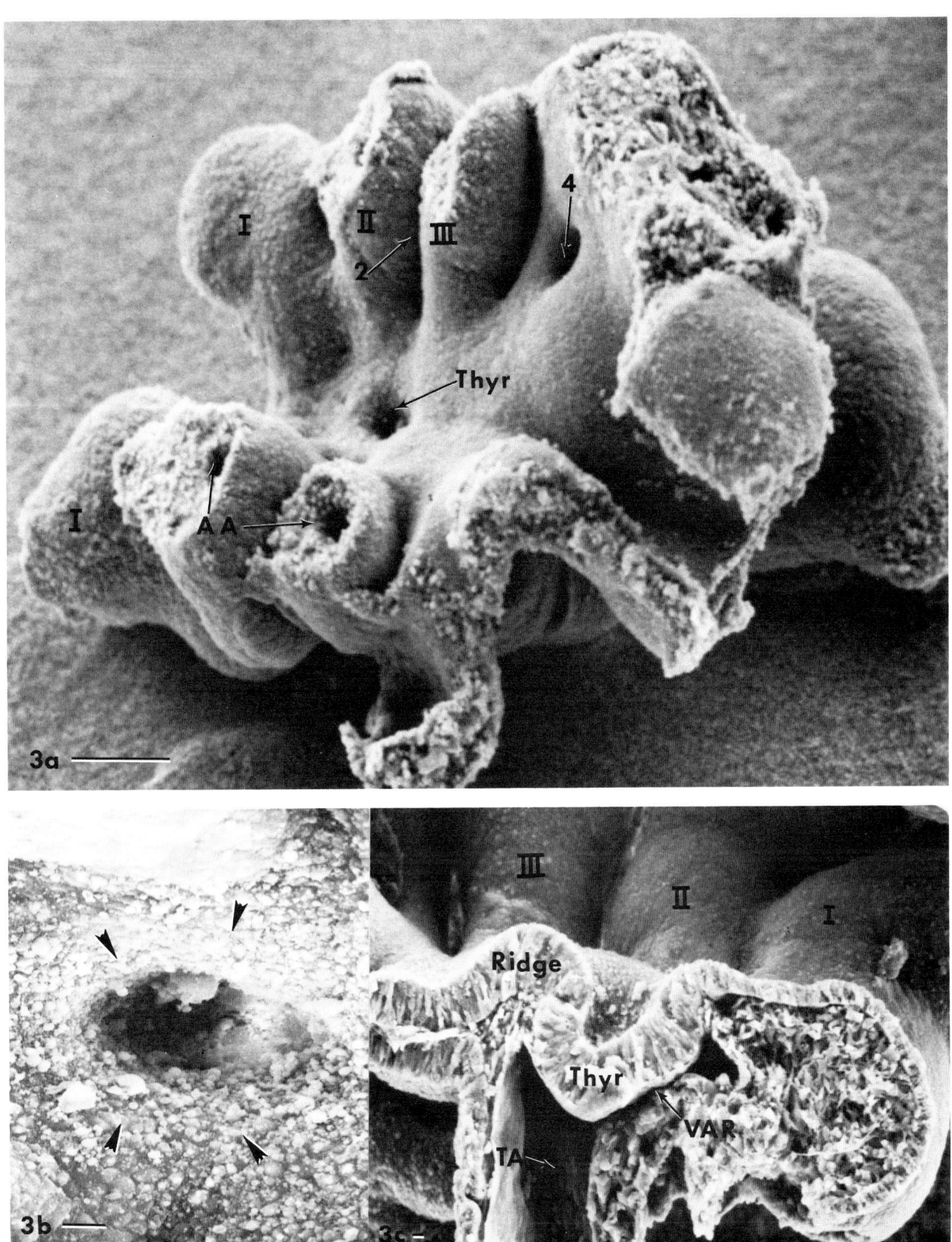

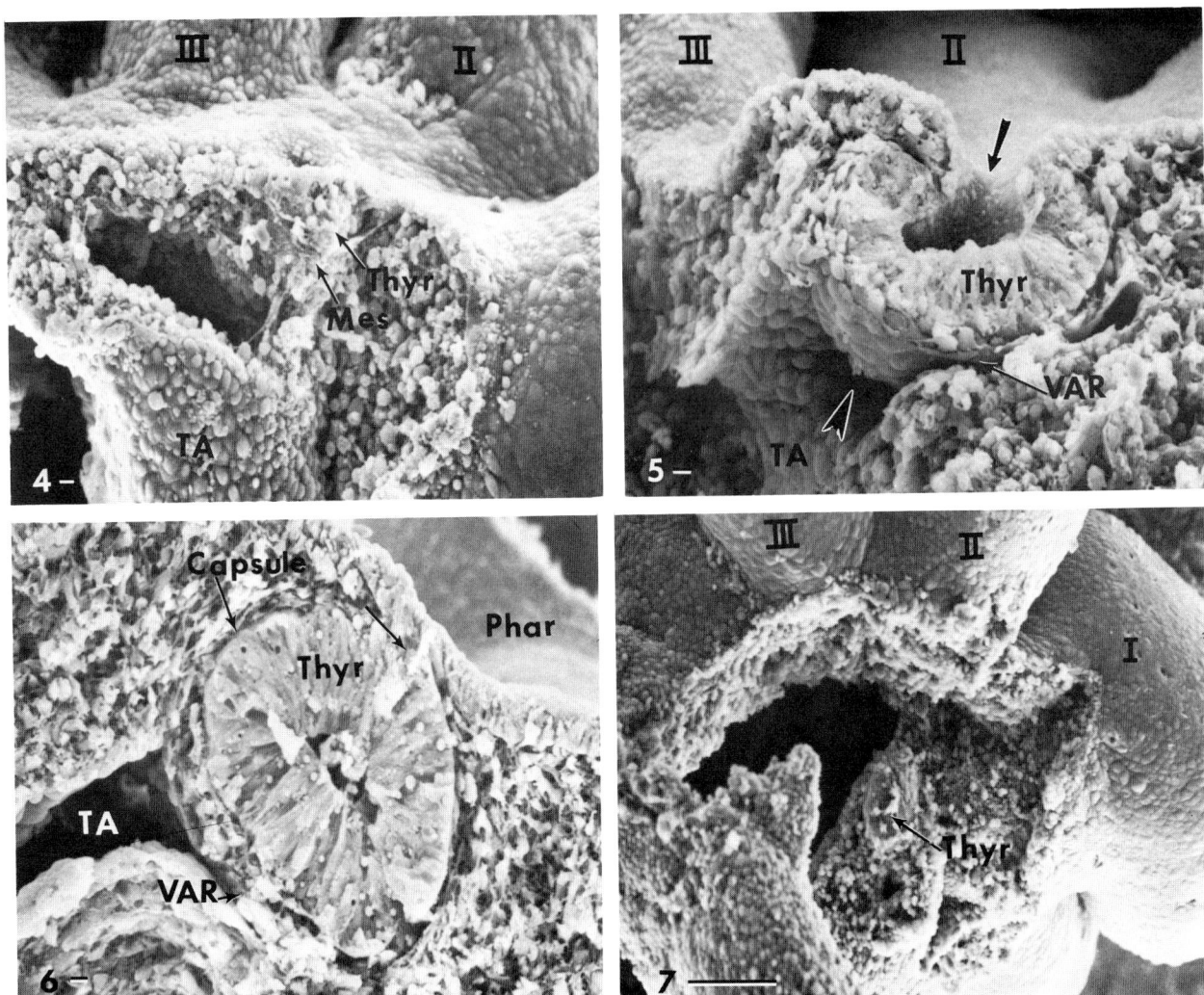

**Figure 4.** Pharyngeal region of an early 11 day mouse, cleaved to the right of the midline. The thyroid (Thyr) is cup-shaped and more symmetrical than in the chick. It is also surrounded by mesenchyme (Mes) rather than being in close association with the aortic blood vessels. The fracture passed along the surface of the truncus arteriosus (TA). II and III = pharyngeal arches II and III. Bar = 10 μm.

**Figure 5.** Longitudinally cleaved pharyngeal region from a stage 20 chick embryo. The plane of cleavage passed through the right ventral aortic root (VAR) and removed much of the truncus arteriosus (TA). The opening of the left ventral aortic root is visible (arrowhead). Pharyngeal arch III is at the caudal end of the preparation. The opening of the thyroid (Thyr) pit (arrow) is narrower than at earlier stages. A few mesenchyme cells lie between the thyroid and the endothelium. Bar = 10 μm.

**Figure 6.** Longitudinal section of the thyroid primordium in a stage 21 chick embryo. The thyroid vesicle is closed off from the pharynx (Phar) and attached by a short stalk (arrow). Although the layer of mesenchyme cells surrounding the thyroid has formed a capsule, there still is a close association with the truncus arteriosus (TA) and the ventral aortic roots (VAR). The cranial end is to the right. Thyr = thyroid. Bar = 10 μm.

**Figure 7.** Ventral view of the pharyngeal region of a mouse embryo late in the eleventh day of gestation. Part of the ventral wall has been removed to expose the thyroid (Thyr). The teardrop-shaped vesicle is suspended from the pharyngeal epithelium. The bases of pharyngeal arches I (lower jaw), II, and III are visible. Bar = 100 μm.

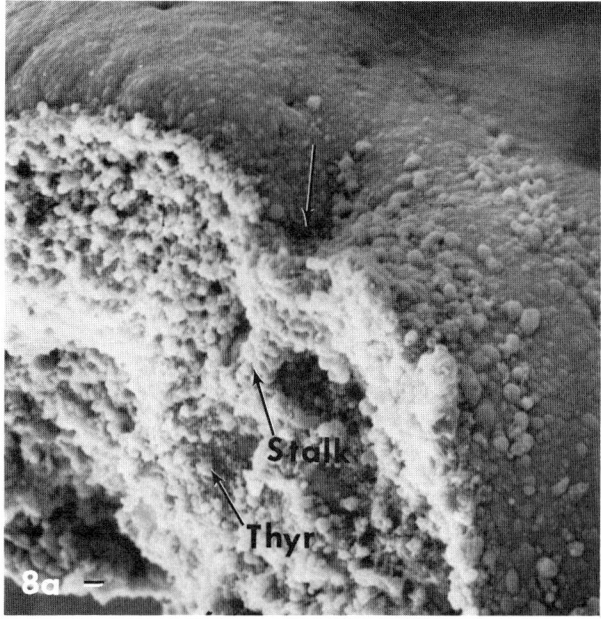
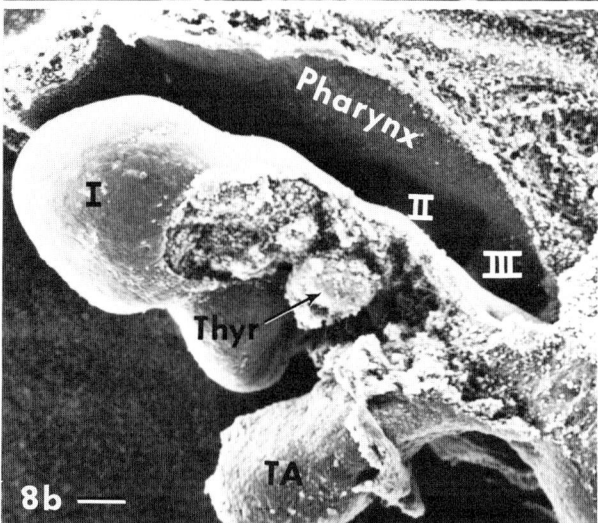
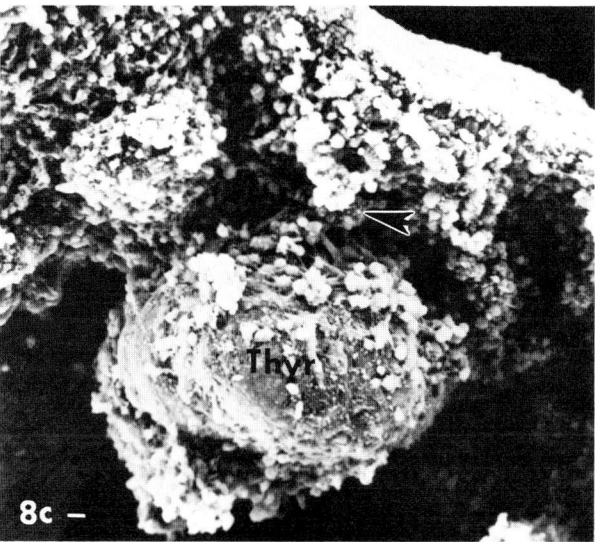

The primordia undergo similar changes in position as in the chick. At 13 days all of the primordia have detached from the pharyngeal endoderm and lie in the mesenchyme close to the surface of either the pharynx or the regressing pharyngeal pouches. The thymi are the lateralmost and caudalmost organs in this group and lie at an angle to the long axis of the neck region (Fig. 15a). The parathyroids lie cranial to the thymi, closer to the lateral tracheal surface. The thyroid is partially obscured by the thymi and can be seen in its entirety only when the thymi are removed (Fig. 15b). The thyroid forms a bar lying perpendicular to the long axis of the trachea and is of fairly uniform diameter along its length. The ends curve slightly dorsolaterally to point toward the parathyroids (Fig. 15c). Thus, the thyroid lies at the level of the fourth pharyngeal arch but has not reached the level of the ultimobranchial bodies. By 14 days, the thymi have moved toward the midline and their caudal ends are close to each other (Fig. 16a). However, they are still separated by the trachea and larynx cranially and the cranial ends lie lateral to the parathyroids (Fig. 16a). The thyroid has grown laterally to reach the parathyroids. The thyroid is not clearly visible unless the thymi are moved out of position (Fig. 16b). The ultimobranchial bodies also lie close to the ends of the thyroid, suggesting that the primordium has moved caudally.

By 15 days, the thymi lie closer together (Fig. 17a) but are still separated along the midline by a thin connective tissue partition (Fig. 17b). Removal of this connective tissue layer results in movement of the thymi closer together and distortion of their shape (Fig. 17a). The thymi have become more oval than previously and are more nearly parallel. The ends of the thyroid are enlarged as a result of engulfment of the parathyroids (Fig. 17c). The ultimobranchial bodies also invade the thyroid, forming the parafollicular cells and an assortment of other cells types that suggest an origin from pharyngeal epithelium (Wollman & Hilfer, 1977, 1978). The two enlarged ends of the thyroid remain connected by a narrower isthmus. The thyroid wraps around the trachea with the enlarged ends lying on its lateral surface (Fig. 17c). At this time all of these organs lie at the base of the neck. By 16 days

**Figure 8.** Pharyngeal region of stage 24 chick embryos. a. The thyroid (Thyr) is suspended from the pharynx by a long stalk. A depression in the floor of the pharynx (arrow) is all that remains of the former opening. Bar = 10 μm. b. When viewed from the side, the thyroid is seen to be suspended behind the lower jaw (I) and cranial to the truncus arteriosus (TA). Pharyngeal arch III is visible. Bar = 100 μm. c. Most of the capsule surrounding the thyroid has been peeled off to reveal the surface and the base of the stalk (arrowhead). Bar = 10 μm.

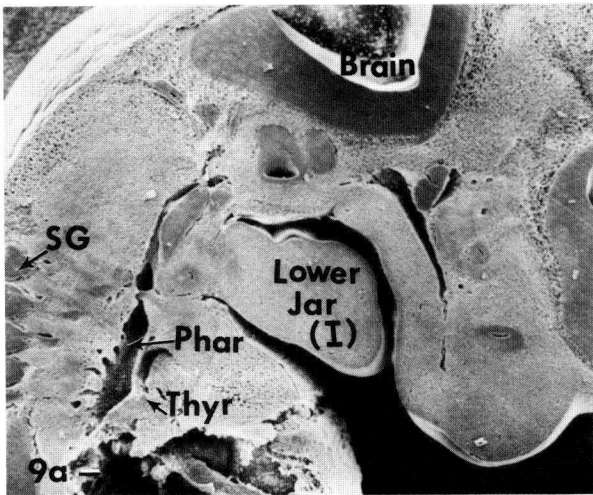
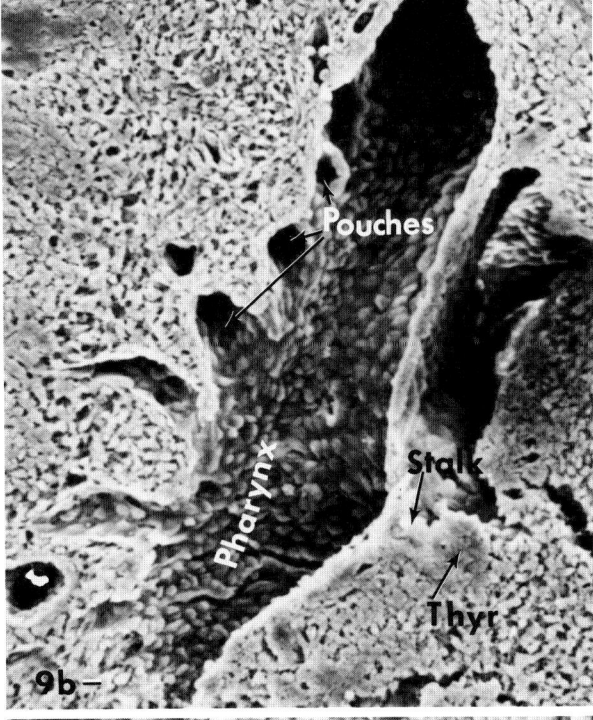
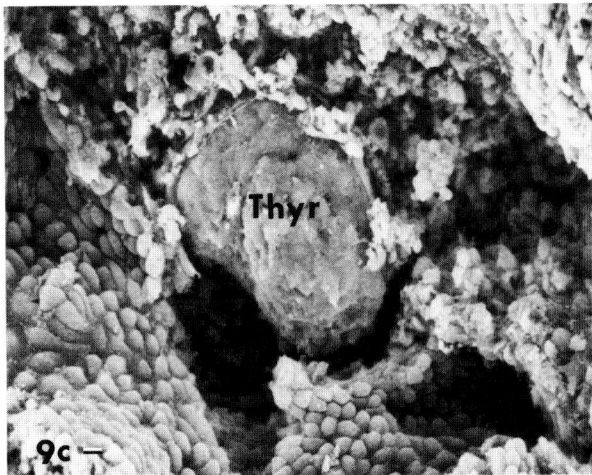

the thymic primorida lie in apposition (Fig. 18a) and will shortly join to make a single median thymus. The thyroid with its incorporated parathyroids and ultimobranchial bodies changes little in shape or position (Fig. 18b).

## Control of Development

Little experimental work has been published on the origin and movement of the pharyngeal endocrine organs. It has been shown that the mesenchyme surrounding these organs has a neural crest origin (Le Douarin, 1980; Bockman & Kirby, 1984). The proximity of the heart-forming region to the site of thyroid formation suggests that a relationship may exist between the two organs. An interesting experiment would be to test whether duplication of the thyroid would occur if the two primordial heart tubes are prevented from fusing. It has been reported that the ends of the thyroid of stage 25 chick embryos are attached to the carotid arteries. Preliminary experiments in which one side was detached gave rise to a single thyroid instead of division into left and right glands (Hilfer, unpublished). These experiments need to be repeated.

Movement of the thyroid and the other endocrine organs appears to be related to the formation of the neck and shoulder region. In the chick, the embryonic pharynx is relatively straight and narrow in comparison to the mouse. The mouse pharynx is relatively flattened and broader at its cranial end than caudally. In both organisms, flexures of the body place this region between the ventrally curved cranial end and the posterior region at the level of the heart. Thus, the thyroid by growing ventrally at its early stages moves toward the heart, undergoing both a ventral and caudal movement. The parathyroids and thymi arise relatively lateral to the ultimobranchial bodies and the medial thyroid. The parathyroids are in closer proximity to the aortic vessels than the thymi, especially in the mouse. At the earlier stages of development, the ventral body is not covered by ectoderm; closure of the amnion ventrally brings in an advancing front of ectodermal cells

**Figure 9.** Mouse embryos at 12 days of gestation. a. Low magnification of a paraffin-embedded embryo that was sectioned to the midline and then prepared for SEM (see Armstrong, 1971 for method). Note the posterior location of the thyroid (Thyr) and the cellularity of the ventral body in comparison to the chick. SG = spinal ganglion. Phar = pharynx. Bar = 100 μm. b. Enlargement of the pharyngeal region of (a). Several pharyngeal pouches are visible to the left. The thyroid is suspended from the pharynx by a short stalk. Bar = 10 μm. c. Ventral view of the thyroid in another embryo. The thyroid is surrounded by mesenchyme; note that it has elongated slightly laterally, but not symmetrically. Bar = 10 μm.

# Development of pharyngeal endocrine organs

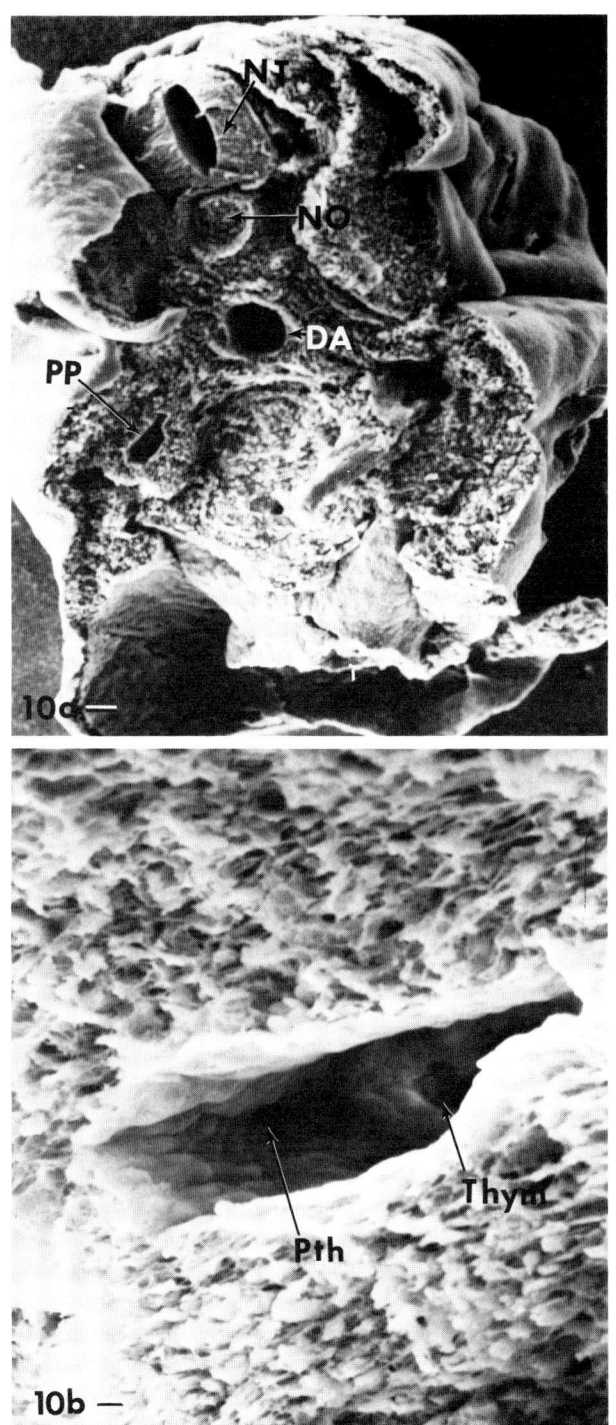

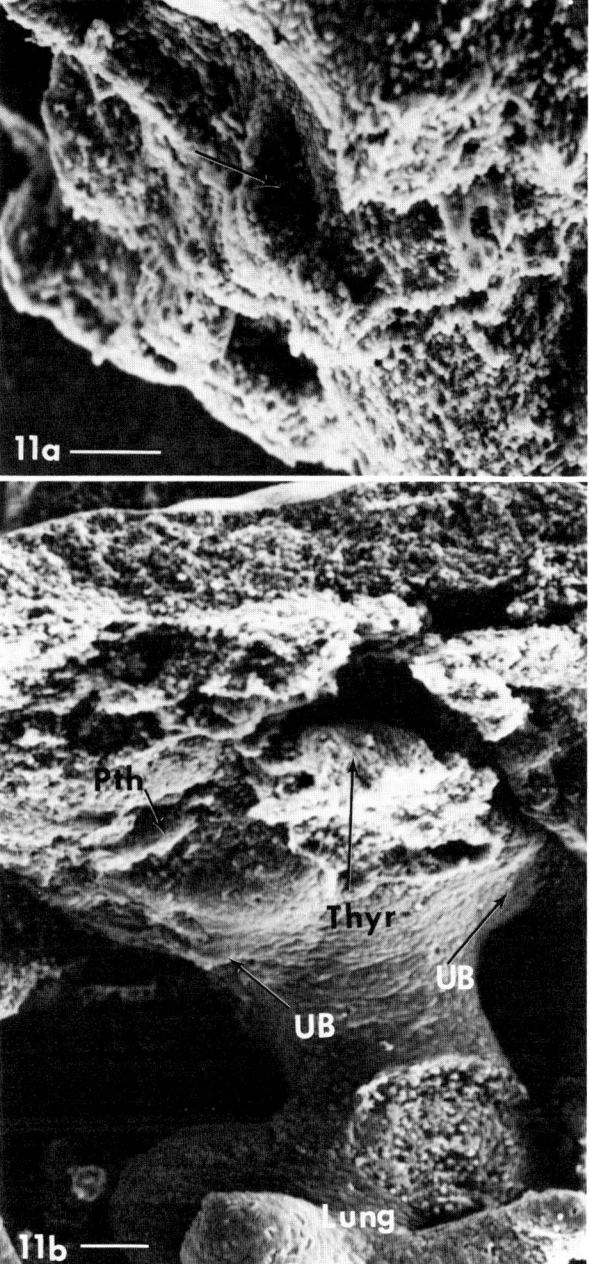

Figure 10. Chick embryo at stage 24, cleaved in cross section. a. Low magnification to show that the cleavage passed through the fourth pharyngeal pouch (PP). DA = dorsal aorta, NO = notochord, NT = neural tube. Bar = 100 μm. b. High magnification of the interior of the pouch. Two depressions in the wall represent the primordia of the thymus (Thym) and parathyroid (Pth). Bar = 10 μm.

Figure 11. Mouse embryos dissected to show the endocrine organs. a. Embryo cleaved through an evagination (arrow) of a pharyngeal pouch. There is no landmark to allow identification of this primordium. b. Embryo from which the ventral mesenchyme has been removed. The ventral pharynx and respiratory primordium are exposed. The thyroid (Thyr) is suspended in the midline and is slightly elongated laterally. The ultimobranchial bodies (UB) form small pouches at the caudal end of the pharynx. A small sac on the right side of the pharynx probably is the parathyroid (Pth). The thymi remain embedded in mesenchyme and are not visible. Bars = 100 μm.

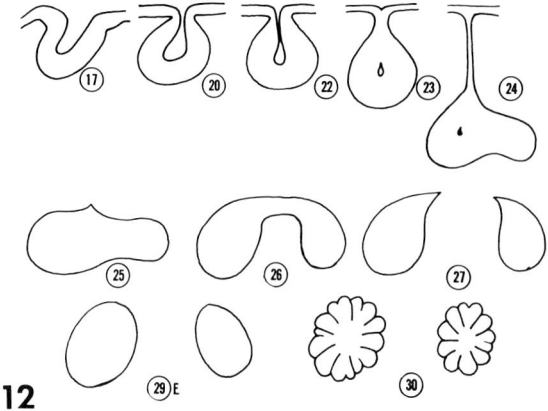

Figure 12. Diagrammatic representation of thyroid development in the chicken embryo from stage 17 (2 1/2 days) to stage 30 (7 days). Stages 17 through 22 show longitudinal sections; the later stages are shown from the ventral aspect. Between stages 17 and 21 the opening between the thyroid evagination and the pharynx closes. A hollow vesicle is formed that becomes filled with cells between stages 22 and 23. The attachment to the pharynx elongates as a stalk during stages 23 and 24. During stage 25 the primordium elongates laterally and detaches from the pharyngeal endoderm. Continued lateral elongation produces a U-shaped body with enlarged ends of unequal size, the left one tending to be smaller. During stage 27 the ends separate from each other and round up to form the left and right thyroids. During stage 30 the solid epithelial primordium is invaded by mesenchyme cells from the capsule and thyroid follicles are formed (Reprinted from Hilfer, 1973).

Figure 13. Stage 27 chick embryos. a. Ventral view of the neck region, opened to expose the paired thyroids (Thyr). The isthmus has broken and the glands are rounded. b. A longitudinal section through the left thyroid (Thyr) which exposed one thymus (Thym) embedded in mesenchyme. An aortic vessel is visible (arrowhead). c. Another longitudinal section through a thyroid that exposed a parathyroid (Pth) and ultimobranchial body (UB). Bars = 100 μm.

to form the ventral body wall. The thymi lie close to this front in the 14 day mouse embryo. It would be interesting to test the possibility that movement of the thymi toward the ventral midline depends upon the formation of the ventral body wall.

## Conclusions

The endocrine organs undergo complex movements to reach their final adult positions. These movements are complicated by the sites of formation, from the pharyngeal pouches in the cases of the thymi and parathyroids. Visualization of the movements is made considerably more understandable when the three-dimensional images of the scanning electron microscope are used. At the present time, the movements of these organs seem to depend upon three mechanisms. They are 1) ventral growth during early stages of development, 2) movement of the ventral aortic vessels and attachment of some of the endocrines to these vessels, and 3) closure of the ventral body wall. Establishment of these influences as the cause of the movements depends upon future experimental verification.

## References

Armstrong PB. (1971). A scanning electron microscope technique for the study of the internal microanatomy of embryos. Microscope 19, 281-284.

Barrington EJW. (1962). Hormones and vertebrate evolution. Experientia 18, 201.

Bockman DE, Kirby ML. (1984). Dependence of thymus development on derivatives of the neural crest. Science 223, 498-500.

Chardard-Ramboult S. (1949). Developpement de la glande thyroide ches la souris pendant la vie intra-uterine. (Development of the thyroid gland of the mouse during uterine life.) C. R. Soc. Biol. (Paris) 143, 40-41.

Crisan C. (1935). Die Entwicklung des thyreo-parathyreo-thymischen Systems der Weissen Maus. (Development of the thyroid-parathyroid-thymic systems of the white mouse.) Z. Anat. Entwicklungsgeschichte. 104, 327-358.

Dudley J. (1942). The development of the ultimobranchial body of the fowl, Gallus domesticus. Am. J. Anat. 71, 65-98.

Hamburger V, Hamilton H. (1951). A series of normal stages in the development of the chick embryo. J. Morphol. 88, 49-92.

Hammond WS. (1954). Origin of thymus in the chick embryo. J. Morphol. 95, 501-521.

Hilfer SR. (1973). Extracellular and intracellular correlates of organ initiation in the embryonic chick thyroid. Amer. Zool. 13, 1023-1038.

Hopkins ML. (1935). Development of the thyroid gland in the chick embryo. J. Morphol. 58, 585-613.

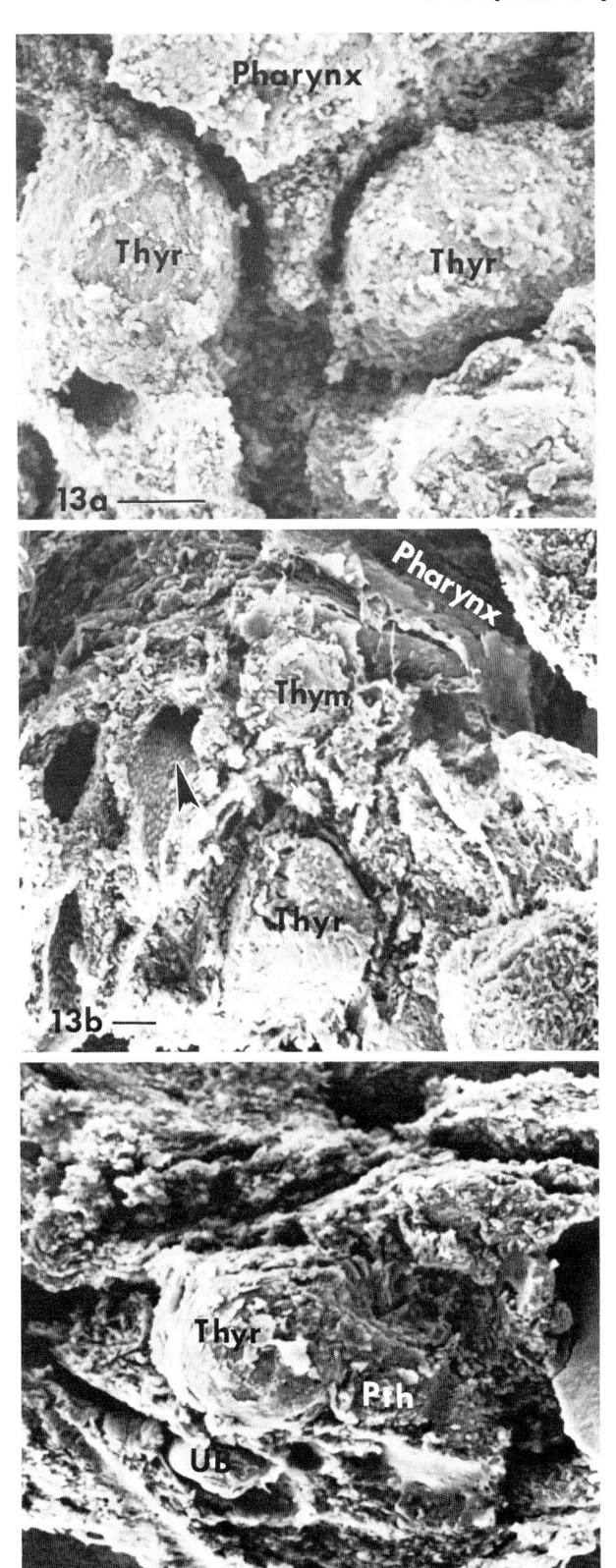
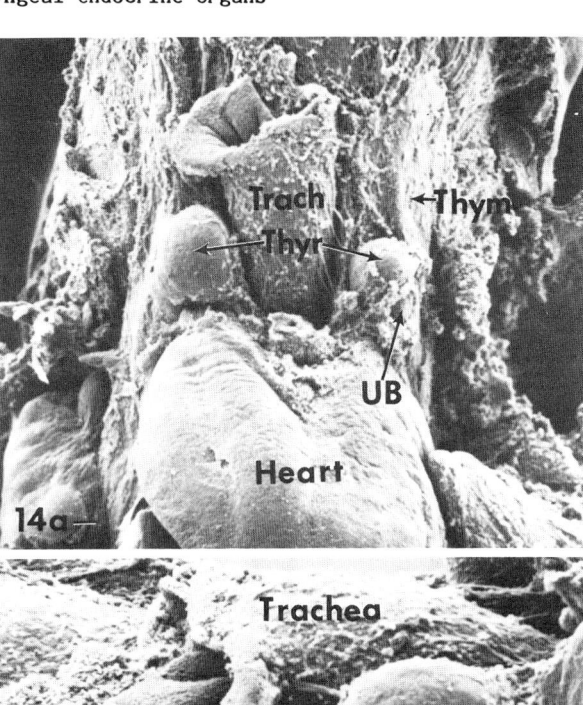
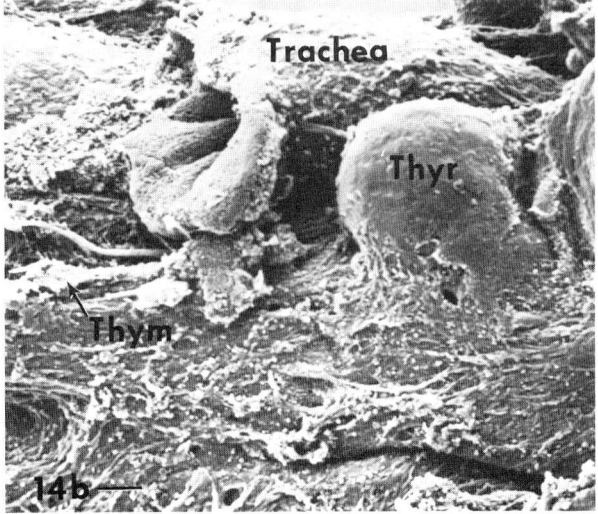

Figure 14. Chick embryo of eight days incubation. a. Ventral view of the neck region. The body wall and subjacent mesenchyme were removed to expose the heart, the severed trachea (Trach), and endocrine organs. The thyroids (Thyr) lie at the base of the neck on either side of the trachea. The left ultimobranchial body (UB) is partially exposed caudal to the thyroid. The thymi (Thym) form cordlike structures along the carotid arteries cranial to the thyroids. The parathyroids either were lost or remained embedded in mesenchyme. b. Same specimen rotated and tilted to show the lateral surface of the right thyroid. The thymus cords are to the left. An ultimobranchial body is barely visible under its mesenchymal capsule. Bars = 100 μm.

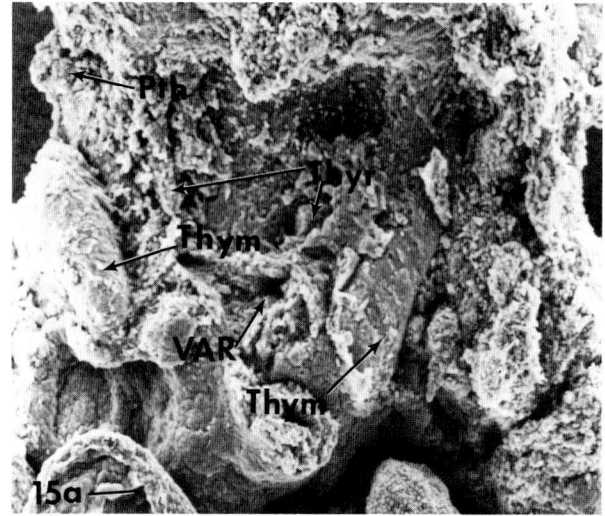
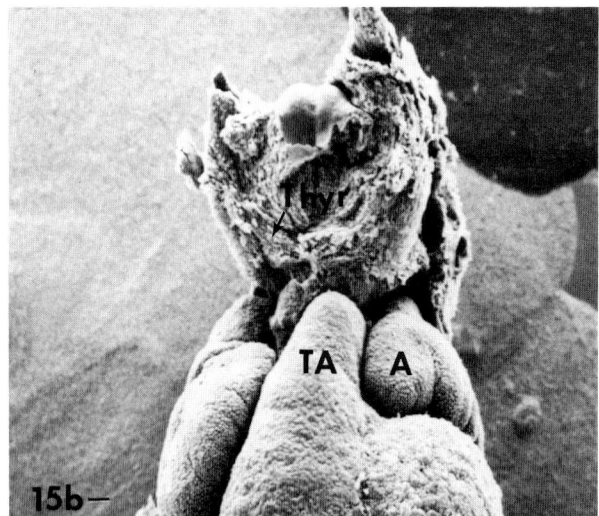
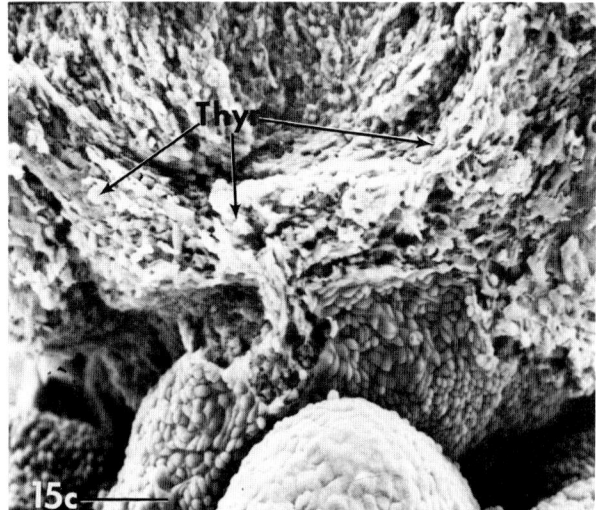

Figure 15. Mouse embryos after 13 days gestation. a. Ventral view of the neck region after removal of the superficial tissue layers. The thymi (Thym) lie at the lateral margins of the pharynx. The ventral aorta has been removed, exposing the bases of the ventral aortic roots (VAR) or common carotids. The isthmus of the thyroid (Thyr) was damaged but the lateral ends remain in place. The right parathyroid (Pth) has been exposed. b. Another embryo from which the superficial lateral and ventral layers, including the thymi, have been removed. The heart, including the truncus arteriosus (TA) and atria (A), was left in place. The thyroid (Thyr) forms a bar across the lower neck. c. Enlargement of the thyroid region of (b). The thyroid is covered by a mesenchymal capsule; its ends curve cranially and dorsally. Bars = 100 µm.

Figure 16. Ventral views of 14 day mouse embryos. a. The ventral body wall was removed, leaving the organs in situ. The glottis (G) and larynx (L) lie between the thymic primordia (Thym). The thymi slant toward each other and partially cover the bar-shaped thyroid (Thyr). A parathyroid (Pth) is visible cranial to the right thymus. The right atrium (A) is visible at the bottom of the figure. b. Another embryo in which the lateral as well as ventral body wall were removed. The thymi are distorted as a result of the dissection but show the relationships among the trachea, thymi and thyroid more clearly. Bars = 100 µm.

Figure 17. Mouse fetuses at 15 days gestation. a. Ventral view of the neck and chest region, after removal of the heart. The base of the truncus arteriosus (TA) remains. The thymic primordia (Thym) lie closer together than at 14 days, partly as a result of the dissection. b. When left in place with their mesenchymal covering, the thymi are separated by a gap. c. Dorsal view to show the parathyroid (Pth) embedded in the lateral thyroid (Thyr) wrapped around the trachea. G = glottis. Bars = 100 µm.

Figure 18. Mouse fetuses at 16 days. a. The thymi (Thym) lie against each other in the ventral midline. The thyroid (Thyr) has changed little from the previous day. b. Ventral view after removal of the thymi. The thyroid covered by a dense capsule forms a compact bar across the trachea. Bars = 100 µm.

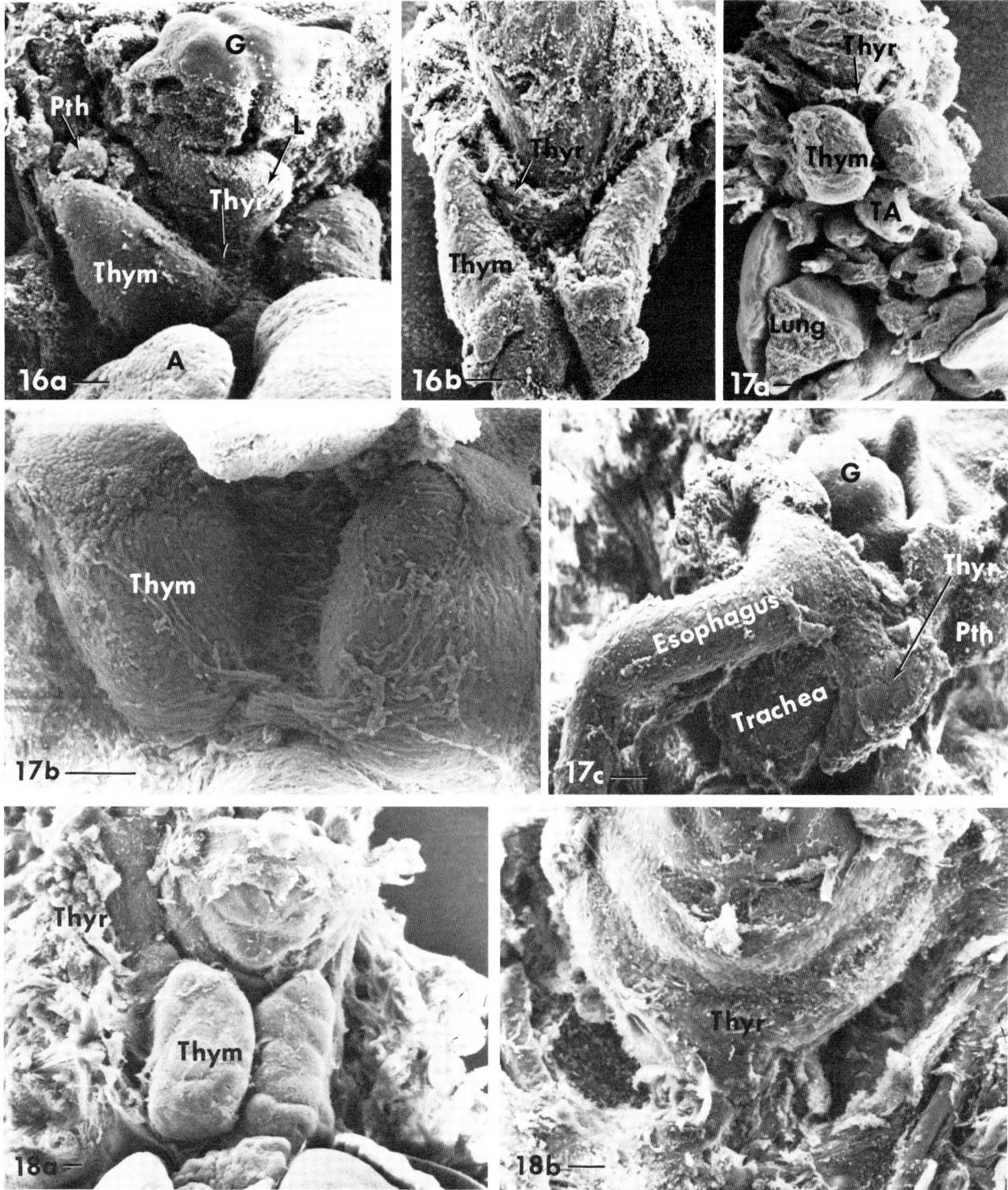

Ishikawa K. (1965). Electron microscopical studies of the ultimobranchial body of the rat in embryologic life, with special emphasis on its fate - The relation to the thyroid tissue and parafollicular cells. Okajamas Fol. Anat. Jap. 41, 313-335.

Le Douarin N. (1980). Migration and differentiation of neural crest cells. Curr. Topics Devel. Biol. 16, 32-85.

Rugh R. (1968). The Mouse, Its Reproduction and Development, Burgess, Minneapolis, 102-207, 251-259.

Schreier JE, Hamilton HL. (1952). An experimental study of the origin of the parathyroid and thymus glands in the chick. J. Exptl. Zool. 119, 165-187.

Shain WG, Hilfer SR, Fonte VG. (1972). Early organogenesis of the embryonic chick thyroid I morphology and biochemistry. Devel. Biol. 28, 202-218.

Theiler K. (1972). The House Mouse. Development and normal stages from fertilization to 4 weeks of age, Springer, Berlin, 53-117.

Tong W, Kerkof P, Chaikoff IL. (1962). Identification of labeled thyroxine and triiodothyronine in $Amphioxus$ treated with $^{131}I$. Biochim. Biophys. Acta. 56, 326.

Venzke WG. (1949). Morphogenesis of the thyroid glands of chicken embryos. Am. J. Vet. Res. 10, 272-287.

Verdun P. (1898). Sur des derives branchiaux du poulet. (On the branchial derivatives of the chicken.) C. R. Soc. Biol. (Paris) 5, 243-244.

Wollman SH, Hilfer SR. (1977). Embryologic origin of various epithelial cell types in the thyroid gland of the rat. Anat. Rec. 189, 467-478.

Wollman SH, Hilfer SR. (1978). Embryologic origin of the various epithelial cell types in the second kind of thyroid follicle in the $C_3H$ mouse. Anat. Rec. 191, 111-112.

### Discussion with Reviewers

**RC Thommes:** Do any of the described glands begin to produce their hormones during the period described?

**Authors:** Using a sensitive assay, thyroxine has been detected in chick thyroid as early as the placode stage (Shain et al., 1972). Secretory granules have been seen by electron microscopy [e.g., Stoeckel ME, Porte A (1970) Origine embryonnaire et differenciation secretoire des cellules a calcitonine (cellules C) dans la thyroide foetale du rat (Embryonic origin and secretory differentiation of the C cells in foetal rat thyroid). Z. Zellforsch. 106, 251-268] and detected with anti-calcitonin antibody [Kameda Y, Shigemoto H, Ikeda A (1980) Development and cytodifferentiation of C cell complexes in dog fetal thyroids. Cell Tissue Res. 206, 403-415] in C-cells shortly after fusion of the ultimobranchial body with the thyroid in several mammalian species. I do not know of any functional studies on the parathyroid during embryonic development.

**JA McAteer:** Are there specializations of the mesenchyme, or epitheliomesenchymal interface that may be linked to the process of thyroid pit formation, and subsequent vesicle separation from the pharyngeal floor? You suggest that a possible interaction may exist between vascular system development and the formation of the thyroid. Does a similar relationship exist for the development of the other pharyngeal endocrines?

**Authors:** These are very interesting developmental questions that need to be answered. As far as we know, they cannot be answered at this time.

Supported by NIH grant #S07RR07115

TRACHEAL MORPHOGENESIS AND FETAL DEVELOPMENT OF THE MUCOCILIARY EPITHELIUM OF THE RAT

James A. McAteer

Department of Anatomy, Indiana University School of Medicine
635 Barnhill Drive, Indianapolis, Indiana 46223

Phone No.: (317) 264 7935.

(Paper received February 17 1984, Completed manuscript received August 19 1984)

## Abstract

In the rat, tracheal development begins at mid-gestation (Day 11) with the formation of the tracheal groove, a longitudinal diverticulum of endodermal epithelium that evaginates from the floor of the pharynx and tubular foregut. At this stage the tracheal groove and developing foregut share a common lumen. Paired primary bronchial buds (lung buds), surrounded by lung bud mesenchyme (splanchnic mesoderm), arise from the caudal end of the tracheal groove. The formation of a longitudinal tracheoesophageal septum divides the combined tracheal groove and developing foregut into two structures, the trachea and esophagus. The trachea and esophagus grow apart, surrounded by independently organized populations of mesenchymal cells. Near this time (Day 12-13), the primary bronchial buds give rise to secondary (lobar) buds of pulmonary epithelium. This establishes the lobar pattern of the right and left lungs and marks the formation of extrapulmonary bronchi.

The development of smooth muscle and cartilage within the tracheal mesenchyme precedes the differentiation of the mucociliary epithelium. Smooth muscle forms transversely oriented fascicles in the dorsal tracheal wall (pars membranacea), while pre-cartilage rings surround the remaining ventral and lateral walls (pars cartilagina). Epithelial differentiation is first evident at Day 17, with the formation of ciliated cells in the epithelium of the pars membranacea. Differentiation in the pars cartilagina trails development in the dorsal epithelium. Cell surface characteristics of the tracheal epithelium indicate that secretory cells differentiate about Day 19-20. The precise time and sequence of differentiation of mature cell types of the tracheal epithelium is yet to be determined. It is clear, however, that the development of the mucociliary epithelium in the rat is not completed at birth, but continues into the neonatal period.

Key Words: Trachea, Lung, Fetal Development, Differentiation, Maturation, Mucociliary Epithelium, Cartilage, Mesenchyme

## Introduction

Study of the tracheobronchial airways is well-suited to morphological analysis of surface features by scanning electron microscopy [1,3,8,14,31]. The extrapulmonary airways and lungs comprise an elaborate epithelial surface in which important regional specializations contribute to respiratory function. The surface epithelium of the trachea and primary airways is a major component, along with submucosal glands, of the pulmonary mucociliary system. This population of ciliated, serous, and mucous cell types contributes to the production and movement of respiratory mucus. As such, the airway epithelium is a principal element of the overall defense mechanism of the respiratory membrane [13]. The maturation of the mucociliary epithelium during the fetal and perinatal period is essential for proper adaptation to respiratory function.

A number of ultrastructural studies have documented aspects of tracheal organogenesis and mucociliary cytodifferentiation in a variety of species [9,10,17,19,22,23,26,30,33,34,36,39]. The majority of these reports have focused on events during specific periods in fetal and neonatal development. The present paper provides a description of tracheal development throughout the fetal period, from the formation of the respiratory primordium, as a derivative of the foregut, to the differentiation of epithelial and mesenchymal cell types during late gestation.

This study was performed to determine the nature, sequence, and timing of major developmental events in the morphogenesis of the trachea and lung primordia. It is a survey of fetal development in one species, the rat. Scanning electron microscopy has been used to demonstrate three-dimensional relationships during the early events of organogenesis and to assess developmental changes in cell surface characteristics during the fetal differentiation of the tracheal mucociliary epithelium.

### Specimen Preparation

Fetuses were removed from the uterus, placed in fluid culture medium F12 and dissected using fine needles. Some specimens were dissected to display structures in situ. For others, the trachea and lungs were excised and processed with the aid of small carriers [29]. Specimens were fixed in 2.5% glutaraldehyde in 0.1 M cacodylate/HCl. Alternatively, intact fetuses were fixed prior to dissection. Following primary fixation, the tissue was fixed in 1% osmium tetroxide in 0.1 M cacodylate/HCl. Treatment with 1% aqueous thiocarbohydrazide was used to enhance osmium binding [25]. Fixed specimens were dehydrated in a graded series of ethanols. Some specimens were cryofractured in liquid nitrogen-frozen absolute ethanol [18]. All specimens were critical-point dried from liquid $CO_2$ [2] and coated with gold-palladium. Specimens for light microscopy were fixed as above and embedded in epoxy resin. One-micrometer-thick sections were prepared and stained with toluidine blue.

## Trachea Formation and Early Events in Lung Development

Tracheal development begins at mid-gestation (Day 11; term 21 days), with the formation of the tracheal groove (also tracheoesophageal or laryngotracheal groove) in the floor of the pharynx and foregut [30,41] (Figs. 1,2). The tracheal groove is a longitudinal outpocketing of endodermal epithelium projecting ventrally into the surrounding splanchnic mesoderm [40]. At this stage in development the primitive foregut and the tracheal groove share a common lumen (Figs. 1c,d; 2). With further development, a longitudinal tracheoesophageal septum forms, which divides the lumen and separates the esophagus (dorsal) from the trachea (ventral) (Figs. 2b,c). The trachea and esophagus subsequently grow apart, separated by interposed tracheal and esophageal mesenchyme (Fig. 4a).

Paired primary bronchial buds (lung buds) emerge from the caudal-most aspect of the tracheal groove (Figs. 1,2). This initial step in lung formation occurs while the tracheal groove is still patent throughout its length. Closure of the tracheal groove begins first caudally, then advances toward the pharynx. During this period, the lungs are simple epithelial pouches which branch directly from the primitive trachea. Discrete mainstem bronchi do not form until secondary bronchial buds (lobar buds) develop and establish the lobar pattern of right and left lungs. From the time of initial formation, and throughout subsequent development, the lungs are not equal in size. The larger right lung develops four lobes (cranial, middle, caudal, accessory), while the left lung forms a single large lobe [27].

At the stage of secondary bronchial bud formation (Day 12), the right lung bears modest dorsolateral, ventromedial, and caudal bulges, although separate lobes are not yet apparent (Fig. 3a). By Day 13, the pattern of lobe formation is conspicuous (Figs. 3c,d). The cranial lobe of the right lung projects dorsally, nearly at right angles to the plane of the middle and caudal lobes. The accessory lobe is prominent and projects ventromedially across the pulmonary midline. The left lung remains smaller and shorter than the right.

## Figure 1 caption

This series of correlative light and scanning electron micrographs demonstrates the formation of the tracheal primordium and primary bronchial buds (lung buds) at Day 11 of fetal development. Frames a, c, e, and f are serial sections taken from the same fetus.

a. The trachea and lungs are derived from a ventral epithelial outpocketing of the developing foregut, termed the tracheal groove. This section shows the tracheal groove (TG) within the floor of the pharynx (P) at the level of the fourth pharyngeal pouch (IV). Dorsal aorta (DA). (bar 100 micrometers).

b. This fetus was sectioned just posterior to the level of the pharyngeal pouches. The tracheal groove (arrowhead) appears as a narrow outpocketing from the floor of the pharynx (P). Heart (H); Neural tube (N); Dorsal aorta (DA). (bar 100 micrometers).

c. This section shows the tracheal groove (TG) as it is found caudal to the pharynx, at the level of the tubular foregut. With continued development, the walls of the combined foregut and tracheal groove will fuse (arrowheads) to form the tracheoesophageal septum. The TE septum will separate the developing trachea from the esophagus (E). Dorsal aorta (DA), Pleural cavity (PC). (bar 100 micrometers).

d. This specimen, likewise sectioned through the tubular foregut, shows the tracheal groove (TG) and foregut (developing esophagus) (E) surrounded by mesenchyme (MES). In this caudally directed view, the lung buds (LB) are seen projecting into the primitive pleural cavity (PC). Developing esophagus (E); Dorsal aorta (DA). (bar 50 micrometers).

e. This section shows the right primary bronchial bud (PB) branching from the tracheal groove (TG). This demonstrates that the lung buds branch from the caudal aspect of the tracheal groove prior to the formation of an independent trachea. Developing esophagus (E); Dorsal aorta (DA). (bar 100 micrometers).

f. A section at the level of the primary bronchial buds (PB) shows the developing lung surrounded by the primitive pleural cavity (arrowhead). The esophagus (E) rests in the midline. Lung bud mesenchyme (MES); Dorsal aorta (DA). (bar 50 micrometers).

g. The primary bronchial buds (PB) are simple unbranched epithelial pockets surrounded by lung mesenchyme (MES). The lung buds sit within the developing pleural cavity (PC) and are connected to the body wall by dorsal (DM) and ventral (VM) mesenteries. Esophagus (E). (bar 100 micrometer).

h. The epithelium (EP) of the primary bronchial buds is pseudostratified columnar. Pleural cavity (PC). (bar 20 micrometers).

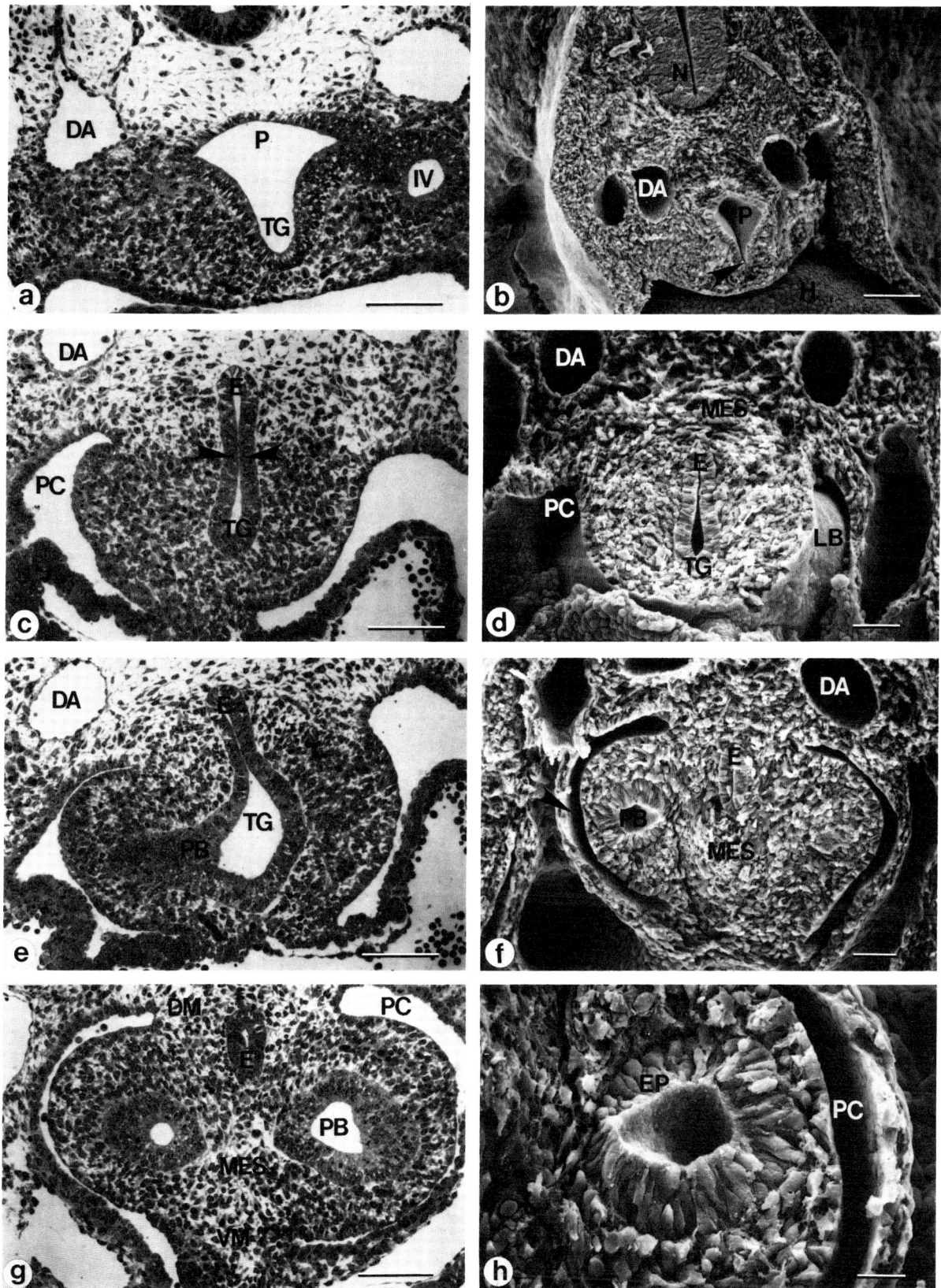

Figure 1. Origin of the trachea and lung buds in the fetal rat.

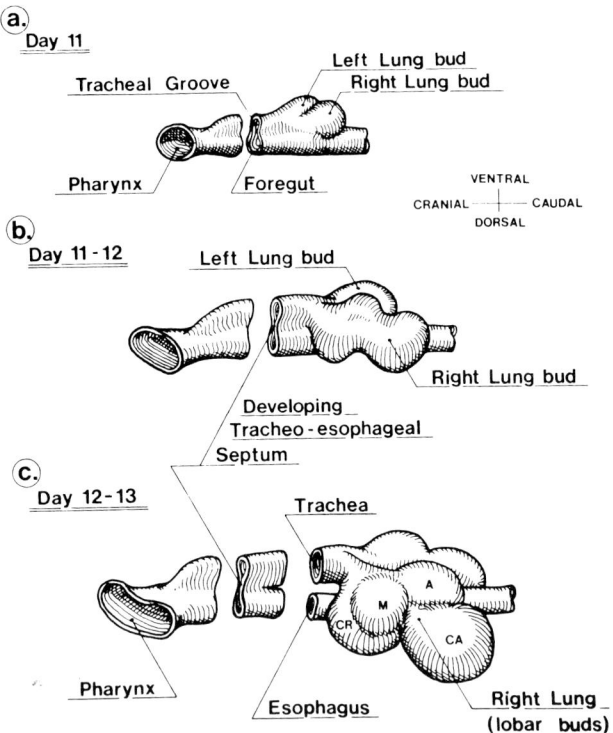

Figure 2. Early events in tracheopulmonary development.

This illustration demonstrates the formation of the trachea and epithelial lung buds as derivatives of the developing foregut. These figures show only epithelial structures. The mesenchyme which surrounds the foregut, trachea, and lung buds is not shown.

a. Development of the trachea and lungs begins at Day 11 with the formation of the tracheal groove, a longitudinal outpocketing from the floor of the pharynx and ventral side of the tubular foregut. Paired lung buds (primary bronchial buds) form at the caudal end of the tracheal groove.

b. During Day 11 to 12 the lateral walls of the combined tracheal groove and developing foregut oppose one another, fuse and form a longitudinal septum. The tracheoesophageal septum divides the lumen into dorsal and ventral portions. The lung buds remain as simple, unbranched epithelial pouches.

c. With continued development (Day 12-13) the newly formed trachea and esophagus grow apart. This formation and separation of the independent trachea and esophagus appears to occur first caudally. The cranial trachea and developing larynx do not become independent of the pharynx until about Day 13-14. During Day 12-13 the lungs form secondary (lobar) buds. Each epithelial bud at this stage of development represents a separate pulmonary lobe (cranial, CR; middle, M; caudal, CA; accessory, A) which persists throughout subsequent development.

Further development of pulmonary lobes is characterized by progressive branching of the bronchial tree and positional changes of the growing lobes in relation to surrounding organs and the developing pleural cavity. At Day 14, the cranial, middle, and caudal lobes of the right lung have a concave ventral surface and are aligned in essentially the same plane (Figs. 3e,f). The accessory lobe projects even further ventromedially. The separation between lobes remains distinct. With continued growth and development, the divisions between lobes will be diminished to narrow fissures. By Day 15, mainstem bronchi are very distinct (Fig. 3g).

### Figure 3 caption

This series of scanning electron micrographs shows the formation of pulmonary lobes during early lung development (Days 12-16) in the fetal rat.

a. At Day 12, the primary bronchial buds give rise to secondary (lobar) buds. In the right lung, lobar buds are apparent as slight ventromedial, dorsolateral, and caudal bulges (arrowheads). (bar 50 micrometers).

b. In this 12-day specimen, a portion of lung mesenchyme has been dissected free to expose the underlying primordial accessory lobe of the right lung (A). The basal surface of the epithelium (inset) is covered only by connective tissue fibers and a few adherent mesenchymal cells (arrow). (bar 100 micrometers).

c. By Day 13, the cranial (CR), middle (M), caudal (CA), and accessory (A) lobes of the right lung are distinct. (bar 100 micrometers).

d. This lateral view of the 13-day lung shows the cranial lobe (CR) projecting dorsally from the plane occupied by the middle (M) and caudal (CA) lobes. Accessory lobe (A). (bar 100 micrometers).

e. At Day 14, the cranial (CR), middle (M), and caudal (CA) lobes of the right lung are concave at their ventral surface. The accessory lobe (A) projects ventromedially across the midline. (bar 100 micrometers).

f. This dorsolateral view of the 14-day lung shows that the cranial lobe (CR) no longer projects dorsally, but is aligned with the middle (M) and caudal (CA) lobes. (bar 100 micrometers).

g. By Day 15, the right lung shows a more concave ventral surface. Mainstem bronchi (B) are clearly visible, branching from the trachea (T). Cranial (CR), middle (M), caudal (CA), accessory (A) lobes. (bar 100 micrometers).

h. At Day 16, the individual lobes of the lung have a more smoothly contoured surface. Divisions between lobes are deep fissures (arrowhead). Each lobe is highly lobulated. In this specimen, the accessory lobe (asterisk) has been removed. Cranial (CR), middle (M), caudal (CA) lobes. (bar 500 micrometers).

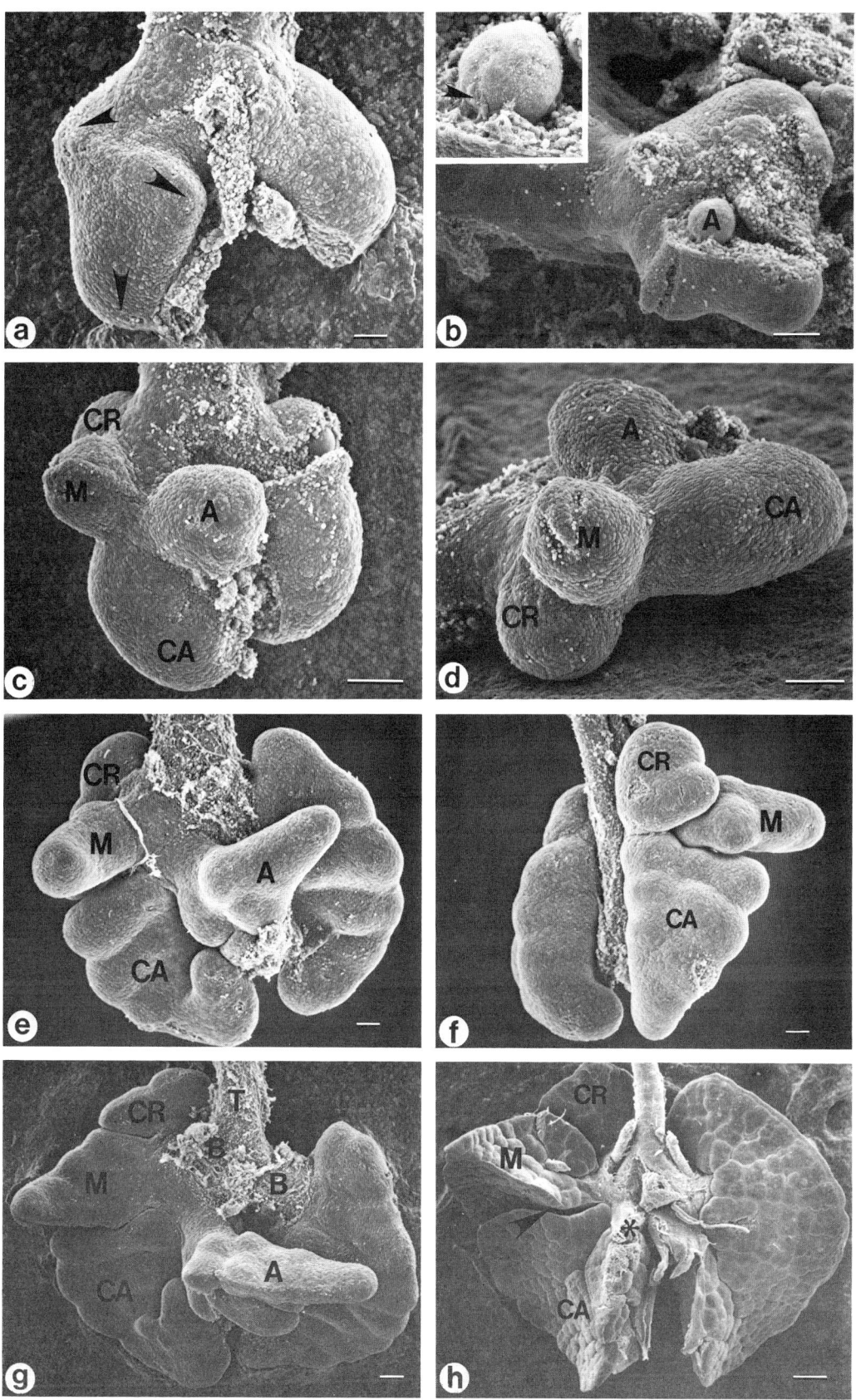

Figure 3. Lung development in the fetal rat.

The surface of the lung shows the contour of the large pulmonary epithelial buds that form the glandular lung parenchyma. By Day 16, each lobe shows a highly lobulated pattern at its surface, indicative of the extensive epithelial branching that is characteristic of the glandular phase of lung development (Fig. 3h).

Following the closure of the tracheal groove and subsequent separation of the trachea from the foregut (esophagus), tracheal development is characterized by differentiation of both the mesenchyme and epithelium. The formation of smooth muscle in the dorsal tracheal wall precedes the development of cartilage rings in the ventral and lateral mesenchyme (Fig. 4). This differentiation of mesenchymally derived elements occurs in advance of epithelial cytodifferentiation.

### Chronology of Tracheal Development in the Rat

#### Day 11

The tracheal groove forms in the midline floor of the posterior pharynx and foregut (Fig. 1). Primary bronchial buds (lung buds) develop at the caudal end of the tracheal groove. The lung buds are simple, paired epithelial pockets surrounded by mesenchyme. Lung buds are present before the trachea has separated from the developing esophagus. The mesenchymal cells that surround the tracheal groove appear to be randomly oriented. The mesenchyme adjacent to the esophageal (dorsal) portion of the lumen is less cellular than that surrounding the primordial trachea and bronchial buds (ventral).

As development proceeds, the lateral walls of the combined tracheal groove and foregut oppose one another and fuse, forming the longitudinal tracheoesophageal (TE) septum (Fig. 2). The TE septum divides the lumen into two portions, the trachea (ventral) and esophagus (dorsal). The newly formed trachea and esophagus grow apart (Day 11-12), separated by proliferating mesenchyme (Fig. 4a).

#### Day 12

The TE septum separates the trachea from the esophagus but may persist at, and near, the pharynx (Fig. 2). The trachea and esophagus grow farther apart. The surrounding mesenchyme shows the first signs of organization (Fig. 4a). Mesenchymal cells organize in a circumferential orientation around the trachea and around the esophagus.

#### Day 13

The trachea and esophagus are independent throughout their length, thus the two can readily be dissected free of one another. The tracheal mesenchyme is less cellular ventrally than dorsally. The dorsal mesenchyme is somewhat condensed, showing little intercellular space.

Lung development has proceeded so that secondary bronchial buds are fully formed (Figs. 3c,d). Each secondary bud represents a primordial pulmonary lobe (cranial, middle, caudal, accessory). With the formation of secondary (lobar) buds, the lungs now possess a glandular structure. True bronchi can now be defined as the short epithelial segments that connect the trachea with the secondary bronchial buds (presumptive pulmonary parenchyma).

#### Day 14

The dorsal tracheal mesenchyme shows increased condensation in the region immediately subjacent to the epithelium (Fig. 4b). This organization of the mesenchyme interposed between the trachea and esophagus represents an early stage in the development of tracheal smooth muscle.

---

### Figure 4 caption

This set of light micrographs demonstrates the morphogenesis of the tracheal wall and the differentiation of smooth muscle and cartilage within the tracheal mesenchyme. All frames show cross sections taken from the middle portion of the trachea.

a. This field shows the trachea (T) and esophagus (E) at Day 12. Following the formation of the tracheoesophageal septum the trachea and esophagus grow apart, separated by mesenchymal cells derived from splanchnic mesoderm. Mesenchymal cells (MES) already show some degree of organization, in concentric array, surrounding the trachea. The tracheal epithelium is clearly pseudostratified columnar. (bar 50 micrometers).

b. By Day 14, a region of condensed mesenchyme can be observed (SM) on the dorsal side of the trachea immediately subjacent to the epithelium (EP). This is a first indication of the formation of smooth muscle within the dorsal tracheal wall. (bar 50 micrometers).

c. At Day 16, cellular organization within the ventral and lateral mesenchyme (C) gives a first indication of cartilage ring formation in the tracheal wall. In this section not all regions of the tracheal mesenchyme (MES) show precartilage development. Smooth muscle (SM) in the dorsal wall forms a distinct bundle. The dorsal epithelium (EP) now forms longitudinal ridges each raised above a core of fibroblastic mesenchyme (arrowheads). (bar 50 micrometers).

d. By Day 17, the development of cartilage rings (C) is more advanced. Epithelial folds (arrowheads) of the dorsal wall (pars membranacea) are very prominent. Smooth muscle (SM). (bar 50 micrometers).

e. At Day 18, hyaline matrix surrounds the developing chondrocytes of a developing cartilage ring (C). The epithelium (EP) of the pars membranacea is highly folded. Smooth muscle (SM). (bar 50 micrometers).

f. By Day 19, mesenchymal elements of the tracheal wall are well developed. In this field, a cartilage ring (C) with abundant hyaline matrix (arrowhead) is seen enclosed by a fibroblastic perichondrium (P). Smooth muscle (SM) of the dorsal wall forms a compact bundle. A well-defined region of fibroblastic mesenchyme (MES) separates cartilage and smooth muscle from the overlying epithelium. (bar 50 micrometers).

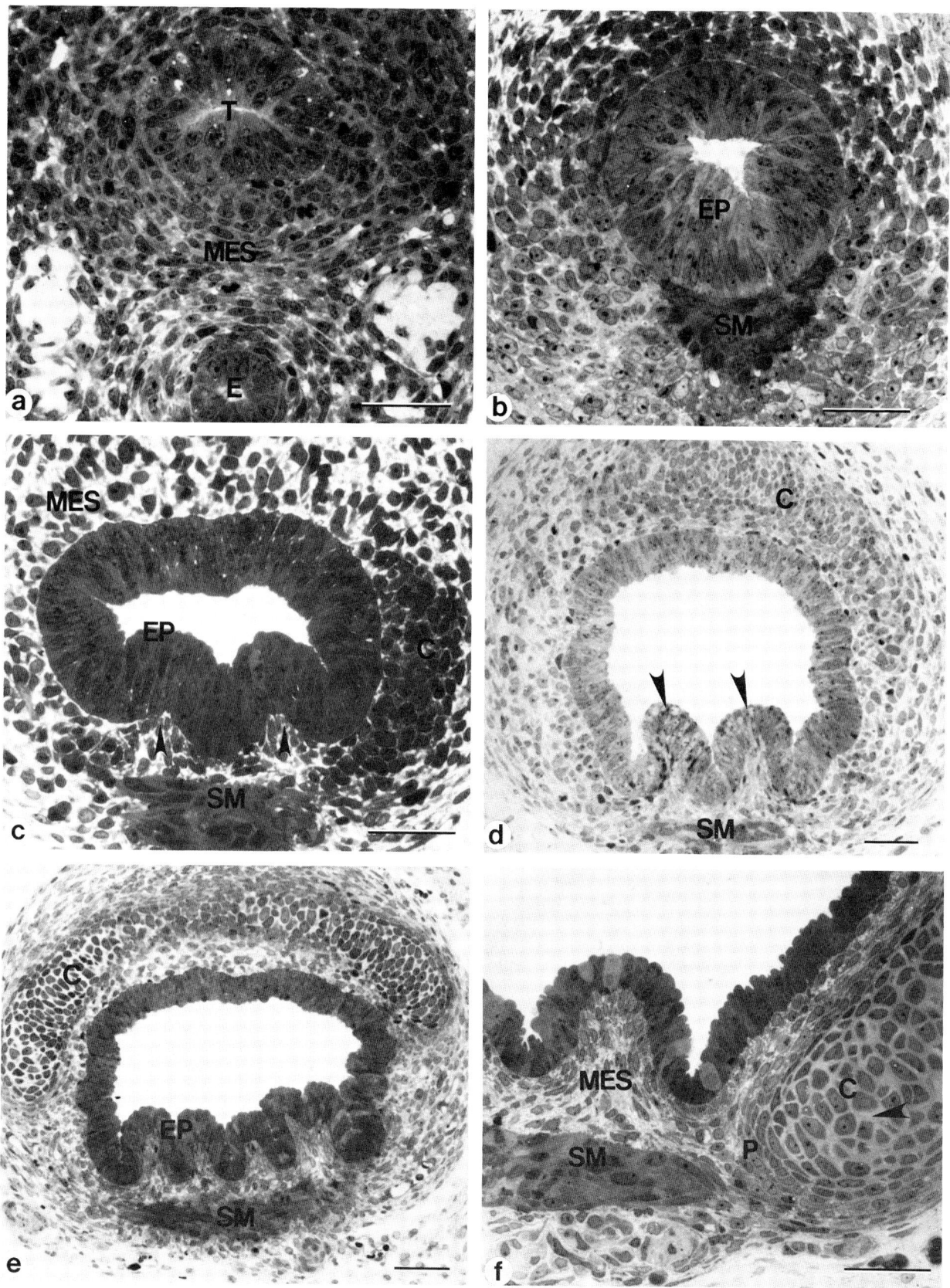

Figure 4. Development of the tracheal wall (Days 12-19).

Day 15
Tracheal form in cross-section is suggestive of the adult pattern. The dorsal wall is flattened to concave at the basal epithelial surface. The remainder of the tracheal wall forms a simple arch.

Day 16
Precartilage formation begins within the mesenchyme (Fig. 4c). Chondrogenic foci occur subjacent to the epithelium along the ventral and lateral tracheal walls. These areas appear as focal accumulations of mesenchymal cells. There is, as yet, no apparent deposition of cartilage matrix. The dorsal tracheal wall has epithelial folds. Smooth muscle within the dorsal mesenchyme forms a distinct, compact bundle.

Day 17
The dorsal tracheal wall is very irregular (Fig. 4d). Epithelial folds, each with a mesenchymal core, project into the lumen. Some cells of the subjacent submucosa bear fibroblastic morphology. The smooth muscle of the dorsal wall now forms discrete fascicles oriented perpendicular to the longitudinal axis of the trachea. Cartilage ring formation has advanced. Cellular organization is evident, with closely spaced polygonal chondroblasts surrounded by several concentric layers of elongated fibroblastic cells. Cartilage matrix is not yet apparent (by light microscopy). Ciliated cells are the first mature epithelial cell type to differentiate.

Day 18
The dorsal tracheal wall is highly folded (Fig. 4e). Cartilage ring formation is well advanced. Tracheal wall microanatomy now closely resembles the pattern of adult anatomy. The trachea has two clearly defined regions, the dorsal pars membranacea and the remaining pars cartilagina. Chondrocytes are now separated by hyaline matrix. Fibroblastic cells ensheath the cartilage rings to form a fetal perichondrium. Cartilage rings are separated from the subjacent epithelium by a distinct region of fibroblastic mesenchyme (primitive submucosal connective tissue).

Days 19-21
The trachea undergoes considerable growth. Cartilage rings show increased matrix deposition and formation of a distinct perichondrium. Principal features of tracheal wall structure established at Day 18 persist through parturition. Major changes involve events in cytodifferentiation of the tracheal epithelium (Figs. 5,6).

Differentiation of the Mucociliary Epithelium

The tracheobronchial epithelium of the adult rat is pseudostratified columnar, consisting of at least seven ultrastructurally identifiable cell types which include ciliated, serous secretory, goblet, basal, brush, intermediate (uncommitted), and enteroendocrine cells [6,15,16,21,28,35]. In addition, the epithelium is populated by cells of the immune system such as wandering lymphocytes and focal accumulations of lymphoid tissue (lymphoepithelium) [5]. Also, mucous secretory glands derived from the epithelium, with ducts emptying into the tracheal lumen, are found within the submucosa [11,22].

### Figure 5 caption

These scanning electron micrographs demonstrate developmental changes in the surface characteristics of the tracheal epithelium.

a. This micrograph shows the open cut end of the trachea at Day 14. The tracheal epithelium (TE) is surrounded by mesenchyme (MES). (bar 10 micrometers).
b. At 14 days, the tracheal epithelium is undifferentiated, and individual apical cell boundaries are not distinct. Each cell bears microvilli (MV) and a solitary primary cilium (arrowheads). (bar 1 micrometer).
c. This micrograph shows 16-day trachea that has been cryofractured. The obvious furrow (arrowhead) and longitudinal ridges in the tracheal epithelium (EP) are characteristic of the dorsal pars membranacea. Tracheal mesenchyme (MES). (bar 10 micrometers)
d. At 16 days, the apex of tracheal epithelial cells begin to bulge into the lumen. The cells are still undifferentiated and bear both microvilli and a primary cilium (arrowheads). (bar 1 micrometer).
e. By Day 18, the pars membranacea of the tracheal wall possesses very distinct longitudinal ridges separated by deep furrows. The epithelium is heterogeneous and ciliated cells are now present (arrowheads) (first evident at 17 days). (bar 10 micrometers).
f. This micrograph shows a cell undergoing ciliogenesis in the pars membranacea of 18-day trachea. The cell possesses cilia of various lengths. The button of cytoplasm (arrowhead) at the apex is a characteristic of a cell in an early stage of cilia formation. Surrounding cells (E) are studded with short microvilli. By Day 18, many of the cells that line the pars membranacea no longer bear primary cilia. (bar 1 micrometer).
g. This field shows the cut edge of the pars cartilagina of 18-day trachea. Well-formed cartilage rings (C), separated by fibroblastic mesenchyme (MES) lie subjacent to the epithelium (TE). (bar 10 micrometers).
h. Epithelial differentiation in the pars cartilagina lags behind the pars membranacea. These cells from the ventral epithelium at 18 days are still undifferentiated. Most of the cells still possess primary cilia (arrowheads). (bar 5 micrometers).

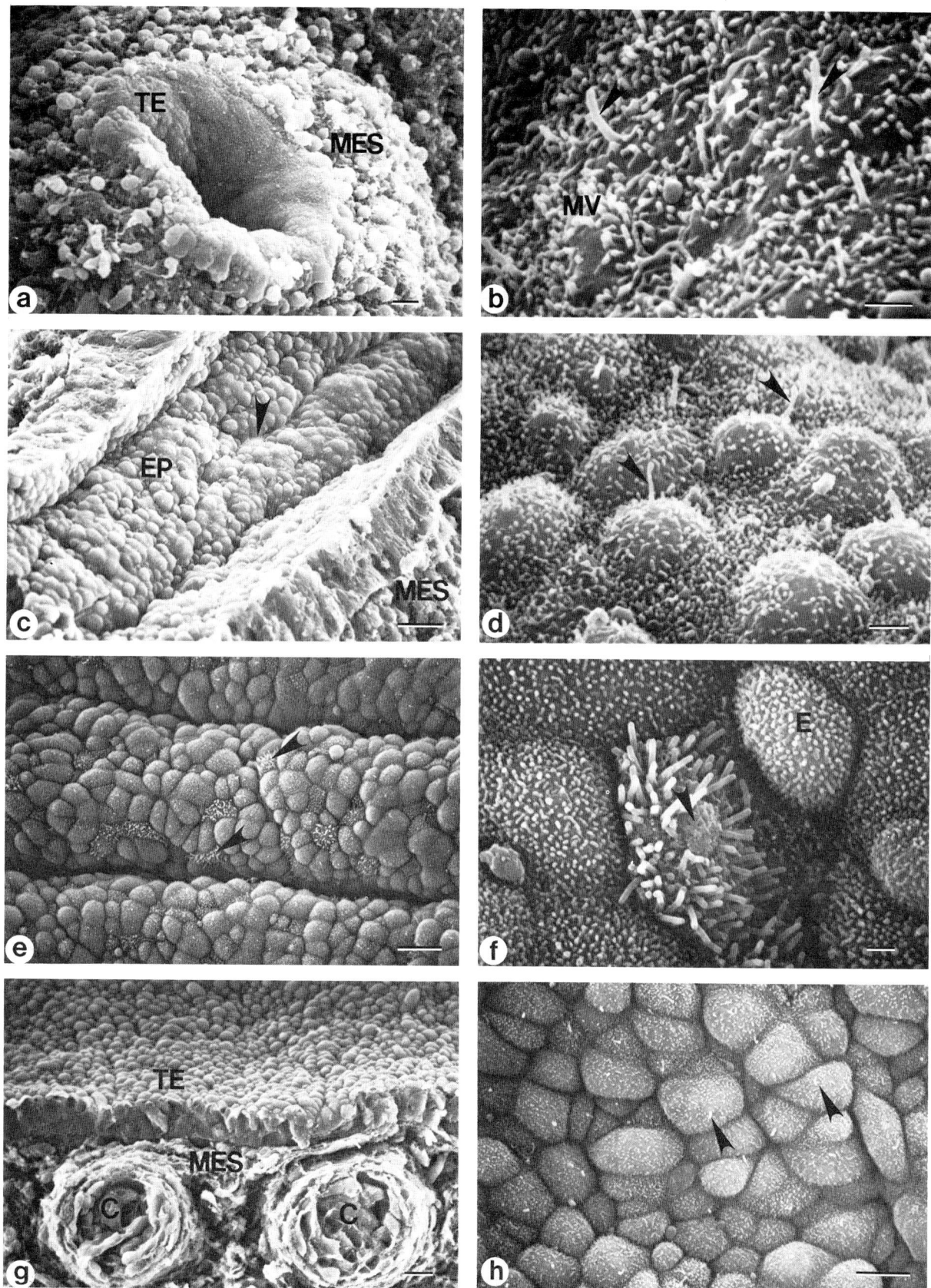

Figure 5. Surface features of the developing tracheal epithelium (Days 14-18).

Important species differences are recognized in airway structure and the identity and distribution of mature cell types [6,32,35]. It is also understood that tracheobronchial development proceeds at different rates in various species, and the gestational age at which epithelial cell types differentiate is species-dependent. For example, the tracheal epithelium of man [7,22] and primates [33] is thought to be structurally advanced at birth, whereas in the mouse [24], rabbit [20,23,34], and rat [9,22,26,36,38], differentiation and maturation appear to continue into the neonatal period.

In the rat, mature cell types of the extrapulmonary airways begin to differentiate during the last several days of fetal development. The precise time of appearance of many of these cell types has not been determined, nor has the sequence of differentiation, particularly regarding the emergence of secretory cell types, been established.

Day 17 marks the emergence of the ciliated cell as the first differentiated cell type identifiable by scanning electron microscopy. Prior to this time, the epithelium appears as a relatively homogeneous population of undifferentiated cells. As discussed previously, the trachea is derived at Day 11 from the endodermal epithelium of the foregut and pharynx. At the stage of tracheal groove formation, the primordial tracheal epithelium is already pseudostratified (Fig. 1). The epithelium is thin and may appear in areas to be simple, but adjacent cells display the staggered nuclear organization (light microscopy) characteristic of pseudostratified morphology. The epithelium of the primary bronchial buds (lung buds) is also pseudostratified.

By Day 12, after the trachea and esophagus have separated, the tracheal epithelium is thicker and appears unquestionably pseudo-stratified (Fig. 4a). The esophagus, destined to form a stratified squamous epithelium, bears a simple columnar epithelium. The tracheal epithelium, through Day 14, forms a relatively regular luminal surface (Fig. 5a,b). The cells have numerous long microvilli, and each cell bears a primary cilium. The boundaries between individual cells are indistinct.

Structure suggestive of developmental changes in the epithelium is first identifiable on Day 16 (Figs. 5c,d). Although no mature cell types are as yet identifiable, the lumen is now distinctly irregular in form due to the presence of cells with rounded apices. At this age, nearly all cells still bear a single, prominent primary cilium.

Ciliated cells first appear in the tracheal epithelium at Day 17-18 (Figs. 5e,f). They occur in the dorsal tracheal wall (pars membranacea) where mucosal folds lie opposite the developing trachealis muscle (Figs. 4d,e). Epithelial development in the pars membranacea region is clearly advanced over the remainder of the trachea. At Day 18 the cells of the ventral epithelium still appear to be undifferentiated (Figs. 5g,h). Ciliated cells do not occur opposite the pars cartilagina until about Day 19. Even as late in development as Day 20, the epithelium of the pars cartilagina has far fewer ciliated cells than the pars membranacea (Figs. 6a,b).

At Day 18, cells undergoing ciliogenesis are the only differentiating cell type positively identifiable by surface characteristics (Figs. 5e,f). However, the epithelium is heterogeneous and also includes non-ciliated cells, many without primary cilia. Most cells have a uniform population of short microvilli, although some cells have small regions of smooth apical surface. These latter cells become more abundant with time. By Day 19-20, cells are present which possess a bulging, rounded, often irregular apex completely devoid of microvilli (Figs. 6a,c). At this time, other cells bear multiple cytoplasmic blebs at their surface (Fig. 6d). These cells, and cells which possess a protruding apical segment attached by a narrow stalk (Fig. 6e), resemble the mucous secretory cells of neonatal and adult airway [3,26,38]. Another cell type observed during this late fetal period possesses microplicae at its apical surface (Figs. 6e,f).

---

#### Figure 6 caption

These scanning electron micrographs show the surface morphology of tracheal epithelial cell types undergoing differentiation during late fetal development.

a. Numerous ciliated cells occur in the pars membranacea among a heterogeneous population of nonciliated cells. (bar 10 micrometers).
b. The epithelium of the pars cartilagina still possesses only a few ciliated cells. Most cells remain undifferentiated. (bar 10 micrometers).
c. This field shows a pair of cells undergoing ciliogenesis. Each has a button of apical cytoplasm (arrowhead) which is positioned among cilia of various lengths. Also in the field is a typical ciliated cell (C) and a number of nonciliated cells (E) covered with microvilli. (bar 1 micrometer).
d. Secretory cells are identifiable by surface morphology at Day 19-20. This field shows several cells that may be mucous secretory cells. Two cells (MC) have a smooth apical surface devoid of microvilli. A third cell (arrowhead) has irregular globular structures at its apex. (bar 2.5 micrometers).
e. This field shows a heterogeneous epithelium including cells with microvilli (E), one with microplicae (M), and a cell with an apical protrusion (MC) connected below by a narrow stalk (arrowhead). This apical protrusion may represent the released secretory product of a mucous cell. (bar 1 micrometer).
f. Cells with microplicae (arrowhead) are observed within the developing tracheal epithelium beginning about Day 18-19. This cell has an elaborate surface with ridges of folded apical membrane (arrowhead). Adjacent cells (E) are covered with typical microvilli. (bar 1 micrometer).

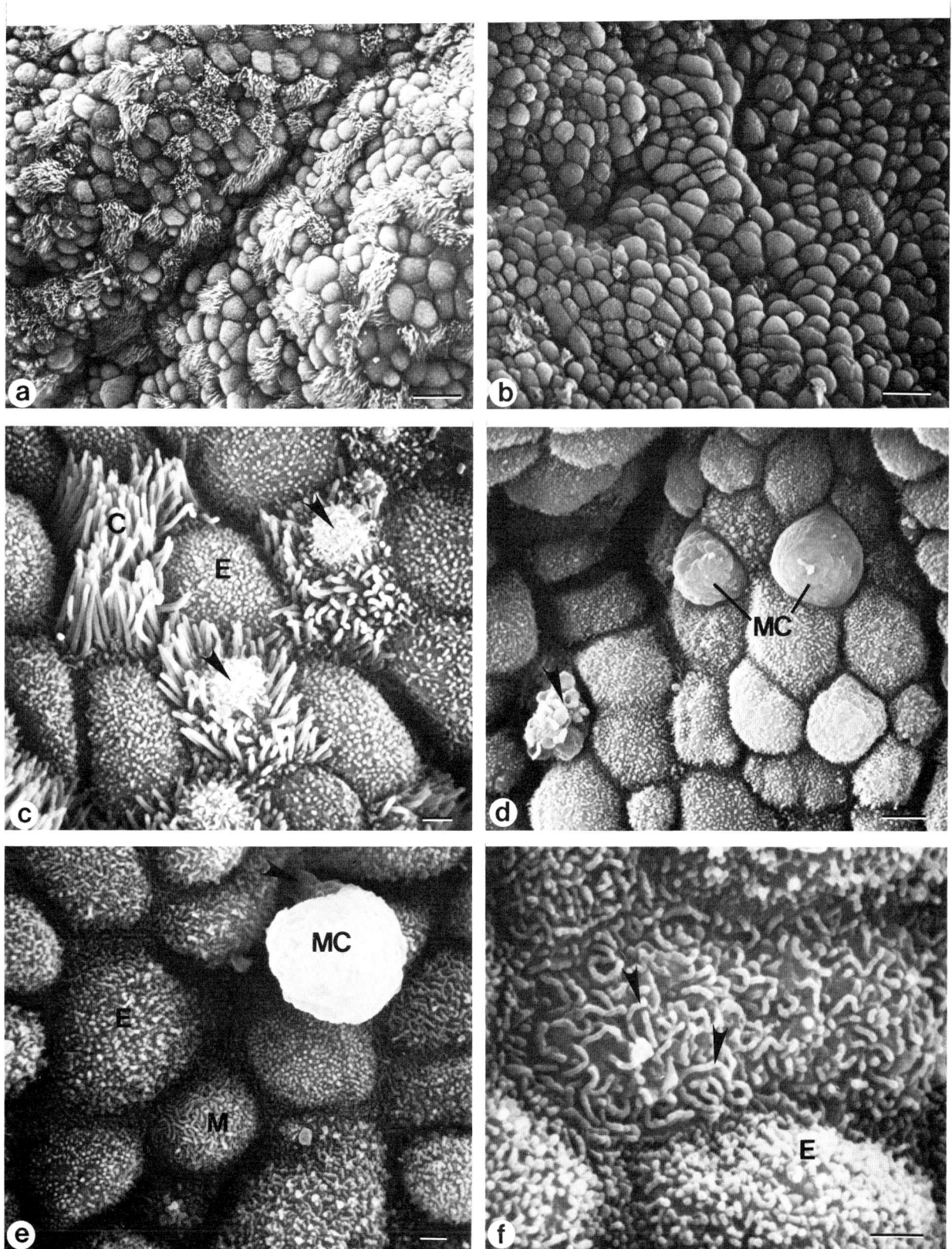

Figure 6. Tracheal mucociliary epithelium, Day 20.

The developmental significance of microplicae in the tracheal epithelium is unknown. They may represent a transient surface feature of the differentiation process, since cells showing microplicae have not been reported in adult trachea. Similar cells have, however, been observed within the nasal epithelium of the adult rat [3]. Microplicae, although not considered a common cell surface feature, are recognized as a characteristic of cells in a variety of organs including cornea, esophagus [4] and renal collecting duct [12].

The time of appearance of the remaining mature cell types of the mucociliary epithelium has not been determined. Investigators have, however, identified brush, intermediate, and serous cell types at Day 21 [22]. Although serous cells are understood to be present at birth [9,22,26,36], it is not known when the rat develops its adult complement of secretory cell types, including the goblet cell.

The tracheal epithelium of the rat is not mature at birth. During the neonatal period, important changes take place in the morphology and distribution of cell types [9,22,26,36,38]. For example, it has been reported that the number and histochemical staining properties of secretory cell types change during the first two weeks after birth [36]. The total number of secretory cell types apparently decreases, while there is an increase in the number of cells which stain for acidic mucosubstances. Other recognized changes occurring during the neonatal period include a substantial increase in the number of ciliated cells, and the formation of submucosal mucous glands [22]. The neonatal period is also the time during which the Clara cell of the distal airways becomes fully differentiated. Clara cells have been identified within the intrapulmonary airways of fetal rats; however, these cells do not show mature ultrastructure [22]. Maturation does not occur until the first several weeks after birth [37]. Similar observations have been made in a variety of species [34].

## Acknowledgements

This study was supported by grant 5-S07-RR-5371 awarded by the Division of Research Resources, National Institutes of Health, Public Health Service, U.S. Department of Health and Human Services. The author wishes to thank Alan Plantz for typing the manuscript. Special thanks to Philip Blomgren for his technical assistance, and to Dr. Andrew Evan for his counsel throughout the study.

## References

1. Alexander I, Ritchie BC, Maloney JE, Hunter CR. (1975) Epithelial surfaces of the trachea and principal bronchi in the rat. Thorax 30, 171-177.
2. Anderson TF. (1951) Technique for the preservation of three-dimensional structure in preparing specimens for the electron microscope. Trans. N.Y. Acad. Sci. 13, 130-134.
3. Andrews P. (1974) A scanning electron microscopic study of the extrapulmonary respiratory tract. Am. J. Anat. 139, 399-424.
4. Andrews PM. (1976) Microplicae: Characteristic ridge-like folds of the plasmalemma. J. Cell Biol. 68, 420-429.
5. Bienenstock J. (1981) Bronchus-associated lymphoid tissue, in: Cellular Biology of the Lung, Cumming G and Bonsignore G (eds.), Plenum, NY, 225-238.
6. Breeze RG, Wheeldon EB. (1977) The cells of the pulmonary airways. Am. Rev. Resp. Dis. 116, 705-777.
7. Bucher U, Reid L. (1961) Development of the mucus-secreting elements in human lung. Thorax 16, 219-225.
8. Castleman WL, Dungworth DL, Tyler WS. (1974) Intrapulmonary airway morphology in three species of monkey: A correlated scanning and transmission electron microscopic study. Am. J. Anat. 142, 107-122.
9. Cireli E. (1965) Elektronenmikroskopische Analyse der prä- und postnatalen Differenzierung des Epithels der oberen Luftwege der Ratte. Zeit. Mikroskop. Anat. Forsch. 74, 132-178.
10. Cutz E, Chan W, Sonstegard KS. (1978) Identification of neuro-epithelial bodies in rabbit fetal lungs by scanning electron microscopy: A correlative light, transmission and scanning electron microscopic study. Anat. Rec. 192, 459-466.
11. DeHaller R. (1969) Development of mucus-secreting elements, in: The Anatomy of the Developing Lung, Emery J (ed.), Tadworth, England, 94-115.
12. Evan A, Huser J, Bengele HH, Alexander EA. (1980) The effect of alterations in dietary potassium on collecting system morphology in the rat. Lab. Invest. 42, 668-675.
13. Green GM, Jakab GJ, Low RB, Davis GS. (1977) Defense mechanisms of the respiratory membrane. Am. Rev. Resp. Dis. 115, 479-514.
14. Greenwood MF, Holland P. (1972) The mammalian respiratory tract surface: A scanning electron microscopic study. Lab. Invest. 27, 296-304.
15. Hage E. (1973) Electron microscopic identification of several types of endocrine cells in the bronchial epithelium of human foetuses. Zeit. Zellforsch. 141, 401-412.
16. Hansell MM, Moretti RL. (1969) Ultrastructure of the mouse tracheal epithelium. J. Morphol. 128, 159-170.

17. Herbst R, Multier-Lajous AM, Grabitz K, Hoder D, Kober HJ, Blumcke S. (1979) Rasterelektronenmikroskopische Untersuchungen der Lungenepithelien des Kaninchens während der Ontogenese. Zeit. Mikroskop. Anat. Forsch. 93, 736-750.
18. Humphreys WJ, Spurlock BO, Johnson JS. (1974) Critical point drying of ethanol-infiltrated cryofractured biological specimens for scanning electron microscopy. Scanning Electron Microsc. 1974: 275-282.
19. Hung KS. (1982) Development of neuroepithelial bodies in pre- and postnatal mouse lungs: Scanning electron microscopic study. Anat. Rec. 203, 285-291.
20. Hyde DM, Plopper CG, Kass PH, Alley JL. (1983) Estimation of cell numbers and volumes of bronchiolar epithelium during rabbit lung maturation. Am. J. Anat. 167, 359-370.
21. Jeffery PK, Reid L. (1975) New observations of rat airway epithelium: A quantitative and electron microscopic study. J. Anat. 120, 295-320.
22. Jeffery PK, Reid LM. (1977) Ultrastructure of airway epithelium and submucosal gland during development, in: Lung Biology in Health and Disease, Hodson WA (ed.), Vol. 6, Marcel Dekker, NY, 87-134.
23. Kanda T, Hilding D. (1968) Development of respiratory tract cilia in foetal rabbits. Electron microscopic investigation. Acta Oto-laryngol. 65, 611-624.
24. Kawamata S, Fujita H. (1983) Fine structural aspects of the development and aging of the tracheal epithelium of mice. Arch. Histol. Jap. 46, 355-372.
25. Kelley RO, Dekker RAF, Bluemink JG. (1975) Thiocarbohydrazide-mediated osmium binding: a technique for protecting soft biological specimens in the scanning electron microscope, in: Principles and Techniques of Scanning Electron Microscopy, Hayat, MA (ed.), Vol. 6, Van Nostrand Rinehold Co., New York, 34-44.
26. Kober HJ. (1975) Die lumenseitige Oberfläche der Rattentrachea während der Ontogenese. Zeit. Mikroskop. Anat. Forsch. 89, 399-409.
27. Liebich HG. (1974) Die segmentale Gliederung der Lunge der weissen Ratte (Rattus norvegicus). Anat. Histol. Embryol. 3, 243-249.
28. Marin ML, Lane BP, Gordon RE, Drummond E. (1979) Ultrastructure of rat tracheal epithelium. Lung 156, 223-236.
29. McAteer JA, Evan AP, Douglas WHJ. (1983) A method for processing free-floating cultured cells for scanning electron microscopy. J. Tissue Cult. Methods 7, 85-88.
30. Morse DE, Gattone VH, McCann P. (1979) Surface features of the developing rat lung. Scanning Electron Microsc. 1979; III: 899-904.
31. Nowell JA, Tyler WS. (1971) Scanning electron microscopy of the surface morphology of mammalian lungs. Am. Rev. Resp. Dis. 103, 313-328.
32. Pavelka M, Ronge HR, Stockinger G. (1976) Vergleichende Untersuchungen am Trachealepithel verschiedener Säuger. Acta Anat. 94, 262-282.
33. Plopper CG, Alley JL. (1983) Differentiation of tracheal epithelium in the fetal Rhesus monkey. Anat. Rec. 205, 155A.
34. Plopper CG, Alley JL, Serabjit-Singh CJ, Philpot RM. (1983) Cytodifferentiation of the nonciliated bronchiolar epithelial (Clara) cell during rabbit lung maturation: An ultrastructural and morphometric study. Am. J. Anat. 167, 329-357.
35. Plopper CG, Mariassy AT, Wilson DW, Alley JL, Nishio SJ, Nettesheim P. (1983) Comparison of nonciliated tracheal epithelial cells in six mammalian species: ultrastructure and population densities. Exp. Lung Res. 5, 281-294.
36. Sannes PL. (1982) The early postnatal development of pulmonary airway epithelium in the rat. Anat. Rec. 202, 167A.
37. Smith P, Heath D, Mossavi H. (1974) The Clara cell. Thorax 29, 147-163.
38. Smolich JJ, Stratford BF, Maloney JE, Ritchie BC. (1976) Postnatal development of the epithelium of larynx and trachea in the rat: Scanning electron microscopy. J. Anat. 124, 657-673.
39. Sorokin S. (1968) Reconstructions of centriole formation and ciliogenesis in mammalian lungs. J. Cell Sci. 3, 207-233.
40. Tuchmann-Duplessis H, Haegel P. (1974) Illustrated Human Embryology, Vol. 2, Organogenesis, Springer-Verlag, NY, 44-49.
41. Witschi E. (1962) Development: Rat, in: Growth, Altman L and Dittmer DJ (eds.), Fed. Am. Soc. Exp. Biol., Washington DC, 304-314.

## Discussion with Reviewers

*Dr. E. Cutz:* What cell types are the cells with surface microplicae?
*Author:* To my knowledge, microplicae have not been observed in adult tracheal epithelium. Microplicae, during fetal development, may be a transient feature representing a stage in the differentiation of another cell type.
*Dr. E. Cutz:* Were any brush cells with short "brushy" microvilli observed?
*Author:* I did not observe brush cells typical of those reported to be within the bronchial epithelium and in the alveolar region of some species.
*Dr. D.L. Luchtel:* The presence of epithelial cells with a single cilium for a short period of time during fetal development is interesting. Can the author speculate about the functional significance of such cells?
*Author:* Primary cilia appear to be very common in occurrence, both in fetal development, and as organelles of many adult cell types. The

function of the primary cilium in the fetal airways is unknown. I am aware of no experimental study in which the function of primary cilia during the fetal period has been investigated.

Dr. S.R. Hilfer: Has any experimental work been done to test the role of hormones and growth factors on differentiation of tracheal cell types?

Author: This is an area of recent, current investigation in my laboratory. Unfortunately, I have only some preliminary information to offer. I have cultured fetal rat tracheal explants in serum-free media, in the complete absence of hormones. Under these conditions the tracheal epithelium continues to differentiate and still exhibits dorsal (pars membranacea) epithelial differentiation in advance of the pars cartilagina. I, as yet, have not determined if the rate of differentiation or the differentiation of specific cell types is influenced by hormones or growth factors.

Dr. S.R. Hilfer: Is there any evidence that the earlier differentiation of dorsal vs. ventral epithelium is related to differentiation of the mesenchyme into muscle?

Author: I have no experimental evidence to support this idea. I do feel, however, that epithelio-mesenchymal developmental events within the pars membranacea deserve attention.

Dr. C.G. Plopper: When and where do submucosal glands form during tracheal development?

Author: I have read reports in the literature which indicate that submucosal glands are absent from the rat. However, I have observed, by SEM, what I believe to be gland pits (containing mucous) in trachea of two-week-old neonates. These structures are limited in distribution to the dorsal wall (pars membranacea) of the proximal trachea.

Dr. C.G. Plopper: Is there a species difference in gland formation?

Author: To my knowledge, the perinatal morphogenesis of tracheal glands has not been studied.

Dr. C.G. Plopper: What are the species differences in the sequence of tracheal groove formation and bronchial outgrowth?

Author: This, also, is an aspect of development of the respiratory system that has not been investigated.

# DEVELOPMENT OF TERMINAL BUDS IN THE FETAL MOUSE LUNG

S. R. Hilfer

Department of Biology, Temple University, 12th and Norris Streets
Philadelphia, PA  19122

Phone No. 215 - 787-8863

(Paper received February 18 1983, Complete manuscript received July 21 1983)

## Abstract

Development of the lungs in mammals begins with the formation of pharyngeal buds that undergo repeated branching to establish the bronchial tree. Late in fetal life the most superficial buds begin to form side branches that will develop into the respiratory surfaces of the lung; the alveolar ducts, alveolar sacs, and alveoli. The bronchial ciliated cells and Clara cells first appear on the seventeenth day of gestation in the mouse. In the respiratory region, a few lamellar bodies are found by 16 days but identifiable type II pneumocytes and extracellular surfactant are not found until 18 days. Flattened type I pneumocytes do not form until after birth, on day 19 or 20. The epithelial branches of the lung are surrounded by mesenchyme and covered by a capsule. At 16 days, the branches are separated by a compact mesenchyme that is highly cellular. By 18 days, the mesenchymal space begins to be reduced. By 19 days, the air passages expand and the mesenchyme forms a thin layer between branches. Little experimental work has been done on the role of the mesenchyme in differentiation of the respiratory surfaces. Since branching of the embryonic lung is influenced by the kind of mesenchyme surrounding the epithelium at early stages, it is likely that mesenchymal control is exerted at later stages in the fetus.

Key words: Lung development; Bronchus, cell types; Respiratory passages, development; Surfactant; Pneumocytes, types I & II; Extracellular matrix; Blood vessels; Clara cells; Branching pattern, Lung; Alveolar surfaces.

## Introduction

In mammals, the respiratory system develops from paired endodermal buds in the floor of the caudal pharynx. The outgrowth of these buds produces the left and right primary bronchi. The trachea forms by horizontal splitting of the pharynx cranial to the bronchial buds (Spooner & Wessells, 1970). The epithelial lining of the lung is established by repeated branching of the bud at the tip of each primary bronchus. The number of major subdivisions, or lobes, that comprise each lung is determined by the number of branches that are formed by the initial bronchial bud. In the mouse, the right lung characteristically has four major lobes whereas the left lung consists of only one small lobe. (Fig. 1).

Development of the lung generally is divided into four periods or phases based upon distinctive patterns of branching, cellular organization, and appearance of specialized cell types. In the mouse, the initial, or embryonic, phase lasts to gestation day 16 or 17. There is a rather abrupt transition to the canalicular phase, which lasts until day 18 or 19. A relatively abrupt change on day 19 results in the beginning of the terminal sac phase; a more gradual change after birth results in the alveolar phase, which lasts throughout childhood. After a brief description of the embryonic period, this discussion will focus on the changes in branching pattern, lobular organization, and cell specialization that occur during the last three phases.

During early development, a modified dichotomous branching pattern produces an enlarging tree-like structure in each lobe. Each bud divides to produce two or three new branches. These branches divide again to form several orders of bronchi, starting at the base, or hilum, of the lobe and progressing toward the margin (Fig. 1). The last branches, or terminal buds, give rise to the respiratory surfaces. The bronchi act as air passages to conduct the air into and away from the respiratory endings. This early branching pattern is called centrifugal because it progresses from the center to the periphery of the developing lobe. Late in fetal life, a second branching pattern becomes superimposed upon the embryonic centrifugal pattern. This latter pattern is designated centripetal because it appears to be initiated near the outer limits of the expanding lung and to progress toward the center. However, it occurs only in the terminal portions that form the respiratory surfaces. The initiation of the centripetal branching and the appearance of the specialized cell types of the lung termini are the subjects of this discussion. The descriptions are based upon the development of the CD1 mouse, which has a gestation time of 19 days. All illustrations are of the dorsal lobe of the right lung. Additional information on branching, especially in humans can be obtained from several good reviews (e.g., Boyden, 1975;

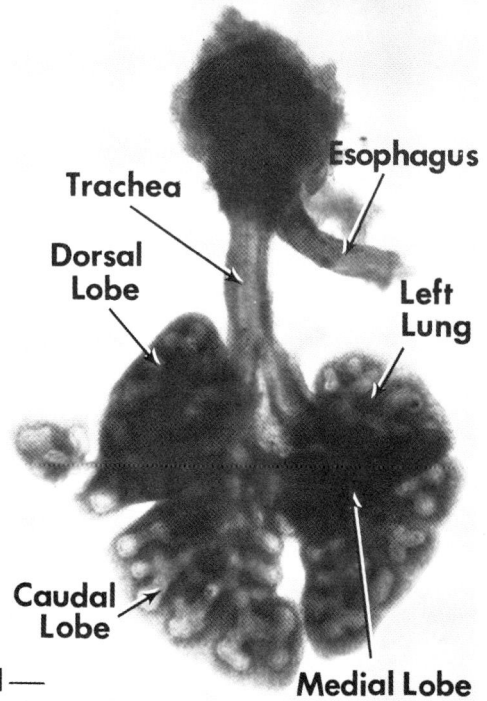

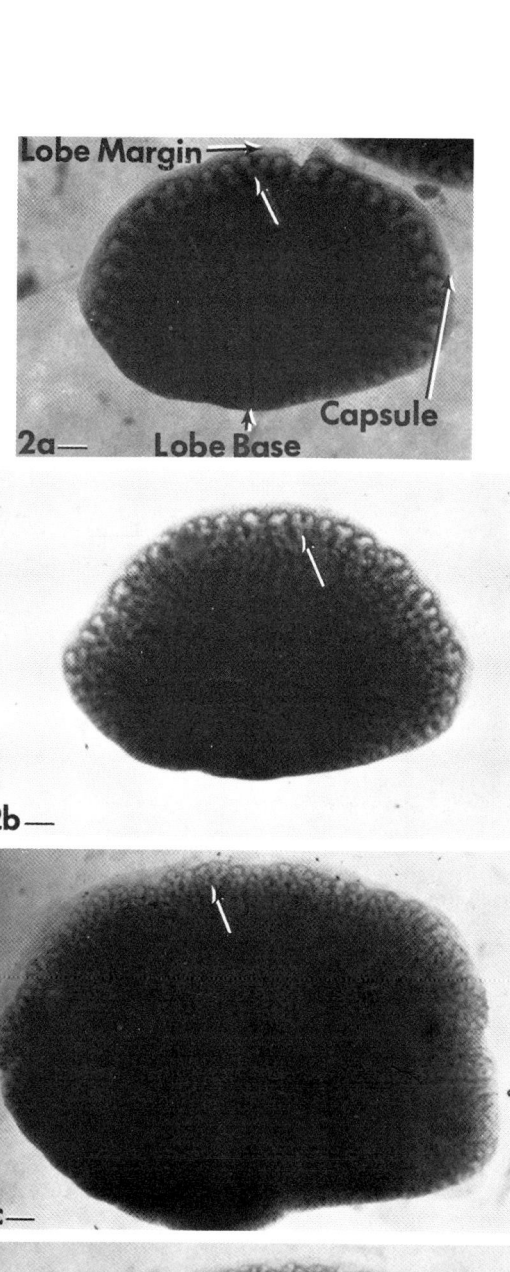

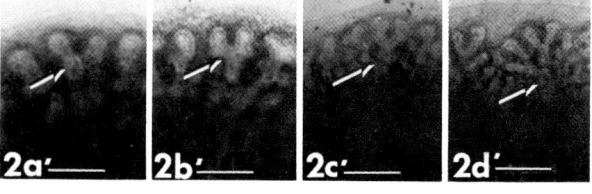

Fig. 1. Ventral view of the respiratory system of a 14 day mouse fetus. Bar = 200 µm. The left lung consists of one lobe. The right lung has four lobes, one of which is hidden in this view. The dorsal cranial lobe overlaps the larger caudal lobe. A smaller medial ventral lobe grows across to the left side of the embryo. Each lobe has branched several times to produce several orders of bronchial passageways.

Fig. 2. Development of the dorsal lobe of the right lung at a) 15; b) 16; c) 17 and d) 18 days gestation. Bar = 200 µm. The lobe increases greatly in size over this three day period. The terminal buds are visible along the thinner outer margin of the lobe. The arrow marks the same bud at low and high magnification at each age. At 15 (a') and 16 (b') days, the buds are relatively large and the branches short. At 17 days (c') the channels are narrower and at 18 days (d') the side buds have elongated. The mesodermal capsule that covers the epithelial tree extends as a thin sheet beyond the terminal buds.

Hutchins et al., 1981; Perelman et al., 1981; Sorokin, 1965). The description of specialized lung cells is especially good in Kuhn (1982).

### Branching Pattern

The pattern of the terminal branches can be seen at the margin of a living lung when examined with a dissecting microscope (Fig. 2). Growth of the branches is three-dimensional; terminal buds also are present at the broad surfaces where the lobe is too thick to allow their visualization. The lung is covered by a mesodermal capsule and mesenchymal cells derived from the mesoderm penetrate deep into the lung between the epithelial branches and surround them. The pattern of branching changes with development. At 15 and 16 days of gestation, the buds are relatively large and exhibit a fairly reproducible pattern of branching from one lung to the next at the same age (Figs. 2a, b). Each bud forms two or three new buds at each branch point (Figs. 2a', b'). Each branching tends to produce a larger bud that is oriented toward the surface and one or two smaller buds that are deflected toward the sides. These smaller buds tend to branch more slowly than the larger peripheral bud, so that each segment of the developing tree consists of a main branch with smaller, less branched side shoots. The appearance of the terminal branches begins to change on the seventeenth day (Fig. 2c). The branches become narrower (Fig. 2d) and the side branches more complex (Figs. 2c', d'). Small sacs begin to appear on the surfaces of the preexisting branches. By 19 days, the small branches are so numerous that it is difficult to visualize the branching pattern in the living lobe.

The three-dimensional pattern of the branching is easier to visualize if the mesenchyme is removed from the surface of the epithelial component. The mesoderm can be loosened by treating the lobe for a few minutes with ice-cold 2% trypsin in balanced salt solution. By light microscopy, the buds appear hollow (Fig. 3a). By scanning electron microscopy the orderly placement of the peripheral bud versus the deeper ones becomes apparent (Fig. 3b). When all of the mesenchyme is cleaned from the terminal portion, the three-dimensional nature of the branching of individual terminal buds can be seen (Fig. 3c),

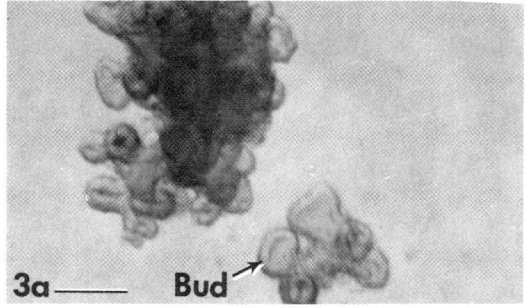

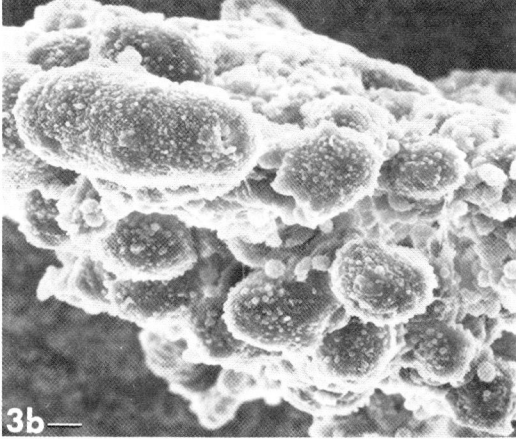

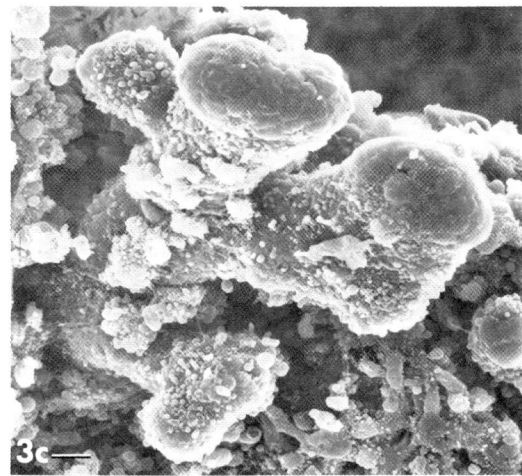

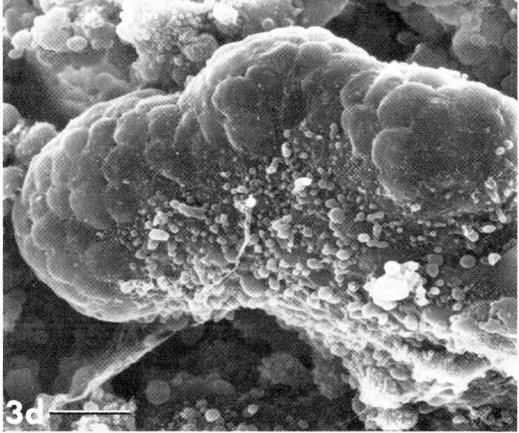

Fig. 3. Terminal parts of the epithelial tree that have been separated from the investing capsule after treatment with trypsin. a. 15 days, light micrograph. The buds are hollow sacs. Bar = 100 µm. b. 16 days, low magnification SEM to show the ordered arrangement of the terminal bud and its more proximal branches. Bar = 10 µm. c. 17 days, the branching pattern of a single terminal bud. Bar = 10 µm. d. 17 days, the surface of a bud that is forming three subdivisions. Bar = 10 µm.

as well as the initial stages in bud formation (Fig. 3d).

### Lobular Organization

The relationship between the epithelial component and the investing mesenchyme is visible in lobes that have been cleaved in various planes. The dorsal lobe of the lung has a convex outer surface that lies against the rib cage and a concave inner surface that rests against the other lobes. It is thick at its dorsal attachment point to the other lobes and thin at its ventral margin. When cut along its dorsoventral axis, perpendicular to the broad surface, it appears as a wedge (Fig. 4a). At 16 days, the mesenchyme has a dense, character with little extracellular space. The bronchi lie toward the center of the lobe and the periphery contains many buds along both the concave and convex surfaces. All of the space between the buds is filled with a highly cellular connective tissue (Fig. 4b). The ventral margin of the lobe contains a single layer of regularly spaced terminal buds (Fig. 4c). There is little change in organization at 17 days (Fig. 5a). The lobe appears highly cellular and the channels of the bronchi are not pronounced (Fig. 5b). Except for the major blood vessel entering the lobe (Fig. 5a), the blood supply forms only narrow channels (Fig. 5c). The tip of the lobe is still occupied by buds that have relatively large lumina (Fig. 5d).

At 18 days a dramatic change occurs in the appearance of the lobe (Fig. 6a). The lumina of the airways become distended and a vascular supply lies parallel to the major bronchi and penetrates to the level of the terminal branches (Fig. 6b). The terminal branches form an elaborate tree with many small side branches. Sections perpendicular to the broad surfaces show that similar changes occur along these surfaces as well (Fig. 6c). At 19 days these changes are even more pronounced (Fig. 7a). The airways are expanded and the branched tree extends under the broad surfaces (Fig. 7b) as well as along the ventral margin of the lobe (Fig. 7c). Narrow vascular channels follow the terminal branches to the periphery of the lobe (Fig. 7d). The lung has a much more open appearance. Shortly after birth the openness of the lung again changes dramatically (Fig. 8a, b). The smaller bronchi terminate at an extensive system of small chambers. These chambers have a complex architecture, with smaller sacs, the alveoli, communicating with slightly larger ones (alveolar sacs and ducts), which connect with respiratory bronchioles.

### Cell Types

The adult lung has been described as containing more than 40 distinct cell types. Not all of these are found in the fetus and neonate, the subject of this discussion. Only the most common cell types of the terminal bronchial and respiratory portions of the fetus will be considered here. Up to 16 days gestation, the epithelium of the respiratory

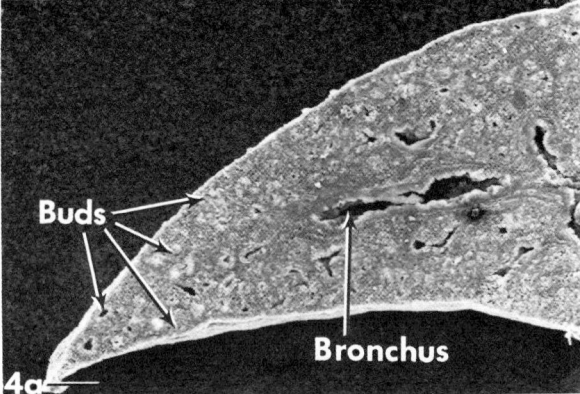
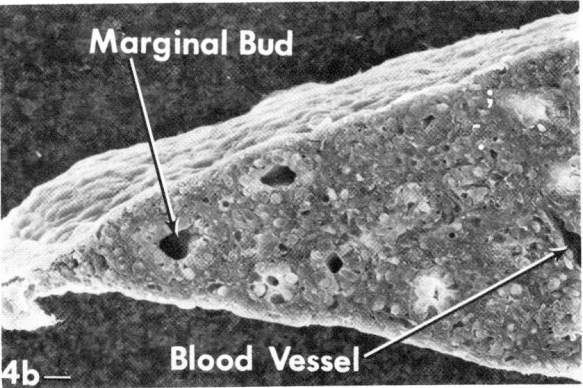
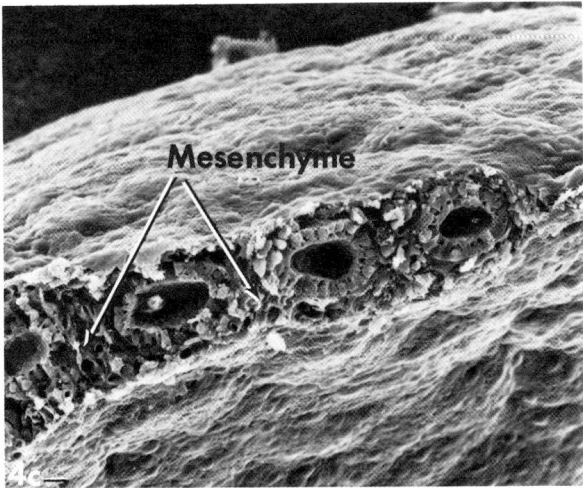

**Fig. 4.** Structure of the lung at 16 days. a. The dorsal lobe was cut on a plane from the base to the thin outer edge and perpendicular to the broad surfaces. The outer curvature fits against the ribs and the concave inner surface lies against the caudal lobe. Note the larger bronchi near the base and center and the smaller bronchi peripheral to these. Many terminal buds line both curved surfaces. Bar = 100 μm. b. The tip region of (a). The marginal bud is larger than those more proximal. The epithelial portion is surrounded by a dense mesenchyme. Bar = 10 μm. c. Part of a lobe cut through the buds along the thin margin. Note the regular spacing of the terminal buds and the packing of mesenchyme between them. Bar = 10 μm.

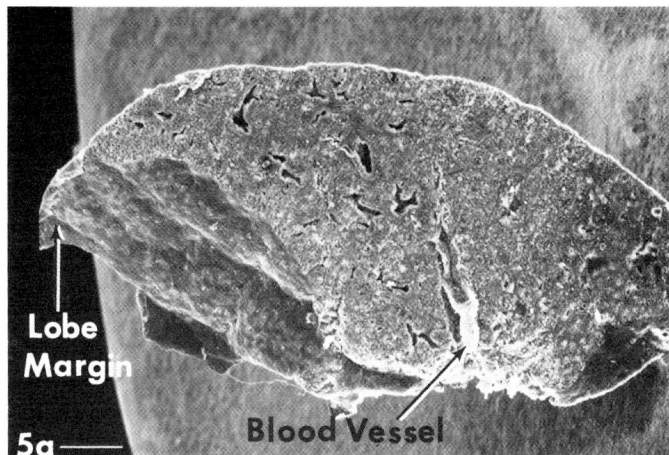
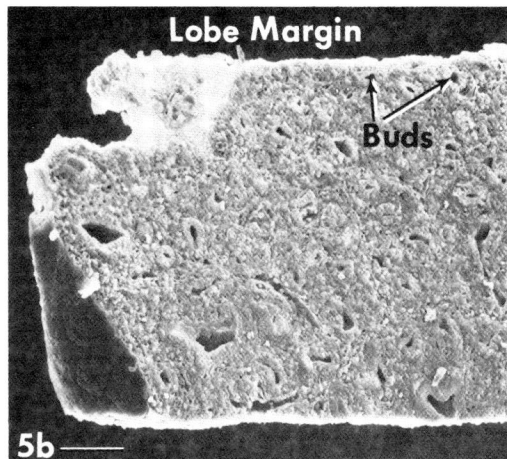
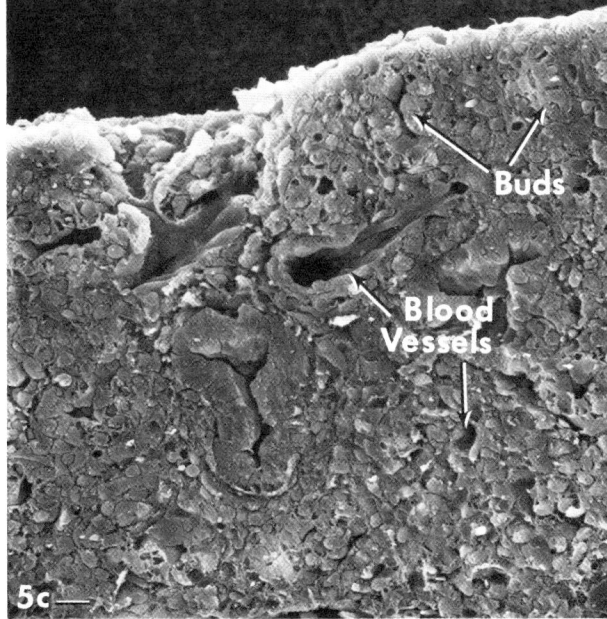
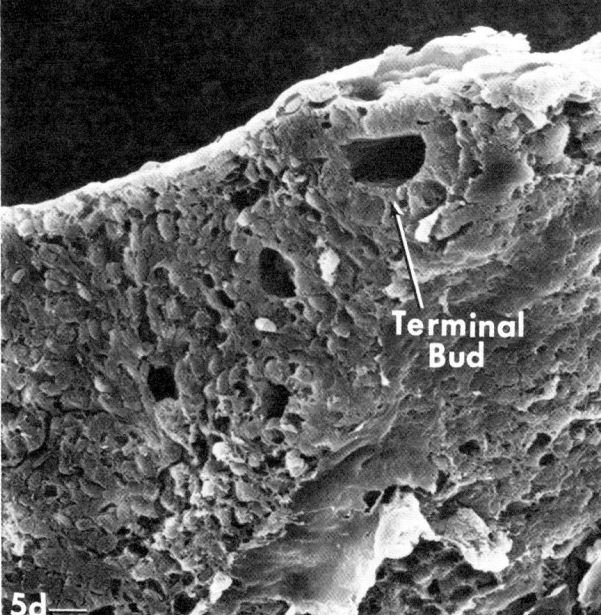

Fig. 5. Structure of the lung at 17 days. An oblique cut was made through the dorsal lobe near one end. The thickened base and thin margin of the lobe are evident. A large blood vessel enters from the base, close to the larger bronchi. Bar = 200 µm. b. A segment of the lobe cut tangential to the convex surface, to expose the many terminal buds beneath the surface. The mesenchyme is densely packed around the epithelial branches. Bar = 200 µm. c. Small portion of the margin of a lobe. Small blood vessels lie between the epithelial branches. Bar = 10 µm. d. Tip of a lobe. The terminal bud is larger than the deeper ones. Bar = 10 µm.

---

buds and the bronchial tubes cannot be distinguished on the basis of shape or organelle content. In all regions of the lung, the cells are columnar with few microvilli and a single central apical cilium (Fig. 9a). These cells have been designated primitive epithelial cells and are thought to be the precursors of all of the specialized epithelial cells of later stages (Ten Have-Opbroek, 1981). At later stages two major (ciliated and Clara) and several minor (neuroepithlial, brush, and mucous) cell types have been identified in the bronchial epithelium. Most of the bronchial surface is covered by cells possessing many cilia interspersed among dome-shaped Clara cells. The first indication of ciliated cells in scanning electron micrographs is the presence of thick, stubby processes among the finer microvilli (Fig. 9b). These processes elongate to form 40 or more motile cilia per cell apex (Fig. 9c). These Clara cells are first distinguishable by their dome-shaped apices with small protrusions (Fig. 9b). Later, the apices spread and the surfaces appear smoother (Fig. 9c). These cells contain secretion droplets that are thought to be serous (Kuhn, 1982). Of the minor components, the most prominent is the neuroepithelial cell type (Fig. 9d). The apices form irregular bulges and the cells either are interspersed between and partly covered by the

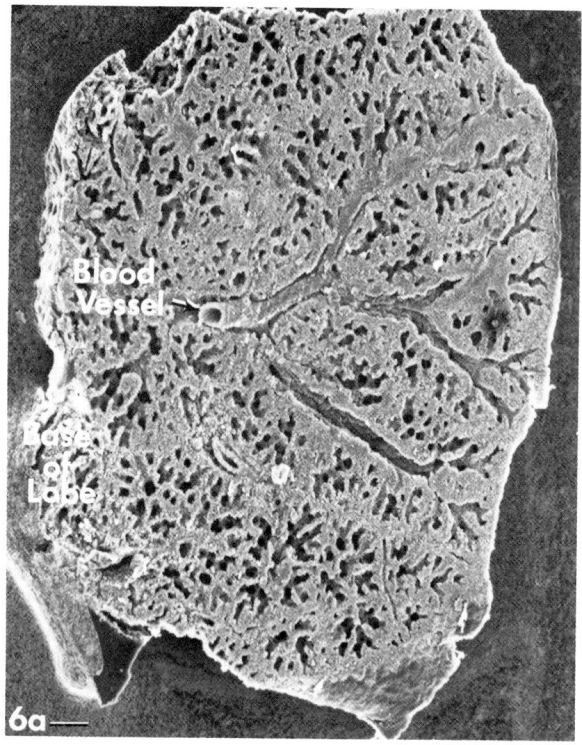
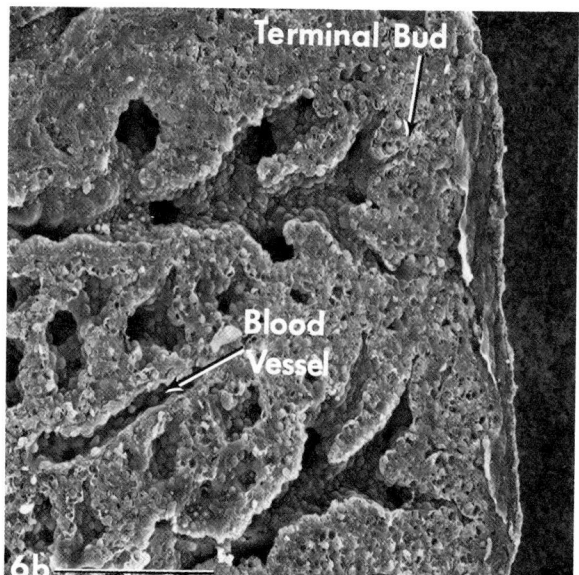
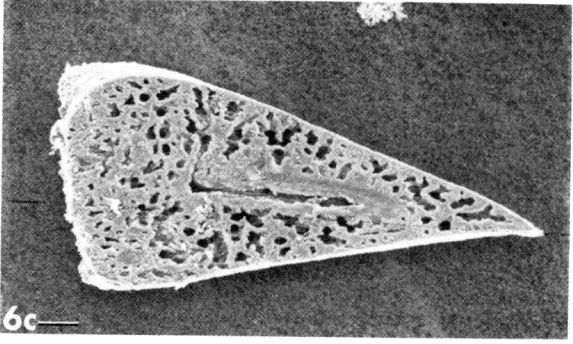

bulging Clara cells or the cells form patches called neuroepithelial bodies. Neuroepithelial bodies are often located at bifurcations of the airways (Hung, 1982). Occasionally a cell apex possessing many coarse microvilli is found among the Clara cells (Fig. 9e). These cells are called brush cells and are of unknown function (Kuhn, 1982). Brush cells also have been described in the respiratory passages. Individual mucous, or goblet, cells have been reported in the terminal bronchial passageways of the adult mouse and in fetuses of other species, but we have not found them in the mouse fetus. These cells possess long tongues of mucous protruding from their apical surfaces. The mucous and serous secretions of the goblet and Clara cells trap particles that are swept toward the outside by the action of the ciliated cells. The neuroepithelial cells, which contain the neurohormone, serotonin, are thought to play a regulatory role in the lung. In most species, the cells are associated with autonomic nerve termini (Kuhn, 1982).

Whereas the epithelial cells of the airways remain cuboidal when they specialize, the cells of the respiratory portion become shorter. Two cell types appear. The first to differentiate is the cuboidal or polygonal pneumocyte type II, which is recognized by the large protrusions of its apical surface (Fig. 9f). In transmission electron micrographs, the protrusions are seen to contain lamellar bodies (see below). These cells produce the surfactant that coats the apical surface of the respiratory epithelium and

Fig. 6. Structure of the lung at 18 days. Bar = 100 μm. a. The section passed through the lobe tangential to the convex surface. The cut end of the major blood vessel protrudes near the base of the lobe. The extensive bronchial tree and terminal branches are visible. The lumina of the bronchi are distended. b. Portions of two terminal (respiratory) branches. A larger blood vessel is between the two branches. The airways are distended but the mesenchyme still has the dense appearance of the younger lungs. c. Section cut perpendicular to (a). The airways are distended from the center almost to the surface of the lobe.

Fig. 7. Structure of the lung at 19 days. a. Oblique section of the dorsal lobe, cut through the base toward the thin margin. Ends of large blood vessels are toward the center of the lobe. The distended airways form an elaborate branched tree. Bar = 100 μm. b. Section cut on a plane perpendicular to the broad surfaces and from base to margin. Distension of the passageways extends to the surface branches. Bar = 100 μm. c. Enlargement of (a) near the margin of the lobe, showing a set of terminal branches. Not as much mesenchyme appears to be between the passageways as at earlier stages. Bar = 20 μm. d. Terminal buds near the margin of the lobe. The diameter of the buds is approximately half that at 17 days. Small blood vessels lie between the branches. The epithelial cells show the effects of extraction of their glycogen. Bar = 20 μm.

Development of Respiratory Branches

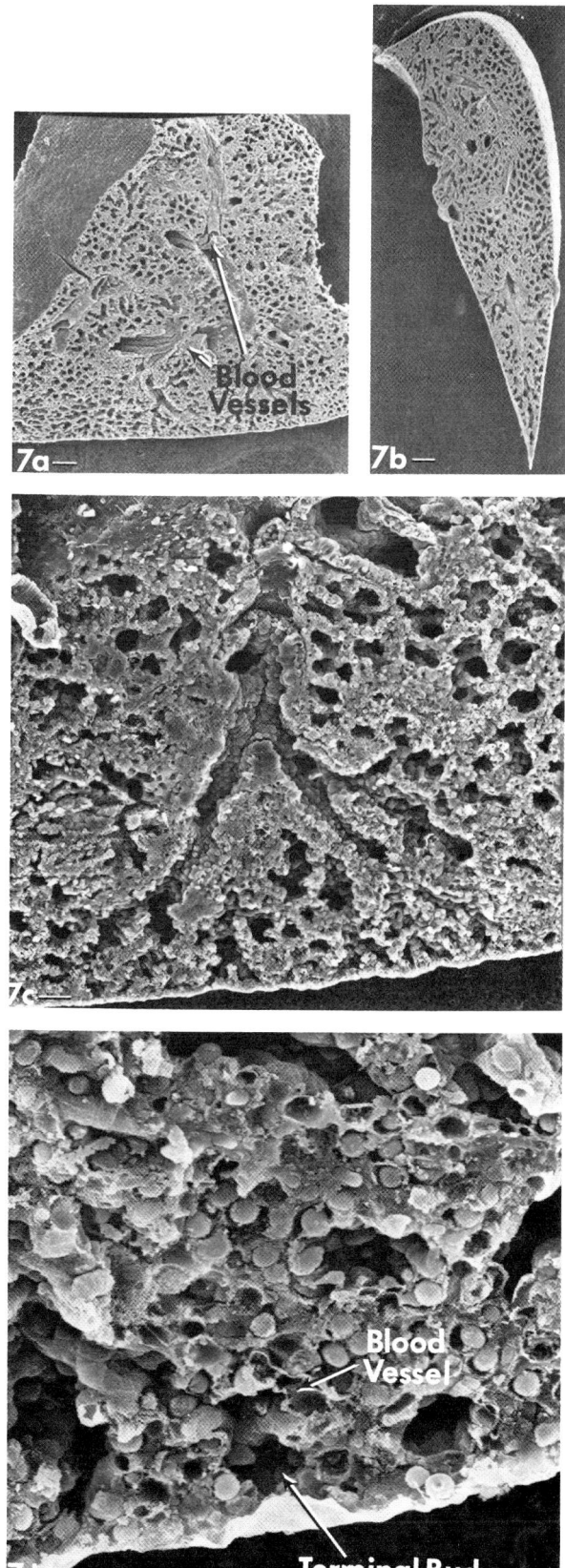
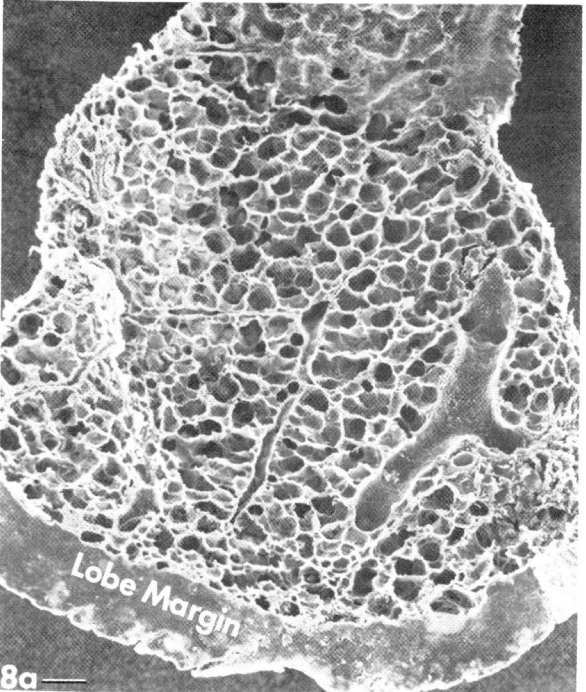
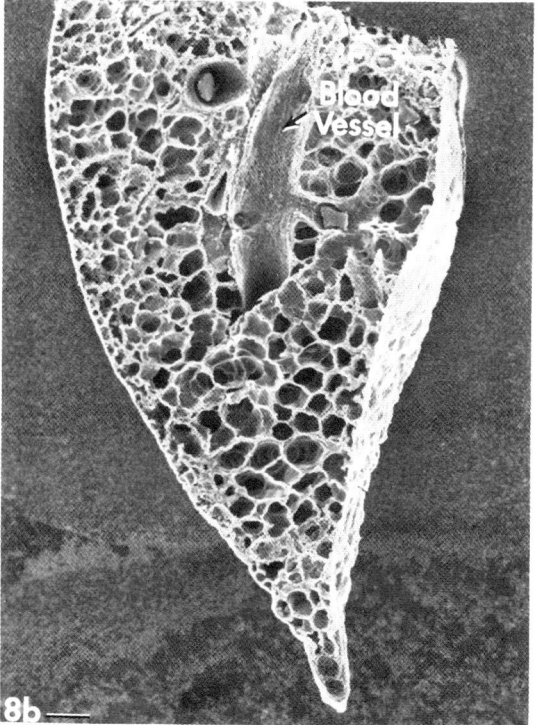

Fig. 8. Structure of the lung two days after birth. Bar = 100 μm. a. Segment of a lobe cut parallel to, but not passing through, the margin of the lobe. b. Lobe cut perpendicular to (a). The lung consists of larger central airways and blood vessels and many small sacs that are packed close together. The sacs under the capsule are the smallest.

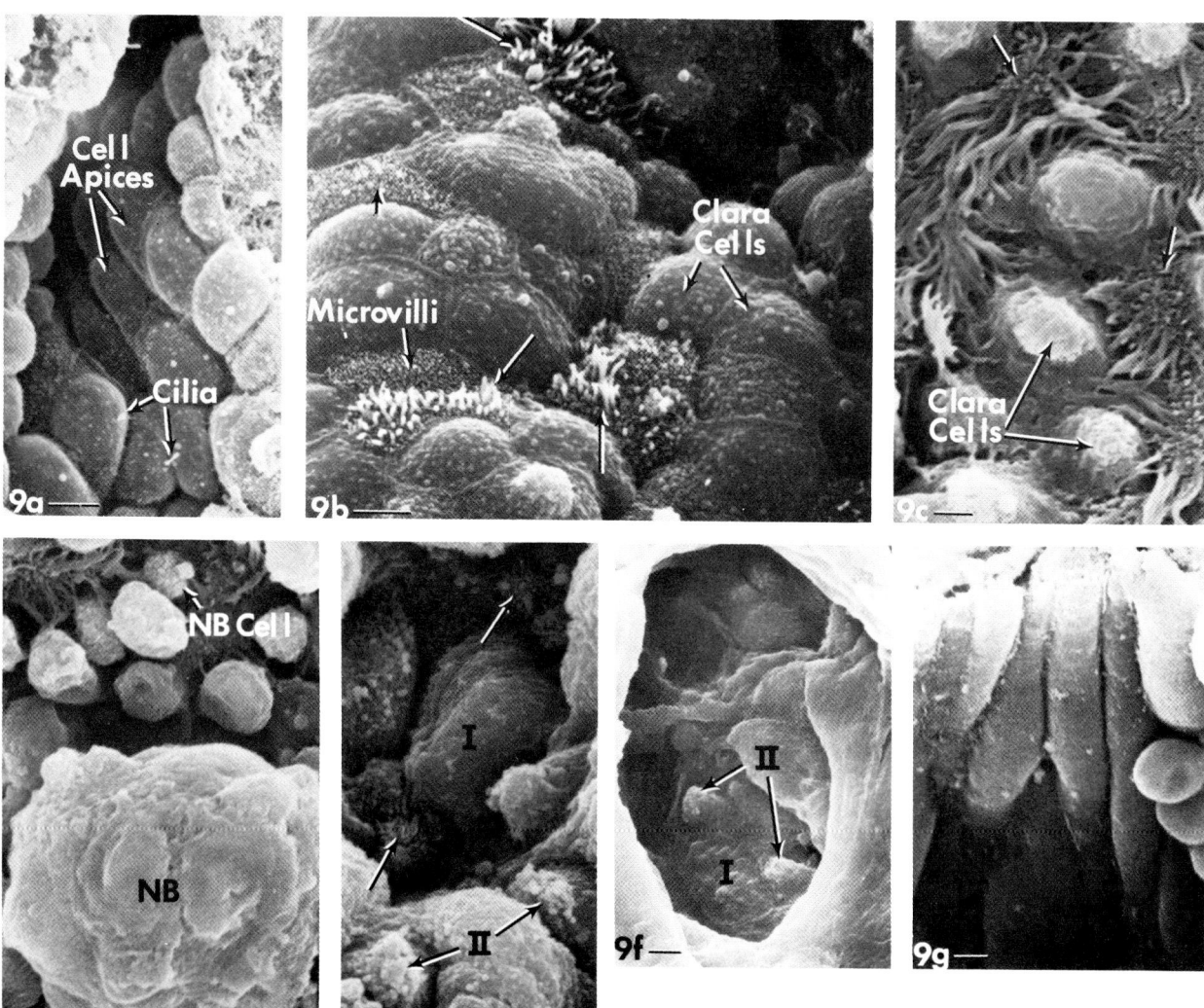

Fig. 9. Lung cell types. Bar = 2 μm. a. Unspecialized cells of the early lung, from a terminal branch at 16 days gestation. Cell apices have one short cilium and microvilli. b. Several stages in the formation of ciliated cells (arrows) in a bronchus of an 18-day fetus. Note the microvilli on the cell apices. The Clara cells are marked by a marginal ridge. c. Fully developed ciliated cells (arrows) in a two- day neonate. The Clara cells are flattened with a secretory bulge. d. Neuroepithelial body (NB) surrounded by Clara cells in a bronchus of a two-day neonate. The blebbed apex may mark a single neuroepithelial cell. e. Brush cells (arrows) in the respiratory region of a 19-day fetus. Type I (I) and type II (II) pneumocytes also are visible. f. Pneumocytes in a two-day neonate. The type II cells (II) have blebbed apices; the type I cells (I) are surrounded by a ridge. g. Endothelial cells of a medium-sized blood vessel in a two-day neonate. The cells are elongated in the direction of blood flow.

lowers surface tension. Presence of surfactant is necessary to prevent collapse of the alveoli during expiration (Perelman et al, 1981). The other respiratory cell type, the pneumocyte type I develops close to and after birth. The cells become almost as flat as capillary endothelial cells and cover a large surface area as compared with the other epithelial cell types (Fig. 9f). The cell bases become intimately associated with capillaries.

The connective tissue contains a variety of cell types that are difficult to distinguish on the basis of scanning electron microscopy. Many of these cells remain unspecialized until after birth. Cartilage and smooth muscle, for instance, develop relatively late and it is not clear how early nerve cells make junctions with the neuroepithelial bodies. A prominent cell type of the connective tissue is the vascular endothelial cell. In the large blood vessels, the cells are elongate and highly flattened (Fig. 9g). The small vessels in the peripheral regions of the lung have linings that resemble the respiraatory epithelium, but the diameter of the vessels is smaller than that of the respiratory passages.

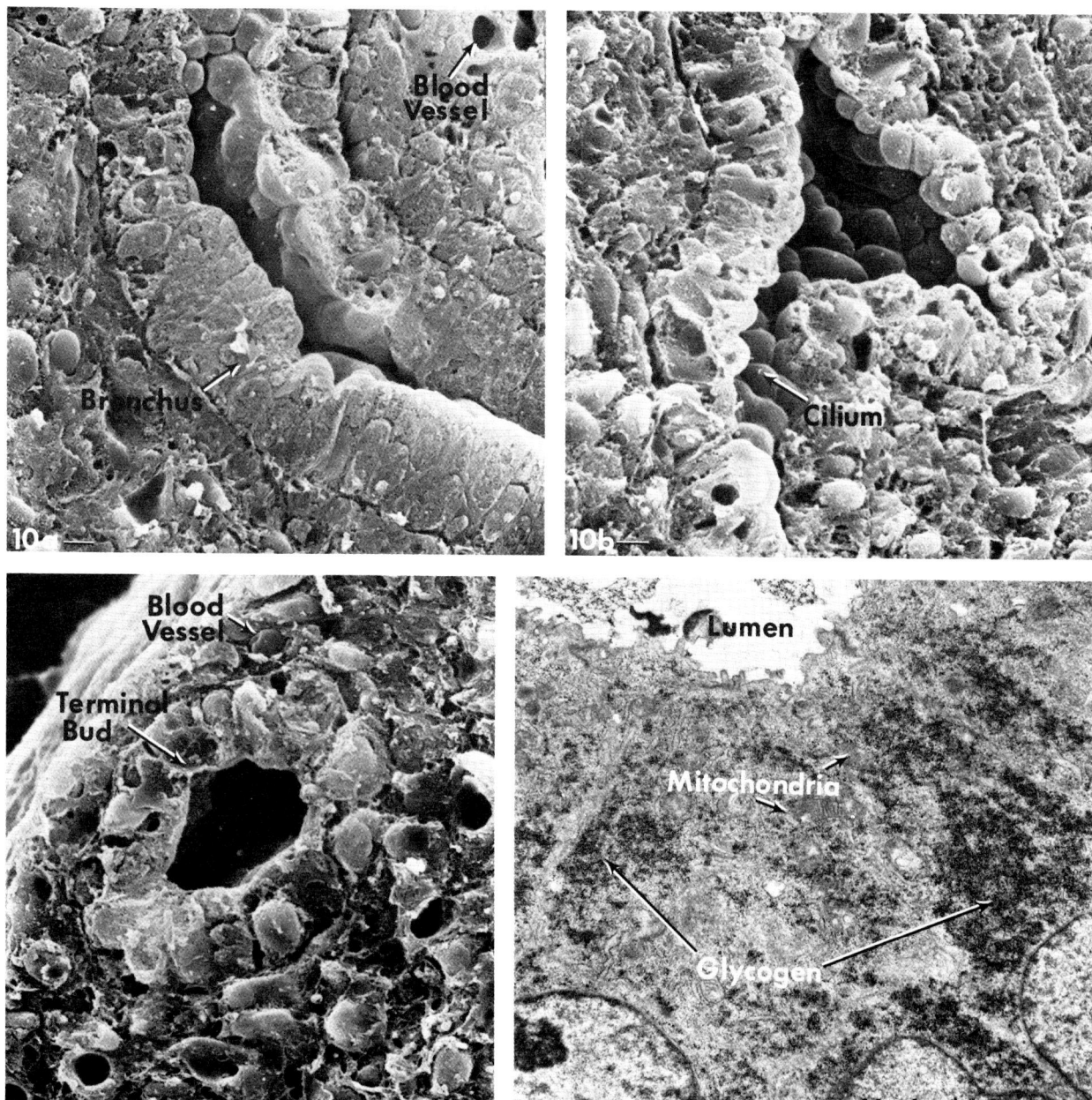

Fig. 10. Cellular organization at 16 days. Bar = 2 μm. a. Small portion of a bronchus. The cells are columnar and their slightly bulging apices possess short microvilli. The airway is surrounded by a dense mesenchyme that contains small blood passageways. b. Branch of a terminal bud. The cells are columnar and have the same bulging apices with microvilli that are seen in the bronchi. The single, central cilium appears as a larger protrusion than the microvilli. c. A terminal bud, showing cellular structure similar to the deeper branches. d. Transmission electron micrograph of a small part of a terminal bud. The cells contain large masses of glycogen. Procedures that fix glycogen do not allow enough contrast to visualize the cytoplasmic organelles, such as mitochondria, Golgi region, and endoplasmic reticulum, all of which are present in the apical cytoplasm.

## Cellular Changes

At 16 days, the organization of the epithelial cells is similar in the bronchi, the proximal terminal branches, and the terminal buds (Fig. 10a,b,c). All consist of low columnar epithelia. The apical cell surfaces are relatively smooth, with many of them possessing a single central cilium. Many of the cells also possess short microvilli (Fig. 10d). Small blood vessels lie near the epithelial passages, but few make contact with the basal cell surfaces (Figs. 10a, c). The cells of the terminal regions contain large glycogen deposits (Fig. 10d) that diminish after the cells specialize. At 17 days, the first cells with many cilia begin to appear in the bronchial passages (Figs. 11a). The cilia begin as short projections from the surfaces of a few of the cells. The bulging apices of the bronchiolar cells (Clara cells) precede the accumulation of secretion droplets in these cells. The cells of the terminal buds are covered with short microvilli and an occasional protruding droplet (Fig. 11b). A few lamellar bodies are present in an occasional cell of the terminal branches at 16 and 17 days (Fig. 11c). These inclusions and the protruding droplets probably mark the first pneumocyte type II cells (surfactant producing cells). The cells are still low

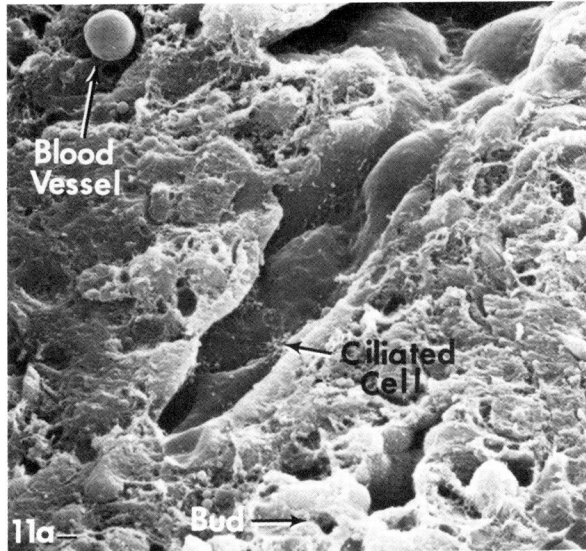
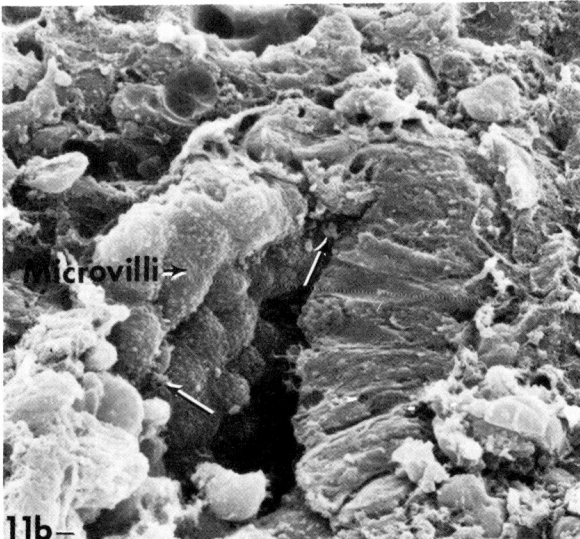
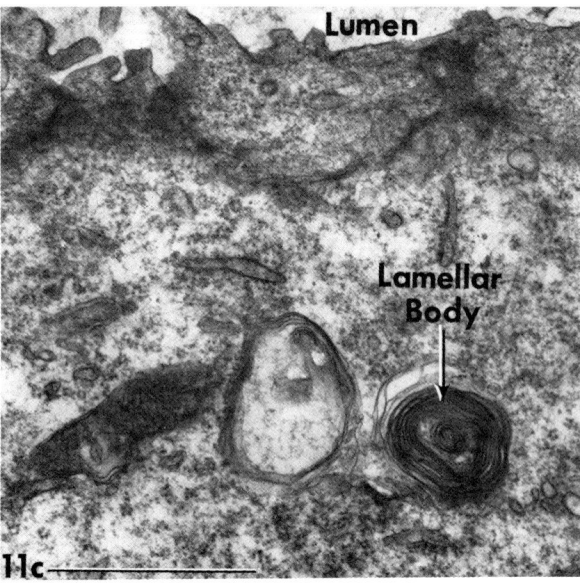

Fig. 11. Cellular organization at 17 days. Bar = 1 µm. a. Small portion of a bronchus, showing the early development of ciliated cells. These cells have short cilia and lie in depressions between the bulging Clara cells. b. Part of a terminal branch near the margin of the lobe. The columnar cells are covered with short microvilli. Small masses (arrows) on the surfaces of the cells probably are the earliest masses of surfactant that are seen in transmission micrographs. c. Transmission micrograph of a cell apex in the terminal region near the margin of the lobe. Two lamellar bodies lie in the apical cytoplasm.

Fig. 12. Cellular organization at 18 days. Bar = 1 µm. a. Part of a bronchus, showing the increase in the number of ciliated cells. The cilia tend to circle the cell apex in contrast to the single central cilium of the bulging Clara cells. The newest ciliated cells have short cilia. The cells with small apical protrusions that are tucked between the Clara cells probably belong to neuroepithelial bodies b. Apices of Clara and ciliated cells. The Clara cell cytoplasm contains rough endoplasmic reticulum and secretion vesicles and has an uneven surface. The ciliated cells contain many basal bodies and microvilli. c. Small portions of a terminal branching point. Many of the cells have masses protruding from their apices. The vesicular nature of some of these (arrows) suggests that this represents secretion of surfactant. d. Apex of a type II pneumocyte. The cytoplasm contains several lamellar bodies that are cut in different planes. These bodies consist of cylindrical layers of membrane that are concentric.

# Development of Respiratory Branches

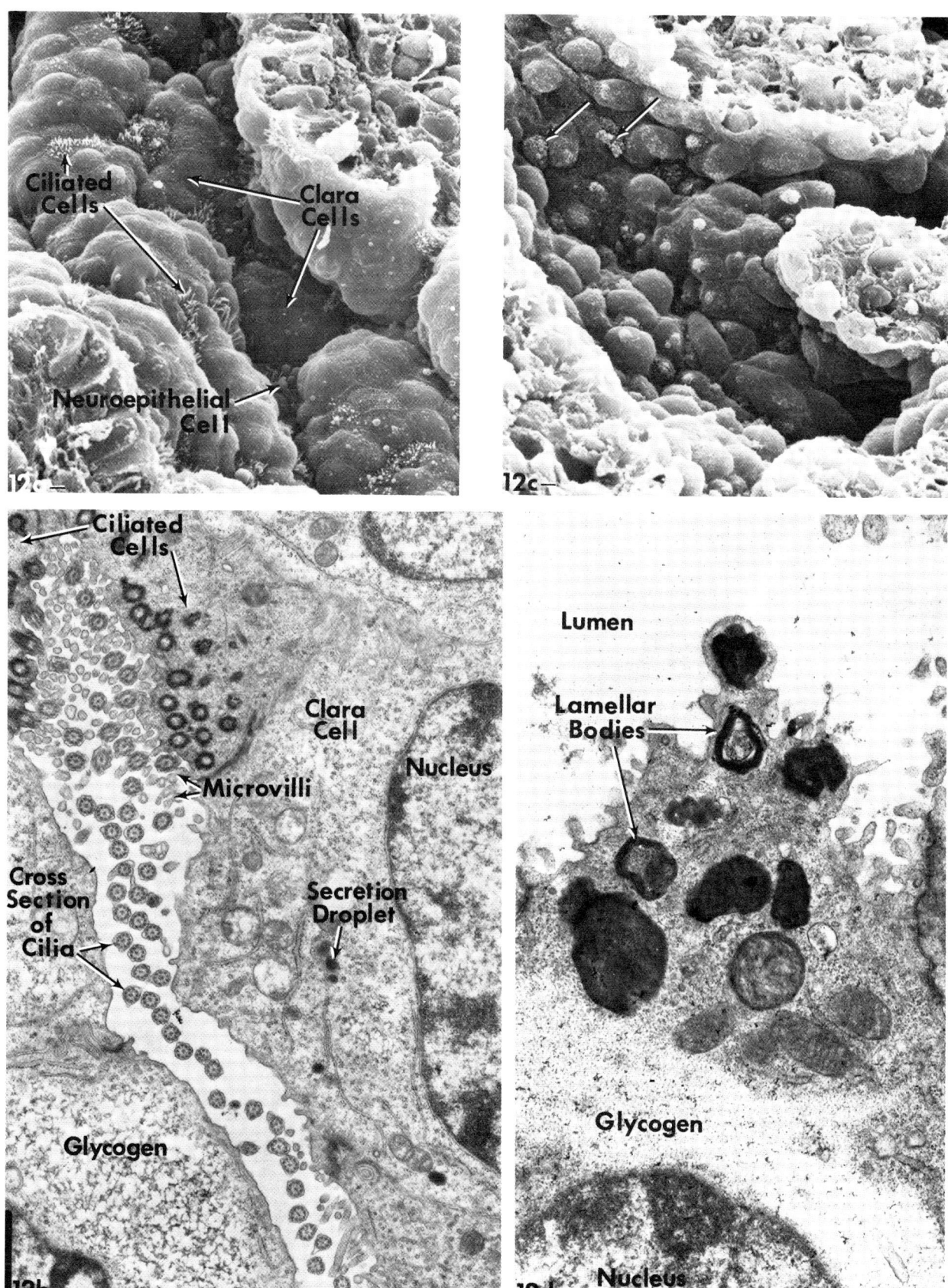

For figure legend see the facing page.

columnar throughout the epithelial branches.

At 18 days, the bronchi contain larger numbers of ciliated cells (Fig. 12a) that form small patches between the Clara cells (Fig. 12b). Small clusters of cells with protruding droplets that are partially covered by these cells are the first neurosecretory cells of neuroepithelial bodies (Fig 12a). In the terminal branches, large numbers of cells have accumulations of droplets at their apices (Fig. 12c). These appear to be too numerous for all to be the type II pneumocytes that are found in transmission electron micrographs (Fig. 12d). The cells still contain large glycogen deposits, which frequently are lost during tissue processing, leaving empty spaces in the cells (Fig. 12d). The terminal epithelial cells are still relatively tall, having a low columnar or cuboidal shape.

By 19 days, some of the cells in the terminal branches are flattened and masses of extracellular material coat the cell apices (Fig. 13a). In alveolar sacs that are close to the surface, taller cells with highly blebbed apices are interspersed among larger, flat cells (Fig. 13b). Transmission electron micrographs show the terminal branches to consist of sacs containing type II pneumocytes with many lamellar bodies (Fig. 13c) and flattened cells. These cells have not reached the highly spread state that occurs after birth, however. The bronchial passageways contain many ciliated cells, flattened Clara cells with blebbed surfaces and neuroepithelial bodies (Fig. 13d). The highly flattened type I pneumocytes make their appearance after birth (Fig. 14a). Many of the terminal portions still are not alveoli but are classified as alveolar sacs. Additional pouching of their walls will produce the true alveoli. The sacs are closely packed to produce the honeycomb impression of the lobe at low magnification (Figs. 8a, b). The alveolar sacs consist of a mixture of tall type II cells, spread type I cells, and other cells that are of intermediate morphology (Fig. 14b). The bronchial surfaces are highly ciliated (Fig. 14c).

### Experimental Studies

Generation of the branching pattern has been studied carefully in several mammalian species (see Boyden, 1975; Hutchins et al., 1981; Ten Have-Opbroek, 1981 for recent examples). Antibodies have been developed against lung cells which have been used to trace the differentiation of Clara cells (Nathrath et al., 1981) and type II pneumocytes (Ten Have-Opbroek, 1979). However, little is known about the way in which the branching pattern and cell differentiation are controlled during the later stages of lung development. A factor has been extracted from lung fibroblasts in cell culture that hastens maturation of type II pneumocytes when injected into fetuses (Smith, 1979). Addition of hormones to the medium of lung organ cultures or injection into fetuses also accelerates the formation of surfactant (see Perelmen et al., 1981). Surfactant is necessary for the function of the respiratory surfaces. Its presence lowers surface tension and permits more efficient gas exchange. The primary medical problem faced in premature birth of a normal fetus is respiratory distress syndrome. This condition is a result of immaturity of the cells in the respiratory passageways and the inability to synthesize sufficient amounts of surfactant.

Development of the lung at early stages has been thoroughly investigated (see Goldin, 1980, for an introduction to the excellent work from Alescio's and Wessells' laboratories). It is known that the capsule of the proximal airways has different properties from that of the terminal bronchial buds. Bronchial mesenchyme induces supernumerary bud formation when transplanted to the tracheal surface and tracheal mesenchyme suppresses branching of the early bronchial buds. The difference between the capsule at the base vs. the periphery of the lung may be a difference in the composition of the matrix. There is a clear difference in the orientation of collagen fibrils in the two regions and it has been reported that mesenchymal cells make direct contact with the epithelial surface where buds are forming (Bluemink et al., 1976). Furthermore, interference with synthesis of either collagen or proteoglycan will suppress bud formation and subsequent branching (Spooner & Faubion, 1980). It seems likely that similar controls exist during the terminal phases of lung development.

### Conclusion

The development of an organ that undergoes three-dimensional branching is difficult to visualize from two-dimensional sections. In the past, the tedious process of three-dimensional reconstruction from light microscopic sections has been used to study the formation of branches. This method suffers from inherent inaccuracies because it is difficult to align the sections and because details of cell structure and surface topology are lost through both lack of resolution and inability to model fine detail. The higher resolution and three-dimensional images of scanning electron microscopy are ideal for studying branched organs such as the lung. Thick slices of the organ allow the morphology of the lumina to be examined and developmental changes to be assessed. These changes include the pattern of branching, the first appearance of specialized cells, the distribution of specialized cells, and the relationship of the epithelium with the surrounding connective tissues.

### Acknowledgements

Much of the original work in this paper would not have been possible without the exceptional skills of Joyce W. Brown, whose contribution I gratefully acknowledge. Thanks also go to Michael Czeredarczuk for his highly competent assistance, to Dr. Robert L. Searls for his help with writing the manuscript, and to Jo-Ann Felder for her skills with the word

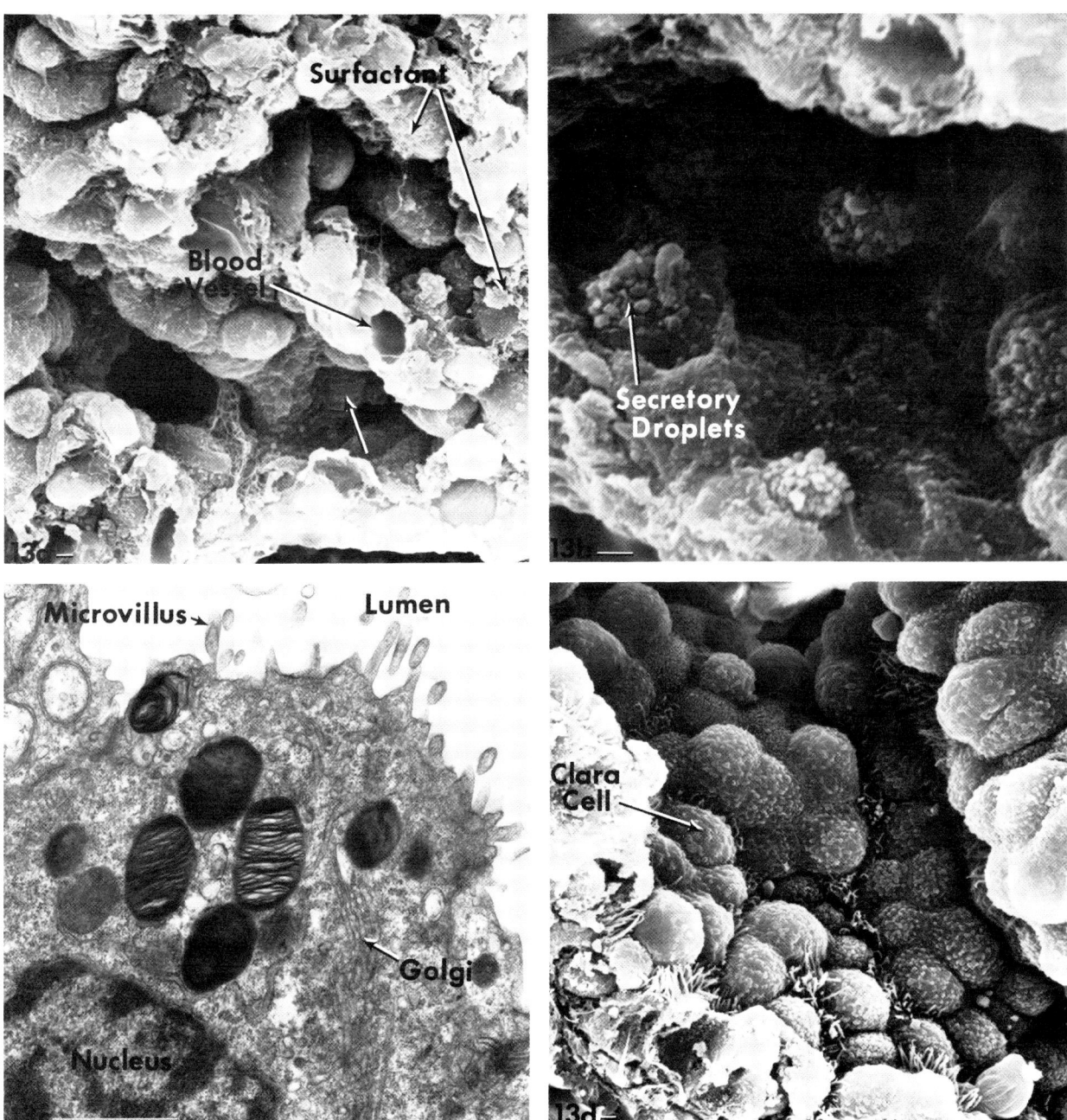

Fig. 13. Cellular organization at 19 days. Bar = 1 μm. a. Terminal buds at the margin of a lobe. The epithelial cells are cuboidal or shorter than broad. The apical surfaces of some cells appear serrated (arrow); all have microvilli. Clumps of surfactant are at the cell surfaces and in the lumen. Small blood vessels lie in the thin partitions between the respiratory branches. b. Respiratory surface just proximal to the terminal bud. Large accumulations of secretory material are at the cell apices. c. Transmission micrograph of a type II pneumocyte. The cell contains large lamellar bodies. The lumenal surface is covered with microvilli. d. Bronchial epithelium. Ciliated cells tend to form rows between the larger Clara cells. Neuroepithelial cells are still covered by the Clara cells.

processor. Supported by DHHS Research Grant HL 28330.

## References

Bluemink JG, van Maurik P and Lawson KA. (1976). Intimate cell contacts at the epithelial/mesenchymal interface in embryonic mouse lung. J. Ultrastruct. Res. 55, 257-270.

Boyden EA. (1975). Development of the human lung. Practice of Pediatrics, Vol. 4, Harper and Row, Hagerstown, MD, Ch. 64, 1-17.

Goldin GV. (1980). Towards a mechanism for morphogenesis in epithelio-mesenchymal organs. Quart. Rev. Biol. 55, 251-265.

Hung K-S. (1982). Development of neuroepithelial bodies in pre- and postnatal mouse lungs: Scanning electron microscopic study. Anat. Rec. 203, 285-291.

Hutchins GM, Haupt HM and Moore W. (1981). A proposed mechanism for the early development of the human tracheobronchial tree. Anat. Rec. 201, 635-640.

Kuhn C. (1982). The cytology of the lung: Ultrastructure of the respiratory epithelium and extracellular lining layers. in: Lung Development: Biological and Clinical Perspectives, PM Farrell (ed.), Academic, New York 27-55.

Nathrath WBJ, Rowlatt C and Detheridge F. (1981). Localization of organ-specific antigens in mouse lung by light and electron microscopy. J. Histochem. Cytochem. 29, 1365-1371.

Perelman RH, Engle MJ and Farrell PM. (1981). Perspectives on fetal lung development. Lung 159, 53-80.

Smith BT. (1979). Lung maturation in the fetal rat: Acceleration by injection of fibroblast-pneumocyte factor. Science 204, 1094-1095.

Sorokin S. (1965). Recent work on developing lungs, in: Organogenesis. R.L. DeHaan and H. Ursprung (eds.), Holt, Rinehart and Winston, New York, 467-491.

Spooner BS and Faubion JM. (1980). Collagen involvement in branching morphogenesis of embryonic lung and salivary gland. Devel. Biol. 77, 84-102.

Spooner BS and Wessells NK. (1970). Mammalian lung development: Interactions in primordium formation and bronchial morphogenesis. J. Exp. Zool. 175, 445-454.

Ten Have-Opbroek AAW. (1979). Immunological study of lung development in the mouse embryo. II. First appearance of the great alveolar cell, as shown by immunofluorescence microscopy. Dev. Biol. 69, 408-423.

Ten Have-Opbroek AAW. (1981). The development of the lung in mammals: An analysis of concepts and findings. Am. J. Anat. 162, 201-219.

## Discussion with Reviewers

**D. L. Luchtel:** What is the functional significance of the single central cilium of the Clara cell? How common is this? Do such cilia also appear on the differentiating type II cell?

**Author:** It has been known for some time that lung cells of the bronchial and respiratory regions possess single cilia before they differentiate [see Sorokin, S. (1968) Reconstructions of centriole formation and ciliogenesis in mammalian lungs. J. Cell Sci. 3, 207-230]. These cilia appear short and of irregular diameter and probably are nonmotile. They are lost from most of the cells when differentiation commences. Although cilia of similiar shape in other organs are thought to have a sensory function, their role in undifferentiated lung cells is not known.

**D. L. Luchtel:** What was the fixation protocol for this study?

**Author:** Lung lobes were removed from fetuses and neonates and rapidly immersed in 2.0% glutaraldehyde in 0.2 M phosphate buffer at pH 7.4. The fixed tissues were washed several times in phosphate buffer and samples for scanning microscopy were cleaved in this solution. Some samples for transmission microscopy were en bloc stained with uranyl acetate. All tissues were postfixed in 1.0% osmium tetroxide in phosphate buffer. Samples for transmission microscopy were dehydrated with ethanol, embedded in Aralite, and sections were stained with uranyl acetate and lead citrate. Samples for scanning microscopy were dehydrated with acetone, critical point dried in $CO_2$, and sputter coated with gold.

---

Fig. 14. Cellular organization 2 days after birth. Bar = 2 µm. a. Near the surface showing alveolar duct and sacs. Several pouches of the alveolar sacs are the beginnings of alveoli. The large, spread cells are type I pneumocytes; a few type II pneumocytes are identified by their surface blebs. Small blood vessels lie in the thin walls between the alveolar sacs. b. Transmission micrograph of an alveolar sac. The type II cell is surrounded by and partly covered by the type I cells. Surfactant in the form of a collection of lamellar bodies forms a mass at the type II cell surface. c. Bronchial epithelium. More ciliated cells are present than before birth. The cilia tend to circle the apex with short microvilli toward the center. The particulate material is surfactant.

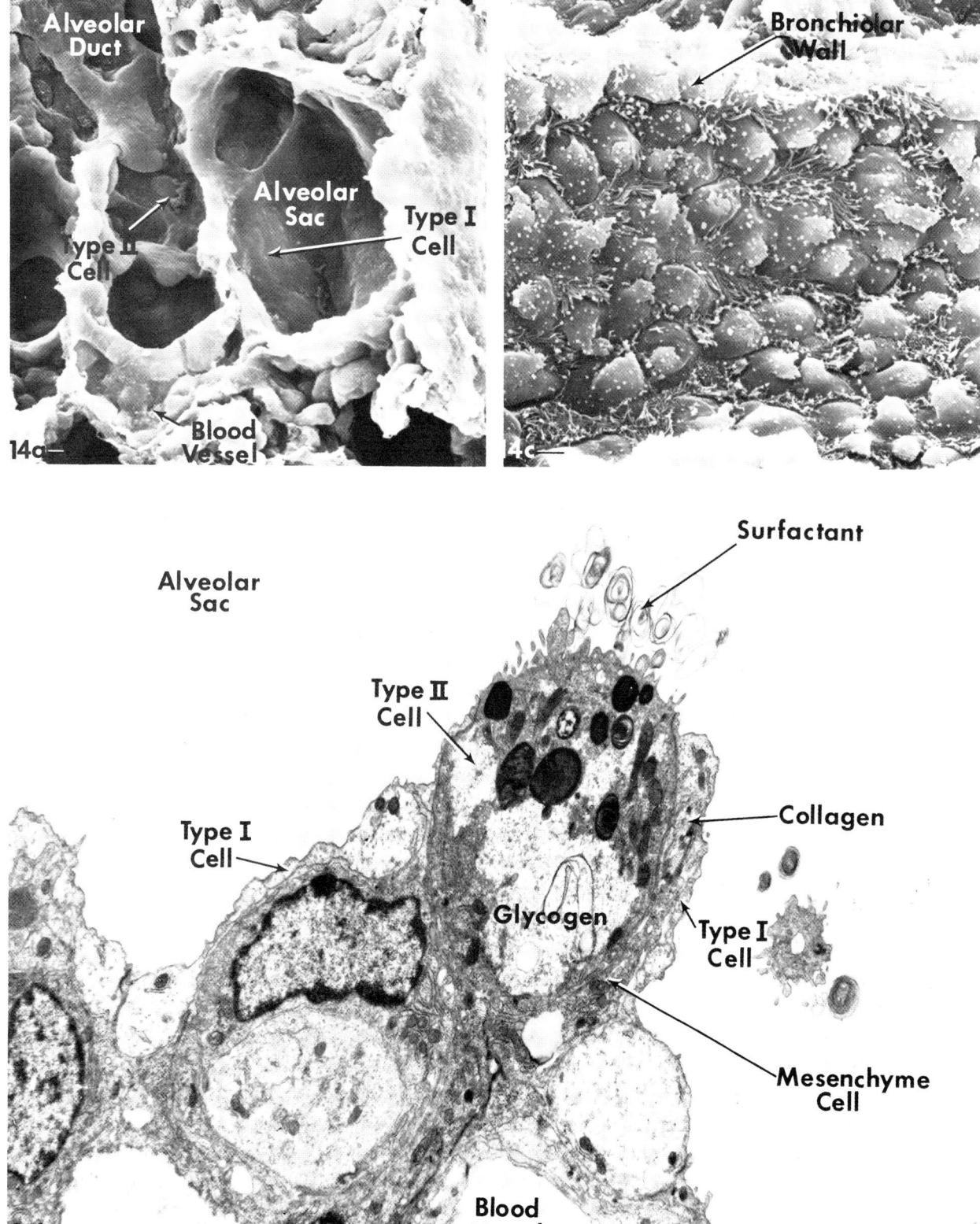

For figure legend see the facing page.

Additional discussion with reviewers of the paper "Preparation of Embryonic tissue for SEM" by R. E. Waterman continued from page 354.

T. Pexieder: Can you summarize the essential differences between the preparation of embryonic and adult tissues for SEM?
Author: The two types of tissue are quite similar, although there are few, if any, studies in which the parameters of preparation of embryonic and adult are compared directly. The preparation of fetal tissue is perhaps more similar to adult samples than very early embryonic stages. Factors to be considered with these younger stages center around their small size, rapidly changing intercellular and cell-matrix relationships, and the relative lack of extracellular matrix components which tend to stabilize many adult tissues. The experience of many investigators further suggests that the rapidly changing osmotic, cell size, and other factors affecting "good" ultrastructural preservation of embryonic tissues make control of chemical fixation more demanding than with corresponding adult tissue. Procedures for drying and rendering the sample conductive appear quite similar relative to the two classes of tissue. Exposing specific, small regions to view at desired orientations is more often a greater problem in embryos than in adult tissues which may usually be dissected or fractured in a more random way.

H.D. Geissinger: Who coined the term "filopodia"? Where do they occur, and what is their presumptive function?
Author: The term "filopodium" is a descriptive term of combined Latin and Greek origin applied to a long, slender, "thread-like" cytoplasmic projection which extends from the surface of many cell types under a variety of conditions. They contain bundles of cytoplasmic filaments, are dynamic structures, and have been associated with "exploratory behavior", cell-substrate interactions, and possibly cellular locomotion. As with similar descriptive terms applied to other types of cellular protrusion (i.e., "bleb", "lobopodium", "lamellipodium", etc.), it is not clear that the shape of the extension can be directly correlated with a single function in all cases. The use of such terms is also not consistent in the literature; hence, one investigator's "bleb" may be another person's "lobopodium". Discussion of the terminology of filopodia may be found in many references dealing with cell motility (e.g., Albrecht-Buehler, G. 1976 The function of filopodia in spreading 3T3 mouse fibroblasts. In: Cell Motility. Book A. Motility, muscle and non-muscle cells. (Goldman, R., Pollard, T. and Rosenbaum, J., eds.) Cold Spring Harbor Conferences on Cell Proliferation. Vol. 3, pp. 247-264).

T. Pexieder: What are the principal differences between preparation of embryonic tissues for TEM and SEM?
Author: Apart from the obvious need for the embedding, sectioning and staining of thin sections for TEM examination, and the drying, fracturing or dissection of intact samples, and producing a conductive surface for SEM examination, the initial stabilization of structure via physical means or chemical fixation and removal of water from the tissue (dehydration) are the preparatory steps which may be most directly compared. The literature indicates that most investigators employ chemical fixation or freeze-drying protocols for SEM which are essentially identical to those for TEM. As indicated in the text, there is some evidence that the osmolarity of fixatives for SEM should perhaps be slightly lower than those used for TEM, but this may not hold true for all cases.

T. Pexieder: Can you sketch the perspectives of further developments in the field of preparation of embryonic tissues for SEM?
Author: It is essential that continued efforts be made to improve the preservation of the overall size and shape of SEM samples, both embryonic and adult. This should provide the opportunity for obtaining more useful quantitative data, as well as providing a more accurate representation of specimen architecture. Increased attention will be paid to preserving and identifying components of the extracellular matrix and should prove useful in leading to a more complete understanding of the morphogenetic roles of these materials. It is hoped that the number of studies examining the parameters of specimen preparation in a detailed and carefully controlled way will increase thus providing more adequate information which will lead to improvements in technique. Such information should also provide a mechanism for more rational assessment of quality and will hopefully lead to improved standards. The increased application of immunologic and cytochemical labeling techniques, perhaps in combination with the use of the x-ray, backscattered electron, and other detection modes, can be expected to provide useful information of developmental significance; especially if the sensitivity and resolution of these methods and equipment continue to improve.

ASPECTS OF LIVER AND GUT DEVELOPMENT IN THE CHICK

J. Overton* and R. Meyer

Department of Biology, University of Chicago, Chicago, Illinois 60637
and Department of Pathology, St. Paul-Ramsey Medical Center, St. Paul,
Minnesota 55101

(Paper received January 16, 1984, Completed manuscript received April 30, 1984)

## Abstract

The development of two organs which form as major derivatives of the endoderm, the liver and the gut, have been followed in early stages of their formation in the chick, using scanning electron microscopy. Morphogenetic events occur in a precisely coordinated manner. We have indicated briefly in each case some of the ways these phenomena are currently being investigated. The liver forms as an outgrowth of the foregut at the level of the vitelline veins. Endoderm invades the blood vessels and forms the hepatic cords, while the liver sinusoids are of mesodermal derivation and come from the endothelium. The development of the gut also involves endodermal and mesenchymal interaction as the villi form in a geometrically regular pattern. During these processes there are changes in amount and spatial organization of cell surface junctions in epithelial cell membranes and changes in organization of extracellular materials.

KEY WORDS: Gut, Villus, Collagen, Pattern formation, Liver, Bile canaliculus, Hepatocyte, Sinusoid, Gap junction, Tight junction

*Address for correspondence:
Department of Biology, University of Chicago
1103 East 57th Street,
Chicago, Illinois 60637
Phone No. (312) 962-8946

## Introduction

A close examination of the conformational changes which occur during organogenesis raises questions concerning the mechanisms by which groups of cells change in a coordinated manner. Classical experiments have shown that cells can influence the developmental path of their neighbors, but we are still largely ignorant of the mechanisms by which this is achieved. Cells may make contact directly, or may interact by way of a complex matrix.

We will examine the normal development of the chick liver, and indicate briefly some ways in which hepatic cell contacts are currently being used to study morphogenetic problems. We will also consider cell matrix interactions in connection with the developing duodenal villi in the chick. Suggestions for further reading are given at the end of the paper.

## The Liver

We will first examine the cellular organization of the mature chick liver then proceed to illustrate a few stages involved in achieving this organization. The mature liver functions as both an endocrine and exocrine organ. These functions are reflected in its cellular organization. The hepatocytes are short (10 µm long) wedge shaped cells arranged spirally around the bile canaliculus (Figure 1). The bile canaliculus is formed by apical specializations of the hepatocytes which form microvilli on this cell surface. The bile produced by the liver is transported via the bile canaliculus to bile ducts and to the gall bladder where it is stored. When a meal is consumed bile is released from the gall bladder into the duodenum and aids in the digestion of fats. At the base of the hepatocytes are blood sinusoids lined by endothelial cells. These sinusoids provide for filtration of the blood and absorption of nutrients which are stored and later released. Drug and hormone levels in the blood are also regulated by degradation of these agents which occurs in the liver.

As the liver develops a complex series of cell interactions takes place. The first appearance of the liver primordium occurs at 22 somites (stage 7, or roughly 48 hrs of incubation).

An endodermal evagination of the margin of the anterior intestinal portal occurs where the paired omphalomesenteric veins enter the embryo. The hepatic evagination divides into primordia which proliferate, and together with the surrounding mesenchymal tissue, differentiate into hepatic cords and sinusoids. The hepatic primordia enlarge and surround the ductus venosus which eventually disappears as it is invaded and fragmented by the proliferating hepatic cords. Hepatocytes are derived from the fore gut endoderm, and the sinusoidal endothelium comes from the ductus venosus and omphalomesenteric veins. In Figure 2 the liver from a chick embryo of about 3½ days of incubation has been cracked open revealing the hollow passage of the ductus venosus lined by endothelial cells. Cavities of several forming sinusoids are evident (s) where endothelial cells (e) and hepatocytes (L) are beginning to become arranged in their mature conformation. The hepatocytes of the 3½ day chick liver are rapidly proliferating and protrude into the ductus venosus (Figure 3) and surrounding mesenchymal tissue (not illustrated). The newly emerging

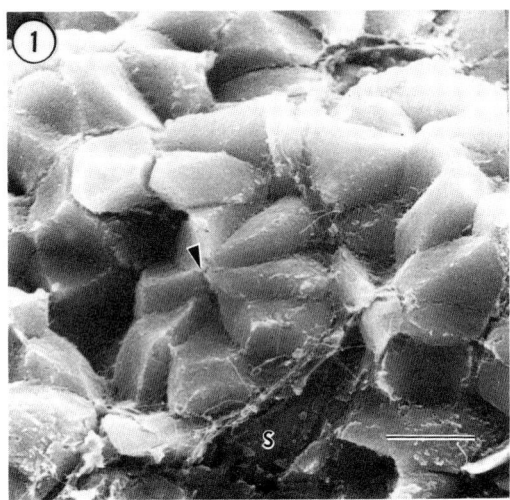

Fig. 1. This illustrates the mature organization of the chick liver. Hepatocytes are arranged spirally around the bile canaliculus (arrowhead) and surrounded at their bases by blood sinusoids (s); bar, 10 μm.

Fig. 2. This is stage 21 chick embryo liver (about 3½ days incubation) looking at the surface of the ductus venosus and the sinusoids forming (s) from it. The sinusoids are lined by endothelial cells (e) and surround forming hepatic plates or cords (L); bar, 30 μm.

Fig. 3. This is also a stage 21 chick embryo liver showing the hepatocytes arranged in hepatic buds (b) surrounded again by blood sinusoids (s); bar, 50 μm.

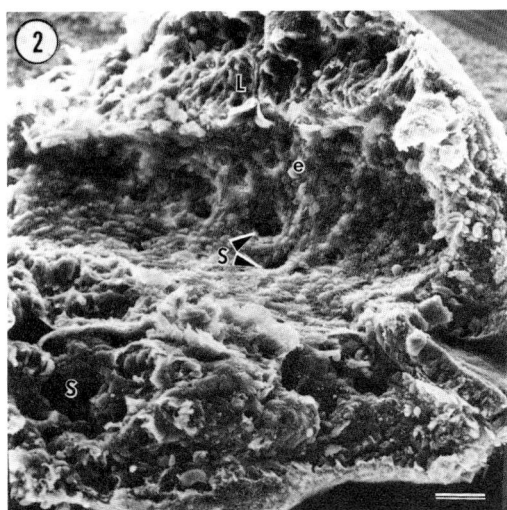

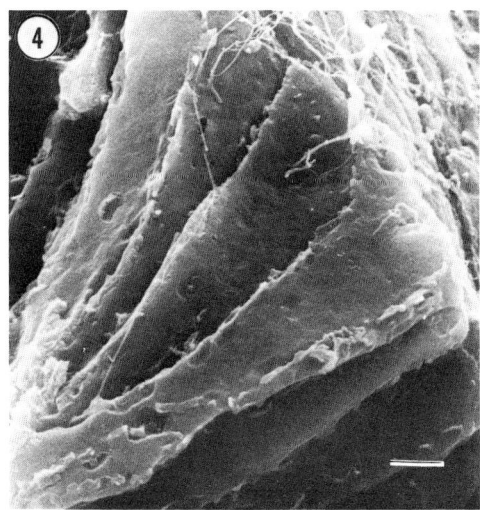

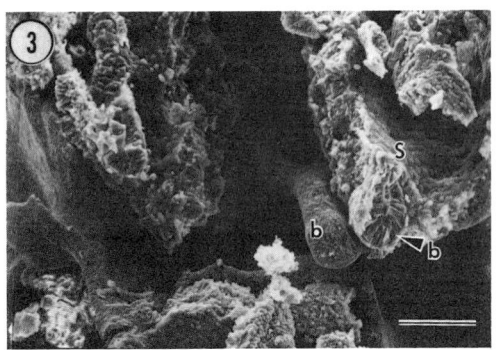

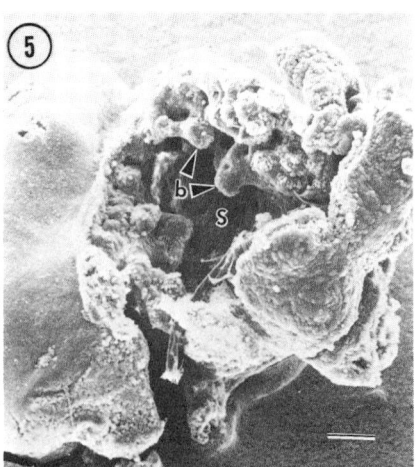

Fig. 4. The hepatocytes found in hepatic buds have a distinctive shape and are elongate cells of about 20 μm in length; bar, 1 μm.

Fig. 5. This liver is from a stage 24 chick (about 4½ days incubation) and illustrates a large central blood sinusoid (s) with emerging hepatic buds projecting into it (b); bar, 100 μm.

tissue is arranged on the form of hepatic buds (Figure 3) whose cells are elongate and about 20 μm in length (Figure 4). In the early embryonic liver large blood sinusoids are seen (Figure 5) with hepatic buds projecting into them. However, as maturation occurs the tissue becomes more compact with smaller and more numerous sinusoids (Figures 6 and 7). Thus the surface area of the blood sinusoids is increased allowing more efficient blood filtration and nutrient absorption. Hepatic buds are present only in the very immature embryonic liver. In liver of older embryonic stages, the hepatocytes are arranged in what has classically been called hepatic cords

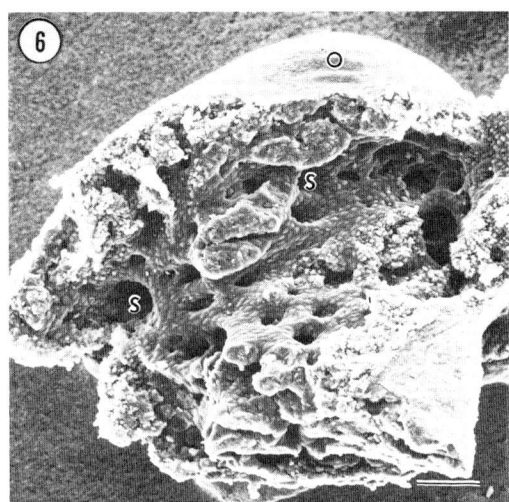

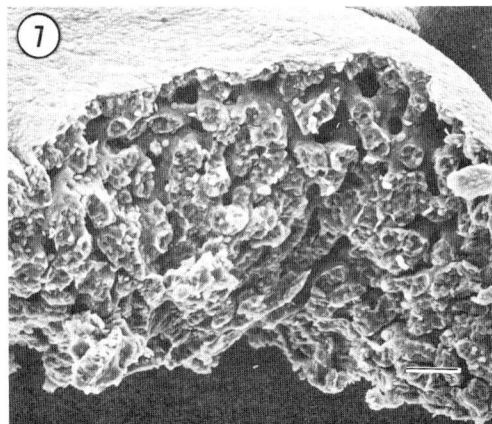

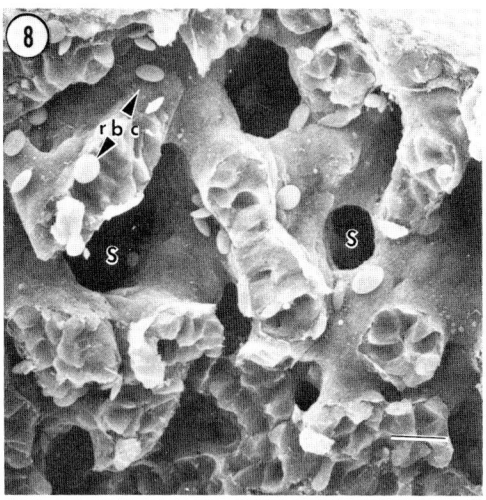

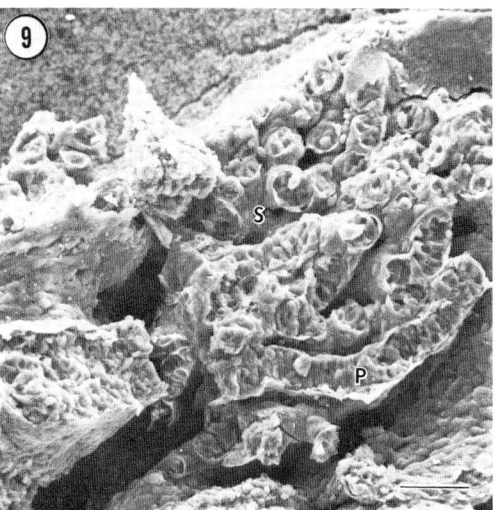

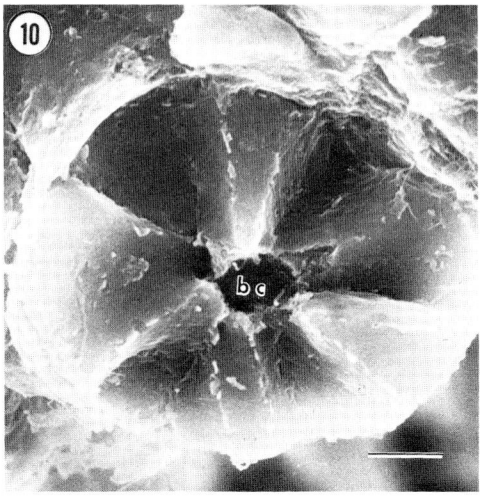

Fig. 6. This stage 25 chick embryo liver has smaller blood sinusoids (s), O, outside surface; bar, 100 μm.

Fig. 7. By embryonic stage 28 the chick liver (about 6 days incubation) is more compact than at earlier stages and the blood sinusoids are smaller and more numerous; bar, 60 μm.

Fig. 8. A stage 28 chick embryo liver is illustrated here with the hepatocytes arranged in cords surrounded by blood sinusoids (s) and red blood cells (rbc); bar, 20 μm.

Fig. 9. This also is a stage 28 chick liver with the hepatocytes viewed from a slightly different view than in Fig. 8. Here they appear to be arranged in plates (P) or sheets of cells surrounded by sinusoids (s); bar, 50 μm.

Fig. 10. This liver is from an 8 day old embryo and illustrates the shape of cells in the hepatic plates or cords. The cells are about 10 μm in length and wedge shaped with their apical cell surfaces modified to form the bile canaliculus (bc); bar, 3 μm.

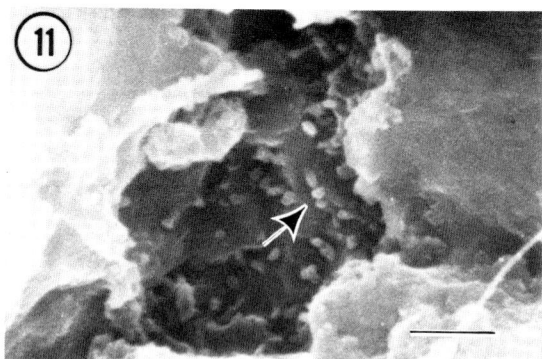

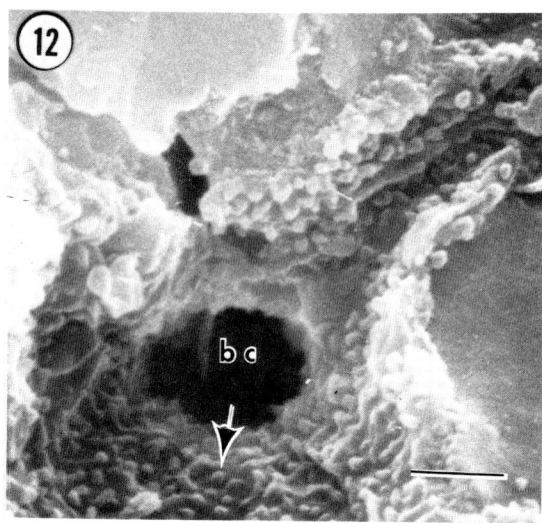

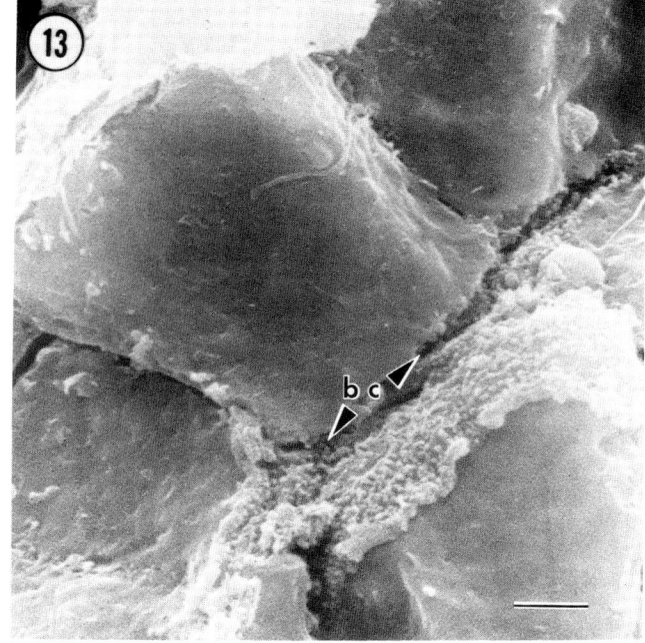

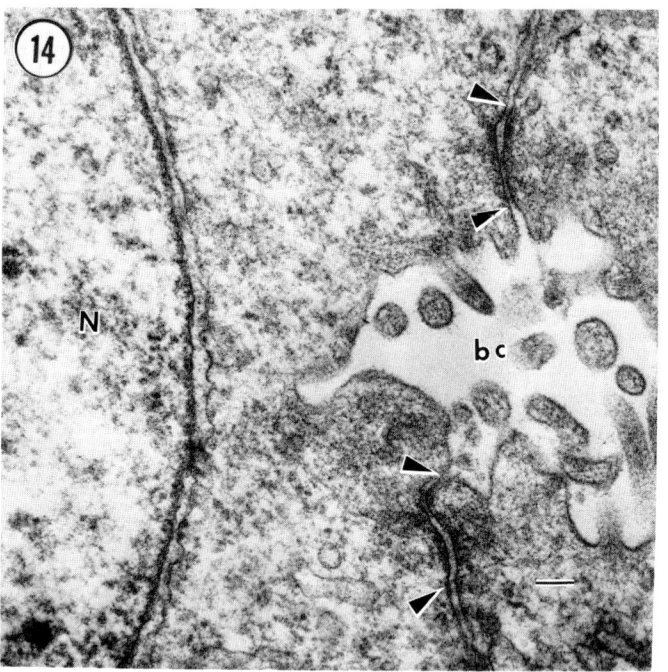

Fig. 11. This is an inside view of a bile canaliculus of a stage 28 embryo liver showing a few short microvilli (arrow); bar, 3 μm.

Fig. 12. By embryonic stage 32, the bile canalicular surface (bc) is more elaborate containing more microvilli (arrow) as is seen here; bar, 1 μm.

Fig. 13. The bile canalicular surface (bc) at embryonic stage 39 shows the numerous apical cell surface modifications of the hepatocytes forming an almost mature in appearance bile canaliculus with abundant microvilli; bar, 2 μm.

Fig. 14. Freeze-fracture micrograph of stage 28 embryo liver illustrating a loosely woven tight junction complex (arrows), lining the bile canaliculus (bc); N, nucleus; bar, 0.1 μm.

(Figure 8) in which the hepatic cells are surrounded by blood sinusoids (s) containing red blood cells (rbc). Yet depending upon how the tissue is cracked and viewed the hepatocytes can also appear to be arranged in plates (p) or sheets of cells surrounded by blood sinusoids (s) (Figure 9). As previously mentioned the mature hepatocyte is about 10 μm in length and wedge shaped (Figure 10). Thus a major change in cell shape between the hepatocytes in buds (Figure 4) and those in plates or cords (Figure 20) occurs.

The apical cell surface of the hepatocyte is elaborated to form the lining of the bile canaliculus which becomes modified during development. The number of microvilli increases as the liver matures (Figures 11, 12, and 13). The lateral hepatocyte surface also changes during development. Conspicuous differences in the amount and distribution of intercellular junctions occur. Several types of intercellular junctions are present on hepatocyte cell surfaces. Adherens junctions and desmosomes which function in cell adhesion are best seen with thin section

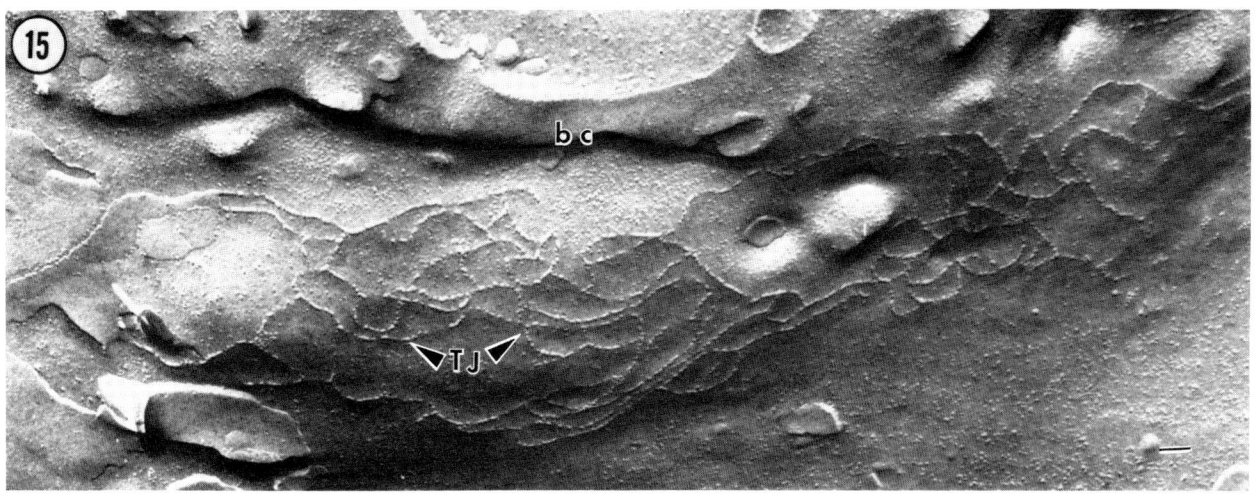

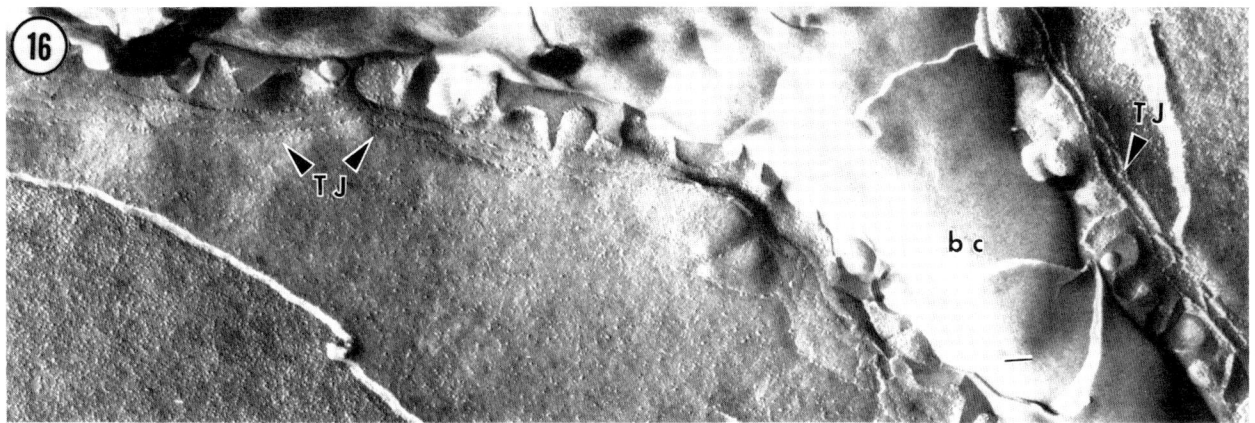

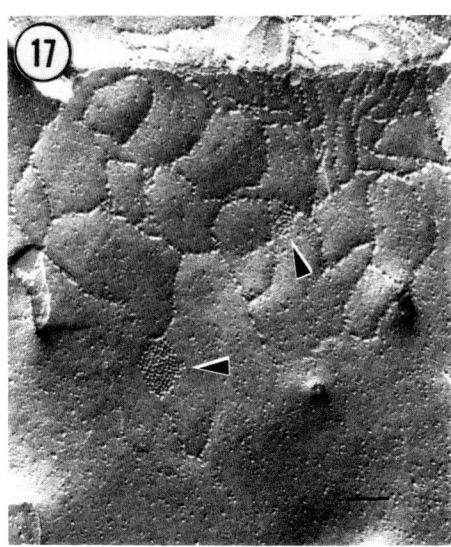

Fig. 15. Freeze-fracture micrograph of liver from a hatched chick illustrating a tight junction complex (TJ) lining the bile canaliculus (bc) where the tight junction strands run roughly parallel to each other; bar, 0.1 μm.

Fig. 16. Stage 21 (about 3 days) chick duodenum. TJ, network of tight junction strands adjacent to the gut lumen; f, fragment of linear tight junction; bar, 0.1 μm.

Fig. 17. Freeze-fracture micrograph of stage 28 embryo liver showing gap junctions intercalated into tight junctions, arrows; bar, 0.1 μm.

transmission electron microscopy and occur in the liver of the early embryo near the apical surface of the cell as part of the junctional complex (Figure 14, arrows) just as in the mature tissue.

In the liver of young chick embryos viewed by using the freeze-fracture technique and transmission electron microscopy the inside of the plasma membrane is seen and membrane specializations such as tight and gap junctions are observed. Tight junctions in the mature tissue provide a barrier at the apical surface of the cell separating the bile canaliculus from the lateral cell interspace. Tight junctions also function in establishing cell polarity by limiting the lateral movement of other membrane components from the apical membrane to the basal membrane and vice versa. Tight junction conformation changes during development going from irregular isolated strands or macular configurations to a wide loose network (Figure 15, TJ) at later stages. In the mature chick liver the tight junction strands become more extensive and regularly arranged forming a junctional complex lining the bile canalicular surface (Figure 16, TJ, bc).

Fig. 18. Stage 21 (about 3 days) chick duodenum. nTJ, network of tight junction strands adjacent to the gut lumen; f, fragment of linear tight junction; bar, 0.1 μm.

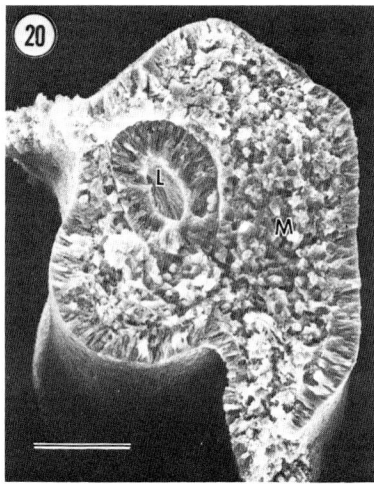

Fig. 20. Stage 28 (about 6 days). Cross section of the gut; L, gut lumen; M, mesenchyme; bar, 5 μm.
Fig. 21. Chick gut at 12 days. Arrow, ridge which has begun to show flexures; bar, 50 μm.

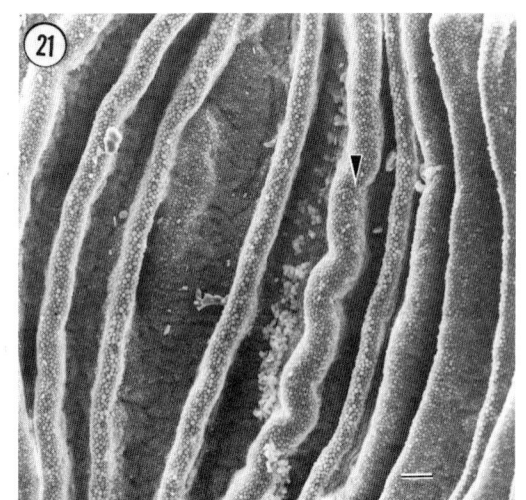

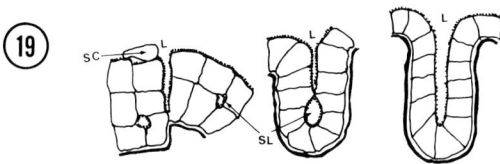

Fig. 19. Diagram of changes in intestinal epithelium as the stratified condition becomes columnar (based on Madara et al., 1981). L, lumen of the gut; SL, secondary lumen which will fuse with the lumen; SC, apical cell which is sloughed. The removal of apical cells by sloughing and the fusion of secondary lumina contribute to this developmental change.

---

Gap junctions provide intercellular communication and thus may play a vital role in cell differentiation and controlled growth by allowing signals (stimulatory or inhibitory) to pass from one cell to its neighbors. Gap junctions are most numerous in early stages of liver development near the time when morphogenesis of hepatic cords is taking place. They are dispersed over the entire lateral cell surface at this time and are characteristically included within the tight junctional network (Figure 17, arrows). Later they are largely separate from the tight junction complex and tend to be confined to the apical half of the lateral cell surface.

Microscopic studies of liver development indicate in some detail the changes in conformation which epithelial and mesenchymal cells undergo during morphogenesis as well as emphasizing differences within contacting surfaces. One aspect of this intricate process which has been studied experimentally in tissue culture is the capability of hepatocytes to adhere to each other. The plasma membrane of lateral cell surfaces not only has morphological features in the form of cell junctions which are related to adhesion as well as other functions, but also possesses integral membrane proteins associated with cell adhesion, liver cell adhesion molecules (L-CAM), which have recently been localized with the junctional complex. Hepatic cells will adhere and form liver cords in culture and this process is inhibited by antibody to L-CAM. These sorts of findings give us no real understanding as yet of morphogenetic mechanisms, but they do enable us to begin to visualize morphogenesis as resulting from a changing temporal and spatial molecular organization on the cell surface.

## The Gut

The absorptive function of the gut is promoted by the large surface area of the lumen. This area is increased in lower vertebrates by formation of straight or folded ridges, and in birds and mammals by numerous finger-like projections or villi. The single layer of epithelial cells covering these contours contains absorptive cells with a brush border of microvilli which further increases the luminal surface.

The early epithelium of the gut from which the liver primordium develops is stratified. Cells toward the basal side which contribute to the liver, have a very distinct cell surface morphology in terms of amount and distribution of

Aspects of Liver and Gut Development in the Chick

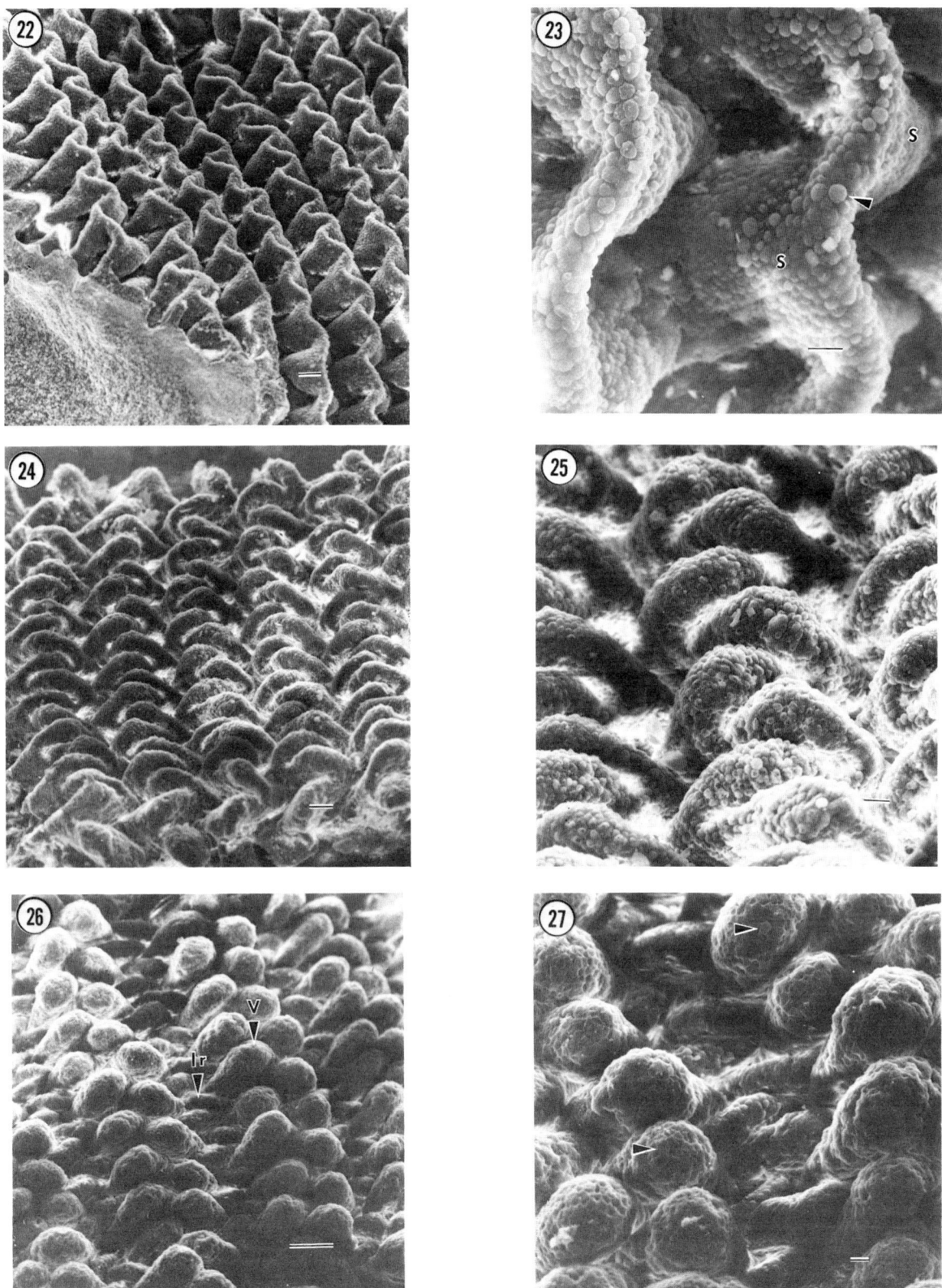

Chick gut: Fig. 22. at 15 days; bar, 50 µm. Fig. 23. at 15 days. S, swelling where villus will form; arrow, rounded cell surface; bar, 10 µm. Fig. 24. at 18 days; bar, 50 µm. Fig. 25. at 18 days; bar 10 µm. Fig. 26. at 20 days. V, villus; lr, less advanced ridge; bar, 100 µm. Fig. 27. Same as Fig. 26. Arrow, angular flattened cell surfaces; bar, 10 µm.

cell junctions when compared to cells of the gut proper, particularly those cells bordering the lumen. Thus from their earliest inception these two cell types differ. In both liver and gut, gap junctions are sparse, but tight junctions, while sparse in the liver, form an extensive network on the lateral cell surface of epithelial cells adjacent to the lumen (Figure 18, TJ). Fragmented linear tight junctions are also characteristically aligned parallel to the long axis of these polarized cells as seen in Figure 18 (f) and small tight junctions strands occur occasionally in deeper layers. This stratified primitive epithelium will transform into the simple columnar lining of the intestinal lumen covering the villi by formation of secondary lumina surrounded by junctions which then fuse with the primary one (Figure 19). In addition, some cells will be lost from the surface.

As development proceeds, the round or oval lumen of the gut (Figure 20, L) becomes modified as the folds of the first 3 to 4 previllus ridges, which start anteriorly and run the length of the intestine, form at about 8 days in the chick. By 11 days there are an average of 8 previllus ridges and as the gut increases in diameter, new ranks of ridges appear between those already formed. As a consequence, ridges of different stages of development may be seen in the same preparation as illustrated in Figure 21. In this view of the luminal surface of the 12-day duodenum the largest ridge (Figure 21, arrow) has begun to show irregular flexures. By day 15 (Figures 22 and 23) a rather regular series of zigzag flexures are characteristic. The apex of each of these flexures then becomes transformed into a villus in the course of the next several days. When these ridges are viewed from an angle which permits one to see the grooves between them, regular swellings where the future villi will form are apparent (Figure 23, S). By 18 days, the ridge has become rounded and subdivision has begun (Figures 24 and 25) and by day 20 villar protrusions are evident (Figure 26, V). At the same time, ranks of ridges which are less advanced lie between the rows of villi (Figure 26, lr). Since the definitive villi arise at each bend of the zigzag ridge (Figure 23), when they finally form, they appear as double rows (Figures 26, 27). By 20 days, the columnar epithelial cells covering the villus have well developed junctional complexes, and an extensive cytoskeleton. Alkaline phosphatase activity, characteristic of the functional condition, also increases rapidly at this time. Thus the formation of a specialized cytoplasmic organization occurs concurrently with morphogenesis. This is evident in a comparison of cell shape at 12 and 15 days (Figures 21 and 23, arrows) and 20 days (Figure 27, arrows). At earlier stages the free cell surface bulges outward in a rounded protrusion, while by 20 days, the cell surface is flattened and cell boundaries are angular. In Figure 26, note that cells in the less advanced ridges still have a rounded contour. At the earlier stage, the apical cytoplasm lacks a regular organization (Figure 28, arrows) and only a narrow apical layer of microfibrils is present, while later, the typical brush border with its extensive array of microfibrils which occupy the

Fig. 28. Apical region from 11-day intestinal epithelium; arrows, apical microfibrils; bar, 5 μm.

Fig. 29. Apical region from 20-day intestinal epithelium; arrows, apical microfibrils; bar, 5 μm.

Fig. 30. Chick gut at 15 days with mesenchymal cells exposed (see Tsai and Overton, 1976, for methods); rm, rounded cell oriented along ridge; fm, flattened mesenchyme cells in trough, m, location where villus will form; bar, 10 μm.

Fig. 31. Collagen matrix, 12 days. Collagen is largely a random meshwork at the longitudinal ridge stage (see Tsai and Overton, 1976, for methods); bar, 10 μm.

Fig. 32. Collagen matrix, 15 days; bar, 10 μm.

Fig. 33. Collagen matrix, 20 days; bar, 10 μm.

Fig. 34. Collagen matrix of villus, 20 days; bar, 10 μm.

apical surface displacing cytoplasmic organelles (Figure 29, arrows) has formed.

The development of this geometrically regular pattern of villus formation can not only be followed in surface view, but the epithelium and basal lamina can be removed enzymatically revealing the mesenchyme. Alternatively, cells can be digested completely away, leaving the collagenous components of the matrix. Figure 30 shows the organization of mesenchyme cells in the 15 day chick. Cells at the apex of the ridges are somewhat rounded and tend to be elongated parallel to the ridge, while cells in the troughs are flattened. At each flexure where we know that a villus will develop, an accumulation of rounded mesenchymal cells which in early stages appear to have little pattern in their distribution (compare Figures 20 and 30) become spatially organized as the villus pattern develops.

When the collagen matrix is viewed at successive stages after removal of cells (Figures 31, 32 and 33) it is evident that an early random network (Figure 31) changes with development as the zigzag ridges form (Figures 32 and 33) and finally becomes highly oriented between the villi, though organization of collagen in the villus proper is random (Figures 33, 34). Each villus retains two highly oriented basal extensions derived from the original zigzag ridge (Figure 33, 34).

This obvious change in spatial organization of the matrix as the villus pattern develops must both reflect the activity of the cells and have some influence on their behavior. It is clear from studies in vitro that fibroblasts placed on oriented collagen matrices can disrupt the oriented pattern, or when cultured in a collagen mesh, can orient and align collagen fibers into bundles. At the same time, there is much evidence that once aligned, a fibrous substrate tends to orient cells and may control their shape and direct their movement. It is reasonable to suppose that such interactions occur in the course of villus development and contribute in some part to the final pattern which is achieved. We are arriving at an increasingly detailed knowledge of the components of the cell matrix, that is, the collagens, glycoaminoglycans and adhesive factors such as fibronectin,

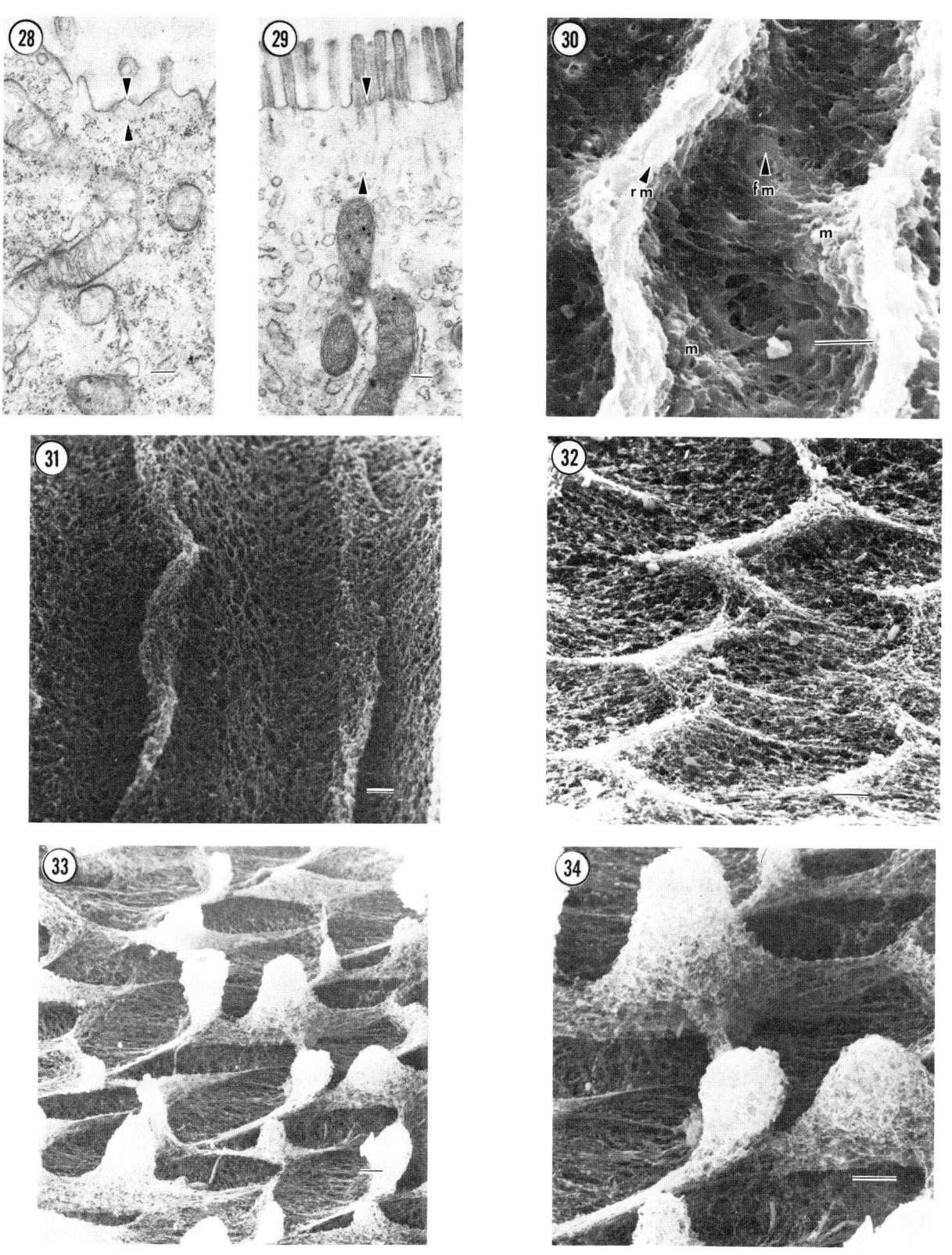

chondronectin and laminin. As more is learned about the time and locations at which these molecules are synthesized and the way they interact with each other and with the cell, we should develop a clearer picture of the controls which operate as epithelial and mesenchymal cells interact during morphogenesis and tissue patterns arise.

## Suggestions for Further Reading

### Development of the Liver

Croisille Y, LeDouarin NM (1965) Development and regeneration of the liver, in: Organogenesis. RL DeHaan, H Urspring (eds), Holt, Rinehart and Winston, NY, 421-466.

Fukuda S (1979) The development of hepatic potency in the endoderm of quail embryos. J. Embryol. Exp. Morphol. 52, 49-62.

Hamilton HL (1952) Lillie's Development of the Chick. Henry Holt & Co., Inc., NY, 224-227, 393-398.

LeDouarin MN (1975) An experimental analysis of liver development. Medical Biol. 53, 427-455.

### Liver Cells in Culture

Bertolotti R, Rutishauser U, Edelman GM (1980) A cell surface molecule involved in aggregation of embryonic liver cells. Proc. Nat. Acad. Sci. 77, 4831-4835.

Edelman GM, Gallin WJ, DeLouvee A, Cunningham BA, Theiry J-P (1983) Early epochal maps of two different cell adhesion molecules. Proc. Nat. Acad. Sci. USA 80, 4384-4388.

Lambiotte M, Vorbrodt A, Benedetti LE (1973) Expression of differentiation of rat foetal hepatocytes in cellular culture under the action of glucocorticoids: Appearance of bile canaliculi. Cell Differen. 2, 43-53.

Meyer R, Overton J (1983) Changes in intercellular junctions. I. Embryonic chick liver development. Devel. Biol. 99, 172-180.

Meyer R, Overton J (1983) Changes in intercellular junctions. II. Modulation in embryonic chick liver in vitro by cytosine arabinoside and dexamethasone. Devel. Biol. 99, 181-187.

Ocklind C, Fursun U, Obrink B (1983) Cell surface localization and tissue distribution of hepatocyte cell-cell adhesion glycoprotein. J. Cell Biol. 96, 1168-1171.

Odon S, Yoshida Noro C, Takeichi M (1983) Calcium-dependent cell-cell adhesion molecules common to hepatocytes and teratocarcinoma stem cells. J. Cell Biol. 97, 944-948.

### Development of the Gut

Burgess DR (1975) Morphogenesis of intestinal villi. II. Mechanism of formation of previllus ridges. J. Embryol. Exp. Morph. 34, 723-740.

Coulombre AJ, Coulombre JL (1958) Intestinal development: Morphogenesis of the villi and musculature. J. Embryol. Exp. Morph. 6, 402-411.

Grey RD (1972) Morphogenesis of intestinal villi. I. Scanning electronmicroscopy of the duodenal epithelium of the developing chick embryo. J. Morph. 137, 193-214.

Hinni JB, Watterson RL (1963) Modified development of the duodenum of chick embryos hypophysectomized by partial decapitation. J. Morph 113, 381-425.

Madara JL, Neutra MR, Trier JS (1981) Junctional complexes in fetal rat small intestine during morphogenesis. Devel. Biol. 86, 170-178.

Moog F (1981) The lining of the small intestine. Scientific American, November, 154-176.

Tsai L-J, Overton J (1976) The relation between villus formation and the pattern of extracellular fibers as seen by scanning microscopy. Devel. Biol. 52, 61-73.

### Cell Matrix Interactions

Bellows CG, Melcher AH, Aubin JE (1981) Contraction and organization of collagen gels by cells cultured from peridontal ligament, gingiva and bone suggest functional differences between cell types. J. Cell Sci. 560, 229-314.

Greenburg G, Hay E (1982) Epithelial cells suspended in collagen gels can lose polarity and express characteristics of migrating mesenchymal cells. J. Cell Biol. 95, 333-339.

Hay E, ed. (1981) Cell Biology of the Extracellular Matrix, Plenum Publishing Co., NY 379-405.

Kleinman H, Klebe RJ, Martin G (1981) Role of collagenous matrices in the adhesion and growth of cells. J. Cell Biol. 88, 472-485.

Tseng SCC, Savion N, Gospodarowicz D, Stern R (1983) Modulation of collagen synthesis by a growth factor and by the extracellular matrix. J. Cell Biol. 97, 803-809.

Yamada KM, Olden K, Hahn L-H (1980) Cell surface protein and cell interactions. in: The Cell Surface: Mediator of Developmental Processes. Eds. Subtelny S and Wessels NK. Academic Press, NY, 43-79.

### Discussion with Reviewers

<u>T. Pexieder</u>: Is there any difference in embryo and/or adult liver anatomy and histology between chick and human?

<u>Authors</u>: The general features of liver structure, function and development are similar throughout the vertebrates. However, there may be differences such as the degree to which the liver functions as a hematopoietic organ, the site of hematopoiesis, or the extent to which the interlobular connective tissue is developed. In the human foetus haematopoiesis is prominent in the mesodermal stroma, but by the time of birth it has been reduced to small foci. In the chick, in contrast, hematopoiesis is less important in the liver, and it is intravascular rather than extravascular.

<u>T. Pexieder</u>: Is the chick embryonic liver a haematopoietic organ as in other species? If yes, have you seen in your SEM micrographs any signs of hematopoiesis?

<u>Authors</u>: Blood flowing through the vascular system of the chick in early stages of incubation contains erythrocytes which originate exclusively in the yolk sac. However, between days 6 and 8 hematopoiesis begins in the liver, when some of the endothelial cells of the sinusoids round up and form erythroblasts which are released into the sinusoids. We have not identified unambiguous evidence of hematopoiesis using SEM.

<u>C. Erickson</u>: Do you know if villus development is dependent upon changes in the underlying mesenchyme or vice versa?

<u>Authors</u>: A number of developmental mechanisms have been proposed to account for formation of villi, which include localized cell proliferation or degeneration in the epidermis, local condensation of mesenchyme through migration and elongation of villi associated with sprouting blood vessels. Also, muscle contraction has been considered as a possible contributing factor. However, these possibilities have not been critically evaluated.

HIGHLIGHTS OF CRANIOFACIAL MORPHOGENESIS IN MAMMALIAN EMBRYOS,
AS REVEALED BY SCANNING ELECTRON MICROSCOPY

K.K. Sulik

Department of Anatomy
University of North Carolina
at Chapel Hill
School of Medicine
Chapel Hill, North Carolina
27514

G.C. Schoenwolf*

Department of Anatomy
University of Utah
School of Medicine
Salt Lake City, Utah
84132

(Paper received August 05, 1985: manuscript received October 11, 1985)

## Abstract

The craniofacial region has a complex developmental history. Some of the major morphogenetic events of this region are described for the mouse embryo, as revealed by the use of scanning electron microscopy. The three germ layers, the primitive building blocks of the embryo, are delineated during gastrulation. During neurulation, the outer ectodermal layer gives rise to the nervous system. Several other events occur concomitantly with neurulation. These include regional subdivision of the mesoderm, body folding, turning, and formation of the heart and gut. The craniofacial region, as well as a number of other regions or organ systems, undergoes major developmental events during the phase of embryogenesis called organogenesis. The visceral (branchial) apparatus develops around the primitive oral cavity and pharyngeal region of the foregut, and gives rise to many of the structures of the head and neck. The face develops from the first visceral arches and derivatives of the frontonasal prominence. The tongue also originates from tissue prominences derived from the visceral arches. The oral and nasal cavities are initially continuous broadly with one another. As a result of partitioning, paired nasal cavities develop and become largely separated from the underlying oral cavity. Scanning electron microscopy reveals much about the morphological events occurring during formation of the craniofacial region. However, emphasis in future studies should be placed on determining the underlying cellular and molecular mechanisms causing these events.

KEY WORDS: Craniofacial Development, Gastrulation, Human Embryos, Mouse Embryos, Neural Crest Cells, Neurulation, Visceral Apparatus

*Address for correspondence:
Department of Anatomy
University of Utah, School of Medicine
Salt Lake City, Utah 84132
Phone No. (801) 581-6453

## Introduction

Fertilization occurs within the oviducts of mammals and a cascade of developmental events are quickly set in motion. The single-celled zygote (Greek: zygotós, yoked), resulting from the union of the egg and sperm, undergoes a series of rapid mitotic divisions (i.e., cleavages) to form a solid ball of cells called the morula (Latin: morus, mulberry). The morula transforms into the blastocyst (Greek: blastos, germ; kystis, bladder) with formation of an internal cavity called the blastocele (blastos- + Greek: koilos, hollow). Groups of cells produced during cleavage first become specialized overtly with formation of the blastocyst. The outermost cells of the blastocyst flatten as the trophoblast (Greek: trophē, nourishment + -blastos). This layer contributes to the extraembryonic membranes supporting and protecting the embryo, as well as aiding in its nourishment and in the elimination of its waste products through interactions with maternal tissues. The innermost cells of the blastocyst concentrate at one pole of the blastocyst as the inner cell mass. These cells form embryonic tissues, chiefly if not exclusively. The scanning electron microscopic aspects of fertilization, cleavage, and blastocyst formation are illustrated by Chávez (1984).

During subsequent development, as the extraembryonic membranes form, the cells of the inner cell mass become rearranged into two layers: a thickened epiblast (Greek: epi-, upon + -blastos) and a flattened hypoblast (Greek: hypo, under + -blastos). (Some authors name these two layers differently but such variations in terminology need not concern us here.) The details of their formation and the surrounding membranes is beyond the scope of this paper. However, it is important to point out that the epiblast and hypoblast together constitute a disk-like structure called the bilaminar blastoderm (blastos- + Greek: derma, skin). Also, it should be noted that the epiblast (dorsal) side of the blastoderm is roofed by the amnion (Greek: amnion, lamb; the membrane around the fetus), an extraembryonic membrane that encases the embryo in fluid. The amnion (and other extraembryonic membranes more intimately associated with the uterus) must be removed to expose the blastoderm for examination with the scanning electron microscope.

Following the appearance of the epiblast and hypoblast, the blastoderm undergoes a series of fascinating morphogenetic events that transforms it from a flattened, two-dimensional plate to a robust, three-dimensional embryo. Some of the major events involved in this transformation will be discussed briefly below (for additional details and suggestions for further reading see Schoenwolf, 1983, and Johnston and Sulik, 1984).

The main purpose of this paper is to describe, by the use of scanning electron microscopy (SEM), some of the major morphogenetic (Greek: morphé, form; genesis, birth or origin) events involved normally in formation of the craniofacial region of mammalian embryos. Although the mouse embryo will be used most frequently as an example some human embryos are pictured as well. Emphasis will be on what normal developmental events occur, rather than how such events occur, because SEM reveals considerably more about the former than the latter.

## From Bilaminar Blastoderm to Embryo

The bilaminar blastoderm becomes trilaminar during gastrulation (Greek: gaster, stomach or belly; the term gastrulation is used because formation of the archenteron or gut is associated with this phase of development). Gastrulation is an interesting process in which certain epiblast cells migrate toward the midline, accumulate there as a longitudinal thickening called the primitive streak, then leave this streak by diving inward to one of two places (Figures 1-2). (1) Some epiblast cells migrate into the hypoblast to form endoderm (Greek: endon, within + -derma) that will ultimately line the gut. Hence, this endoderm is called embryonic (or intraembryonic), whereas the hypoblast is extraembryonic because it is displaced centrifugally by the embryonic endoderm and contributes to certain extraembryonic membranes. (2) Other epiblast cells migrate between the epiblast and forming endodermal layer as the mesoderm (Greek: mesos, middle + -derma). The remainder of the epiblast cells remain on the dorsal surface of the blastoderm as the ectoderm (Greek: ektos, outside + -derma). The blastoderm becomes trilaminar with delineation of the ectoderm, mesoderm, and endoderm--the so-called primary germ layers (Figure 3). Note that the ectoderm and endoderm are epithelial sheets in which cells abut one another closely with little intervening space (Figures 3-5). In contrast, the mesodermal cells are joined to one another more loosely and large intercellular spaces abound (Figures 3, 6-8). Thus, the mesoderm is organized as a mesenchymal cell population.

While cell migration from the primitive streak is occurring in the caudal half of the blastoderm, neurulation (Greek: neuron, nerve), another phase of embryogenesis, begins within the cranial half. Neurulation results in formation of the nervous system as its name implies. In 7-day mouse embryos (the day of fertilization is considered gestational day 0), most of the ectoderm within the cranial half of the blastoderm is thickened as the neural plate (Figures 1, 3, 9). This thickening is due to the presence of elongated (columnar) cells rather than cellular stratification. The midline of the neural plate is indicated by a shallow furrow (Figure 1) overlying a structure called the notochordal plate (Figure 9). Although the notochordal plate is now continuous with flattened cells constituting the endoderm, it will eventually separate from the endoderm and roll up into a solid longitudinal rod called the notochord (see Figure 18).

---

Figure 1. Extraembryonic membranes have been dissected away to reveal the dorsal (epiblast) surface of a 7-day mouse blastoderm. The innermost membrane that completely enclosed the epiblast previously was the amnion; its cut edge is indicated by arrowheads. The neural plate is forming from the ectoderm in the cranial half of the blastoderm while cell migration from the primitive streak is occurring more caudally. The midline furrowed region of the neural plate (arrow) marks the position of the underlying notochordal plate. The location of the primitive streak is indicated by an asterisk. Reproduced with permission from: Sulik and Johnston (1982). Bar = 100 μm.

Figure 2. Transverse slice through the primitive streak region of a mouse blastoderm similar to that shown in Figure 1. The directions of cell migrations from the epiblast (EP) through the primitive streak (PS) to form the mesoderm (M) are indicated by arrows. E, endoderm. Bar = 10 μm.

Figure 3. Transverse slice through the lateral portion of the neural plate region of a mouse blastoderm similar to that shown in Figure 1. The ectodermal cells of the neural plate (NP) are elongated dorsoventrally and constitute a pseudostratified columnar epithelium. E, endoderm; M, mesoderm; asterisk, large intercellular space. Reproduced with permission from: Sulik and Johnston (1982). Bar = 10 μm.

Figure 4. Dorsal surface of the neural plate from a mouse blastoderm similar to that shown in Figure 1. Apices of neural plate (ectodermal) cells are shaped irregularly and boundaries between cells are demarcated by short membranous ruffles. Numerous small blebs, short sparse microvilli, and filopodia-like prominences are also visible. Bar = 10 μm.

Figure 5. Ventral surface of a mouse blastoderm from a stage similar to that shown in Figure 1. Endodermal cells are less rounded where they overlie the notochordal plate (N) and cardiac mesoderm (asterisk). Arrow indicates the beginnings of the foregut. Bar = 100 μm.

Figure 6. Ventral surface of a 7-day mouse blastoderm from a stage slightly more advanced than the one shown in Figure 5. The foregut has deepened (arrow). Some of the endoderm has been removed revealing underlying cells. H, developing heart. Numbers 7 and 8 indicate areas of mesodermal cells enlarged in Figures 7 and 8. Arrowhead, midline portion of neural plate. Bar = 50 μm.

Figure 7. Enlargement of head mesodermal cells

SEM of Craniofacial Development

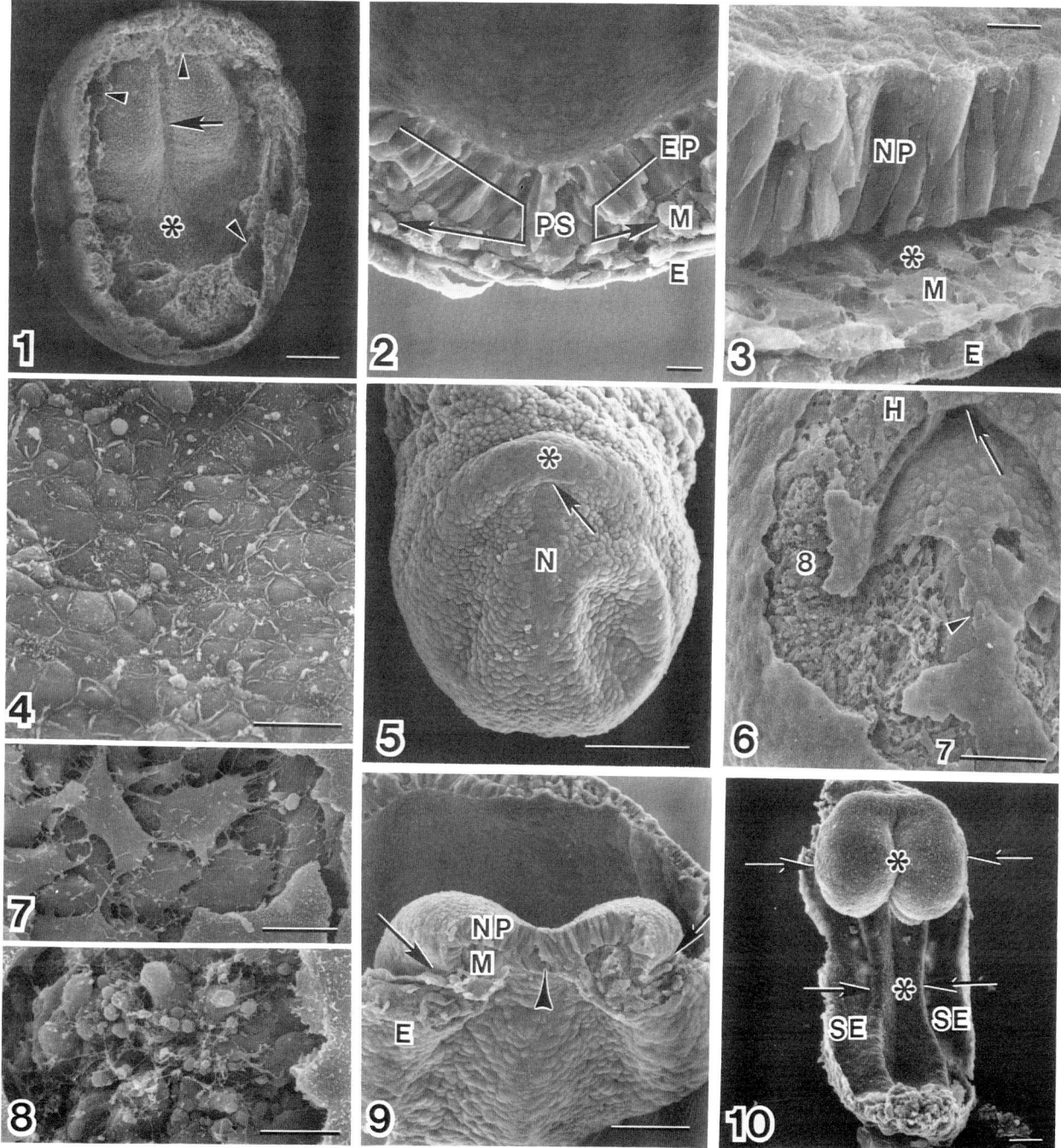

in the area indicated by the number 7 in Figure 6. Bar = 10 μm.
Figure 8. Enlargement of the mesodermal cells in the area indicated by the number 8 in Figure 6. These mesodermal cells, which would have become incorporated into the developing heart at subsequent stages, are morphologically much different from the more medial mesodermal cells. Bar = 10 μm.
Figure 9. Transverse slice through the neural plate (NP) region of a 7-day mouse blastoderm slightly more advanced than the one shown in Figure 1. The notochordal plate (arrowhead) is associated closely with the midline furrowed region of the neural plate. Mesodermal cells (M) occupy the space between the neural plate and endoderm (E). The lateral body folds (arrows) are beginning to delineate the lateral aspects of the body. Bar = 50 μm.
Figure 10. Dorsal surface of an 8-day mouse embryo. The neural groove (asterisks) is flanked by widely separated neural folds (arrows). The neural plate is much broader in the future brain region than in the spinal cord. SE, surface ectoderm. Reproduced with permission from: Sulik and Johnston (1982). Bar = 100 μm.

Shortly after its appearance, the neural plate folds or bends and its lateral edges, the neural folds, elevate above the midline furrow (Figures 10-14). This folding creates a trough called the neural groove. Each neural fold consists of an inner layer of neural plate and an outer layer of surface ectoderm; the latter is composed of flattened (squamous) cells (Figures 13-15). The inner layer contributes to the central nervous system (i.e., the brain and spinal cord), whereas the outer layer contributes to the epithelium (i.e., skin) covering the body. The cranial one-third of the neural plate, the future brain region, is much broader than the remainder of the plate, the future spinal cord (Figures 10, 12). Very early, this broader portion is subdivided into the three principal divisions of the brain: the forebrain, midbrain, and hindbrain (Figure 15).

The mesoderm formed during gastrulation also undergoes regional subdivision. Mesoderm of the cranial region is organized as loosely-packed head mesenchyme (Figure 13). More caudally in the future spinal cord region, mesoderm immediately flanking the neural groove, the paraxial mesoderm, is organized into longitudinal bands (Figure 14) that will form somites, condensed, segmental blocks of tissue. The most lateral mesoderm at this level, called the lateral plate, consists of two plate-like layers separated by a space called the body cavity or coelom.

Closure of the neural groove by fusion of the neural folds across the dorsal midline occurs concomitantly with regional subdivision of the mesoderm. Fusion of the neural folds begins in the cranial spinal cord/caudal hindbrain (i.e., cervical) region (Figure 16-17). Paraxial mesoderm flanking the closed neural tube segments into somites near the time that fusion occurs (Figures 16-18). Fusion of the neural folds is preceded by the appearance of a transitional zone between the surface ectoderm and neural plate of each neural fold (Figure 19). Cells within this zone display numerous protrusions consisting in the mouse of slender filopodia (Latin: _filum_, thread + Greek: _pous_, foot) and broad flattened lamellipodia (Latin: _lamina_, plate or leaf + -_pous_). Although the distribution, times of appearance, and morphological features of these protrusions have been described in detail (Waterman, 1976), the precise developmental role of the protrusions is still unknown. The extreme cranial and caudal regions of the neural groove are the last areas to undergo fusion so small pore-like structures (called neuropores) transiently occupy these areas (see Figures 37, 53).

As the neural folds elevate in the future brain regions of mouse embryos, cells located precisely at the junction between the surface ectoderm and neural plate on each side leave the fold (Figures 20, 21). These cells, called neural crest cells, differentiate into a diversity of cell types such as nerve cells, glial (nerve supporting) cells, pigment cells, cartilage cells, bone cells, skeletal muscle sheath cells, and smooth muscle cells (reviewed by LeDouarin, 1982). The details of neural crest cell formation and migration within the head region of rodent embryos are provided by Nichols (1981) and Tan and Morriss-Kay (1985).

Neural crest cells first appear in the trunk (spinal cord region) after the neural groove has closed to form the neural tube. Hence, trunk neural crest cells are seen first in the roof of the neural tube and then in the space subjacent to the overlying surface epithelium. Removal of the surface epithelium exposes neural crest cells as they migrate laterad from the roof of the neural tube in the sub-surface epithelial space (Figures 22-24). Migrating neural crest cells display numerous filopodia, lamellipodia, and blebs (Figure 23). Initially, the cells show an essentially random orientation within the roof of the neural tube, but as migration begins they become aligned perpendicularly to the long axis of the spinal cord, especially as they traverse the somites (Figure 24).

A number of other important developmental events occur concomitantly with neurulation. For example, folding of the blastoderm (i.e., body folding) occurs laterally, cranially, and caudally

---

Figure 11. Frontolateral view of the future brain portion of the bending neural plate from the embryo shown in Figure 10. The transition between the neural ectoderm of the neural plate and surface ectoderm is indicated by arrows. The heart is developing deep in the area indicated by an asterisk. Reproduced with permission from: Johnston and Sulik (1984). Bar = 100 μm.

Figure 12. Dorsal surface of an 8-day mouse embryo more advanced than that shown in Figure 10. B, future brain region of the neural plate; SC, future spinal cord region of the neural plate. The caudal end of the neural groove flares laterad as the sinus rhomboidalis. Persisting remnants of the primitive streak (arrow) occupy the caudal part of the sinus rhomboidalis. Numbers 13 and 14 indicate levels similar to those shown in Figures 13 and 14. Reproduced with permission from: Sadler (1985). Bar = 100 μm.

Figure 13. Transverse slice through a level similar to that indicated by the number 13 in Figure 12. NP, neural plate; SE, surface ectoderm; asterisk, neural fold; arrow, neural groove; N, notochordal plate; HM, head mesenchyme; FG, foregut. Bar = 10 μm.

Figure 14. Transverse slice through a level similar to that indicated by the number 14 in Figure 12. The lateral walls of the neural groove are elevated fully at this level and fusion of the neural folds (asterisks) would have occurred shortly. C, coelom; E, endoderm; LP, lateral plate mesoderm; NG, neural groove; P, paraxial mesoderm; SE, surface ectoderm. Reproduced with permission from: Sadler (1985). Bar = 50 μm.

Figure 15. Enlargement of the frontolateral region of a mouse embryo similar to that shown in Figure 12. FB, forebrain; MB, midbrain; HB, hindbrain; H, heart (covering layers removed partially); SE, surface epithelium; arrow, opening of foregut (cranial intestinal portal). Bar = 100 μm.

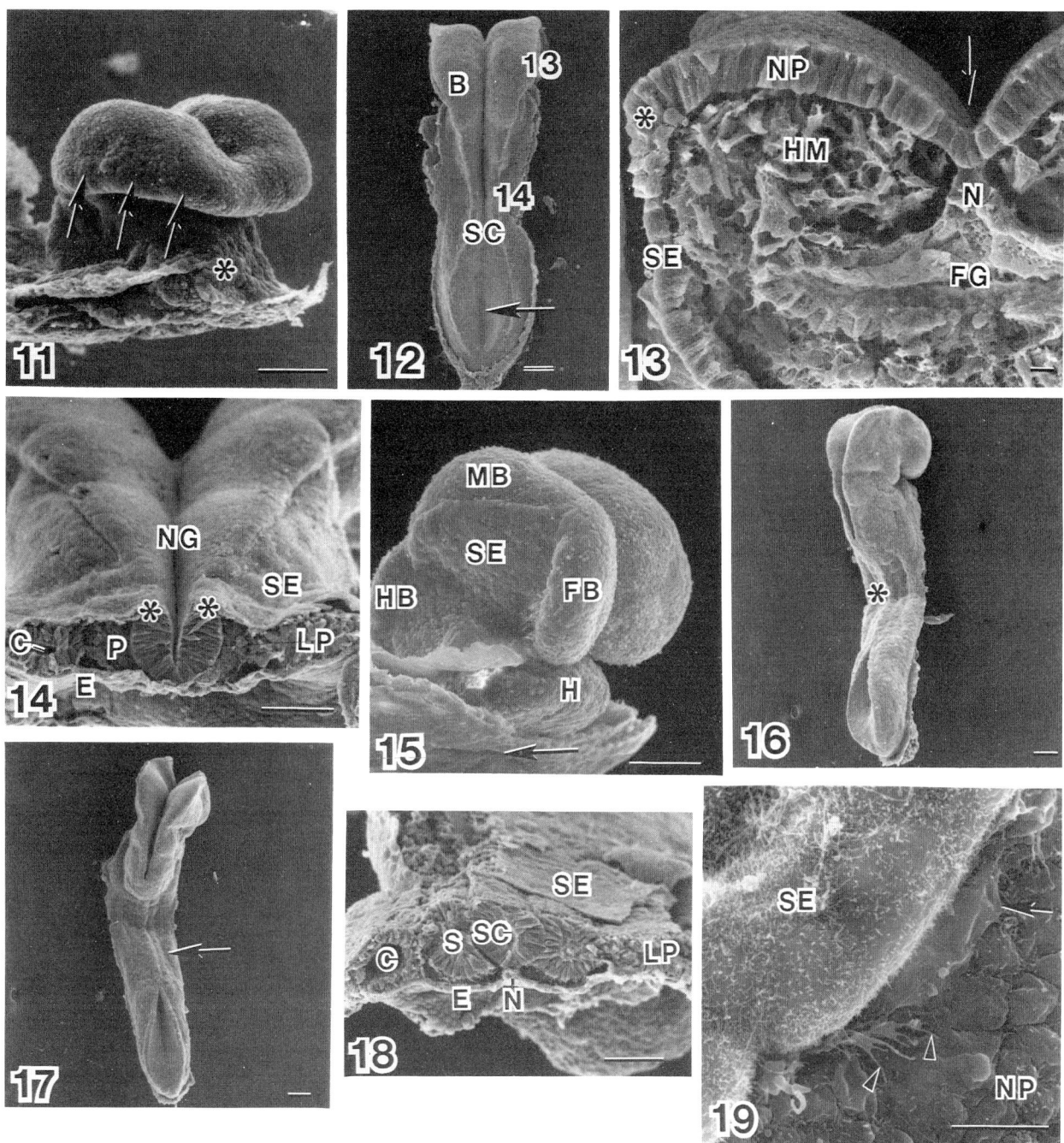

Figure 16. Dorsolateral view of an 8-day mouse embryo. The neural groove has closed in the caudal hindbrain/cranial spinal cord levels (asterisk). Bar = 100 μm.

Figure 17. Dorsal view of the mouse embryo shown in Figure 16. The neural groove is open throughout much of the brain and caudal spinal cord (sinus rhomboidalis). Arrow indicates a level similar to that shown in Figure 18. Reproduced with permission from: Sadler (1985). Bar = 100 μm.

Figure 18. Transverse slice through a level similar to that shown by the arrow in Figure 17. The surface ectoderm (SE) has been removed largely. SC, spinal cord; S, somite; N, notochord; E, endoderm; LP, lateral plate mesoderm; C, coelom. Reproduced with permission from: Sadler (1985). Bar = 50 μm.

Figure 19. Transitional zone between the surface ectoderm (SE) and neural plate (NP) components of a neural fold from the brain region of an 8-day mouse embryo. Delicate filopodia (arrowheads) and broad lamellipodia (arrow) extend from the transitional zone toward the approaching contralateral fold (not shown). Bar = 10 μm.

to separate the developing embryo from its extraembryonic membranes. Such foldings, called the lateral (Figure 9), cranial or head (Figures 25, 26), and caudal or tail (Figures 25, 26) body folds, respectively, create what is known as the vertebrate tube-within-a-tube body plan. In essence, the vertebrate embryo consists of an outer epithelial (ectodermal) tube facing the external environment, and an inner epithelial (endodermal) tube. The latter tube, the gut, takes in and digests nutritive substances and excretes undigested waste products. Formation of the gut involves all three types of body folds. The foregut is formed principally through the action of the cranial fold, the midgut, by the paired lateral folds, and the hindgut, by the caudal fold (Figures 25, 26).

A unique process similar to body folding occurs between 8 and 9 days of gestation in the mouse. During this process, called turning, the embryo is inverted. At the beginning of 8 days of gestation the embryo is shaped like a "U" (Figure 10) with its head and tail projecting dorsad and forming the two limbs of the "U". Turning has occurred by 9 days such that the embryo is distinctly "C"-shaped at this time (Figure 27). Now, the head and tail again form the two limbs of the "C", but they project ventrad rather than dorsad. Note that approximately one-half of the "C"-shaped embryo consists of the head and neck. The remainder consists of the trunk (beginning at the level of the caudal aspect of the heart bulge) and a short tail.

In addition to body folding and turning, the heart forms during neurulation. It is initially a crescent-shaped rudiment located just cranial to the incipient foregut (Figures 5, 6). Due to cranial and lateral body folding, the heart becomes situated more caudally with formation of the foregut, and the caudal limbs of the crescent-shaped rudiment fuse together as a broad midline tube or sac (Figure 25). This primitive tube elongates craniocaudally, thins, and loops upon itself during subsequent development (Figure 26), and is eventually partitioned by ingrowths from its walls into four separate chambers. Morphogenesis of the heart tube is described in detail by Hay et al. (1984).

## Craniofacial Morphogenesis

### Overview

The craniofacial region has a complex developmental history. Major events in its formation occur during the phase of embryogenesis called organogenesis. Our approach will be first to describe the visceral (branchial or pharyngeal) apparatus, major building blocks of the head and neck. We will then discuss development of the face, including that of the external ears and eyes, and partitioning of the oral and nasal cavities.

### Development of the Visceral Apparatus

The visceral or branchial (Greek: branchia, gills; structures associated with gill slits of lower vertebrates) arches are a series of epithelia-covered, mesenchymal bars, which arch around the lateral and ventral aspects of the primitive oral cavity and pharyngeal portion of the foregut. The visceral arches are delineated from one another externally by ectodermal grooves (Figures 27, 28, 32) and internally by pharyngeal pouches (Figures 31, 32, 33). The arches develop temporally in a cranial to caudal sequence. Initially, the first arch appears on each side as a straight bar of tissue located caudal to the ectoderm-lined primitive oral cavity or stomodeum (Figure 27). Later, at the dorsal end of the arch, a growth center called the maxillary prominence develops (see Figure 54). Growth of this prominence above the ventral aspect of the arch (termed the mandibular prominence) causes the first arch to look somewhat like an inverted "U" (see Figures 42, 48). Each limb of the "U" is important in development of the face with the cranial limb, the maxillary prominence,

---

Figure 20. Enlargement of the frontolateral region of a mouse embryo similar to that shown in Figure 12. The surface ectoderm has been removed partially to reveal underlying neural crest cells (asterisks). H, heart (covering layers removed partially). Box indicates area enlarged in Figure 21. Bar = 50 μm.

Figure 21. Enlargement of neural crest cells from the boxed area shown in Figure 20. Micrograph was rotated about 45° clockwise. Bar = 10 μm.

Figure 22. Dorsal surface of the recently formed neural tube from the future spinal cord region of a 8-day mouse embryo. The surface ectoderm (SE) has been removed partially. Neural crest cells are migrating laterad from the roof of the neural tube toward the paraxial mesoderm (PM). Box indicates area enlarged in Figure 23. Bar = 50 μm.

Figure 23. Enlargement of the boxed area in Figure 22. Neural crest cells display blebs, filopodia, and lamellipodia. Bar = 5 μm.

Figure 24. Dorsolateral surface of the future spinal cord region from a 9-day mouse embryo slightly more advanced than the one shown in Figure 22. The surface ectoderm (SE) has been removed partially. Somites (S) are forming within the paraxial mesoderm. Arrows, neural crest cells. Bar = 50 μm.

Figure 25. Ventral view of an 8-day mouse embryo similar to that shown in Figure 12. The foregut and hindgut have formed through the action of the cranial and caudal body folds, respectively. The now open midgut (MG) will be closed eventually through the action of the lateral body folds. FB, forebrain region; H, heart (covering layers removed partially); arrow, the foregut (cranial intestinal portal); arrowhead, opening of the hindgut (caudal intestinal portal). Bar = 100 μm.

Figure 26. Ventral view of an 8-day mouse embryo similar to that shown in Figure 17. The heart (H) is beginning to loop (covering layers removed partially). The foregut and hindgut have formed through the action of the cranial and caudal body folds, respectively. The now open midgut (MG) will be closed eventually through the action of the lateral body folds. Arrow, cranial intestinal portal; arrowhead, caudal intestinal portal; HG, floor of the hindgut. The neural groove is still

SEM of Craniofacial Development

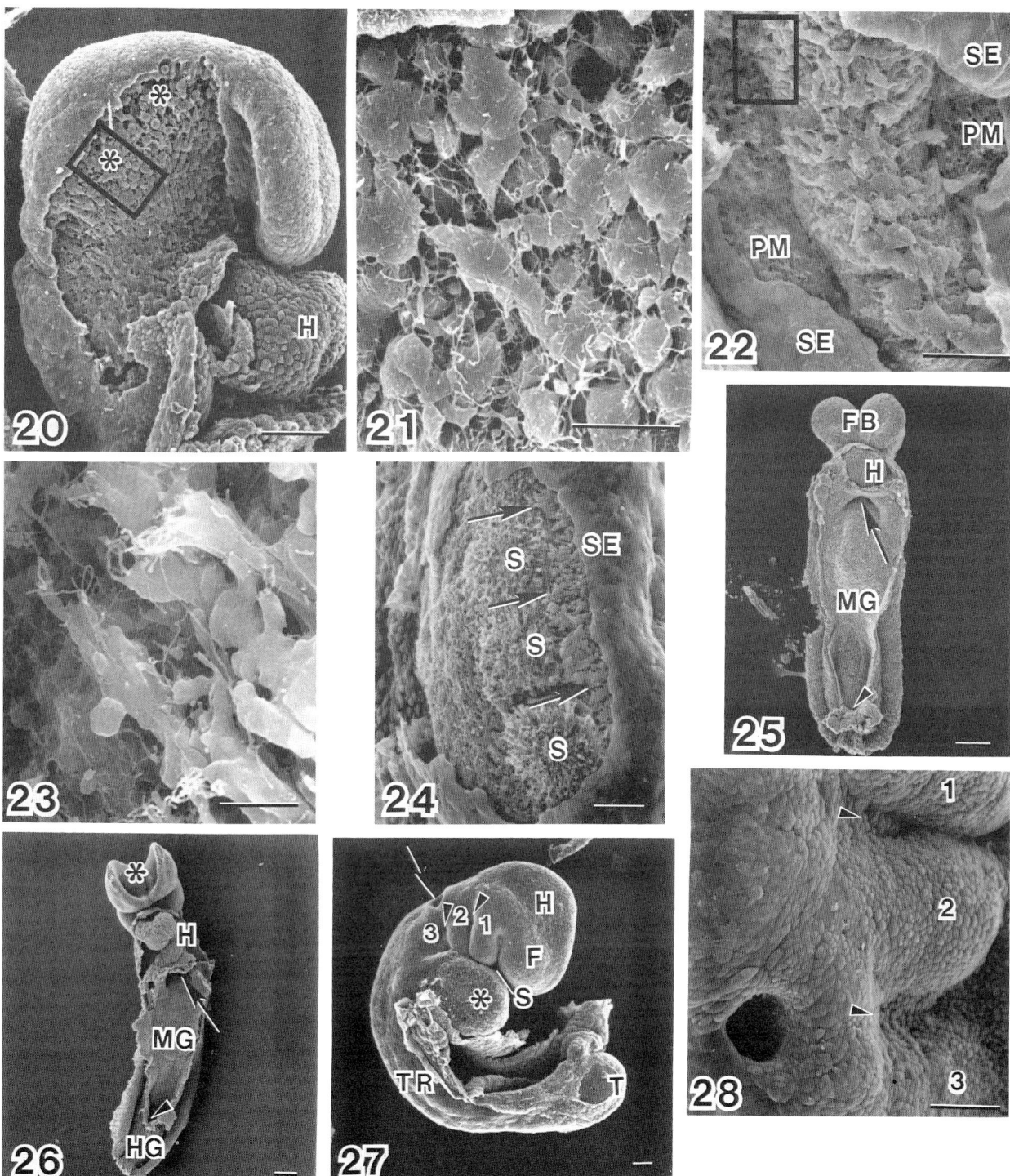

open throughout most of the brain region (asterisk). Reproduced with permission from: Sadler (1985). Bar = 100 μm.

Figure 27. Lateral view of a 9-day mouse embryo. The neural groove has closed completely and the body has become "C"-shaped through turning. H, head; TR, trunk; T, tail; asterisk, location of the heart; arrow, auditory (otic) pit; 1-3, visceral arches 1-3; F, frontonasal prominence; S, stomodeum; arrowheads, first and second visceral grooves. Bar = 100 μm.

Figure 28. Enlargement of the upper left portion of Figure 27; the enlargement has been rotated 90° counterclockwise. The cells lying dorsal to the auditory (otic) pit are less rounded than those lying cranial, caudal, and ventral to the pit. 1-3, visceral arches 1-3; arrowheads, first and second visceral grooves. Bar = 50 μm.

contributing to the upper jaw, and the caudal limb, the mandibular prominence, to the lower jaw (see below). The mandibular prominence is separated from the second visceral arch by the ectoderm-lined first visceral groove, and the second arch is separated from the third arch by the second visceral groove. Six pairs of visceral arches develop in mammalian embryos, but the fifth are rudimentary. It is difficult to discriminate the fourth-sixth arches from one another externally because their boundaries are indistinct.

The auditory (or otic) pit (or vesicle) is located dorsal to the second visceral groove (Figures 28, 29). Earlier, this pit began as a flattened ectodermal thickening, the auditory placode. This placode forms as a result of a tissue interaction (induction). The hindbrain and adjacent head mesenchyme induce the overlying surface ectodermal cells to remain elongated. Subsequent invagination or infolding of the placode forms the pit, continuous with the adjacent surface ectoderm. Later, the pit will constrict free from the surface ectoderm forming a closed vesicle, which will differentiate into the entire inner ear. For details of this process see Streeter (1907) and Waterman (1938).

As mentioned above, the visceral arches are demarcated from one another both externally and internally. To view internal boundaries with scanning electron microscopy, intact embryos must be cut into slices (Figures 30-36). The initially simple structure of a visceral arch can be appreciated readily in such slices. Each arch consists of a mesenchymal core (derived from primitive streak mesoderm and neural crest ecto-mesenchyme) covered by two epithelial sheets: an internal sheet of pharyngeal endoderm and an external sheet of surface ectoderm (Figures 30-32). Pharyngeal pouches expand laterad between adjacent arches (see especially Figures 31, 32, 35, 36) directly opposite correspondingly numbered visceral grooves. (Details on development of the pharyngeal pouches are given by Hilfer and Brown, 1984.) The area of apposition between the ectoderm of each visceral groove and the endoderm of each pharyngeal pouch is called a closing plate (branchial membrane). Such plates may rupture temporarily (see Kaufmann et al., 1981; Mangold et al., 1981; and Waterman, 1985, for details of closing plate formation and rupture) forming clefts, which correspond to the gill slits in lower forms.

As the visceral apparatus is developing, an area of apposition forms between gut endoderm and stomodeal ectoderm at the cranial end of the foregut. This structure, the oral (buccopharyngeal, pharyngeal, or oropharyngeal) membrane, quickly becomes perforated (Figures 31 and 33), largely through active cell rearrangements and detachments of intercellular junctions, but also through localized cell death. The oral membrane is eventually removed completely allowing continuity of the primitive oral cavity and foregut (Figures 34-36) (See Waterman, 1977; Waterman and Schoenwolf, 1980; Watanabe et al., 1984; and Waterman and Kao, 1982, for details).

The stomodeal ectoderm and foregut endoderm are continuous around this opening and form the lining of the mouth and oropharynx.

Development of the Face and Related Structures

The face develops from the five facial prominences that surround the stomodeum: the frontonasal prominence and the paired maxillary and mandibular prominences. Recall that the maxillary and mandibular prominences are subdivisions of the first visceral arches (Figures 27, 37). The frontonasal prominence appearing with closure of the cranial neuropore surrounds the developing forebrain (Figure 37). The ventral forebrain is responsible for the induction of nasal (olfactory) placodes, thickened areas of surface ectoderm on the ventrolateral aspects of the frontonasal prominence (Figure 38). As these placodes invaginate or fold inward to form the nasal pits, the surrounding tissue forms a raised ridge or nasal prominence (Figure 38, 39). Those prominences on the outer rim of the placodes are the lateral nasal prominences, whereas those closer to the midline are the medial nasal prominences. The portion of the frontonasal prominence that is not involved in the formation of the nasal prominences is now referred to as the frontal prominence. Note that each nasal prominence consists of a mesenchymal core, derived principally from neural crest cells, and an epithelial covering of surface ectoderm (Figure 40). Initially, the cavity of each nasal pit is continuous broadly with the stomodeum (Figures 38, 39, 40).

Formation of the face involves the union of various facial prominences by way of two distinct developmental events--merging or fusion. During merging, two adjacent prominences are united by an upheaval of the floor of the valley separating them. This upheaval could result from rapid cell proliferation within the floor of the valley, active migration of cells into this area, or accumulation of extracellular materials. The overlying epithelium remains intact as mergence occurs. The union of certain other facial prominences (namely the union of the medial nasal prominence with the maxillary and lateral nasal prominence on each side, or the union of the secondary palatal shelves; see below) involves fusion. Fusion differs from merging in that the free ends of adjacent prominences are brought into contact and a double-layered epithelial sheet is trapped between the mesenchymal cores of the two prominences. This double-layered sheet is removed subsequently allowing the underlying mesenchymal cells to intermingle. The epithelium is removed subsequently either by cell death or by its transformation into mesenchymal cells.

---

Figure 29. Slice through an auditory (otic) pit similar to that shown in Figure 28. SE, surface epithelium; HB, hindbrain. Bar = 10 μm.
Figure 30. Lateral view of a frontal slice through an embryo similar to that shown in Figure 27. S, stomodeum; 1-3, visceral arches 1-3; arrowheads, first and second visceral grooves; asterisk, location of the heart. Reproduced with permission from: Johnston and Sulik (1980). Bar = 100 μm.

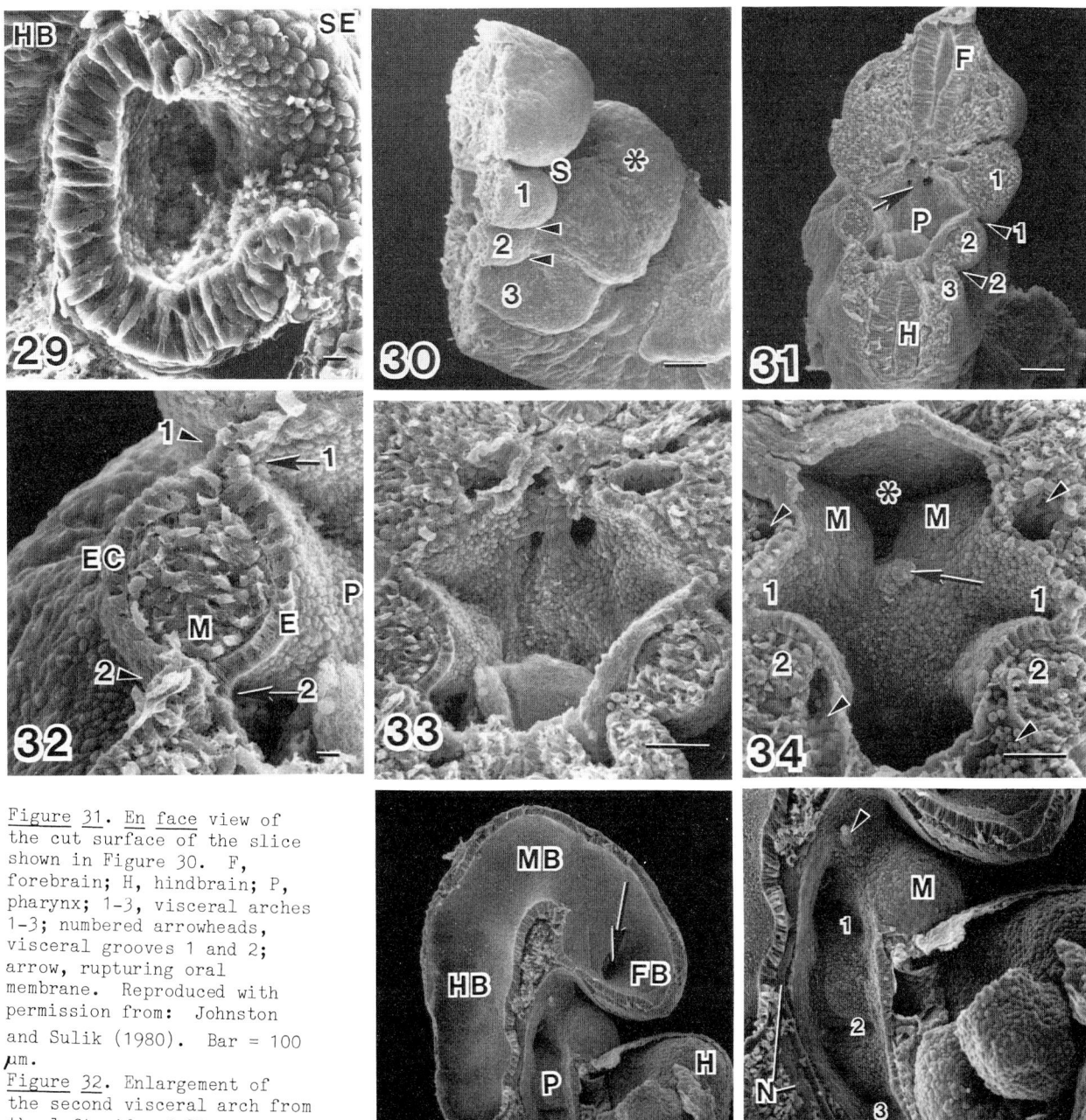

Figure 31. En face view of the cut surface of the slice shown in Figure 30. F, forebrain; H, hindbrain; P, pharynx; 1-3, visceral arches 1-3; numbered arrowheads, visceral grooves 1 and 2; arrow, rupturing oral membrane. Reproduced with permission from: Johnston and Sulik (1980). Bar = 100 μm.

Figure 32. Enlargement of the second visceral arch from the left side of Figure 31. EC, ectoderm; E, endoderm; M, mesenchymal core; P, pharynx; numbered arrowheads, visceral grooves 1 and 2; numbered arrows, pharyngeal pouches 1 and 2. The first and second closing plates lie between the correspondingly numbered pouches and grooves. Bar = 10 μm.

Figure 33. Enlargement of the pharyngeal region of Figure 31. Bar = 50 μm.

Figure 34. Enlargement of the pharyngeal region from a frontal slice similar to that shown in Figure 31. Rupture of the oral membrane is largely completed forming the opening between the stomodeum and pharynx (asterisk). Arrow, remnant of the oral membrane; arrowheads, aortic arches; M, mandibular prominence of first visceral arches; 1, first pharyngeal pouches; 2, second visceral arches. Bar = 50 μm.

Figure 35. Sagittal slice through a 9-day mouse embryo. FB, forebrain; MB, midbrain; HB, hindbrain; H, heart (covering layers partially removed); P, pharynx; arrow, optic stalk. Bar = 100 μm.

Figure 36. Enlargement of the pharyngeal region of Figure 35. M, mandibular prominence of first visceral arch; N, notochord; 1-3, first-third pharyngeal pouches; arrowhead, remnant of oral membrane in mouth opening. Bar = 100 μm.

The upper jaw is formed from four prominences: the paired medial nasal prominences and the paired maxillary prominences. The medial nasal prominence on each side fuses laterally with the adjacent maxillary prominence and lateral nasal prominence (Figures 41-44). This fusion separates the cavities of the nasal pits and stomodeum from one another and forms the external (or anterior) nares or nostrils. The paired medial nasal prominences then gradually merge with one another across the midline forming the philtral region of the upper lip, the intermaxillary segment of the upper jaw, and the tip of the nose. Concomitantly, the maxillary prominences enlarge considerably to form the cheeks and maxillae. Formation of the upper jaw and cheeks of mouse and human embryos involves similar events (cf. Figures 41-44 with Figures 45-47).

The nose forms from five facial prominences. The bridge derives from the frontal prominence along with the forehead (Figure 44). The crest and tip of the nose derive from the merged medial nasal prominences, whereas the sides (alae) derive from lateral nasal prominences (Figures 45, 46). Merging also occurs lateral to the nose between the lateral nasal prominence and maxillary prominence on each side (Figure 45). At this juncture on each side, an epithelial cord of cells, which extends from the medial corner of the eye to the external naris, sinks inward where it cavitates (i.e., undergoes canalization) to form the nasolacrimal duct.

The lower jaw has a much simpler developmental history than the upper jaw or nose. It derives from paired mandibular prominences which merge across the midline (Figures 38, 39, 41, 43, 44).

As the face is forming from the facial prominences, a series of auricular hillocks appear around the first visceral groove (Figures 48-50). There are generally considered to be 6 hillocks on each side, three derived from the mandibular prominence of the first visceral arch and three from the second arch. These hillocks give rise to the external ear or pinna (Figure 51). The first visceral groove on each side forms the external auditory meatus (external ear canal), the first pharyngeal pouch forms the tympanic (middle ear) cavity and the tubotympanic recess (Eustachian tube), and the first closing plate (which is later infiltrated with mesenchyme derived from neural crest cells) forms the tympanic membrane (ear drum). The middle ear ossicles and their ligaments are derived from cartilages (originating from neural crest cells) of the first and second visceral arches.

Formation of the eyes begins before closure of the neural groove in the future forebrain region and precedes formation of the face. As the neural folds of the forebrain region are elevated toward the dorsal midline, a pair of depressions form, one associated with each neural fold (Figure 52). These depressions, the optic sulci, constitute the first rudiments of the eyes. With subsequent development, the neural folds of the future forebrain region approach one another (Figure 53), eventually obliterating the cranial neuropore, and the optic sulci expand laterad as the optic vesicles (Figures 53-55). Each optic vesicle retains a connection to the forebrain (diencephalon level) called the optic stalk.

Each optic vesicle undergoes a process of invagination or inpocketing after maximal evagination occurs, resulting in a two-layered structure called the optic cup (Figures 56, 57). The thinner of the two layers of the optic cup forms the pigmented epithelium (pigmented retina) of the eye and the thicker layer, the neural (nervous) retina and iris.

The evaginating optic vesicle (and the invaginating optic cup), induces the overlying surface epithelium to thicken as the lens placode (Figure 56). The lens placode invaginates concurrently with invagination of the optic vesicle forming a lens pit or vesicle (Figure 57). The lens pit is initially continuous broadly with the adjacent surface epithelium, but as invagination of the pit continues, the region of continuity narrows (Figures 58) and is eventually obliterated. Now, a second embryonic induction occurs, this time between the lens vesicle and overlying surface epithelium. As a result of this, the surface epithelium differentiates into the epithelium of the cornea. The corneal stroma is derived from mesenchymal neural crest cells, which invade the space between the lens and overlying epithelium (Figures 59, 60). The cells composing the walls of the spherical lens undergo regional changes as the cornea is forming. The outermost cells thin as lens epithelium, while the innermost cells elongate as lens fibers (Figure 59). The cavity of the lens vesicle is obliterated ultimately as the innermost cells continue to elongate. Futher details on eye development can be found in Hilfer (1983).

---

Figure 37. Frontal view of the head of a 9-day mouse embryo. The anterior neuropore (arrow) has just closed. M, mandibular prominences of first visceral arches; X, maxillary prominences of first visceral arches; F, frontonasal prominence; S, stomodeum; asterisk, area from which the heart was removed. Bar = 100 μm.

Figure 38. Frontal view of the head of a 10-day mouse embryo. F, frontonasal prominence; N, nasal placodes; S, stomodeum; M, mandibular prominence of first visceral arch; X, maxillary prominence of first visceral arch; 2, second visceral arch. Reproduced with permission from: Sulik (1983). Bar = 100 μm.

Figure 39. Frontal view of the head of a 10-day mouse embryo more advanced than that shown in Figure 38. F, frontal prominence; M, medial nasal prominence; L, lateral nasal prominence; MAN, mandibular prominence of first visceral arch; X, maxillary prominence of first visceral arch; S, stomodeum; arrow, Rathke's pouch (rudiment of the anterior pituitary gland; forms by evagination of the stomodeum); arrowheads, nasal pits. Reproduced with permission from: Sulik and Johnston (1982). Bar = 200 μm.

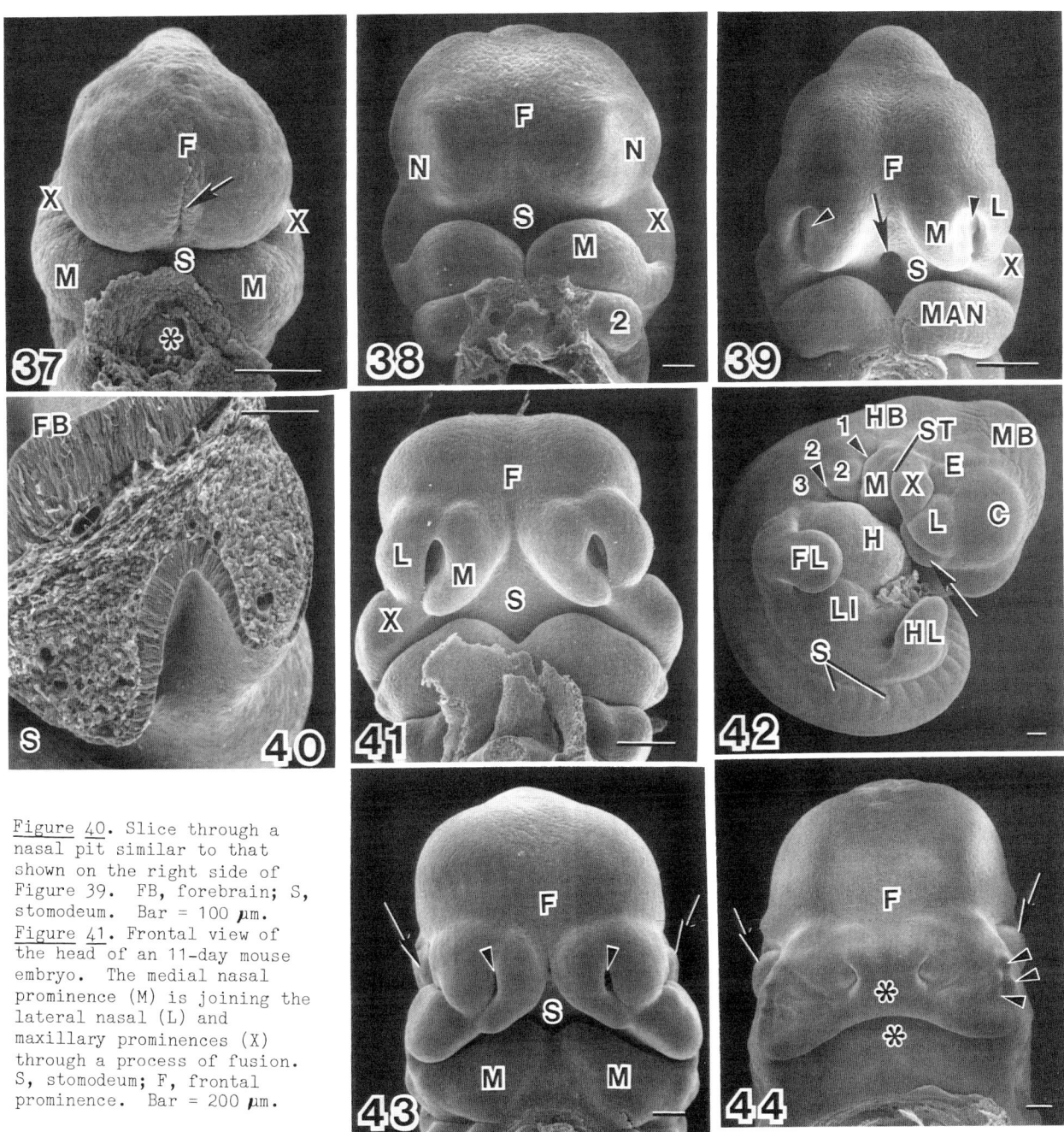

Figure 40. Slice through a nasal pit similar to that shown on the right side of Figure 39. FB, forebrain; S, stomodeum. Bar = 100 μm.

Figure 41. Frontal view of the head of an 11-day mouse embryo. The medial nasal prominence (M) is joining the lateral nasal (L) and maxillary prominences (X) through a process of fusion. S, stomodeum; F, frontal prominence. Bar = 200 μm.

Figure 42. Lateral view of a mouse embryo similar to that shown in Figure 41. Merging is occurring between the maxillary prominences (X) and lateral nasal prominences (L). The positions of the eye (E), cerebral hemisphere of the forebrain (C), midbrain (MB), hindbrain (HB), heart (H), liver (LI), and somites (S) are indicated. FL, forelimb bud; HL, hindlimb bud; M, mandibular prominence of the first visceral arch; ST, stomodeum; arrow, medial nasal prominence; 2-3, second and third visceral arches; numbered arrowheads, first and second visceral grooves. Bar = 200 μm.

Figure 43. Frontal view of the head of an 11-day mouse embryo more advanced than that shown in Figure 41. The eyes (arrows) are positioned farther rostrally. The mandibular prominences (M) are merging in the midline and the external nares (arrowheads) are now clearly separated from the stomodeum (S). F, frontal prominence. Bar = 200 μm.

Figure 44. Frontal view of the head of a 12-day mouse embryo. Arrows, eyes; arrowheads, whisker papillae; F, frontal prominence. The two medial nasal prominences have merged together in the midline as have the two mandibular prominences (asterisks). Bar = 200 μm.

Recall that each optic vesicle is originally connected to the forebrain via an optic stalk (Figures 54, 55). The optic stalk invaginates like the optic vesicle, creating a channel bounded by the now double-layered optic stalk. This channel allows blood vessels to grow from the head toward and into the cavity of the developing eye (Figure 60), providing for gas exchange, nutritive supply, and waste removal as rapid growth of this structure occurs. Also, optic nerve fibers grow through the wall of the channel from the neural retina to the diencephalon.

The developing eyes are bounded superiorly and inferiorly during late in utero life by folds of surface epithelium constituting the rudiments of the eyelids (Figure 61). These two folds grow toward one another and fuse, temporarily covering the cornea completely (Figure 62). Considerable alterations occur in cell surface morphology near and along the raphe demarcating the site of transient fusion (Figure 63).

Development and Partitioning of the Oral and Nasal Cavities

The ventral ends of the visceral arches form the floor of the mouth and give rise to rudiments (i.e., glossal processes) contributing to the tongue (Figure 64). The anterior two-thirds or body of the tongue derives from principally two large swellings, the lateral lingual swellings or distal tongue buds (Figures 65, 66). (A rudimentary median tongue bud or tuberculum impar also forms between the two lateral lingual swellings but is later overgrown contributing little or nothing to the definitive tongue.) These lateral swellings originate from the mandibular prominences of the first visceral arches. The posterior one-third or the root of the tongue derives from part of the copula or hypobranchial eminence, a swelling formed by the second-fourth visceral arches (Figures 65, 66). A depression, the foramen caecum, lies between the body and root of the tongue (Figures 65, 66) marking the point of origin of the thyroid gland. The latter originates as a vesicular evagination from the floor of the pharynx between the second and third visceral arches (Figure 64). Later, it migrates down into the root of the neck. In addition to contributing to the tongue, the ventral ends of the visceral arches (that is, the posterior part of the hypobranchial eminence) form the epiglottis and arytenoid swellings, bordering the glottis, the opening of the pharynx into the larynx (Figure 67). The surfaces of the derivatives of the ventral ends of the visceral arches quickly acquire their distinctive morphological features (Figures 68-70).

The primitive oral and nasal cavities are initially broadly continuous with one another (Figure 39). They begin to separate during formation of the upper jaw, when fusion occurs on each side of the midline between the medial nasal prominences and the maxillary and lateral nasal prominences (Figures 41, 43, 71). This separation is only temporary, since internal connections are soon re-established between the oral and nasal cavities. Internally, the oral and nasal cavities are separated on each side by a thin membrane called the oronasal or bucconasal membrane, which quickly ruptures (Figures 72, 73). Rupture of this paired membrane establishes the internal (posterior) nares (or choanae) (Figure 74). The internal nares of the adult (i.e., the definitive choanae) are located at the back of the nasopharynx after formation of the palate (see below). Hence, the first internal nares are called the primitive choanae to distinguish them from the later ones.

The oral and nasal cavities are separated by formation of the hard and soft palates and uvula. The definitive palate consists of a primary and a secondary palate. The secondary is derived from medial outgrowths of the maxillary prominences termed palatal shelves (Figure 71), whereas the primary palate is an anteromedian derivative of the medial nasal prominences (Figure 74). As the tongue develops, the secondary palatal shelves become oriented in a vertical plane (Figures 74, 75), lateral to it. The shelves soon rotate 90° mediad to assume a horizontal position above the tongue (Figures 76-78), and approach one another to fuse across the midline. Fusion begins about one-third of the distance from the anterior aspect of the shelves and then extends anteriorly and posteriorly (Figures 77, 79). Details of this fusion are given by Waterman et al. (1973) and Waterman and Meller (1974). As the secondary palatal shelves are approaching one another, the nasal septum grows downward to meet them (Figures 75, 78). Fusion then occurs between the nasal septum, primary palate, and secondary palatal shelves forming paired nasal cavities, separated (except posteriorly via the definitive choanae) from the oral cavity. The areas of fusion between the contralateral secondary palatal shelves and primary palate are indicated by raphes (Figures 80-82).

Coda

In this brief overview, we have described some of the major morphogenetic events in formation of the craniofacial region as revealed by scanning electron microscopy. It is obvious that although our knowledge of what happens morphologically in sculpturing this complex region is fairly complete, we know very little about biochemical changes or cellular and molecular mechanisms responsible for the observed morphological events. Determining how morphogenesis occurs remains the challenge for future studies. It seems certain that the rudiments of the craniofacial region will serve as important model systems in meeting this important and stimulating challenge.

Acknowledgments

We would like to thank Ms. Deborah Dehart for her excellent technical assistance, Dr. M. C. Johnston for his support and teaching, and Ms. Melinda Engelhardt for her secretarial assistance.

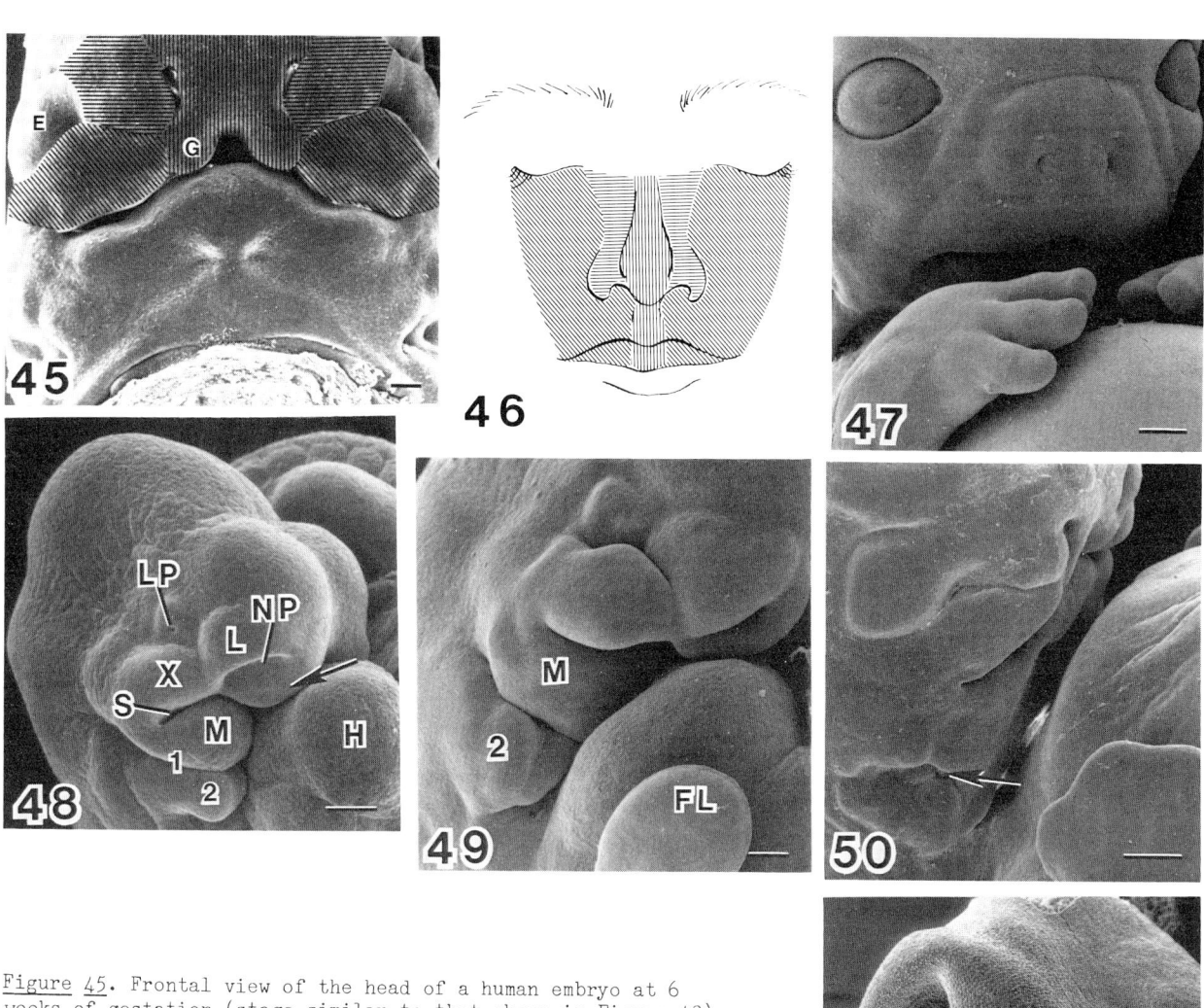

Figure 45. Frontal view of the head of a human embryo at 6 weeks of gestation (stage similar to that shown in Figure 43). Shading has been used to demarcate facial prominences associated with the upper jaw (vertical shading, medial nasal prominences; horizontal shading, lateral nasal prominences; oblique shading, maxillary prominences of first visceral arches). E, eye; G, globular process of medial nasal prominence. Reproduced with permission from: Sulik and Johnston (1982). Bar = 200 μm.

Figure 46. Diagram of the contributions of the shaded facial prominences shown in Figure 45 to the adult face. Reproduced with permission from: Sulik and Johnston (1982).

Figure 47. Frontal view of the head of a human embryo at 8 weeks of gestation. Formation of the face is largely completed. Bar = 500 μm.

Figure 48. Lateral view of the head of a mouse embryo at a stage similar to that shown in Figure 42. LP, lens pit; NP, nasal pit; L, lateral nasal prominence; arrow, medial nasal prominence; M, mandibular prominence of first visceral arch; X, maxillary prominence of first visceral arch; S, stomodeum; 1, first visceral groove; 2, second visceral arch; H, location of heart. Reproduced with permission from: Sulik and Johnston (1982). Bar = 200 μm.

Figure 49. Lateral view of the head of a mouse embryo at 11 days of gestation (embryo similar to that shown in Figure 43). The external ear (pinna) is forming from hillocks derived from the first (i.e., mandibular prominence, M) and second (2) visceral arches. FL, forelimb bud. Bar = 200 μm.

Figure 50. Lateral view of the head of a human embryo at 6 weeks of gestation (embryo similar to that shown in Figure 45). The individual hillocks that will form the external ear can be identified both cranial and caudal to the first visceral groove (arrow). Bar = 500 μm.

Figure 51. Developing external ear from a human embryo at 8 weeks of gestation. Bar = 200 μm.

## References

Chávez DJ. (1984). Mammaliam preimplantation embryogenesis. Scanning Electron Microsc. 1984; II:729-735.

Hay DH, Markwald RR, Fitzharris TP. (1984). Selected views of early heart development by scanning electron microscopy. Scanning Electron Microsc. 1984; IV:1983-1993.

Hilfer SR. (1983). Development of the eye of the chick embryo. Scanning Electron Microsc. 1983; III:1353-1369.

Hilfer SR, Brown JW. (1984). The development of pharyngeal endocrine organs in mouse and chick embryos. Scanning Electron Microsc. 1984; IV:2009-2022.

Johnston MC, Sulik KK. (1980). Development of the face and oral cavity. In: Orban's Oral Histology and Embryology. Bhaskar SN (Ed). Mosby Co., St. Louis, MO. pp. 1-23.

Johnston MC, Sulik KK. (1984). Embryology of the head and neck. In: Pediatric Plastic Surgery. Serafin D, Georgiade NG. (Eds). Mosby Co., St. Louis, MO. pp. 184-215.

Kaufmann P, Leisten H, Mangold U. (1981). Die keimenbogenentwicklung bei ratte und maus. I. Zur entwicklung von sinus cervicalis und operculum. Acta Anat. $\underline{110}$, 7-22.

LeDouarin N. (1982). The Neural Crest. Cambridge University Press, Cambridge. 259 pp.

Mangold U, Dörr A, Kaufmann P. (1981). Die kiemenbogenentwicklung bei ratte und maus. II. Zur existenz von kiemensplaten. Acta Anat. $\underline{110}$, 23-34.

Nichols DH. (1981). Neural crest formation in the head of the mouse embryo as observed using a new histological technique. J. Embryol. Exp. Morph. $\underline{64}$, 105-120.

Sadler TW. (1985). Langman's Medical Embryology. Williams and Wilkins, Baltimore, MD. 409 pp.

Schoenwolf GC. (1983). The chick epiblast: A model for examining epithelial morphogenesis. Scanning Electron Microsc. 1983; III:1371-1385.

Streeter GL. (1907). On the development of the membranous labyrinth and the acoustic and facial nerves in the human embryo. Amer. J. Anat. $\underline{6}$, 139-165.

Sulik KK. (1983). Sequence of developmental alterations following acute ethanol exposure in mice: Craniofacial features of the fetal alcohol syndrome. Amer. J. Anat. $\underline{166}$, 257-269.

Sulik KK, Johnston MC. (1982). Embryonic origin of holoprosencephaly: Interrelationship of the developing brain and face. Scanning Electron Microsc. 1982; I:309-322.

Tan SS, Morriss-Kay G. (1985). The development and distribution of cranial neural crest in the rat embryo. Cell Tissue Res. $\underline{240}$, 403-416.

Watanabe K, Sasake F, Takahama H. (1984). The ultrastructure of oral (buccopharyngeal) membrane formation and rupture in the anuran embryo. Anat. Rec. $\underline{210}$, 513-524.

Waterman AJ. (1938). The development of the inner ear rudiment of the rabbit embryo in a foreign environment. Amer. J. Anat. $\underline{63}$, 161-219.

Figure 52. Dorsal view of the cranial neural folds from a mouse embryo similar to that shown in Figure 15. Arrows indicate the optic sulci. FB, forebrain; MB, midbrain. Reproduced with permission from: Sulik (1983). Bar = 50 μm.

Figure 53. Frontal view of the forebrain region from a 9-day mouse embryo slightly younger than that shown in Figure 37. Arrow indicates the closing cranial neuropore. The optic vesicles lie just beneath the surface ectoderm in the areas indicated by asterisks. Reproduced with permission from: Sulik (1983). Bar = 100 μm.

Figure 54. Frontolateral view of a slice through the forebrain region from a mouse embryo similar to that shown in Figure 53. Arrows, optic stalks; OV, optic vesicle; M, mandibular prominence of first visceral arch; X, maxillary prominence of first visceral arch; 2, second visceral arch. Reproduced with permission from: Johnston and Sulik (1984). Bar = 100 μm.

Figure 55. Enlargement of the sliced optic vesicle and stalk from the embryo shown in Figure 54. Bar = 50 μm.

Figure 56. Slice through the early optic cup from a 10-day mouse embryo. PE, pigmented epithelium; NR, neural retina; arrow, lens placode. Bar = 50 μm.

Figure 57. Slice through the late optic cup/lens vesicle from an 11-day mouse embryo. PE, pigmented epithelium; NR, neural retina; LV, lens vesicle; arrow, surface ectoderm that forms the epithelium of the cornea. Reproduced with permission from: Sadler (1985). Bar = 50 μm.

Figure 58. View of the surface epithelium partially covering and directly continuous with the lens vesicle from a mouse embryo at a stage similar to that shown in Figures 48 and 57. Arrow, opening into the lens vesicle. Bar = 50 μm.

Figure 59. Slice through the late optic cup/lens vesicle from a 13-day mouse embryo. Arrowheads, pigmented epithelium; NR, neural retina; LF, lens fibers; E, lens epithelium; arrow, surface ectoderm that forms the epithelium of the cornea. Reproduced with permission from: Sadler (1985). Bar = 50 μm.

Figure 60. Slice through a developing eye from a mouse embryo similar to that shown in Figure 59. Blood vessels (arrow) have extended through the optic fissure to reach the cavity of the optic cup. LV, lens vesicle. Bar = 100 μm.

Figure 61. Surface view of a developing eye from a human embryo similar to that shown in Figure 47. Arrows indicate the rudiments of the upper and lower eyelids. C, cornea. Bar = 100 μm.

Figure 62. Surface view of the fused eyelids from a human embryo at 10 weeks of gestation. Bar = 200 μm.

Figure 63. Enlargement of the raphe between the fused upper and lower eyelids from the embryo shown in Figure 62. Bar = 100 μm.

SEM of Craniofacial Development

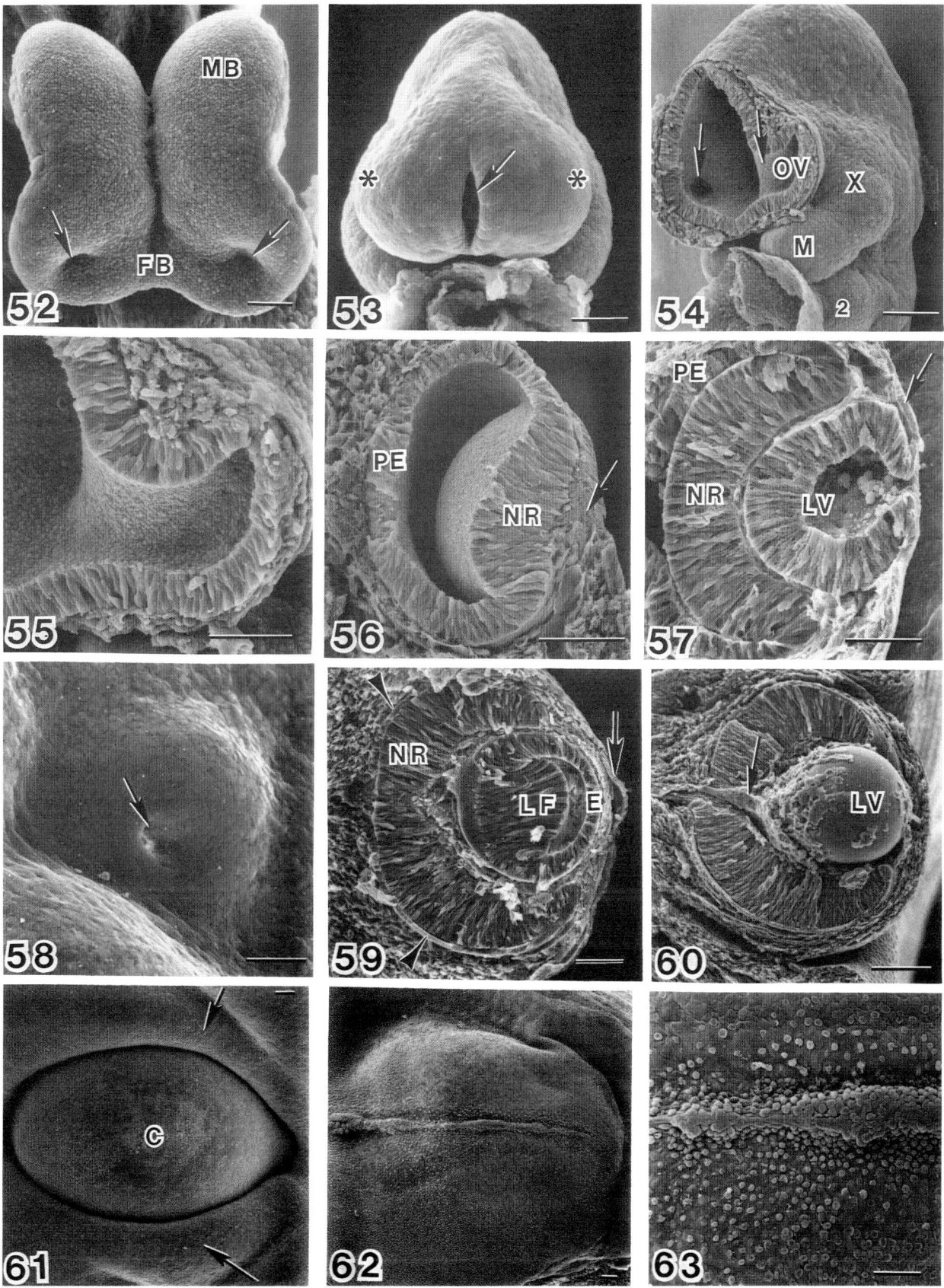

Waterman RE. (1976). Topographical changes along the neural fold associated with neurulation in the hamster and mouse. Amer. J. Anat. 146, 151-172.

Waterman RE. (1977). Ultrastructure of oral (buccopharyngeal) membrane formation and rupture in the hamster embryo. Dev. Biol. 58, 219-229.

Waterman RE. (1985). Formation and perforation of closing plates in the chick embryo. Anat. Rec. 211, 450-457.

Waterman RE, Kao R. (1982). Formation of the mouth opening in the zebrafish embryo. Scanning Electron Microsc. 1982; III:1249-1257.

Waterman RE, Meller SM. (1974). Alterations on the epithelial surface of human palatal shelves prior to and during fusion: A scanning electron microscopic study. Anat. Rec. 180, 111-136.

Waterman RE, Ross LM, Meller SM. (1973). Alterations in the epitheial surface of A/Jax mouse palatal shelves prior to and during palatal fusion: A scanning electron microscopic study. Anat. Rec. 176, 361-376.

Waterman RE, Schoenwolf GC. (1980). The ultrastructure of oral (buccopharyngeal) membrane formation and rupture in the chick embryo. Anat. Rec. 197, 441-470.

Figure 64. Dorsal view of the floor of the pharynx from a 10-day mouse embryo. M, mandibular prominences of first visceral arches; 2, 3, second and third visceral arches; T, location of the thyroid rudiment beneath the endodermal floor of the pharynx; arrowheads, aortic arch vessels. Reproduced with permission from: Johnston and Sulik (1984). Bar = 100 μm.

Figure 65. Dorsal view of the floor of the pharynx from an 11-day mouse embryo. Asterisks, lateral lingual swellings; arrow, foramen caecum. Reproduced with permission from: Johnston and Sulik (1984). Bar = 100 μm.

Figure 66. Dorsal view of the floor of the pharynx from a 12-day mouse embryo. M, mandibular prominences of first visceral arches; arrow, foramen caecum. Reproduced with permission from: Johnston and Sulik (1984). Bar = 100 μm.

Figure 67. Dorsal view of the floor of the oral cavity from a human embryo at 10 weeks of gestation. E, epiglottis; arrow, glottis. Bar = 500 μm.

Figure 68. Enlargement of the anterior portion of the tongue from the specimen shown in Figure 67. Bar = 100 μm.

Figure 69. Enlargement of the middle lateral portion of the tongue from the specimen shown in Figure 67. Bar = 100 μm.

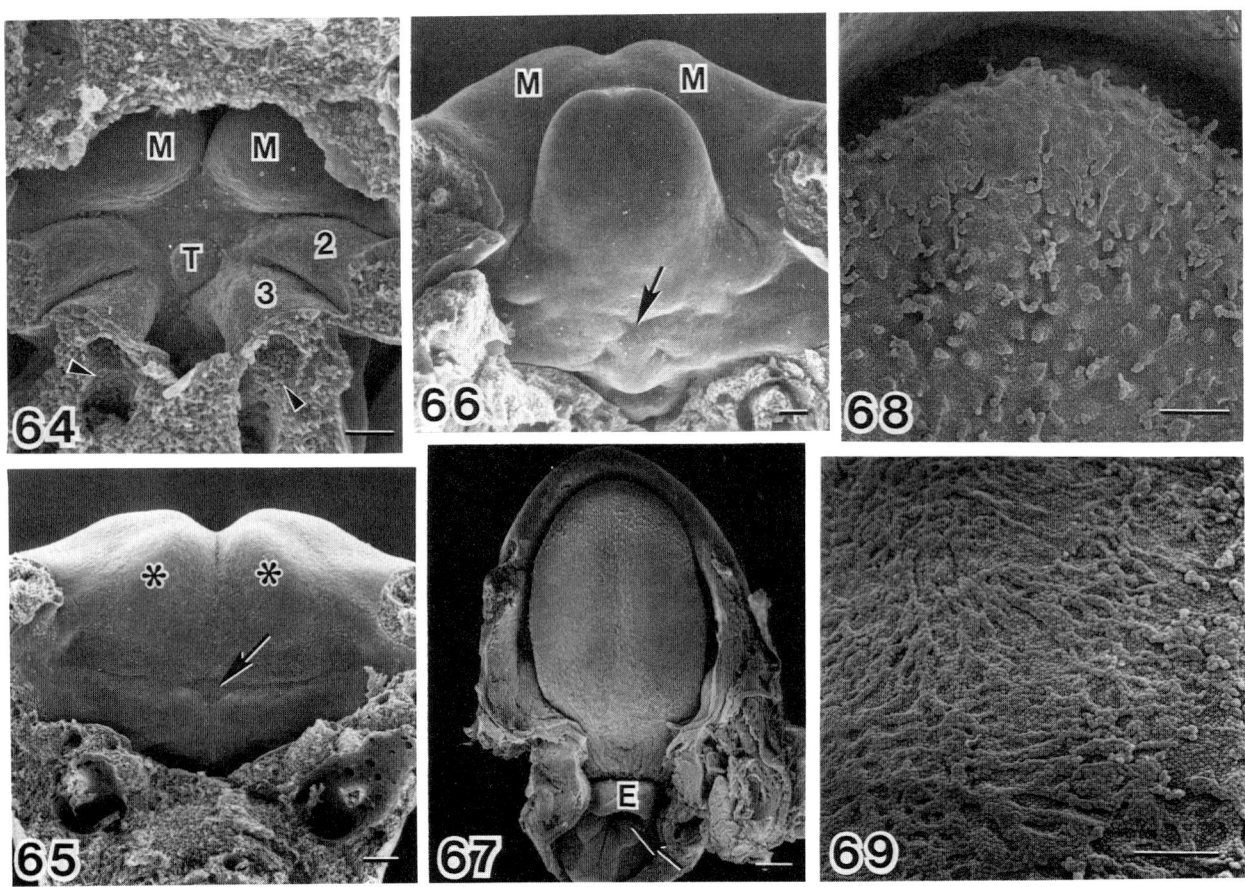

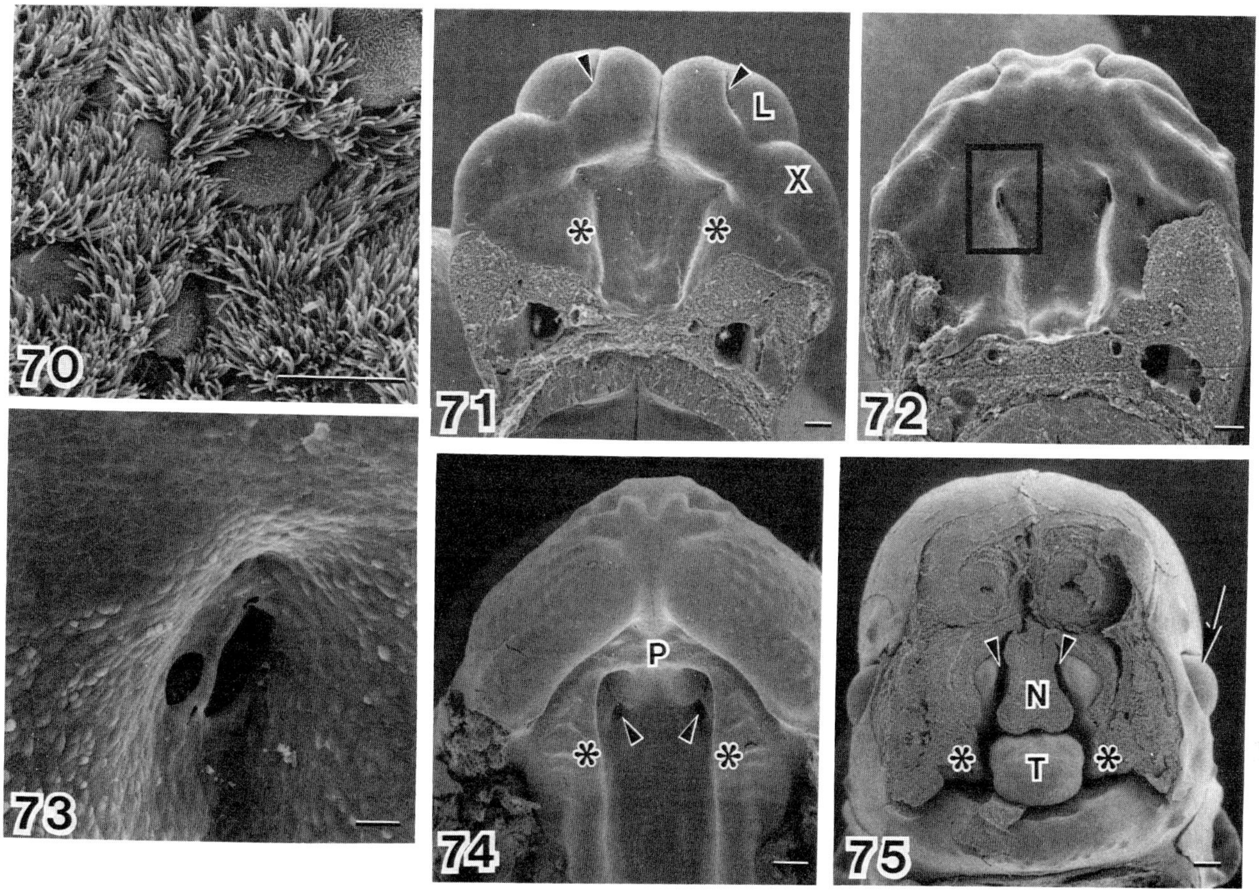

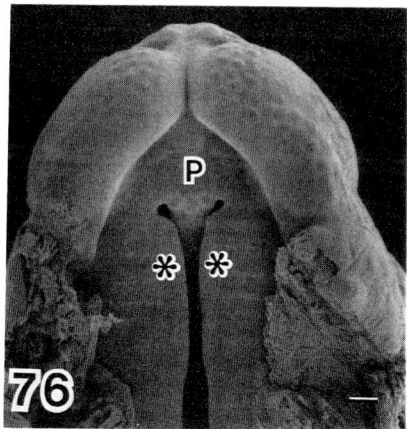

Figure 70. Enlargement of the epithelium covering the glottis from the specimen shown in Figure 67. Bar = 10 μm.

Figure 71. Developing upper jaw viewed from the oral cavity from an 11-day mouse embryo. L, lateral nasal prominence; X, maxillary prominence; asterisks, secondary palatal shelves; arrowheads, external nares. Reproduced with permission from: Johnston and Sulik (1980). Bar = 100 μm.

Figure 72. Developing upper jaw viewed from the oral cavity from a 12-day mouse embryo. Boxed area is enlarged in Figure 73. Reproduced with permission from: Johnston and Sulik (1980). Bar = 100 μm.

Figure 73. Boxed area from Figure 72. The oronasal membrane is rupturing. Reproduced with permission from: Johnston and Sulik (1980). Bar = 20 μm.

Figure 74. Developing upper jaw viewed from the oral cavity from a 14-day mouse embryo. Arrowheads, primitive choanae; asterisks, secondary palatal shelves; P, primary palate. Bar = 200 μm.

Figure 75. Frontal slice through the head of a mouse embryo similar to that shown in Figure 74. Arrow, eye; T, tongue; N, nasal septum; arrowheads, nasal cavities; asterisks, secondary palatal shelves. Bar = 200 μm.

Figure 76. Developing upper jaw viewed from the oral cavity from a 15-day mouse embryo. P, primary palate; asterisks, secondary palatal shelves. Bar = 200 μm.

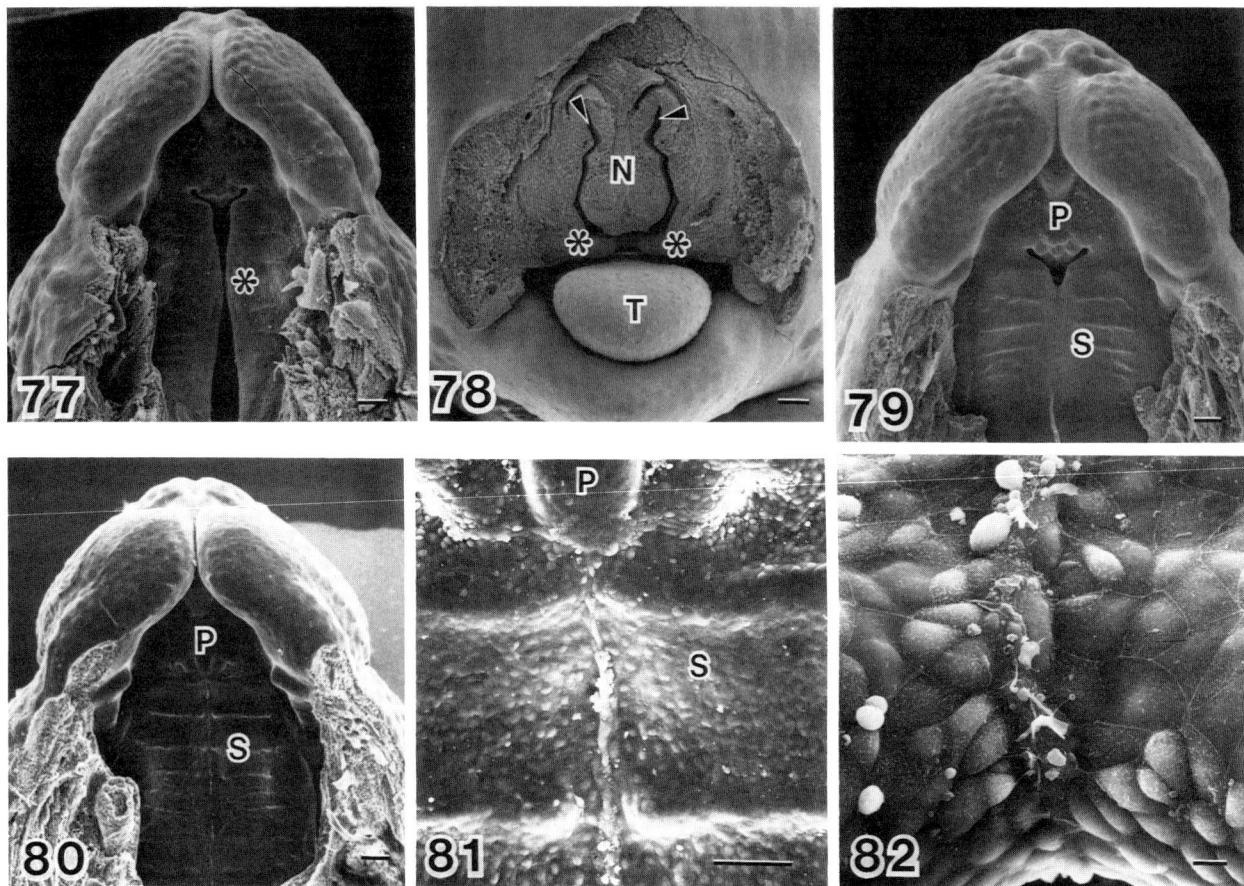

Figure 77. Developing upper jaw viewed from the oral cavity from a 14-day mouse embryo more advanced than that shown in Figure 76. Secondary palatal shelves are about to fuse in the area indicated by the asterisk. Bar = 200 μm.

Figure 78. Frontal slice through the head of a mouse embryo similar to that shown in Figure 77. T, tongue; N, nasal septum; arrowheads, nasal cavities; asterisks, secondary palatal shelves. Bar = 200 μm.

Figure 79. Developing upper jaw viewed from the oral cavity from a 15-day mouse embryo. P, primary palate; S, secondary palate. Bar = 200 μm.

Figure 80. Developing upper jaw viewed from the oral cavity from a 15-day mouse embryo more advanced than that shown in Figure 79. P, primary palate; S, secondary palate. Bar = 200 μm.

Figure 81. Enlargement of the fusion line near the anterior end of the secondary palate. P, primary palate; S, secondary palate. Bar = 50 μm.

Figure 82. Enlargement of the fusion line near the posterior end of the secondary palate. Bar = 10 μm.

## Discussion with Reviewers

R.E. Waterman: Would you please discuss how the cranial somitomeres (as described by Steve Meier) relate to the visceral arches and facial prominences?

Authors: Somitomeres are segmental units of paraxial mesoderm consisting of circular domains of radially arranged cells, and have been observed in many vertebrate embryos by Meier and colleagues. Eight pairs of somitomeres form in tandem in the cranial region of mouse embryos (Meier S and Tam PL. (1982) Metameric pattern development in the embryonic axis of the mouse. I. Differentiation of the cranial segments. Differentiation 21, 95-108). The first pair (along with the prechordal mesoderm) underlies the forebrain, the second and third pairs underlie the midbrain, and the fourth-seventh pairs underlie the hindbrain; the eighth pair forms the first pair of somites.

The precise relationships between the somitomeres and visceral arches/facial prominences is unknown. Noden has shown in the chick (where a similar pattern of somitomeres exists) that somitomeres 1-3 and 5 contribute to the extrinsic eye muscles, somitomere 4 to first arch muscles, somitomere 6 to second arch muscles, and somitomere 7 to third arch muscles. The remaining muscles of the head and neck originate from the myotomes of the somites (Noden DM. (1983) The embryonic origins of avian cephalic and cervical muscles and associated connective tissue. Amer. J. Anat. 168, 257-276).

EARLY DEVELOPMENT OF THE VERTEBRATE LIMB: AN INTRODUCTION TO MORPHOGENETIC TISSUE
INTERACTIONS USING SCANNING ELECTRON MICROSCOPY

Robert O. Kelley

Department of Anatomy
University of New Mexico School of Medicine
Albuquerque, New Mexico 87131
Phone no.: (505) 277-5555

(Paper received August 21 1984, Completed manuscript received May 15 1985)

## Abstract

The developing limbs of most vertebrates serve as a model system for studies of morphogenesis, pattern formation, cell and tissue interactions and cell differentiation. Mesoderm in the flank of the embryo induces overlying ectoderm to form a thickened, stratified or pseudo-stratified epithelium which becomes the highly specialized apical ectodermal ridge. In turn, the apical ridge specifies individual limb parts (first from structures proximal to the body axis, then to more distal components) and is required for those elements to form. If the ridge is removed, subsequent limb development ceases and no further limb parts appear. The series of ectodermal-mesodermal interactions is poorly understood at the molecular level, but scanning electron microscopy permits the visualization of tissues and cells which participate in this remarkable process of morphogenesis and differentiation. This paper is intended to serve as an introduction for the student beginning an investigation into the multiple, integrated biological processes which culminate in the establishment of a normal vertebrate limb.

Key words: Ectoderm; mesoderm; apical ectodermal ridge; mesenchyme; matrix; cartilage, chondrogenesis; induction; tissue interactions; intercellular communication.

## Introduction

Studies of the development of limbs in vertebrate embryos have attracted a great deal of recent attention in laboratories throughout the world. The reason for this interest is that the developing vertebrate limb involves many fundamental mechanisms which are central to the development and differentiation of other organ systems within the embryo. These mechanisms include epithelial-mesenchymal interactions; cell-cell interactions; intercellular communication; differentiation of cartilage, bone, and vascular structures; and the factors which regulate proliferation and cell death. Taken together, over a precise period of developmental time within the embryo, these forces shape the limb and ultimately give rise to functional structures recognized as wings, arms and legs.

In addition, these developmental events are remarkably similar when compared between reptiles, birds and mammals. In this regard, it is reasonably clear that all amniote embryos develop along a central evolutionary theme. It is this general theme which we will focus upon in this paper. A series of scanning electron micrographs is used to illustrate the major events which occur during early limb development using the mammalian embryo as a model. Each specimen has been chemically preserved, dried by the critical point method, and coated with a conductive metal alloy. The student is expected to use this paper as an introduction into a model morphogenetic system. Selected readings are listed in the reference section of the paper in addition to specific references of papers in the research literature which document important discoveries regarding the mechanisms of limb development.

## Early Shape of the Embryo

As a mammalian embryo develops the tissues which will eventually become the skin of the embryo and the underlying musculature and connective tissues, the embryo also begins to acquire a unique specific shape. Figure 1 shows a mouse embryo early in development. The head is beginning to form as is the flank (side) of the embryo. Along the flank, individual somites can be seen, in addition to an elevation of tissue adjacent to the somites which is the early fore limb bud. The bud (Figure 2) consists of two parts. First, the outermost covering is ectoderm, a tissue layer which is continuous over the surface of the embryo. Immediately beneath the ectoderm is a layer of undifferentiated mesoderm. Some of the cells in the mesoderm form an embryonic connective tissue which is called mesenchyme.

As the embryo continues to change shape and increase in length, the buds which will form the upper extremities (fore legs) develop somewhat earlier than the bud which will form the hind legs of the organism. Approximately one developmental stage separates the growth and differentiation of the fore from the hind limb bud. As can be seen in Figure 3, the lower limb bud appears as a slight swelling along the caudal region of the embryo slightly forward of the developing tail bud.

### Polarity of the Early Limb Bud

At this early stage in the development of limb buds, polarity may already be seen. The anterior border of the limb bud is marked in Figure 4 as is the posterior border. To help in orientation, in an adult human, the anterior border of the embryonic limb will eventually correspond to the top of the shoulder, whereas the posterior border will develop into the armpit (axilla). Furthermore, there is a dorsal surface of ectoderm illustrated in Figure 4, as well as a ventral ectodermal surface. In addition, as the limb bud continues to grow into a more cylindrical structure, a proximo-distal polarity is apparent. The proximal boundary is closest to the midline of the embryo. The distal boundary would be farthest from the midline of the embryo (where the fingers and toes will eventually be formed).

### Mesodermal-Ectodermal Interactions

One of the more interesting and least understood features of early limb development begins to occur during the stage of early bud formation depicted in Figure 5. We know from previous experimental studies that the mesoderm of the limb bud has two origins. One is the tissue which can be seen as part of individual somites adjacent to the developing bud. The second point of origin is in the mesodermal tissue of the body (the somatopleure). Both tissues contribute to the mesenchyme underlying the covering ectoderm of the limb bud. These two tissues can be seen more clearly in Figure 6 in which a portion of the ectoderm has been pulled back to reveal the underlying mesoderm (arrows, Fig. 6). It is important to note that there is no change in the contour of the surface of the ectoderm at this developmental stage. However, a very important event is occurring as a result of some unknown type of communication from the underlying mesoderm. This mesodermal "induction" will eventually result in the appearance of a ridge (see Figure 10) which develops along the distal border of the developing limb bud. More will be said regarding this ridge in a moment.

However, it is important to note that during this early developmental period, proliferation of underlying mesoderm results in the development of a hemispherical limb bud and, in addition, the first of two known tissue interactions begins to induce the future apical ridge by action of the underlying mesoderm. The student should remember that this interaction is required for the apical ridge to appear and, once the thickened epithelium is in place, its maintenance is dependent upon the continued action of the underlying mesoderm in some, as yet unknown, manner.

It is of further interest to note that, because of the suspected dual origin of the mesoderm, these cells are developing along two very different developmental lines, although they have a remarkably homogeneous structural appearance at this stage. The future muscle of the limb appears to be of somite origin, whereas tendons and chondrogenic tissues develop from cells derived from the somatic (flank) mesoderm. Although it is not possible at this early stage to see differentiation into muscle or cartilage, experimental evidence clearly documents that this differentiation is occurring.

The structural nature of the interaction between the mesoderm and the ectoderm is interesting in that neither layer actually touches free cell surface. Inductive communication occurs through a highly structured matrix of extracellular macromolecules. Figure 7 reveals the structural relationships of the ectoderm and its position adjacent to mesoderm before the thickened ridge develops. At higher magnification in Figure 8, it can be seen that the mesodermal cells reside in an extracellular matrix. This matrix consists of collagen, proteoglycan and other matrix molecules (for example, fibronectin). The matrix enmeshes adjacent cells and provides an in vivo culture matrix for the further development and differentiation of mesenchymal cells in the core, subridge, and dorsal and ventral compartments of developing limb mesoderm. In addition, it can be seen in Figure 9 that an intimate relationship exists between the mesoderm and the overlying ectoderm. It is curious that no direct cell to cell contact occurs at this boundary which is so active as an inductive interface. However, Figure 9 reveals that matrix molecules span the interface between the basal surface of the ectoderm and the opposing surfaces of mesodermal cells without permitting the plasma membranes of individual cells to contact.

### The Apical Ectodermal Ridge

As a result of mesodermal induction, an apical ectodermal ridge forms on the distal boundary of the developing bud (Figure 10). A great deal of experimentation in birds and mammals has revealed that the presence of the apical ridge is responsible for the establishment of all parts of the limb. Experimentally, if the ridge is removed or damaged, limb development ceases and no further limb parts are developed.

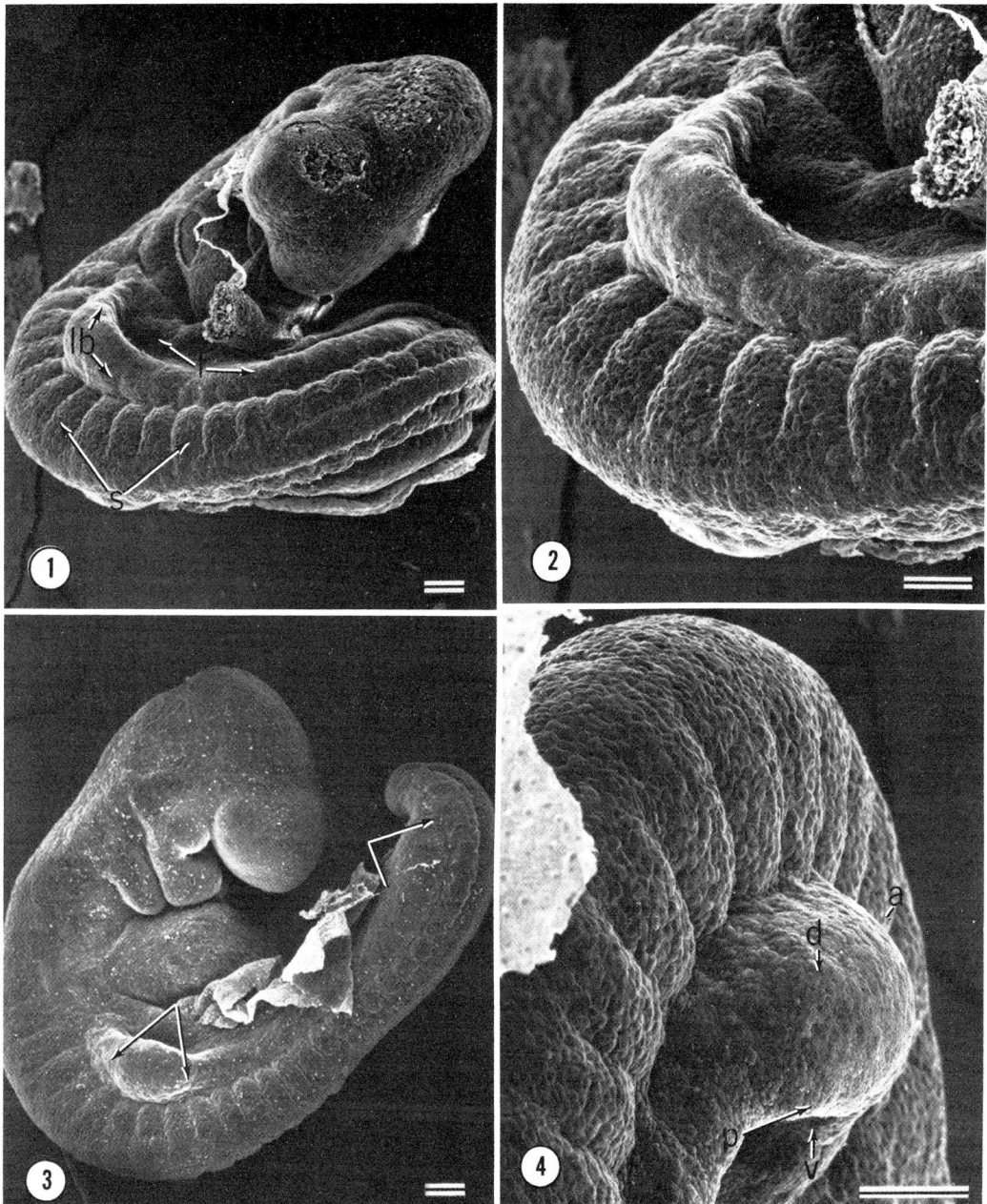

Figure 1: Mouse embryo showing morphological relationships of somites(s), flank(f), and limb buds (lb). The amnion and other extraembryonic membranes have been dissected away to reveal the contours of the embryo. Bar = 100 μm.

Figure 2: Higher magnification reveals the early limb bud to be an elevation on the embryonic flank ventrolateral to the somites. As development continues, the bud will acquire a more cylindrical shape which will be molded by proliferatin, migration, differentiation and death of cells into a definitive limb. Bar = 100 μm.

Figure 3: The fore and hind limb buds are visible on this hamster embryo. The hind limb is delayed approximately one developmental stage from that exhibited by the fore limb, a phenomenon associated with the general cranial to caudal gradient exhibited by all vertebrate embryos. Bar = 100 μm.

Figure 4: As the limb bud enlarges, it develops polarity in relation to other body parts and axes. The dorsal (d) and ventral (v) surfaces are continous with ectoderm covering the somites and the flank, respectively. The border closest to the head is anterior (a), wheras that closest to the tail is posterior (p). The base of the bud is proximal wheras the tip is distal to the longitudinal axis of the embryo's body. Bar = 100 μm.

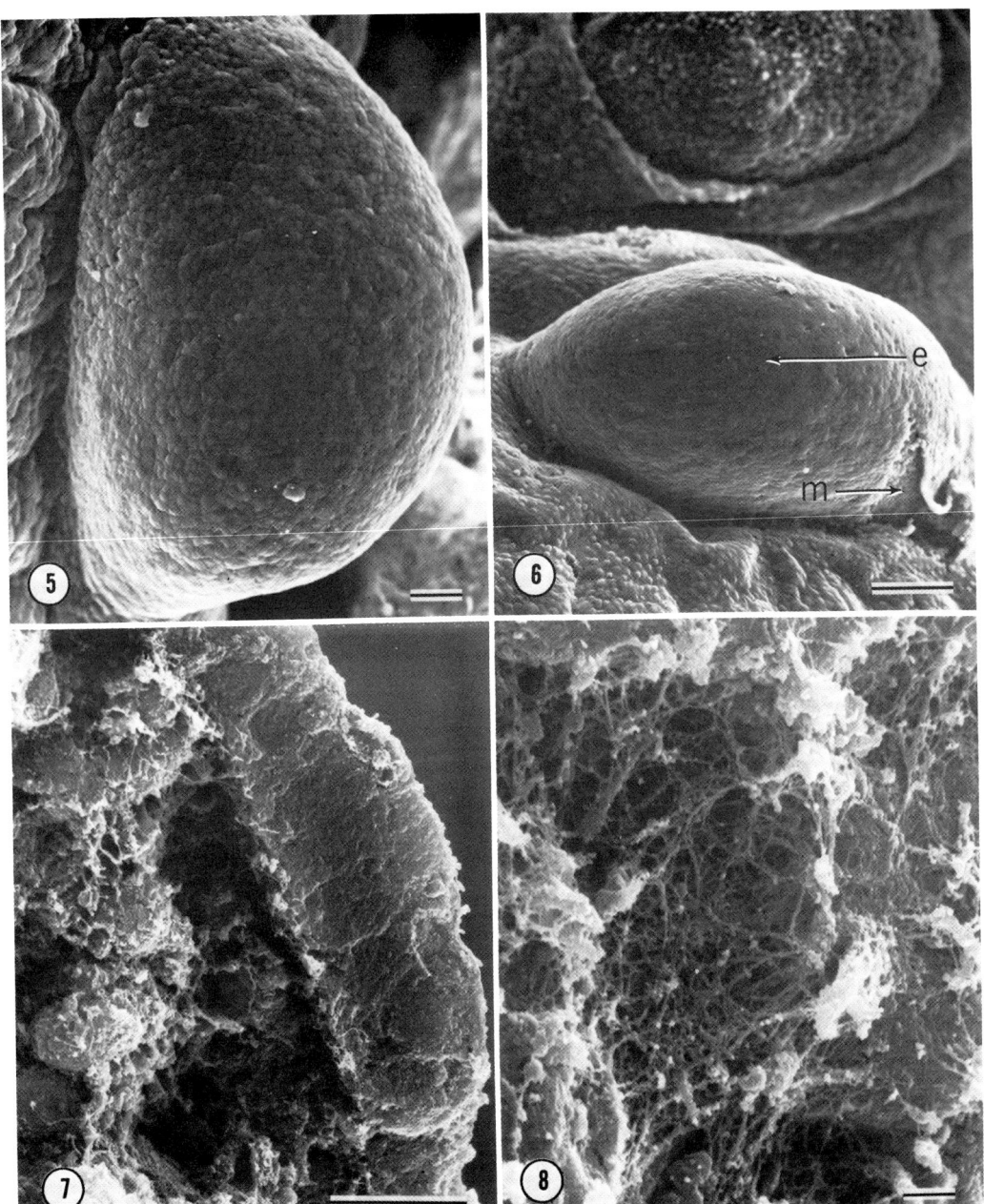

Figure 5: It is important to note that the early limb bud acquires a considerable mass of mesodermal tissue before the appearance of the apical ridge (see figure 10 for reference). This mesoderm will differentiate to form muscle, cartilage, bone and blood vessels in the limb. Bar = 20 μm.

Figure 6: This micrograph shows a limb bud which has been partially dissected to reveal the mesoderm (m) beneath the covering ectoderm (e). Limb mesenchymal tissue comes from both somite and flank mesoderm. The somite-derived cells differenticate into muscle, wheras the tendons and cartilage of the limb develop from mesoderm of the flank (technically, the somatopleure of the embryo). Bar = 50 μm.

Figure 7: In this dissected limb bud, the mesoderm is seen to be closely opposed to the covering ectoderm. Inductive interactions transmit signals from the mesoderm to the overlying ectoderm which initiates the appearance and maintenance of the apical ridge. The precise nature of this inductive process is not understood, although it is clear that direct cell to cell contact between mesoderm and ectoderm does not occur. Consequently signals must be mediated through the extracellular matrix. Bar = 10 μm.

Figure 8: The scanning electron micrscope provides a unique image of extracellular materials in the mesoderm, although it does not facilitate recognition of the exact molecular species present in the matrix. However, other techniques reveal the cob-web-like material to consist of collagen, proteoglycan macromolecules, glycosaminoglycans, and adhesive molecules, e.g. fibronectin. Bar = 1 μm.

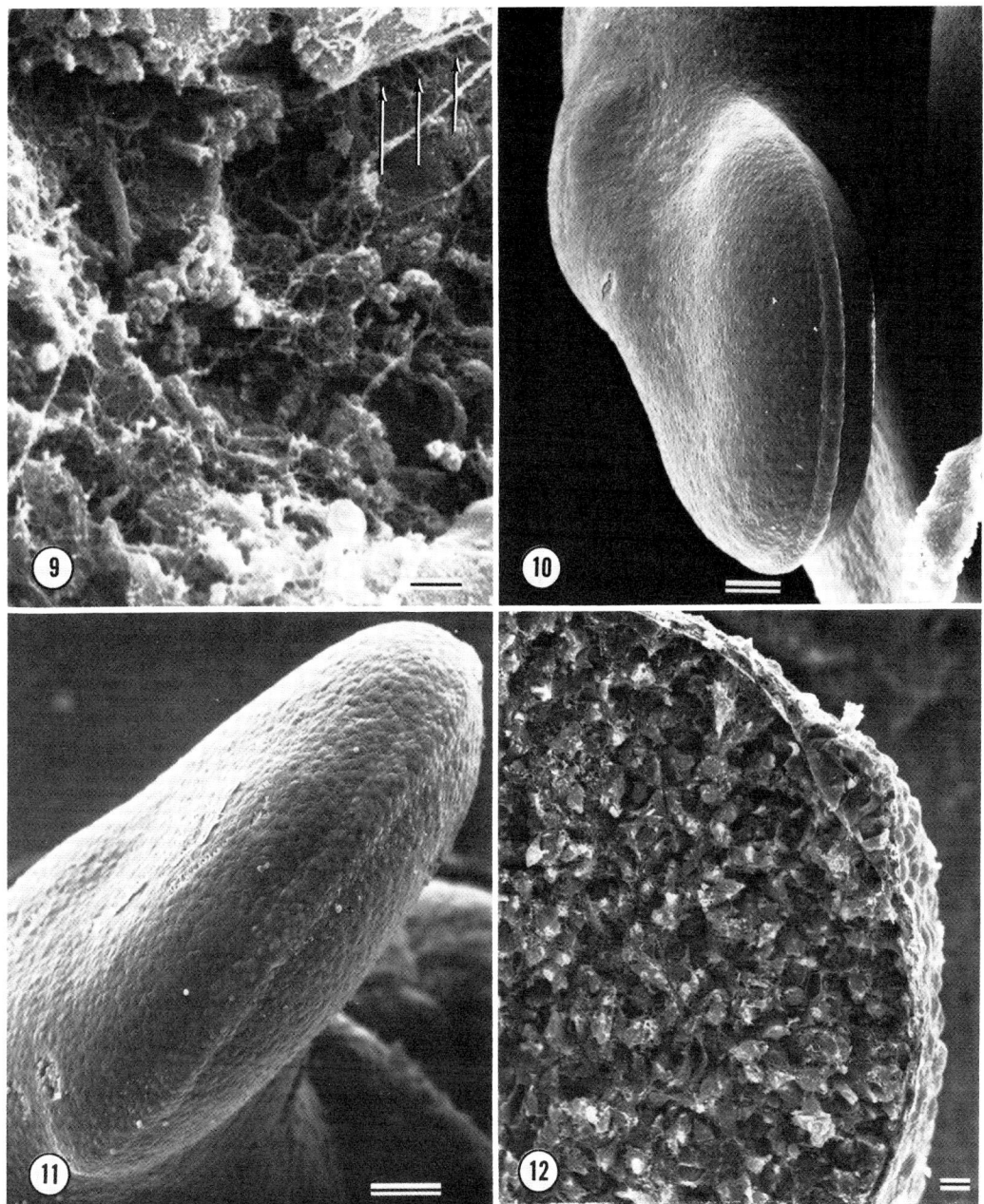

Figure 9: As noted, no direct intercellular contact occurs between the ectoderm and mesoderm during the period of ridge induction. However, it can be seen in this image that the extracellular matrix spans the interface (arrows) between interacting embryonic tissues. Bar = 1 μm.

Figure 10: As development continues, the limb bud elongates and exhibits a thickened ridge of ectoderm along its distal border (the "apical ectrodermal ridge"). This structure is required for the specification of the parts of the limb. If the ridge is removed, limb development ceases and no subsequent limb parts will form. Bar = 100 μm.

Figure 11: Since the apical ridge somehow signals underlying mesoderm, it is important for the borders of cells at the surface of the embryo to be tightly sealed to facilitate directional flow of information into the mesodermal compartment. As can be seen in this micrograph, individual cells, both in the apical ridge and the non-ridge ectroderm, are tightly joined at their apical surfaces, creating a cobble-stone appearance at the embryo's surface. Bar = 100 μm.

Figure 12: This dissected limb bud illustates the structual relationships of the apical ridge and non-ridge ectroderm to the underlying, as yet undifferentiated, mesoderm. A vascular plexus is beginning to form at this stage to provide nutrients to all regions of limb mesenchyme. Bar = 10 μm.

In addition, the ridge is responsible for the growth and proliferation of cells immediately subjacent to the ridge. This region has been termed the "progress zone" (see Figures 16 and 17) and it is a region of mesoderm in which cell proliferation participates in limb outgrowth. Figure 11 shows that the outer boundary of the limb bud is tightly sealed from leakage either of amniotic fluid or other extracellular fluids into the mesoderm across the ectoderm. In addition, the tightly joined ectoderm also prohibits materials which are secreted within the ectoderm in finding their way out to the extraembryonic environment. Indeed, materials which are synthesized and secreted by ectoderm have their way into the subjacent mesoderm facilitated as a result of the tight intercellular junctions which form the cobblestone pattern of apical cell surfaces within both ridge and non-ridge ectoderm.

As the apical ridge thickens, it is possible to dissect specimens in such a way as to reveal the ectodermal-mesodermal interface at regions of ridge as well as non-ridge ectoderm. Figure 12 illustrates such a dissected limb bud showing both the ridge, the non-ridge ectoderm and the subjacent mesoderm. It is noted that even at this stage of limb development, the mesoderm is structurally homogeneous, whereas the ridge is readily distinguished from the non-ridge ectoderm. Further dissection of ectoderm (Figure 13) reveals the interface between the mesoderm and the ectoderm where inductive interactions occur. Higher magnification (Figure 14) shows that the surface of mesodermal cells immediately beneath the apical ridge exhibits a labyrinth of cellular projections which intertwine with one another immediately beneath the basal lamina which is attached to the apical ridge. Studies with transmission electron microscopy reveal that these cellular projections often contact each other to form intercellular junctions with adjacent cells as well as with cells some distance away in the mesoderm. This observation is illustrated better in Figure 15 where it can be seen that cell processes contact both adjacent and more distal cells, and form communicating junctions which may provide a structural mechanism for cell-cell signalling during development. The molecular nature of such alleged signals is unknown. However, the diameter of channels which form within communicating junctions would permit only ions and small molecular weight metabolites to be transferred between cells coupled by such structures.

### Early Limb Vasculature

As the apical ridge establishes the compartments which shape the limb, additional structures appear which provide mechanisms not only for cellular nutrition, but potentially for shaping the limb as a result of the mediation of specific informational signals. This component is the developing vasculature. Figure 16 shows a small capillary partially dissected to reveal a red cell contained within. The capillary system is well developed within the developing limb and specific variations in fluid dynamics are observed within the developing limb. In addition, Figure 17 reveals the zone of proliferation (the "progress zone") which is composed of some 8-10 cell layers between the boundary of the apical ectoderm and the subjacent vascular sinus. This zone is a special region of cell proliferation with little, if any, cell differentiation occurring. It is interesting to question the regulation of the relationship between cell proliferation and cell differentiation at this stage, because no new structural development occurs. However, cell growth and cell division in all parts of the limb results in a gradual elongation of the developing limb.

### The Hand and Foot Plate

As limb development nears the period where the apical ridge ceases to dictate limb parts, the hand and foot plate appears, displaying the primordia of future digits (Fig. 18). Again, the polarity of the limb is retained, as well as similar polarity in the five distal digits which will develop further into fingers and toes. Several features are noted at this stage. First, there is a gradual demise of the apical ridge. Since all limb parts have now been specified, the apical ridge undergoes cellular death and subsequently regresses to become indistinguishable from embryonic skin. In addition, it is possible to observe the appearance of morphogenetically regulated cell death, which will ultimately give rise to the spaces between digits. The indentations at the distal boundary of the limb in Figure 18 mark the beginning of this phenomena. Furthermore, the cartilages which will contribute to the blastema of individual digits are in evidence within elevated regions radiating from the developing wrist region.

By dissecting the hand plate at this stage in development, it can be seen that individual mesenchymal cells in the core of the hand plate are surrounded by an elaborate extracellular matrix (Fig. 19). However, Figure 20 reveals that cells in the digital core, as well as the differentiating radius and ulna, are beginning to show surface characteristics of cartilage. Individual mesenchymal cells aggregate together, form short intercellular projections, and begin to elaborate cartilage-specific matrix containing proteoglycan molecules rich in chondroitin and other sulfated glycosaminoglycans. Figure 21 also shows that many of the cells that are peripheral to regions of developing cartilage retain the elaborate arrangement of intercellular projections which contribute to cell-cell communication.

By focussing one's attention on the region of distal indentation (which is contributing to the developing interdigital zones), it can be seen that the extracellular matrix in these regions is also changing. Collagen fibers are disassembling from the overlying ectoderm (Fig.

# Early Limb Development

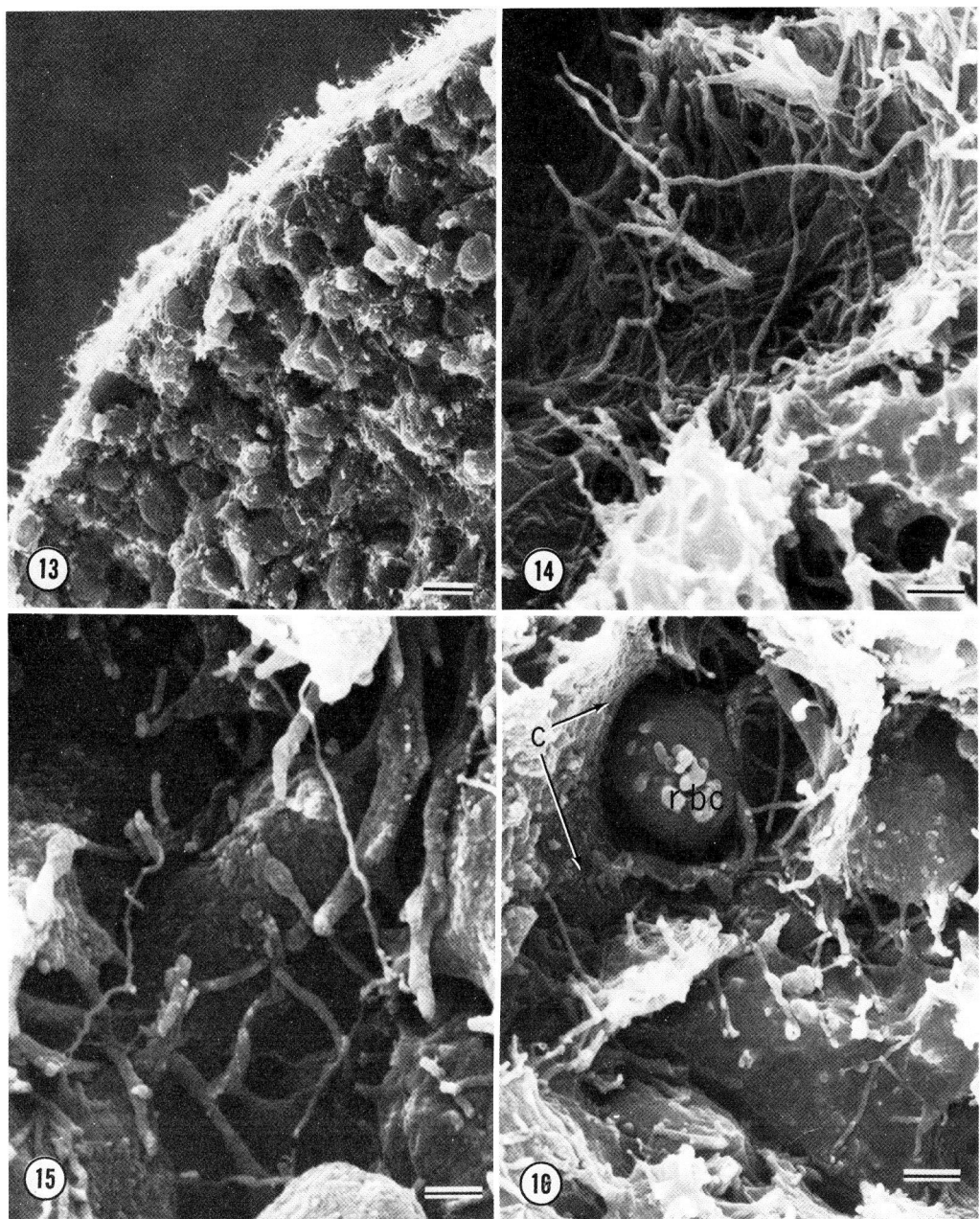

Figure 13: When the apical ridge is removed, the scanning electron microscope permits imaging of mesenchymal surfaces adjacent to inductive and non-inductive areas. This micrograph reveals the area beneath the ridge. Hundreds of slender, cellular projections from mesodermal cells are seen at the epithelial-mesenchymal interface. Bar = 10 μm.

Figure 14: Higher magnification of the mesodermal surface beneath the apical ridge reveals the interwoven nature of cellular processes. These projections contact each other and adjacent cell bodies, and are observed to penetrate several cell diameters into the mesoderm further from the apical ridge. It is probable that intercellular signalling is mediated by communicating junctions which form at sites of contact between cells. Bar = 2 μm.

Figure 15: It is important to recognize that contacts between cells in the limb mesoderm provide the basis for a signalling network which may play a role in organizing regions of tissues with morphogenetic significance and in coordinating differentation of cells into limb structures. This micrograph further illustrates the interconnections in this mesenchymal network. Bar = 2 μm.

Figure 16: This micrograph of a dissected region of mesoderm reveals a capillary (c) containing a red blood cell (rbc). It is not known whether the developing vasculature participates in limb morphogenesis beyond providing nutrients to cells and tissues. Bar = 2 μm.

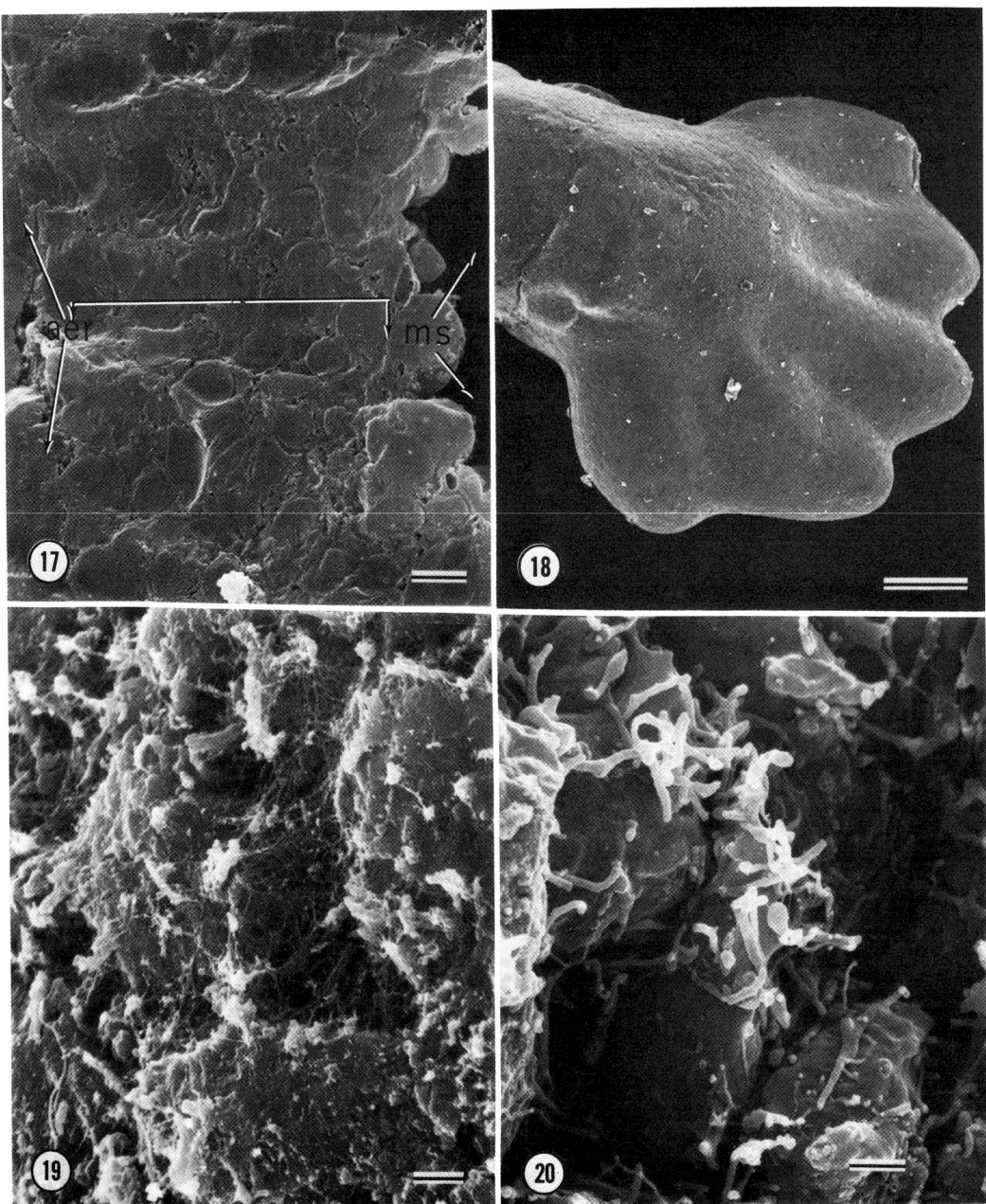

Figure 17: The region of mesoderm between the marginal vascular sinus (ms) of the limb and the apical ridge (aer) is known as the progress zone (brackets). This region does not undergo cellular differentiation as long as the apical ridge is present. However, cell proliferation (mitosis) is active which contributes to the outward growth of the limb. This specimen was prepared by freezing the embryo, then fracturing the tissue with a razor blade. Bar = 2 μm.

Figure 18: As development contiunes, distal structures on the limb begin to form. This is a micrograph of the developing hand plate from a human embryo. The five digits are beginning to take shape, as are the contours of the future interdigital spaces. A combination of digital outgrowth and interdigital cell death will shape the hand to its definitive form. Bar = 10 μm.

Figure 19: When the hand plate is dissected open, it is apparent that mesenchymal cells are aggregating to form cartilage. This micrograph reveals the matrix surrounding these cells as differentiation of cartilage begins. In the embryo, all the bony components of limbs are formed first as cartilage. Bar = 1μm.

Figure 20: With the matix chemically removed, the surfaces of cartilage cells can be seen. This image shows that cells are closely opposed (aggregated) and are losing the long processes observed earlier in development. Short projections, characteristic of chondrocytes (cartilage cells), are now observed. Bar = 2 μm.

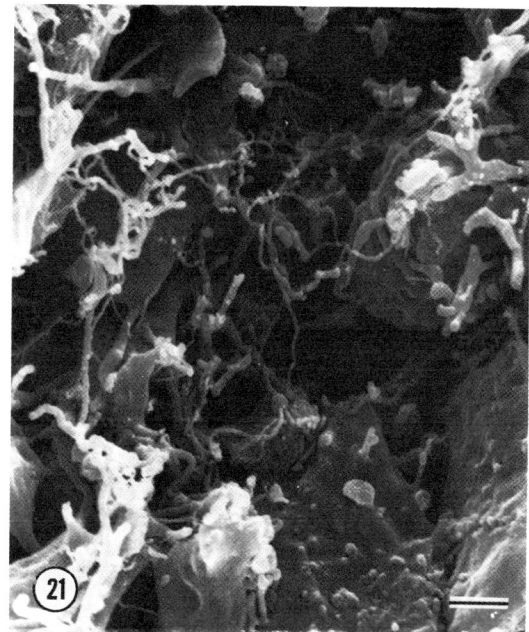
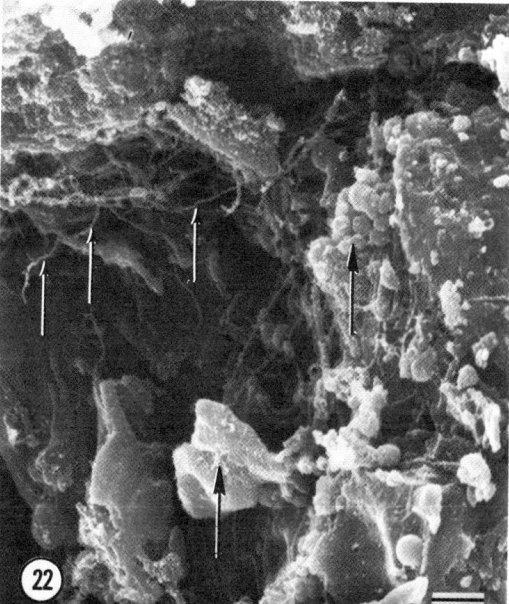
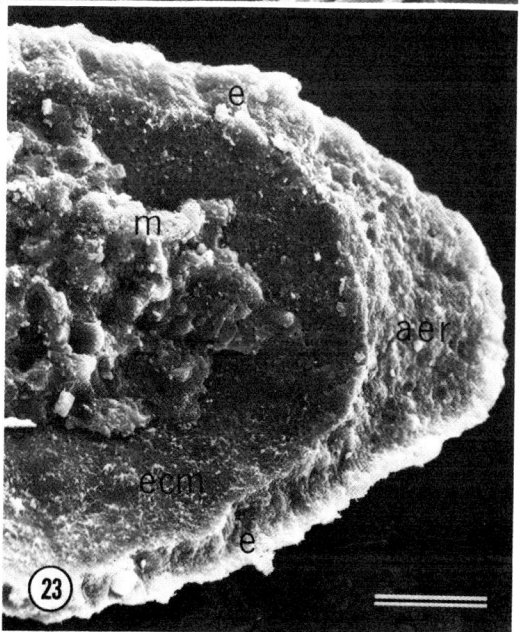

Figure 21  Mesenchymal cells in the hand plate which are located away from chondrogenic regions retain the long, interwoven projections characteristic of earlier, undifferentiated mesenchyme. Some of these cells will form muscle tissue, whereas others will undergo senescence and cell death.  Bar = 2 µm

Figure 22  Regions of the limb experiencing cell death (e.g. the interdigital spaces) exhibit reorganization and deterioration of both matrix elements and cells. In this micrograph, collagen fibrils beneath the ectoderm are visible (small arrows) as is cellular debris form necrotic tissue (large arrows).
Bar = 2 µm

Figure 23  A dissected limb bud which shows the principal cells and structures which interact to form the definitive limb: apical ridge (aer); mesoderm (m); non-ridge ectoderm (e); and the interposed extracellular matrix (ecm).  Bar = 10 µm

22), creating a disorganized pattern, and cellular death is in evidence. One must assume that adjacent living cells engulf dead elements by the process of phagocytosis and, indeed, there is evidence from other microscopic methods to verify this suggestion. As cell death proceeds, the overlying ectoderm will indent further to contribute to the developing interdigital space. Concomitantly, continued growth of the digit will result in the final pattern characteristic of the mammalian hand or foot plate.

## Summary

In summary, it can be seen that the developing vertebrate limb serves as a model morphogenetic system (Figure 23). This experimental model requires epithelial mesenchymal interactions, as evidenced by the mutual inductions between subjacent mesoderm and the overlying apical ectodermal ridge mediated through extracellular matrix. In addition, it is clear that for subsequent organization and development of mesoderm, cell contacts must be formed between mesoderm cells which facilitate intercellular signaling, exchange of nutrients and, potentially, the exchange of informational macromolecules. Furthermore, as development proceeds, the dual origin of mesoderm expresses itself in the formation of muscle and of cartilage and tendons. Tissue differentiation ultimately results in the expression of the various bony parts of the limb as well as the associated muscular and vascular compartments.

Finally, limb development is remarkably similar in reptiles, birds and mammals (although significant variants appear in amphibians). Since many of these embryos lend themselves readily to experimentation, we should be able to design experiments which will help us to understand more of the genetic control of limb morphogenesis. By so doing, one hopes that investigators will be able to define further those factors which control and regulate the shape of the vertebrate limb.

## Suggested Reading

### Books and Review Articles:

1. <u>Limb Development and Regeneration</u> Part A. (1983) ed., J.F. Fallon and A.I. Caplan; Alan R. Liss, Inc., New York.

2. <u>Limb Development and Regeneration</u> Part B. (1983) ed., R.O. Kelley, P.F. Goetinck and J.A. McCabe; Alan R. Liss, Inc., New York.

3. Kelley, R.O., J.F. Fallon (1981) <u>The Developing Limb</u>: an analysis of interacting cells and tissues in a model morphogenetic system in <u>Morphogenesis and Pattern Formation</u>. Ed., T.G. Connally, B. Carlson and L. Brinkley; Raven Press, New York, pp. 49-85.

4. Zwilling, E. (1961) Limb Morphogenesis in <u>Advances in Morphogenesis</u>. Ed., M. Abercrombie and J. Bracket; Academic Press, New York, pp. 302-330.

### Selected References in Journals:

1. Chevallier, A., Kieny, M., Mauger, A. (1977) Limb-somite relationships: origin of the limb musculature. J. Embryol. Exp. Morph. $\underline{41}$:245-258.

2. Fallon, J.F., Saunders, J.W. Jr. (1968) <u>In vitro</u> analysis of the control of cell death in a zone of prospective necrosis in the chick wing bud. Devel. Biol. $\underline{18}$:553-570.

3. Hornbruch, A., Wolpert, L. (1970) Cell division in the early growth and morphogenesis of the chick limb. Nature (London) $\underline{226}$:764-766.

4. Kelley, R.O., Fallon, J.F. (1976) Ultrastructural analysis of the apical ectodermal ridge during vertebrate limb morphogenesis. I. The human forelimb with special reference to gap junctions. Develop. Biol. $\underline{51}$:241-256.

5. Saunders, J.W. Jr. (1948) The proximo-distal sequence of origin of the parts of the chick wing and the role of the ectoderm. J. Exp. Zool. $\underline{108}$:363-403.

6. Wolpert, L. (1971) Positional information and pattern formation. Curr. Topics Dev. Biol. $\underline{6}$:183-224.

Editor's Note: All of the reviewers' concerns were appropriately addressed by text changes hence there is no Discussion with Reviewers.

PREPARATION OF EMBRYONIC TISSUES FOR SEM

R. E. Waterman

Department of Anatomy
University of New Mexico
School of Medicine
Albuquerque, NM 87131

## Abstract

Examination of the SEM literature regarding embryonic stages of a variety of animal species reveals that similar results may be obtained using a range of preparative techniques, but that images of significantly different quality may also be obtained by different investigators studying similar tissues processed by presumably identical methods. This paper provides a discussion of basic procedures which may be used to obtain "good" quality images of embryonic stages of a variety of representative species including insect, marine invertebrate, amphibian, avian and mammalian forms commonly used in developmental biology. It is hoped that such information will serve as an initial starting point for those with little or no previous experience in preparing embryonic specimens for SEM examination. In addition to comments regarding fixation, dehydration, drying and preparation of a conductive surface, representative procedures for procuring, handling and removal of extraembryonic layers are also mentioned. Some of the more common types of artifacts are also considered in an attempt to aid in interpretation of micrographs.

KEY WORDS: Embryonic Tissue, Fixation, Insect, Marine Invertebrate, Amphibian, Avian, Mammalian, Scanning Electron Microscopy

## Introduction

The SEM is a valuable tool for visualizing the complex topography of developing animal embryos. The increased resolution obtainable compared with that of the light microscope also makes it an effective instrument for quantitative study of a variety of developmental problems. Aspects of embryonic tissues commonly examined include the shape and surface features of individual cells, the changing three dimensional relationships between cells in rapidly developing embryos, and the relationship between cells and their extracellular environment during cell migration, tissue formation, and organogenesis.

A large literature dealing with SEM observations of embryonic and fetal tissues has evolved since publication of studies in the early 1970's stressing the value of critical point- or freeze-drying methods for reproducibly preserving the delicate structure of embryos. Although the current range of SEM techniques used for the study of soft biological tissues is potentially available for examination of embryos, the vast majority of the investigations of embryonic tissue to date involve the use of the SEM in the secondary electron mode. Very few studies have been reported using other modes (e.g., X-ray, backscattered electrons, cathodoluminesence). The SEM literature regarding vertebrate embryos has been reviewed by Waterman[1,2], and pertinent bibliographies and reviews of special topics may be found in other publications[3]; particularly these SEM symposia proceedings.

SEM examination of embryos usually involves the following sequence of steps: procurement of samples, fixation, drying, preparation of a conductive surface, and viewing in the SEM. Embryonic samples are subject to the same general problems of specimen preparation as other biological tissues, although extra care in handling, or special methods for obtaining young embryos are often required. Artifacts may be introduced at any stage and an ability to recognize such artifacts and a basis for devising means to eliminate them are important aspects in evaluating the resulting images.

The purpose of this tutorial is to provide

procedures and rationale which, in the author's experience or by reference to the available literature, can be used to provide useful and "good" quality images with currently available SEM's operated in the secondary electron mode. In view of the large number of studies and the wide range of types of embryos and embryonic stages which can be examined, no attempt is made to present a complete review of the literature. Instead, only representative protocols and species are presented to illustrate typical problems and approaches. Since the more advanced, or fetal, stages of developing organisms pose problems quite similar to "adult" tissues, emphasis in this tutorial is placed on the earlier developmental stages. It is hoped that the information provided will serve as an initial starting point which can be expected to yield useful results for those having limited experience in preparing embryos for SEM examination. A wide range of acceptable protocols are possible, but need to be determined for each species and for each developmental stage. Each method has advantages and disadvantages, and the choice of acceptable disadvantages should be dictated by the problem and the end result desired. It is essential that each problem or tissue be examined by a number of different techniques to provide a more complete understanding of the nature of the resulting artifacts and how these relate to the in vivo situation. References to more detailed or extensive studies are provided to aid in exploring modifications of the basic protocols presented.

As in all laboratory work with flammable, volatile, toxic and other potentially hazardous materials and equipment, safety must be considered. Detailed discussion of specific safety consideration in the SEM laboratory may be found in references[4,5].

## Procurement of Embryonic Tissues

Techniques for obtaining and handling embryos vary greatly. Embryos of certain species may be available for a limited time during the year, or may be obtained only with difficulty by fertilization in the laboratory. Early stages of many forms must be recovered from within the genital tract of the female, and most embryos are surrounded by one or more protective cellular or extracellular layers ("membranes") which must be removed to allow visualization of the embryonic surface. This is particularly true of the earliest stages of embryonic development. Certain of these protective coverings may be mechanically removed (e.g., with forceps or needles) or rinsed from the cells with a stream of buffered salt solution or fixative. Others require chemical or enzymatic treatment to dissolve or digest these materials; and possible damage to the cells of the embryo must be considered.

Because of the variety of extraembryonic coverings, no single procedure suffices for all situations. Specific information may be found in the literature or in reviews of useful techniques[6,7]. Since many experimental situations necessitate manipulation of embryos prior to fixation, however, the recipes of several isotonic, balanced salt solutions are presented in Table 1 which can be utilized for this purpose.

Table 1. Recipes for useful dissection solutions.

Synthetic Seawater (Woods Hole MBL)

| | |
|---|---|
| NaCl | 24.72 gm |
| KCl | 0.67 gm |
| $CaCl_2 \cdot 2H_2O$ | 1.36 gm |
| $MgCl_2 \cdot 6H_2O$ | 4.66 gm |
| $MgSO_4 \cdot 7H_2O$ | 6.29 gm |
| $NaHCO_3$ | 0.18 gm (add after all other salts are dissolved) |
| Distilled $H_2O$ | To make 1 liter volume |

Amphibian Ringer

| | |
|---|---|
| $CaCl_2$ | 120 mg |
| KCl | 140 mg |
| $NaHCO_3$ | 200 mg |
| NaCl | 6500 mg |
| Distilled $H_2O$ | 1 liter |
| Osmolarity about 240 | |

Mammalian Ringer

| | |
|---|---|
| $H_2O$ to make | 200.0 ml |
| NaCl | 1.8 gm |
| KCl | 0.84 gm |
| $CaCl_2 \cdot 6H_2O$ | 0.48 gm |
| $NaHCO_3$ | 0.1 gm |
| glucose | 0.1 gm |
| $MgCl_2 \cdot 6H_2O$ | 0.05 gm |

Insect Ringer (Pringle)

| | |
|---|---|
| NaCl | 9.00 gm |
| KCl | 0.20 gm |
| Glucose (dextrose) | 4.00 gm |
| $CaCl_2$ | 0.27 gm |
| Distilled $H_2O$ | 1 liter |
| Adjust pH to 7.2 with 1 M $NaHCO_3$ | |

Buffered Chick Ringer's Solution (Physiological Saline-Avian)

For 100 mls of Ringers solution:

| | |
|---|---|
| NaCl | 0.9 gm |
| KCl | 0.42 gm |
| $CaCl_2$ | 0.024 gm |
| $Na_2HPO_4$ (anhydrous) | 0.060 gm |
| or $Na_2HPO_4 \cdot 12H_2O$ | 0.08 gm |
| $KH_2PO_4$ | 0.020 gm |
| $NaHCO_3$ | 0.055 gm |

The temperature of the solutions should probably approximate that of the tissue in vivo for most routine morphologic studies. Commercial culture media or balanced salt solutions are also available. Manipulations of short duration may often be performed in a simple isotonic NaCl solution, but should be no longer than necessary since there is some evidence that surface changes may occur rapidly in certain tissues even in "physiologic" holding solutions[8].

## Handling

Several procedures have been described for handling and transporting small and delicate embryos through the phases of processing. Adherence of gametes and preimplantation embryos to various substrates such as glass coverslips or millipore filters[9], or enclosure in small containers constructed of filter or lens paper[10], empty ant pupae cases[11], or more complex mechanical design[12-14], have all been utilized. Construction of carriers or baskets is perhaps best left to the inventiveness of the investigator. Essential features of any such container include the use of materials (plastic, metal, etc.) which: 1) will not be dissolved by the solvents used, 2) will allow adequate and rapid exchange of fluid throughout the chamber while retaining the samples, 3) will hold a convenient number of specimens in identifiable locations, and 4) will conveniently fit into the drying chamber used. Multiple containers, or a single container with several compartments, are useful to maximize the efficiency of the drying step, but some means of identifying the individual compartments must be made, and access to the containers for introduction of samples and removal of the dried specimens must be such as to avoid mechanical damage.

A convenient technique for use with eggs, early cleavage embryos, or very small embryos is to affix them to coverslips coated with poly(L)lysine. Several cationic proteins, including poly(L)lysine, have been used to enhance the attachment of cultured cells to a variety of substrata[15]. These molecules, when absorbed onto the substrate, are thought to act as ligands which interact with negatively charged groups on the cell surface. As described by Mazia et al.[16]: a) Dissolve 0.1% poly-L-lysine (80,000-100,000 Daltons) in distilled water, b) coat the surface of a coverslip (12 mm, round slips conveniently fit the specimen stubs of several SEM instruments) by dipping or place a drop of poly(L)lysine solution onto the center of the glass and allow it to disperse into a thin layer; and c) wash the coverslip with running water. Embryos may either be allowed to settle onto the poly(L)lysine-coated surface then fixed and processed, or fixed in glutaraldehyde prior to being placed onto the poly(L)lysine. A slightly higher percentage of specimens may be retained if they are allowed to attach to the poly-L-lysine coated surface prior to fixation, although the rapidity and extent of the adhesion reaction may cause deformation of the adhering cells due to spreading and flattening on the substrate.

## Fixation

Except for a few special cases, hydrated embryonic tissue must be processed to allow the specimen to be scanned by the beam in the vacuum of the column of the SEM. Two general methods for stabilizing tissue are freeze-drying (or freeze-substitution) and chemical fixation. Of these, chemical fixation followed by dehydration and drying is by far the most commonly used method.

The aim of chemical (or any) fixation is to stabilize cell structure so that it can withstand the rather drastic forces during subsequent dehydration, drying, and interaction with the electron beam. The structure of cells and extracellular materials is inevitably altered during the fixation process, and since the complex chemistry of fixation is imperfectly known, what is judged to be the "best" fixation is a compromise. In general, criteria for "good" fixation include: 1) preservation of cellular morphology as close to the in vivo state as possible (usually as monitored by light microscopy) (Figs. 1, 2); 2) the dimensions of fixed structures should be similar to that expected from data obtained by other means; and 3) a partially subjective judgement as to what "things should look like." (e.g.,

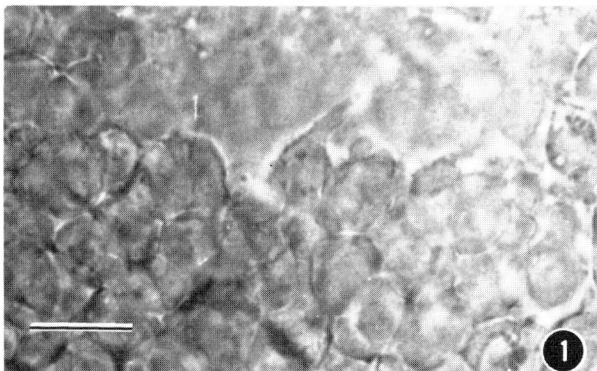

Fig. 1. Cells at anterior end of the embryonic shield in an intact, living zebrafish (Brachydanio rerio) embryo exhibit filopodia which contact the overlying ectoderm. Bar=100 μm.

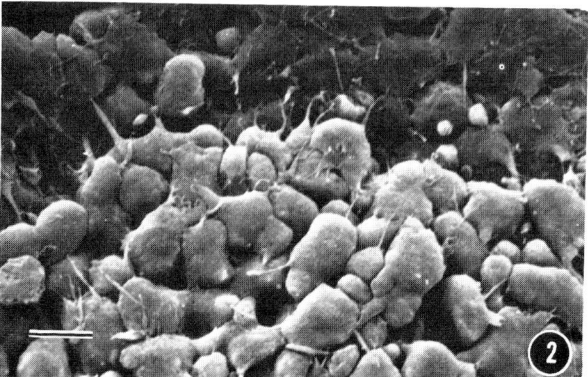

Fig. 2. Region of zebrafish embryo similar to that in Fig. 1 viewed from inferior surface with SEM. Bar=10 μm.

membranes should not have large holes). The latter is often difficult to assess, and an appreciation for what structures should probably look like is best achieved by comparing the results of a number of different preparative procedures, although this is too seldom done or reported. Another criterion useful for assessing fixation for SEM is to examine the ultrastructure of specimens embedded, sectioned and viewed in the transmission electron microscope (TEM) after being scanned.

A fixative solution normally consists of a fixative (e.g., glutaraldehyde or osmium tetroxide) in a suitable vehicle (usually a buffer and/or various salts or other additives)[17]. Since detectable secondary electrons are limited to a small distance beneath the surface, the main requirement for fixation of SEM specimens might seem to be preservation of the surface only. Preservation of the cytoplasm and other components of the tissue are also important, however, to aid in resisting the forces encountered during subsequent processing. Fixation of the internal portions of larger embryos or pieces of embryos also permits viewing of internal surfaces exposed by fracturing or dissecting the specimen prior to, or after drying, and the amount of information obtainable from a single specimen can be extended further by examination of thin sections cut from scanned specimens following embedding and sectioning.

Although many fixatives used to obtain adequate preservation for light microscopy might also yield adequate preservations for some types of SEM examination, most current investigations utilize a fixation protocol which has been demonstrated to provide "good" fixation of a particular tissue at the ultrastructural level. It should be mentioned, however, that because of the considerable shrinkage produced during the drying process, and since SEM techniques are not presently capable of routinely resolving many details of subcellular fine structure, the permissible range of variation in fixative procedure may be greater for SEM than for TEM preparations.

Historically, osmium tetroxide was shown to be one of the best light microscopic fixatives for preserving cell ultrastructure (particularly membranes) and was a common fixative used during the early development of fixatives for transmission electron microscopy[18]. In the early 1960's, Sabatini and coworkers published data comparing the efficacy of several aldehydes as ultrastructural fixatives[19,20], and strongly recommended glutaraldehyde as a primary fixative to be followed by secondary fixation with $OsO_4$. Other aldehydes (e.g., formaldehyde, acrolein) have also been used as fixatives. Formaldehyde, freshly prepared from paraformaldehyde[21], penetrates more rapidly than glutaraldehyde as a fixative. Mixtures of one or more aldehydes were subsequently demonstrated to provide fixation better than that obtained by use of any single component. This was particularly true for a mixture of paraformaldehyde and glutaraldehyde[22,23]. The addition of other components (e.g., trinitro-compounds)[24] or combinations of aldehyde and osmium are also used.

## Factors affecting fixation

A large number of factors are involved during the fixation process. Because of this, it is impossible to accurately calculate the ideal fixative or protocol for each tissue, or for each developmental stage, and the choice of a fixative for a particular purpose is often a matter of trial and error. Some of the general factors of the fixative to be considered include: penetration, pH, ionic strength, ionic composition, temperature and time of fixation, and osmolarity. Detailed discussion of these factors and recipes for solutions may be found in a number of sources[13,17,25,26]. Only brief mention of several points relating to fixation as presented here as a guide to better understanding the theoretical basis for the protocols most commonly used in SEM studies.

*Penetration.* An ideal fixative should stabilize tissue components immediately, causing minimal changes in structure. The rate of stabilization or "fixation" depends on many factors, including the speed of penetration into the tissue and the rate of chemical stabilization of molecular structure. The rate at which a particular fixative penetrates depends in some cases on the manner in which the tissue is exposed to the fixative. Small embryos or samples may be adequately fixed by immersion into the fixative solution. If dissection of the embryo is necessary prior to complete fixation, it should be accomplished as rapidly as possible in an isotonic solution at appropriate temperature, or during the initial stages of fixation. The specimen should be constantly immersed to prevent air drying or distortion of delicate structures by surface tension forces. Larger fetuses may perhaps require perfusion of the fixative via the vasculature to achieve rapid dissemination of the fixative to the internal regions of the specimen. The inherent rate of penetration of fixatives varies greatly, and is one of the underlying rationales for using combinations of fixatives; a slowly penetrating plus a rapidly penetrating fixative. Other factors such as the characteristics of the fixative vehicle, temperature, and concentrations of the fixative, also may affect the speed of penetration. It must be remembered, however, that the rate of penetration of particular fixative does not always correlate directly with its rate of chemical stabilization of various cellular and extracellular components.

*pH.* Most fixatives used for ultrastructural preservation are "non-coagulant" fixatives (e.g., acrolein, formaldehyde, glutaraldehyde, osmium tetroxide) and rely in part on cross-linking between proteins as a mechanism of stabilization of structure. The isoelectric pH of a protein depends on the relative numbers of acidic and basic groups within the molecule. Proteins exhibit minimal solubility and viscosity, lowest osmotic pressure and the least swelling at their isoelectric pH. Small changes in the pH of the fixative may lead to large changes within cells, and the pH of the fixative should therefore remain as close as possible to the normal pH of the tissue in an attempt to minimize changes in protein structure. The pH of different proteins and in different regions of an organism may differ, however, and some variation in optimal pH values may exist for special purposes or tissues. In practice, there appears to be little evidence that the pH of a fixative is critical from pH 6.5-8.0 for most tissues[17]. Most fixatives are therefore adjusted to near neutrality since the average pH of most cells and tissues is approximately 7.4.

Vehicles. The primary components of most fixative vehicles are buffers. Buffer solutions contain a weak acid and a base and its salt. Buffers are normally present in biological tissues and tend to resist changes in hydrogen ion concentration (pH) when exposed to strong acids or bases. The most common buffers used in fixatives for electron microscopy are phosphate and cacodylate. Others, such as veronal-acetate, or s-collidine, are also used; usually for special purposes or historic reasons. Two important aspects to consider in choosing a buffer are: 1) its buffering capacity over a desired range of pH values; and 2) possible effects of the buffer on the structure of the tissue. An ideal buffer would exactly match the pH, osmolarity and ionic constitution of the living tissue[17].

Osmolarity. Considerations of osmolarity largely revolve around effects of shrinkage or swelling of tissues and cells. Theoretically, hypotonic solutions may cause swelling, hypertonic solutions shrinkage, of cells and tissues. The osmolarity of the fixative is therefore usually adjusted to be isosmotic with the cells being fixed. The osmolarity of a fixative is normally adjusted by varying the concentration of the buffer (not the fixative agent), or by the addition of ionic (e.g., $NaCl_2$, $CaCl_2$) or non-ionic (e.g., glucose, sucrose) compounds. A useful guide for making solutions of various osmolarity for a wide range of tissues can be found in the lab manual of Millonig[26]. It is extremely difficult to achieve isosmotic conditions in practice, since the osmolarity of the tissues is difficult to determine accurately, and since the action of the fixative with cell components may radically alter the osmotic characteristics of cells. For example, membranes become rather freely permeable to macromolecules after $OsO_4$ fixation[27], whereas they remain less freely permeable following exposure to glutaraldehyde or formaldehyde[28]. The fixing agent may also contribute to the osmolarity of the fixative solution. Experience in preparing tissue for TEM examination indicates that a slightly hypertonic fixative solution is often less damaging than a hypotonic one; particularly in minimizing swelling. It has also been suggested, however, that an isotonic solution is more critical for SEM than for TEM preparation, if the fixation of surface features is the principal goal[13].

Ionic composition. The primary effect to be expected if the ionic composition of the fixative solution differs from that of the natural environment of the cells is artifacts due to extraction of materials or precipitation of material resulting from interaction with the fixative[29]. Since these occur before the tissue is fully stabilized, these considerations are most important when dealing with fixatives which penetrate slowly (e.g., $OsO_4$) or react (e.g., formaldehyde) slowly[17]. Calcium chloride is often added to fixative solutions since calcium ions appear to aid in stabilizing cellular membranes.

Buffers

Phosphate buffers. These are the most "physiological" buffers since they approximate certain components of tissue fluids and are non-toxic to living cells in culture[17]. They are therefore especially good for use with slowly penetrating and/or reacting fixatives. Temperature changes have little effect on the pH of phosphate buffers. They are non-toxic, and are stable for weeks at 4°C if sugars have not been added and growth of microorganisms is monitored. Most of the phosphate buffers are based on the formula of Sorenson.

Cacodylate buffers. Proposed as buffers for electron microscopy by Sabatini et al. in 1963[19], cacodylate buffers are easy to prepare, stable and do not promote growth of microorganisms. However, they do contain arsenic and have an unpleasant smell.

Veronal-acetate buffers. These buffers contain a barbituate, and they should be handled and exposed of with care. The Michaelis veronal-acetate buffers were used with osmium fixatives during the early days of transmission electron microscopy since they were thought to cover the physiological range of pH at constant ionic strength. It was subsequently shown, however, that veronal-acetate is a poor buffer in the range pH 5.2-7.5[30], and should not be used with aldehydes since sodium veronal reacts with aldehydes to yield a solution with no buffering capacity in the physiological range[31]. They can be used to buffer $OsO_4$ to be used as a secondary fixative (following primary aldehyde fixation), and can be used with uranyl acetate during "en bloc" staining since it will not result in precipitates as do phosphate or cacodylate buffers[17].

Fixative agents

Osmium tetroxide. Osmium tetroxide ($OsO_4$) is a volatile substance and dissolves rather slowly in water. Since reduced osmium imparts high contrast to osmiophilic components of the cell, it was extensively used as both a fixative and electron dense stain for TEM. The main disadvantage of $OsO_4$ is that it penetrates and reacts with tissues so slowly that changes in structure may occur prior to stabilization. The ionic composition of the buffer and fixative solution must therefore be chosen to minimize these effects. Osmium tetroxide, however, largely destroys the osmotic properties of cell membranes, thus making the osmolarity of the fixative solution of less importance. Primary fixation with $OsO_4$ alone has been almost entirely discontinued for "routine" specimen preparation, and $OsO_4$ is currently used in combination with glutaraldehyde as a primary fixative, or as a secondary (post-) fixative.

Aldehydes. Aldehydes have largely superceded the use of $OsO_4$ as a primary fixative. The most common fixation protocols involve initial fixation with glutaraldehyde or a combination of (para) formaldehyde and glutaraldehyde followed by secondary fixation with $OsO_4$.

Glutaraldehyde is a dialdehyde which rapidly stabilizes structures by cross-linking proteins thus minimizing the opportunity for extraction by the buffer. Glutaraldehyde fixation does not destroy the osmotic properties of cells[28] and more care must be taken to control the osmolarity of the fixative solution as well as of the medium following exposure to glutaraldehyde alone. Since the final oxidation product of glutaraldehyde is glutamic acid, some workers initially recommended storing glutaraldehyde over barium carbonate since barium glutamate is insoluble[25]. However, glutaraldehyde should be stored at 4ºC and barium carbonate should not be added. Polymerization of glutaraldehyde results in a yellow solution and can occur in the presence of even small amounts of water. Solutions of glutaraldehyde also deteriorate rather rapidly at room temperature and in the presence of oxygen, or at alkaline pH. It may gel at concentrations of 50% or more, although 25% solutions may remain stable for long periods at 4ºC. Polymerization decreases its ability as a cross-linking agent and the purity of commercial preparation may vary. Techniques for purifying glutaraldehyde solutions may be found in Glauert[17]. Glutaraldehyde does not preserve lipids to a significant degree, hence another reason for combining with $OsO_4$ either in the primary fixative or as a secondary fixative. Penetration of glutaraldehyde may be monitored by observing the pale yellow color (due to formation of Schiff-positive bases when the dialdehyde reacts with basic amino groups[25]), and by the degree of firmness of the fixing tissue.

Formaldehyde has been, and still is, used as an important fixative for light microscopy. It was also used during the early days of electron microscopy, but was not adopted because of poor results[31]. This was presumably due to the presence of methanol (11-16%) in commercial solutions of formaldehyde[32], and fixation can be improved with methanol-free formaldehyde prepared fresh from powdered paraformaldehyde (trioxy-methylene)[21]. Like glutaraldehyde, formaldehyde also cross-links proteins, but to a lesser degree; presumably since formaldehyde is a monoaldehyde. The reaction is also slow and reversible so that tissue should not be washed or stored in buffer or other solutions prior to further processing. The permeability of cellular membranes is also not entirely abolished by formaldehyde, and therefore the buffer characteristics are important as with glutaraldehyde. The principle advantage of formaldehyde is its faster rate of penetration than either glutaraldehyde or osmium. It can be used to preserve tissue for SEM, and is often used for preserving human embryonic and fetal tissue in a hospital setting. It is currently more often used, however, in combination with other fixatives.

Acrolein (acrylic aldehyde) was introduced as an ultrastructural fixative by Luft in 1959[33]. The principle advantage of this unsaturated monoaldehyde seems to be its rapid rate of penetration, and it is usually used in combination with other aldehydes[34,35]. It polymerizes on exposure to light, air and certain chemicals, and should be stored in tightly stoppered clean, brown glassware at 4ºC to retard polymerization which can be initiated by small amounts of impurities. Commercial preparations usually contain an inhibitor of oxidation (e.g., hydroquinone) which can be removed by distillation of the acrolein prior to use, but there is little evidence that purified acrolein yields better fixation[17]. Acrolein is a powerful lachrymosing agent, and should be used with great care in a fume hood or in other situations where adequate ventilation exists.

Combined fixatives. Glutaraldehyde-osmium: Combinations of fixing agents have been developed in an attempt to maximize the advantages and/or minimize the disadvantages of the component fixing agents. Simultaneous use of glutaraldehyde and osmium was shown to yield good results for TEM fixation by Trump and Bulger in 1966[36], and has been applied to certain embryonic tissues. Since these two fixatives react with each other, the mixture is not stable, and should be made up immediately prior to use.

Glutaraldehyde-(para)formaldehyde: The use of formaldehyde to rapidly penetrate and partially stabilize structure which can be more permanently fixed by the more slowly penetrating, but better cross-linking, glutaraldehyde was used by Karnovsky[22], and as originally described or in diluted form[23], is widely used as a primary fixative for both TEM and SEM. The original fixative solution was considered to be highly hypertonic (approximately 2,010 mosmol), and hence it was subsequently suggested that this be diluted to a more appropriate osmolarity with buffer. These calculations were made with the assumption that glutaraldehyde was osmotically active, however, which may not be the case[17,37-39].

Aldehyde fixatives containing trinitro compounds: Addition of trinitrophenol (picric acid) to (para)formaldehyde fixatives was used by several workers to fix a variety of tissues, and addition of a variety of other trinitro compounds (all potentially explosive!) to the basic Karnovsky fixative was reported by Ito and Karnovsky in 1968[24]. These mixtures yield results judged to be as good, or better, than other fixatives for a variety of tissues, and can be used for large samples because of rapid penetration. The following variation of this fixative (PAFG) containing picric acid is used frequently in the author's laboratory to fix a wide range of embryonic tissue:

a. Add 20 gm paraformaldehyde to 150 ml of doubly filtered saturated aqueous picric acid (trinitrophenol) solution.

b. Heat to near boiling and add several drops of (0.05-0.1 N, 2-4%) NaOH to clear the solution.

c. Place solution into a graduate cylinder and bring volume to 1,000 ml with phosphate buffer. To make buffer, dissolve 3.31 gm $NaH_2PO_4 \cdot H_2O$ and 33.77 gm $Na_2HPO_4 \cdot 7H_2O$ in 1 liter distilled water.

d. Osmolarity of this solution (PAF) is 900 mOsM; pH 7.3. Solution is stable for a long time.

e. To make final fixative (PAFG), add 3% glutaraldehyde to the PAF solution just before use (12 ml of 25% glutaraldehyde per 88 ml of PAF).

Some shrinkage may result, particularly from prolonged exposure, and the author's experience indicates that tissue must be thoroughly washed prior to secondary fixation with osmium to prevent precipitation artifacts.

---

Figs. 3-8 illustrate cardiac jelly of stage 17-19 chick hearts fixed in different ways. (Courtesy of Dr. R. Markwald).

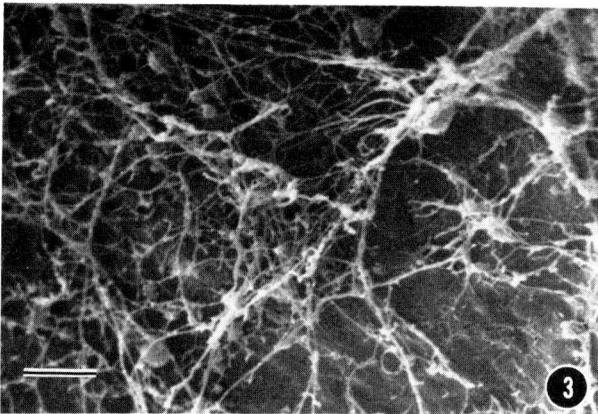

Fig. 3. SEM of matrix fixed in glutaraldehyde without CPCL. Only microfibrils and large globules (complexed glycoprotein?) are visible. Bar=1 μm.

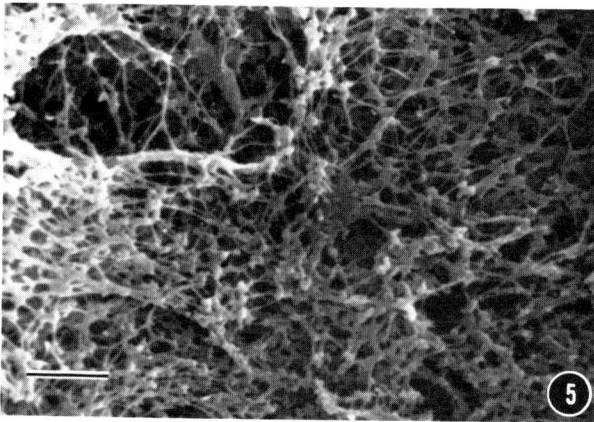

Fig. 5. A continuum of pleomorphic materials ("CPCL-dependent matrix") in addition to microfibrils appears in cardiac jelly fixed with glutaraldehyde + CPCL. Bar=1 μm.

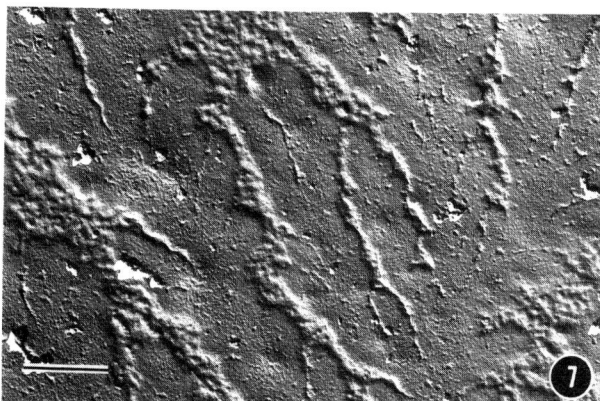

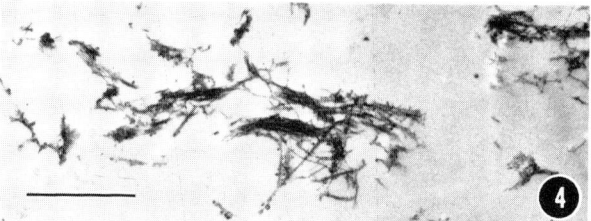

Fig. 4. TEM of matrix fixed in glutaraldehyde without CPCL. Less material is preserved than when CPCL is present (cf., Fig. 6). Bar=1 μm.

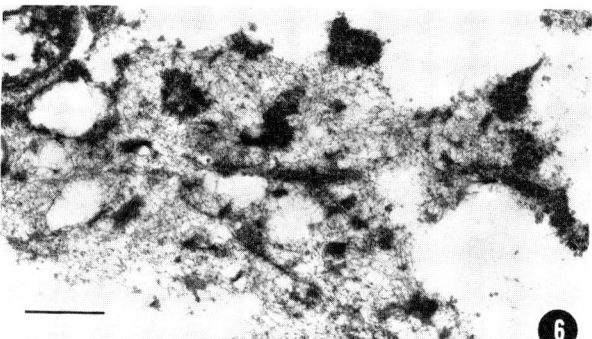

Fig. 6. TEM reveals CPCL-dependent matrix is composed of amorphous dense material (incorporates $^3$H-fucose), 30-35 nm granules (chondroitin sulfate-protein) and 3 nm filaments (probably hyaluronate). Bar=0.5 μm.

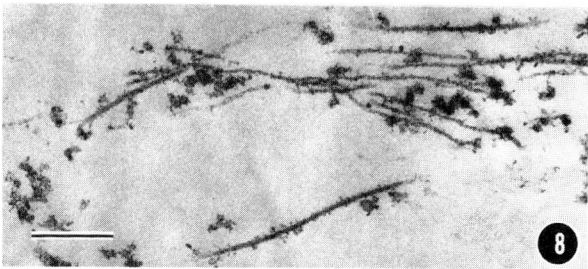

Fig. 8. TEM of matrix fixed in glutaraldehyde + 3% tannic acid in 0.1 M phosphate buffer (pH 7.2). More matrix material is visible than after glutaraldehyde only, but may reflect lead-mordanting properties of tannic acid and not increased preservation of material. Bar=0.25 μm.

Fig. 7. TEM of carbon-platinum replica of unfixed, frozen (liquid $N_2$) externalized cardiac jelly. As in SEM preparations (Fig. 5), material fills some intercellular spaces and surrounds microfibrils. Bar=1 μm.

Other additives: Many of the initial TEM fixatives were designed primarily to preserve cellular membranes and intracellular structure and do not preserve components of extracellular coats or the extracellular matrix. With increased interest in the extracellular matrices generated by their roles in intercellular communication and embryogenesis, several compounds have been added to routine EM fixative solutions to enhance preservation of such materials which contain significant amount of polysaccharides and are not usually preserved to a significant extent by either osmium or aldehyde fixatives alone[40]. Several compounds which have been extensively used in embryonic tissues, include: alcian blue, lanthanum, ruthenium red, tannic acid and cetyl pyridinium chloride. Precipitation of extracellular material by such additives may obscure cell surface features, however, and the identity of the preserved material can not be made with confidence from SEM data alone.

An example of the usefulness of these additives is a series of combined SEM, TEM and biochemical studies to determine the composition of cardiac jelly during heart development published by Markwald and co-workers[41,42]. Among the additives used in these studies is cetylpyridinium chloride (CPCL; CPC), a cationic, long chain carbon quartenary ammonium compound which ionically interacts with glycosaminoglycans (GAG) causing reduced solvation of both substances. CPCL added to glutaraldehyde is a useful additive for retaining water soluble matrix moieties not conjugated by aldehydes. Glutaraldehyde (1-2%) containing 0.25-1.0% CPCL in 0.065 M cacodylate buffer is a particularly useful fixative for matrices rich in hyaluronate. If the matrix contains little hyaluronate, however, the addition of CPCL to the primary fixative will tend to condense or "complex" persisting charged matrical components (proteoglycans, acidic glycoproteins) into somewhat amorphous appearing electron dense strands. To a lesser extent, the same "complexing" of components occurs in some matrices if other charged substances are used; e.g., ruthenium red or alcian blue. Thus, if the matrix is known to contain little hyaluronate or water soluble material, CPCL should probably not be added to the fixative. Cetylpyridinium chloride is also a mild detergent and at concentration of 0.5% or greater will produce some undesirable alteration of cell morphology, although such damage seems minimal at concentrations below 1%. Unfortunately, as yet, no fixative appears to retain water soluble matrix moieties and simultaneously give excellent cellular detail, thus necessitating a compromise between matrix preservation and cellular preservation. Acceptable results may usually be obtained by using 0.25% CPCL in 2% glutaraldehyde in 0.065 M cacodylate buffer pH 7.0 at room temperature for 2 hrs, but a range of concentrations should probably be examined for each tissue studied. The beneficial results obtainable in CPCL in preserving matrices containing free hyaluronate (as opposed to hyaluronate incorporated or internalized into a macromolecular aggregate such as chondroitin sulfate proteoglycans) are illustrated in Figure 3-8. These micrographs illustrate the cell free embryonic cardiac jelly matrix located near the myocardium of stage 17-19 chick embryos. Similar results have been obtained using embryonic matrix associated with cephalic neural crest cells, as well as a number of non-embryonic tissues.

## Dehydration and Drying

Hydrated tissue must be dried in order to be observed in the SEM under "normal" circumstances, and it is this step of the preparation procedure which probably accounts for the most obvious distortion artifacts in most SEM specimens. Because water plays an integral role in the macromolecular architecture of cellular and extracellular components, from the point of view of preservation of structure, the success of all drying techniques ultimately depend on the degree to which the distorting forces generated during removal of this water can be minimized. Such forces at the cellular and subcellular levels are primarily due to considerations of surface tension.

The simplest of the several possible drying procedures is to allow the water to evaporate ("air-drying"). However, the distortion produced by evaporation of water, or even of solvents with lower surface tensions following chemical fixation to harden the tissue and replacement of water with these solvents, is so great that air drying is not adequate for routine studies, and is only to be considered for special problems when other techniques are not possible (Fig. 9). The two techniques for removal of water which are most commonly used are freeze-drying (FD) and critical point drying (CPD). Of these, CPD is used more extensively.

### CPD

This technique was introduced by Anderson[43,44] to improve the preservation of microorganisms for electron microscopy, and has been widely applied to embryonic tissues since the early 1970's. Many critical point drying apparatuses are commercially available, and the theoretical and practical considerations of the technique have been recently reviewed[45].

Fluids which undergo a phase change from liquid to gas at the critical point are termed "transitional fluids". Although water is such a fluid, the critical temperature and pressure of water are too high to be practicable and the water within the tissue must be replaced by another transitional fluid with more appropriate critical point characteristics for CPD. Transitional fluids most commonly used are $CO_2$ and certain fluorocarbons ("Freons")[46]. Unfortunately, both of these transitional fluids are not completely miscible with water, and hence the water within the tissue must be first replaced by an "intermediate fluid" (or dehydrating agent) which is miscible with the transitional fluid used. More than one intermediate fluid

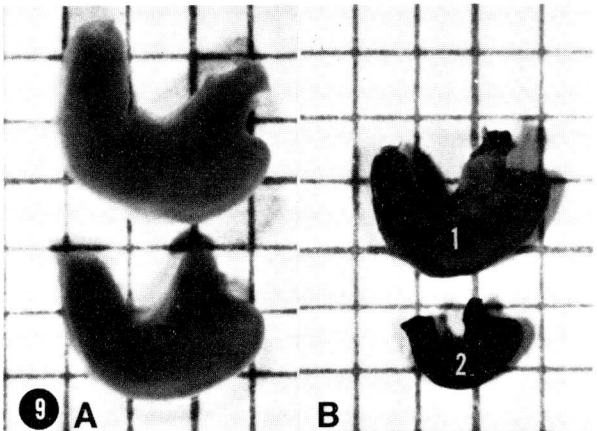

Fig. 9. Photomicrographs of fetal human stomachs comparing air- and critical point-drying. A: Stomachs 1 and 2 during initial PAFG fixation. B: Same stomachs following drying. Number 1: post-fixed in $OsO_4$, dehydrated in ethanol, infiltrated with isoamyl acetate, CPD from $CO_2$. Number 2: post-fixed in $OsO_4$, air dried from acetone. Photographed against a 1 mm square grid.

may be used if the initial dehydrating fluid is not completely miscible in the transition fluid [45].

For most CPD protocols, the tissue is fixed and then dehydrated by passing through increasing concentration (e.g., 30, 50, 70, 95, 100%) of ethanol, methanol, or acetone which replaces the water within the tissue. All appear to yield essentially identical results, although some differences in dimensional changes during dehydration in various fluids have been documented[47]. Other dehydrating agents have been used[17,48,49], but are not routine; especially for embryonic tissues. A major effect of dehydration with these solvents on the composition of the tissue involves extraction of lipids[50] or other components. Extraction of lipids can be minimized by osmication, the use of acetone, and avoiding prolonged exposure to the dehydrating agents[17]. Anderson[43] originally reported that liquid $CO_2$ is not miscible with ethanol, and used a second intermediate fluid, amyl acetate, which is miscible in both ethanol and liquid $CO_2$. Although amyl acetate is still used occasionally, and can provide a sensitive olfactory clue to the complete purging of the chamber and tissue with liquid $CO_2$, it is toxic and tissue may be passed from absolute ethanol into liquid $CO_2$ with no adverse effects[51]. The $CO_2$ used should not contain water, and care should be taken not to inadvertantly allow the tissue to air dry during the process. The standard CPD protocol for tissue in 100% ethanol followed in the author's laboratory involves three, 3 minute purges of the chamber of the CPD apparatus with liquid $CO_2$, alternated with 3 minute intervals with the exhaust valve closed. The chamber is sealed and heated following the third flushing. This has been determined empirically, and yields reproducible results. However, whether all the ethanol is purged from the tissue after this procedure has not been measured.

For critical point drying using one of the fluorocarbons, tissue is fixed and dehydrated in ethanol as above, then, although ethanol is miscible with any of the freons used, the ethanol is replaced with Freon TF (Freon 113)[45,46]. Freon TF is liquid at room temperature, nearly non-toxic, non-flammable, and repeated flushing of the chamber to completely remove it prior to CPD need not be done since it has properties similar to Freon 13 which is generally used as the transitional fluid for critical point drying. A small amount of Freon TF can therefore be carried over when the tissue is placed into the drying chamber and will act as a transitional fluid with it's own critical point. The chamber does not have to be thoroughly flushed with Freon 13 since Freon TF need not be completely removed, saving time and expense. Because Freon TF is readily miscible with $CO_2$, it may be used as an additional intermediate fluid between ethanol (acetone) and $CO_2$, and may aid in preventing evaporative drying since an amount sufficient to cover specimens may be transferred into the chamber of the drying apparatus. Freon TF is quite volatile, however, and transfer of the specimen and filling of the chamber must be done without delay. Critical point drying from either $CO_2$ or Freon 13 yield similar results, although some data suggest CPD with Freon 13 may yield somewhat less shrinkage[47]. The use of fluorocarbons may entail adverse environmental considerations, however, and the freons are generally more expensive than $CO_2$. Dried specimens should be stored in a dessicator or other dry environment between manipulations to prevent uptake of moisture. Although critical point drying is rapid, and a great improvement over air drying in preserving delicate surface features, considerable overall shrinkage of CPD specimens is routinely observed[1,47].

FD

The applications and theory of FD for SEM have been extensively discussed[52,53,54]. Less shrinkage is observed after FD than CPD, but the process is more time consuming and has not yet been used extensively to prepare embryonic specimens for SEM which makes detailed comparison of these methods difficult at present. A major disadvantage of the freeze drying procedures for many studies is the disruption of subcellular structures by ice crystals which usually form during freezing and which may reduce the usefulness of TEM observations of scanned specimens. A number of methods have been described for minimizing the amount of damage due to ice crystals[54], including FD from Freon 113[55], and this technique offers distinct advantages for the future. It is interesting, however, that SEM images resulting from studies comparing CPD and FD tissues many times appear remarkably similar.

## Mounting

Once dried, specimens must be mounted on stubs for viewing. This involves not only affixing the specimen to the stub with a suitable adhesive, but also necessitates establishment of a conductive contact between the specimen and the stub since most biological tissues are poor conductors of heat and electric charge. The manipulations may be accomplished separately by using a non-conducting glue or tape as the adhesive followed by creation of a conductive contact, or simultaneously by using a conductive paint or colloidal solution as both an adhesive and conductive medium. If the specimens have been previously affixed to a substrate (e.g., millipore filter, poly(L)lysine-coated cover slip, etc), this substrate may be simply affixed to the stub and a conductive path between stub and specimen then created with conductive paint. Any thin object such as a wooden splinter, wire, pine needle or cactus spine may be used to accomplish this with minimum risk to the specimen.

Placing specimens on strips of double-sided adhesive tape permits better control over the orientation of the specimen, and allows more possibilities for dissection. The strips of tape should be as small as possible to reduce movement of specimens as a result of cracking of the tape during subsequent metal coating (Fig. 10). Cracking of the tape can also be minimized by painting as much of the unused surface of the tape as possible with conductive paint. This also usually provides a more uniform background in the resulting micrographs. Specimens too small to permit handling by forceps may be simply scattered over the adhesive surface, or may be transferred by means of a small triangular sliver of double-adhesive tape attached to a small wooden stick if more control over the orientation is required. This is best done under a dissecting microscope. Care must be taken to avoid mechanical damage during removal of the dried specimens from the CPD- or FD-apparatus and their placement on stubs. If only conductive paint is used as an adhesive, caution must be taken that the solvent used does not permit the paint to infiltrate the specimen to such an extent that important surfaces are obscured. This is especially true for very small embryos which may totally disappear beneath a sea of paint.

## Preparing a Conductive Surface

Since most biological tissues are poor conductors, the surfaces of specimens must usually be rendered conductive prior to viewing to prevent charging and thermal damage to the specimen during interaction with the primary electron beam. This is most commonly accomplished by depositing a thin layer of heavy metal over the sample surface by evaporation in a vacuum or by cathode sputtering[56,57]. In addition, since most biological tissues are composed of elements which are not particularly good sources of secondary electrons, a surface layer of heavy atoms also provides more secondary electrons, and hence, a better signal than can be normally obtained from uncoated biological specimens. A variety of heavy metals may be used, although gold or gold-palladium alloy is generally preferred. Carbon may also be used but generally does not allow viewing at increased magnification with higher accelerating voltages and is usually only used prior to coating with a heavy metal in studies involving secondary electron detection. The main disadvantage of both coating techniques relates to thermal damage during coating. Heating may be minimized by using several short periods of coating interspersed with cooling-off periods, mechanical shielding devices, or by the use of triode sputter coaters[13].

Other techniques have been devised for improving the conductivity of the sample which do not require coating with metal (although they do not necessarily preclude it). These have been recently reviewed by Murphy[58]. One of the original, and the basis of many subsequent modifications, is the O-T-O method of Kelley et al.[59,60] which involves impregnation of the tissue with excess osmium using thiocarbohydrozide (TCH) as a bridging molecule prior to drying. Such methods may not only obviate the necessity for metal coating, but also increase the conductivity throughout the bulk of the sample, and may permit dissection and viewing of the sample without the necessity to recoat the freshly exposed surface. Embryonic tissues treated with the O-T-O procedure may become very brittle, however, and have a tendency to shatter. Impregnation with osmium or other heavy metals may also decrease charging of coated specimens by providing some degree of surface conductivity in parts of the specimen (e.g., cavities and irregularities) which may not be heavily coated in intact specimens by either vacuum evaporation or sputter coating[61].

## Viewing and Photographing

Improper use of the SEM instrument may negate even the best and most careful preparation of the sample. A discussion of the factors involved in operating the SEM and photographic recording and printing of images is clearly not the intent of this tutorial, but since the point of preparing specimens is to allow the gathering of useful data from them, the use of the appropriate accelerating voltage, focus, depth of focus, photographic exposure, and other parameters of the instrument are important in determining the amount of type of data obtained. Damage or distortion of delicate surface projections such as microvilli, cilia or of long, slender cellular processes and extracellular strands by an excessive accelerating voltage have been documented and may create a source of frustrating artifact. While information content is the most important feature of SEM images, there is no reason why esthetic considerations should not also be considered. For example, specimens can be mounted on the stub and oriented within the SEM in such a way as to minimize the inclusion distracting debris or substrate in the background of microphoto-

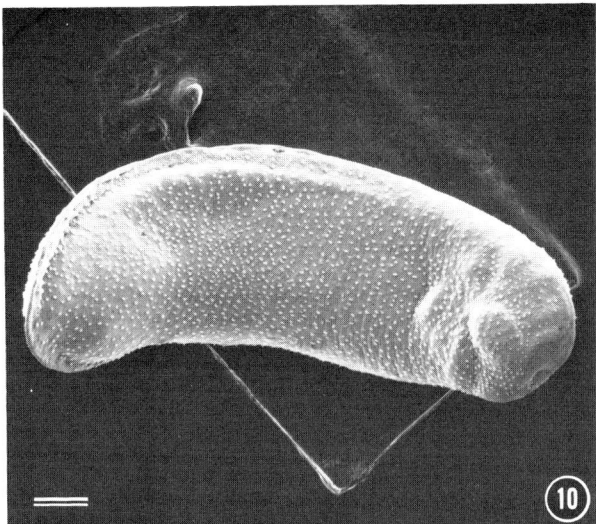
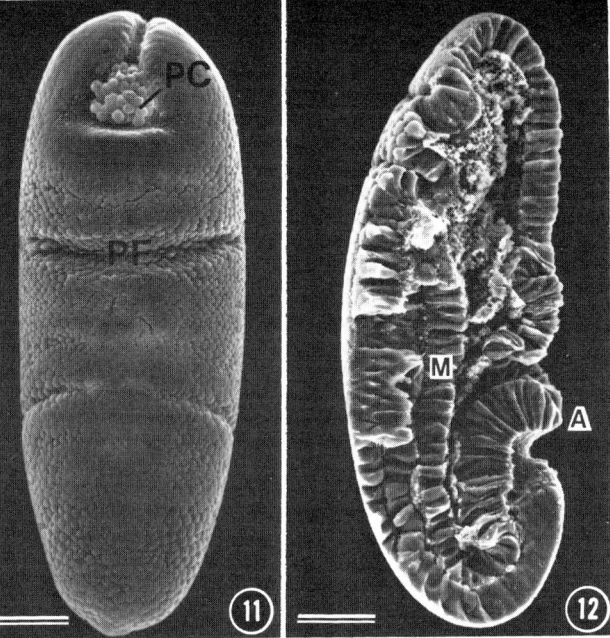

Fig. 10. Xenopus laevis embryo mounted on aluminum stub with strip of double-sticky tape and silver paint. PAFG-fixed, CPD, coated with Al. Bar=150 μm.
Figs. 11-12 illustrate Drosophila embryos at approximately 4 hours of development at 25°C processed as described in text (courtesy Dr. A.P. Mahowald).

Fig. 11. Dorsal view of intact embryo. Pole cells (PC) in posterior midgut invagination. Proctodeal fold (PF). Bar=50 μm.
Fig. 12. Embryo cleaved in mid-sagittal plane. Mesodermal tube ventrally split, appearing as two layers (M). Anterior midgut invagination (A). Bar=50 μm.

graphs (Figs. 11-12). If multiple specimens are mounted on a single stub, they should be mounted far enough apart to decrease the possibility that regions of a particular specimen may be obscured at certain angles of view by an adjacent specimen. A small amount of attention to such details, as well as proper photographic processing of images to assure the best contrast, and to avoid disconcerting scratches, dust, etc., can improve the esthetic impact of images without sacrificing their information content.

## Representative Protocols and Species

### Insects

The insect embryo offers the possibility for studying a wide range of developmental phenomena. Of particular interest is the study of the genetic control of development. The "fruitfly", Drosophila melanogaster, has proven to be a commonly studied insect in this regard since its introduction as a laboratory animal by T.H. Morgan in 1911. Basic techniques for procurement and handling of embryos may be found in many sources [62].

The Drosophila embryo is surrounded by two protective envelopes which must be removed to expose the surface of the embryo. Traditional methods for fixation of Drosophila embryos have necessitated puncture of these envelopes with needles or use of hot fixatives to obtain penetration through the inner, vitelline membrane[63]. The hydrophobic vitelline membrane is impermeable to aqueous solutions by virtue of a waxy coating which is fully established soon after oviposition[64]. Organic solvents, such as ether or chloroform, can penetrate this barrier, but kill the embryo. However, a method utilizing organic solvents saturated with glutaraldehyde has been used to permit adequate penetration of the membrane by the fixative. Zalokar[65] showed that treatment of dechorionated eggs with heptane for short periods does not adversely effect their development, nor their ultrastructure. This "phase-partition" fixation is similar to fixation with vapor. The organic solvent (heptane) which is immiscible with water, penetrates the hydrophobic vitelline membrane and carries the fixative (glutaraldehyde) into contact with the aqueous environment of the embryo[66]. The use of this technique of fixation to examine early stages of Drosophila development with the SEM has been elegantly demonstrated by Mahowald and Turner[67-69] (Figs. 11, 12).

a. The outer layer (chorion) is removed by soaking embryos 1-2 minutes in a 30% Chlorox solution; the active ingredient of which is sodium hypochlorite (5.25% by wgt).

b. Wash well in distilled $H_2O$.

c. Immerse embryos for 1 minute in heptane saturated with glutaraldehyde. Prepare by mixing equal parts heptane and 50% biological grade glutaraldehyde. Shake for 1 minute and allow the heptane phase to separate.

d. Transfer embryos to the trialdehyde fixative of Kalt-Tandler[34].

e. Allow embryos to float on the fixative until the heptane has evaporated, then submerge the embryos.

f. Remove the vitelline membrane with sharpened tungsten needles, and allow the embryos to fix "overnight".

g. Wash briefly in 0.2M sucrose in 0.1M sodium cacodylate buffer (pH 7.2).

h. Post-fix in 1% $OsO_4$ in 0.1M cacodylate buffer for 1-2 hours.

i. Rinse in distilled water.

j. Dehydrate in graded ethanol series, and CPD.

Internal details can be revealed by dissecting the embryo with tungsten needles while in the fixative. Embryos may be sonicated for 30 seconds - 1 minute in an ultrasonic cleaner prior to osmication to remove loose yolk which might contaminate the fractured surface.

Marine Invertebrates

The embryos of many marine invertebrates provide useful systems for descriptive, biochemical and experimental studies in cell biology and early embryology. This is particularly true of the echinoderms; about which a staggering amount has been written. Many techniques for maintaining and handling various echinoderms and their gametes have been published over the last century[70]. One of the most widely used techniques for inducing release of gametes in sea urchins, sand dollars or starfish is to inject 0.5 M Kl into the coelom at various points around the peristomial membrane. The amount injected varies with the size of the animal, but is usually about 1-10 ml. Eggs are collected in sea water which is usually sterilized by passing through a millipore filter. Eggs may then be passed through several layers of cheesecloth or other mesh to remove debris, and washed (to remove coelomic fluid which may effect fertilization adversely) by suspending the eggs in a large volume of sea water, allowing them to settle, decanting the sea water and resuspending the eggs several times. Sperm have a short span of activity in solution and are therefore collected "dry" and not suspended in sea water until needed. They can be covered and stored in the refrigerator; usually remaining viable for a day or more. Other techniques for obtaining gamete release include electric shock, and injection of other compounds. Removal of gonads is also possible but destroys the animal. Fertilization can be accomplished by suspending a small amount (one drop or less) of viable sperm in about 10 ml of fresh sea water and adding a drop or two of this suspension to the eggs. The tendency in the beginning is to add too many sperm, and experience will dictate the correct proportions.

To examine the topography of the egg surface, one or more extracellular coats will have to be removed. Jelly coats of some species can be removed by simply passing eggs through cheesecloth or fine mesh. More tenacious coats may be removed by suspending eggs in acidified sea water. An example of this sequential removal of several extraneous layers from starfish eggs is illustrated in Figures 13-18. Unlike sea urchin eggs, starfish eggs are obtained from the ovary still surrounded by follicle cells. These can be dislodged by washing the eggs in calcium-free artificial sea water but the underlying jelly and vitelline layers remain. Exposing eggs to 1% protease (a broad spectrum proteolytic enzyme derived from Saccheromyces) in calcium-free sea water for 2-10 minutes will remove these layers. Lesser treatment removes only jelly and longer treatment strips both layers and exposes a uniform field of microvilli.

Alternatively, 1.0 M urea can be used after the initial washes in calcium-free sea water, although thorough substitution of the surrounding medium with urea will not remove the vitelline layer unless mechanical shearing is introduced; e.g., by sucking the eggs up and down in a fine glass pipette. Glycine does not substitute for urea in terms of a satisfactory egg surface for SEM since microvilli are massively damaged. On the other hand, microvilli are probably longer after urea-treatment than when the hyaline layer is still present. The number of eggs which can be handled this way is small but the resulting image of microvilli on the egg surface is superior to images after protease treatment. Normal early development will occur following either treatment if the eggs have been fertilized, although since there is little or no intercellular cement in early starfish embryos under normal conditions, the blastomeres fail to remain together as cleavage proceeds in the absence of vitelline layer. The eggs shown in Figs. 13-18 were fixed in 1% glutaraldehyde in 90% filtered sea water (or calcium-free artificial sea water, pH 8.0) in test tubes for 30 minutes, post-fixed with 1% $OsO_4$ in 0.4 M sodium acetate (pH 6.0) for another 30 minutes, then dehydrated in ethanol (holding in 70% ethanol for up to several weeks) prior to passage into 2,2-dimethoxypropane (which combines with $H_2O$ to form methanol and acetone) followed by acetone and CPD from $CO_2$. Eggs were pipetted into truncated BEEM capsules capped with nylon mesh for all manipulations subsequent to infiltration with 70% ethanol. Dried specimens were sprinkled onto silver painted stubs.

To examine the membrane events occurring during fertilization in the sea urchin, the vitelline membrane must be removed at an early stage. The vitelline envelope can usually be thoroughly removed only after it is allowed to lift after fertilization. However, due to its hardening within a few minutes of fertilization, removal is virtually impossible unless the hardening reaction is impeded. Several methods for preventing the formation, or removal, of fertilization membranes are avialable[71-73]. The following procedure has been used by Eddy and Shapiro[74,75] to examine membrane events during fertilization in the sea urchin, Strongylocentrotus purpuratus.

a. Gametes were obtained by injection of 1-3 ml of 0.5 M KCl intracoelomically. Eggs were passed through cheesecloth, then washed extensively. Sperm were collected dry and diluted just prior to insemination. All sea water used was passed through 0.5 μm millipore filters (millipore-filtered sea water; MSW) and all handling of gametes accomplished at 12°C.

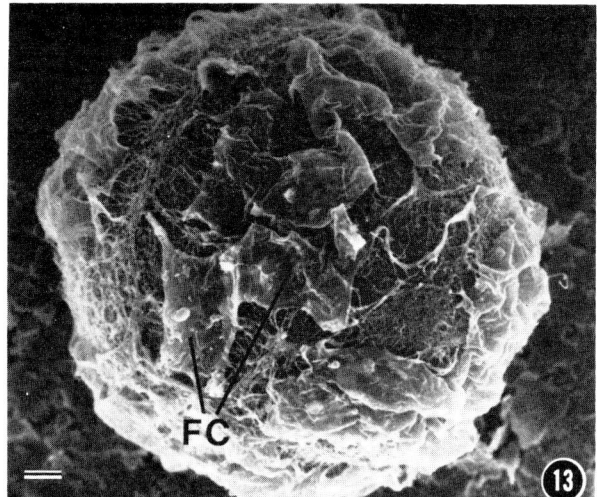

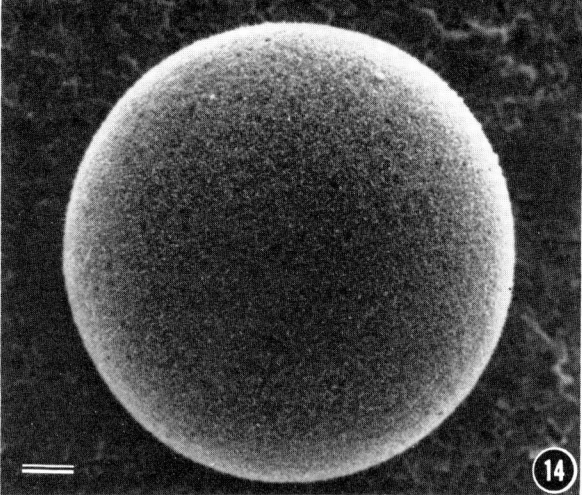

Figs. 13-18 illustrate sequential removal of extracellular layers to expose surface of starfish embryo (courtesy Dr. T. Schroeder).

Fig. 13. Ovarian egg surrounded by follicle cells (FC), jelly coat and vitelline layer. Bar = 10μm.

Fig. 14. Egg after removal of all coats. Bar=10 μm.

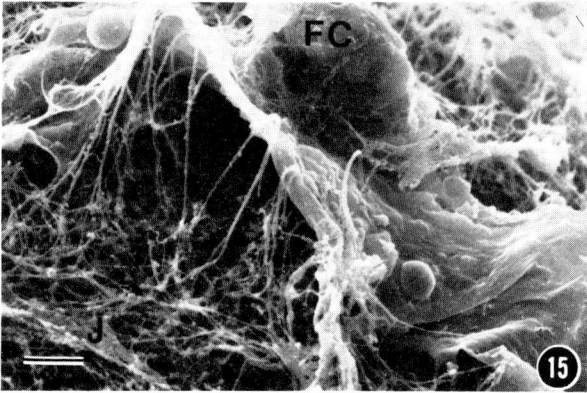

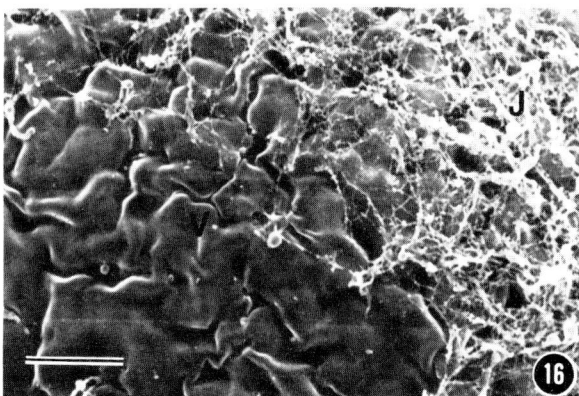

Fig. 15. Follicle cells (FC) and jelly coat (J) of ovarian egg at higher magnification. Bar= 2 μm.

Fig. 16. Vitelline membrane (V) and portions of jelly coat (J) remain after treatment with $Ca^{++}$-free, artificial seawater to dislodge follicle cells and brief exposure to 1% protease. Bar=5 μm.

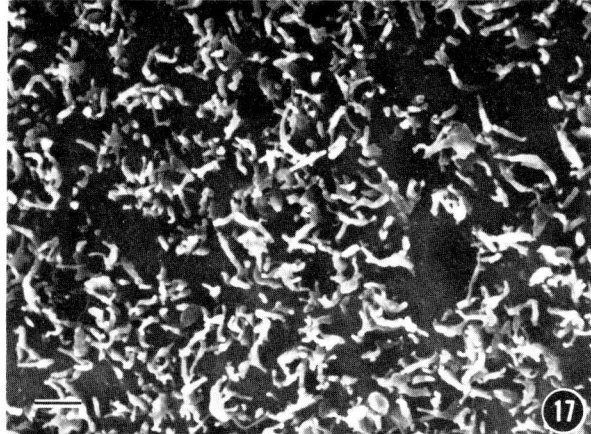

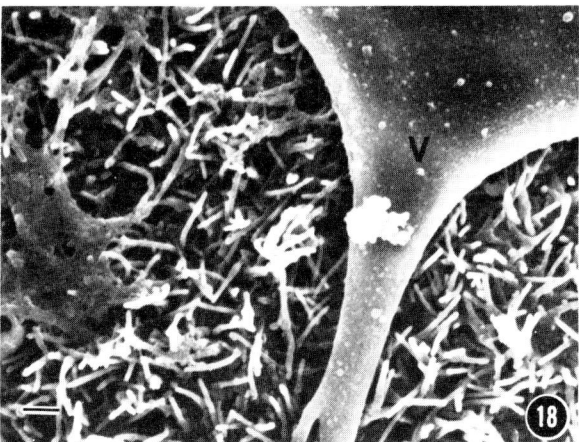

Fig. 17. Complete removal of both jelly and vitelline layers after longer protease digestion exposes microvilli at egg surface. Bar=10 μm.

Fig. 18. Vitelline envelope is partially removed following exposure to 1.0 M urea. Microvilli appear more numerous and regular than after protease treatment (cf., Fig. 17). V=vitelline membrane. Bar=1 μm.

b. The jelly coat was removed by suspending eggs in sea water at pH 4.5 for 2 minutes.

c. The vitelline layer was removed either by washing eggs in dithiothreitol (DTT) or with trypsin. Eggs were suspended in 5 mM DTT in MSW at pH 9 for 2-3 minutes, and then decanted into a large excess of fresh MSW and allowed to settle. This was repeated several times. DTT is a disulfide reducing agent which readily solublizes newly formed vitelline membrane and consequently prevents it from lifting off the egg surface upon fertilization as it should[76]. Eggs incubated in TPCK-trypsin in MSW at a final concentration of 2.5 mg per ml for 20 minutes, were then washed 2 times in MSW[77].

d. Fixation was accomplished by adding 0.5 ml of eggs in sea water to 0.5 ml of a solution of 4% glutaraldehyde and 0.01 M cacodylate buffer (pH 7.8) in 80% sea water for one hour at 12°C. They were then rinsed in cacodylate-buffered sea water and post-fixed in a solution of one part 4% $OsO_4$ and three parts sea water containing a 0.01 M cacodylate buffer for 2-3 hours at room temperature. Ruthenium red (1.0 mg/ml) was included in all solutions in some cases. Eggs were then washed repeatedly in distilled water, dehydrated in ethanol, and critical point dried. For CPD, eggs were placed in 20 µm mesh Nytex containers in Freon TF, placed in the chamber of a CPD apparatus, infiltrated with Freon 13 and dried. Dried eggs were fixed to stubs with silver paste, coated with gold and viewed.

Removal of the fertilization membrane allows visualization of the interaction between the egg surface and sperm and membrane changes accompanying the discharge of cortical granules (Fig. 19) which were obscured in most early SEM studies by presence of the vitelline membrane[78]. Variations of this technique have been used by Schatten and Schatten[79], who used a 3 minute exposure to isosmotic urea (1 M) with trace amounts of calcium chloride $CaCl_2$ (0.1 mM) to stabilize membranes. This treatment removes the vitelline membrane as well as initiates discharge of cortical granules. Eggs were fixed onto glass slips coated with poly(L)-lysine (0.1%) as described by Mazia et al.[80] After fixation for 20 minutes in 5% glutaraldehyde and sea water at pH 7.8, samples were dehydrated in ethanol, and dried by CPD from Freon 13. Specimens were coated with platinum and carbon.

The passage of sperm into the egg has also been studied by viewing the process from the inside of the egg[81,82]. Eggs and sperm of S. purpuratus were obtained as described above. Unfertilized eggs were fixed to polylysine-coated glass plates, dilute sperm was added and specimens were rapidly transferred to fixative (5% glutaraldehyde in 80% sea water (pH 8.2) for 30 minutes). Two methods were employed to view events from the inner aspect of the membrane. Eggs were either fertilized in suspension, homogenized by a single stroke of a teflon pestle, and the resulting fragments placed on polylysine-coated glasses and fixed (all within 45 seconds), or the fertilization membrane was removed from fertilized eggs as soon as the fertilization membrane became visible by exposure to 0.1% mercaptoethyl gluconamide, 10 mM ethylene diamine tetraacetic acid (EDTA) in sea water (pH 8.2) followed by three passages through bolting silk. Several minutes after fertilization, they were placed on polylysine-coated slips, washed two times in homogenization buffer and the tops of the cells sheared off by means of a squirt of this buffer. The eggs were poured rather than pipetted since they become sticky. Remnants of the egg cortex were then fixed with 0.5 M KCl, 2 mM $MgCl_2$, 2 mM EGTA, 1% acrolein, and 2.5% glutaraldehyde (pH 7.5) for 15 minutes, dehydrated in ethanol and critical point dried with Freon 13.

The morphology of blastula and gastrula stages of several echinoderms has also been examined. Good preservation of cell structure has been obtained when embryos are fixed in glutaraldehyde (1-2.5%) in sea water (pH 8.0) or in a buffer vehicle isotonic with sea water, then either dehydrated in ethanol and CPD or postfixed in buffered $OsO_4$ prior to ethanol dehydration and CPD with either Freon 13 or $CO_2$[83-85]. After mounting, embryos may be randomly fractured with a razor blade or other instrument to allow visualization of such interior structures as the forming mesenchyme and gut.

The outer egg capsules of Ilyanessa or Limnea can be removed into forceps and scissors, but additional material of a gelatinous nature still obscures the egg surface. Much of the material surrounding Ilyanessa eggs can be digested away by a 10 minute treatment with 0.1% amylase followed by 5 minute treatment with 1.0% protease. Many mollusks such as oysters shed individual eggs with discrete coats resembling vitelline envelopes, and it is possible that thiol-reducing agents would successfully remove such extraneous coats as in the sea urchin, although much less information is presently available regarding the preparation of molluscan eggs and embryos for SEM examination.

Embryonic stages of the squid, Loligo pealei, have been examined by Arnold et al.[86] Embryos for SEM were fixed in 2% glutaraldehyde in MSW buffered to pH 7.4 with collidine-HCl (osmolarity about 900), for 1.5 hours. Embryos were rinsed three times in MSW buffered to pH 7.4 with collidine-HCl, then post-fixed in 1% $OsO_4$ in MSW (adjusted to 900 mos and pH 7.4) for 1 hour. Embryos were dehydrated in acetone, CPD with $CO_2$ and coated with gold by vacuum evaporation. Cracks in the surface of these yolk-laden embryos were routinely seen, but comparison of TEM and SEM observations showed good correlation of surface features involved in formation of the funnel or siphon.

Amphibians

The amphibian embryo has been a long standing favorite of developmental biologists because of its accessability and amenability to surgical manipulations. In recent years, SEM observations have been used to confirm and extend classical concepts of early development, including fertilization, cleavage and gastrulation[2]. Techniques for obtaining and staging various anurans (e.g., Rana, Xenopus) and urodeles (e.g., Ambystoma, Triturus, Taricha) may be easily found in the literature of experimental embryology. The extracellular layers surrounding the early embryos of some species may be removed mechanically with fine forceps or these may be chemically removed; e.g., in a solution of 0.6% sodium thioglycolate and 0.25% crude trypsin[87].

Fixation of the yolk-laden cells of amphibian embryos for ultrastructural examination has been a recognized problem. The following trialdehyde mixture containing dimethyl sulfoxide (DMSO)[88] was introduced in 1971 by Kalt and Tandler[34] to improve overall preservation of ultrastructure in pregastrula amphibian embryos. This fixative contains 3% glutaraldehyde, 2% formaldehyde, 1% acrolein and 2.5% DMSO in 0.1 M cacodylate-HCl buffer plus 0.001 M $CaCl_2$. Recommended fixation time was 16 hours at room temperature followed by several, 15 minute rinses with 0.1 M cacodylate buffer in 0.1 M sucrose, and postfixation in 2% $OsO_4$, 0.1 M cacodylate buffer containing 0.2 M sucrose.

The DMSO seemed to reduce shrinkage and increase the penetration of the primary fixative. Other fixatives have also been used with good results, and older embryos may be sectioned to enhance fixative penetration.

In a recent LM, TEM and SEM study of neural crest cell migration in Ambystoma mexicanum by Lofberg and Ahlfors[89], embryos were removed from their capsules and anesthetized in a balanced salt solution containing a few crystals of MS222 (Sandoz) per 10 ml of solution, then fixed in 1.5% glutaraldehyde plus 1.5% paraformaldehyde in 0.1 M phosphate (or cacodylate) buffer (pH 7.4) for 3-5 hours at room temperature or overnight at 4°C. Specimens were washed well in buffer at which time the epidermis was mechanically removed from some embryos to expose the underlying neural crest cells and extracellular matrix, then post-fixed in 1% $OsO_4$ in 0.1 M buffer. They were then dehydrated in a graded ethanol series and transferred to Freon TF prior to CPD from $CO_2$. Dried specimens were coated with gold in a diode sputter apparatus and viewed. Resulting cellular preservation in the SEM images was similar to that in embryos fixed by other standard fixatives. Small holes or defects in the cell membranes are frequently seen in cells of early amphibian embryos fixed in a number of ways in spite of the fact that very fine filopodia and overall cell shape and arrangement are preserved[90-91].

An interesting technique for exposing the microarchitecture of the collagenous laminae of the basement lamella in frog larvae has been described by Overton[92] and extended to chick embryonic dermis[93]. Whole specimens, or selected portions, were fixed in 70% ethanol for 30 minute (or stored in alcohol until processed further), then transferred (gradually) to water. They were then incubated in 1.0% trypsin (Grand Island; 1:250) in Hank's solution at pH 7.8 for 30 minutes at 37°C to remove the epidermis. Denuded specimens were rinsed with distilled water, fixed routinely in diluted Karnovsky's fixative[22], rinsed, post-fixed in 1% $OsO_4$, then dehydrated in ethanol (and amyl acetate) and CPD. Some distortion was observed in fiber size, but the overall pattern of fiber arrangement was preserved.

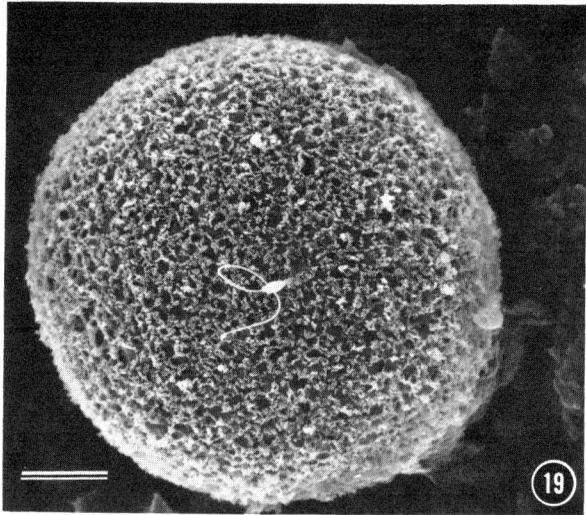

Fig. 19. Mosaic morphology of DTT-treated sea urchin egg 1 minute after insemination (courtesy Drs. E.M. Eddy and B.M. Shapiro). Bar=10 μm.

Fig. 20. Stereopair illustrating initial stage of sequential dissection of cranial region of stage 11 chick embryo. Ectoderm removed from ventral surface of head with small piece of adhesive tape. Bar = 50μm.

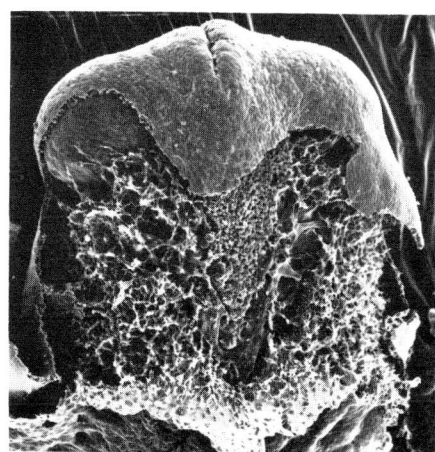 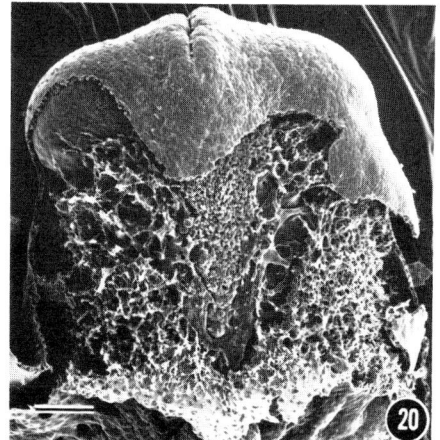

## Avian

Because of the popularity of the chick embryo in developmental biology, a large variety of techniques for preparing chick tissues for microscopic examination are available in the literature[2]. For routine SEM observation, the choice of fixative and drying protocols do not differ significantly from those for other tissues; i.e., fixation in buffered glutaraldehyde or an aldehyde mixture followed by post-osmication, dehydration in ethanol or acetone and CPD with Freon or $CO_2$. The following protocol used by Schoenwolf[94] is an example of a procedure which in the author's experience yields good SEM results.

For observation of early stages, fertile eggs at appropriate stages of incubation[95] may be carefully cracked and opened into a dish containing isotonic saline; holding the egg close to, or partially immersed in the saline. The blastoderm is cut from the yolk by grasping the egg just outside the area of the blastoderm and cutting around the blastoderm with sharp scissors. The embryo and overlying vitelline membrane may then be transferred to a fixation dish, such as a small petri dish, and floated in a small amount of saline to allow for orientation and spreading of the blastoderm. Excess saline is then pipetted off and fresh fixative pipetted onto the embryo.

Essentially similar results are obtained when embryos are fixed initially with glutaraldehyde (1-2.5%) in either 0.1 M phosphate or cacodylate buffer, or with one-half diluted Karnovsky's[22] glutaraldehyde-paraformaldehyde mixture in the same buffer. They are usually also post-osmicated (after washing with 0.1 M buffer) prior to dehydration in ethanol or acetone and CPD.

The early chick embryo often seems more sensitive to osmotic or drying artifacts than some other species. Several studies of the effect of various fixatives or pre-fixation environments on the resulting surface topography of chick embryonic cells suggest that the fixative vehicle should be approximately isotonic for chick tissue, with an aldehyde concentration of no greater than 2% added[96].

Embryos may be easily fractured or dissected to reveal migrating cell populations and epithelial surfaces[97,98] (Figs. 20, 21). In a study of epiblast expansion[99] pieces of vitelline membrane, with or without associated cells, were attached to round glass discs using a method described by Vial and Porter[100]. The discs were coated with 3% gelatin, then chilled at 4°C for at least 10 minutes. Samples were then placed onto the discs which were submerged in Hank's solution and fixed at 37°C in dilute Karnovsky's fixative for 30 minutes to 1 hour until firmly attached. Samples were held down with glass rings which also formed a well for the fixative. Samples were then post-osmicated, dehydrated in ethanol and CPD from $CO_2$.

## Mammalian

To examine fertilization (Fig. 22) and early preimplantation stages of mammalian development in vitro, mature oocytes may be obtained by puncture of ovarian follicles and the surrounding granulosa cells and zona pellucida removed mechanically or enzymatically[2,101,102]. Ovulated eggs and blastocysts may also be recovered from the oviduct (or uterus) by flushing with isotonic saline or fixative[103]. Following attachment and implantation within the uterine wall, embryos must be dissected from maternal and/or extra-embryonic tissues for examination.

Fixation of mammalian embryos is usually accomplished by glutaraldehyde with or without subsequent $OsO_4$ protocols, or by initial fixation with an aldehyde mixture such as Karnovsky's followed by post-fixation with $OsO_4$[104]. The following protocol has been used by Phillips et al.[101] to examine fertilization of rat oocytes in vitro.

1. Ovaries were removed from proestrus rats and rinsed in HEPES-buffered Ringer's solution (pH 7.2).

2. Contents of the largest follicles were expelled into approximately 0.5 ml of mammalian Ringer's solution. Ovulated oocyte-cumulus complexes were recovered from the ampullae of estrus rats.

3. Preovulatory complexes were incubated for 5 minutes in 0.5% trypsin (1:250 Difco) followed by 1.6 mM ATP; both in HEPES-buffered saline. The mechanism of action of the trypsin-ATP sequence is not clear, but was determined empirically.

3. Post ovulatory oocyte-cumulus complexes were exposed to hyaluronidase (1 mg/ml ovine testicular hyaluronidase; Sigma; type II, 420 units/mg in saline), prior to trypsin-ATP treatment. Hyaluronidase effectively removes the post-ovulatory cumulus, but is effective in removing the cumulus of preovulatory ova which are only a few hours prior to ovulation.

4. When the zonae appeared to lyse as observed under a dissecting microscope (usually 1 minute), the complexes were fixed in 0.1 M collidine-buffered glutaraldehyde (4%), pH 7.4, for 1-4 hours.

5. The fixed complexes were rinsed in saline and allowed to settle for 10-20 minutes in a humid environment onto a cover slip pre-soaked for 10 or more minutes in 0.1 to 0.2% aqueous poly-L-lysine (Sigma).

6. Coverslips with attached complexes were then dehydrated in either acetone or ethanol and CPD using $CO_2$.

7. Dried specimens were sputter coated with gold (fairly heavily since oocyte is in contact with the substrate only over a small area).

A variation of this technique was used by Ducibella et al.[105] who permitted embryos to settle onto poly-L-lysine coated silver metal membranes. The filters were placed on a small Buchner funnel and rinsed with 0.1 M cacodylate buffer with suction. Embryos were then absorbed onto the filters with gentle suction, and the filters rapidly immersed in buffer, dehydrated with acetone or ethanol, CPD with $CO_2$ and coated with gold:platinum in a sputter coater.

The following protocol for examining implanting and gastrulating rodent embryos was described by Enders and coworkers[106,107].

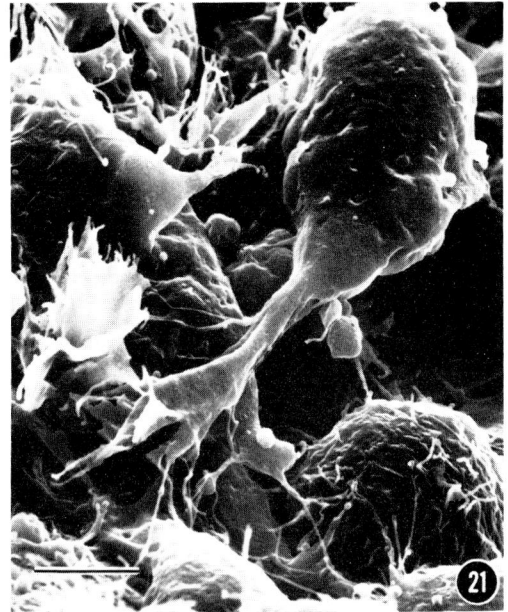

Fig. 21. Presumptive mesenchyme cell in stage 3 chick embryo (courtesy Dr. J.-P. Revel). Bar-5 μm.

Fig. 22. Hamster egg and associated sperm. Cumulus cells removed by hyaluronidase. Zona pellucida removed mechanically following fixation in 2.5% glutaraldehyde. Mounted on poly(L)lysine-coated coverslip and CPD (courtesy Dr. D.M. Phillips). Bar=5 μm.

1. Rat and mice on days 5-7 of pregnancy (day 1 = day of finding vaginal plug) were anesthetized with ether and 1% Pontamine Blue in saline (0.25 ml for rats; 0.2 ml for mice) was injected into the femoral vein to localize implantation sites.

2. 15 minutes later, the animals were perfused via the abdominal aorta with 2% glutaraldehyde, 2.5% formaldehyde in 0.1 M phosphate buffer, pH 7.3.

3. The hardened uteri were removed by laparotomy and the implantation sites, identified by their blue coloration, trimmed and placed into fresh fixative.

4. The antimesometrial muscle and mesometrial third of the uterus were sliced away at right angles to the mesometrium. This exposes the endometrium and cut edge of the luminal epithelium which appears at this stage as a thin, straight line.

5. The two halves of the uterus were gently separated to reveal the implantation chamber which either remains intact or is fractured open, depending on technique and degree of adhesion between the embryo and uterine wall.

6. After about 90 minutes of fixation, the tissue was rinsed overnight in cold 0.1 M phosphate buffer.

7. Post-fix in 2% $OsO_4$ in phosphate buffer (90 minutes), wash, and dehydrate in ethanol.

8. Specimens in this study were placed in absolute acetone (overnight or longer) then CPD with $CO_2$, coated with chromium.

Later stages of embryonic and fetal development may be examined by removing the uterine horns and carefully dissecting the embryo from surrounding uterine tissue and extraembyronic membranes while immersed in fixative. Similar procedures will also suffice for preserving human embryonic material following removal for medical reasons, although initial fixation in the glutaraldehyde-formaldehyde mixture containing picric acid (dinitrophenol) described by Ito and Karnovsky[24] is routinely used in this lab, followed by osmication, ethanol dehydration and CPD from $CO_2$. This fixative penetrates larger samples rapidly and hardens the tissue which can be subsequently dissected to expose desired structures[108].

Additional Techniques for Maximizing Information from Single Specimens

In addition to viewing the surfaces of intact specimens, various procedures have been devised for exposing deeper areas of the specimen for examination in the SEM or to allow correlation of SEM images with other types of examination. The simplest of these procedures involves cutting or fracture of specimens before or after drying[109]. This often allows precise dissection and exposure of specific structures if done prior to drying, but often results in a more unpredictable result, particularly following drying which may make the specimen brittle or fragile.

Better control over the plane of section may be obtained by supporting embryos on a substrate such as solidified albumin prior to cutting with a razor blade[110], or by sectioning paraffin-embedded specimens until the desired level and plane are reached, after which the paraffin can be removed from the unsectioned portion by passage through several changes of xylene or toluene (after first trimming away as much excess paraffin as possible) prior to transfer to several changes of 100% ethanol and CPD[111,112]. Images obtained from such preparations are valuable for overall orientation of internal features which may be correlated with structures seen in the associated sections whicy may be subjected to the usual histological staining procedures. Distortion resulting from the embedding, sectioning and removal of paraffin is usually evident, however, and damage to the cells at the surface cut by the razor or microtome blade normally prevents detailed analysis of their structure. Removal of epoxy resins from embedded specimens and examination of the exposed surfaces has also been reported[113], although changes in surface ultrastructure may occur.

A smoother fracture surface results if the specimen is either infiltrated with resin[114] or fractured while frozen. A simple and useful procedure of the latter type known as "ethanolic cryofracture" has been introduced by Humphreys et al.[115] Fixed specimens are dehydrated to absolute ethanol, and placed in small cylinders of parafilm while immersed in absolute ethanol. The ends of the cylinders containing the tissue and ethanol are sealed by crimping with a hemostat or other instrument and the cylinder is then submerged in liquid nitrogen. The frozen cylinder and enclosed tissue is placed on a metal surface under liquid nitrogen and fractured with a pre-cooled razor blade held by a forceps and struck sharply with a hammer or other implement. The resulting pieces of tissue may then be transferred to absolute ethanol and allowed to warm to room temperature, after which they are CPD in the usual way. The smooth, often shiny, fractured surfaces of tissue pieces are easily recognized and can be oriented appropriately when mounting. The plane of fracture follows an unpredictable course, however, and tends to pass through cells rather than between them (Fig. 23) (as in cracking or tearing of dried specimens), often making examination of surface features difficult.

Dissection of dried specimens can be done prior to mounting them on the viewing stub, or after placing them on the adhesive (tape or paint), depending on the size of the embryo and the desired result. Slightly more control over the fracture and decreased likelihood of loss or damage to the fractured portions of smaller tissues is possible if the specimen is first affixed to the specimen stub and then fractured, with the pieces allowed to remain close together. This often permits examination of the complementary surfaces of the fracture surface, a possibility wich may be lost if the fractured pieces must be transferred some distance from the site of the fracture.

Another dissection technique which can be very useful for removing sheets of cells or portions of embryos is to affix a specimen to the stub, gently bring a piece of double-sided adhesive tape into contact with a portion of the specimen surface, then lift the portion of the specimen adhering to the tape away (Fig. 20). Complementary surfaces may be viewed by simply inverting the tape and mounting it adjacent to the remainder of the specimen. Sequential removal of cells, recoating and photography of the newly exposed surfaces can provide a significant amount of information relative to the organization of an embryo, especially if a series of stereopairs of micrographs is recorded.

Dissection of the specimen within the column of the SEM by means of a micro-manipulator is also possible. The details of this procedure, and the limitations imposed by commercially available equipment, have been reviewed[116]. Charging of the surface exposed by fracture of the specimen may be reduced or eliminated by impregnation of the specimen with heavy metal by means of the O-T-O or related techniques[50].

Scanned specimens may be infiltrated with resin and sectioned for light or TEM[117]. Except for special cases, this is most readily accomplished by simply placing the specimen into the unpolymerized resin, or into a solvent such as propylene oxide, then through a graded series of resin dilutions and polymerized in the standard manner. Sections cut from scanned specimens usually differ slightly in their staining characteristics at both the light and TEM levels, but can provide extremely useful data regarding the identity of surface features. Theoretically, it should be possible to identify the same structure in both the SEM and in sections, although this is often tedious and seldom done. Procedures for carefully marking the feature of interest, usually by placing small tears or holes around it, or by its relationship to plastic spheres or other particles placed on the specimen, have been reported. A scanned specimen may also be embedded in paraffin and sectioned for histologic staining, although this has not been frequently reported.

A number of techniques for producing casts of the vascular system have been devised and applied to various adult tissues[118,119]. These are potentially very useful in the study of developing embryonic and fetal vascular systems, but have received relatively little use. The circulatory system of chick embryos has been studied by Dollinger and Armstrong[120] who perfused with 2% $KNO_3$ via the vitelline vein until the cessation of heartbeat, at which point replication medium was injected at the bifurcation of the right vitelline artery. Batson's corrosion compound was judged to yield better results than the methacrylate used. Following hardening of the injected material, embryos were macerated using 10% KOH for 24 hours. The resulting casts were washed thoroughly with water, allowed to air dry, mounted on stubs, coated with gold in a vacuum evaporator and viewed.

## Artifacts

Since all steps of specimen preparation for SEM examination can produce some change in the state of the tissue from that occurring in vivo, the usefulness of SEM images for discerning something about the tissue in its pre-fixation state depends on the ability of the observer to recognize such changes, or artifacts, to determine how they arose and to formulate criteria for determining what constitutes an acceptable degree of artifact for a given purpose[53,121]. Some types of artifacts which are frequently encountered are briefly mentioned.

Mechanical damage due to abrasion, stretching, or tearing during dissection usually results in gouging or flattening of the surface and is generally easy to recognize. Mechanical damage may also result from overcrowding of specimens during dehydration or result from lack of care in handling specimens during transfer or mounting.

Shrinkage artifacts usually result in a wrinkled appearance of the surface or collapse of a portion of a sample. This is particularly obvious over hollow structures with thin walls. Flattening of surface features, such as microvilli, cilia or larger cytoplasmic laminae or filopodia may also result from inadequate drying. Small cracks in cell surfaces may represent drying artifacts, but may also result from thermal damage during the coating process. Surface irregularities may result from exposure to an improper osmotic environment prior to or during fixation[39]. They may also represent secretion products, normal cytoplasmic protrusions, or contamination by debris or microorganisms. The identity of such structures often requires correlated in vivo observations and/or sectioning of scanned specimens. Small areas of cell lysis resulting from mechanical or osmotic damage may also create surface features which are then stabilized by subsequent fixation. Small blebs, or "blister" created by fixation with glutaraldehyde may be stabilized by subsequent osmication[122].

Discontinuities ("pits") in the cell surface suggest inadequate fixation, but may represent real structures, and this must be established by sectioning the specimen. Treatment of cultured cells with 1% tannic acid in cacodylate buffer at pH 7.2 for 30 minutes following sequential glutaraldehyde and osmium tetroxide fixation has been reported to reduce the number of small surface defects normally seen after CPD[123], but this has not yet been extensively examined in embryonic specimens.

Charging may result from inadequate contact between specimen and stub, inadequate preparation of a conductive surface or mechanical damage to the specimen due to jarring during transfer of the specimen to and from storage and the column of the SEM. Defects in a surface metal coat may also result from swelling of the dried specimen by uptake of $H_2O$ from the environment; either natural humidity or due to prolonged breathing of the investigator directly onto the specimen.

Precipitation artifacts may result from a large number of sources, and are often difficult to identify. Filamentous meshworks and flocculent material over the surface often result from fixation of proteinaceous or other components of the natural environment. Examples of this are the contaminate which occurs in body cavities (Fig. 24) or at the surface of rodent embryos fixed and processed with the amnion intact. To avoid the presence of these precipitates, extraembryonic membranes should be removed or at least torn, and cavities opened, prior to or during initial fixation to allow ready access of solutions. Crystalline precipitates commonly result from interactions between components of the buffer or fixative solutions and the specimen, between buffers, or between fixatives if multiple solutions are used. These types of precipitation artifacts may be reduced or eliminated by careful cleaning or washing of the surface prior to fixation and adequate washing of the specimen between steps during processing. Some precipitates (or dust) may be carefully removed from the dried specimen by

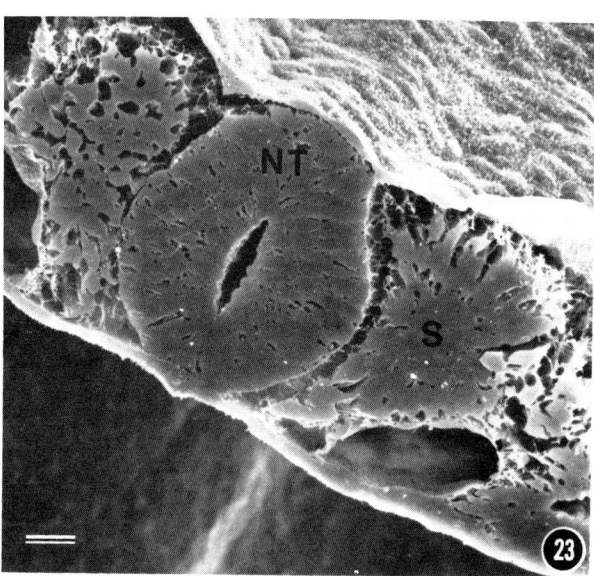

Fig. 23. Cross-section through midbody region of hamster embryo following ethanolic cryofracture. Neural tube (NT). Somite (S). Bar=10 μm.

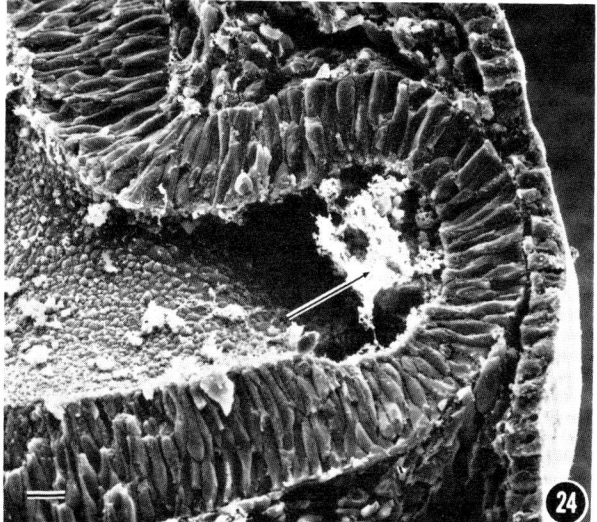

Fig. 24. Precipitate (arrow) obscures cell surfaces in lumen of optic vesicle of chick embryo fractured following CPD. Fracture plane tends to expose lateral cell surface. Bar=10 μm.

gentle blowing of the surface with a dry gas, or mechanically with fine forceps, adhesive tape or by electrostatic forces. This type of surface cleaning is best attempted prior to deposition of a layer of heavy metal over the surface since removal of material from the surface of coated specimens may result in charging of the sample due to breaks in the conductive layer.

Acknowledgements

The author wishes to thank the following individuals for providing their comments and micrographs on short notice: Drs. E.M. Eddy, T. Schroeder, W.J. Humphreys, Jean-Paul Revel, A.P. Mahowald, R.R. Markwald and D.M. Philips. Dr. R.E. Waterman is the recipient of a PHS R.C.D.A. #DE00013.

References

1. Waterman, R.E. 1974 Embryonic and fetal tissues of vertebrates. In: Principles and Techniques of Scanning Electron Microscopy. Biological Applications. (Hayat, M.A., ed.) Van Nostrand Reinhold Co., N.Y., Vol. 2, Chap. 8, pp. 93-110.
2. Waterman, R.E. 1979 Embryonic and foetal tissues of vertebrates. In: Biomedical Research Applications of Scanning Electron Microscopy. (Hodges, G.M. and Hallowes, R.C., eds.) Academic Press, London, Vol. 1, Chap. 1, pp. 1-125.
3. Herbst, R. 1978 Bibliographic service. Scan. 1: 250-257.
4. Thurston, E.L. 1978 Health and safety hazards in the SEM laboratory: update 1978. SEM/1978/Vol. II. SEM, Inc., AMF O'Hare, IL 60666, pp. 849-853.
5. Humphreys, W.J. 1977 Health and safety hazards in the SEM laboratory. SEM/1977/ Vol.I, IIT Research Institute, Chicago, IL 60616, pp. 537-544.
6. Wilt, F.H. and Wessells, N.K., eds. 1967 Methods in Developmental Biology. Thomas Y. Cromwell Co., N.Y., pp. 1-813.
7. Johnson, L.G. and Volpe, E.P., eds. 1973 Patterns and Experiments in Developmental Biology. Wm. C. Brown Co., Dubuque, IA, pp. 1-255.
8. Peine, C.J. and Low, F.N. 1975 Scanning electron microscopy of cardiac endothelium of the dog. Am. J. Anat. 142: 137-158.
9. Flechon, J.-E., Bergstrom, S., Jaszczak, S. and Hofez, E.S.E. 1975 Techniques for critical point drying of gametes and embryos. SEM/1975, IIT Research Institute, Chicago, IL 60616, pp. 325-331.
10. Tamarin, A. and Boyde, A. 1976 Three-dimensional anatomy of the eight-day mouse conceptus: A study of scanning electron microscopy. J. Embryol. Exp. Morph. 36: 575-596.
11. Calarco, P.G. and Epstein, C.J. 1973 Cell surface changes during preimplantation and development in the mouse. Devel. Biol. 32: 208-213.
12. Rostgaard, J. and Christensen, P. 1975 A multipurpose specimen-carrier for handling small biological objects through critical point drying. J. Microsc. 105: 107-113.
13. Hayat, M.A. 1978 Introduction to Biological Scanning Electron Microscopy. University Park Press, Baltimore, MD, pp. 1-323.
14. Crossley, A. 1976 A versatile multi-specimen holder for processing and critical point drying of materials for examination in the scanning electron microscope (SEM). J. Microsc. 108: 349-352.
15. Grinnell, F. 1978 Cellular adhesiveness and extracellular substrata. Int. Rev. Cytol. 53: 65-144.
16. Mazia, D., Schatten, G. and Sale, W. 1975 Adhesion of cells to surfaces coated with poly-lysine. J. Cell Biol. 66: 198-200.
17. Glauert, A.M. 1974 Fixation, dehydration and embedding of biological specimens. In: Practical Methods in Electron Microscopy. (Glauert, A.M., ed.) North-Holland Pub. Co., Amsterdam, Vol. 3, Part 1, pp. 1-207.
18. Palade, G.E. 1952 A study of fixation for electron microscopy. J. exp. Med. 95: 285-298.
19. Sabatini, D.D., Bensch, K. and Barrnett, R.J. 1963 Cytochemistry and electron microscopy. The preservation of cellular ultrastructure and enzymatic activity by aldehyde fixation. J. Cell Biol. 17: 19-58.
20. Sabatini, D.D., Miller, F. and Barrnett, R.J. 1964 Aldehyde fixation for morphological and enzyme histochemical studies with the electron microscope. J. Histochem. Cytochem. 12: 57-71.
21. Robertson, J.D., Bodenheimer, T.S. and Stage, D.E. 1963 The ultrastructure of Mauthner cell synapses and nodes in goldfish brain. J. Cell Biol. 19: 159-199.
22. Karnovsky, M.J. 1965 A formaldehyde-glutaraldehyde fixative of high osmolarity for use in electron microscopy. J. Cell Biol. 137A-138A.
23. Karnovsky, M.J. 1967 The ultrastructural basis of capillary permeability studied with peroxidase as a tracer. J. Cell Biol. 35: 213-236.
24. Ito, S. and Karnovsky, M.J. 1968 Formaldehyde-glutaraldehyde fixatives containing trinitro compounds. J. Cell Biol. 39: 168a-169a.
25. Hayat, M.A. 1970 Principles and Techniques of Electron Microscopy. Biological Applications. Vol. 1, Van Nostrand Reinhold Co., N.Y., pp. 5-107.
26. Millonig, G. 1976 Laboratory-Manual of Biological Electron Microscopy. Mario Saviolo. Vercelli, Italy, pp. 1-67.
27. Tormey, J. McD. 1965 Artifactual localization of ferritin in the ciliary epithelium in vitro. J. Cell Biol. 25(2): 1-7.
28. Jard, S., Bourguet, J., Carasso, N. and Favard, P. 1966 Action de divers fixateurs sur la perméabilité et l'ultrastructure de la vessie de grenoville. J. Microscopie 5: 31-50.
29. Schiff, R.I. and Gennaro, J.F. (Jr.) 1979 The role of the buffer in the fixation of biological specimens for transmission and scanning electron microscopy. Scan. 2: 135-148.

30. Bennett, H.S. and Luft, J.H. 1959 S-collidine as a basis for buffering tissues. J. Biophys. Biochem. Cytol. 6: 113-114.
31. Holt, S.J. and Hicks, R.M. 1961 Studies on formalin fixation for electron microscopy and cytochemical staining purposes. J. Biophys. Biochem. Cytol. 11: 31-45.
32. Pease, D.C. 1964 Histological Techniques for Electron Microscopy. (2nd ed.) Academic Press, N.Y., Chap. 3, pp. 34-81.
33. Luft, J.H. 1959 The use of acrolein as a fixative for light and electron microscopy. Anat. Rec. 133: 305.
34. Kalt, M.R. and Tandler, B. 1971 A study of fixation of early amphibian embryos for electron microscopy. J. Ultrastr. Res. 36: 633-645.
35. Stehr, C.M. and Hawkes, J.W. 1979 The comparative ultrastructure of the egg membrane and associated pore structures in the starry flounder, Platichthys stellus (Pallas), and pink salmon, Oncorhynchus gorbuscha (Walbaum). Cell Tiss. Res. 202: 347-356.
36. Trump, B.F. and Bulger, R.E. 1966 New ultrastructural characteristics of cells fixed in a glutaraldehyde-osmium mixture. Lab. Invest. 15: 368-379.
37. Bone, Q. and Denton, E.J. 1971 The osmotic effects of electron microscopic fixatives. J. Cell Biol. 49: 571-581.
38. Bone, Q. and Ryan, K.P. 1972 Osmolarity of osmium textroxide and glutaraldehyde fixatives. Hist. J. 4: 331-
39. Arborgh, B., Bell, P., Brunk, U. and Collins, V.P. 1976 The osmotic effect of glutaraldehyde during fixation. A transmission electron microscopy, scanning electron microscopy and cytochemical study. J. Ultrastr. Res. 56: 339-350.
40. Luft, J.H. 1976 The structure and properties of the cell coat. Int. Rev. Cytol. 45: 291-382.
41. Markwald, R.R., Fitzharris, T.P. and Bernanke, D.H. 1979 Morphologic recognition of complex carbohydrates in embryonic cardiac extracellular matrix. J. Histochem. Cytochem. 27: 1171-1173.
42. Markwald, R.R., Fitzharris, T.P., Bank, H. and Bernanke, D.H. 1978 Structural analyses on the matrical organization of glycosaminoglycans in developing endocardial cushions. Devel. Biol. 62: 292-316.
43. Anderson, T.F. 1951 Techniques for the preservation of three-dimensional structure in preparing specimens for the electron microscope. Trans. N.Y. Acad. Sci. 13: 130-133.
44. Anderson, T.F. 1956 Electron microscopy of micro-organisms. In: Physical Techniques in Biological Research. (Oster, G. and Pollister, A.W., eds.) Academic Press, N.Y., Chap. 5, pp. 177-240.
45. Cohen, A.L. 1974 Critical point drying. In: Principles and Techniques of Scanning Electron Microscopy. Biological Applications. Vol. 1, (Hayat, M.A., ed.) Van Nostrand Reinhold Co., N.Y., Chap. 2, pp. 44-112.
46. Cohen, A.L., Marlow, D.P. and Garner, G.E. 1968 A rapid critical point method using fluorocarbons ("freons") as intermediate and transitional fluids. J. Microscopie 7: 331-342.
47. Boyde, A. and Maconnachie, E. 1979 Volume changes during preparation of mouse embryonic tissue for scanning electron microscopy. Scan. 2: 149-163.
48. Maser, M.D. and Trimble, J.J. (III) 1977 Rapid chemical dehydration of biologic samples for scanning electron microscopy using 2,2-dimethoxypropane. J. Histochem. Cytochem. 25: 247-251.
49. Christ, B., Jacob, H.J. and Jacob, M. 1977 Experimental analysis of the origins of the wing musculature in avian embryos. Anat. Embryol. 150: 171-186.
50. Stein, O. and Stein, Y. 1971 Light and electron microscopic radioautography of lipids: Techniques and biological applications. Adv. Lipid Res. 9: 1-72.
51. DeBault, L.E. 1973 A critical point drying technique for scanning electron microscopy of tissue culture cells grown on plastic substratum. SEM/1973, IIT Research Institute, Chicago, IL 60616, pp. 317-324.
52. Boyde, A. 1978 Pros and cons of critical point drying and freeze drying for SEM. SEM/1978/Vol. II, SEM, Inc., AMF O'Hare, IL 60666, pp. 303-314.
53. Rebhun, L.I. 1972 Freeze-substitution and freeze-drying. In: Principles and Techniques of Electron Microscopy. Biological Applications. (Hayat, M.A., ed.) Van Nostrand Reinhold Co., N.Y., Vol. 2, Chap. 1, pp. 1-49.
54. Echlin, P. 1978 Low-temperature biological scanning electron microscopy. In: Advanced Techniques in Biological Electron Microscopy. (Koehler, J.K., ed.) Springer-Verlag, Berlin, Vol. 2, pp. 89-122.
55. Boyde, A. and Maconnachie, E. 1979 Freon 113 freeze-drying for scanning electron microscopy. Scan. 2: 164-166.
56. DeNee, P.B. and Walker, E.R. 1975 Specimen coating techniques for the SEM - A comparative study. SEM/1975, IIT Research Institutes, Chicago, IL 60616, pp. 225-232.
57. Echlin, P. 1975 Sputter-coating techniques for scanning electron microscopy. SEM/1975. IIT Research Institute, Chicago, IL 60616, pp. 218-224.
58. Murphy, J.A. 1978 Non-coating techniques to render biological specimens conductive. SEM/1978/Vol. II, SEM, Inc., AMF O'Hare, IL 60666, pp. 175-193.
59. Kelley, R.O., Dekker, R.A.F. and Bluemink, J.G. 1973 Ligand-mediated osmium binding: Its application in coating biological specimens for scanning electron microscopy. J. Ultrastr. Res. 45: 254-258.
60. Kelley, R.O., Dekker, R.A.F. and Bluemink, J.G. 1975 Thiocarbohydrazine-mediated osmium binding: A technique for protecting soft biological specimens in the scanning electron microscope. In: Principles and Techniques of Scanning Electron Microscopy. Biological Applications. (Hayat, M.A., ed.) Van Nostrand Reinhold Co., N.Y., Vol. 4, Chap. 2, pp. 34-44.
61. de Slobodrian, M.-L., de Estable-Puig, R.F. and Estable-Puig, J.F. 1977 Metallic deposi-

tion in specimens presenting cavities using the sputter coater. J. Microsc. 112: 365-369.

62. Doane, W.W. 1967 Drosophila. In: Methods in Developmental Biology. (Wilt, F.H. and Wessells, N.K., eds.) Thomas Y. Crowell Co. N.Y., pp. 219-244.

63. Rickall, W.L. 1976 Cytoplasmic continuity between embryonic cells and the primitive yolk sac during early gastrulation in Drosophila melanogaster. Devel. Biol. 49: 304-310.

64. Limbourg, B. and Zalokar, M. 1973 Permeabilization of Drosophila eggs. Devel. Biol. 35: 382-387.

65. Zalokar, M. 1971 Fixation of Drosophila eggs without pricking. Dros. Inform. Serv. 47: 128-129.

66. Zalokar, M. and Enk, I. 1977 Phase-partition fixation and staining of Drosophila eggs. Stain Tech. 52: 89-95.

67. Mahowald, A.P. and Turner, F.R. 1978 Scanning electron microscopy of Drosophila embryos. SEM/1978/Vol. II, SEM Inc., AMF O'Hare Chicago, IL 60666, pp. 11-19.

68. Turner, F.R. and Mahowald, A.P. 1976 Scanning electron microscopy of Drosophila embryogenesis. I. The structure of the egg envelopes and formation of the cellular blastoderm. Devel. Biol. 50: 95-108.

69. Turner, F.R. and Mahowald, A.P. 1977 Scanning electron microscopy of Drosophila embryogenesis. II. Gastrulation and segmentation. Devel. Biol. 57: 403-416.

70. Hinegardner, R.T. 1967 Echinoderms. In: Methods in Developmental Biology. (Wilt, F.H. and Wessells, N.K. eds.) Thomas Y. Crowell Co., N.Y., pp. 139-155.

71. Berg, W.E. 1967 Some experimental techniques for eggs and embryos of marine invertebrates. In: Methods in Developmental Biology. (Wilt, F.H. and Wessells, N.K., eds.) Thomas Y. Crowell Co., N.Y., pp. 767-776.

72. Schroeder, T.E. 1979 Surface area change at fertilization: Resorption of the mosaic membrane. Devel. Biol. 70: 306-326.

73. Schroeder, T.E. 1978 Microvilli on sea urchin eggs: a second burst of elongation. Devel. Biol. 64: 342-346.

74. Eddy, E.M. and Shapiro, B.M. 1976 Changes in the topography of the sea urchin egg after fertilization. J. Cell Biol. 71: 35-48.

75. Eddy, E.M. and Shapiro, B.M. 1979 Membrane events of fertilization in the sea urchin. SEM/1979/Vol. II, SEM, Inc., AMF O'Hare Chicago, IL 60666, pp. 287-298.

76. Epel, D., Weaver, A.M. and Mazia, D. 1970 Methods of the vitelline membrane of sea urchin eggs. I. Use of dithiothreitol (Cleland's reagent). Exp. Cell. Res. 61: 64-68.

77. Epel, D. 1970 Methods for removal of the vitelline membrane of sea urchin eggs. II. Controlled exposure to trypsin to eliminate post fertilization clumping of the embryos. Exp. Cell Res. 61: 69-70.

78. Tegner, M.J. and Epel, D. 1976 Scanning electron microscope studies of sea urchin fertilization. 1. Eggs with vitelline layers. J. Exp. Zool. 197: 31-58.

79. Schatten, G. and Schatten, H. 1979 Sperm-egg membrane fusions and interactions in denuded sea urchin eggs. SEM/1979/Vol. II, AMF O'Hare, Chicago, IL 60666, pp. 299-306.

80. Mazia, D., Schatten, G. and Sale, W. 1975 Adhesion of cells to surfaces coated with polylysine. Applications to electron microscopy. J. Cell Biol. 66: 198-200.

81. Schatten, G. and Mazia, D. 1976 The surface events at fertilization: The movements of the spermatozoon through the sea urchin egg surface and the roles of the surface layers. J. Supramolec. Struct. 5: 343-369.

82. Schatten, G. and Mazia, D. 1976 The penetration of the spermatozoon through the sea urchin egg surface at fertilization. Expl. Cell Res. 98: 325-337.

83. McClay, D.R. and Marchase, R.B. 1979 Separation of ectoderm and endoderm from sea urchin pluteus larvae and demonstration of germ layers-specific antigens. Devel. Biol. 71: 289-296.

84. Crawford, B.J. and Chia, F.S. 1978 Coelomic pouch formation in the starfish Pisaster ochraceus (Echinodermata: Asteroidea). J. Morph. 157: 99-120.

85. Holland, N.D. 1976 The fine structure of the embryo during the gastrula stage of Comanthus japonica (Echinodermata: Crinoidea). Tiss. Cell 8: 491-510.

86. Arnold, J.M., Williams-Arnold, L.D. and Peters, V. 1978 Fusion of tissue masses in embryogenesis. A scanning electron microscope and transmission electron microscope study of funnel development in the squid Loligo pealei. Devel. Biol. 65: 155-170.

87. Okamoto, M. 1972 A method for the removal of the jelly and vitelline membrane from the embryos of Xenopus laevis. Devel., Growth, Diff. 14: 37-41.

88. Sandborn, E.B., Makita, T. and Lin, K-.N. 1969 The use of dimethyl sulfoxide as an accelerator in the fixation of tissue for ultrastructural and cytochemical studies and in freeze etching of cells. Anat. Rec. 163: 255.

89. Lofberg, J. and Ahlfors, K. 1978 Extracellular matrix organization and early neural crest cell migration in the axototl embryo. Zoon 6: 87-101.

90. Karfunkel, P. 1977 SEM analysis of amphibian mesodermal migration. Roux's Arch. Devel. Biol. 181: 31-40.

91. Keller, R.E. and Schoenwolf, G.C. 1977 An SEM study of cellular morphology, contact, and arrangement, as related to gastrulation in Xenopus laevis. Roux's Arch. Devel. Biol. 182: 165-186.

92. Overton, J. 1976 Scanning microscopy of collagen in the basement lamella of normal and regenerating frog tadpoles. J. Morph. 150: 805-824.

93. Tsai, L-.J. and Overton, J. 1976 The realtion between villus formation and the pattern of extracellular fibers as seen by scanning microscopy. Devel. Biol. 52: 61-73.

94. Schoenwolf, G.C. 1979 Observations on closure of the neuropore in the chick embryo. Am. J. Anat. 155: 445-466.

95. Hamburger, V. and Hamilton, H.L. 1951 A series of normal stages in the development of the chick embryo. J. Morph. 88: 49-92.
96. Litke, L.L. and Low, F.N. 1977 Fixative tonicity for scanning electron microscopy of delicate chick embryos. Am. J. Anat. 148: 121-127.
97. Revel, J.-P. and Solursh, M. 1978 Ultrastructure of primary mesenchyme in chick and rat embryos. SEM/1978/Vol. II., SEM, Inc., AMF O'Hare, Chicago, IL 60666, pp. 1041-1046.
98. Wakely, J. and England, M.A. 1977 Scanning electron microscopy (SEM) of the chick embryo primitive streak. Differentiation 7: 181-186.
99. Chernoff, E.A.G. and Overton, J. 1977 Scanning electron microscopy of chick epiblast expansion on the vitelline membrane. Cell-substrate interactions. Devel. Biol. 57: 33-46.
100. Vial, J. and Porter, K.R. 1975 Scanning microscopy of dissociated cells. J. Cell Biol. 67: 345-360.
101. Phillips, D.M., Shalgi, R., Kraicer, P. and Segal, S.J. 1978 The rat oocyte-cumulus complex during ovulation and fertilization as seen with the SEM. SEM/1978/II., SEM, Inc., AMF O'Hare, Chicago, IL 60666, pp. 1113-1122.
102. Flechon, J.-E., Bergstrom, S., Jaszczak, S. and Hafez, E.S.E. 1975 Techniques for critical point drying of gametes and embryos. SEM/1975. IIT Research Institute, Chicago, IL 60616, PP. 325-331.
103. Shalgi, R. and Sherman, M.I. 1979 Scanning electron microscopy of the surface of normal and implantation - delayed mouse blastocysts during development in vitro. J. Expl. Zool. 210: 69-80.
104. Gould, K.G. 1973 Preparation of mammalian gametes and reproductive tract tissues for scanning electron microscopy. Fert. Steril. 24: 448-456.
105. Ducibulla, T., Ubena, T., Karnovsky, M.J. and Anderson, E. 1977 Changes in cell surface and cortical cytoplasmic organization during early embryogenesis in the preimplantation mouse embryo. J. Cell Biol. 74: 153-167.
106. Enders, A.C. 1975 The implantation chamber, blastocyst and blastocyst imprint in the rat: A scanning electron microscope study. Anat. Rec. 182: 137-150.
107. Enders, A.C., Gwen, R.L. and Schlafke, S. 1978 Differentiation and migration of endoderm in the rat and mouse at implantation. Anat. Rec. 190: 65-78.
108. Waterman, R.E. and Meller, S.M. 1974. Alterations in the epithelial surface of human palatal shelves prior to and during fusion: A Scanning electron microscopic study. Anat. Rec. 180: 111-136.
109. Boyde, A. 1975 A method for the preparation of cell surfaces hidden within bulk tissue for examination in the SEM. SEM/1975 IIT Research Institute, Chicago, IL 60616, pp. 295-303.
110. Ukeshima, A. 1976 A technique for making the cut surface of the avian embryo for scanning electron microscopy. J. Electr. Microsc. 25: 49-50.
111. Armstrong, P.B. 1971 A scanning electron microscope technique for study of the internal microanatomy of embryos. Microscope 19: 281-284.
112. Armstrong, P.B. and Parenti, D. 1973 Scanning electron microscopy of the chick embryo. Devel. Biol. 33: 457-462.
113. Steffens, W.L. 1977 A method for the removal of epoxy resins from tissue in preparation for scanning electron microscopy. J. Microsc. 113: 95-99.
114. Boyde, A. 1974 Freezing, freeze-fracturing and freeze drying in biological specimen preparation for the SEM. SEM/1974 IIT Res. Inst., Chicago, IL 60616, pp. 1043-1046.
115. Humphreys, W.J., Spurlock, B.O. and Johnson, J.S. 1978 Critical-point drying of cryofractured specimens. In: Principles and Techniques of Scanning Electron Microscopy. Biological Applications. (Hayat, M.A., ed.) Van Nostrand Reinhold Co., N.Y., Vol. 6, Chap. 5, pp. 136-158.
116. Pawley, J.B, Hayes, T.L. and Nowell, J.A. 1975 Microdissection. In: Principles and Techniques of Scanning Electron Microscopy. Biological Applications. (Hayat, M.A., ed.) Van Nostrand Reinhold Co., N.Y., Vol. 3, Chap. 2, pp. 17-44.
117. Meller, S.M., Coppe, M.R., Ito, S. and Waterman, R.E. 1973 Transmission electron microscopy of critical point dried tissue after observation in the scanning electron microscope. Anat. Rec. 176: 245-252.
118. Murakami, T. 1978 Methyl methacrylate injection replica method. In: Principles and Techniques of Scanning Electron Microscopy. Biological Applications. (Hayat, M.A., ed.) Van Nostrand Reinhold Co., N.Y., Vol. 6, Chap. 6, pp. 159-169.
119. Gannon, B.J. 1978 Vascular casting. In: Principles and Techniques of Scanning Electron Microscopy. Biological Applications. (Hayat, M.A., ed.) Van Nostrand Reinhold Co., N.Y., Vol. 6, Chap. 7, pp. 170-193.
120. Dollinger, R.K. and Armstrong, P. 1974 Scanning microscopy of injection replicas of the chick embryo circulatory system. J. Microsc. 102: 179-186.
121. Clark, J.M. and Glagov, S. 1976 Evaluation and publication of scanning electron micrographs. Sci. 192: 1360-1361.
122. Shelton, E. and Mowczko, W.E. 1978 Membrane blisters: A fixation artifact. A study in fixation for scanning electron microscopy. Scan. 1: 166-173.
123. Mannweiler, K., Baigent, C.L., Rutter, G., Andresen, I., Neumayer, U. and Hohenberg, H. 1979 The galloylglucose mordanting effect as postfixative for tissue culture cells during SEM studies. Mikroskopie (Wien) 35: 127-132.

## Discussion with Reviewers

E.M. Eddy: Have you observed variability in fixation and artifacts with different batches of glutaraldehyde of presumably different degrees of purity?

Author: Although much has been discussed, and some written, regarding the effects of various impurities, concentrations, and suppliers of glutaraldehyde on the fixation of tissue for TEM, I have not personally compared these factors on fixation of embryonic tissue for SEM. I would refer the interested reader to text references 13, 25 and 32 for a more detailed discussion of glutaraldehyde as a fixative.

H.D. Geissinger: You state that the rate of penetration does not always correspond to the rate of chemical stabilization of the fixatives. In which fixative(s) commonly used for SEM of embryonic tissues (ectoderm, endoderm, mesoderm) is this the case?
Author: The statement to which you refer was intended to indicate a general property of different chemical fixatives. It was not intended to apply specifically to different germ layers of early embryonic stages, and is probably more relevant to the size of the sample. For the earliest stages of development, differences which might result from the factors which effect the rate of penetration (molecular weight of fixative, number and type of reactive sites, solubility, etc.) are probably not significant, whereas for larger, chemically more complex embryonic stages or fetal tissues, the size of the sample may be an important consideration for achieving uniform fixation throughout; especially if immersion fixation is used. The aldehydes, osmium and other chemical fixatives used for stabilization of embryonic tissues for SEM are known to vary in their rates and depth of penetration into both sample systems (ex., gelatin) and various biological tissues. For details, the reader is referred to more extensive works dealing with various aspects of fixation (text references 17, 25, 32).

E.M. Eddy: Can you comment on the advantages and disadvantage of dimethoxypropane as a dehydrating agent?
Author: The use of acidified 2,2-dimethoxy-propane (DMP) has been used to dehydrate specimens for TEM, and to a limited degree for dehydrating SEM tissues prior to CPD. Maser and Trimble (text reference 48) reported that a variety of conventionally fixed cells and tissues showed "no difference" in SEM appearance when dehydrated in DMP for periods from 5 minutes to 30 days prior to CPD from $CO_2$ when compared with similar specimens dehydrated routinely in ethanol or acetone series. Among the advantages listed by these authors were: 1) dehydration of both large and small samples with DMP is more rapid than with ethanol or acetone (attributed to the fact that physical exchange of $H_2O$ with ethanol or acetone is required whereas combination of DMP with water produces methanol and acetone directly); 2) the products of the dehydration reaction, methanol and acetone, as well as DMP, are miscible liquid $CO_2$; 3) virtually all the $H_2O$ is removed if excess DMP is used; 4) the endothermic nature of the reaction and the short time required for dehydration may minimize extraction of components such as lipids; and 5) DMP is less expensive. Comparison of direct measurements made of embryonic mouse limb exposed to DMP or ethanol-acetone dehydration by Boyde and Maconnachie (text reference 47), however, suggest that DMP results in severe and rapid shrinkage which is even greater than that seen with acetone and methanol mixtures. More data are clearly required regarding the use of DMP for this purpose.

T. Pexieder: On what evidence is your statement that incomplete removal of Freon 113 (TF) does not disturb the CPD procedure and will not result in artifacts based?
Author: This is based on the work of Cohen (cf., text reference 45), and the model developed by Pawley and Dole (Pawley, J. and Dole, S. 1976 A totally automatic critical point dryer. SEM/1976 (Part 1). IIT Research Institute, Chicago, IL. Johari, O. (ed.), pp. 287-294).

T. Pexieder: What minimal and maximal thickness of the metal layer would you recommend? Can you inform the readers about the different ways by which a given thickness can be obtained?
Author: The range of acceptable thickness of the deposited metal layer will depend largely on the required operating conditions of the SEM, the degree of resolution required, and the contour of the sample. The minimum thickness is that which prevents charging of the sample and provides adequate electron emission for good signal properties. A coating of approximately 10 nm has been recommended for examination at high resolution and a slightly thicker coat (ca., 20-50 nm) at lower resolution. The thickness of the layer deposited may be determined and controlled empirically by measurement of the metal film in thin sections of coated specimens examined in the TEM. The theoretical thickness of metal deposited in a sputter-coater may be calculated from the parameters of the unit, or may be measured more accurately by such methods as a quartz-crystal, thin film monitor, in which changes in vibration of the coated crystal reflect the mass of the layer deposited, or optical methods such as multi-beam interferometry. A useful discussion of these topics may be found in text reference 13.

---

For additional discussion see page 292.

## SUBJECT INDEX

Acephalic embryos, 157
Acrosomal process, 1
Acrosomal vesicle, 1
Acrosome, 1
Acrosome reaction, 1
Actin, 1, 23, 97
Activation, 37
Active transport, 37
Adenosine triphosphate, 129
Adherens junctions, 293
Adrenaline, 191
Adrenergic neurons, 117
African clawed frog embryos, 55, 157, 221
Air passages or airways, 263, 277
Aldehydes, 331
Alkaline phosphatase, 293
Alveolar ducts, 277
Alveolar phase of lung development, 277
Alveolar sacs, 277
Alveolar surfaces, 277
Alveoli, 277
Amacrine cells, 129
Ambystoma mexicanum embryos, 55, 221
Amnion, 249, 303
Amorphous material, 117
Amphibian embryos, 55, 157, 221
Amphioxus, 249
Ampulla, 229
Anaphase, 81, 97
Animal pole, 45, 55
Anterior intestinal portal, 81, 293
Anteroposterior axis, 45
Antrum of follicle, 37
Aneural embryos, 157
Anuran embryos, 55, 145, 157, 221
Aortae, 181, 229, 249
Aortic arches, 181, 249
Aortic vestibule, 181
Aorticopulmonary septum, 181
Apical ectodermal ridge, 321
Archenteron, 55
Area opaca, 67
Area pellucida, 81, 67
Arms, 321
Artifacts, 331
Arytenoid swelling, 303
Ascending limb of kidney tubule, 229
Asexual buds, 13
Atherogenous diet, 191
Atria, 181, 191,
Atrioventricular canal, 117, 181
Atrioventricular cushions, 191
Auditory pits, 303
Auditory placodes, 303
Auditory vesicles, 303
Auricular hillocks, 303
Autogenous venous grafts, 191
Autophagic vesicles, 157
Axial organs or structures, 157, 169
Axilla or armpit, 321
Axis, 1, 45
Axolotl embryos, 55, 221
Basal body, 1
Basal cells, 263
Basal laminae, 67, 81, 97, 129, 157, 293, 321
Basal membrane, 67
Basement membranes, 117

$\beta$-D-xyloside, 97
Beaded threads, 81, 97
Bicuspid valve, 181
Bilaminar blastoderm, 45, 303
Bilateral furrows of neural plate, 81, 97
Bile, 293
Bile canaliculus, 293
Bile ducts, 293
Bindin, 1
Bipolar cells, 129
Blastemal cells, 117
Blastocoel, 1, 37, 303
Blastocoel roof, 55
Blastocyst, 37, 81, 303
Blastoderm, 45, 303
Blastodisc, 45
Blastomeres, 37, 45, 157, 181
Blastoporal cells, 97
Blastoporal groove, 55
Blastoporal lip, 55
Blastoporal pigment line, 55
Blastopore, 45, 55
Blastula, 1, 45
Blastulation, 37
Blebs, 45, 75, 81, 277, 303
Blood vessels, 181, 191, 229, 249, 277, 293, 321
Body cavity, 181, 303
Body folding, 81, 303
Body of tongue, 303
Bone, 117, 303, 321
Bottle cells, 55, 67, 75, 81, 97
Bowman's capsule, 229
Bowman's space, 229
Bow region of lens, 145
Brain, 81, 129, 303
Branchial apparatus, 303
Branchial membranes, 303
Branching or branching pattern, 169, 277
Branchydania rerio embryos, 45
Bronchial buds, 277
Bronchial mesenchyme, 277
Bronchial tree, 263, 277
Bronchus, 277
Brush border, 293
Brush cells, 263, 277
Buccopharyngeal membrane, 303
Buffers, 331
Bulbar septae, 181
Bulboventricular fold, 181
Bulbus cordis, 181
Calcium, 67, 97, 129
Calcium agonists and antagonists, 169
Calcium-dependent contraction, 169
Calcium ions, 1
Calmodulin, 129, 169
Canalicular phase of lung development, 277
Canalization, 303
Capillaries and endothelial cells, 229, 277, 321
Capsule, 249, 277
Cardiac jelly, 117, 181
Cardiac septation, 181
Cardiogenesis, 181
Carotid arteries, 249
Cartilage, 117, 169, 263, 277, 303, 321
Cartilage matrix, 263
Cartilage rings, 263
Caudal or tail body fold, 303

Cavitation, 81
Cell adhesion, 81, 293
Cell cycle, 1, 97, 145
Cell-cell signaling, 321
Cell death, 81, 129, 157, 303, 321
Cell divisions or proliferation, 1, 221, 303, 321
Cell elongation or palisading, 81, 129
Cell-free space, 117
Cell intermingling, 81
Cell migration or movement, 67, 75, 81, 117, 169, 181, 221, 303
Cell polarity, 23, 293
Cell rearrangement, 221, 303
Cell shape and cell shape changes, 75, 129, 145, 169, 181
Cell specialization, 277
Cellular adhesivity, 169
Cellular projections, 321
Central nervous system, 81, 97, 303
Central periblast, 45
Centrioles, 1
Cetylpyridinium chloride, 117, 129
Charging of specimens, 331
Cheeks, 303
Chick or avian or bird embryos, 67, 75, 81, 97, 117, 129, 169, 181, 191, 221, 249, 293, 321, 331
Cholesterol, 191
Cholinergic neurons, 117
Chondrogenesis, 169, 263, 277, 321
Chondroitin sulfate, 169, 321
Chondronectin, 293
Chordamesoderm, 55
Choroid coat, 129
Choroid fissure, 129
Choroquine, 23
Chromosomes, 23
Ciliated cells, 263, 277
Ciliogenesis, 263
Cilia, 1, 13, 81, 97, 229, 263, 277
Clamp ischemia, 191
Clara cells, 263, 277
Clathrin-coated vesicles, 1
Cleavage, 37, 81, 303
Cloaca, 221, 229
Close junctions, 45
Closing plates, 303
Cnidaria, 13
Cnidocyte cell types, 13
Coated vesicles, 97
Coelom, 81, 181, 221, 303
Colchicine, 23, 97
Colchicine cells or figures, 97
Collagen, 67, 117, 129, 157, 169, 181, 221, 277, 293, 321
Collecting duct of kidney tubules, 229
Combined fixatives, 331
Common carotid arteries, 249
Concanavalin A, 23, 97
Cones of retina, 129
Congenital heart defects, 181, 191
Connecting segment of kidney tubule, 229
Connective tissue, 117, 249, 277, 303
Conotruncus, 191
Contact guidance, 67
Contractile ring, 1
Conus ligament, 181
Convergence, 55
Convergence of the neural folds, 81, 97
Copula, 303

Cornea, 117, 129, 277, 303
Corona radiata, 37
Coronary blood vessels, 181
Correlative microscopy, 331
Cortex, 1, 23, 45, 229
Cortical granules, 1, 23
Cranial or head body fold, 303
Craniofacial region, 303
Critical point drying, 331
Cryofracture, 13
Cumulus fluid, 37
Cutting or fracture of specimens, 331
Cytochalasins, 1, 23, 97, 169, 191
Cytodifferentiation, 263
Cytokinesis, 1, 45
Cytosegresomes, 191
Cytoskeleton, 23, 169
Deep cells, 55
Deeper blastomeres, 45
Definitive endoblast, 67
Dehydration, 331
Dermamyotome, 81, 169
Dermis, 169
Desmosomes, 37, 45, 97, 293
Determination, 157
6-Diazo-5-Oxo-L-Norleucine, 117
Dibutyrylcyclic AMP, 23
Diencephalon, 129, 303
Differential growth, 81
Differentiation, 117, 157, 263, 321
Digestive cell types, 13
Digital core, 321
Digits, 321
Dissection of specimens, 331
Distal tongue bud, 303
Distal tubule of kidney tubule, 229
Dog embryos, 191, 229
Dorsal aortae, 229
Dorsal marginal zone, 55
Drying of specimens, 331
Ductus venosus, 293
Duodenum, 293
Ear drum, 303
Ectoblast, 67
Ectoderm, 13, 55, 67, 81, 117, 169, 221, 249, 303, 321
Ectodermal-mesodermal interactions, 321
Egg, 1, 13, 23, 303
Egg activation, 37
Egg cortex, 1, 23
Egg cylinder, 37
Egg maturation, 23
Egg membrane, 37
Elevation of the neural folds, 81, 97
Emboly, 81
Embryonic axis, 1
Embryonic or initial phase of lung development, 277
Embryonic polarity, 23
Embryonic shield, 45
Endocardial cushions, 117, 181
Endocardium, 181, 191
Endocrine gland or organ, 249, 293
Endoderm, 13, 37, 45, 55, 67, 81, 169, 181, 249, 263, 277
Endodermal-mesenchymal interactions, 293
Endoplasmic reticulum, 229
Endostyle, 249
Endothelium, 181, 229, 293

Endotoxin lipid A, 191
Endotoxin treatment, 191
Enhanced preservation of extracellular materials, 331
Enteroendocrine cells, 263
Enveloping layer, 45
Epiblast, 45
Epiboly, 45, 55, 81
Epicardium, 181, 191
Epidermis, 45
Epiglottis, 303
Epithelial branching, 117
Epithelial fusion, 181
Epithelial-mesenchymal interactions, 321
Epithelial-mesenchymal transformation, 81, 169, 303
Epitheliomuscular cell types, 13
Epithelium, 75, 81, 263, 277, 293, 303
Esophagus, 263
Eustachian tube, 303
Evagination, 303
Exocrine organ, 293
Extension, 55
External auditory meatus, 303
External carotid arteries, 249
External ear canals, 303
External ears, 303
External nares, 303
Extracellular matrix or materials, 67, 97, 117, 129, 169, 181, 221, 277, 293, 303, 321
Extraembryonic membranes, 67
Extrinsic factors in bending of neural plate, 97
Eyes, 129, 303
Eyelids, 303
Face, 303
Facial prominences, 303
Fate maps, 67
Female pronucleus, 1
Fenestrae, 229
Fertilization, 1, 23, 37, 45, 303
Fertilization coat, 1
Fertilization cone, 1
Fibrils, 67
Fibroblasts, 293
Fibronectin, 67, 97, 117, 169, 181, 293, 321
Filaments, 117
Filopodia, 45, 67, 75, 81, 191, 221, 303
Fingers, 321
First meiotic division, 23
First polar body, 23
Fish embryos, 45, 331
Fixation, 13, 23, 37, 45, 97, 117, 145, 157, 191, 263, 331
Fixative ionic composition, 331
Fixative ionic strength, 331
Fixative osmolarity, 331
Fixative penetration, 331
Fixative pH, 331
Fixative temperature, 331
Fixative tonicity, 191
Fixative vehicles, 331
Flask cells, 55, 67, 75, 81, 97
Flask-shaped cells of neuroepithelium, 81, 97
Foot plate, 321
Foot processes, 229
Foramen caecum, 303
Foramen primum, 181
Foramen secundum, 181
Forebrain, 129, 303

Foregut, 81, 181, 263, 293, 303
Forehead, 303
Fragmentation, 37
Free fatty acids injection, 191
Freeze fracture, 293
Freeze-substitution, 331
Freeze drying, 331
Frog embryos, 55, 145, 157, 221, 331
Frontal prominence, 303
Frontonasal prominence, 293
Fusion, 81, 181, 303
Fundus, 129
Gall bladder, 293
Gametes, 1, 23, 37
Ganglionic cells, 13, 129
Gap junctions, 37, 97, 145, 293
Gastrodermal cells, 13
Gastrozooids, 13
Gastrula, 1, 45, 55
Gastrulation, 1, 45, 55, 67, 75, 157, 303
Germinal vesicle breakdown, 23
Germinative zone, 145
Germ layers, 13, 37, 45, 55, 67, 75, 81, 97, 117, 157, 169, 181, 221, 249, 263, 277, 293, 303, 321
Germ ring, 45
Gill slits, 303
Glandular cell types, 13
Glandular stroma, 117
Glial cells, 129, 303
Glomerular endothelium, 229
Glomerular visceral epithelium, 229
Glomerulogenesis, 229
Glomerulus, 229
Glossal processes, 303
Glycoconjugates, 129
Glycogen, 277
Glycoproteins, 67, 97, 117, 129, 181
Glycosaminoglycan lyases, 169
Glycosaminoglycans, 67, 97, 117, 293, 321
Goblet cells, 263
Granules, 13, 117
Gray crescent, 23
Gut, 81, 293, 303
Hagfish, 221
Handling tissues, 331
Hand plate, 321
Head, 303
Head fold of the body, 81
Head mesenchyme, 97, 303
Head process, 81
Heart, 117, 181, 191, 249, 303
Heart tubes, 181
Heavy meromyosin, 97
Hemodynamics, 191
Hensen's node, 67, 81, 169
Heparin sulfate, 117
Hepatic buds, 293
Hepatic cords, 293
Hepatic evagination, 293
Hepatic primordia, 293
Hepatocytes, 293
Hilum of lung, 277
Hindbrain, 303
Horizontal cells, 129
Human embryos, 145, 191, 303, 331
Hyaluronate or hyaluronic acid, 97, 117, 181
Hyaluronidase, 97, 221
Hydrozoa, 13
Hyperplasia, 81
Hypertropy, 81

Hypoblast, 45, 67, 75, 81, 303
Hypobranchial eminence, 303
Hypovitaminosis A, 191
Hypovitaminosis C, 191
Hypoxia, 191
Incorporation cone, 37
Induction, 81, 97, 129, 157, 221, 303, 321
Inner cell mass, 37, 303
Inner ear, 303
Inner plexiform layer, 129
Insect embryos, 331
Insemination, 1
Integral membrane proteins, 293
Interatrial septum, 181
Intercalated cells, 229
Intercellular communication, 293, 321
Intercellular junctions, 37, 45, 75, 81, 97, 145, 169, 293, 303, 321
Intercellular openings, 191
Interkinetic nuclear migration, 81, 97, 129
Interlocking devices of lens fibers, 145
Intermaxillary segment of upper jaw, 303
Intermediate cells, 263
Intermediate mesoblast, 221
Internal blastoporal lip, 55
Internal nares, 303
Interstitial cell types, 13
Interventricular septum, 181
Intestine, 293
Intramembrane particles, 145
Intrinsic factors in bending of neural plate, 97
Invagination, 55, 67, 169, 303
Invertebrate embryos, 1, 13, 331
In vitro, 23, 37, 67
Involution, 55
Iodine, 249
Iodothyronines, 249
Ionophore, 97, 129, 169
Ions, 321
Iris, 303
Ischemia, 191
Isthmus, 249
Jelly coat, 1
Keeshond strain of dogs, 191
Kidneys, 221, 229
Lamellar bodies, 277
Lamelliform protrusions, 55
Lamellipodia, 45, 75, 81, 191, 221, 303
Lamina lucida, 117
Laminin, 117, 293
Larva, 1, 13
Laryngotracheal groove, 263
Larynx, 249
Lateral body folds, 81, 303
Lateral lingual swellings, 303
Lateral nasal prominences, 303
Lateral plate mesoderm, 181, 221, 303
Legs, 321
Length of fixation, 331
Lens capsule, 129
Lens cortex, 145
Lens crystallins, 129
Lens differentiation, 145
Lens epithelial cells, 97
Lens epithelium, 129, 145, 303
Lens fibers, 129, 145, 303
Lens pits, 129, 303
Lens placodes, 97
Lens pores, 129

Lens primordia, 129
Lens vesicles, 97, 129, 303
Lenses, 129, 145
Limbs, 117, 321
Lipid droplets, 229
Liver, 293
Liver cell adhesion molecules, 293
Lobes of lungs, 263, 277
Lobopodia, 45
Loop of Henle, 229
Looping of heart, 181
Loop-tail embryos, 97
Lower jaw, 303
Lung bud mesenchyme, 263
Lung buds, 263
Lungs, 263, 277
Lymphocytes, 263
Lymphoepithelium, 263
Lymphoid tissue, 263
Lysosomes, 157, 229
Macrophages, 191
Macula densa, 229
Magnesium, 67
Mainstem bronchi, 263
Major calyces, 229
Male pronucleus, 1, 37
Mammalian embryos, 23, 37, 145, 157, 229, 263, 277, 321, 331
Mandibular prominences, 303
Marginal folds, 191
Marginal periblast, 45
Marginal zone, 55
Marine invertebrate embryos, 331
Maturation, 23
Maxillae, 303
Maxillary prominences, 303
Medial nasal prominences, 303
Median or midline furrow of neural plate, 81, 97, 303
Median tongue bud, 303
Medulla, 229
Medullary cord, 81, 97
Medusae, 13
Meiosis, 23, 37
Meiotic spindle, 23
Melanoblasts, 117
Membranous interventricular septum, 181
Merging of facial processes, 303
Mesangium, 229
Mesenchymal cells, 263, 293
Mesenchymal-epithelial transformations, 81, 169
Mesenchyme, 75, 81, 97, 117, 129, 169, 229, 249, 263, 277, 293, 303, 321
Mesoblast, 75
Mesoderm, 45, 55, 67, 81, 169, 181, 277, 293, 303, 321
Mesodermal mantle, 55
Mesogleal layer, 13
Mesonephric ducts, 81, 229
Mesonephric glomeruli, 229
Mesonephric nephrons, 229
Mesonephric tubules, 81, 221
Mesonephric vesicles, 229
Mesonephroi, 221, 229
Metamorphosis, 13
Metanephric blastema, 229
Metanephric buds, 229
Metanephric ducts, 229
Metanephrogenic tissue, 229

Metanephroi, 229
Metaphase, 81, 97
Methodology, 331
Mexican axolotl, 55
Microappendages, 191
Microcephalic embryos, 157
Microenvironment, 117
Microfibrils, 293
Microfilaments, 1, 23, 97, 129, 157, 169
Microfolds, 97
Microplicae, 191, 229, 263
Microtubule organizing centers, 1
Microtubules, 1, 81, 97, 157, 169
Microvilli, 1, 23, 67, 75, 97, 191, 229, 263, 277, 293
Microvilliomuscular cells, 13
Midbodies, 81, 97
Midbrain, 129, 303
Middle ear cavity, 303
Middle ear ossicles, 303
Minor calyces, 229
Mitochondria, 157, 229
Mitosis and mitotic cycle, 1, 81, 97, 129, 191
Mitral valve, 181
Monozygotic twinning, 37
Morphogenetic events, 97, 181, 293, 303
Morphogenetic forces, 169
Morphogenetic movements, 45, 55, 181
Morphogenetic processes, 81
Morula, 1, 37, 81, 303
Mouse embryos, 23, 37, 191, 249, 321, 331
Mouse fetuses, 277
Mounting tissues for scanning electron microscopy, 331
Mucociliary epithelium, 263
Mucous cells, 13, 277
Mucus, 263, 277
Mueller cells, 129
Muscle, 321
Muscular interventricular septum, 181
Musculature, 321
Myoblast, 157
Myocardium, 181, 191
Myosin, 97
Nasal cavities, 303
Nasal epithelium, 263
Nasal pits, 303
Nasal placodes, 303
Nasal septum, 303
Nasolacrimal ducts, 303
Nasopharynx, 303
Neck, 249
Nematoblast cell types, 13
Nephrogenesis, 229
Nephron, 229
Nervous system, 303
Neural crest cell pathways, 117
Neural crest cells, 81, 97, 117, 129, 249, 303
Neural ectoderm, 81, 97
Neural folds, 81, 97, 117, 157, 303
Neural groove, 81, 97, 303
Neural keel, 45
Neural plate, 81, 97, 157, 169, 303
Neural retina, 129
Neural tube, 81, 97, 117, 129, 157, 169, 221
Neural tube fluid, 81
Neuraminidase, 67
Neuroblasts, 129
Neurocele, 97

Neuroepithelial bodies, 277
Neuroepithelial cells, 97, 277
Neuroepithelium, 97, 169
Neurohormone, 277
Neurons, 117, 129
Neuropores, 23, 81, 97, 129, 303
Neurulation, 81, 97, 303
Neurulation overlap zone, 81
Nile blue sulfate, 221
Nocodazole, 97, 169
Nose, 303
Nostrils, 303
Notochord, 45, 55, 67, 81, 97, 157, 169, 303
Notochordal plate, 303
Notochordal sheath, 157
Notochord-defective embryos, 157
Notochordless embryos, 157
Nuclear envelope, 1
Nuclei, 45
Occlusion of neurocele, 81
Oel strain of mice, 97
Olfactory pits, 303
Olfactory placodes, 303
Omphalomesenteric veins, 293
Oocytes, 23, 37
Oogenesis, 23
Oolemma, 23
Optic cups, 129, 303
Optic nerves, 303
Optic primordia, 129
Optic stalks, 129, 303
Optic sulci, 303
Optic tectum, 129
Optic vesicles, 129, 303
Optocoel, 129
Oral membrane, 303
Organogenesis, 129, 169, 191, 263, 293, 303
Organ rudiments, 81
Oronasal membranes, 303
Oropharyngeal membrane, 303
Osmium tetroxide, 331
Osmolarity, 191, 331
Ostium primum, 181
Otic pits, 303
Otic placodes, 97, 303
Otic vesicles, 303
Outer plexiform layer, 129
Outer segments of rods and cones, 129
Outflow tract of heart, 181
Ova, 23
Oviducts, 303
Ovulation, 37
Palatal shelves, 81, 303
Palate, 303
Pancreatic rudiment, 97
Papaverine, 97, 169
Parafollicular cells, 249
Parathyroid, 249
Paraxial mesoblast, 169
Paraxial mesoderm, 157, 303
Paraxial microtubules, 97
Pars cartilagina of trachea, 263
Pars convoluta, 293
Pars membranacea of trachea, 263
Pars rectae, 229
Partitioning of heart, 181
Paternal genome, 1
Pattern formation, 293, 321
Pedicels, 229

Pennaria tiarella embryos, 13
Perfusion fixation, 191, 331
Periblast, 45
Perichondrium, 263
Periderm, 129
Perinotochordal matrix, 169
Perinotochordal sheath, 169
Perivitelline space, 1
Phagocyte, 191
Phagocytosis, 321
Pharyngeal apparatus, 303
Pharyngeal arches, 249
Pharyngeal buds, 277
Pharyngeal endoderm, 117
Pharyngeal membranes, 303
Pharyngeal pouches, 249
Pharynx, 249, 263, 277, 303
Philtral region of upper lip, 303
Photography, 331
Photoreceptors, 129
Pigment cells, 303
Pigmented epithelium, 129, 303
Pigmented retina, 303
Pinnae, 303
Pinocytotic vesicles, 191, 229
Planulae, 13
Plasmalemmae or plasma membranes, 1, 23, 37, 97, 293, 321
Pneumocytes types I and II, 277
Podocytes, 229
Polar bodies, 23, 37
Polarity, 23, 321
Poly-L-Lysine, 331
Polyp, 13
Polysaccharides, 129
Posterior cardinal vein, 221
Postinvolution mesoderm, 55
Prechordal plate mesoderm, 97
Pregerminative zone, 145
Preinvolution mesoderm, 55
Preparation of a conductive surface, 331
Preparation techniques, 331
Primary bronchi, 277
Primary bronchial buds, 263
Primary germ layers, 55, 81
Primary interventricular foramen, 181
Primary mesenchyme, 75
Primary nephric duct, 221
Primary neurulation, 97
Primary palate, 303
Primary tubular heart, 181
Primitive choanae, 303
Primitive groove, 67, 81
Primitive knot, 67, 81
Primitive node, 67
Primitive oral cavity, 303
Primitive pit, 67, 81
Primitive ridges, 81
Primitive streak, 67, 75, 81, 117, 303
Primordial germ cells, 67
Principal cells, 229
Proamnion, 67, 81
Processing techniques, 331
Procuring tissues, 331
Progress zone of limb bud, 321
Pronephric ducts, 221, 229
Pronephric tubules, 229
Pronephroi, 221, 229
Prophase, 1, 81, 97

Prospective mesoderm, 55
Prostaglandin, 191
Proteins, 129
Proteoglycans, 97, 129, 277, 321
Protrusions, 55
Proximal tubule of kidney tubules, 229
Pseudostratified epithelium, 67, 81, 129, 263
Pulmonary arteries, 181
Pulmonary parenchyma, 263
Punctate type junctions, 45
Radial cell columns, 145
Radial interdigitation, 55
Radius, 321
Rat embryos, 75, 145, 191, 263, 303, 331
Rat fetuses, 263
Red blood cells, 293
Removal of extraembryonic layers, 331
Renal capsules, 229
Renal collecting ducts, 263
Renal corpuscles, 229
Renal vesicles, 229
Respiration, 263
Respiratory bronchioles, 277
Respiratory distress syndrome, 277
Respiratory passages, 277
Respiratory surfaces, 277
Retinae, 129, 145
Retinal discs, 129
Retinal pigmented epithelium, 129
Retraction fibers, 75
Ringer solution, 331
Rods of retina, 129
Root of tongue, 303
Rough endoplasmic reticulum, 229
Rudiments, 97
Ruffles, 81, 191
Ruffling, 75
Salamander embryos, 55, 221
Sclerotome, 81, 169
Sea urchin embryos, 1
Secondary bronchial buds, 263
Secondary neurulation, 97
Secondary palatal shelves, 303
Secondary palate, 81, 303
Second meiotic division, 37
Second polar body, 37
Secretory cells, 263
Segmental plate mesoderm, 81, 169
Segmental tubules, 221
Segmentation, 169
Semilunar valves, 181, 191
Sensory cells, 13
Septation of heart, 181
Septum intermedium, 181
Septum primum, 181
Serotonin, 277
Serous cells, 263
Shoulders, 321
Sialic acid, 67
Simple tubular heart, 181
Sinusoid, 293
Sinus rhomboidalis, 97
Sinus venosus, 181
Skeletal muscles, 169
Skeletal muscle sheath cells, 303
Skin, 169, 303, 321
Small molecular weight metabolites, 321
Smooth muscle, 263, 277, 303
Somatic mesoderm, 81, 321

Somatopleure, 81, 321
Somites, 55, 81, 97, 157, 169, 191, 221, 303, 321
Somitogenesis, 157, 169
Somitomeres, 169
Sperm, 1, 13, 303
Sperm aster, 1
Spermatozoa, 1, 37
Sperm head, 1
Sperm mid-piece, 1
Sperm nucleus, 1
Sperm tail, 1
Spherical cells of neuroepithelium, 81, 97
Spinal cord, 81, 169, 303
Spinal ganglia, 117
Spindle-shaped cells of neuroepithelium, 81, 97
Spiral ridges, 181
Splanchnic mesoderm, 81, 181, 263
Splanchnopleure, 81
Steinberg's solution, 221
Stomodeum, 303
Streptomyces hyaluronidase, 97, 117
Subblastodermic space, 81
Subblastoporal endodermal cells, 55
Subcephalic pocket, 81
Subclavian arteries, 249
Submandibular gland, 97
Submucosal glands, 263
Submucosa of trachea, 263
Superficial epithelium, 55
Sublastoporal endodermal cells, 55
Supranotochordal cells of neural plate, 97
Surface ectoderm or epithelium, 81, 97, 117, 303
Surface polarity, 23
Surfactant, 277
Sympathoblasts, 117
Syncytium, 45
Syngamy, 1
Synthetic seawater, 331
Tail, 303
Tail bud, 67, 81, 97, 321
Taricha torosa embryos, 97
Teleost embryos, 45
Telophase, 81
Telophase bridges, 97
Tendons, 321
Terminal bars, 97
Terminal buds, 277
Terminal sac phase of lung development, 277
Testicular hyaluronidase, 117
Thin limbs of kidney tubules, 229
Thymus, 249
Thyroid, 169, 249
Thyroid gland, 303
Thyroid placode, 249
Thyroid rudiment, 97
Thyroid vesicle, 249
Tight junctions, 37, 75, 97, 293
Tissue interactions, 157, 321
Toes, 321
Tongue, 303
Totipotency, 37
Trachea, 249, 263, 277
Tracheal groove, 263
Tracheoesophageal groove, 263
Tracheoesophageal septum, 263
Transitional zone, 145
Trifluoperazine, 169
Trilaminar blastoderm, 303
Trophoblast, 37, 303

Truncus arteriosus, 181
Trunk, 303
Trypsin, 97
Tuberculum impar, 303
Tube-within-a-tube vertebrate body plan, 81, 303
Tubotympanic recess, 303
Tubulin, 23
Tubulogenesis, 229
Tunicamycin, 129
Turning, 303
Tympanic cavities, 303
Tympanic membranes, 303
Ulna, 321
Ultimobranchial body, 249
Upper jaw, 303
Ureteric buds, 229
Ureters, 221, 229
Uriniferous tubules, 221, 229
Urochordate, 249
Urodela embryos, 1, 55, 221
Urogenital morphogenesis, 221
Urogenital ridge, 229
Uterine cells, 37
Uterus, 37
UV irradiation, 157
Uvula, 303
Vascular endothelial cell, 277
Vasculature, 321
Vegetal pole, 45, 55
Venous autografts, 191
Ventral aortae, 249
Ventral body wall, 249
Ventricle, 181
Verapamil, 169
Vertebral column, 169
Vertebrate tube-within-a-tube body plan, 81, 303
Villi, 293
Vinblastine sulfate, 97
Visceral apparatus, 303
Visceral arches, 303
Visceral epithelium, 229
Visceral grooves, 229
Visceral pouches, 303
Vital dye, 221
Vitamin A, 97
Vitelline layers, 1
Vitelline membrane, 67
Vitelline veins, 293
Vitreous cavity, 129
Wing, 321
Wedge-shaped cells of neuroepithelium, 97
Wolffian duct, 229
Wrist, 321
Xenopus laevis embryos, 55, 157, 221
Yolk, 45, 67, 81
Yolk cytoplasmic layer, 45
Yolk gel layer, 45
Yolk plug, 55
Yolk sac, 67
Zebrafish embryos, 45
Zona pellucida, 37
Zygotes, 1, 45, 303

AUTHOR INDEX

| | |
|---|---|
| Apkarian, R.P. | 13 |
| Blomgren, P.M. | 229 |
| Bolender, D.L. | 117 |
| Brauer, P.R. | 117 |
| Brown, J.W. | 249 |
| Chávez, D.J. | 37 |
| Chen, D.Y. | 23 |
| Chernoff, E.A.G. | 169 |
| England, M.A. | 67 |
| Evan, A.P. | 229 |
| Fitzharris, T.P. | 181 |
| Gattone, II, V.H. | 229 |
| Hay, D.A. | 181 |
| Hilfer, S.R. | 129, 249, 277 |
| Hotchkiss, A.E. | 13 |
| Jurand, A. | 157 |
| Keller, R. | 55 |
| Kelley, R.O. | 321 |
| Kuszak, J.R. | 145 |
| Longo, F.J. | 23 |
| Lundmark, C. | 55 |
| Macsai, M.S. | 145 |
| Malacinski, G.M. | 157 |
| Markwald, R.R. | 117, 181 |
| Martin, V.J. | 13 |
| McAteer, J.A. | 263 |
| Meyer, R. | 293 |
| Overton, J. | 293 |
| Pexieder, T. | 191 |
| Poole, T.J. | 221 |
| Rae, J.L. | 145 |
| Revel, J.P. | 75 |
| Schatten, G. | 1 |
| Schatten, H. | 1 |
| Schoenwolf, G.C. | 81, 97, 303 |
| Seliger, W.G. | 117 |
| Shih, J. | 55 |
| Solursh, M. | 75 |
| Steinberg, M.S. | 221 |
| Sulik, K.K. | 303 |
| Thomas, R.G. | 45 |
| Tibbetts, P. | 55 |
| Waterman, R.E. | 45, 331 |
| Youn, B.W. | 157 |

# LIST OF REVIEWERS

Following list presents some of the reviewers who kindly helped us with the reviewing of papers in this monograph (when they were prepared for their initial publication in **Scanning Electron Microscopy**).

| | |
|---|---|
| Armstrong, P.B. | University of California, Davis, CA |
| Barber, V.C. | Memorial University, St.John's, NF, Canada |
| Beebe, D.C. | Uniform.Svc.Univ.Hlth.Sci., Bethesda, MD |
| Bell, P.B. | University of Oklahoma, Norman, OK |
| Bellairs, R. | University College London, U.K.. |
| Bernanke, D.H. | Mass.Gen.Hosp. Burns Inst., Boston, MA |
| Brick, I. | New York Univ. - Biology, New York, NY |
| Brunk, U. | University Hospital, Linkoping, Sweden |
| Calarco, P.G. | University of California, San Francisco, CA |
| Carlson, B.M. | University of Michigan, Ann Arbor, MI |
| Carr, K.E. | Queen's Univ. Med. Bio. Ct., Belfast, U.K. |
| Challice, C.E. | University of Calgary, Calgary, Alt., Canada |
| Clark, E.B. | Univ. Iowa Hospital, Iowa City, IA |
| Collins, V.P. | Karolinska Sjukhuset, Stockholm, Sweden |
| Corliss, C.E. | Univ.Tenn.Ctr.Health Sci., Memphis, TN |
| Cutz, E. | Hosp. for Sick Children, Toronto, Ont., Canada |
| Eddy, E.M. | N.I.H.-Inst.Env.Hlth.Sci., Res.Tri.Park, NC |
| England, M.A. | University of Leicester, Leicester, U.K. |
| Erickson, C.A. | University of California, Davis, CA |
| Fallon, J.F. | University of Wisconsin, Madison, WI |
| Farrell, P.M. | Univ. Wisconsin Hospital, Madison, WI |
| Fitzharris, T.P. | Organon Teknika Corp., Charleston, SC |
| Flickinger, R.A. | State University New York, Buffalo, NY |
| Geissinger, H.D. | University of Guelph, Ont., Canada |
| Goeringer, G.C. | Georgetown Univ.Sch.Med., Washington, DC |
| Hilfer, S.R. | Temple University, Philadelphia, PA |
| Hirakawa, R. | Saitama Med.Sch. Anatomy, Saitama, Japan |
| Jacobson, A.G. | University of Texas, Austin, TX |
| Kelley, R.O. | Univ. New Mexico Sch.Med., Albuquerque, NM |
| Kenemans, P. | Univ. Nijmegen Med., Netherlands |
| Kuwabara, T. | National Institute Health, Bethesda, MD |
| Lee, H.-Y. | Rutgers State University, Camden, NJ |
| Luchtel, D.L. | University of Washington, Seattle, WA |
| Manasek, F.J. | University of Chicago, Chicago, IL |
| Markwald, R.R. | Medical Coll. Wisconsin, Milwaukee, WI |
| Massover, W.H. | New Jersey Medical School, Newark, NJ |
| McAteer, J.A. | Indiana Univ. School Med., Indianapolis, IN |
| Meier, S.P. | University of Texas, Austin, TX |
| Moran, D.J. | State University New York, New Paltz, NY |
| Morse, D.E. | Medical College of Ohio, Toledo, OH |
| Motta, P. | Univ. "La Sapienza" Anat., Rome, Italy |
| Nagele, R.G. | Rutgers Med.Sch. UMDNJ, Piscataway, NJ |
| Nicosia, S.V. | Univ. South Florida Med., Tampa, FL |
| O'Rahilly, R. | Univ.Calif. Primate Res.Ctr., Davis, CA |
| Overton, J.H. | University of Chicago, Chicago, IL |
| Perelman, R.H. | Univ. Wisconsin Hospital, Madison, WI |
| Pexieder, T. | Univ. Lausanne, Switzerland |
| Plopper, C.G. | University of California, Davis, CA |
| Runner, M.N. | University of Colorado, Boulder, CO |
| Sadler, T.W. | University North Carolina, Chapel Hill, NC |
| Schatten, G.P. | Florida State University, Tallahassee, FL |
| Schoenwolf, G.C. | Univ. Utah Sch. Med., Salt Lake City, UT |
| Shimada, Y. | Chiba Univ. Sch. Medicine, Chiba, Japan |
| Solursh, M. | University of Iowa, Iowa City, IA |
| Spooner, B.S. | Kansas State University, Manhattan, KS |
| Stocum, D.L. | University of Illinois, Urbana, IL |
| Streeten, B.W. | State Univ. NY Medical Ct., Syracuse, NY |
| Tamarin, A. | University of Washington, Seattle, WA |
| Thommes, R.C. | DePaul University, Chicago, IL |
| Tosney, K.W. | Yale Univ. School Med., New Haven, CT |
| Unakar, N.J. | Oakland University, Rochester, MI |
| Van Praagh, R. | Children's Hosp. Med. Ct., Boston, MA |
| Wakely, J. | University of Leicester, Leicester, U.K. |
| Waterman, R.E. | Univ. New Mexico Sch.Med., Albuquerque, NM |
| Westfall, J.A. | Kansas State University, Manhattan, KS |
| Wilborn, W.H. | Univ.S.Alabama Coll. Med., Mobile, AL |
| Wolbarsht, M.L. | Duke University Med. Ctr., Durham, NC |

# SEM PUBLICATIONS

### (i) International Journal "Scanning Electron Microscopy"† (ISSN: 0586-5581)

SEM publishes the quarterly international journal "Scanning Electron Microscopy". Each issue (four issues in a year) is hardbound and contains over 40 papers and 400 pages covering all aspects of techniques, instrumentation, theory and applications of SEM and **related** techniques. Each paper is published after a thorough peer review and contains the unique feature "Discussion with Reviewers" which provides added insight about the paper. The micrograph reproduction quality is superb.

Complete sets of Tables of Contents are separately available at $1.00 **per issue** ($4.00 per year).

**Prices for SEM/1980, 1981, 1982, 1983, 1984, 1985 or 1986 each:**

|  | U.S. Delivery | Elsewhere |
|---|---|---|
| ☐ Complete Set (Parts 1-4) | $109.00 | $119.00 |
| ☐ any 2 different parts | $ 84.00 | $ 87.50 |
| ☐ any 1 part, also for 1979 | $ 52.00 | $ 55.00 |
| **SEM/1978 & 1979** (2 parts in 1978, 3 parts in 1979): | | |
| ☐ 1979(I+II) (Part III out of print) | $ 65.00 | $ 71.50 |
| ☐ 1978/I (physical) | $ 37.00 | $ 41.00 |
| ☐ 1978/II (biological) | $ 40.50 | $ 44.50 |
| ☐ 1978/2 part set | $ 67.50 | $ 74.00 |

### (ii) International Journal Food Microstructure† (ISSN: 0730-5419)

Started in 1982 this semi-annual journal contains papers on all aspects of microstructure and microanalysis of foods, feeds, and their ingredients. There are 18 papers each in the two issues of Volume 4 (1985) covering meat, dairy products, plant foods, fats, etc. The subscription price **per year** is $50.00 (U.S. delivery) and $55.00 (elsewhere by surface mail) or $70.00 (outside U.S. by air mail). A major subject index of all food-related papers published by SEM is available on request.

### (iii) Proceedings of "Pfefferkorn Conferences" Series (hardbound):

Started in 1982, this series is devoted to an in-depth coverage of basic topics related to electron microscopy.

**1st – 1982: Electron Beam Interactions with Solids for Microscopy, Microanalysis and Microlithography**
Edited by David F. Kyser, Dale E. Newbury, Heinz Niedrig, Ryuichi Shimizu  (ISBN: 0-931288-30-4)
31 papers (372 pages)  *Price $51.00 (U.S. delivery) and $54.00 (elsewhere).*

**\*2nd – 1983: The Science of Biological Specimen Preparation for Microscopy and Microanalysis**
Edited by Jean Paul Revel, Tudor Barnard, Geoffrey H. Haggis  (ISBN: 0-931288-32-0)
28 papers (246 pages)  *Price $40.00 (U.S. delivery) and $43.00 (elsewhere).*

**3rd – 1984: Electron Optical Systems for Microscopy, Microanalysis, and Microlithography**
Edited by John J. Hren, Friedrich A. Lenz, Eric Munro, Peter B. Sewell  (ISBN: 0-931288-34-7)
27 papers (272 pages)  *Price $44.00 (U.S. delivery) and $47.00 (elsewhere).*

### (iv) Special Publications Derived from SEM Volumes:

**1. The Integument: Scanning Electron Microscopy in Skin Biology** — A soft cover book with 31 papers from SEM/1978 to 1985 (294 pages).  *Price $34.00 (U.S. delivery), $37.00 (elsewhere).*

**\*2. Preparation of Biological Specimens for Scanning Electron Microscopy** — A soft cover book with 22 papers from SEM/ 1978 to 1984 (352 pages).  *Price $32.00 (U.S. delivery), $35.00 (elsewhere).*

**\*3. SEM of Cells in Culture** — Soft cover book with 30 papers from SEM/1978 to 1983 (320 pages).
*Price $29.00 (U.S. delivery), $32.00 (elsewhere).*

**\*4. Basic Methods in Biological X-ray Microanalysis** — A soft cover book with 20 papers from SEM/1979 to 1982 (284 pages).  *Price $22.00 (U.S. delivery), $25.00 (elsewhere).*

**\*5. Ultrastructural Effects of Radiation on Tissues & Cells** — A soft cover book with 17 papers from SEM/1981 and 1982.  *Price $18.00 (U.S. delivery), $20.50 (elsewhere).*

**\*6. Cell Surface Labeling** — A soft cover booklet with 10 papers from SEM/1979.  *Price $10.00.*

**7. Clinical Applications of the SEM** — A soft cover booklet with 3 papers from SEM/1980.  *Price $5.00.*

**8. Brain Ventricular Surfaces** — A soft cover booklet with 18 papers from SEM/1978.  *Price $10.00.*

**9. Studies of Food Microstructure** — A hardbound book with 36 papers from SEM/1979, 1980 and 1981.
*Price $49.00 (U.S. delivery), $52.00 (elsewhere).*

---

†Reprints (or copies) of papers from SEM publications are available at $5.00 each.
*Special price for six books marked with an asterisk $99.00 for U.S. delivery and $109.00 (elsewhere).

Library of Davidson College